Lecture Notes in Computer Science

**Lecture Notes in Artificial Intelligence** **16597**

Founding Editor

Jörg Siekmann

The series Lecture Notes in Artificial Intelligence (LNAI) was established in 1988 as a topical subseries of LNCS devoted to artificial intelligence.

The series publishes state-of-the-art research results at a high level. As with the LNCS mother series, the mission of the series is to serve the international R & D community by providing an invaluable service, mainly focused on the publication of conference and workshop proceedings and postproceedings.

Raymond Chi-Wing Wong · Hanghang Tong ·
Hua Lu · James Kwok · Flora Salim ·
Yuanfeng Song · Man Lung Yiu
Editors

# Advances in Knowledge Discovery and Data Mining

30th Pacific-Asia Conference
on Knowledge Discovery and Data Mining, PAKDD 2026
Hong Kong, China, June 9–12, 2026
Proceedings, Part I

Springer

*Editors*
Raymond Chi-Wing Wong
Hong Kong University of Science and Technology
Hong Kong, China

Hua Lu
Aalborg University
Copenhagen, Denmark

Flora Salim
University of New South Wales
Sydney, NSW, Australia

Man Lung Yiu
Hong Kong Polytechnic University
Hong Kong, China

Hanghang Tong
University of Illinois Urbana-Champaign
Urbana, IL, USA

James Kwok
Hong Kong University of Science and Technology
Hong Kong, China

Yuanfeng Song
ByteDance
Shanghai, China

ISSN 0302-9743 ISSN 1611-3349 (electronic)
Lecture Notes in Artificial Intelligence
ISBN 978-981-92-1299-6 ISBN 978-981-92-1300-9 (eBook)
https://doi.org/10.1007/978-981-92-1300-9

LNCS Sublibrary: SL7 – Artificial Intelligence



This Springer imprint is published by the registered company Springer Nature Singapore Pte Ltd.
The registered company address is: 152 Beach Road, #21-01/04 Gateway East, Singapore 189721, Singapore

# Foreword

On behalf of the Organizing Committee, we are delighted to welcome all participants to the 30th Pacific-Asia Conference on Knowledge Discovery and Data Mining (PAKDD 2026). Since the first PAKDD conference in 1997, PAKDD has established itself as one of the leading international conferences on data mining and knowledge discovery. PAKDD provides an international forum for researchers and industry practitioners to share their new ideas, original research results, and practical development experiences across all areas of data mining. This year, after its previous edition in Hong Kong (2001), PAKDD was held for the second time in the fascinating city of Hong Kong, during June 9–12, 2026.

We extend our sincere gratitude to the researchers who submitted their research work to the PAKDD 2026 conference. The main conference of PAKDD 2026 had three tracks, namely Main Track, Survey Track, and Special Track on LLMs for Data Science. Following the initial practice of PAKDD 2025, this year, we organized a Special Session on Data Science Foundations and Applications (DSFA). The conference program was further enriched with an AI forum (involving 11 invited speakers), ten high-quality tutorials, six workshops on cutting-edge topics, and a doctoral consortium. We would like to express our sincere thanks to all authors for their efforts in research, as well as in preparing high-quality presentations. We also express our appreciation to all the collaborators and sponsors for their support, trust, and cooperation. We were honored to have five distinguished keynote speakers joining the conference: Zhi-Hua Zhou (Nanjing University), Geoff Webb (Monash University), Xin Luna Dong (Meta Reality Labs), Xia "Ben" Hu (Shanghai AI Laboratory), and Qingsong Wen (Squirrel AI Learning), each with high reputations in their respective areas. Moreover, we had three award talks, namely (1) PAKDD Distinguished Research Contributions Award Talk (by Jian Pei (Duke University)), (2) PAKDD Most Influential Paper Award Talk (by Truyen Tran (Deakin University)), and (3) PAKDD Early Career Research Award Talk (by Guansong Pang (Singapore Management University)). We enjoyed their participation and talks, which made the conference one of the best academic platforms for knowledge discovery and data mining. We also included a panel session, moderated by Flora Salim (University of New South Wales), involving the following panelists: Geoff Webb (Monash University), Jian Pei (Duke University), Cyrus Shahabi (University of Southern California), and Xia "Ben" Hu (Shanghai AI Laboratory).

We would like to express our sincere gratitude for the contributions of the members and reviewers in the Senior Program Committee and the Program Committee, led by program chairs Raymond Wong, Hanghang Tong, and Hua Lu, and the track chairs including Elena Baralis and Guandong Xu on the Survey Track, and Carl Yang and Xiang Li on the Special Track on LLMs for Data Science. It is through their untiring efforts that the conference had an excellent technical program. We are also thankful to the other Organizing Committee members: including the Chairs of the DSFA Special Session: Guansong Pang, Xuyun Zhang, and Xin Wang; Workshop Chairs Cheng Long,

Jieming Shi, and Yuxuan Liang; Tutorial Chairs Jianliang Xu, Gao Cong, and Sinno Jialin Pan; Doctoral Consortium Chairs Xiao Huang and Byron Choi; Publicity Chairs Chen Ma, Yixiang Fang, and Huiping Cao; Industrial Chairs Xiaoli Li, Jason Zhang, and James Cheng; Proceedings Chairs Yuanfeng Song and Man Lung Yiu; Finance Chair Xin Huang and Yuchen Li; Web and Content Chairs Wenqi Fan and Chenhao Ma; Registration Chairs Haibo Hu and Harry Chan; and Sponsorship Chairs Yangqiu Song, Amelia Jing Li, Wei Bi, Mi-Yen Yeh, and Lu Hou. We would like to particularly thank the Local Arrangement Team, Hong Va Leong and Sibo Wang. The conference would not have been successful without their dedication, commitment, and efforts. We are also grateful to the PAKDD Steering Committee for their guidance, the hosting organizations, and all our sponsors for continuously providing institutional and financial support to PAKDD 2026.

Finally, we hope that you found our program interesting and thought-provoking and will preserve a cherished memory of the conference and Hong Kong. Once again, our warmest welcome to you all!

June 2026

James Kwok
Flora Salim

# Preface

It is our great pleasure to present the 30th Pacific-Asia Conference on Knowledge Discovery and Data Mining (PAKDD 2026) as Program Committee Chairs. PAKDD is one of the longest-established and leading international conferences in the areas of data mining and knowledge discovery. It provides an international peer-reviewed forum for presenting and sharing new ideas, original research results, and practical development experiences from all data science areas, including data mining, data warehousing, machine learning, artificial intelligence, databases, statistics, knowledge engineering, visualization, decision-making systems, and emerging applications.

The PAKDD 2026 technical program featured three tracks, the Main Track, the Survey Track, and a Special Track on Large Language Models in Data Science. The Main Track continued its tradition of being the premier forum for the presentation of research results and experience reports on knowledge discovery, data science, and machine learning. The Survey Track was introduced in 2025 for the first time, to promote the dissemination of insightful survey papers: survey papers are intended to provide a structured synthesis of a particular topic in the area of data mining, including but not limited to theoretical foundations of mining, inference, and learning, big data technologies, as well as security, privacy, and integrity, for the perusal of junior researchers and of experts from other research fields. As the rapid advancements in Large Language Models (LLMs) have opened new avenues for innovation and research across various domains, particularly in the field of data science, we introduced the Special Track on LLMs, which aimed to explore the transformative potential of LLMs for data science, bringing together researchers, practitioners, and industry experts to discuss the latest developments, challenges, and opportunities in this rapidly growing area.

This year's conference was highly competitive. The three tracks received 728 submissions with 552 to the Main Track, 39 to the Survey Track, and 137 to the Special Track on LLMs. The selection process was competitive, rigorous, and thorough, involving 532 PC and 206 Senior PC (SPC) members. Each submission was double-blindly reviewed by at least 3 reviewers with many receiving 4 or more reviews, followed by a discussion, and then further evaluated by a designated SPC member, the Track chairs, and the PC program co-chairs. The final selection considered multiple factors such as novelty, technical depth, experimental validation, and relevance to the conference. We accepted 139 papers in the Main Track, 12 in the Survey Track, and 33 in the LLM Track, thus having an acceptance rate of approximately 25% for the main conference. Following the practice of PAKDD 2025, in PAKDD 2026, considering the high quality of some of the rejected submissions, we decided to have a Special Session on Data Science – Foundations and Applications (DSFA), to which we recommended 125 of the original submissions. 281 of all submissions accepted to the main conference program and to the DSFA Special Session were presented orally during the conference. We would like to thank the Chairs of the Survey Track (Elena Baralis and Guandong Xu) and the Chairs of the LLM Track (Carl Yang and Xiang Li), the SPC members, PC members, and external reviewers, together with the Chairs of the DSFA special session (Guansong Pang, Xuyun Zhang, and Xin Wang), for their invaluable efforts in ensuring a timely, fair, and highly effective paper review and selection procedure. We formed the PAKDD 2026 Best Paper Selection Committee, involving all PC co-chairs, all chairs of the Survey Track and the LLM Track and two new members, namely De-Nian Yang and Xintao Wu, for selecting the best paper award, the best student paper award, and the best survey paper award. Their diligence, dedication, and expertise helped produce the high-quality program for PAKDD 2026.

In addition to the exciting technical presentations, our program also featured five outstanding keynotes, a panel session, three award talks, the AI forum, ten high-quality tutorials, six workshops, and a doctoral consortium. We hope our PAKDD 2026 program provided a wonderful journey for researchers, users, practitioners, and developers.

June 2026

Raymond Chi-Wing Wong
Hanghang Tong
Hua Lu

# Organization

## Organizing Committee

### General Chairs

| | |
|---|---|
| James Kwok | Hong Kong University of Science and Technology, China |
| Flora Salim | University of New South Wales, Australia |

### Program Committee Chairs

| | |
|---|---|
| Raymond Wong | Hong Kong University of Science and Technology, China |
| Hanghang Tong | University of Illinois Urbana-Champaign, USA |
| Hua Lu | Aalborg University, Denmark |

### Sponsorship Chairs

| | |
|---|---|
| Yangqiu Song | Hong Kong University of Science and Technology, China |
| Amelia Jing Li | Hong Kong Polytechnic University, China |
| Wei Bi | Tencent AI Lab, China |
| Mi-Yen Yeh | Academia Sinica, Taiwan |
| Lu Hou | Huawei, China |

### Best Paper Selection Committee

| | |
|---|---|
| Raymond Wong | Hong Kong University of Science and Technology, China |
| Hanghang Tong | University of Illinois Urbana-Champaign, USA |
| Hua Lu | Aalborg University, Denmark |
| Elena Baralis | Politecnico di Torino, Italy |
| Guandong Xu | Education University of Hong Kong, China |
| Carl Yang | Emory University, USA |
| Xiang Li | East China Normal University, China |
| De-Nian Yang | Institute of Information Science, Academia Sinica, Taiwan |
| Xintao Wu | University of Arkansas, USA |

## Local Arrangement Chairs

| | |
|---|---|
| Hong Va Leong | Hong Kong Polytechnic University, China |
| Sibo Wang | Chinese University of Hong Kong, China |

## Industrial Track Chairs

| | |
|---|---|
| Xiaoli Li | Agency for Science, Technology and Research, Singapore |
| Jason Zhang | Hong Kong Polytechnic University, China |
| James Cheng | Chinese University of Hong Kong, China |

## Survey/Vision Track Chairs

| | |
|---|---|
| Elena Baralis | Politecnico di Torino, Italy |
| Guandong Xu | Education University of Hong Kong, China |

## Special Track on Large Language Models in Data Science Chairs

| | |
|---|---|
| Carl Yang | Emory University, USA |
| Xiang Li | East China Normal University, China |

## Special Session on Data Science: Foundations and Applications (DSFA) Chairs

| | |
|---|---|
| Guansong Pang | Singapore Management University, Singapore |
| Xuyun Zhang | Macquarie University, Australia |
| Xin Wang | University of Calgary, Canada |

## Tutorial Chairs

| | |
|---|---|
| Jianliang Xu | Hong Kong Baptist University, China |
| Gao Cong | Nanyang Technological University, Singapore |
| Sinno Jialin Pan | Chinese University of Hong Kong, China |

## Workshop Chairs

| | |
|---|---|
| Cheng Long | Nanyang Technological University, Singapore |
| Jieming Shi | Hong Kong Polytechnic University, China |
| Yuxuan Liang | Hong Kong University of Science and Technology (Guangzhou), China |

### Doctoral Consortium Chairs

| | |
|---|---|
| Xiao Huang | Hong Kong Polytechnic University, China |
| Byron Choi | Hong Kong Baptist University, China |

### Publicity Chairs

| | |
|---|---|
| Chen Ma | City University of Hong Kong, China |
| Yixiang Fang | Chinese University of Hong Kong (Shenzhen), China |
| Huiping Cao | New Mexico State University, USA |

### Financial Chairs

| | |
|---|---|
| Xin Huang | Hong Kong Baptist University, China |
| Yuchen Li | Singapore Management University, Singapore |

### Proceedings Chairs

| | |
|---|---|
| Yuanfeng Song | ByteDance, China |
| Man Lung Yiu | Hong Kong Polytechnic University, China |

### Web Masters

| | |
|---|---|
| Wenqi Fan | Hong Kong Polytechnic University, China |
| Chenhao Ma | Chinese University of Hong Kong (Shenzhen), China |

### Registration Chairs

| | |
|---|---|
| Haibo Hu | Hong Kong Polytechnic University, China |
| Harry Chan | University of Sheffield, UK |

## PAKDD Steering Committee

### Chair

| | |
|---|---|
| Longbing Cao | University of Technology Sydney, Australia |

**Vice Chair**

| | |
|---|---|
| Gill Dobbie | University of Auckland, New Zealand |

**Treasurer**

| | |
|---|---|
| Longbing Cao | University of Technology Sydney, Australia |

**Working Group**

| | |
|---|---|
| Longbing Cao (Chair) | University of Technology Sydney, Australia |
| Ming-Syan Chen (Senior Award Chair) | National Taiwan University, Taiwan |
| Gill Dobbie (Vice Chair and Treasurer) | University of Auckland, New Zealand |
| Hady W. Lauw (Diversity, Equity, and Inclusion Chair) | Singapore Management University, Singapore |
| Jaideep Srivastava (Strategy Chair) | University of Minnesota, USA |
| Raymond Chi-Wing Wong (Junior Award Chair) | Hong Kong University of Science and Technology, China |
| Min-Ling Zhang (Engagement Chair) | Southeast University, China |

**Members**

| | |
|---|---|
| Ramesh Agrawal | Jawaharlal Nehru University, India |
| Gill Dobbie | University of Auckland, New Zealand |
| Joao Gama | University of Porto, Portugal |
| Zhiguo Gong | University of Macau, China |
| Hisashi Kashima | Kyoto University, Japan |
| Hady W. Lauw | Singapore Management University, Singapore |
| Jae-Gil Lee | Korea Advanced Institute of Science & Technology, South Korea |
| Dinh Phung | Monash University, Australia |
| Kyuseok Shim | Seoul National University, South Korea |
| Geoff Webb | Monash University, Australia |
| Raymond Chi-Wing Wong | Hong Kong University of Science and Technology, China |
| De-Nian Yang | Institute of Information Science, Academia Sinica, Taiwan |
| Xintao Wu | University of Arkansas, USA |

### Life Members

| | |
|---|---|
| Longbing Cao | University of Technology Sydney, Australia |
| Ming-Syan Chen | National Taiwan University, Taiwan |
| David Cheung | University of Hong Kong, China |
| Joshua Z. Huang | Shenzhen Institutes of Advanced Technology, Chinese Academy of Sciences, China |
| Masaru Kitsuregawa | Tokyo University, Japan |
| Rao Kotagiri | University of Melbourne, Australia |
| Ee-Peng Lim | Singapore Management University, Singapore |
| Huan Liu | Arizona State University, USA |
| Hiroshi Motoda | AFOSR/AOARD and Osaka University, Japan |
| Jian Pei | Simon Fraser University, Canada |
| P. Krishna Reddy | International Institute of Information Technology, Hyderabad, India |
| Jaideep Srivastava | University of Minnesota, USA |
| Thanaruk Theeramunkong | Thammasat University, Thailand |
| Tu-Bao Ho | Japan Advanced Institute of Science and Technology, Japan |
| Vincent S. Tseng | National Yang Ming Chiao Tung University, Taiwan |
| Takashi Washio | Osaka University, Japan |
| Kyu-Young Whang | Korea Advanced Institute of Science & Technology, South Korea |
| Graham Williams | Australian National University, Australia |
| Chengqi Zhang | University of Technology Sydney, Australia |
| Ning Zhong | Maebashi Institute of Technology, Japan |
| Zhi-Hua Zhou | Nanjing University, China |

### Past Members

| | |
|---|---|
| Arbee L. P. Chen | Asia University, Taiwan |
| Hongjun Lu | Hong Kong University of Science and Technology, China |
| Takao Terano | Tokyo Institute of Technology, Japan |

## Senior Program Committee

| | |
|---|---|
| Albert Bifet | Université Paris-Saclay, France |
| Amparo Alonso-Betanzos | University of A Coruña-CITIC, Spain |
| Anirban Dasgupta | IIT Gandhinagar, India |

| | |
|---|---|
| Arianna Dagliati | University of Pavia, Italy |
| Aris Anagnostopoulos | Sapienza University of Rome, Italy |
| Binbin Hu | Ant Group, China |
| Bo Han | Hong Kong Baptist University, China/RIKEN, Japan |
| Bo Yuan | University of Queensland, Australia |
| Branislav Kveton | Adobe Research, USA |
| Byung Suk Lee | University of Vermont, USA |
| Chang Tang | China University of Geosciences, Wuhan, China |
| Chang-Tien Lu | Virginia Tech, USA |
| Chao Chen | RMIT University, Australia |
| Chao Huang | University of Hong Kong, China |
| Chao Lan | University of Oklahoma, USA |
| Chen Gong | Nanjing University of Science and Technology, China |
| Chen Wang | Huazhong University of Science and Technology, China |
| Chen Zhao | Baylor University, USA |
| Cheng Wang | Amazon, USA |
| Chenhao Ma | Chinese University of Hong Kong (Shenzhen), China |
| Christan Grant | University of Florida, USA |
| Chuxu Zhang | University of Connecticut, USA |
| Da Yan | Indiana University Bloomington, USA |
| De-Nian Yang | Academia Sinica, Taiwan |
| Depeng Xu | University of North Carolina at Charlotte, USA |
| Di Jin | Tianjin University, China |
| Dongjin Song | University of Connecticut, USA |
| Dongxiao He | Tianjin University, China |
| Dongxiao Yu | Shandong University, China |
| Eirini Ntoutsi | University of the Bundeswehr Munich, Germany |
| Elio Masciari | University of Naples, Italy |
| Elisa Bertino | Purdue University, USA |
| Fabrizio Marozzo | Università della Calabria, Italy |
| Feiping Nie | Northwestern Polytechnical University, China |
| Feng Chen | University of Texas at Dallas, USA |
| Feng Xia | RMIT University, Australia |
| Fenglong Ma | Pennsylvania State University, USA |
| Fu-lai Chung | Hong Kong Polytechnic University, China |
| Gan Sun | South China University of Technology, China |
| Gang Li | Deakin Univeristy, Australia |
| Giancarlo Sperlì | University of Naples Federico II, Italy |

| | |
|---|---|
| Mingbo Zhao | Donghua University, China |
| Mingming Gong | University of Melbourne, Australia |
| Min-Ling Zhang | Southeast University, China |
| Mirco Nanni | CNR-ISTI Pisa, Italy |
| Mi-Yen Yeh | Academia Sinica, Taiwan |
| Muhammad Aamir Cheema | Monash University, Australia |
| Nan Ye | University of Queensland, Australia |
| Neil Shah | Snap Inc., USA |
| Nizar Bouguila | Concordia University, Canada |
| Noseong Park | KAIST, South Korea |
| P. Krishna Reddy | IIIT Hyderabad, India |
| Panagiotis Papapetrou | Stockholm University, Sweden |
| Pang-Ning Tan | Michigan State University, USA |
| Paul Honeine | University of Rouen Normandy, France |
| Paulo Cortez | University of Minho, Portugal |
| Pengpeng Zhao | Soochow University, China |
| Qi Yu | Rochester Institute of Technology, USA |
| Qiang Cheng | University of Kentucky, USA |
| Qinghua Hu | Tianjin University, China |
| Quanming Yao | Tsinghua, China |
| Rajeev Raman | University of Leicester, UK |
| Rajiv Khanna | Purdue University, USA |
| Ranga Raju Vatsavai | North Carolina State University, USA |
| Ricardo J. G. B. Campello | University of Southern Denmark, Denmark |
| Riccardo Guidotti | University of Pisa, Italy |
| Sang-Wook Kim | Hanyang University, South Korea |
| Sen Wang | University of Queensland, Australia |
| Shirui Pan | Griffith University, Australia |
| Shui Yu | University of Technology Sydney, Australia |
| Shuo Wang | Shanghai Jiao Tong University, China |
| Siyu Huang | Clemson University, USA |
| Slawomir Nowaczyk | Halmstad University, Sweden |
| Suhang Wang | Pennsylvania State University, USA |
| Sung-Bae Cho | Yonsei University, South Korea |
| Tingting Wang | RMIT University, Australia |
| Tong Chen | University of Queensland, Australia |
| Tru Hoang Cao | UTHealth, USA |
| Tsuyoshi Ide | IBM Research, T. J. Watson Research Center, USA |
| Tzung-Pei Hong | National University of Kaohsiung, Taiwan |
| Uday Kiran Rage | University of Aizu, Japan |
| Wei Fan | University of Oxford, UK |

| | |
|---|---|
| Wei Ju | Sichuan University, China |
| Wei Emma Zhang | University of Adelaide, Australia |
| Weiqing Wang | Monash University, Australia |
| Wei-Shinn Ku | Auburn University, USA |
| Wenbin Zhang | Florida International University, USA |
| Wenqi Fan | Hong Kong Polytechnic University, China |
| Wenqing Lin | Tencent, China |
| Xia Ning | Ohio State University, USA |
| Xiang Deng | Harbin Institute of Technology (Shenzhen), China |
| Xiangyu Zhao | City University of Hong Kong, China |
| Xiao Huang | Hong Kong Polytechnic University, China |
| Xiaoning Liu | RMIT University, Australia |
| Xiaowei Jia | University of Pittsburgh, USA |
| Xin Cao | University of New South Wales, Australia |
| Xindong Wu | Hefei University of Technology, China |
| Xintao Wu | University of Arkansas, USA |
| Xun Zhou | Harbin Institute of Technology (Shenzhen), China |
| Xuyun Zhang | Macquarie University, Australia |
| Yanchang Zhao | CSIRO, Australia |
| Yanchao Tan | Fuzhou University, China |
| Yang Cao | Institute of Science Tokyo, Japan |
| Yibing Zhan | JD Explore Academy, China |
| Yinbin Miao | Xidian University, China |
| Ying Li | AwanTunai, Indonesia |
| Yingxia Shao | Beijing University of Posts and Telecommunications, China |
| Yipeng Zhou | Macquarie University, Australia |
| Yiqun Xie | University of Maryland, USA |
| Yongkai Wu | Clemson University, USA |
| Yu Zhang | Southern University of Science and Technology, China |
| Yuan Luo | Northwestern University, USA |
| Yue Xu | Queensland University of Technology, Australia |
| Yufei Han | Inria, France |
| Yu-Feng Li | Nanjing University, China |
| Yunjun Gao | Zhejiang University, China |
| Yunlong Yu | Zhejiang University, China |
| Yunyun Wang | Nanjing University of Posts and Telecommunications, China |
| Yuxuan Liang | Hong Kong University of Science and Technology (Guangzhou), China |

| | |
|---|---|
| Zhao Kang | University of Electronic Science and Technology of China, China |
| Zheng Wang | University of Leeds, UK |
| Zhengming Ding | Tulane University, USA |
| Zhengzhang Chen | NEC Laboratories America, USA |
| Zhong Ji | Tianjin University, China |
| Zhongfei Zhang | Binghamton University, USA |
| Zimu Zhou | City University of Hong Kong, China |
| Ziwei Zhu | George Mason University, USA |
| Ziyu Jia | Chinese Academy of Sciences, China |

## Program Committee

| | |
|---|---|
| Abdur Rehman Khan | Queensland University of Technology, Australia |
| Abheek Dhingra | University of the Bundeswehr Munich, Germany |
| Akhlaque Ahmad | Indiana University Bloomington, USA |
| Akihiro Yamaguchi | Toshiba Corporation; Kyushu University, Japan |
| Akira Sakai | Fujitsu Laboratories Ltd., Japan |
| Alex Memory | Johns Hopkins University Applied Physics Laboratory, USA |
| Alfredo Cuzzocrea | University of Calabria, Italy |
| Amir Gandomi | University of Technology, Sydney, Australia |
| Andrzej Skowron | University of Warsaw, Poland |
| Ankit Sharma | University of Minnesota, USA |
| Anne Denton | North Dakota State University, USA |
| Antonela Tommasel | ISISTAN, CONICET-UNICEN, Argentina |
| Anushka Vidanage | Australian National University, Australia |
| Arindam Pal | TechSoftX, Australia |
| Aris Gkoulalas-Divanis | Merative, USA |
| Arnab Sharma | Paderborn University, Germany |
| Aryya Gangopadhyay | University of Maryland, Baltimore County, USA |
| Ashwin Viswanathan Kannan | Amazon, USA |
| Ashwini Kodipalli | RV University, India |
| Atsuhiro Takasu | National Institute of Informatics, Japan |
| Augusto A. Santos | Instituto de Telecomunicações, Portugal |
| Bang An | University of Akron, USA |
| Baodi Liu | China University of Petroleum (East China), China |
| Beibei Li | Chongqing University, China |
| Bijay Prasad Jaysawal | University of Southampton, UK |
| Bijaya Adhikari | University of Iowa, USA |

| | |
|---|---|
| Dai Hai Nguyen | Hokkaido University, Japan |
| Dalin Zhang | Hangzhou Dianzi University, China |
| Dan Simovici | University of Massachusetts Boston, USA |
| Daokun Zhang | Monash University, Australia |
| De-Chuan Zhan | Nanjing University, China |
| Deguang Kong | Yahoo Research, USA |
| Delvin Ce Zhang | Singapore Management University, Singapore |
| Derui Lyu | University of Science and Technology of China, China |
| Dharmalingam Muthusamy | Bharathiar University, India |
| Dong Li | Baylor University, USA |
| Dong-Dong Wu | University of Tokyo, Japan |
| Dongqi Fu | Meta, USA |
| Dongyuan Li | Tokyo Institute of Technology, Japan |
| Dugang Liu | Shenzhen University, China |
| Edouard Fouché | Karlsruhe Institute of Technology, Germany |
| Eduardo Ogasawara | Federal Center for Technological Education of Rio de Janeiro, Brazil |
| Ekta Gujral | University of California, Riverside, USA |
| Eliana Pastor | Politecnico di Torino, Italy |
| Enea Vincenzo Napolitano | University of Naples Federico II, Italy |
| Erhu He | University of Pittsburgh, USA |
| Esra Akbas | Georgia State University, USA |
| Fali Wang | Penn State University, USA |
| Fan Wu | Central South University, China |
| Fanglan Chen | Virginia Tech, USA |
| Fei Zhao | University of Alabama at Birmingham, USA |
| Felipe R. Azúa | University of Edinburgh, UK |
| Feng Xiao | Chinese University of Hong Kong (Shenzhen), China |
| Fengxin Li | Renmin University of China |
| Florent Masseglia | Inria, France |
| Francesco Marcelloni | University of Pisa, Italy |
| Francesco Spinnato | University of Pisa, Italy |
| Fuqiang Yu | Qilu University of Technology, China |
| Fuyuan Cao | Shanxi University, China |
| Giovanna Nicora | University of Pavia, Italy |
| Giuseppe Albi | University of Pavia, Italy |
| Guangyan Huang | Deakin University, Australia |
| Guimu Guo | Rowan University, USA |
| Guixian Zhang | China University of Mining and Technology, China |

| | |
|---|---|
| Guodu Xiang | Hefei University of Technology, China |
| Guoyin Wang | Chongqing University of Posts and Telecommunications, China |
| Guozhong Li | King Abdullah University of Science and Technology, Saudi Arabia |
| Hadi Talal Jaafar | University of Birmingham, UK |
| Haidong Kang | Northeastern University, China |
| Haifang Li | Baidu Inc., China |
| Hailiang Zhao | Zhejiang Univerisity, China |
| Haiquan Chen | California State University, Sacramento, USA |
| Hanchen Wang | University of New South Wales, Australia |
| Hang Yu | Shanghai University, China |
| Hao Xue | University of New South Wales, Australia |
| Hao Zhen | University of Georgia, USA |
| Hao Zheng | Central South University, China |
| Haomin Yu | Aalborg University, Denmark |
| Haoran Duan | Wuhan University, China |
| Haoran Luo | Nanyang Technological University, Singapore |
| He Kong | Jilin University, China |
| He Li | Xidian University, China |
| Heitor M. Gomes | Victoria University of Wellington, New Zealand |
| Hezhe Qiao | Singapore Managment University, Singapore |
| Hoang H. Nguyen | University of Tennessee at Chattanooga, USA |
| Hongchao Qin | Beijing Institute of Technology, China |
| Hongliang Zhang | Nanjing University of Science and Technology, China |
| Hongtao Wang | North China Electric Power University, China |
| Hongzu Su | University of Electronic Science and Technology of China, China |
| Huaiyuan Yao | Arizona State University, USA |
| Huan Li | Zhejiang University, China |
| Huan Wang | University of Wollongong, Australia |
| Huan Zhang | Zhengzhou University, China |
| Huaxiang Zhang | Shandong Normal University, China |
| Hui Fang | Shanghai University of Finance and Economics, China |
| Huifang Ma | Northwest Normal University, China |
| Huiping Cao | New Mexico State University, USA |
| Huizhong Wang | Chinese University of Hong Kong (Shenzhen), China |
| Ickjai Lee | James Cook University, Australia, Australia |
| Jae Oh Woo | Samsung SDS America, USA |

| | |
|---|---|
| Jahnavi A. Kachhia | meta, USA |
| Ja-Hwung Su | National University of Kaohsiung, Taiwan |
| Jean Paul Barddal | Pontifical Catholic University of Paraná, Brazil |
| Jean-Paul Calbimonte | University of Applied Sciences and Arts of Western Switzerland HES-SO, Switzerland |
| Jeongwhan Choi | KAIST, South Korea |
| Jeremiah D. Deng | University of Otago, New Zealand |
| Jia Chen | University of California, Riverside, USA |
| Jiahao Li | Guangdong University of Technology, China |
| Jialiang Li | Roskilde University, Denmark |
| Jiancheng Lv | Sichuan University, China |
| Jianfei Zhang | Université de Sherbrooke, Canada |
| Jiangneng Li | Nanyang Technological University, Singapore |
| Jianwu Wang | University of Maryland, Baltimore County, USA |
| Jianxin Li | Beihang University, China |
| Jian-Yun Nie | Université de Montréal, Canada |
| Jianzhong Qi | University of Melbourne, Australia |
| Jiaqi Zhao | China University of Mining and Technology, China |
| Jiatan Huang | University of Connecticut, USA |
| Jiaxin Jiang | National University of Singapore, Singapore |
| Jiaxing Xu | National University of Singapore, Singapore |
| Jiayan Guo | Peking University, China |
| Jiazheng Li | University of Connecticut, USA |
| Jie Li | RMIT University, Australia |
| Jie Wang | Southwest Jiaotong University, China |
| Jie Yang | University of Technology Sydney, Australia |
| Jierong Li | Yunnan University, China |
| Jilin Hu | Aalborg University, Denmark |
| Jimmy Xiangji Huang | York University, Canada |
| Jin Liu | Shanghai Maritime University, China |
| Jin Lu | University of Georgia, USA |
| Jinbin Huang | Shenzhen University, China |
| Jinghui Yuan | Northwestern Polytechnical University, China |
| Jinhua Guo | Southern University of Science and Technology, China |
| Jin-Hui Wu | Nanjing University, China |
| Jinjun Chen | Swinburne University of Technology, Australia |
| Jinpeng Chen | Beijing University of Posts and Telecommunications, China |
| Jinyi Liu | Tianjin University, China |

| | |
|---|---|
| Jipeng Guo | Beijing University of Chemical Technology, China |
| Jiuding Duan | Allianz Global Investors, Japan Co., Japan |
| Jonathan A Zhu | Carnegie Mellon University, USA |
| Jože Rožanec | Jožef Stefan Institute, Slovenia |
| Jun Huang | Anhui University of Technology, China |
| Jun Wang | National University of Defense Technology, China |
| Jun Wu | Michigan State University, USA |
| Jungeun Kim | Inha University, South Korea |
| Junjie Yao | East China Normal University, China |
| Jun-Ki Min | Korea University of Technology and Education, South Korea |
| Junlin Wu | Washington University in St. Louis, USA |
| Jun-Zhao Du | Xidian University, China |
| K. Selcuk Candan | Arizona State University, USA |
| Kai Guo | Sichuan University, China |
| Kai Zhao | Aalborg University, Denmark |
| Kaiqiang Yu | Nanjing University, China |
| Kaiqun Fu | South Dakota State University, USA |
| Kaixin Wang | Beijing University of Technology, China |
| Kang Xu | Nanjing University of Posts and Telecommunications, China |
| Karim Tabia | CRIL, Artois University, France |
| Keith Burghardt | University of North Carolina at Charlotte, USA |
| Kejia Chen | Nanjing University of Posts and Telecommunications, China |
| Kewen Liao | Deakin University, Australia |
| Ki Yong Lee | Sookmyung Women's University, South Korea |
| Kia Teymourian | University of Texas at Austin, USA |
| Ki-Hoon Lee | Kwangwoon University, South Korea |
| Kijung Shin | KAIST, South Korea |
| Kishlay Jha | University of Iowa, USA |
| Kohei Hatano | Kyushu University/RIKEN AIP, Japan |
| Koji Maruhashi | Fujitsu Research, Fujitsu Limited, Japan |
| Kok-Leong Ong | RMIT University, Australia |
| Krishnamoorthy Sivakumar | Washington State University, USA |
| Krisztian Buza | Budapest University of Economics and Business, Hungary |
| Kuan-Ting Lai | National Taipei University of Technology, Taiwan |
| Kun Fang | Hong Kong Polytechnic University, China |
| Kunpeng Liu | Portland State University, USA |

| | |
|---|---|
| Kunpeng Xu | University of Sherbrooke, Canada |
| Kun-Ta Chuang | National Cheng Kung University, Taiwan |
| Kun-Yang Yu | Nanjing University, China |
| Lam Ngo | RMIT University, Australia |
| Lecheng Zheng | Virginia Tech, USA |
| Lei Chen | Tsinghua University, China |
| Lei Duan | Sichuan University, China |
| Lei Gu | Nanjing University of Posts and Telecommunications, China |
| Lei Li | Hong Kong University of Science and Technology (Guangzhou), China |
| Lei Zhang | Nanjing Normal University, China |
| Li Hu | Hong Kong Polytechnic University, China |
| Li Tang | Yunnan University, China |
| Liang Hong | Wuhan University, China |
| Liang Li | National University of Defense Technology, China |
| Liang Qu | Edith Cowan University, Australia |
| Libin Wang | Northwestern Polytechnical University, China |
| Lihui Liu | Wayne State University, USA |
| Lijing Wang | New Jersey Institute of Technology, USA |
| Likang Wu | University of Science and Technology of China, China |
| Lin Zhang | IDEA.EDU, USA |
| Ling Ding | Tianjin University, China |
| Ling Zheng | Fujian Provincial Maternity and Children's Hospital, China |
| Lingwei Chen | Rochester Institute of Technology, USA |
| Lingyang Chu | McMaster University, Canada |
| Liu Xichao | Yunnan University, China |
| Liwei Deng | Aalborg University, Denmark |
| Lizhong Ding | Beijing Institute of Technology, China |
| Lo Pang-Yun Ting | National Cheng Kung University, Taiwan |
| Longxu Sun | Hong Kong Baptist University, China |
| Lu Jiang | Dalian Maritime University, China |
| Luca Cagliero | Politecnico di Torino, Italy |
| Lujun Li | Hong Kong University of Science and Technology, China |
| Lulwah AlKulaib | Kuwait University, Kuwait |
| Luyuan Yang | University of Oklahoma, USA |
| Luzhi Wang | Dalian Maritime University, China |
| Lyu Xu | Nanyang Technological University, Singapore |

| | |
|---|---|
| Lyuheng Yuan | Indiana University Bloomington, USA |
| Maciej Grzenda | Warsaw University of Technology, Poland |
| Mandar Chaudhary | eBay, USA |
| Man-Sheng Chen | Sun Yat-sen University, China |
| Maolin Wang | City University of Hong Kong, China |
| Marc Plantevit | EPITA, France |
| Marcellin M Atemkeng Teufack | Rhodes University, China |
| Maria Luisa Sapino | Università di Torino, Italy |
| Marzena Kryszkiewicz | Warsaw University of Technology, Poland |
| Masahiro Nakano | NTT Communication Science Laboratories, Japan |
| Matej Mihelčić | University of Zagreb, Croatia |
| Matjaž Gams | Jožef Stefan Institute, Slovenia |
| Melike Yildiz Aktas | Virginia Tech, USA |
| Meng Wang | Tongji University, China |
| Mengdi Huai | Iowa Sate University, USA |
| Menghan Wang | eBay, USA |
| Miao Hu | University of Southampton, UK |
| Miguel Rodriguez | Google, USA |
| Milos Radovanovic | University of Novi Sad, Serbia |
| Milos Savic | University of Novi Sad, Serbia |
| Min Gao | Chongqing University, China |
| Ming Yang | Nanjing University, China |
| Minglai Shao | Tianjin University, China |
| Minglong Lei | Beijing University of Technology, China |
| Mingyu Yang | Hong Kong University of Science and Technology (Guangzhou), China |
| Mingyue Cheng | University of Science and Technology of China, China |
| Mingzhong Wang | University of the Sunshine Coast, China |
| Minh-Hao Van | University of Arkansas, USA |
| Minhua Lin | Pennsylvania State University, USA |
| Min-Te Sun | National Central University, Taiwan |
| Mohamad El-Hajj | MacEwan University, Canada |
| Mohammad M. Masud | United Arab Emirates University, United Arab Emirates |
| Morteza Saberi | University of Technology Sydney, Australia |
| Moshe Eliasof | University of Cambridge, UK |
| Mrinal Das | IIT Palakkad, India |
| Muhammad Abulaish | South Asian University, India |
| Muhammad Daffa Robani | AwanTunai, Indonesia |
| Nannan Wu | Tianjin University, China |
| Nassim Belmecheri | Simula Research Laboratory, Norway |

| | |
|---|---|
| Nathan S. Gavenski | King's College London, UK |
| Nian Liu | National University of Singapore, Singapore |
| Ning Liu | Shandong University, China |
| Ou Wu | Tianjin University, China |
| P. Radha Krishna | National Institute of Technology, Warangal, India |
| Panagiotis Liakos | Athens University of Economics and Business, Greece |
| Paolo Garza | Politecnico di Torino, Italy |
| Pedro G. Ferreira | University of Porto, Portugal |
| Peer Kröger | Christian Albrechts University Kiel, Germany |
| Peiliang Zhang | Wuhan University of Technology, China |
| Peng Li | Dalian University of Technology, China |
| Peng Qian | Zhejiang University, China |
| Pengfei He | University of Washington, USA |
| Pengpeng Zhao | Soochow University, China |
| Pengyang Wang | University of Macau, China |
| Pinhan Fu | Shanxi University, China |
| Pranjal Shankhdhar | Meta Platforms, USA |
| Praveen Kumar Donta | Stockholm University, Sweden |
| Primoz Kocbek | University of Maribor, Slovenia |
| Purushottam Panta | University of Kentucky, USA |
| Qian Li | Curtin University, Australia |
| Qian Sun | Hong Kong University of Science and Technology, China |
| Qiang Gao | Southwestern University of Finance and Economics, China |
| Qiang Tang | Luxembourg Institute of Science and Technology, Luxembourg |
| Qiange Wang | National University of Singapore, Singapore |
| Qianying Ren | University of Connecticut, USA |
| Qiao Xue | Nanjing University of Aeronautics and Astronautics, China |
| Qiheng Sun | Hong Kong Polytechnic University, China |
| Qing Liao | Harbin Institute of Technology (Shenzhen), Singapore |
| Qing Liu | Zhejiang University, China |
| Qing Tian | Nanjing University of Information Science and Technology, Singapore |
| Qingxiang Liu | Chinese Academy of Sciences, China |
| Qiqi Wang | Nankai University, China |
| Qisen Cheng | Samsung Display America Lab, USA |
| Qixu Gong | New Mexico State University, USA |
| Qizhou Wang | Monash University, Australia |

| | |
|---|---|
| Quangui Zhang | Liaoning Technical University, China |
| Qun Liu | Louisiana State University, USA |
| Rajatsubhra Chakraborty | University of North Carolina at Charlotte, USA |
| Ravi Chhabra | Nvidia, USA |
| Renchi Yang | Hong Kong Baptist University, China |
| Renqiang Luo | Jilin University, China |
| Robert J. Churchill | Georgetown University, USA |
| Rong Du | Taiyuan University of Technology, China |
| Rosalba Giugno | University of Verona, Italy |
| Ru Wang | Shandong Normal University, China |
| Ruiqi Zheng | University of Queensland, Australia |
| Ruliang Xiao | Fujian Normal University, China |
| Rundong Zuo | Hong Kong Baptist University, China |
| Runlong Yu | University of Alabama, USA |
| Sadok Ben Yahia | Southern Denmark University, Denmark |
| Senzhang Wang | Central South University, China |
| Shafiq Alam | Massey University Auckland, New Zealand |
| Shangqing Zhao | University of Oklahoma, Oklahoma |
| Shao-Yuan Li | Nanjing University of Aeronautics and Astronautics, China |
| Shengchao Chen | University of Technology Sydney, Australia |
| Shengda Zhuo | Jinan University, China |
| Shengkun Wang | Virginia Tech, USA |
| Shen-Shyang Ho | Rowan University, USA |
| Shichao Zhang | Central South University, China |
| Shifei Ding | China University of Mining and Technology, China |
| Shiyu Yang | Guangzhou University, China |
| Shi-Yu Tian | Nanjing University, China |
| Shridhar B Devamane | Amrita Vishwa Vidyapeetham, India |
| Shu Wang | Chinese University of Hong Kong (Shenzhen), China |
| Shuai Li | University of Cambridge, UK |
| Shubham Jain | Visa Research, USA |
| Shuichi Miyazawa | Graduate University for Advanced Studies, Japan |
| Shuo Li | Guangxi University, China |
| Shuo Yu | Dalian University of Technology, China |
| Siddhartha Bhattacharyya | University of Illinois Chicago, USA |
| Simone Monaco | Politecnico di Torino, Italy |
| Siqi Liang | Amazon, USA |
| Siqing Li | Unversity of New South Wales, Australia |

| | |
|---|---|
| Siru Zhong | Hong Kong University of Science and Technology (Guangzhou), China |
| Siyang Liu | Nanjing University, China |
| Song Wu | Aalborg University, Denmark |
| Sungsu Lim | Chungnam National University, South Korea |
| Tadashi Nomoto | National Institute of Japanese Literature, Japan |
| Tai Le Quy | University of Koblenz, Germany |
| Tao Chen | China University of Geosciences, Wuhan, China |
| Tao Meng | Central South University of Forestry and Technology, China |
| Tao Wang | University of Chinese Academy of Sciences, China |
| Tetsuya Yoshida | Nara Women's University, Japan |
| Thi Huyen Nguyen | L3S Research Center, Germany |
| Thilina N Ranbaduge | Australian National University, Australia |
| Tianrui Li | Southwest Jiaotong University, China |
| Tiantian Liu | Aalborg University, Denmark |
| Tianyi Li | Aalborg University, Denmark |
| Tiejin Chen | Arizona State University, USA |
| Tim Oates | University of Maryland, Baltimore County, USA |
| Tong Chen | University of Queensland, Australia |
| Trung-Hoang Le | Nanyang Technological University, Singapore |
| Tsunenori Mine | Kyushu University, Japan |
| Tuan Le | New Mexico State University, USA |
| Tung Kieu | Aalborg University, Denmark |
| Vahid Taslimitehrani | PhysioSigns Inc., USA |
| Vana Kalogeraki | Athens University of Economics and Business, Greece |
| Wanfu Gao | Jilin University, China |
| Wang Lizhen | Yunnan University, China |
| Waqas Jamil | Bournemouth University, UK |
| Wei Chen | Beihang University, China |
| Wei Wang | Fudan University, China |
| Wei Ye | Tongji University, China |
| Wei Emma Zhang | University of Adelaide, Australia |
| Weifeng Liu | China University of Petroleum, China |
| Weiwei Jiang | Beijing University of Posts and Telecommunications, China |
| Wenbo Shang | Hong Kong Baptist University, China |
| Wen-Chao Hu | Nanjing University, China |
| Wenda Zhang | Walmart Inc., USA |
| Wengen Li | Tongji University, China |

| | |
|---|---|
| Xinyu Su | University of Melbourne, Australia |
| Xinyu Wu | Baylor University, USA |
| Xinyue Sun | Harbin Institute of Technology, China |
| Xiuling Wang | Stevens Institute of Technology, USA |
| Xiwen Chen | Clemson University, USA |
| Xu Zhang | Macquarie University, Australia |
| Xuliang Zhu | Shanghai Jiao Tong University, China |
| Xuyun Zhang | Macquarie University, Australia |
| Yan Cao | University of Texas at Dallas, USA |
| Yan Ru Pei | NVIDIA, USA |
| Yanchao Liu | Communication University of China, China |
| Yanchao Tan | Fuzhou University, China |
| Yang Cao | Great Bay University/Tsinghua University, China |
| Yang-Sae Moon | Kangwon National University, South Korea |
| Yanhao Wang | East China Normal University, China |
| Yanwei Yu | Ocean University of China, China |
| Yaqian Zhang | University of Waikato, New Zealand |
| Yi Zhang | University of Queensland, Australia |
| Yibing Zhan | JD Explore Academy, China |
| Yifan Wang | University of Hawaii, USA |
| Yifeng Gao | University of Texas Rio Grande Valley, USA |
| Yi-Hui Chen | Chang Gung University, Taiwan |
| Yijun Lin | University of Minnesota, USA |
| Yimin Wen | Wuyi University, China |
| Yiming Tang | Hefei University of Technology, China |
| Yin Xian | Hong Kong Baptist University, China |
| Yingjie Wang | Nanyang Technological University, Singapore |
| Yingli Zhou | Chinese University of Hong Kong (Shenzhen), China |
| Yiran Li | University of Toronto, Canada |
| Yi-Shin Chen | National Tsing Hua University, Taiwan |
| Yixin Liu | Monash University, Australia |
| Yongcan Huang | University of Georgia, USA |
| Yongjie Wang | Nanyang Technological University, Singapore |
| Yongjing Hao | Suzhou University of Science and Technology, China |
| Yongmin Yoo | Macquarie University, Australia |
| Yongpan Sheng | Southwest University, China |
| Yongyu Wang | Michigan Technological University, USA |
| Yu Li | Hangzhou Dianzi University, China |
| Yu Liu | Beijing Jiaotong University, China |
| Yuanbo Xu | Jilin University, China |

| | |
|---|---|
| Yuanyuan Xu | University of New South Wales, Australia |
| Yuanyuan Zhu | Wuhan University, China |
| Yuchen Fang | University of Electronic Science and Technology of China, China |
| Yucong Dai | Clemson University, USA |
| Yudong Zhang | University of Science and Technology of China, China |
| Yue Deng | Michigan State University, USA |
| Yue Kou | Northeastern University, China |
| Yue Pang | Hong Kong University of Science and Technology, China |
| Yue Zhang | Chinese University of Hong Kong (Shenzhen), China |
| Yuefeng Li | Queensland University of Technology, Australia |
| Yue-Shi Lee | Ming Chuan University, Taiwan |
| Yuexuan An | Hohai University, China |
| Yufei Han | Inria, France |
| Yujie Wang | Hefei University of Technology, China |
| Yukun Yuan | University of Tennessee at Chattanooga, USA |
| Yulong Gu | Bytedance, China |
| Yun Li | Nanjing University of Posts and Telecommunications, China |
| Yuncheng Hua | University of New South Wales, Australia |
| Yunjun Gao | Zhejiang University, China |
| Yuping Wang | Xidian University, China |
| Yuqing Sun | Shandong University, China |
| Yuqing Zhou | George Mason University, USA |
| Yushan Jiang | University of Connecticut, USA |
| Yutong Feng | Hong Kong University of Science and Technology (Guangzhou), China |
| Yuwen Liu | China University of Petroleum (East China), China |
| Yuxiang Wang | Hangzhou Dianzi University, China |
| Yuyang Luo | Emory University, USA |
| Zee Hen Tang | Academia Sinica, Taiwan |
| Zeyu Wang | Zhejiang University of Technology, China |
| Zhanke Zhou | Hong Kong Baptist University, China |
| Zhaohong Deng | Jiangnan University, China |
| Zhen Wu | Nanjing University, China |
| Zhenfeng He | National University of Singapore, Singapore |
| Zheng Liu | Nanjing University of Posts and Telecommunications, China |
| Zhichao Huang | JD Intelligent Cities Research, China |

## External Reviewers

| | |
|---|---|
| Congning Ni | Vanderbilt University Medical Center, USA |
| Cristiano Mesquita Garcia | Instituto Federal de Santa Catarina, Brazil |
| Dazhi Fu | Chinese University of Hong Kong (Shenzhen), China |
| Enea Vincenzo Napolitano | University of Naples Federico II, Italy |
| Eric Inae | University of Notre Dame, USA |
| Fangwei Ye | Nanjing University of Aeronautics and Astronautics, China |
| Feng Xiao | Chinese University of Hong Kong (Shenzhen), China |
| Fuhong Song | Guizhou University of Finance and Economics, China |
| Gözde Tataroglu Özbulak | University of Lausanne, Switzerland |
| Hadi Al-Kadhimi | University of Birmingham, UK |
| Hao Yuan | Northeastern University, China |
| Hoang Hai Nguyen | University of Manitoba, Canada |
| Huan Wang | University of Wollongong, Australia |
| Huy-Son Nguyen | TU Delft, Netherlands |
| Hy Dang | University of Notre Dame, USA |
| Ichsan Budiman | University of Birmingham, UK |
| Jaime Hieu Do | Singapore Management University, Singapore |
| Jianzhe Li | East China Normal University, China |
| Jiaqi Zhao | China University of Mining and Technology, China |
| Kan Guo | Beijing University of Technology, China |
| Kexin Meng | Zhengzhou University, China |
| Liqun Deng | China University of Petroleum (East China), China |
| Lorenzo Peracchio | University of Pavia, Italy |
| Lorenzo Corso | University of Pavia, Italy |
| Ludovico Ferreri | University of Pavia, Italy |
| Lyuheng Yuan | Indiana University Bloomington, USA |
| Manuel Tognon | Verona University, Italy |
| Marco De Luca | Politecnico di Torino, Italy |
| Matteo Berta | Politecnico di Torino, Italy |
| Miguel Rodriguez | Google, USA |
| Mong Yuan Sim | Adelaide University, Australia |
| Moshe Eliasof | University of Cambridge, UK |
| Nan Zhang | East China Normal University, China |
| Padipat Sitkrongwong | National Institute of Informatics, Japan |
| Paolo Garza | Politecnico di Torino, Italy |
| Peiliang Zhang | Wuhan University of Technology, China |

| | |
|---|---|
| Pengpeng Qiao | Institute of Science Tokyo, Japan |
| Qi Wen | East China Normal University, China |
| Qing Tian | Nanjing University of Information Science and Technology, China |
| Qingyuan Song | Vanderbilt University, USA |
| Quanjiang Guo | University of Electronic Science and Technology of China, China |
| Renxuan Hou | Hong Kong Polytechnic University, China |
| Ritabrata Chakraborty | Manipal University Jaipur, India |
| Ruimin Dai | Clemson University, USA |
| Saranya Maneeroj | Chulalongkorn University, Thailand |
| Show-Jane Yen | Ming Chuan University, Taiwan |
| Siqing Li | University of New South Wales, Australia |
| Tingzheng Jia | Beijing University of Technology, China |
| Tung Doan | Hanoi University of Science and Technology, Vietnam |
| Vyshnavi Gutta | Georgia Institute of Technology, USA |
| Wajid Manzoor | Rowan University, USA |
| Weixin Liu | Vanderbilt University, USA |
| Wengang Guo | Tongji University, China |
| Wenke Du | Cornell University, USA |
| Xiaoyang Li | Adelaide University, Australia |
| Xin Ai | Northeastern University, China |
| Xuanting Xie | University of Electronic Science and Technology of China, China |
| Yanwei Xu | Peking University, China |
| Yijun Lin | University of Minnesota, USA |
| Ying Li | Northeastern University, China |
| Yiyang Zhang | East China Normal University, China |
| Yongmin Yoo | Macquarie University, Australia |
| Yue Deng | Michigan State University, USA |
| Yuhong Dai | Shenzhen University, China |
| Yunhe Feng | University of North Texas, USA |
| Yushan Jiang | University of Connecticut, USA |
| Yutong Ye | Beihang University, China |
| Zhen Wu | East China Normal University, China |
| Zhiyu Chen | Chinese University of Hong Kong (Shenzhen), China |
| Zifeng Zhou | Jinan University, China |

## Sponsors

### Exclusive Diamond Sponsor

王寬誠教育基金會
K. C. WONG EDUCATION FOUNDATION

### Gold Sponsors

香港科技大學
THE HONG KONG UNIVERSITY OF SCIENCE AND TECHNOLOGY

AsianArk

### Silver Sponsor

THE HONG KONG POLYTECHNIC UNIVERSITY
香港理工大學

**Bronze Sponsors**

香港浸會大學
HONG KONG BAPTIST UNIVERSITY

香港教育大學
The Education University of Hong Kong

香港中文大學（深圳）
The Chinese University of Hong Kong, Shenzhen

**Publication**

Springer

# Contents

# Engineering Multi-Agent Systems

# Estimating Subgraph Importance with Structural Prior Domain Knowledge

Changhyun Kim[1], Seunghwan An[2], and Jong-June Jeon[1(✉)]

[1] University of Seoul, Seoul, South Korea
{ch.kim,jj.jeon}@uos.ac.kr
[2] Incheon National University, Incheon, South Korea
sh.an@inu.ac.kr

**Abstract.** We propose a subgraph importance estimation method for pretrained Graph Neural Networks (GNNs) on graph-level tasks, formulated as a linear Group Lasso regression problem in the embedding space. Our method effectively leverages prior domain knowledge of graph substructures, while remaining independent of the specific form of the output layer or readout function used in the GNN architecture, and it does not require access to ground-truth target labels. Experiments on real-world graph datasets demonstrate that our method consistently outperforms existing baselines in subgraph importance estimation. Furthermore, we extend our method to identify important nodes within the graph.

**Keywords:** Subgraph importance estimation · Graph Neural Networks · Linear Group Lasso regression · Embedding space · Graph mining

## 1 Introduction

Graph Neural Networks (GNNs) have emerged as a specialized model for analyzing graph-structured data due to their ability to effectively capture both node features and the connectivity relationships between nodes via edges [10,14,24]. Specifically, the post-hoc approaches providing an interpretation of graphs via fitted node feature matrices have attracted considerable attention due to their simplicity and independence from specific model architectures.

In this context, GraphLIME [10] and STGraphLIME [21] propose local explanation methods that identify influential nodes or subgraphs based on the Hilbert-Schmidt Independence Criterion (HSIC) Lasso [26]. SubgraphX [29] extends the Shapley value [12] to the graph domain by computing the contributions of connected subgraphs, thereby providing subgraph-level explanations for GNN predictions. The aforementioned models retrospectively estimate node importance by quantifying changes in the GNN's outputs with respect to variations in graph-level features.

C. Kim, S. An—Both authors contributed equally to the paper.


R. C.-W. Wong et al. (Eds.): PAKDD 2026, LNAI 16597, pp. 3–15, 2026.
https://doi.org/10.1007/978-981-92-1300-9_1

However, existing explanation methods for GNN models are limited in their ability to incorporate prior knowledge, resulting in the frequent identification of important subgraphs that are not accepted by domain experts. For instance, subgraphs such as ecological clusters, functional components in engineered systems, and functional groups or motifs in molecular structures are known to determine key graph-level properties [5,29]. This observation indicates that subgraph-level explanations can provide a more domain-aligned understanding of model behavior. Furthermore, the permutation-invariant readout function in GNNs often makes the contribution of each node embedding to the final network output identical, which hinders the development of gradient-based methods.

To address these two issues of post-hoc GNN explanation methods, we propose a simple and effective domain-aligned important subgraph identification method based on a structural sparsity prior, *Group Lasso* [30]. Our method can incorporate domain-specific structural knowledge through the predefined groups. In addition, since it is independent of the GNN architecture (such as the readout function and output layer), it can be easily applied to various GNN models, aiming to evaluate the importance of substructures of a graph. To evaluate our method's effectiveness, we employ 13 real-world graph datasets focusing on molecular property prediction, comparing the results against ground-truth where available. Our method is referred to as eXplainable Embedding-Space Group Lasso regression for GNN (**eXEL**[1]).

## 2 Related Works

Existing approaches for GNN explanation and importance estimation can be categorized into four groups: surrogate-, decomposition-, perturbation-, and gradient-based methods. Surrogate methods fit simple proxy models to the local decision boundary of a GNN; for example, GraphLIME [10] applies kernel regression with HSIC Lasso on a node's neighborhood to identify important features, RelEx [32] trains a Graph Convolution Network (GCN) surrogate on sampled subgraphs to learn explanation masks, and PGM-Explainer [22] captures dependencies between features and predictions using a probabilistic graphical model. Decomposition methods redistribute the model's output to input components. LRP [3] propagates relevance scores through layers, ExcitationBP [16] redistributes relevance via conditional likelihood, and GNN-LRP [20] attributes importance along multi-hop walks via higher-order Taylor expansion. Perturbation-based and gradient-based methods estimate importance by modifying inputs or analyzing gradients. Perturbation methods such as GNNExplainer [28], PGExplainer [13], GraphMask [19], and SubgraphX [29] assess the sensitivity of predictions to feature or edge changes. Gradient- and feature-based methods, including Saliency Analysis, Guided Backpropagation, CAM, and Grad-CAM [3,16], estimate importance directly from gradients or feature maps.

[1] Our implementation code and Appendix are available at:https://github.com/Nempet9398/graph_xai.

# 3 Preliminaries

**Notations.** Consider a graph consisting of $n$ nodes, denoted by $G = (X, A)$, where $X$ is the node feature matrix whose rows represent node attributes, and $A$ is the adjacency matrix. A labeled dataset is given by $\mathcal{D} = \{(G^{(i)}, y^{(i)})\}_{i=1}^{N} = \{(X^{(i)}, A^{(i)}, y^{(i)})\}_{i=1}^{N}$. To represent a subgraph of $G$, we introduce $\mathcal{P}_G$, an $m$-partition of the node index set $\{1, 2, \ldots, n\}$, and let $I_s \in \mathcal{P}_G$ $(s = 1, 2, \ldots, m)$ denote the index set of nodes belonging to the $s$th subgraph.

**Graph Neural Networks.** Let the composition of the embedding layer and the graph convolution function be denoted by $f$, which outputs the embedding matrix of a given graph $G$ lying in $\mathbb{R}^d$ as follows: $f : \mathbb{R}^{n \times p} \times \{0, 1\}^{n \times n} \mapsto \mathbb{R}^{n \times d}$. Here, the first dimension, corresponding to the number of nodes, remains invariant. The readout function, denoted as $h$, produces a $d$-dimensional graph-level representation by applying a row-wise permutation-invariant aggregation to the output of $f$, such as a column-wise mean, sum, or maximum (we will demonstrate in Sect. 5 that our proposed method is applicable to all these cases). Lastly, the final output layer $h_{\text{out}} : \mathbb{R}^d \mapsto \mathcal{Y}$ determines the form of the output space $\mathcal{Y}$ (e.g., class probabilities for classification tasks). Then, the GNN model is trained by the following objective:

$$\min_{h_{\text{out}}, f} \quad \frac{1}{N} \sum_{i=1}^{N} \mathcal{L}\Big(h_{\text{out}}\big((h \circ f)(X^{(i)}, A^{(i)})\big), y^{(i)}\Big), \tag{1}$$

where $\mathcal{L} : \mathcal{Y} \times \mathcal{Y} \mapsto \mathbb{R}^+$ is the loss function corresponding to the target task (e.g., cross-entropy loss for classification), and $y^{(i)} \in \mathcal{Y}$ denotes the target label.

# 4 Proposal: eXEL

Our goal is to develop an importance estimation method that assigns continuous importance scores to all subgraphs for a GNN model, which is pretrained according to (1). We introduce the concept of subgraph-level sparsity and define importance as the norm of the subgraph feature.

## 4.1 Motivation

In GNN, the permutation-invariant readout function is widely used to aggregate node features into a fixed-size graph-level representation. While this readout function preserves the isomorphic property of graphs, the node contributions are pooled, which hinders the explanation of graphs on the node level. Figure 1 (a) and (b)

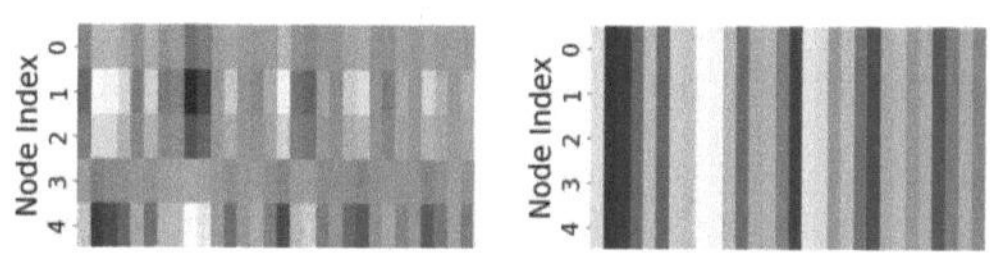


(a) Node embeddings (b) Gradients

**Fig. 1.** (a): Node embedding matrix. (b): Gradient values with respect to the node embeddings.

show the node embedding matrix and the gradient of the node embeddings. Since all gradient vectors for each node embedding are equal, it is difficult to directly apply gradient-based attribution methods.

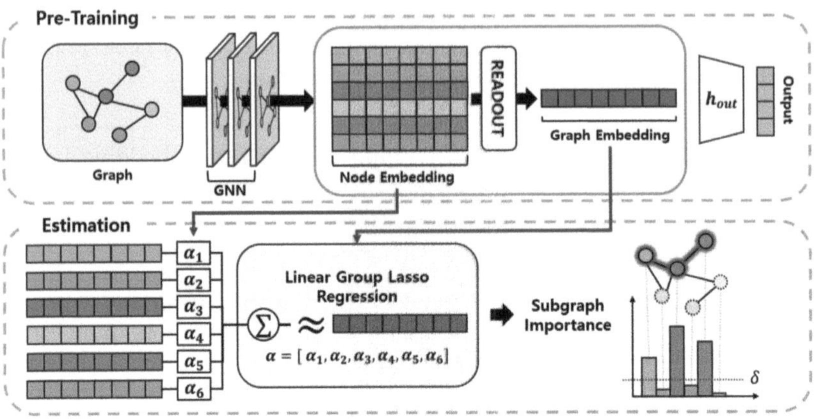


**Fig. 2.** Overall process of our proposed subgraph importance estimation method.

Regardless of the pretraining task (i.e., independent of the choice of the output layer $h_{\text{out}}$ or the output space $\mathcal{Y}$), the graph-level embedding, $(h \circ f)(X, A)$, serves as the *optimal* $d$-dimensional representation of the given graph, in the sense that the pretrained GNN model achieves its best performance on the graph-level task using this embedding vector. In addition, since the readout function aggregates node embeddings along each embedding dimension, a mapping exists for the same graph-level representations when the number of nodes is large. If the prior knowledge restricts a partition of subgraphs, $\mathcal{P}_G$, the same graph-level representation within the constraint can be reasonably interpreted in the context of a specific data domain.

Thus, our approach is guided by the motivation that the explanation for a graph should be as structurally simple as possible under the prior domain knowledge $\mathcal{P}_G$. Let $\alpha = (\alpha_1, \cdots, \alpha_n)^\top$ and a subvector of $\alpha$ be $\alpha_s = (\alpha_k : k \in I_s \in \mathcal{P}_G)$, then our ideal objective function is given by $\max_{\alpha \in \mathbb{R}^n} \sum_{s=1}^{m} \mathbb{I}(\|\alpha_s\| = 0)$ with the constraint $(h \circ f)(X, A) = \alpha^\top f(X, A)$. Since this constraint is generally infeasible for non-linear readout functions, we set our objective function by

$$\max_{\alpha \in \mathbb{R}^n} \sum_{s=1}^{m} \mathbb{I}(\|\alpha_s\| = 0) \quad \text{subject to } \left\|(h \circ f)(X, A) - \alpha^\top f(X, A)\right\| < \epsilon, \tag{2}$$

where $\mathbb{I}(\cdot)$ is indicator function, $\|\cdot\|$ denotes the $L_2$ norm, and $\epsilon$ is a positive tuning parameter. The constraint in (2) can be interpreted as a well-posed regression problem when $d > n$, where $d$ and $n$ correspond to the number of samples and the number of covariates in linear regression framework, respectively (in our experiments, we set $d = 256$, and the average number of nodes was 26;

see the Appendix for detailed dataset descriptions). In the evaluation, we used mean and sum readout functions, which are commonly employed for molecular graph datasets [4], as well as the max operation for a dimension-wise nonlinear readout function.

## 4.2 Subgraph Importance Estimation

Since (2) is a nonconvex combinatorial problem, we relax this objective function using the Group Lasso penalty and define, for each graph, the corresponding optimization problem as follows:

$$\min_{\alpha \in \mathbb{R}^n} \quad \left\| (h \circ f)(X, A) - \alpha^\top f(X, A) \right\|^2 + \lambda \sum_{s=1}^{m} \|\alpha_s\|, \tag{3}$$

where $\lambda > 0$ is a hyperparameter that controls the degree of sparsity. From (3), we can observe the following advantages:

1. Domain knowledge of graph substructures is incorporated into the sparsity penalty by grouping nodes that belong to each known subgraph unit.
2. Our method is independent of the GNN architecture, including the readout function, output layer, and target space.

**Assumption 1.** *The loss function $\mathcal{L}$ used for the pre-training of a GNN model is $K$-Lipschitz with respect to its first argument if, for all graph-level embedding vectors $\mathbf{z}_1, \mathbf{z}_2 \in \mathbb{R}^d$ and $y \in \mathcal{Y}$, $\left|\mathcal{L}(\mathbf{z}_1, y) - \mathcal{L}(\mathbf{z}_2, y)\right| \leq K\|\mathbf{z}_1 - \mathbf{z}_2\|$, where $K > 0$.*

Without loss of generality, the loss function $\mathcal{L}$ in (1) can be rewritten for a fixed GNN model as $\mathcal{L}((h \circ f)(X, A), y)$, where the output layer $h_{\text{out}}$ is considered to be absorbed into the loss function $\mathcal{L}$. Note that many loss functions commonly used in graph-level tasks (e.g., cross-entropy [17,23,27,31], hinge loss [7]) satisfy Assumption 1. Then, under Assumption 1,

$$\left|\mathcal{L}((h \circ f)(X, A), y) - \mathcal{L}(\alpha^\top f(X, A), y)\right|^2 \leq K^2 \left\|(h \circ f)(X, A) - \alpha^\top f(X, A)\right\|^2. \tag{4}$$

Since our optimization problem (3) is a well-posed linear regression problem with a large number of nodes, it can yield group coefficients $\alpha_s$ that produce nearly identical graph-level representations. Moreover, reliable $\alpha_s$ can be obtained under Assumption 1, as (4) ensures similar loss values (note that the right-hand side of (4) corresponds to the first term in our objective (3)). Therefore, our proposed method can structurally select subgraphs that yield the same model prediction, guided by prior domain knowledge.

After solving the optimization problem in (3), we transform the resulting score vector $\alpha$ to obtain the estimate of subgraph importance $\beta = (\beta_1, \cdots, \beta_m)$:

$$\beta_s := \frac{1}{|I_s|} \sum_{l \in I_s} \alpha_l, \quad \text{for } s = 1, 2, \cdots, m,$$

which serves as an averaging step to ensure consistent importance values among nodes within the same subgraph. Then, we define the set of important subgraphs $\mathcal{S}$ in graph $G$ as those whose absolute scores exceed a threshold $\delta > 0$, which filters out subgraphs with negligible weights (in practice, we set $\delta = 10^{-5}$):

$$\mathcal{S} := \{s \in \{1, 2, \cdots, m\} \mid \delta < |\beta_s|\}.$$

The overall process of our proposed method is illustrated in Fig. 2.

## 5 Experiments

We evaluate eXEL by addressing the following research questions:

1. **Estimation performance:** Can eXEL accurately identify important subgraphs?
2. **Ablation study:** Does the Group Lasso penalty effectively incorporate domain-specific prior knowledge?
3. **Model analysis:** How does reconstructing the original graph-level embedding contribute to importance estimation?
4. **Scalability:** Can eXEL be extended to node-level importance estimation?

### 5.1 Overview

**Datasets.** Following prior studies [28,29], we evaluate our method on molecular datasets that exhibit interpretable subgraph structures such as functional groups. We use `MUTAG` [15], `Benzene(BENZ)`, `Alkane`, `Fluoride(FLUOR)` [1,18], `Ames` [9], `BACE`, and `BBBP` [8], where atomic features represent node attributes. Motif structures (i.e., predefined groups) are extracted using BRICS decomposition [5] and tree decomposition [33]. Among these, `Benzene`, `Alkane`, and `Fluoride` provide ground-truth annotations for subgraph-level importance. Each dataset is randomly split into 80% training and 20% test, and the optimal $\lambda$ is selected via 4-fold cross-validation.

**Baselines.** Following prior work [29], we train four widely used GNN architectures: GCN, GAT, GIN, and GraphSAGE (SAGE). We also experiment with three types of readout functions (mean, sum, and max). Each GNN architecture is trained using the cross-entropy loss, which is Lipschitz continuous (i.e., Assumption 1), and the following baseline explanation methods are applied to the fitted GNNs: PGExplainer (PGExp) [13], GNNExplainer (GNNExp) [28], GradCAM [16], GraphMask [19], Guided Backpropagation (GBP) [16], Saliency Maps (SA) [16], and SubgraphX [29]. Each method generates importance scores over nodes and averages them to provide a subgraph-level explanation. Further details about the baseline models and configurations are provided in the Appendix.

**Evaluation.** We adopt a *fidelity metric* to measure the influence of the identified subgraphs on model predictions. Extending the metric proposed by [2], we compute the drop in $F_1$-score as $\texttt{Fidelity}_{F_1} = F_1(G) - F_1(G \setminus S)$, where $S$ is the selected subset of nodes, $F_1(G)$ denotes the $F_1$-score on the original graph, and $F_1(G \setminus S)$ is obtained by zeroing out the features of nodes in $S$. A higher fidelity score indicates that the identified subset of nodes has a stronger influence, i.e., greater importance. All fidelity scores are averaged over different readout functions.

## 5.2 Main Results

**Estimation Performance.** As in [29], to ensure a fair comparison, we control the sparsity of the explanation by selecting subgraphs to achieve total coverage slightly below 70%. Table 1 presents the subgraph-level experiment results. Our

**Table 1.** Subgraph-level $\texttt{Fidelity}_{F_1}$ scores. The best result is highlighted with color and the second-best is underlined. Higher is better.

| model | data | eXEL | PGExp | GNNExp | GradCAM | GraphMask | GBP | SA | SubgraphX |
|---|---|---|---|---|---|---|---|---|---|
| GCN | MUTAG | $.158_{\pm.140}$ | $.000_{\pm.094}$ | $.151_{\pm.122}$ | $.001_{\pm.068}$ | $.009_{\pm.075}$ | $.031_{\pm.181}$ | $.020_{\pm.121}$ | $.243_{\pm.297}$ |
| | alkane | $.325_{\pm.138}$ | $.258_{\pm.192}$ | $.277_{\pm.180}$ | $.321_{\pm.191}$ | $.097_{\pm.094}$ | $.133_{\pm.127}$ | $.225_{\pm.111}$ | $.080_{\pm.086}$ |
| | ames | $.155_{\pm.155}$ | $.024_{\pm.044}$ | $.082_{\pm.057}$ | $.009_{\pm.039}$ | $.037_{\pm.042}$ | $.026_{\pm.032}$ | $.044_{\pm.070}$ | $.194_{\pm.298}$ |
| | bace | $.243_{\pm.038}$ | $.088_{\pm.051}$ | $.165_{\pm.057}$ | $.136_{\pm.058}$ | $.071_{\pm.049}$ | $.116_{\pm.053}$ | $.119_{\pm.070}$ | $.108_{\pm.062}$ |
| | bbbp | $.090_{\pm.041}$ | $.041_{\pm.030}$ | $.069_{\pm.026}$ | $.044_{\pm.054}$ | $.043_{\pm.034}$ | $.054_{\pm.034}$ | $.068_{\pm.028}$ | $.055_{\pm.045}$ |
| | BENZ | $.196_{\pm.112}$ | $.076_{\pm.130}$ | $.156_{\pm.044}$ | $.075_{\pm.060}$ | $.073_{\pm.031}$ | $.147_{\pm.080}$ | $.095_{\pm.056}$ | $.112_{\pm.034}$ |
| | FLUOR | $.323_{\pm.281}$ | $.171_{\pm.303}$ | $.142_{\pm.183}$ | $.122_{\pm.084}$ | $.105_{\pm.120}$ | $.217_{\pm.185}$ | $.218_{\pm.316}$ | $.104_{\pm.101}$ |
| GAT | MUTAG | $.138_{\pm.152}$ | $.006_{\pm.059}$ | $.044_{\pm.063}$ | $.016_{\pm.052}$ | $.034_{\pm.042}$ | $.058_{\pm.058}$ | $.035_{\pm.198}$ | $.142_{\pm.100}$ |
| | alkane | $.455_{\pm.198}$ | $.164_{\pm.169}$ | $.336_{\pm.149}$ | $.409_{\pm.149}$ | $.144_{\pm.099}$ | $.482_{\pm.269}$ | $.073_{\pm.074}$ | $.073_{\pm.306}$ |
| | ames | $.168_{\pm.046}$ | $.074_{\pm.053}$ | $.146_{\pm.014}$ | $.035_{\pm.019}$ | $.072_{\pm.041}$ | $.100_{\pm.053}$ | $.099_{\pm.046}$ | $.014_{\pm.039}$ |
| | bace | $.247_{\pm.055}$ | $.121_{\pm.107}$ | $.222_{\pm.075}$ | $.123_{\pm.080}$ | $.158_{\pm.068}$ | $.188_{\pm.107}$ | $.149_{\pm.062}$ | $.131_{\pm.073}$ |
| | bbbp | $.079_{\pm.036}$ | $.036_{\pm.025}$ | $.068_{\pm.027}$ | $.026_{\pm.017}$ | $.049_{\pm.037}$ | $.043_{\pm.025}$ | $.043_{\pm.027}$ | $.043_{\pm.027}$ |
| | BENZ | $.578_{\pm.144}$ | $.062_{\pm.050}$ | $.163_{\pm.049}$ | $.045_{\pm.054}$ | $.105_{\pm.076}$ | $.166_{\pm.049}$ | $.158_{\pm.058}$ | $.558_{\pm.144}$ |
| | FLUOR | $.587_{\pm.182}$ | $.177_{\pm.232}$ | $.316_{\pm.135}$ | $.317_{\pm.232}$ | $.356_{\pm.099}$ | $.517_{\pm.233}$ | $.244_{\pm.216}$ | $.538_{\pm.182}$ |
| GIN | MUTAG | $.139_{\pm.175}$ | $.005_{\pm.027}$ | $.010_{\pm.162}$ | $.095_{\pm.210}$ | $.006_{\pm.090}$ | $.013_{\pm.047}$ | $.074_{\pm.138}$ | $.141_{\pm.251}$ |
| | alkane | $.406_{\pm.370}$ | $.158_{\pm.325}$ | $.147_{\pm.174}$ | $.424_{\pm.372}$ | $.112_{\pm.078}$ | $.246_{\pm.241}$ | $.381_{\pm.245}$ | $.041_{\pm.065}$ |
| | ames | $.242_{\pm.060}$ | $.062_{\pm.045}$ | $.109_{\pm.050}$ | $.144_{\pm.086}$ | $.103_{\pm.041}$ | $.100_{\pm.054}$ | $.176_{\pm.089}$ | $.174_{\pm.112}$ |
| | bace | $.228_{\pm.081}$ | $.089_{\pm.064}$ | $.206_{\pm.066}$ | $.218_{\pm.064}$ | $.112_{\pm.060}$ | $.185_{\pm.063}$ | $.216_{\pm.109}$ | $.162_{\pm.073}$ |
| | bbbp | $.111_{\pm.047}$ | $.032_{\pm.023}$ | $.061_{\pm.016}$ | $.032_{\pm.021}$ | $.047_{\pm.017}$ | $.032_{\pm.021}$ | $.036_{\pm.026}$ | $.034_{\pm.035}$ |
| | BENZ | $.388_{\pm.095}$ | $.057_{\pm.066}$ | $.160_{\pm.087}$ | $.206_{\pm.162}$ | $.075_{\pm.016}$ | $.152_{\pm.088}$ | $.331_{\pm.088}$ | $.168_{\pm.053}$ |
| | FLUOR | $.596_{\pm.190}$ | $.227_{\pm.284}$ | $.399_{\pm.194}$ | $.598_{\pm.212}$ | $.216_{\pm.115}$ | $.439_{\pm.275}$ | $.598_{\pm.321}$ | $.146_{\pm.180}$ |
| SAGE | MUTAG | $.200_{\pm.253}$ | $.012_{\pm.039}$ | $.075_{\pm.091}$ | $.091_{\pm.205}$ | $.031_{\pm.041}$ | $.038_{\pm.067}$ | $.060_{\pm.089}$ | $.258_{\pm.312}$ |
| | alkane | $.587_{\pm.277}$ | $.127_{\pm.204}$ | $.244_{\pm.125}$ | $.532_{\pm.357}$ | $.112_{\pm.057}$ | $.294_{\pm.245}$ | $.352_{\pm.195}$ | $.078_{\pm.065}$ |
| | ames | $.265_{\pm.089}$ | $.052_{\pm.027}$ | $.160_{\pm.073}$ | $.139_{\pm.105}$ | $.098_{\pm.098}$ | $.147_{\pm.091}$ | $.212_{\pm.068}$ | $.212_{\pm.069}$ |
| | bace | $.240_{\pm.083}$ | $.100_{\pm.047}$ | $.198_{\pm.083}$ | $.192_{\pm.065}$ | $.132_{\pm.080}$ | $.217_{\pm.090}$ | $.160_{\pm.049}$ | $.160_{\pm.130}$ |
| | bbbp | $.115_{\pm.076}$ | $.044_{\pm.018}$ | $.060_{\pm.032}$ | $.027_{\pm.022}$ | $.037_{\pm.039}$ | $.061_{\pm.050}$ | $.026_{\pm.046}$ | $.047_{\pm.017}$ |
| | BENZ | $.370_{\pm.143}$ | $.053_{\pm.053}$ | $.161_{\pm.085}$ | $.150_{\pm.150}$ | $.065_{\pm.024}$ | $.161_{\pm.127}$ | $.281_{\pm.159}$ | $.149_{\pm.052}$ |
| | FLUOR | $.665_{\pm.218}$ | $.239_{\pm.265}$ | $.433_{\pm.162}$ | $.459_{\pm.311}$ | $.227_{\pm.085}$ | $.480_{\pm.224}$ | $.571_{\pm.266}$ | $.206_{\pm.130}$ |

method ranks first in 20 out of 28 settings (71.4%; 4 GNN models × 7 datasets) and, notably, achieves either first or second place in *all* scenarios. **These results show that our method outperforms existing baselines by effectively identifying important subgraphs.** SubgraphX shows the next best performance.

**Ablation Study: Group Lasso vs Lasso.** To further evaluate whether the application of the Group Lasso penalty effectively integrates substructure domain knowledge, we conduct an ablation study by replacing the Group Lasso penalty with a simple Lasso penalty, $\lambda\|\alpha\|_1$, where $\|\cdot\|_1$ denotes the $L_1$-norm. This variant, denoted as $\text{eXEL}_{\text{node}}$, no longer leverages substructure domain knowledge and computes subgraph-level importance by averaging node-level importance.

The ablation results in Table 2 show that using the Group Lasso penalty achieves higher fidelity metric scores than using the simple Lasso penalty in almost all settings (82.1%, 23 out of 28 scenarios). **This implies that the Group Lasso penalty effectively leverages prior domain knowledge when optimizing the score vector $\alpha$.** This supports the advantage of eXEL, where domain knowledge of graph substructures is incorporated into the sparsity penalty by grouping nodes that belong to each predefined subgraph unit.

**Table 2.** Ablation study: Effect of the Group Lasso penalty on subgraph-level $\texttt{Fidelity}_{\text{F}_1}$ scores. 'Improve(%)' denotes the relative performance difference between the two penalties, and the color entries highlight positive improvements.

| model | penalty | MUTAG | alkane | ames | bace | bbbp | BENZ | FLUOR |
|---|---|---|---|---|---|---|---|---|
| GCN | Lasso | $.115_{\pm.133}$ | $.473_{\pm.163}$ | $.148_{\pm.057}$ | $.209_{\pm.046}$ | $.091_{\pm.034}$ | $.180_{\pm.140}$ | $.217_{\pm.216}$ |
| | GroupLasso | $.158_{\pm.140}$ | $.325_{\pm.138}$ | $.155_{\pm.155}$ | $.243_{\pm.038}$ | $.090_{\pm.041}$ | $.196_{\pm.112}$ | $.323_{\pm.281}$ |
| | Improve(%) | **+37.4** | −31.3 | **+4.7** | **+16.3** | −1.1 | **+8.9** | **+48.8** |
| GAT | Lasso | $.106_{\pm.100}$ | $.411_{\pm.106}$ | $.152_{\pm.046}$ | $.225_{\pm.073}$ | $.088_{\pm.024}$ | $.538_{\pm.169}$ | $.484_{\pm.158}$ |
| | GroupLasso | $.138_{\pm.152}$ | $.455_{\pm.198}$ | $.168_{\pm.046}$ | $.247_{\pm.055}$ | $.079_{\pm.036}$ | $.578_{\pm.144}$ | $.587_{\pm.182}$ |
| | Improve(%) | **+30.2** | **+10.7** | **+10.5** | **+9.8** | −10.2 | **+7.4** | **+21.3** |
| GIN | Lasso | $.085_{\pm.125}$ | $.321_{\pm.248}$ | $.206_{\pm.092}$ | $.205_{\pm.071}$ | $.113_{\pm.078}$ | $.352_{\pm.143}$ | $.576_{\pm.271}$ |
| | GroupLasso | $.139_{\pm.175}$ | $.406_{\pm.370}$ | $.242_{\pm.060}$ | $.228_{\pm.081}$ | $.111_{\pm.047}$ | $.388_{\pm.095}$ | $.596_{\pm.190}$ |
| | Improve(%) | **+63.5** | **+26.5** | **+17.5** | **+11.2** | −1.8 | **+10.2** | **+3.5** |
| SAGE | Lasso | $.181_{\pm.221}$ | $.379_{\pm.281}$ | $.204_{\pm.079}$ | $.248_{\pm.081}$ | $.113_{\pm.035}$ | $.256_{\pm.176}$ | $.501_{\pm.248}$ |
| | GroupLasso | $.200_{\pm.253}$ | $.587_{\pm.277}$ | $.265_{\pm.089}$ | $.240_{\pm.083}$ | $.115_{\pm.076}$ | $.370_{\pm.143}$ | $.665_{\pm.218}$ |
| | Improve(%) | **+10.5** | **+54.9** | **+29.9** | −3.2 | **+1.8** | **+44.5** | **+32.7** |

**Model Analysis: Reconstruction Strategy.** Building on the previous ablation study results, we further investigate the role of the reconstruction loss through our variant, $\text{eXEL}_{\text{node}}$. When ground-truth labels are available in binary form, we treat the importance estimation task as binary classification using ROC-AUC and PR-AUC as evaluation metrics. To ensure that pre-trained GNN models capture meaningful graph structures, we evaluate only those that achieve high fidelity with the ground-truth important subgraphs. Table 3 reports the average scores across the datasets with ground-truth labels, where our method consistently achieves the best performance in both metrics. This demonstrates that **even without prior domain knowledge (i.e., the Group Lasso penalty), our reconstruction strategy can *partially* identify the ground-truth important subgraphs.**

**Table 3.** Evaluation of importance estimation against ground-truth subgraph importance. The best result is highlighted with color and the second-best is underlined

| model | PR-AUC | ROC-AUC |
|---|---|---|
| GNNExp | $0.574_{\pm 0.038}$ | $0.596_{\pm 0.022}$ |
| PGExp | $0.404_{\pm 0.074}$ | $0.350_{\pm 0.284}$ |
| GradCAM | $0.478_{\pm 0.102}$ | $0.526_{\pm 0.237}$ |
| SA | <u>$0.619_{\pm 0.107}$</u> | <u>$0.689_{\pm 0.200}$</u> |
| GraphMask | $0.386_{\pm 0.016}$ | $0.524_{\pm 0.010}$ |
| GBP | $0.487_{\pm 0.116}$ | $0.520_{\pm 0.106}$ |
| $\text{eXEL}_{\text{node}}$ | $0.677_{\pm 0.086}$ | $0.744_{\pm 0.102}$ |

However, this partial identification occurs because the reconstruction strategy alone cannot identify *connected* subgraph-level structures, which domain experts consider semantically valid. We visualize the identified important node subsets in Fig. 3. As shown, $\text{eXEL}_{\text{node}}$, unlike eXEL, fails to identify important node subsets that form connected subgraph units. This implies that the Group Lasso penalty reinforces the reconstruction strategy by ensuring the identification of semantically valid subgraphs.

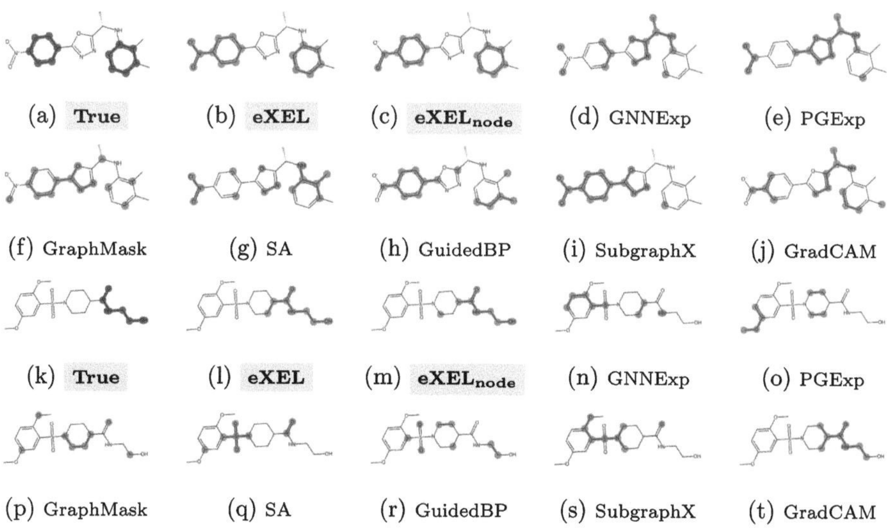

(a) **True** (b) **eXEL** (c) $\textbf{eXEL}_{\textbf{node}}$ (d) GNNExp (e) PGExp
(f) GraphMask (g) SA (h) GuidedBP (i) SubgraphX (j) GradCAM
(k) **True** (l) **eXEL** (m) $\textbf{eXEL}_{\textbf{node}}$ (n) GNNExp (o) PGExp
(p) GraphMask (q) SA (r) GuidedBP (s) SubgraphX (t) GradCAM

**Fig. 3.** Identified important node subsets (top two rows: `Benzene`; bottom two rows: `Fluoride`). The tasks are to determine whether a molecule contains a benzene ring, a fluoride group, or a carbonyl group.

**Scalability: Node-Level Estimation.** We evaluate our variant, $\text{eXEL}_{\text{node}}$, to verify whether our method can be extended to node-level importance estimation. We use IMDB-BINARY (IMDB), ENZYMES, Protein (PROTE) [15], and UPFD [6] for real-world graphs, and BAMultiShapes (MULTI) and BA2-Motif (MOTIF) for synthetic ones. For datasets with available node features, we use the original features; for those without, we adopt one-hot encoding of node degrees [25].

As in the graph-level experiments, sparsity is similarly controlled by retaining only the top 30% of nodes based on the importance scores from each method. Table 4 presents the node-level experiment results, where the scores are averaged across different readout functions. For the fidelity metric, our method ranks first in 12 out of 24 settings (50%, 4 GNN models × 6 datasets), and also achieves either first or second place in 22 scenarios (91.7%). This result demonstrates that **our variant, $\text{eXEL}_{\text{node}}$, can not only be extended to node-level importance estimation but also consistently achieves notable performance compared to other baselines.** On synthetic datasets (MOTIF and MULTI), SubgraphX generally performs better than other methods, whereas our $\text{eXEL}_{\text{node}}$ demonstrates strong performance on real-world datasets.

**Table 4.** Node-level $\text{Fidelity}_{\text{F}_1}$ scores. The best result is highlighted with color and the second-best is underlined. Higher is better.

| model | data | $\text{eXEL}_{\text{node}}$ | GNNExp | GradCAM | GraphMask | GBP | SA | SubgraphX | PGExp |
|---|---|---|---|---|---|---|---|---|---|
| GCN | ENZYME | $.150_{\pm.079}$ | $.083_{\pm.078}$ | $.113_{\pm.096}$ | $.049_{\pm.063}$ | $.109_{\pm.097}$ | $.147_{\pm.075}$ | $.072_{\pm.113}$ | $.066_{\pm.110}$ |
| | IMDB | $.021_{\pm.055}$ | $.024_{\pm.034}$ | $.008_{\pm.039}$ | $.010_{\pm.033}$ | $.006_{\pm.046}$ | $.000_{\pm.049}$ | $.012_{\pm.039}$ | $.013_{\pm.034}$ |
| | PROTE | $.103_{\pm.089}$ | $.061_{\pm.057}$ | $.059_{\pm.053}$ | $.061_{\pm.057}$ | $.055_{\pm.066}$ | $.054_{\pm.069}$ | $.089_{\pm.077}$ | $.007_{\pm.051}$ |
| | UPFD | $.100_{\pm.052}$ | $.007_{\pm.023}$ | $.032_{\pm.026}$ | $.019_{\pm.020}$ | $.024_{\pm.024}$ | $.010_{\pm.018}$ | $.012_{\pm.031}$ | $.002_{\pm.013}$ |
| | MOTIF | $.659_{\pm.232}$ | $.226_{\pm.097}$ | $.243_{\pm.156}$ | $.244_{\pm.142}$ | $.490_{\pm.328}$ | $.363_{\pm.325}$ | $.791_{\pm.212}$ | $.429_{\pm.360}$ |
| | MULTI | $.695_{\pm.219}$ | $.255_{\pm.107}$ | $.237_{\pm.136}$ | $.269_{\pm.156}$ | $.452_{\pm.286}$ | $.355_{\pm.335}$ | $.864_{\pm.157}$ | $.466_{\pm.353}$ |
| GAT | ENZYME | $.180_{\pm.085}$ | $.138_{\pm.086}$ | $.126_{\pm.080}$ | $.151_{\pm.050}$ | $.151_{\pm.076}$ | $.162_{\pm.085}$ | $.124_{\pm.067}$ | $.144_{\pm.077}$ |
| | IMDB | $.061_{\pm.043}$ | $.039_{\pm.037}$ | $.029_{\pm.038}$ | $.033_{\pm.045}$ | $.020_{\pm.057}$ | $.043_{\pm.058}$ | $.032_{\pm.045}$ | $.030_{\pm.044}$ |
| | PROTE | $.104_{\pm.074}$ | $.078_{\pm.069}$ | $.045_{\pm.044}$ | $.077_{\pm.076}$ | $.077_{\pm.070}$ | $.064_{\pm.065}$ | $.092_{\pm.105}$ | $.084_{\pm.080}$ |
| | UPFD | $.146_{\pm.082}$ | $.034_{\pm.026}$ | $.086_{\pm.078}$ | $.020_{\pm.024}$ | $.030_{\pm.017}$ | $.077_{\pm.070}$ | $.019_{\pm.036}$ | $.018_{\pm.031}$ |
| | MOTIF | $.312_{\pm.137}$ | $.203_{\pm.070}$ | $.110_{\pm.182}$ | $.170_{\pm.052}$ | $.143_{\pm.136}$ | $.162_{\pm.161}$ | $.406_{\pm.226}$ | $.164_{\pm.237}$ |
| | MULTI | $.309_{\pm.069}$ | $.207_{\pm.067}$ | $.131_{\pm.186}$ | $.156_{\pm.053}$ | $.144_{\pm.139}$ | $.206_{\pm.213}$ | $.424_{\pm.239}$ | $.149_{\pm.239}$ |
| GIN | ENZYME | $.176_{\pm.076}$ | $.091_{\pm.093}$ | $.177_{\pm.094}$ | $.093_{\pm.077}$ | $.174_{\pm.086}$ | $.153_{\pm.067}$ | $.116_{\pm.091}$ | $.108_{\pm.108}$ |
| | IMDB | $.022_{\pm.055}$ | $.014_{\pm.018}$ | $.008_{\pm.047}$ | $.003_{\pm.026}$ | $.018_{\pm.041}$ | $.019_{\pm.046}$ | $.023_{\pm.048}$ | $.028_{\pm.023}$ |
| | PROTE | $.035_{\pm.041}$ | $.017_{\pm.045}$ | $.010_{\pm.043}$ | $.009_{\pm.051}$ | $.020_{\pm.038}$ | $.042_{\pm.082}$ | $.028_{\pm.050}$ | $.000_{\pm.037}$ |
| | UPFD | $.129_{\pm.095}$ | $.005_{\pm.031}$ | $.149_{\pm.106}$ | $.000_{\pm.016}$ | $.029_{\pm.035}$ | $.005_{\pm.036}$ | $.006_{\pm.029}$ | $.001_{\pm.020}$ |
| | MOTIF | $.800_{\pm.125}$ | $.164_{\pm.056}$ | $.564_{\pm.245}$ | $.130_{\pm.029}$ | $.587_{\pm.391}$ | $.539_{\pm.297}$ | $.734_{\pm.069}$ | $.006_{\pm.010}$ |
| | MULTI | $.793_{\pm.137}$ | $.166_{\pm.071}$ | $.541_{\pm.268}$ | $.138_{\pm.039}$ | $.596_{\pm.401}$ | $.456_{\pm.297}$ | $.731_{\pm.082}$ | $.004_{\pm.008}$ |
| SAGE | ENZYME | $.117_{\pm.090}$ | $.101_{\pm.091}$ | $.112_{\pm.115}$ | $.078_{\pm.095}$ | $.183_{\pm.073}$ | $.132_{\pm.088}$ | $.076_{\pm.076}$ | $.100_{\pm.087}$ |
| | IMDB | $.044_{\pm.068}$ | $.054_{\pm.079}$ | $.009_{\pm.073}$ | $.043_{\pm.069}$ | $.042_{\pm.068}$ | $.027_{\pm.068}$ | $.001_{\pm.071}$ | $.042_{\pm.085}$ |
| | PROTE | $.095_{\pm.043}$ | $.092_{\pm.083}$ | $.024_{\pm.040}$ | $.053_{\pm.039}$ | $.064_{\pm.068}$ | $.057_{\pm.097}$ | $.066_{\pm.056}$ | $.073_{\pm.071}$ |
| | UPFD | $.157_{\pm.070}$ | $.041_{\pm.030}$ | $.036_{\pm.034}$ | $.031_{\pm.020}$ | $.036_{\pm.049}$ | $.062_{\pm.048}$ | $.042_{\pm.046}$ | $.029_{\pm.031}$ |
| | MOTIF | $.382_{\pm.096}$ | $.270_{\pm.139}$ | $.103_{\pm.090}$ | $.200_{\pm.119}$ | $.313_{\pm.273}$ | $.323_{\pm.185}$ | $.816_{\pm.169}$ | $.323_{\pm.293}$ |
| | MULTI | $.394_{\pm.086}$ | $.304_{\pm.157}$ | $.126_{\pm.092}$ | $.195_{\pm.112}$ | $.232_{\pm.217}$ | $.473_{\pm.265}$ | $.809_{\pm.165}$ | $.265_{\pm.241}$ |

## 6 Conclusion and Limitations

In this paper, we introduced eXEL, a subgraph importance estimation method for pretrained GNNs, formulated as a linear Group Lasso regression problem in the embedding space. Our method is applicable regardless of the output layer type, the readout function, or the ground-truth target label. Through extensive experiments, we demonstrate that our method consistently outperforms existing approaches while effectively leveraging prior domain knowledge of graph substructures through the Group Lasso penalty.

However, our method also has the following limitations. First, it reconstructs the graph-level embedding vector, implying that it is applicable only to GNN models equipped with a readout function. Furthermore, since we adopt the Group Lasso penalty, our method can identify only partitioned subgraphs. As a future direction, it can be extended to incorporate tree-structured hierarchical subgraph domain knowledge by employing the tree-guided Group Lasso [11]. On the other hand, although the simple Lasso penalty does not require domain prior knowledge, applying it to the final GNN layer may not be appropriate, as it treats nodes independently even though the embeddings have already aggregated information from neighboring nodes through the message-passing mechanism. Therefore, we plan to enhance our method to a hybrid approach that applies the simple Lasso penalty to earlier GNN layers and the Group Lasso penalty to the later layers.

**Acknowledgements.** This study was supported by the National Research Foundation of Korea (NRF) grant funded by the Ministry of Science and ICT (MSIT) (No. RS-2022-NR068754, No. RS-2025-23525436). The authors acknowledge the Urban Big data and AI Institute of the University of Seoul supercomputing resources (http://ubai.uos.ac.kr) made available for conducting the research reported in this paper.

## References

1. Agarwal, C., Queen, O., Lakkaraju, H., Zitnik, M.: Evaluating explainability for graph neural networks. Sci. Data **10** (2022)
2. Amara, K., Ying, R., Zhang, Z., Han, Z., Shan, Y., Brandes, U., Schemm, S., Zhang, C.: Graphframex: Towards systematic evaluation of explainability methods for graph neural networks (2024)
3. Baldassarre, F., Azizpour, H.: Explainability techniques for graph convolutional networks. arXiv preprint arXiv:1905.13686 (2019)
4. Buterez, D., Janet, J.P., Kiddle, S.J., Oglic, D., Liò, P.: Graph neural networks with adaptive readouts. Adv. Neural. Inf. Process. Syst. **35**, 19746–19758 (2022)
5. Degen, J., Wegscheid-Gerlach, C., Zaliani, A., Rarey, M.: On the art of compiling and using 'drug-like' chemical fragment spaces. ChemMedChem **3**(10), 1503–1507 (2008)
6. Dou, Y., Shu, K., Xia, C., Yu, P.S., Sun, L.: User preference-aware fake news detection. In: Proceedings of the 44th International ACM SIGIR Conference on Research and Development in Information Retrieval, (2021)

7. Duarte, G.J., Pereira, T.A., do Nascimento, E.J.F., Mesquita, D.P.P., Junior, A.H.S.: How do loss functions impact the performance of graph neural networks? Anais do 15. Congresso Brasileiro de Inteligência Computacional (2021)
8. Hu, W., et al.: Open graph benchmark: datasets for machine learning on graphs. Adv. Neural. Inf. Process. Syst. **33**, 22118–22133 (2020)
9. Huang, K., et al.: Therapeutics data commons: machine learning datasets and tasks for drug discovery and development. In: NeurIPS Datasets and Benchmarks (2021)
10. Huang, Q., Yamada, M., Tian, Y., Singh, D., Yin, D., Chang, Y.: Graphlime: local interpretable model explanations for graph neural networks. IEEE Trans. Knowl. Data Eng. **35**, 6968–6972 (2020)
11. Kim, S., Xing, E.P.: Tree-guided group lasso for multi-response regression with structured sparsity, with an application to eQTL mapping. Ann. Appl. Stat. **6**(3), 1095–1117 (2012)
12. Kuhn, H.W., Tucker, A.W., Dresher, M., Wolfe, P., Luce, R.D., Bohnenblust, H.F.: Contributions to the theory of games (1953)
13. Luo, D., et al.: Parameterized explainer for graph neural network. In: NeurIPS. NIPS '20 (2020)
14. Mika, G.P., Bouzeghoub, A., Wegrzyn-Wolska, K., Neggaz, Y.M.: Hgexplainer: explainable heterogeneous graph neural network. In: IEEE International Conference on Web Intelligence and Intelligent Agent Technology, pp. 221–229 (2023)
15. Morris, C., Kriege, N.M., Bause, F., Kersting, K., Mutzel, P., Neumann, M.: Tudataset: a collection of benchmark datasets for learning with graphs. ArXiv arxiv:2007.08663 (2020)
16. Pope, P.E., Kolouri, S., Rostami, M., Martin, C.E., Hoffmann, H.: Explainability methods for graph convolutional neural networks. In: IEEE/CVF Conference on Computer Vision and Pattern Recognition (CVPR, pp. 10764–10773 (2019)
17. Rong, Y., et al.: Self-supervised graph transformer on large-scale molecular data. Adv. Neural. Inf. Process. Syst. **33**, 12559–12571 (2020)
18. Sánchez-Lengeling, B., et al.: Evaluating attribution for graph neural networks. In: Neural Information Processing Systems (2020)
19. Schlichtkrull, M.S., Cao, N.D., Titov, I.: Interpreting graph neural networks for NLP with differentiable edge masking. In: ICLR 2021 (2021)
20. Schnake, T., et al.: Higher-order explanations of graph neural networks via relevant walks. IEEE Trans. Pattern Anal. Mach. Intell. **44**, 7581–7596 (2020)
21. Toyokuni, A., Yamada, M.: Structural explanations for graph neural networks using HSIC. ArXiv (2023)
22. Vu, M., Thai, M.T.: PGM-explainer: probabilistic graphical model explanations for graph neural networks. Adv. Neural. Inf. Process. Syst. **33**, 12225–12235 (2020)
23. Wang, T., et al.: Mogonet integrates multi-omics data using graph convolutional networks allowing patient classification and biomarker identification. Nat. Commun. **12** (2021)
24. Wu, Z., Pan, S., Chen, F., Long, G., Zhang, C., Yu, P.S.: A comprehensive survey on graph neural networks. IEEE Trans. Neural Netw. Learn. Syst. **32**, 4–24 (2019)
25. Xu, K., Hu, W., Leskovec, J., Jegelka, S.: How powerful are graph neural networks? ArXiv (2018)
26. Yamada, M., Jitkrittum, W., Sigal, L., Xing, E.P., Sugiyama, M.: High-dimensional feature selection by feature-wise kernelized lasso. Neural Comput. **26**, 185–207 (2012)
27. Yao, L., Mao, C., Luo, Y.: Graph convolutional networks for text classification. In: Proceedings of the AAAI Conference on Artificial Intelligence, vol. 33, pp. 7370–7377 (2019)

28. Ying, R., Bourgeois, D., You, J., Zitnik, M., Leskovec, J.: Gnnexplainer: generating explanations for graph neural networks. Adv. Neural. Inf. Process. Syst. **32**, 9240–9251 (2019)
29. Yuan, H., Yu, H., Wang, J., Li, K., Ji, S.: On explainability of graph neural networks via subgraph explorations. In: International Conference on Machine Learning (2021)
30. Yuan, M., Lin, Y.: Model selection and estimation in regression with grouped variables. J. R. Stat. Soc. Ser. B (Stat. Methodol.) **68** (2006)
31. Zhang, D., Chen, J., Lu, X.: Blockchain phishing scam detection via multi-channel graph classification. In: Dai, H.-N., Liu, X., Luo, D.X., Xiao, J., Chen, X. (eds.) BlockSys 2021. CCIS, vol. 1490, pp. 241–256. Springer, Singapore (2021). https://doi.org/10.1007/978-981-16-7993-3_19
32. Zhang, Y., DeFazio, D., Ramesh, A.: Relex: a model-agnostic relational model explainer. In: Proceedings of the 2021 AAAI/ACM Conference on AI, Ethics, and Society (2020)
33. Zhang, Z., Liu, Q., Wang, H., Lu, C., Lee, C.K.: Motif-based graph self-supervised learning for molecular property prediction. In: NeurIPS (2021)

# A Hierarchy Tree Data Structure for Behavior-Based User Segment Representation

Yang Liu(✉), Xuejiao Kang, Sathya Iyer, Idris Malik, Ruixuan Li, Juan Wang, Xinchen Lu, Xiangxue Zhao, Dayong Wang, Menghan Liu, Isaac Liu, Feng Liang, and Yinzhe Yu

Meta, Menlo Park, California 94025, USA
{yliu9,xuejiao,sath,idrismalik,ruixuan,juanw,xinchenlu,sherryz, dayongwang,menghanliu,iliu,liangfeng,yinzheyu}@meta.com

**Abstract.** User attributes are critical in modern recommendation systems, especially for alleviating cold-start challenges and enhancing experiences for new or infrequent users. Integrating diverse categorical attributes, such as demographics and interests, in a scalable and behavior-aware manner remains challenging. We present Behavior-based User Segmentation (*BUS*), a tree-based framework that hierarchically partitions users guided by product-specific engagement behaviors. During tree construction, a greedy algorithm optimizes Normalized Discounted Cumulative Gain (NDCG) to ensure behavioral representativeness between marginal and active users. A novel *regress* operator filters and aggregates irrelevant user attributes using loss signals at each iteration. The *BUS* tree enables aggregation across leaf and internal nodes to identify representative user segments and generate popular content and interaction patterns. To improve fairness and robustness, *BUS* incorporates connection-based segmentation via social graphs, combining individual and social behavioral patterns. Deployed at industrial scale, *BUS* serves billions of users daily, achieving significant improvements in metrics such as music ranking and email notifications delivery time. To our knowledge, *BUS* is the first list-wise learning-to-rank framework for tree-based recommendation that integrates diverse categorical attributes with semantic interpretability at industrial scale.

**Keywords:** user segmentation · demographic-based recommendation · learning-to-rank · behavioral patterns · retrieval

## 1 Introduction

User attributes provide structured representations of user-related signals that guide personalization in modern recommendation systems. These attributes include relatively static demographic and technical features (e.g., age, gender, device type) as well as dynamic behavioral and interest-based representations (e.g., preferred content categories or long-term engagement topics). Together,


R. C.-W. Wong et al. (Eds.): PAKDD 2026, LNAI 16597, pp. 16–30, 2026.
https://doi.org/10.1007/978-981-92-1300-9_2

they capture both stable and evolving aspects of user profiles, offering a rich context for modeling preferences and improving personalization [1].

Demographic-based recommendation, one of the earliest approaches, assumes that users with similar demographic characteristics tend to have similar preferences [2]. This method is transparent, interpretable, and effective for new or infrequent users, addressing the cold-start problem and enabling easily explainable recommendations compared to complex deep learning-based models [3].

However, attribute-based recommendation faces several limitations. First, it often provides limited personalization and may amplify bias, leading to repetitive and homogeneous recommendations [4]. Second, it largely depends on empirically derived user attributes that are specific to certain product contexts, making them difficult to generalize across products with distinct market positions or objectives. Third, categorical features often exhibit high cardinality and dependency, posing challenges for efficient processing, fairness, and scalability. Scaling user attribute-based recommendation to support diverse products and billions of users poses substantial challenges in both complexity and scalability. The large and heterogeneous user base of each product requires highly efficient algorithms capable of processing massive volumes of behavioral and attribute data. When multiple attributes are directly combined or aggregated (Fig. 1), controlling user segment size becomes difficult, and skewed demographic distributions can lead to biased or unrepresentative segments. Furthermore, the dynamic nature of user behavior—with evolving demographic profiles and shifting preferences—necessitates adaptive learning mechanisms that continuously update attribute-based models to capture the latest user - product engagement pattern. Notably, deep learning approaches such as neural collaborative filtering and graph neural networks require sufficient interaction history to learn meaningful representations, leaving cold-start and marginal users underserved.

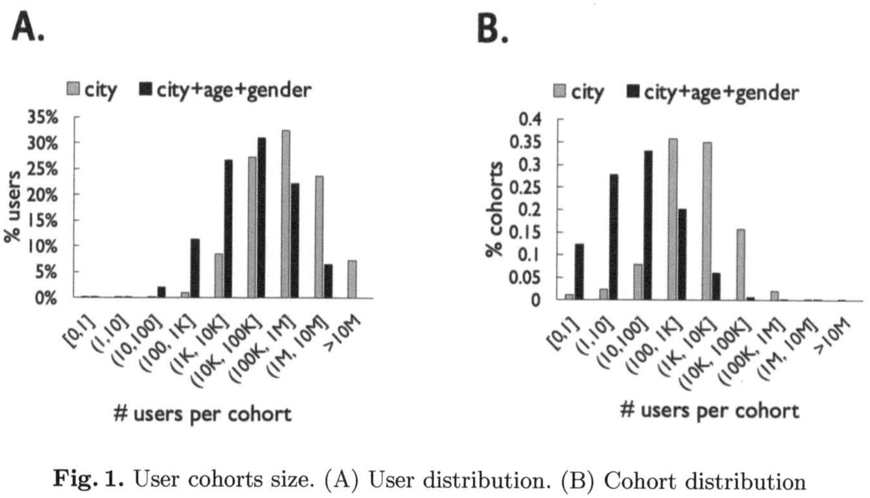


**Fig. 1.** User cohorts size. (A) User distribution. (B) Cohort distribution

In this paper, we reformulate user attribute-based recommendation as a listwise learning-to-rank problem, leveraging diverse categorical features to construct targeted user segments that transfer knowledge from active users to new or infrequent (marginal) users. To address this, we propose a novel tree-based data structure, *BUS*, as the core of our approach. We describe the construction of the *BUS* tree, including attribute selection and aggregation through the *regress* operator. Additionally, we extend *BUS* with connection-based segments derived from the social graph to enhance recommendation quality and ecosystem effects. Finally, we demonstrate *BUS*-based recommendation in two real-world applications: candidate retrieval for music ranking and delivery timing optimization in email notifications.

Our main contributions are as follows:

- We introduce a novel tree-based data structure, *BUS*, to represent user attribute segments, and productionize the data structure and training algorithm in SQL.
- We propose a listwise learning-to-rank solution leveraging *BUS* to generate product-specific segments and enable *BUS*-based candidate retrieval for recommendation.
- We evaluate our approach on production traffic serving billions of users daily and demonstrate substantial improvements in key metrics across multiple applications.

## 2 Related Work

Demographic-based user targeting and behavior modeling are widely adopted strategies in large-scale social media, search, and e-commerce platforms. Incorporating demographic attributes into recommendation systems enhances accuracy and preserves users' local experiences [5]. Beyond serving as ranking features or filters, demographic attributes can also support retrieval for cold-start users [3]. However, prior studies often examine only fixed orderings of attributes without a principled learning algorithm to weight their contributions. Attribute selection is frequently driven by heuristics or experimentation rather than systematic, data-driven methods.

Most demographic attributes are categorical and can be encoded via integer encoding, one-hot encoding, or entity embeddings for deep learning [6]. While embeddings improve memory efficiency and capture intrinsic feature properties [7,8], they reduce interpretability, complicating debugging and signal quality assessment. Recently, Large Language Models (LLMs) have been applied to convert structured user attributes into textual descriptions, which are then embedded and clustered into user groups using methods such as K-Means [9–11]. These approaches, however, largely ignore user–product interactions and rely on heuristic clustering without optimizing cluster granularity or downstream ranking quality at scale.

Beyond flat encodings, hierarchical and tree-based structures have been explored to capture richer user representations. HieRec [28] constructs a hierarchical user interest tree for news recommendation, capturing fine-grained interest representations. Tree-based indexing has also been explored for efficient retrieval, with joint optimization of tree-based index structures and deep models shown to improve recommendation efficiency [29]. The role of user segmentation granularity in recommendation performance has been examined [31], highlighting the importance of segment design. Other approaches model user behavior through attention mechanisms [30] or leverage graph neural networks with metapath aggregation for heterogeneous settings [32]. However, these methods typically require sufficient interaction history and focus on active users rather than cold-start scenarios.

Listwise learning-to-rank (LTR) methods directly optimize ranking quality by considering the entire list of items for each query as a training instance, rather than evaluating items individually or in pairs. It has been extensively studied in information retrieval and recommendation systems, often outperforming pointwise and pairwise approaches [12,13]. Algorithms such as ListNet [14], RankCosine [15], ListMLE [13], SQL-Rank [16], and GFN4Rec [17] improve ranking quality but are typically limited to small candidate sets. Applying listwise approaches to large candidate pools, especially for cold-start users, remains largely unexplored. Content-aware listwise collaborative filtering addresses new items [18], but effective solutions for new or infrequent users are still needed.

# 3 Methodology

## 3.1 Problem Overview

We formulate user attribute-based recommendation as an optimization problem to identify an optimal set of user segments defined by categorical attributes (e.g., age, gender, location, interest). Since active users usually provide sufficient behavioral signals, our goal is to ensure that aggregated active-user behaviors can effectively represent marginal users. Let $E_{u,p}$ denote the engagement behavior of user $u$ with product $p$, represented as a ranked list of items the user has interacted with. For active users in segment $s$, we compute the aggregated top-$K$ behavior $\hat{E}_{AU_s,p}$ by ranking items based on their popularity within the segment. We define a listwise loss for each product $p$ as:

$$L_{T_p} = -\sum_{s \in T_p} \sum_{u \in MU_s} f(E_{u,p}, \hat{E}_{AU_{s,p}})$$

where $f$ is the listwise function (e.g., NDCG) to measure how well the aggregated active-user behavior $\hat{E}_{AU_s,p}$ predicts marginal user $u$'s actual engagement $E_{u,p}$. The objective is to construct a segment tree $T'_p$ that minimizes this loss:

$$T'_p = \arg\min_{T_p} L_{T_p}.$$

## 3.2 User Segment Tree

Marginal users typically exhibit limited product engagement activity and weak graph connections, making demographic attributes the primary modeling signals. However, traditional encoding methods (e.g., one-hot, label encoding) lose semantic meaning and perform poorly with high-cardinality features. To address this, we construct a hierarchical tree-based structure that preserves semantic dependencies among categorical attributes (e.g., *San Francisco* city nested under *US* country). Each user is mapped to a unique leaf node based on their attribute path, enabling semantically coherent segmentation such as $global \rightarrow US \rightarrow 30s \rightarrow California \rightarrow San\_Francisco$.

We initialize the $BUS$ tree with root node *global* containing all users in universe $U$, and iteratively grow tree $T$ following Algorithm 1. At each level $i$, we consider eligible attribute types $Attrs = \tau_1, \tau_2, ..., \tau_n$ and generate staging child nodes for each attribute $\tau$. Hierarchical dependencies (e.g., country before city) can be enforced for semantic consistency. For each staging node $s_{i_k}$, we aggregate top-K product behaviors $\hat{E}_{AU_{s_{i_k}},p}$ from active users (e.g., top 100 clicked items), forming a predicted ranking list. The list is compared with marginal users' actual engagement lists using normalized discounted cumulative gain (NDCG@K) [19] to compute the node loss.

$$L_{s_{i_k}} = \sum_{u \in MU_{s_{i_k}}} -NDCG_u@K = \sum_{u \in MU_{s_{i_k}}} -\frac{DCG_u}{IDCG_u}$$

We assign zero relevance when the predicted list fails to cover the user's engaged items.

**Algorithm 1.** *BUS* Tree Construction

Initialize attribute list $Attrs$, user universe $U = AU_{global} \cup MU_{global}$, and root node $S_{global}$.
Compute global top-$K$ behaviors $\hat{E}_{AU_{global},p}$ and initialize loss $L = \{-\text{NDCG}(E_{u,p}, \hat{E}_{AU_{global},p}) : u \in MU_{global}\}$.
**while** $|Attrs| > 0$ **do**
  **for** each attribute $\tau \in Attrs$ **do**
    **for** each leaf node $s_{i-1}$ in $T$ **do**
      Split $s_{i-1}$ by $\tau$ into child nodes $s_i = \{s_{i_1}, ..., s_{i_n}\}$.
      **for** each $s_{i_k}$ **do**
        Compute top-$K$ behaviors $\hat{E}_{AU_{s_{i_k}},p}$ and loss $L_{s_{i_k}} = \sum_{u \in MU_{s_{i_k}}} -\text{NDCG}(E_{u,p}, \hat{E}_{AU_{s_{i_k}},p})$.
        Compute inherited loss $\hat{L}_{s_{i_k}}$ from parent $s_{i-1}$.
        **if** $\sum L_{s_{i_k}} > \omega \sum \hat{L}_{s_{i_k}}$ or $|AU_{s_{i_k}}| < \mu$ **then**
          Replace $i_k$ with "*regress*" and append $\hat{L}_{s_{i_k}}$ to $L$.
        **else**
          Append $L_{s_{i_k}}$ to $L$.
        **end if**
      **end for**
      Record total loss $L_\tau = \sum_{s \in s_i} L_s$.
    **end for**
  **end for**
  Select $\hat{\tau} = \arg\min_\tau L_\tau$, grow $T$ with its child nodes, and remove $\hat{\tau}$ from $Attrs$.
**end while**

Because NDCG is non-differentiable, we adopt a greedy optimization strategy to ensure monotonically improved ranking quality during tree construction. Specifically, a *regress* operator is introduced to compare each node's loss $L_{s_{i_k}}$ with its inherited parent loss $\hat{L}_{s_{i_k}}$. A staging node is retained only if its loss is no greater than its parent's; otherwise, it is replaced by a *regress* node that reverts to the parent's aggregate behavior (Fig. 2). The *regress* tree nodes have a clear real-world interpretation. For instance, as shown in Fig. 2B, the *regress* nodes derived from the city attribute correspond to all $US$ users not residing in San Francisco or New York; their behavior is represented using the aggregate activity patterns of all $US$ users.

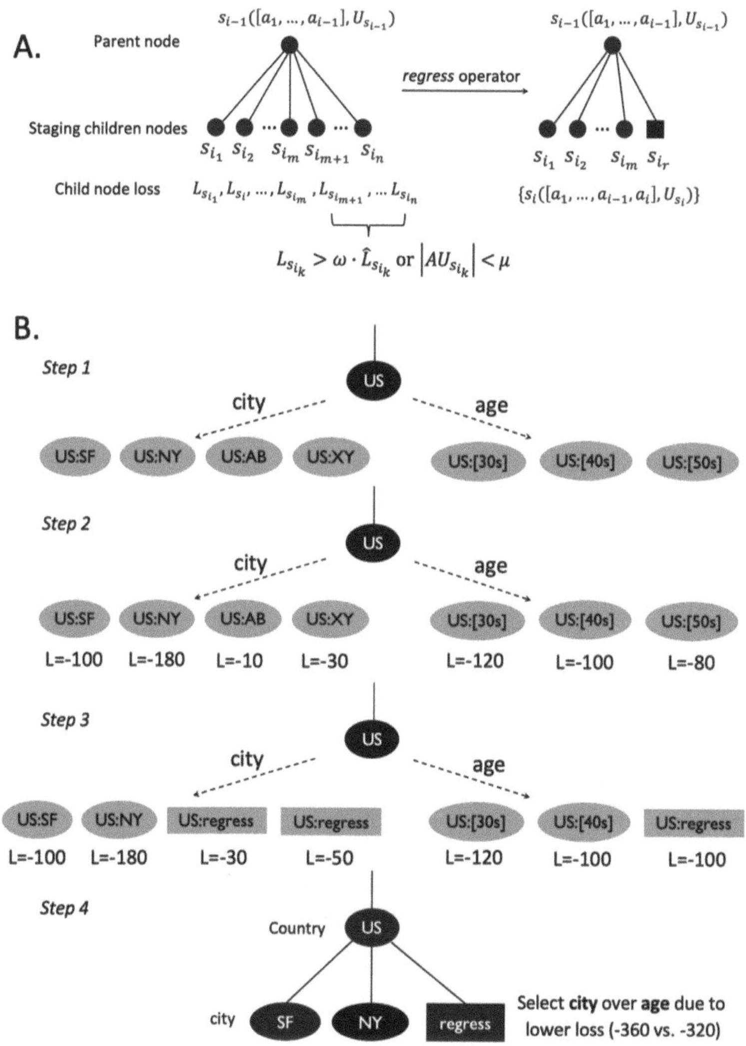


**Fig. 2.** *BUS* tree construction: (A) The *regress* operator generates a regress node, shown as a square; (B) Example of user attribute selection in *BUS* tree construction

Two hyperparameters, $\omega$ (default 1.0) and the minimum active user count $\mu$, are introduced in the training configuration to regulate tree growth. The threshold $\omega$ controls the strictness of the regress operator: lower values (e.g., $\omega < 1.0$) allow more aggressive splitting, while $\omega \geq 1.0$ ensures strictly monotonic loss reduction (Lemma 1). In our experiments, we set $\omega = 1.0$ and use the minimum active user count $\mu$ to control the segment size explicitly.

At each level, the attribute type with the minimum total loss $L_\tau$ is selected for expansion, yielding a hierarchically optimized and interpretable user segmentation iteratively.

**Lemma 1:** The overall loss during *BUS* tree construction decreases monotonically when $\omega \geq 1.0$.

**Proof:** For each staging child node, two cases arise. If the node's loss exceeds its parent's, the *regress* operator replaces it with the parent's aggregate behavior, ensuring the child's loss equals the parent's. If the node's loss is equal to or lower than the parent's, it is retained. In both cases, each child node has loss no greater than its parent, guaranteeing monotonic decrease in overall loss.

Each user is assigned to exactly one leaf node, satisfying the MECE principle, and the resulting tree has uniform depth. The algorithm is practical to implement and scalable. Its time complexity is $O(m^2 \cdot n \cdot K \cdot C)$, where $m$ is the number of attribute types, $n$ the number of marginal users, $K$ the ranking depth used for NDCG calculation, and $C$ the cost of computing top-K behaviors. Training billions of users with 10–15 attributes typically requires 40–50 Batch Compute Units (BCUs), far less than deep learning embedding methods which typically require thousands of BCUs.

### 3.3 *BUS*-Based Recommendation

To generate retrieval candidates, each leaf node traverses upward to the first non-*regress* ancestor, aggregating activity behaviors from all users under that ancestor, including descendants. The *BUS* tree is transformed into $T'$ by collapsing regress attributes, and for each user, internal nodes are prioritized by distance to the leaf (Fig. 3A). Aggregated behaviors from these priority nodes are used individually or combined as the retrieval source for users assigned to the leaf. Nodes in *BUS*-based recommendation are not mutually exclusive and a user contributes to multiple segments simultaneously. For example, the behavior of a user in *global* → *US* → *regress* → 30*s* segment contributes to *global*, *global* → *US*, and *global* → *US* → 30*s* nodes. This property distinguishes *BUS*-based recommendation from standard clustering approaches, where users typically influence only their own cluster, while preserving interpretability and flexibility in candidate generation.

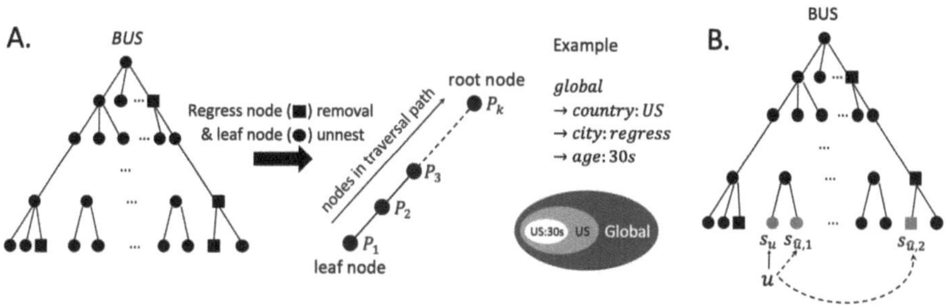


**Fig. 3.** *BUS*-based recommendation. (A) A constructed *BUS* tree undergoes *regress* node removal and leaf node unnest. (B) Connection aware *BUS*-based recommendation where content retrieved from user's own segment and connection segments.

### 3.4 Connection-Aware *BUS*-Based Recommendation

In product scenarios with a large number of content candidates, relying solely on a user's own segment may limit personalization and raise fairness concerns. To mitigate this, we leverage users' social graph to generate connection segments and combine content from both the user's own segment ($s_u$) and the connected users' segments ($s_{\hat{u},1}$, $s_{\hat{u},2}$,...) as retrieval sources (Fig. 3B). The social graph $G = (U, E)$ is constructed from the platform's friendship network, where $U$ represents all users and an edge $(u, u') \in E$ exists if users $u$ and $u'$ are mutual friends on the platform. For each user $u$, let $C_u = \{u' \in U \mid (u', u) \in E\}$ and $S_{\hat{u}} = \bigcup_{u' \in C_u} s_{u'}$ denote the connection segments. To reduce computation, we filter long-tailed segments using a percentile threshold $\phi$ (default 0.1) and round retained segment weights to 0.1. For example, connection distributions of $\{62\%, 38\%\}$ and $\{61\%, 39\%\}$ are both rounded to $\{0.6, 0.4\}$, enabling shared computation. Aggregated behaviors from weighted connection segments are then combined with the user's own segment to produce the final connection-aware *BUS*-based recommendation, improving robustness, personalization, and fairness.

# 4 Applications and Experiments

## 4.1 System Overview

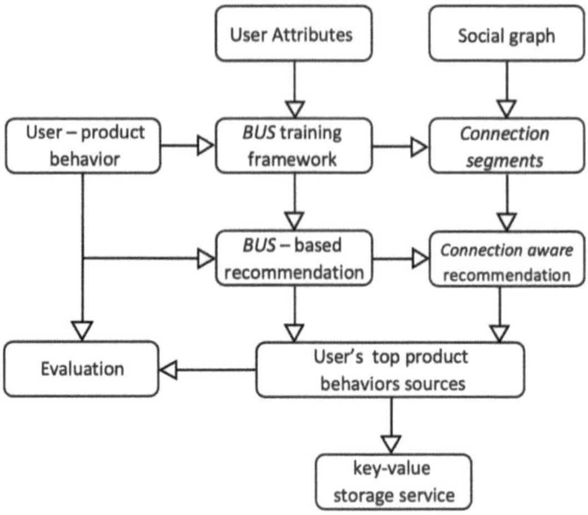


**Fig. 4.** System overview of *BUS*-based recommendation.

To support *BUS* tree training and recommendation at billion-user scale, we implement the *BUS* data structure in SQL, executed via a distributed query engine (Presto [20]) within a workflow management service [21]. We streamlined the *BUS* training framework as a series of APIs to generate Dataswarm SQL codes that use the user attributes table and the user-product interaction behavior table to construct the *BUS* tree iteratively in the workflow. We utilized relational database tables to model the tree node conceptual objects and optimized the SQL implementation by random sampling of up to 100K active users per staging segment in the computation of $\hat{E}_{AU_{s,p}}$ and storing the inherited reward to avoid duplicated computation.

The *BUS* tree is periodically rebuilt using the latest user-product engagement data to maintain the long-term effectiveness of *BUS*-based recommendations. For recommendation, top-K behaviors/contents are retrieved from each segment and optionally combined with connection segments. Popular behaviors are stored in Hive tables and cached in a distributed key-value store [22] for fast real-time serving (Fig. 4).

## 4.2 Training and Serving

We evaluated *BUS*-based recommendation in two personalization use cases: music ranking [23] and email notifications [24]. In music ranking, millions of music and artist candidates were ranked to generate the top 100 IDs per *BUS* segment, serving as one dedicated retrieval source. Prior work shows strong correlations between users' music habits and demographics [25,26]. For email notifications, we ranked 24-hour buckets per segment and use the segment's activity pattern to optimize delivery timing [24,27].

We defined active users as monthly active users on the target product (MAU), and marginal users as non-MAU but with activity on the product in the next 7 days. We set $K = 100$ and $K = 24$ for NDCG in music ranking and email notifications, respectively, with $\omega = 1.0$ and 11 categorical features. Numerical demographics were bucketized, and missing features assigned a value of '*NULL*'. The *BUS* tree is rebuilt weekly to incorporate the latest behaviors.

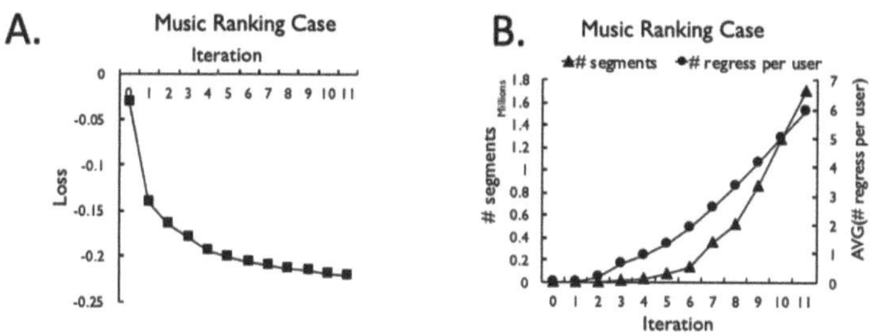


**Fig. 5.** Offline evaluation. (A) Overall loss in the *BUS* tree construction. (B) The number of segments and *regress* operators in the *BUS* tree construction.

Consistent with Lemma 1, overall loss decreases monotonically during tree growth (Fig. 5A), and *regress* nodes are more frequent at lower levels (Fig. 5B), replacing 50% of attributes on average. Periodic rebuilding dynamically changes segment grouping, balancing stability and plasticity in recommendations [4]. The *BUS*-based recommendation sources are updated on a daily basis to provide the most recent segment's product behaviors or content candidates. In connection-aware recommendation, segment percentile threshold $\phi = 0.1$ filters connection segments and popular candidates are ranked by a utility function:

$$U_{u,c} = P_{c,s_u} + \sum_{s \in S_{\hat{u}}} w_{s,u} \cdot P_{c,s}$$

where $P_{c,s}$ is candidate popularity in segment $s$, $s_u$ is the user's own segment, $S_{\hat{u}}$ denotes the user's connection segments, and $w_{s,u}$ is the weight of connection segment $s$ (e.g., the proportion of the user's friends belonging to that segment).

### 4.3 Offline Evaluation

Deep learning approaches such as neural collaborative filtering, graph neural networks, and embedding-based methods require sufficient interaction history to learn meaningful user representations. For marginal users with limited or no engagement data, these methods cannot generate reliable embeddings, making direct comparison infeasible. We focus on evaluating retrieval sources for marginal users, as our approach specifically targets this cold-start scenario where only categorical attributes are available. This complements rather than replaces deep learning methods for active users. Accordingly, our comparison is between $BUS$-based recommendation and one-hot encoding-based aggregation.

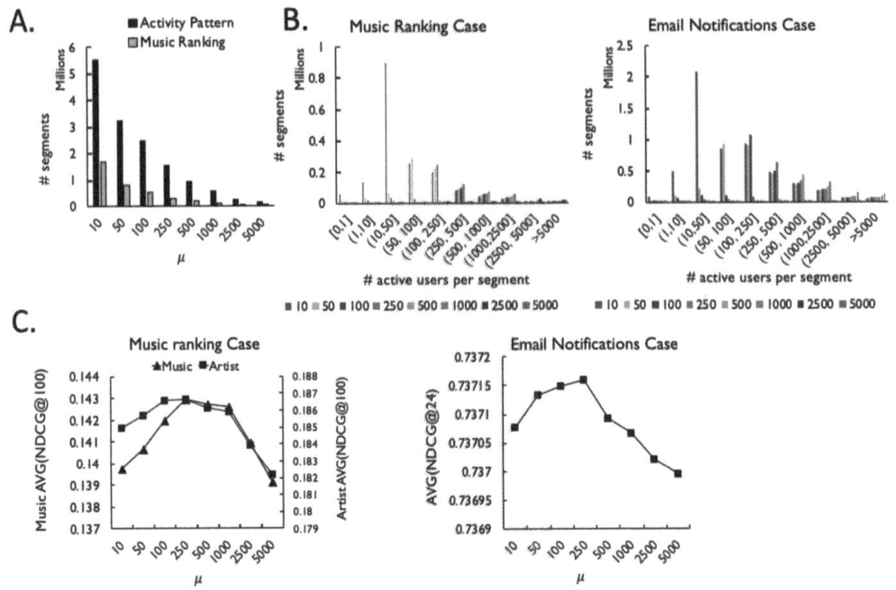


**Fig. 6.** Effect of segment size control in the $BUS$ tree construction. (A) Number of leaf node segments in $BUS$ trees built with varying minimum active user thresholds ($\mu$). (B) Size distribution of $BUS$ leaf node segments across different buckets for varying $\mu$ values. (C) NDCG scores for $BUS$ trees built with varying $\mu$.

We constructed 8 $BUS$ trees with minimum active user thresholds ($\mu$) from 10 to 5,000 and evaluated ranking quality by computing NDCG using marginal users' actual engagement behaviors in the subsequent 7 days as ground truth. Increasing $\mu$ reduced segment number and size (Figs. 6AB). Both use cases achieved the highest ranking quality at $\mu = 250$ (Fig. 6C), outperforming one-hot aggregation (Table 1), suggesting that suitable segment size control mitigates overfitting. We also observed different product use cases exhibited distinct attribute ordering in $BUS$ trees, and $\mu$ influenced this ordering.

In the music ranking use case, we further implemented the connection-aware $BUS$-based recommendation, generating 400–500 million distinct connection segments shared across billions of users. We observed that for 70% of users, their own segment is not overlapped with their connection segments. This approach further boosted the music NDCG@100 from 0.1429 (Table 1) to 0.150, with an additional computational cost of approximately 500–600 BCUs.

**Table 1.** Ranking quality of *BUS*-based vs. one-hot aggregation

| Case | Music Ranking | | | Email Notifications | |
|---|---|---|---|---|---|
| Metric | music NDCG@100 | artist NDCG@100 | #Segments | NDCG@24 | #Segments |
| *BUS* | 0.1429 | 0.1867 | 303K | 0.7372 | 1.7M |
| City | 0.1104 | 0.1536 | 195K | 0.7355 | 195K |
| City + Age + Gender | 0.1325 | 0.1770 | 5M | 0.7365 | 5M |

## 4.4 Online Experiments

We conducted two 30-day online experiments to assess the impact of *BUS*-based recommendation. The analysis is based on the aggregated results from the last 7 days of each experiment. In the email notifications, we evaluated an ensemble approach combining *BUS*-based segment's activity patterns with the pointwise learning-based personalized prediction method [24] (our baseline) to optimize email notification delivery times. We observed that the improved user activity pattern leads to statistically significantly ($p < 0.05$) increases in both email click-through rates and daily user app engagement (Table 2). Both marginal and active users became more active in the treatment, which suggests that the *BUS* user segment can serve as a powerful collaborative filtering mechanism by capturing and disseminating cross-behavioral signals.

**Table 2.** Online experiments of *BUS*-based recommendation.

| Experiment/Metric | Overall | Active | Marginal |
|---|---|---|---|
| **Email notifications** (*BUS* **vs. no-***BUS*) | | | |
| Click-through rate | +2.95% | +3.10% | +2.36% |
| Daily active users | +0.057% | +0.021% | +0.330% |
| **Music ranking** (*BUS* **vs. no-***BUS*) | | | |
| Daily active producers | +0.126% | +0.11% | +0.91% |
| **Music ranking (Connection-aware** *BUS* vs. *BUS*) | | | |
| Daily active producers | +0.045% | +0.043% | +0.086% |
| Shared rate | +0.16% | +0.13% | +0.58% |

In music ranking, we evaluated the integration of *BUS*-based recommendation as an additional retrieval source to provide the top 50 music IDs and artist IDs per segment. We compared this source with existing retrieval sources, which serve as our baselines and are based on various deep learning and collaborative filtering methods. Our results showed that *BUS*-based recommendation leads to a statistically significant increase in content producers activities for both active and marginal producers (Table 2). Although we saw an increase in production, the content shared rate among users' friends was not significantly improved. To improve recommendation fairness and diversity, we introduced connection-aware *BUS*-based recommendation and compared them with *BUS*-based recommendation derived solely from the user's own segment. This approach not only further boosts producer activity but also leads to a statistically significant increase in content sharing among friends, demonstrating a positive ecosystem impact.

## 5 Conclusion

We propose a hierarchical tree structure, *BUS*, representing diverse user attributes to generate interpretable user segments and enhance attribute-based recommendations across multiple product scenarios. By combining list-wise learning-to-rank objectives with tree-based segmentation, *BUS* produces semantically coherent user groups. Industrial-scale experiments in music ranking and email notifications show substantial improvements in key metrics, demonstrating efficiency and effectiveness. The *BUS* framework is highly extensible and infrastructure-friendly, supporting diverse objectives and entity types beyond users. To our knowledge, this is the first large-scale deployment of a list-wise learning-to-rank framework leveraging tree-based recommendation with diverse entity attributes. Further studies are needed to clarify the optimal values of $\omega$ and $\phi$, and the strategy for combining *BUS* with embedding methods for users transitioning from marginal to active status.

Looking forward, integrating LLMs with *BUS* offers promising opportunities. Each segment provides an explicit, semantically consistent audience, enabling audience-level prompts for content summarization, creative generation, and insight extraction. Segment semantics can also be incorporated into ranking systems, bridging audience representation with content optimization. Thus, *BUS* provides a unified foundation for recommendation, ranking, and generative AI applications.

**Acknowledgments.** We thank Yuankai Ge, Weizhe Shi, and partner teams for support and contributions.

**Author Contributions.** Conceptualization and methodology: Y. Liu; experiments and analysis: Y. Liu, J. Kang, S. Iyer, I. Malik, R. Li, J. Wang, M. Liu, X. Lu, X. Zhao; supervision and administration: D. Wang, I. Liu, W. Shi, F. Liang, Y. Yu; manuscript preparation: Y. Liu. All authors reviewed and approved the final manuscript.

## References

1. Adomavicius, G., Tuzhilin, A.: Toward the next generation of recommender systems: a survey of the state-of-the-art and possible extensions. IEEE Trans. Knowl. Data Eng. **17**(6), 734–749 (2005)
2. Pazzani, M.J.: A framework for collaborative, content-based and demographic filtering. Artif. Intell. Rev. **13**, 393–408 (1999)
3. Al-Shamri, M.Y.H.: User profiling approaches for demographic recommender systems. Knowl.-Based Syst. **100**, 175–187 (2016)

4. Fayyaz, Z., Ebrahimian, M., Nawara, D., Ibrahim, A., Kashef, R.: Recommendation systems: algorithms, challenges, metrics, and business opportunities. Appl. Sci. **10**(21), 7748 (2020)
5. Gon, M.: Local experiences on instagram: social media data as source of evidence for experience design. J. Destinat. Mark. Manag. **19**, 100435 (2021)
6. Hancock, J.T., Khoshgoftaar, T.M.: Survey on categorical data for neural networks. J. Big Data **7**(1), 28 (2020)
7. Guo, C., Berkhahn, F.: Entity embeddings of categorical variables. arXiv preprint arXiv:1604.06737 (2016)
8. Zhang, Y., Zheng, X., Helbich, M., Chen, N., Chen, Z.: City2vec: urban knowledge discovery based on population mobile network. Sustain. Cities Soc. **85**, 104000 (2022)
9. Sun, S., et al.: Random silicon sampling: simulating human sub-population opinion using a large language model based on group-level demographic information. arXiv preprint arXiv:2402.18144 (2024)
10. Li, Y., Liu, Y., Yu, M.: Consumer segmentation with large language models. J. Retail. Consum. Serv. **82**, 104078 (2025)
11. Tissera, M., Asanka, P., Rajapakse, R.: Enhancing customer segmentation using large language models (LLMs) and deterministic, independent-of-corpus embeddings (DICE). In: 2024 4th International Conference on Advanced Research in Computing (ICARC), pp. 73–78. IEEE (2024)
12. Chapelle, O., Chang, Y., Liu, T.-Y.: Future directions in learning to rank. In: Proceedings of the Learning to Rank Challenge, pp. 91–100. PMLR (2011)
13. Xia, F., Liu, T.-Y., Wang, J., Zhang, W., Li, H.: Listwise approach to learning to rank: theory and algorithm. In: Proceedings of the 25th International Conference on Machine Learning, pp. 1192–1199 (2008)
14. Cao, Z., Qin, T., Liu, T.-Y., Tsai, M.-F., Li, H.: Learning to rank: from pairwise approach to listwise approach. In: Proceedings of the 24th International Conference on Machine Learning, pp. 129–136 (2007)
15. Qin, T., Zhang, X.-D., Tsai, M.-F., Wang, D.-S., Liu, T.-Y., Li, H.: Query-level loss functions for information retrieval. Inf. Process. Manag. **44**(2), 838–855 (2008)
16. Wu, L., Hsieh, C.-J., Sharpnack, J.: SQL-rank: a listwise approach to collaborative ranking. In: International Conference on Machine Learning, pp. 5315–5324. PMLR (2018)
17. Liu, S., et al.: Generative flow network for listwise recommendation. In: Proceedings of the 29th ACM Conference on Knowledge Discovery and Data Mining, pp. 1524–1534 (2023)
18. Ravanifard, R., Mirzaei, A., Buntine, W., Safayani, M.: Content-aware listwise collaborative filtering. Neurocomputing **461**, 479–493 (2021)
19. Järvelin, K., Kekäläinen, J.: Cumulated gain-based evaluation of IR techniques. ACM Trans. Inf. Syst. (TOIS) **20**(4), 422–446 (2002)
20. Sethi, R., et al.: Presto: SQL on everything. In: IEEE 35th International Conference on Data Engineering (ICDE), pp. 1802–1813. IEEE (2019)
21. Alex: Data engineering at Meta: High-Level Overview of the internal tech stack. https://medium.com/@AnalyticsAtMeta/data-engineering-at-meta-high-level-overview-of-the-internal-tech-stack-a200460a44fe
22. Chen, G.J., et al.: Realtime data processing at facebook. In: Proceedings of the 2016 International Conference on Management of Data, pp. 1087–1098 (2016)
23. Afchar, D., Melchiorre, A., Schedl, M., Hennequin, R., Epure, E., Moussallam, M.: Explainability in music recommender systems. AI Mag. **43**(2), 190–208 (2022)

24. Liu, Y., et al.: Personalized execution time optimization for the scheduled jobs. arXiv preprint arXiv:2203.06158 (2022)
25. Cheng, Z., Shen, J.: Just-for-me: an adaptive personalization system for location-aware social music recommendation. In: Proceedings of International Conference on Multimedia Retrieval, pp. 185–192 (2014)
26. Krismayer, T., Schedl, M., Knees, P., Rabiser, R.: Predicting user demographics from music listening information. Multimedia Tools Appl. **78**(3), 2897–2920 (2019)
27. Liu, Y., Chen, Z., Virochsiri, K., Wang, J., Wu, J., Liang, F.: Reinforcement learning-based product delivery frequency control. In: Proceedings of the AAAI Conference on Artificial Intelligence, pp. 15355–15361 (2021)
28. Qi, T., et al.: HieRec: hierarchical user interest modeling for personalized news recommendation. In: Proceedings 59th Annual Meeting of the ACL and 11th International Joint Conference on Natural Language Processing, vol. 1: Long Papers, pp. 5446–5456 (2021)
29. Zhu, H., et al.: Joint optimization of tree-based index and deep model for recommender systems. In: Proceedings of 33rd International Conference Neural Information Processing Systems (NeurIPS), vol. 32 (2019)
30. Zhou, C., et al.: ATRank: an attention-based user behavior modeling framework for recommendation. In: Proceedings of 32nd AAAI Conference on Artificial Intelligence, vol. 32, no. 1 (2018)
31. Erdem, E.E., Orman, G.K.: On the role of user segmentation in recommender systems' performances. Procedia Comput. Sci. **225**, 2333–2342 (2023)
32. Fu, X., Zhang, J., Meng, Z., King, I.: MAGNN: Metapath aggregated graph neural network for heterogeneous graph embedding. In: Proceedings of WWW 2020, pp. 2331–2341 (2020)

# S$^2$tory: Story Spine Distillation for Movie Script Summarization

Mingzhe Lu[1,2], Yanbing Liu[1,2], Qihao Wang[1,2], Jiarui Zhang[1,2], Jiayue Wu[1,2], Yue Hu[1,2], Yunpeng Li[1,2(✉)], and Yangyan Xu[3]

[1] Institute of Information Engineering, Chinese Academy of Sciences, Beijing, China
liyunpeng@iie.ac.cn

[2] School of Cyber Security, University of Chinese Academy of Sciences, Beijing, China

[3] HiThink Research, Hangzhou, China

**Abstract.** Movie scripts pose a fundamental challenge for automatic summarization due to their non-linear, cross-cut narrative structure, which makes surface-level saliency methods ineffective at preserving core story progression. To address this, we introduce S$^2$tory (Story Spine Distillation), a narratology-grounded framework that leverages character development trajectories to identify plot nuclei, the essential events that drive the narrative forward, while filtering out peripheral satellite events that merely enrich atmosphere or emotion. Our Narrative Expert Agent (NEAgent) performs theory-constrained reasoning, whose distilled knowledge conditions a small model to identify plot nuclei. Another model then uses these plot nuclei to generate the summary. Experiments on the MovieSum dataset demonstrate state-of-the-art semantic fidelity at approximately 3.5× compression, and zero-shot evaluation on BookSum confirms strong out-of-domain generalization. Human evaluation further validates that narratological theory provides an indispensable foundation for modeling complex, non-linear narratives.

**Keywords:** Screenplay · Summarization · Narratology

## 1 Introduction

Large language models (LLMs) have achieved remarkable progress in text understanding. However, their performance declines on long-form and structurally complex narratives, especially in maintaining the core storyline. This limitation becomes particularly evident in movie script summarization, which poses unique challenges [17] beyond traditional text summarization.

Long-form summarization typically adopts a two-stage process: shortening the source, then generating the summary [13]. Among recent approaches in movie script summarization, MENSA [18] selects scenes based on saliency estimation, while DiscoGraMS [6] constructs a graph over characters and dialogues to model cross-scene coherence. Despite their effectiveness, these approaches remain largely data-driven and rely on lexical or shallow structural patterns.


R. C.-W. Wong et al. (Eds.): PAKDD 2026, LNAI 16597, pp. 31–42, 2026.
https://doi.org/10.1007/978-981-92-1300-9_3

Movie scripts are shaped by symbolic narrative structures, a foundational element of screenwriting, and effective summarization must recover this underlying logic. Capturing these structures requires reasoning about how events function within the story rather than how they appear in text. This gap motivates a model that integrates symbolic representation with neural abstraction to reconstruct the narrative backbone of a movie, an aspect largely overlooked in previous work.

Inspired by Barthesian narrative theory [1], we posit that movie script summarization requires explicit modeling of narrative structure to overcome the limitations of saliency-based approaches. Accordingly, we propose S$^2$tory, a framework that identifies plot nuclei as the essential events forming the core narrative thread and uses them as the foundation for high-fidelity summarization.

In S$^2$tory, a Narrative Expert Agent (NEAgent) follows narratological principles to identify plot nuclei essential to narrative integrity. The NEAgent analyzes the screenplay through character-arc trajectories, using character development to determine the indispensability of plot events. A student model distills this reasoning, and a fine-tuned summarizer performs nuclei-conditioned summarization.

In summary, our main contributions are as follows:

- We propose S$^2$tory, a narratologically-grounded framework that includes a NEAgent for identifying plot nuclei through character-arc trajectories.
- Within S$^2$tory, we introduce a distillation step where a student model learns the reasoning of the NEAgent, and a fine-tuned summarizer performs nuclei-conditioned summarization.
- We achieve state-of-the-art performance on the MovieSum benchmark, and demonstrate out-of-distribution generalization on BookSum through automated validation, human evaluation, and case studies.

# 2 Related Work

## 2.1 Computational Narratology

Computational narratology, rooted in classical narratology [4], aims to operationalize how stories produce meaning. Early approaches focused on symbolic representations, such as story grammars [16], plot units [11], and character-centered scripts [19], yielding interpretable but domain-limited frameworks. The field subsequently evolved into LLM-driven approaches, which can implicitly capture narrative patterns in text [8]. However, despite their proficiency in narrative processing, LLMs lack a deep understanding of underlying relational structures and narrative functions [3]. Recent studies have re-grounded narrative modeling in theory via turning points and events [9], yet they still fail to capture the Barthesian [1] distinction between structural necessity and surface prominence.

### 2.2 Long-Text Summarization

Long-text summarization poses distinct challenges for LLMs. Transformer architectures face quadratic complexity, leading to long-context variants such as Longformer [2], BigBird [21], and LongT5 [7]. Although these models extend context windows, they only partially capture global structure and still treat narratives as flat token sequences, missing the multi-level organization [2] inherent in human storytelling. In movie script summarization, datasets like SummScreen [5] and MovieSum [17] highlight the need to model cross-scene coherence and character-driven progression. Their coherence arises from causal and thematic continuity rather than textual adjacency. Building on this, structure-aware models such as DiscoGraMS [6] and ScreenWriter [12] incorporate discourse or character graphs but remain limited to surface cues-modeling who interacts with whom, but not why those interactions matter narratively.

## 3 Narrative Theoretical Formulation

### 3.1 Narratological Foundation

S$^2$tory is grounded in Barthes' narratological distinction between nuclei and satellites [1]. In a narrative, nuclei are essential events that drive the core progression of the story, while satellites are auxiliary elements that enrich atmosphere or emotion without altering the narrative trajectory.

### 3.2 Narrative World Modeling

In computational narrative modeling, we represent a story world as a quadruple $(\mathcal{C}, \mathcal{E}, \mathcal{S}, \mathcal{R})$ , where $\mathcal{C}$ is the set of characters, $\mathcal{E}$ is the set of events, $\mathcal{S}$ is the space of character states, and $\mathcal{R} \subseteq \mathcal{S} \times \mathcal{S}$ is the set of state transition relations.

Each transition r: $\mathrm{s}_t \rightarrow s_{t+1}$ is an element of $\mathcal{R}$, capturing a character's developmental trajectory. Based on the causal origin of r, we classify it into two types: intrinsic and extrinsic transitions. Intrinsic transitions arise from internal processes such as reflection, cognitive reevaluation, or emotional realization. Extrinsic transitions, by contrast, are triggered by external events $e \in \mathcal{E}$, representing environmental or social influences that compel a character to adapt.

Our approach focuses on identifying events that directly shape character development. We define a subset $e^{\rightarrow s} \subseteq \mathcal{E}$ as the collection of events that causally induce a state transition in at least one character:

$$e^{\rightarrow s} = \{e \in \mathcal{E} \mid \exists r : s_t \rightarrow s_{t+1} \in \mathcal{R}, \mathrm{Dep}(r, e) \wedge s_{t+1} \neq s_t\}, \tag{1}$$

where $\mathrm{Dep}(r, e)$ indicates that the transition r is causally dependent on event e and $s_{t+1} \neq s_t$ ensures that the event results in an actual change in the character's state. This definition operationalizes narrative significance:

- An event is considered narratively relevant if it leads to a detectable change in a character's state.
- Conversely, an event whose removal blocks character growth is crucial, marking a plot turning point.

### 3.3 Modeling Character Dynamics

Each character $c \in \mathcal{C}$ maintains a time-dependent attribute set $A_c^{(t)} = \{(k_j, v_j)\}$ representing evolving properties such as identity, goals, or affiliations.

A state transition $r : s_t \to s_{t+1}$ for character c is said to be event-induced if there exists $e \in \mathcal{E}$ such that Dep$(r, e)$ and $A_c^{(t+1)} \neq A_c^{(t)}$. We classify each such transition by its update type: $\tau(r) \in \{\ominus, \oplus\}$, where $\oplus$ denotes an *increment* (addition of a new attribute) and $\ominus$ a *modification* (replacement of an existing attribute). The state difference associated with r is:

$$\Delta(r) = \left(S^+, S^-\right), \quad A_c^{(t+1)} = \left(A_c^{(t)} \setminus S^-\right) \cup S^+, \tag{2}$$

with $S^+ = A_c^{(t+1)} \setminus A_c^{(t)}$ and $S^- = A_c^{(t)} \setminus A_c^{(t+1)}$.

For example, being knighted is an *increment*($\oplus$), adding a role without erasing history; conversely, moving cities is a *modification*($\ominus$), replacing the old location.

### 3.4 Narrative Nuclei Reasoning

This module implements the nucleus-satellite distinction via counterfactual reasoning over character state trajectories, as formalized in the previous subsections.

*Narrative Context.* To analyze a scene $\mathcal{C}_k$, the model uses the processed prior scenes, the structured character states, and the naturalized current scene $\tilde{\mathcal{C}}_k$ Based on this context, each sentence $u_{k,i}$ in $\tilde{\mathcal{C}}_k$ is evaluated to see if removing it would break character development.

*Reasoning Logic.* Define Cont $(\mathcal{M}_k, \cdot)$ as a predicate that returns 1 if all character state trajectories remain structurally continuous. The nucleus-satellite label is assigned by:

$$\kappa\left(u_{k,i}\right) = \begin{cases} 1, \text{ if} \neg \operatorname{Cont}\left(\mathcal{M}_k, \left(\tilde{\mathcal{C}}_k \setminus \{u_{k,i}\}\right) \cup B_{<k}\right), \\ 0, \text{ otherwise.} \end{cases} \tag{3}$$

here, $\kappa(u_{k,i}) = 1$ indicates a *nucleus*, and 0 a *satellite*. A soft prior from the transition type $\tau(r) \in \{\oplus, \ominus\}$ can modulate the sensitivity of Cont$(\cdot)$, but does not override the continuity test. For instance, given two candidate units in $\tilde{\mathcal{C}}_k$:

(1). $u_{k,1}$: "Leon claims to be a deserter; mutual distrust is established."
(2). $u_{k,2}$: "On the road, they encounter bandits, a plague village, and a rebel checkpoint."

Given $\mathcal{M}_k$ and $B_{<k}$, the system tests whether deletion breaks any character trajectory. For $u_{k,1}$, removal disrupts Leon's trust evolution, causing a critical break in narrative coherence. Consequently, it is classified as a nucleus $u_{k,1}^+$. In contrast, removing $u_{k,2}$ leaves all state transitions intact and coherence preserved, marking it as a satellite $u_{k,2}^-$.

This reasoning process defines narrative indispensability in functional terms, where an event is considered a nucleus if and only if its removal breaks the continuity of character development within the evolving symbolic world.

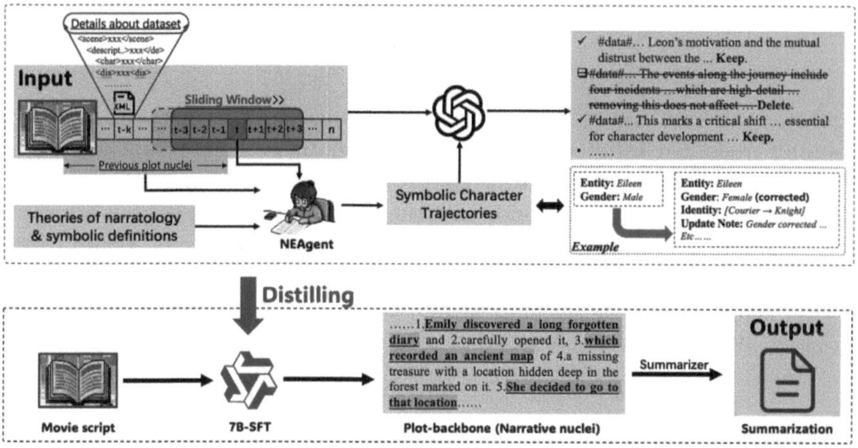


**Fig. 1.** The S$^2$tory framework leverages character-guided reasoning to identify plot nuclei and filter satellites, thereby distilling expert knowledge to effectively condition abstractive summarization.

# 4 Methodology

In S$^2$tory, NEAgent models characters, infers plot nuclei under narrative constraints, and distills this reasoning into a nuclei-conditioned backbone for summarization. The overall process is illustrated in Fig. 1.

## 4.1 NEAgent

Implementing narratological reasoning remains challenging: symbolic models lack openness, while neural models lack narrative data. We address this by designing NEAgent, an In-Context Learning (ICL) agent that integrates narratological principles with dynamic character modeling.

At step $t$, NEAgent processes the current narrative segment $\text{story}_t$ through ICL prompts that operationalize the narrative world model (Sect. 3.2) and character dynamics formalism (Sect. 3.3). The agent combines $\text{story}_t$ with a rolling memory $M_{t-1}$, which stores evolving character states $A_c^{(t)}$ and event-character dependencies $\text{Dep}(r, e)$ (Sect. 3.4). This creates a contextualized narrative representation. Guided by the prompts, NEAgent evaluates each narrative unit $u_{k,i}$ using the nucleus-satellite classification rule in Eq. (2), determining whether its removal would disrupt character trajectory continuity. The prompts explicitly encode narratological principles as actionable instructions, ensuring alignment with the theoretical framework.

## 4.2 Distillation and Summarization

NEAgent offers detailed reasoning at high token cost, hindering large-scale use. To make this reasoning scalable, we record its analytical process as an experience context and construct a structured dataset as follows:

$$\mathcal{D}_{\text{distill}} = \{(x_i, r_i, B_i)\}_{i=1}^{N}, \tag{4}$$

where $x_i$ is the naturalized narrative text of a scene, $r_i$ denotes the symbolic reasoning trace generated by the NEAgent, and $B_i$ is the corresponding set of nuclei identified through trajectory-guided counterfactual reasoning.

The reasoning trace $r_i$ encodes how the agent tracks character-state transitions, redefines goals, and evaluates whether removing an event would disrupt continuity within the evolving narrative structure.

The neural inducer is initialized from a 7B language model and fine-tuned on $\mathcal{D}_{\text{distill}}$. It learns a mapping:

$$f_\theta : (x^n_{\text{shot}}, r^n_{\text{shot}}, b^n_{\text{shot}}, x_i) \rightarrow (r_i, B_i), \tag{5}$$

where each training instance includes few-shot examples $(x^n_{\text{shot}}, r^n_{\text{shot}}, b^n_{\text{shot}})$ and the current input $x_i$, to generate the corresponding reasoning trace $r_i$ and predicted nuclei set $B_i$.

Finally, each backbone $B_i$ predicted by the distilled model is paired with its corresponding gold summary $y_i$ from the dataset, forming a new training set.

$$\mathcal{D}_{\text{sum}} = \{(B_i, y_i)\}_{i=1}^{M}. \tag{6}$$

The summarization model is then trained on $\mathcal{D}_{\text{sum}}$, where the model learns to generate reference summaries conditioned on the distilled backbones.

## 5 Experiment

### 5.1 Setup and Implementation Details

*NEAgent.* The NEAgent was implemented in LangGraph with GPT-4o as the reasoning engine, operating deterministically with temperature set to 0.0. See Appendix A.1 for a comprehensive description of the prompts.

*Reasoning Distillation.* Parameter-efficient fine-tuning (LoRA) was applied to Qwen2.5-7B-Instruct using LLaMA-Factory (32K input, 1K output) on 8xA100-80G GPUs.

*Nuclei-Conditioned Summarization.* Using distilled backbones $B = \{e_k^{\text{nuclei}}\}$ as input, Qwen2.5–0.5B-Instruct was full fine-tuned on $\{B, S^{\text{ref}}\}$.

### 5.2 Dataset and Baselines

*Dataset.* We focus on movie screenplays, which feature XML-style structural formatting. Our experiments use the MovieSum dataset [17], a near-complete superset of existing screenplay corpora. For example, 98% of the MENSA test set [18] is contained within MovieSum. To assess generalization beyond screenplay-style narratives, we also evaluate on BookSum [10], a long-form prose corpus without XML formatting, demonstrating that NEAgent transfers through narratological reasoning rather than dataset-specific pattern learning.

**Table 1.** Results on MovieSum: ROUGE ($R_1$, $R_2$, $R_L$), compression ratio (Comp.), and BERTScore ($BS_P$, $BS_R$, $BS_{F1}$).

| Type | Model | $R_1$ | $R_2$ | $R_L$ | Comp. | $BS_P$ | $BS_R$ | $BS_{F1}$ |
|---|---|---|---|---|---|---|---|---|
| ↑ Extractive | Lead-512 | 10.35 | 1.27 | 9.84 | / | *49.25* | *43.59* | *46.23* |
| | Lead-768 | 14.43 | 1.79 | 13.76 | / | *49.29* | *45.70* | *47.41* |
| | Lead-1024 | 17.93 | 2.24 | 17.15 | / | *49.12* | *46.91* | *47.98* |
| | TextRank | 33.32 | 5.27 | 32.10 | / | *51.46* | *52.47* | *51.85* |
| | FLAN-UL2 | 23.62 | 4.29 | 22.01 | **27.6%** | *52.90* | *49.57* | *50.87* |
| | Vicuna | 16.35 | 3.55 | 15.44 | 55.2% | *48.89* | *48.49* | *47.07* |
| | TextRank+Vicuna | 17.14 | 3.68 | 15.47 | 55.2% | *59.24* | *49.05* | *53.57* |
| | MW-Vicuna | 19.56 | 3.32 | 18.57 | 55.2% | 54.95 | 48.70 | 51.53 |
| Hybrid | **$S^2$tory (Ours)** | **45.98** | **7.93** | **42.45** | 28.4% | 59.18 | **59.36** | **59.23** |
| ↓ Abstractive | LED-Desc. | *44.72* | *9.72* | *42.92* | 55.2% | **59.47** | 58.45 | 58.92 |
| | LED-Heur. | *44.45* | *9.78* | *42.71* | 55.2% | 58.93 | 58.15 | 58.54 |
| | LED-Dialogue | *44.68* | *10.02* | *42.94* | 55.2% | 59.30 | 58.29 | 58.74 |
| | LED | *44.85* | *9.83* | *43.12* | 55.2% | 59.11 | 58.43 | 58.73 |
| | LongT5 | *41.49* | *8.54* | *39.78* | 55.2% | 56.09 | 55.36 | 55.68 |
| | Pegasus-X | *42.42* | *8.16* | *40.63* | 55.2% | 58.81 | 50.56 | 54.36 |

*Baselines.* To ensure comprehensive evaluation, we compare against a diverse set of baselines. These include extractive techniques (e.g., TextRank [14]), instructiontuned LLMs (e.g., Vicuna [22], FLAN-UL2 [20]), and long-context summarization models (e.g., Pegasus-X [15], LongT5 [7], LED [2]). All results are either adopted from prior work or reproduced under the same evaluation protocol as previous studies to ensure fair comparison and consistent evaluation metrics.

### 5.3 Main Results

Our experiments on the MovieSum benchmark show that $S^2$tory effectively combines the strengths of extractive and abstractive summarization. As shown in Table 1, our model achieves exceptional ROUGE scores ($R_1 = 45.98, R_2 = 7.93, R_L = 42.45$), outperforming all extractive baselines by 32–38% while matching the $R_L$ performance of abstractive models.

$S^2$tory also achieves the highest BERTScore recall (59.36) among all methods, indicating superior semantic fidelity to reference summaries. Crucially, it accomplishes this with a compression ratio of only 28.4%—less than half that of abstractive methods (55.2%)—demonstrating its ability to generate concise yet comprehensive summaries of extremely long movie scripts.

These results validate our approach: by strategically integrating extractive precision with abstractive flexibility, $S^2$tory achieves an optimal balance between information coverage, semantic quality, and conciseness that neither purely extractive nor purely abstractive methods can match for long-form content summarization.

# 6 Further Analysis

We further analyze the model from four perspectives: ablation, human evaluation, qualitative visualization, and out-of-domain generalization.

## 6.1 Ablation Study

To assess the impact of character trajectory modeling on nucleus identification, we conduct an ablation study where the trajectory-based profiling module is removed from NEAgent. In the full model, NEAgent constructs structured profiles for each character, capturing evolving goals, identities, and inter-character dependencies, which serve as the foundation for evaluating narrative indispensability. Without this module, the agent directly applies nucleus-satellite reasoning on raw text, losing access to causal continuity in character development.

As shown in Table 2, removing trajectory modeling results in a marked drop in performance, with BERTScore F1 decreasing from 59.23 to 53.69. This confirms that modeling character development trajectories provides essential structural grounding for identifying narrative nuclei and maintaining cross-scene coherence in long-form scripts.

**Table 2.** Ablation study on character trajectory modeling in nucleus reasoning.

| Method Variant | BertScore-P | BertScore-R | BertScore-F1 |
|---|---|---|---|
| NEAgent w/o trajectory profiling | 53.09 | 55.28 | 53.69 |
| NEAgent w/ trajectory profiling | **59.18** | **59.36** | **59.23** |

## 6.2 Human Evaluation

Although the extracted nuclei are intermediate outputs, their quality directly affects summarization performance. We conducted a human evaluation along four narratological dimensions: *indispensability*, *coherence*, *character consistency*, and *satellite reduction*. Each was rated on a five-point scale (1–5) by doctoral students trained in narrative theory. The full evaluation rubric is provided in Appendix A.2.

As shown in Table 3, human ratings consistently exceed automatic scores, particularly in *satellite reduction*, where GPT-4o-mini tends to overvalue descriptive or emotional details. This gap indicates that large models capture surface fluency but struggle with the functional segmentation that defines narrative structure, whereas our nuclei extraction aligns more closely with narratological judgments.

**Table 3.** Evaluation results across four narrative dimensions (1–5 scale).

| Dimension | Auto Metric | Human Metric |
|---|---|---|
| Indispensability | 3.59 | 3.84 |
| Coherence | 3.79 | 3.91 |
| Character Consistency | 3.97 | 4.18 |
| Satellite Reduction | 3.41 | 3.83 |

## 6.3 Out-of-Domain Generalization

To evaluate cross-domain robustness, we directly apply the 7B distilled model to the BookSum corpus without further tuning. BookSum preserves long-form narrative coherence but lacks screenplay formatting, making it suitable for testing whether the model generalizes through narratological reasoning rather than XML-specific cues.

Because BookSum has no annotated nuclei, we adopt an LLM-as-Judge protocol [22], prompting multiple LLMs to assess whether each generated nucleus constitutes a structurally essential event. Each case is labeled as positive or negative, while rejected indicates that the LLM declined to respond due to policy or copyright constraints.

**Table 4.** Out-of-domain evaluation on BookSum using the LLM-as-Judge protocol.

| Evaluator | Positive (%) | Negative (%) | Rejected (%) |
|---|---|---|---|
| GPT-4.1 | 92.45 | 5.45 | 2.10 |
| Qwen3-235B-A22B | 78.34 | 21.24 | 0.42 |
| DeepSeek-R1-671B | 84.71 | 15.55 | 0.28 |
| **Average** | **85.17** | **14.08** | **0.93** |

As shown in Table 4, over 85% of the generated nuclei are judged as narratively essential by large evaluators, demonstrating strong cross-domain generalization. This suggests that NEAgent internalizes character-centric causal reasoning rather than overfitting to screenplay structures, enabling consistent narrative interpretation across diverse text domains.

## 6.4 Case Study

We present a qualitative case study on *Roma(2018)* to illustrate how $S^2$tory preserves narrative rhythm while achieving substantial compression. As shown in Fig. 2, the logarithmic scene-length distributions of the original screenplay (blue) and the generated nuclei (orange) exhibit a highly aligned oscillatory pattern across 85 scenes. Despite significant token reduction, the temporal fluctuations in narrative density are faithfully preserved, indicating that $S^2$tory maintains the underlying rhythmic structure of the film.

This alignment is not coincidental: key emotional or structural peaks – such as the sharp length spike in Scene 79 – are preserved in both versions, reflecting principled modeling of cinematic pacing. Moreover, as demonstrated in the chapterwise distribution for Book-151 (right panel), the generated nuclei maintain a structurally coherent proportionality across chapters. For instance, Fig. 3 increases in relative share from 13.8% to 20.6%, suggesting targeted condensation of less salient content while preserving the prominence of core narrative arcs.

Together, these results show that $S^2$tory does not compress via uniform truncation, but by selectively preserving the narrative pulse – capturing both microlevel rhythmic dynamics and macro-level structural emphasis. This dual fidelity ensures that the distilled narrative remains faithful to the original in terms of dramatic tension and narrative cadence.

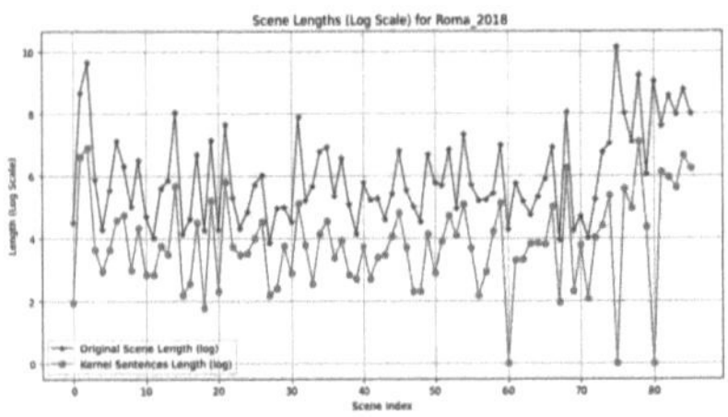


**Fig. 2.** Scene-length distribution: Original (Blue) vs. Nuclei (Orange). (Color figure online)

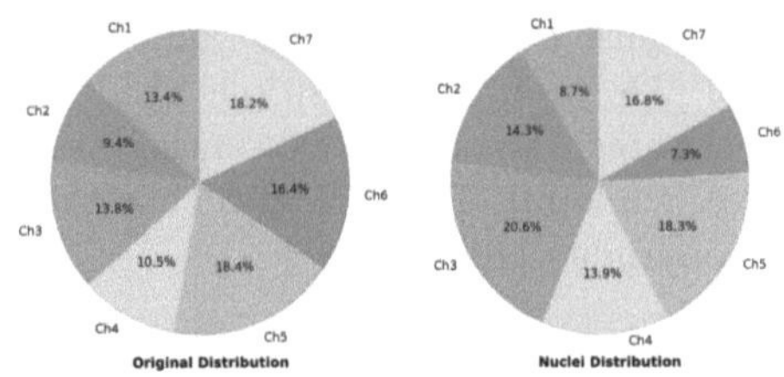


**Fig. 3.** Chapter-wise token distribution: Original (Left) vs. Nuclei (Right).

# 7 Conclusion

We presented $S^2$tory, a narratology-grounded framework for screenplay summarization. A Narrative Expert Agent (NEAgent) reasons over character trajectories to isolate plot nuclei from satellites, with its reasoning distilled into a compact model that conditions an abstractive summarizer. This demonstrates that narratology-guided reasoning provides a principled foundation for long-form story understanding. Future work will refine the causal link between nuclei and character state changes and extend the framework beyond screenplay conventions.

# A Appendix

## A.1 Prompt Details

Due to page constraints, we present a prompt card summarizing key components from our full prompt set. All narrative theories are implemented through carefully designed prompts to guide the NEAgent's reasoning process.

**1. Pronoun Replacement.** Expert linguistics role; identifies coreference clusters with character indices; groups mentions referring to same entities.

**2. Entity Profile Update.** Expert information extraction role; updates structured entity records using coreference clusters; maintains field merge rules.

**3. Narrative Units.** Replaces pronouns for independent readability; preserves sentence structure; segments by standard punctuation.

**4. Counterfactual Analysis.** Evaluates script coherence when removing sentences; assesses 5 dimensions including key information preservation.

**5. Kernel Events.** Based on Barthes' narrative theory; extracts only essential plot-driving elements; outputs micro-drama text ending with STOP token.

**6. Voting Protocol.** LLM-as-judge voting mechanism; selects most consistent events across multiple extraction attempts; resolves conflicts through majority rule.

**7. OOD Verification.** Validates drama miniaturization task; judges if compressed screenplay preserves narrative essence under extreme length constraints.

### A.2 Human Evaluation Criteria

Human annotators assessed the output quality based on the following criteria:

**1. Indispensability (Mainline Necessity).** 1 = irrelevant to main plot; 2=partly related, missing causal links; 3=mostly covers plot, minor gaps; 4=complete and logical; 5=precise and non-redundant.

**2. Coherence.** 1=fragmented; 2 = weak logic, disjoint; 3=generally coherent; 4=smooth and consistent; 5=fully coherent and well-structured.

**3. Character Consistency.** 1 = irrational or contradictory; 2=partly consistent, major jumps; 3=mostly consistent; 4=logical overall; 5=fully consistent with growth.

**4. Satellite Reduction.** 1 = mostly redundant satellites; 2=over 50% satellites; 3=30–40% satellites; 4=10–20% satellites; 5=nearly none, pure mainline.

## References

1. Barthes, R., Duisit, L.: An introduction to the structural analysis of narrative. New Literary Hist. **6**(2), 237–272 (1975)
2. Beltagy, I., Peters, M.E., Cohan, A.: Longformer: the long-document transformer. arXiv preprint arXiv:2004.05150 (2020)
3. Brahman, F.: Modeling Key Narrative Elements for Story Understanding and Generation, Santa Cruz. Ph.D. thesis (2022)
4. Chatman, S.B., Chatman, S.: Story and discourse: Narrative structure in fiction and film. Cornell university press (1978)
5. Chen, M., Chu, Z., Wiseman, S., Gimpel, K.: Summscreen: a dataset for abstractive screenplay summarization. In: Proceedings of the 60th Annual Meeting of the Association for Computational Linguistics, vol. 1: Long Papers, pp. 8602–8615 (2022)

6. Chitale, M.P., Bindal, U., Rajkumar, R.P., Mishra, R.: Discograms: enhancing movie screen-play summarization using movie character-aware discourse graph. In: Proceedings of the 2025 Conference of the Nations of the Americas Chapter of the Association for Computational Linguistics: Human Language Technologies, vol. 2: Short Papers, pp. 954–965 (2025)
7. Guo, M., et al.: Longt5: efficient text-to-text transformer for long sequences. In: Findings of the Association for Computational Linguistics: NAACL 2022, pp. 724–736 (2022)
8. Huang, Z., Zhao, J., Jin, Q.: ECR-chain: advancing generative language models to better emotion-cause reasoners through reasoning chains. arXiv preprint arXiv:2405.10860 (2024)
9. Jiayang, C., Qiu, L., Chan, C., Liu, X., Song, Y., Zhang, Z.: Eventground: narrative reasoning by grounding to eventuality-centric knowledge graphs. In: Proceedings of the 2024 Joint International Conference on Computational Linguistics, Language Resources and Evaluation (LREC-COLING 2024), pp. 6622–6642 (2024)
10. Kryściński, W., Rajani, N., Agarwal, D., Xiong, C., Radev, D.: Booksum: a collection of datasets for long-form narrative summarization. In: Findings of the Association for Computational Linguistics: EMNLP 2022, pp. 6536–6558 (2022)
11. Lehnert, W.G.: Plot units and narrative summarization. Cogn. Sci. **5**(4), 293–331 (1981)
12. Mahon, L., Lapata, M.: Screenwriter: automatic screenplay generation and movie summarisation. arXiv preprint arXiv:2410.19809 (2024)
13. Mei, L., et al.: A survey of context engineering for large language models. arXiv preprint arXiv:2507.13334 (2025)
14. Mihalcea, R., Tarau, P.: Textrank: bringing order into text. In: Proceedings of the 2004 Conference on Empirical Methods in Natural Language Processing, pp. 404–411 (2004)
15. Phang, J., Zhao, Y., Liu, P.J.: Investigating efficiently extending transformers for long input summarization. In: Proceedings of the 2023 Conference on Empirical Methods in Natural Language Processing, pp. 3946–3961 (2023)
16. Rumelhart, D.E.: Notes on a schema for stories. In: Bobrow, D.G., Collins, A. (eds.) Representation and Understanding: Studies in Cognitive Science (1975)
17. Saxena, R., Keller, F.: Moviesum: an abstractive summarization dataset for movie screenplays. In: Findings of the Association for Computational Linguistics: ACL 2024, pp. 4043–4050 (2024)
18. Saxena, R., Keller, F.: Select and summarize: scene saliency for movie script summarization. In: Findings of the Association for Computational Linguistics: NAACL 2024, pp. 3439–3455 (2024)
19. Schank, R.C., Abelson, R.P.: Scripts, plans, goals, and understanding: An inquiry into human knowledge structures. Psychology press (2013)
20. Tay, Y., et al.: Ul2: unifying language learning paradigms. arXiv preprint arXiv:2205.05131 (2022)
21. Zaheer, M., et al.: Big bird: transformers for longer sequences. Adv. Neural. Inf. Process. Syst. **33**, 17283–17297 (2020)
22. Zheng, L., et al.: Judging LLM-as-a-judge with mt-bench and chatbot arena. Adv. Neural. Inf. Process. Syst. **36**, 46595–46623 (2023)

# Spatio-Temporal Context-Aware Web Service QoS Prediction via Contrastive Learning

Xiang Mao[1], Yu Zhang[1], Yiwen Zhang[1(✉)], and Xiaoyao Zheng[2,3]

[1] School of Computer Science and Technology, Anhui University, Hefei, China
e23301174@stu.ahu.edu.cn, zhangyiwen@ahu.edu.cn

[2] Key Laboratory of Intelligent Computing and Signal Processing, Ministry of Education, Anhui University, Hefei, China
zxiaoyao@ahnu.edu.cn

[3] School of Computer and Information, Anhui Normal University, Wuhu, China

**Abstract.** Accurate Quality of Service (QoS) prediction is a major challenge in the field of service recommendation. Existing QoS prediction methods typically predict missing QoS values by analyzing historical call records of users and services. While these methods demonstrate effectiveness, they face limitations in addressing the sparsity of user service call data and fully utilizing contextual information related to service calls. To address these challenges, we propose a spatio-temporal context-aware QoS prediction method via contrastive learning, which effectively integrates temporal and spatial contextual information of users and services. First, we employ a neighborhood selection method that combines spatial and temporal dimensions to construct embeddings for similar users and services. Second, we generate contrastive views based on the similar and original embeddings, leveraging contrastive learning to capture self-supervised signals and alleviate data sparsity. Finally, the resulting embeddings, which incorporate both neighborhood and context information, are input into a multi-layer perceptron (MLP) for QoS prediction. Extensive experiments on real-world datasets demonstrate the effectiveness of our method, particularly in scenarios with extremely sparse training data, resulting in more accurate QoS predictions.

**Keywords:** Web service · QoS prediction · Spatio-temporal context-aware · Contrastive learning

## 1 Introduction

The rapid development of service computing has led to the proliferation of Web services offering similar functionalities [14], making it increasingly challenging for users to identify the most suitable service from a vast range of candidates. Quality of Service (QoS) [7], which reflects non-functional attributes such as response time, throughput, and bandwidth, has thus become a key criterion for


R. C.-W. Wong et al. (Eds.): PAKDD 2026, LNAI 16597, pp. 43–55, 2026.
https://doi.org/10.1007/978-981-92-1300-9_4

service selection and recommendation [30]. However, users can only observe QoS values for a small subset of services, leading to a highly sparse QoS matrix and posing challenges for accurate prediction.

Although various existing methods have explored temporal [27] or spatial [9,28]factors, few have considered both aspects simultaneously [33]. Specifically, existing approaches exhibit the following limitations: 1) even when user and service location information remains unchanged, QoS values can still fluctuate significantly over time, due to variations in the network environment and continuous changes in server load . 2) the service invocation process is complex and involves multiple spatial factors, including direct interactions between clients and servers as well as interactions within the network environment. Moreover, the communication paths of service invocations may traverse multiple subnets and autonomous systems (AS), collectively influencing service performance and QoS. 3) as the number of users and services increases and user behavior preferences become more personalized, users cannot invoke services at every moment but only at specific times. During other periods, invocation values may be empty, leading to increased sparsity of user and service calls across consecutive time slices. This phenomenon not only complicates the analysis of user behavior but also exacerbates the sparsity of the user-service-time matrix, presenting a significant challenge for accurate QoS prediction.

To address these challenges, we propose a spatio-temporal context-aware QoS prediction via contrastive learning (STCL) model. This model considers both temporal and spatial dimensions to identify similar neighborhoods of users and services, leveraging spatio-temporal context. By separately comparing the original embeddings of users and services with their respective similar embeddings, our model effectively captures valuable self-supervised signals, which enhance the representations of both users and services. Finally, the enriched embeddings integrating comprehensive contextual information are input into MLP for QoS prediction. The main contributions of this paper are as follows:

- We propose a novel model called STCL, which integrates a neighborhood selection method that hybridizes temporal volatility and spatial features, effectively constructing embeddings for similar users and services. By integrating neighborhood and rich contextual information, our approach improves the accuracy of QoS prediction.
- To alleviate the issue of data sparsity, we leverage cluster-aware and time-aware to generate contrastive views, and design a multi-scale contrastive learning module, such that the self-supervised signals can be better mined at different scales for QoS prediction.
- Through extensive experimental analysis on real-world web service datasets, our method is shown to outperform existing state-of-the-art approaches, which results in a substantial improvement in prediction accuracy.

## 2 Related Works

### 2.1 QoS Prediction

Collaborative filtering (CF) is a widely used and effective approach for service recommendation, generally divided into neighbor-based and model-based methods. Neighbor-based CF predicts missing values based on user-service similarities [16,17], while model-based methods, such as matrix factorization (MF) [10], which decomposes the user-service interaction matrix into two low-rank representations. Later studies enhance similarity measures and incorporate contextual factors such as geographical location or network conditions [3,5,12], but scalability remains a challenge on large datasets. With deep learning, neural models have been introduced to capture nonlinear user-service interactions for better prediction [24,34,35]. However, data sparsity still hinders effective representation learning. To address this, we propose the STCL model, which integrates contrastive learning into spatio-temporal modeling to enhance feature robustness and improve QoS prediction accuracy.

### 2.2 Contrastive Learning

In recent years, Contrastive Learning (CL) [13,29] has achieved remarkable success in various domains, such as computer vision and natural language processing [8,19,26]. As a self-supervised learning paradigm, CL learns effective representations by contrasting positive and negative pairs, and has been increasingly adopted in recommendation systems to alleviate data sparsity by extracting self-supervised signals from interaction data. For example, SGL [22] enhances recommendation performance through self-supervised graph learning, while Xia et al. [23] integrated CL into a knowledge distillation framework to address dimensional inconsistency in tensor decomposition. In summary, CL has demonstrated strong potential in learning robust representations under data sparsity across various recommendation tasks. This makes it a promising solution for improving the accuracy and robustness of QoS prediction.

## 3 Methodology

In this section, we present the STCL framework as illustrated in Fig. 1.

### 3.1 Feature Representation Module

Quality of Service (QoS) prediction is influenced by user locations and service attributes. Since user-service interactions occur within specific spatial contexts, these spatial characteristics directly affect the observed QoS values. Therefore, incorporating diverse spatial contextual information is crucial for improving prediction accuracy. In addition to user and service IDs, we consider several representative spatial features—including IP address, country, city, latitude, longitude, and autonomous system (AS)—to provide a more informative representation of each entity. These features are first one-hot encoded in the input layer and then mapped to fixed-length dense embedding vectors in the embedding layer, resulting in the user and service feature embeddings as follows:

$$l_u = \Phi(e_u^{ID}, e_u^{IP}, e_u^{country}, e_u^{city}, e_u^{AS}, e_u^{lat}, e_u^{lon}), \tag{1}$$

$$l_s = \Phi(e_s^{ID}, e_s^{IP}, e_s^{country}, e_s^{city}, e_s^{AS}, e_s^{lat}, e_s^{lon}), \tag{2}$$

where $l_u$,$l_s \in \mathbb{R}^{1\times d}$ denote the embedding feature vectors of users and services, respectively. $\Phi(\cdot)$ concatenates the input embeddings and maps them into a fixed-length dense vector.

### 3.2 Cluster-Aware Module

Clustering algorithms [25] are widely used in information retrieval to group similar users based on their learned representation. As shown in module (b) of Fig. 1, we utilize the K-Means algorithm [2] to cluster the location embeddings of users and services. Specifically, the location embedding vectors of users or services are taken as input and iteratively calculated by the clustering algorithm to obtain K cluster centers, that is $\{m_1, m_2,... m_k\}$. The model then partitions the user embeddings into K distinct groups, represented as $\{C_1,C_2,... C_K\}$. The goal is to minimize the sum of squared distances between each embedding vector $x$ and the nearest cluster center $m_i$, thus achieving optimal clustering:

$$\min_{C_1,\ldots,C_K} \sum_{k=1}^{K} \sum_{i\in C_k} (x_i - m_k)^2, \tag{3}$$

where $x_i$ is the $i$th embedding vector and $m_k$ is the center of the $k$th class.

Through this clustering process, we effectively group users with similar location features $l_u$, resulting in a final similar user embedding vector $c_u \in \mathbb{R}^{1\times d}$. Following the same procedure, we derive a service similarity embedding vector $c_s \in \mathbb{R}^{1\times d}$ based on location information $l_s$.

### 3.3 Time-Aware Module

To capture temporal dynamics in user-service interactions, we begin by analyzing the distribution of QoS invocation records over time. Specifically, we compute the average QoS value for each user at a given time step to reflect their overall service experience at that moment. For instance, at time $t$, if user $u$ invokes the services $\{s_1, s_2, s_4\}$ with corresponding QoS values {0.48,0.79,0.54}, the average QoS value for user $u$ at time $t$ is 0.60.

To construct time-aware embeddings that capture changes in user preferences and service performance across different time periods, we use this average QoS value as a temporal representation of the user's behavior and service performance, computed as follows:

$$Avg_{u,t} = \frac{\sum_{s=1}^{\mathcal{N}_{u,t}} r_{u,s,t}}{\mathcal{N}_{u,t}}, \quad Avg_{s,t} = \frac{\sum_{u=1}^{\mathcal{M}_{s,t}} r_{u,s,t}}{\mathcal{M}_{s,t}}, \tag{4}$$

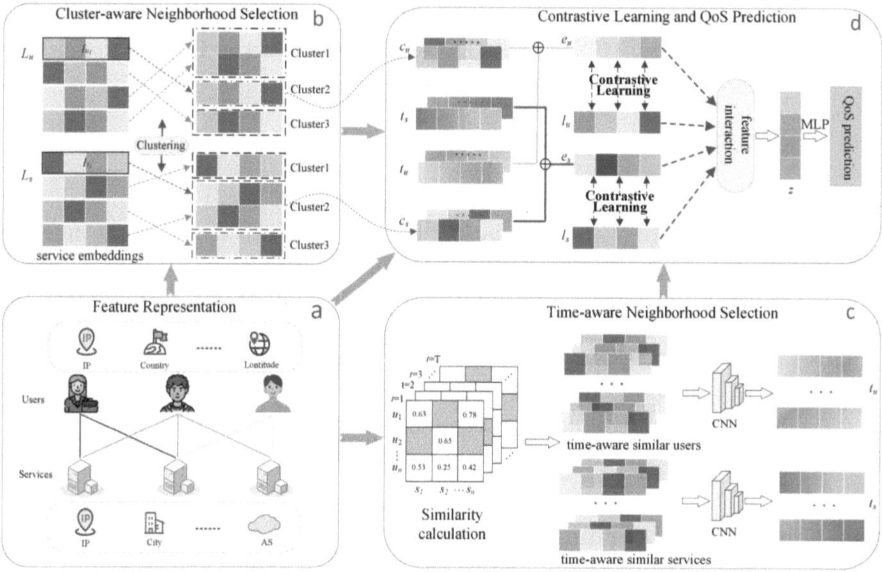


**Fig. 1.** The architecture of our proposed STCL framework: (a) feature extraction, (b) cluster-aware module for spatial features, (c) time-aware module for temporal features, and (d) contrastive learning and QoS prediction.

where $Avg_{u,t}$ denotes the average QoS value of all services invoked by user $u$ at time $t$, and $\mathcal{N}_{u,t}$ is the number of services invoked by user $u$ during time slice $t$. Similarly, $Avg_{s,t}$ represents the average QoS value of service $s$ based on all users who invoked it at time $t$, and $\mathcal{M}_{s,t}$ is the number of users who invoked service $s$ at time $t$. Here, $r_{u,s,t}$ is the actual QoS value observed from the interaction between user $u$ and service $s$ at time $t$. The similarity between users and between services across multiple time slices is computed as follows:

$$Sim_{u,v} = \frac{1}{1 + \sqrt{\sum_{t=1}^{T} |Avg_{u,t} - Avg_{v,t}|^2}}, \tag{5}$$

$$Sim_{s,f} = \frac{1}{1 + \sqrt{\sum_{t=1}^{T} |Avg_{s,t} - Avg_{f,t}|^2}}, \tag{6}$$

where $T$ is the total number of time slices, $Sim_{u,v}$ denotes the similarity between user $u$ and user $v$, $Sim_{s,f}$ represents the similarity between service $s$ and service $f$. A higher similarity value indicates a greater resemblance, which helps to identify the top-$m$ most similar users or services.

For a given user $u$, we first identify its top-$m$ most similar neighbors based on time-aware similarity. Each neighbor is associated with a $d$-dimensional embedding vector, which is retrieved from the trainable embedding layer. These $m$ vectors are stacked row-wise to form a similarity matrix $\mathrm{E}_u \in \mathbb{R}^{m \times d}$. To capture higher-order structural patterns among the neighborhood, we employ a convolutional neural network (CNN) [11]. The initial input to the CNN is denoted as $\mathrm{E}^{(0)} = \mathrm{E}_u$. The convolutional process is defined as:

$$\mathrm{E}_u^{(\ell)} = \sigma(\mathrm{Conv}(\mathrm{E}_u^{(\ell-1)}, W_\ell, B_\ell)), \quad \ell \in \{1, 2, \ldots, L\}, \tag{7}$$

where $\mathrm{Conv}(\cdot)$ denotes the convolution operation, $W_\ell$ and $B_\ell$ represent the weight and bias of the $\ell$-th convolutional layer, respectively. $L$ is the number of convolutional layers, and $\sigma$ is the activation function (e.g., ReLU) [1]. After the final convolutional layer, we apply a flattening operation to compress the output feature map into a fixed-length vector. Specifically, the output $\mathrm{E}_u^{(L)}$ is reshaped to obtain the final representation $t_u \in \mathbb{R}^{1\times d}$. This process ensures that the structural neighborhood information is preserved while generating a compact and trainable embedding. A similar process is applied to the service side to obtain the corresponding context-aware representation $t_s \in \mathbb{R}^{1\times d}$.

### 3.4 Contrastive Learning Module

In real-world QoS prediction scenarios, user-service interaction data is typically sparse and lacks explicit supervisory signals. To address this issue, we introduce a contrastive learning mechanism that enables the model to learn more discriminative representations by comparing different contrastive views derived from spatial and temporal contexts. Specifically, we adopt the InfoNCE loss [4] as our training objective. The loss is defined as:

$$\mathcal{L}(e_a, e_a') = -\sum_{a\in\mathcal{B}} \log \frac{\exp\left(e_a e_a'^{\top}/\tau\right)}{\sum_{b\in\mathcal{B}} \exp\left(e_a e_b^{\top}/\tau\right)}, \tag{8}$$

where $\mathcal{B}$ represents a batch of training samples, and $\langle e_a, e_a' \rangle$ denotes the set of positive pairs, $\{e_b \mid b \in \mathcal{B}, b \neq a\}$ are treated as negative samples, and $\tau$ represents the temperature coefficient.

To construct meaningful contrastive views, we leverage both spatial and temporal contextual information. Specifically, spatial embeddings $c_u$ and $c_s$ are derived from the cluster-aware module, while temporal embeddings $t_u$ and $t_s$ are obtained from the time-aware module. These are then combined through a fusion function to form the final user and service representations:

$$e_u = \mathrm{Agg}(t_u, c_u), \quad e_s = \mathrm{Agg}(t_s, c_s), \tag{9}$$

where $\mathrm{Agg}(\cdot)$ is an element-wise addition operation that combines the time-aware vector $t_u$ with the cluster-aware vector $c_u$, or combines $t_s$ with $c_s$, to form the final user $e_u \in \mathbb{R}^{1\times d}$ and service $e_s \in \mathbb{R}^{1\times d}$ feature vectors.

Based on the fused embeddings $e_u$ and $e_s$, we construct contrastive views to enrich semantic information and enhance representation quality. The contrastive loss is computed from two perspectives: the user view and the service view. In the user view, we maximize the similarity among users with similar spatio-temporal characteristics and minimize it for dissimilar ones; the same applies to services. This mechanism effectively captures intrinsic relationships among user behaviors, spatial contexts, and service properties.

For sample construction, each fused embedding $e_u$ (or $e_s$) and its corresponding location embedding $l_u$ (or $l_s$) form a positive pair, as they represent two complementary views of the same entity. Negative pairs are generated by matching $e_u$ (or $e_s$) with location embeddings of other users (or services) within the same batch, preventing representation collapse and encouraging uniformity in the latent space [21]. The contrastive losses for users and services are computed separately using the InfoNCE loss function as follows:

$$\mathcal{L}(e_u, l_u) = -\sum_{\{u\}\in\mathcal{B}_u} \log \frac{\exp\left(e_u l_u^\top/\tau\right)}{\sum_{i\in\mathcal{B}_u}\exp\left(e_u l_i^\top/\tau\right)}, \tag{10}$$

$$\mathcal{L}(e_s, l_s) = -\sum_{\{s\}\in\mathcal{B}_s} \log \frac{\exp\left(e_s l_s^\top/\tau\right)}{\sum_{j\in\mathcal{B}_s}\exp\left(e_s l_j^\top/\tau\right)}, \tag{11}$$

where $\mathcal{B}_u$ and $\mathcal{B}_s$ denote the sets of user and service embeddings in a training batch. The positive pairs $\langle e_u, l_u\rangle$ and $\langle e_s, l_s\rangle$ capture the interactions of users and services across temporal and spatial contexts. The negative pairs are sampled from other users or services within the same batch, denoted as $\langle e_u, l_i\rangle$ and $\langle e_s, l_j\rangle$, where $l_i$ and $l_j$ are the location embeddings of different users and services, respectively.

Finally, we design a contrastive loss function by constructing two types of contrastive views—based on cluster-aware and time-aware to extract self-supervised signals. This contrastive strategy enhances the expressiveness of user and service representations and helps our model to learn more effective representations for accurate QoS prediction. The final contrastive loss is defined as:

$$\mathcal{L}_{cl} = \mathcal{L}\left(e_u, l_u\right) + \mathcal{L}\left(e_s, l_s\right). \tag{12}$$

### 3.5 QoS Prediction and Model Training

After obtaining the similar user embedding $e_u$, similar service embedding $e_s$, and corresponding location embeddings $l_u$ and $l_s$, we capture nonlinear relationships through a feature interaction mechanism that incorporates spatial context information. Inspired by the highway network [18], a gating mechanism is introduced to control the feature flow between transformed and original representations. For each embedding pair, such as $(e_u, e_s)$, the interaction is defined as:

$$f = G\cdot H + (1-G)\cdot(e_u+e_s), \tag{13}$$

where $H = W_H(e_u+e_s)+b_H$, $G=\sigma(W_G(e_u+e_s)+b_G)$, and $W_H, W_G \in \mathbb{R}^{d\times d}$ are learnable parameters with biases $b_H, b_G \in \mathbb{R}^{1\times d}$. Following the same process, six second-order interactions—$(e_u, e_s)$, $(e_u, l_u)$, $(e_u, l_s)$, $(e_s, l_u)$, $(e_s, l_s)$, and $(l_u, l_s)$—are generated. To retain the most informative relationships, we compute their $L_2$-norms and select the top four with the largest magnitudes:

$$z = [f_{i_1}, f_{i_2}, f_{i_3}, f_{i_4}] \in \mathbb{R}^{1\times 4d}, \tag{14}$$

where $z$ denotes the final aggregated feature representation. The feature vector $z$ is then fed into a Multi-Layer Perceptron (MLP) for QoS prediction:

$$\hat{r}_{u,s,t} = \sigma(W_n \cdot g_{n-1} + b_n), \tag{15}$$

where $g_{n-1}$ is the previous layer's output, and $W_n, b_n$ are learnable parameters.

For model optimization, we employ the $\mathcal{L}_1$ loss, which is less sensitive to outliers and provides stable gradients. Finally, the total loss combines the main and contrastive objectives:

$$\mathcal{L} = \sum_{u,s,t} |r_{u,s,t} - \hat{r}_{u,s,t}| + \lambda\|\theta\| + \gamma\mathcal{L}_{cl}, \tag{16}$$

where $\gamma$ balances the contribution of contrastive learning. This unified training strategy enables STCL to jointly optimize representation learning and QoS prediction in an end-to-end manner.

**Table 1.** Statistics of the WS-Dream dataset.

| User Statistics | Values | Service Statistics | Values |
|---|---|---|---|
| Users | 140 | Services | 1000 |
| Invocation time | 96 | Service invocations | 13,440,000 |
| Users' IP addresses | 114 | Services' IP addresses | 317 |
| Users' countries | 32 | Services' countries | 29 |
| Users' cities | 65 | Services' cities | 161 |
| Users' autonomous systems | 58 | Services' autonomous systems | 180 |
| Users' longitude and latitude | 65 | Services' longitude and latitude | 166 |
| Range of response time (s) | 0–20 | | |

# 4 Experiment and Evaluation

## 4.1 Dataset Details and Evaluation Metrics

To evaluate the model performance, we introduce the WS-Dream dataset [32,34], which contains large-scale interaction records between users and services. The dataset details are summarized in Table 1. Due to computational limitations and dataset constraints, our experiments use the first 96 time slices of the WS-Dream dataset, including the first 140 users and 1,000 services in terms of response time.

The complete dataset contains 13,440,000 possible user-service-time records, among which 1,311,744 QoS values are actually observed. To ensure data quality, records with response times of 0 or 20 are removed, as they indicate either missing invocations or abnormal delays [20].

For experiments, we randomly select subsets of the data with five different matrix densities (1%, 2%, 3%, 4%, and 5%) as training sets, while the remaining samples under each density serve as the test sets. We evaluate model performance using Mean Absolute Error (MAE) and Root Mean Square Error (RMSE) [31], which are widely adopted in QoS prediction. Lower values of both metrics indicate higher prediction accuracy.

### 4.2 Performance Comparison

To verify the effectiveness of the STCL model, we compare our method with traditional collaborative filtering (CF)-based methods and newer methods on the response time dataset. The comparison results are summarized in Table 2. UMEAN [6], UPCC [17], IPCC [16], and UIPCC [31] represent typical memory-based CF methods; PMF [15] is a representative model-based method; PLMF [24] and TRCF [35] are time-aware methods; LDCF [28] considers location factors; and SCATSF [33] integrates spatiotemporal context.

**Table 2.** Performance comparison on RT under different data density settings.

| Methods | MD = 1% | | MD = 2% | | MD = 3% | | MD = 4% | | MD = 5% | |
|---|---|---|---|---|---|---|---|---|---|---|
| | MAE | RMSE | MAE | RMSE | MAE | RMSE | MAE | RMSE | MAE | RMSE |
| UMEAN | 0.9398 | 1.9119 | 0.9115 | 1.8643 | 0.9062 | 1.8415 | 0.8905 | 1.8116 | 0.8571 | 1.7200 |
| UPCC | 0.9400 | 1.8432 | 0.9057 | 1.8277 | 0.8197 | 1.7389 | 0.8112 | 1.7341 | 0.7559 | 1.6335 |
| IPCC | 0.9980 | 2.1579 | 0.9178 | 1.9857 | 0.9398 | 1.9611 | 0.9137 | 1.9226 | 0.8770 | 1.8437 |
| UIPCC | 0.8754 | 1.7434 | 0.8465 | 1.7332 | 0.8170 | 1.7033 | 0.7886 | 1.6883 | 0.7452 | 1.6145 |
| PMF | 0.9893 | 1.8524 | 0.8583 | 1.8045 | 0.8066 | 1.7424 | 0.7918 | 1.6504 | 0.7779 | 1.5807 |
| PLMF | 0.8297 | 1.7013 | 0.7873 | 1.6371 | 0.7671 | 1.5894 | 0.7543 | 1.5094 | 0.6771 | 1.4568 |
| LDCF | 0.7809 | 1.6591 | 0.7027 | 1.5096 | 0.6689 | 1.4140 | 0.6449 | 1.3815 | 0.5924 | 1.2774 |
| SCATSF | 0.6731 | 1.4525 | 0.5554 | 1.4005 | 0.4683 | 1.3211 | 0.4428 | 1.2866 | 0.4386 | 1.2561 |
| TRCF | 0.4231 | 1.4016 | 0.3813 | 1.3727 | 0.3567 | 1.3130 | 0.3493 | 1.2719 | 0.3452 | **1.2316** |
| STCL | **0.3777** | **1.2861** | **0.3552** | **1.2649** | **0.3475** | **1.2550** | **0.3416** | **1.2389** | **0.3387** | 1.2364 |
| Gains | 10.73% | 8.25% | 6.84% | 7.85% | 2.58% | 4.42% | 2.21% | 2.60% | 1.88% | –0.39% |

SCATSF and TRCF demonstrate strong performance among existing approaches. SCATSF effectively leverages spatiotemporal context to learn deep feature representations, while TRCF models temporal dynamics through time-averaged QoS and deviation migration. However, our proposed STCL consistently achieves competitive or superior results across different matrix densities. In particular, under sparse conditions (e.g., 1% density), STCL reduces MAE by 10.73% and RMSE by 8.25% compared to the best baseline. Although its RMSE is slightly higher than the best baseline at 5% density (1.2364 vs. 1.2316), this difference is negligible and reflects the reduced benefit of contrastive learning when sufficient data is available.

**Table 3.** Ablation Experiment

| Methods | MD = 1% | | MD = 2% | | MD = 3% | | MD = 4% | | MD = 5% | |
|---|---|---|---|---|---|---|---|---|---|---|
| | MAE | RMSE | MAE | RMSE | MAE | RMSE | MAE | RMSE | MAE | RMSE |
| w/o Cluster-aware | 0.3840 | 1.2952 | 0.3590 | 1.2681 | 0.3516 | 1.2565 | 0.3435 | 1.2449 | 0.3411 | 1.2406 |
| w/o Time-aware | 0.3827 | 1.2922 | 0.3604 | 1.2795 | 0.3517 | 1.2613 | 0.3460 | 1.2465 | 0.3416 | 1.2429 |
| w/o CL | 0.4062 | 1.3170 | 0.3960 | 1.2911 | 0.3652 | 1.2622 | 0.3546 | 1.2532 | 0.3481 | 1.2487 |
| STCL | **0.3777** | **1.2861** | **0.3552** | **1.2649** | **0.3475** | **1.2550** | **0.3416** | **1.2389** | **0.3387** | **1.2364** |

### 4.3 Ablation Study

To further verify the effectiveness of contrastive learning and the rationality of the contrastive view design, we examine three STCL variants to analyze the impact of key components. In STCL, contrastive views are constructed from cluster-aware $(c_u, c_s)$ and time-aware $(t_u, t_s)$ similarities, which jointly capture user-service relationships. We also test variants using only one similarity type, and the results in Table 3 show that removing either leads to a clear performance drop, confirming their complementary roles.

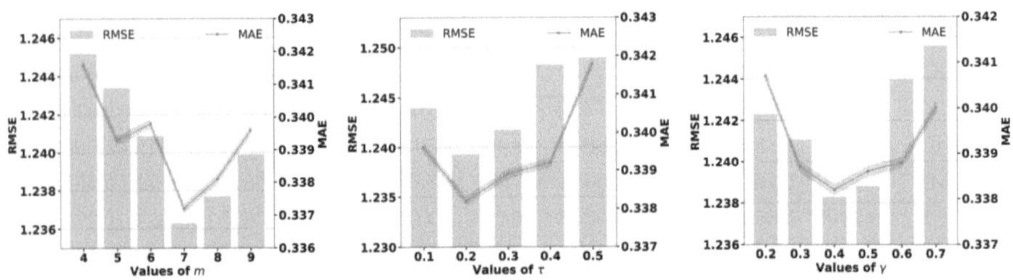


(a) $m$ (b) $\tau$ (c) $\gamma$

**Fig. 2.** Model comparison and parameter sensitivity analysis.

- **w/o Cluster-aware**: Removes $(c_u, c_s)$, keeping only temporal features $(t_u, t_s)$. The decline in accuracy indicates that clustering-based similarities provide critical contextual information beyond temporal features.
- **w/o Time-aware**: Excludes $(t_u, t_s)$, using only cluster-aware features $(c_u, c_s)$. The degraded results demonstrate that temporal dynamics are essential for capturing changing user-service interactions.

- **w/o CL**: Removes the contrastive loss $\mathcal{L}(e_u, l_u)$ and $\mathcal{L}(e_s, l_s)$, preventing the model from leveraging contrastive signals. The performance drop highlights that contrastive learning significantly enhances feature representation and generalization, especially under sparse conditions.

### 4.4 Key Parameters Analysis

**Number of Similar Neighborhoods.** The parameter $m$ denotes the number of similar users or services identified from interaction records. We vary $m$ from 4 to 9 to assess its impact. As shown in Fig. 2(a), the model achieves the lowest MAE and RMSE when $m = 7$. Too small an $m$ limits contextual information, while too large a value introduces redundant neighbors and noise. Hence, $m = 7$ yields the best balance between information diversity and reliability.

**Temperature Coefficient.** The parameter $\tau$ denotes the temperature coefficient controlling negative sample discrimination. As shown in Fig. 2(b), a small $\tau$ overemphasizes hard negatives, while a large one weakens discrimination. We set $\tau = 0.2$, which offers the best balance between generalization and accuracy.

**Weight Coefficient.** The parameter $\gamma$ balances the main loss and contrastive loss in the model by controlling their relative weights in the total loss function, thereby influencing overall performance. To examine its effect, we fixed other parameters and varied $\gamma$. As shown in Fig. 2(c), model performance first improves and then declines as $\gamma$ increases, achieving the best results when $\gamma = 0.4$. This indicates that a proper balance between the main loss and contrastive loss allows the model to effectively capture global data patterns while enhancing local similarity learning through contrastive objectives.

## 5 Conclusion and Fulture Work

In this paper, we propose a spatio-temporal context-aware QoS prediction method via contrastive learning. The method effectively captures rich contextual information of users and services by considering their spatio-temporal backgrounds and utilizing clustering and spatial perception modules. To address the challenge of data sparsity, we introduce a multi-scale contrastive learning module, which extracts self-supervised signals at different scales, significantly improving the accuracy of QoS prediction. In future work, we plan to develop adaptive clustering strategies, explore advanced temporal modeling techniques, and design learnable contrastive view construction mechanisms to further enhance model generalization and robustness.

**Acknowledgments.** This work is supported by the National Natural Science Foundation of China (No.62272001), and the Open Project of the Key Laboratory of Intelligent Computing & Signal Processing of the Ministry of Education (No. 2023A008), and Hefei City Project for Selecting the Best Candidates to Undertake the Key Technology Research (No. 2023SGJ014).

## References


1. Agarap, A.F.: Deep learning using rectified linear units (relu). CoRR (2018)
2. Ahmed, M., Seraj, R., Islam, S.M.S.: The k-means algorithm: a comprehensive survey and performance evaluation. Electronics (2020)
3. Chen, K., Mao, H., Shi, X., Xu, Y., Liu, A.: Trust-aware and location-based collaborative filtering for web service QoS prediction. In: COMPSAC (2017)
4. Chen, T., Kornblith, S., Norouzi, M., Hinton, G.: A simple framework for contrastive learning of visual representations. In: Proceedings of the 37th International Conference on Machine Learning (2020)
5. Fan, X., Hu, Y., Zheng, Z., Wang, Y., Brézillon, P., Chen, W.: Casr-tse: context-aware web services recommendation for modeling weighted temporal-spatial effectiveness. TSC (2017)
6. Fu, Q., Ma, Z., Ben, X., Liu, L.: Using nearest graph QoS prediction method for service QoS prediction. Secur. Commun. Netw. (2018)
7. Ghafouri, S.H., Hashemi, S.M., Hung, P.C.: A survey on web service QoS prediction methods. TSC (2020)
8. Jaiswal, A., Babu, A.R., Zadeh, M.Z., Banerjee, D., Makedon, F.: A survey on contrastive self-supervised learning. Technologies (2021)
9. Jia, Z., Jin, L., Zhang, Y., Liu, C., Li, K., Yang, Y.: Location-aware web service QoS prediction via deep collaborative filtering. TCSS (2022)
10. Koren, Y., Bell, R., Volinsky, C.: Matrix factorization techniques for recommender systems. Computer (2009)
11. Li, Z., Liu, F., Yang, W., Peng, S., Zhou, J.: A survey of convolutional neural networks: analysis, applications, and prospects. TNNLS (2022)
12. Liu, J., Tang, M., Zheng, Z., Liu, X., Lyu, S.: Location-aware and personalized collaborative filtering for web service recommendation. TSC (2015)
13. Liu, X., et al.: Self-supervised learning: generative or contrastive. TKDE (2021)
14. Ma, H., Hu, Z., Li, K., Zhu, H.: Variation-aware cloud service selection via collaborative QoS prediction. TSC (2019)
15. Mnih, A., Salakhutdinov, R.R.: Probabilistic matrix factorization. In: NIPS (2007)
16. Sarwar, B., Karypis, G., Konstan, J., Riedl, J.: Item-based collaborative filtering recommendation algorithms. In: WWW (2001)
17. Shao, L., Zhang, J., Wei, Y., Zhao, J., Xie, B., Mei, H.: Personalized QoS prediction for web services via collaborative filtering. In: ICWS (2007)
18. Srivastava, R.K., Greff, K., Schmidhuber, J.: Highway networks. arXiv preprint arXiv:1505.00387 (2015)
19. Sun, H., Liu, J., Zhang, J.: A survey of contrastive learning in NLP. In: ISAEECE (2022)
20. Tong, E., Niu, W., Liu, J.: A missing QoS prediction approach via time-aware collaborative filtering. TSC (2022)
21. Wang, T., Isola, P.: Understanding contrastive representation learning through alignment and uniformity on the hypersphere. CoRR (2020)
22. Wu, J., et al.: Self-supervised graph learning for recommendation. In: SIGIR (2021)
23. Xia, X., Yin, H., Yu, J., Wang, Q., Xu, G., Nguyen, Q.V.H.: On-device next-item recommendation with self-supervised knowledge distillation. In: SIGIR (2022)
24. Xiong, R., Wang, J., Li, Z., Li, B., Hung, P.C.: Personalized LSTM based matrix factorization for online QoS prediction. In: ICWS (2018)
25. Xu, R., Wunsch, D.: Survey of clustering algorithms. TNN (2005)

26. Yu, J., Yin, H., Xia, X., Chen, T., Li, J., Huang, Z.: Self-supervised learning for recommender systems: a survey. TKDE (2023)
27. Zhang, Y., Yin, C., Lu, Z., Yan, D., Qiu, M., Tang, Q.: Recurrent tensor factorization for time-aware service recommendation. ASC (2019)
28. Zhang, Y., Yin, C., Wu, Q., He, Q., Zhu, H.: Location-aware deep collaborative filtering for service recommendation. TS (2019)
29. Zhang, Y., Zhang, Y., Zhang, Y., Sang, L., Yang, Y.: Unveiling contrastive learning capability of neighborhood aggregation for collaborative filtering. In: SIGIR (2025)
30. Zheng, X., Da Xu, L., Chai, S.: QoS recommendation in cloud services. IEEE Access (2017)
31. Zheng, Z., Ma, H., Lyu, M.R., King, I.: Wsrec: a collaborative filtering based web service recommender system. In: ICWS (2009)
32. Zheng, Z., Zhang, Y., Lyu, M.R.: Investigating QoS of real-world web services. TSC (2012)
33. Zhou, J., Ding, D., Wu, Z., Xiu, Y.: Spatial context-aware time-series forecasting for QoS prediction. TNSM (2023)
34. Zhou, Q., Wu, H., Yue, K., Hsu, C.H.: Spatio-temporal context-aware collaborative QoS prediction. FGCS (2019)
35. Zou, G., Huang, Y., Hu, S., Gan, Y., Zhang, B., Chen, Y.: TRCF: temporal reinforced collaborative filtering for time-aware QoS prediction. TSC (2024)

# ShipTraj-R1: Reinforcing Ship Trajectory Prediction in Large Language Models via Group Relative Policy Optimization

Yang Zhan[1,2], Yunhao Li[3], Chao Zhang[3], Yuxu Lu[2], and Yan Li[3(✉)]

[1] School of Artificial Intelligence, Optics, and Electronics (iOPEN), Northwestern Polytechnical University, Xi'an 710072, China

[2] Department of Logistics and Maritime Studies, The Hong Kong Polytechnic University, Hong Kong 999077, China

[3] State Key Laboratory of Information Engineering in Surveying, Mapping and Remote Sensing, Wuhan University, Wuhan 430079, China

liyanWHU@whu.edu.cn

**Abstract.** Recent advancements in reinforcement fine-tuning have significantly improved the reasoning ability of large language models (LLMs). In particular, methods such as group relative policy optimization (GRPO) have demonstrated strong capabilities across various fields. However, applying LLMs to ship trajectory prediction remains largely unexplored. In this paper, we propose **ShipTraj-R1**, a novel LLM-based framework that reformulates ship trajectory prediction as a text-to-text generation problem. (1) We design a dynamic prompt containing trajectory information about conflicting ships to guide the model to achieve adaptive chain-of-thought (CoT) reasoning. (2) We introduce a comprehensive rule-based reward mechanism to incentivize the reasoning format and prediction accuracy of the model. (3) Our ShipTraj-R1 is reinforced through the GRPO mechanism guided by domain-specific prompts and rewards, and utilizes the Qwen3 as the model backbone. Extensive experimental results on two complex and real-world maritime datasets show that the proposed ShipTraj-R1 achieves the least error compared with state-of-the-art deep learning and LLM-based baselines.

**Keywords:** Ship Trajectory Prediction · Large Language Models · Group Relative Policy Optimization · Reinforcement Fine-Tuning

## 1 Introduction

The surge in global maritime trade and the increasing density of ship traffic in complex waterways have intensified the demand for accurate and robust ship trajectory prediction [9]. Accurate prediction is critical for advancing maritime intelligent transportation services and ensuring navigation safety through effective collision avoidance and congestion management [12,27]. Traditional methods primarily rely on Deep Learning (DL) models, such as Long Short-Term Memory


R. C.-W. Wong et al. (Eds.): PAKDD 2026, LNAI 16597, pp. 56–68, 2026.
https://doi.org/10.1007/978-981-92-1300-9_5

(LSTM) and Convolutional Neural Networks (CNNs), and their hybrids, demonstrating a strong ability to capture sequential and spatio-temporal features.

Recently, Large Language Models (LLMs) has shown great potential in sequence generation and complex reasoning tasks across various domains [5,10]. Leveraging their powerful semantic understanding and inherent reasoning capabilities, researchers have begun exploring LLMs for trajectory prediction. Approaches like LMTraj-SUP [1] and LG-Traj [3] encode continuous coordinates into symbolic tokens, demonstrating the potential of LLMs to generate trajectories autoregressively, analogous to generating natural language sentences.

However, the application of LLMs specifically for ship trajectory prediction remains largely unexplored. Crucially, even the existing LLM-based prediction models are black box regressors that rely on supervised fine-tuning. These methods imitate historical trajectories, which is insufficient. The cost of error in collision avoidance is extremely high, requiring that the complex reasoning process of the prediction model not only refer to historical data but also take into account the status of the surrounding ships, especially those with collision risks.

To address these challenges, we propose ShipTraj-R1, a novel LLM-based prediction framework built upon reinforcement fine-tuning and Chain-of-Thought (CoT) reasoning. Our method fundamentally reformulates ship trajectory prediction as a text-to-text generation problem. The core of ShipTraj-R1 lies in its ability to enforce domain-specific reasoning and precision by employing the Group Relative Policy Optimization (GRPO) mechanism. Specifically, we design a dynamic prompt incorporating trajectory information about conflicting ships to induce adaptive CoT reasoning, which is essential for proactive collision avoidance. Then, the reinforcement process is dynamically guided by a novel rule-based reward function that simultaneously incentivizes the model's structured reasoning format and quantifies coordinate prediction accuracy, ensuring reliability in complex maritime scenarios. Finally, by fine-tuning our model, ShipTraj-R1 can iteratively self-improve its reasoning and prediction capabilities in complex scenarios. To the best of our knowledge, we have for the first time provided a benchmark evaluation of LLMs in the ship trajectory prediction.

Our contributions can be summarized as follows:

- We propose ShipTraj-R1, a novel LLM-based ship trajectory prediction framework. It reformulates the traditional numerical regression task into a text-to-text generation problem with CoT reasoning.
- We design a prompt containing context about conflicting ships, rule-based reward mechanisms and the reinforcement fine-tuning paradigm using GRPO, specifically for ship trajectory prediction.
- We provide a comprehensive benchmark on real-world ship trajectory data in various representative and SOTA models and LLM-based baselines. Extensive experiments demonstrate that ShipTraj-R1 is promising and of great potential for ship trajectory prediction.

# 2 Related Work

## 2.1 Ship Trajectory Prediction

Deep learning (DL) methods have become highly competitive in ship trajectory prediction [9] and are mainly divided into three categories: discriminative, generative, and hybrid methods. Discriminative methods are widely used, learning direct mappings from input to output. LSTM and Gate Recurrent Unit (GRU) can effectively capture long-term dependencies in trajectory sequences and have become mainstream. CNNs are often combined with these temporal models [14] to capture comprehensive spatiotemporal information. Furthermore, graph-based models like Graph Convolutional Neural Networks (GCNN) and Graph Auto-Encoders (GAE), are applied to effectively model the mutual influence and interactive behaviors among individual ships [20]. Generative methods are designed to comprehend data distributions and produce new data samples. This category includes Generative Adversarial Networks (GANs) [13], with specific variants like Time-series GAN and Sequence GAN [24] being highly relevant for time-series analysis. Hybrid methods provide an effective and adaptable approach by combining two or more DL models.

## 2.2 LLM-Based Trajectory Prediction

The application of LLMs for trajectory prediction is an emerging research direction, rapidly expanding due to their powerful semantic and reasoning capabilities [25]. Existing methods can be broadly divided into direct language-based modeling and LLM-based auxiliary prediction. For direct language-based modeling, trajectory data is adapted into textual format for LLMs, enabling them to predict future trajectory coordinate "tokens" akin to generating words. For example, LMTraj-SUP [1] reformulates the task as question-answering. It converts numerical and scene features into structured prompts and utilizes a decimal-aware tokenizer to bridge the gap between continuous values and discrete tokens. Traj-LLM [7] integrates a frozen GPT-2 backbone directly into the prediction pipeline. It encodes multimodal scene context into transformer-compatible token sequences and employs LoRA for fine-tuning, demonstrating strong few-shot adaptability. LLMs also serve several critical auxiliary functions in this domain. Language-guided scene understanding uses language as a complementary modality to enrich spatial context [3]. By encoding semantics like scene images or traffic rules, models achieve more interpretable and socially aware predictions. Language-driven data generation leverages LLMs as generative engines to synthesize diverse and controllable trajectory datasets from high-level textual descriptions [23]. Finally, language-based reasoning and interpretability employs LLMs to explain or justify a model's predictions [15]. This often involves CoT prompting to verbalize the model's decision-making process, enhancing safety and user trust. However, despite this significant progress, the application of LLMs for ship trajectory prediction remains a largely unexplored area.

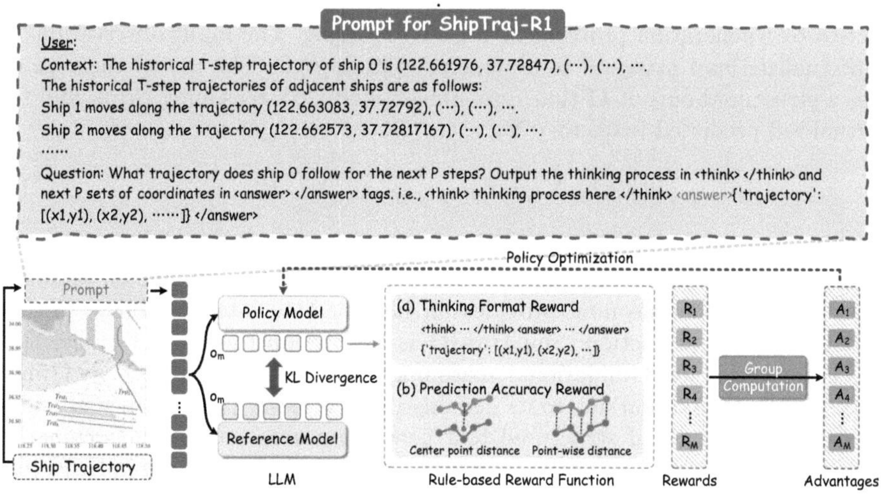


**Fig. 1.** Overview of our proposed ShipTraj-R1 framework.

# 3 Methodology

## 3.1 Problem Definition

The objective of ship trajectory prediction is to forecast a ship's future coordinates given a sequence of its past observations. Formally, the historical trajectory $\mathbf{X}_i$ with length $T_{obs}$ for ship $i$ can be defined as a sequence of spatial points $\mathbf{X}_i = \{(lon_i^t, lat_i^t)|t \in [1, \ldots, T_{obs}]\}$, where $(lon_i^t, lat_i^t)$ is the longitude and latitude at time step $t$. The task is to predict the ground truth future trajectory $\mathbf{Y}_i = \{(lon_i^t, lat_i^t)|t \in [T_{obs}+1, \ldots, T_{obs}+T_{pred}]\}$. The prediction model $\mathcal{F}$ seeks to minimize the error or distance between the predicted trajectory $\hat{\mathbf{Y}}_i = \mathcal{F}(\mathbf{X}_i)$ and the ground truth trajectory $\mathbf{Y}_i$.

## 3.2 ShipTraj-R1 Model

We propose ShipTraj-R1, a specialist LLM for ship trajectory prediction. The overall architecture of ShipTraj-R1 is illustrated in Fig. 1. The framework is built upon two core principles: (1) reformulating the numerical prediction into a text-to-text generation task, and (2) leveraging a powerful, open-source LLM model as the core reasoning agent.

**Task Reformulation and Model.** As defined in Sect. 3.1, existing trajectory prediction is treated as a numerical regression problem. They typically encode continuous GPS coordinates through normalization or specialized schemes, use models with a decoder for numerical prediction, and then apply inverse transformations. In contrast, we instead adopt a "first-reasoning, then-predicting" paradigm. As an LLM-based approach, ShipTraj-R1 reformulates the task as a text-to-text generation problem with CoT reasoning. The input observation $\mathbf{X}_i$ is textualized and attached to a structured prompt $P$. The model then generates a structured output $\mathbf{O}$ that fuses the explicit CoT reasoning *think* and the textualized predicted trajectory $\hat{\mathbf{Y}}_i$:

$$think,\ \hat{\mathbf{Y}}_i = \mathcal{F}_{ShipTraj-R1}\left(\mathbf{X}_i,\ P\right). \tag{1}$$

The text-based nature eliminates the need for normalization/inverse-normalization steps of longitude and latitude, simplifying the model architecture. The explicit reasoning provides an intermediate interpretability process that justifies its prediction. ShipTraj-R1 is built upon the Qwen3 model [22], which serves as the foundational reasoning and prediction backbone. Qwen3 serves as our foundation due to its demonstrated superior performance in mathematical reasoning and structured text generation across various benchmarks, making it highly suitable for the coordinate-level precision required in trajectory prediction.

**Prompt Design Contextualizing Conflicting Ships.** We design a dynamic prompt that explicitly incorporates the contextual information of the trajectories of nearby ships with a significant risk of conflict. In maritime navigation, this is crucial because the trajectory is not purely kinematic but is dominated by collision avoidance logic. As illustrated in Fig. 1, the prompt serves as a cognitive scaffold that guides the model through a "first-reasoning, then-predicting" paradigm. The prompt structure explicitly provides the necessary spatio-temporal context, including the historical trajectory of the target ship and the trajectories of adjacent ships that pose a risk of conflict. Our methodology leverages the inherent capability of LLMs to formulate CoT reasoning without requiring labor-intensive, manually annotated reasoning data. Instead of being constrained to mimic fixed logical patterns, the model is incentivized to autonomously generate an adaptive thinking process that is tailored to the adjacent ship context of each trajectory sample. This comprehensive context is critical for prompting the LLM to perform adaptive social-spatial reasoning.

**Ship Domain Conflict Detection Algorithm:** We establish a quaternion ship domain (QSD) model to conduct ship domain conflict detection [19]. This step assesses the collision risk between the raw adjacent ship and the target ship. The trajectory information of conflicting ships that exceed the risk threshold is provided to the model instead of providing trajectories of all neighboring ships. The QSD model defines a ship's requisite safe water domain, which other ships must avoid to prevent collision risk. The QSD is defined in the ship-centered coordinate system by the four safety distance: $R_{bow}$ (bow), $R_{stern}$ (stern), $R_{starb}$ (starboard), and $R_{port}$ (port). The boundary of the QSD is defined as:

$$\text{QSD} = \{(x,y) | f_k(x,y;H) \leq 1, H = \{R_{fore}, R_{aft}, R_{starb}, R_{port}\}\}, \tag{2}$$

where $k$ determines the shape of the QSD. When $k$=1, its shape is a combined quadrilateral. When $k=2$, its shape is a combined ellipse.

### 3.3 Rule-Based Reward Function

To reinforce the LLM's ability to generate both reliable CoT reasoning and accurate coordinate outputs, we design a comprehensive rule-based reward function, including the thinking format reward and the prediction accuracy reward.

**Thinking Format Reward.** This reward ensures the model's output is correctly structured and reliably parsable for accuracy evaluation. The reward is binary (1 or 0) for compliance. The model's response must strictly adhere to the required structure, separating the reasoning and the final answer using explicit tags, *i.e.*, *<think> the CoT thinking </think> <answer>* ... *</answer>*. Within the ⟨answer⟩⟨/answer⟩ tag, the contained content must be a valid trajectory object. This object must contain the key "trajectory" whose value is a list of exactly $T_{pred}$ predicted coordinate pairs. Furthermore, each coordinate pair must fall within the predefined geographic boundary constraints of the maritime region. Under the above constraints, the trajectory format is considered correct.

**Prediction Accuracy Reward.** This task-specific reward quantifies the error between the predicted trajectory $\hat{\mathbf{Y}}_i$ and the ground truth trajectory $\mathbf{Y}_i$. We use the Vincenty formula to calculate the geodesic distance between coordinates. First, we calculate the center point of both the ground-truth and predicted trajectories by averaging their respective coordinates, *i.e.*, $\mathbf{C}_i$ and $\hat{\mathbf{C}}_i$. A reward of 1 is assigned if their distance is less than 120 m; otherwise, the reward is 0. Then, we need to encourage each predicted trajectory point to achieve high precision. We calculate the distance between the predicted coordinates and the true coordinates for each point. For each distance below the threshold of 90 m, we increment the reward by $\frac{1}{T_{pred}}$.

### 3.4 Reinforcement Fine-Tuning

The ultimate objective of ShipTraj-R1 is to elevate the LLM's ability to produce highly accurate and safety-compliant ship trajectory predictions. We achieve this by optimizing the model using the GRPO algorithm [16], guided by our designed rule-based rewards. For each input $x$, which includes the historical trajectory and adjacent ship context, the old policy model $\pi_{\theta_{old}}$ samples $M$ candidate output completions (CoT reasoning and trajectories) $\{o_m\}_{m=1}^{M}$. Each completion $o_m$ is evaluated by the rule-based reward function to obtain a scalar reward $R_m$. This reward quantifies both the correctness of the reasoning structure and the accuracy of the predicted trajectory. The advantage $\hat{A}_m$ for each output $o_m$ is calculated relative to the average reward of its group:

$$\hat{A}_m = \left(R_m - \text{mean}\{R_1, R_2, \ldots, R_M\}\right)/\text{std}\{R_1, R_2, \ldots, R_M\}. \tag{3}$$

This relative advantage calculation naturally promotes samples that exhibit superior prediction accuracy than the group average, effectively utilizing feedback to steer the learning process. The objective $\mathcal{J}_{\text{GRPO}}(\theta)$ is designed to maximize the expected advantage while ensuring the updated policy $\pi_\theta$ does not deviate excessively from the reference policy $\pi_{\text{ref}}$. The objective is formulated as:

$$\begin{aligned} \mathcal{J}_{GRPO}(\theta) &= \mathbb{E}\{o_m\}_{m=1}^{M} \sim \pi_{\theta_{old}}(O|x) \\ &\left[\frac{1}{M}\sum_{m=1}^{M} \min\left(\rho_\theta^{(m)}\hat{A}_m, \text{clip}(\rho_\theta^{(m)}, 1-\epsilon, 1+\epsilon)\hat{A}_m\right) - \beta D_{KL}(\pi_\theta \,\|\, \pi_{\text{ref}})\right], \end{aligned} \tag{4}$$

where $\rho_\theta^{(m)} = \frac{\pi_\theta(o_m|x)}{\pi_{\theta_{\text{old}}}(o_m|x)}$, $\epsilon$ is the clipping hyperparameter, and $\beta$ controls the KL-divergence regularization strength.

# 4 Experiment

## 4.1 Experimental Settings

**Datasets.** To rigorously evaluate the predictive capability and stability of our proposed ShipTraj-R1 framework, we conducted extensive experiments using real-world ship trajectory data derived from the Automatic Identification System (AIS). Our benchmark dataset focuses on two distinct maritime regions: ChengShanJiao Promontory (CSJP) and CaoFeiDian Port (CFDP). The waterway layouts and ship traffic patterns in CSJP waters are generally more complex than those in CFDP. The AIS data was collected over a two-week period, from July 6 to July 19, 2020, for both areas. Prior to model training, the raw, non-equidistant trajectory data was preprocessed using cubic spline interpolation to ensure uniform time intervals. Specifically, the trajectories were sampled at a 5-s interval per point. The CSJP dataset contains 2,649 complete ship trajectories, including 25,592 GPS points. The CFDP dataset comprises 2,948 ship trajectories, including 34,038 GPS points. The total dataset for each region was partitioned into 90% for training and 5% each for validation and testing.

**Evaluation Metrics.** We employ two widely used evaluation metrics for ship trajectory prediction: the Final Displacement Error (FDE) and the Average Displacement Error (ADE). The FDE measures the $L_2$ distance between the predicted position and the true position solely at the final predicted time step. The ADE measures the mean $L_2$ distance between the predicted position and the true position across all predicted time steps in a trajectory. Mathematically, they are defined as:

$$FDE = \frac{1}{M}\sum_{i=1}^{M}\sqrt{(x_i^P - \hat{x}_i^P)^2 + (y_i^P - \hat{y}_i^P)^2}, \tag{5}$$

$$ADE = \frac{1}{M \times P} \sum_{i=1}^{M} \sum_{t=1}^{P} \sqrt{(x_i^t - \hat{x}_i^t)^2 + (y_i^t - \hat{y}_i^t)^2}, \quad (6)$$

where $M$ is the number of trajectories, $P$ is the number of predicted points, $(x_i^t, y_i^t)$ is the ground truth coordinate, and $(\hat{x}_i^t, \hat{y}_i^t)$ is the predicted coordinate at time step $t$ for trajectory $i$. For both ADE and FDE, a lower value indicates higher prediction accuracy.

**Implementation Details.** For the GRPO training, we perform the RL-based post-training paradigm, implemented via VLM-R1 [17]. We adopt Qwen3-8B [22] as our default language model. The initial learning rate is set to 1e-6, and the weight decay is 1e-2. We set the batch size to 16. Our experiments are conducted on four NVIDIA RTX 4090 GPUs and four L20 GPUs. For our prediction task, we utilized past observed trajectories with length $T_{obs} \in [4, 8]$ (20 s or 40 s) to predict future trajectories with length $T_{pred} \in [1, 2, 3, 4]$ (5 s, 10 s, 15 s, or 20 s). Our LLM-based model can unify prediction tasks under different $T_{obs}$ and $T_{pred}$ conditions and has strong flexibility. However, traditional DL-based methods must train separate models for different combinations of $T_{obs}$ and $T_{pred}$.

**Table 1.** Comparisons with representative DL-based methods, LLM-based methods, and SoTA open-source LLMs.

| Model | Release Year-month | CSJP | | CFDP | |
|---|---|---|---|---|---|
| | | FDE | ADE | FDE | ADE |
| ***DL-based Methods*** | | | | | |
| TBENet TRE'24 [8] | 24-11 | 0.009827 | 8.1000e-04 | 0.005993 | 4.0200e-04 |
| GeoCLSTM CAEE'24 [9] | 24-12 | 0.070173 | 5.4530e-03 | 0.064019 | 4.1890e-03 |
| DBSCAN-GeoCLSTM CAEE'24 [9] | 24-12 | 0.033494 | 2.2640e-03 | 0.030183 | 1.4730e-03 |
| GAT-LSTM JMSE'24 [11] | 24-12 | 0.012828 | 2.7260e-03 | 0.009032 | 1.0520e-03 |
| DGCN-Transformer EAAI'25 [26] | 25-02 | 0.009216 | 8.2400e-04 | 0.006087 | 5.8900e-04 |
| SeqLSTM-U-Net OE'25 [4] | 25-08 | 0.009907 | 1.4360e-03 | 0.007183 | 9.6300e-04 |
| ***LLM-based Methods*** | | | | | |
| LMTraj-SUP CVPR'24 [1] | 24-03 | 0.004734 | 9.2331e-05 | 0.001297 | 5.0249e-05 |
| LG-Traj CVPR'25 [3] | 24-03 | 0.004081 | 7.9923e-05 | 0.001004 | 3.7743e-05 |
| Traj-LLM T-IV'25 [7] | 24-06 | 0.003894 | 4.7202e-05 | 0.000674 | 9.6587e-06 |
| ***Open-source LLMs*** | | | | | |
| Baichuan2-7B [21] | 23-12 | 0.028621 | 2.1598e-04 | 0.003834 | 2.8329e-05 |
| Hunyuan-7B [2] | 25-07 | 0.010099 | 1.5845e-04 | 0.003294 | 1.2795e-05 |
| Llama-3-8B [5] | 24-04 | 0.002671 | 7.0153e-05 | 0.000897 | 9.7625e-06 |
| Qwen2.5-3B [18] | 24-09 | 0.003862 | 4.8339e-05 | 0.000882 | 2.8602e-06 |
| Qwen2.5-7B [18] | 24-09 | 0.003255 | 2.9338e-05 | 0.000498 | 6.1298e-07 |
| DeepSeek-R1 [6] | 25-05 | 0.002303 | 2.2688e-05 | 0.000490 | 6.9395e-07 |
| Qwen3-4B [22] | 25-07 | 0.002350 | 2.7751e-05 | 0.000504 | 7.7631e-07 |
| Qwen3-8B [22] | 25-07 | 0.002293 | 2.0529e-05 | 0.000474 | 6.7403e-07 |
| **ShipTraj-R1-4B** (Ours) | – | 0.001582 | 1.8084e-05 | 0.000386 | 4.9705e-07 |
| **ShipTraj-R1-8B** (Ours) | – | **0.001297** | **1.1547e-05** | **0.000311** | **3.8912e-07** |

## 4.2 Comparison with State-of-Art Methods

**Quantitative Results.** The quantitative results presented in Table 1 establish the superior performance of our proposed ShipTraj-R1 framework across both the complex CSJP and CFDP datasets. Firstly, our ShipTraj-R1 achieves the lowest FDE and ADE across all tested baselines, setting a new state-of-the-art for ship trajectory prediction. The results represent a marked improvement over all competitors. The gain over traditional DL-based models is substantial. This large performance gap validates the fundamental advantage of leveraging Large Language Models over small models for complex maritime scenarios. Furthermore, ShipTraj-R1 significantly outperforms previous SOTA LLM-based trajectory models like Traj-LLM. On the CFDP dataset, our model reduces the FDE from Traj-LLM's 0.000674 to 0.000311. Crucially, by comparing ShipTraj-R1-8B against its base model, Qwen3-8B, we observe a definitive reduction in error. This gain directly validates the efficacy of our designed prompt and reward mechanisms in reinforcing the model's spatial reasoning based on latitude and longitude coordinates through reinforcement fine-tuning. The consistent and pronounced error reduction across both datasets highlights that the ShipTraj-R1 approach is not only highly accurate but also robust in generalizing to diverse and challenging maritime traffic environments.

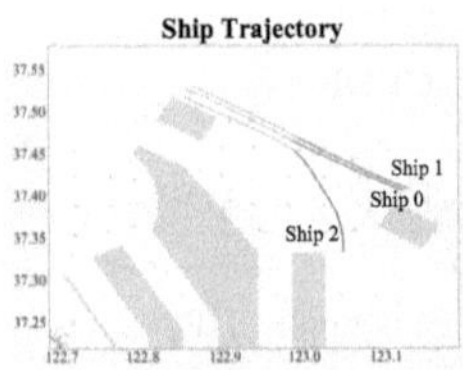


**User:**
Context: The historical 4-step trajectory of ship 0 is (123.0465317,37.43752066), (123.0252457,37.44706119), (123.0039597,37.4566005), (122.9850769,37.4653211). The historical 4-step trajectories of adjacent ships are as follows: Ship 1 moves along the trajectory (123.0533123, 37.43983775), (123.0312538,37.44999125), (123.0112553,37.45980272), (122.9929733,37.46825045). Ship 2 moves along the trajectory (123.0307388, 37.39634613), (123.0204391,37.41284593), (123.0101395,37.42879682), (123.0003548,37.44188219).
Question: What trajectory does ship 0 follow for the next 4 steps? Output the thinking process in <think> </think> and next 4 sets of coordinates in <answer> </answer> tags. i.e., <think> thinking process here </think> <answer>{'trajectory': [(x1,y1), (x2,y2), ······]} </answer>

**ShipTraj-R1's Output**:
<think>I need to figure out the next 4-step trajectory for ship 0 based on the given historical data and the trajectories of ships 1 and 2. Let me analyze the movement of ship 0 first. Let's calculate the differences between consecutive points to see if there's a pattern.
Step 1: Understanding the Movement Pattern ······
Step 2: Calculate the Movement Difference ······
Step 3: Extrapolating the Next Steps. ··· Ship 2 is approaching and there is a risk of collision, we will estimate the next longitude change and latitude change···
Step 4: Predict the next 4 points ······ </think>
<answer>{'trajectory': [(122.9664763, 37.4745879), (122.9479865, 37.4838547), (122.9295861, 37.4914829), (122.911286, 37.5002035)]}</answer>

**Ground Truth:** (122.9685974, 37.47431315), (122.9486847,37.483576555), (122.9301453,37.492294), (122.9119492,37.50073806)

**Fig. 2.** Qualitative examples of our ShipTraj-R1 on the CSJP dataset. In this case, the ShipTraj-R1 adaptively generates CoT reasoning by itself based on the user prompt.

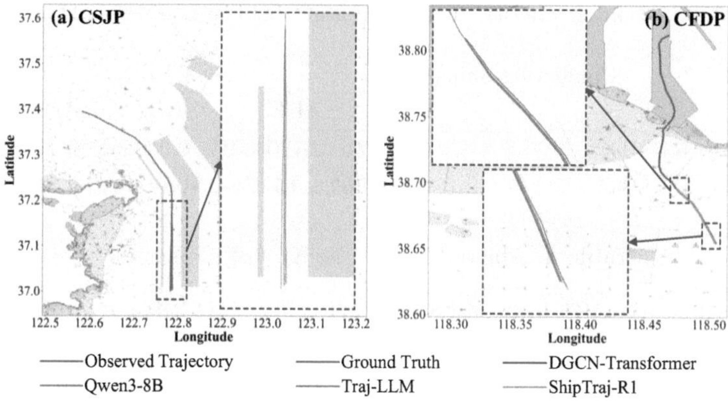


**Fig. 3.** Illustration of predicted ship trajectories from the CSJP and CFDP datasets.

**Qualitative Results.** To visually validate the superiority of ShipTraj-R1 in complex maritime environments, we provide an example in Fig. 2. We have observed that the reasoning process is helpful for physics-based motion analysis and the estimation of the impact of conflicting ships. Figure 3 presents comparative qualitative results against several competitive baselines across the CSJP and CFDP datasets. DGCN-Transfromer [26], Traj-LLM [7], and Qwen3-8B [22] are respectively the most advanced methods among DL-based methods, LLM-based methods, and open-source LLMs. As seen in the zoomed-in views, ShipTraj-R1's predicted trajectory aligns almost perfectly with the Ground Truth trajectory. The qualitative visualization confirms that ShipTraj-R1's reinforcement fine-tuning successfully addresses the limitations of pure geometric and language-based models, yielding predictions that are significantly more accurate, physically compliant, and stable, thereby minimizing the risk of collision or violation in real-world applications.

**Table 2.** Ablation on different fine-tuning methods and chain-of-thought.

| Model | Type | CoT | CSJP | | CFDP | |
|---|---|---|---|---|---|---|
| | | | FDE | ADE | FDE | ADE |
| Baseline-8B | – | – | 0.002293 | 2.0529e-05 | 0.000474 | 6.7403e-07 |
| ShipTraj-R1-8B | SFT | ✗ | 0.001701 | 1.8153e-05 | 0.000412 | 5.1295e-07 |
| ShipTraj-R1-8B | RFT | ✗ | 0.001430 | 1.6216e-05 | 0.000341 | 4.1216e-07 |
| ShipTraj-R1-8B | RFT | ✓ | **0.001297** | **1.1547e-05** | **0.000311** | **3.8912e-07** |

**Table 3.** Ablation on whether the user prompt includes conflicting ship context.

| Model | Conflicting Ships | CSJP | | CFDP | |
|---|---|---|---|---|---|
| | | FDE | ADE | FDE | ADE |
| ShipTraj-R1-8B | ✗ | 0.002756 | 2.7804e-05 | 0.000569 | 6.3881e-07 |
| ShipTraj-R1-8B | ✓ | **0.001297** | **1.1547e-05** | **0.000311** | **3.8912e-07** |

**Table 4.** Ablation on the the KL loss coefficient.

| Model | KL loss coef | CSJP | | CFDP | |
|---|---|---|---|---|---|
| | | FDE | ADE | FDE | ADE |
| ShipTraj-R1-8B | 1e-4 | **0.001297** | **1.1547e-05** | **0.000311** | **3.8912e-07** |
| ShipTraj-R1-8B | 5e-4 | 0.001394 | 1.4580e-05 | 0.000336 | 4.4628e-07 |
| ShipTraj-R1-8B | 1e-3 | 0.002287 | 2.0466e-05 | 0.000519 | 7.3491e-07 |
| ShipTraj-R1-8B | 5e-3 | 0.003650 | 3.6814e-05 | 0.000842 | 2.1277e-06 |

### 4.3 Ablation Studies

**SFT vs. RFT.** We compare the efficacy of our reinforcement fine-tuning (RFT) using GRPO against supervised fine-tuning (SFT), as detailed in Table 2. The RFT-only model (without CoT) significantly outperforms the SFT model. This performance gap validates RFT's superior ability to leverage the rule-based reward signal for enforcing prediction. Furthermore, with CoT reasoning, our ShipTraj-R1 yields the best overall accuracy.

**User Prompt.** We investigate the impact of trajectory information about conflicting ships within the prompt, as shown in Table 3. The omission of this conflict context results in a significant performance drop. This degradation confirms the crucial role of conflicting ship context in guiding the LLM's reasoning.

**KL Loss Coefficient.** The KL divergence regularization term balances the acquisition of new domain knowledge against the retention of the LLM's pre-existing capabilities. As shown in Table 4, the optimal performance is achieved at the lowest coefficient 1e-4. As the coefficient increases, performance degrades.

## 5 Conclusion

In this work, we propose ShipTraj-R1, a novel framework that leverages LLMs with CoT reasoning to achieve robust ship trajectory prediction. (1) Our method reformulates the complex prediction task as an LLM sequence generation problem and utilizes GRPO during reinforcement fine-tuning. (2) Our model can generate an explicit CoT reasoning process to justify its predictions, enhancing interpretability. (3) Unlike traditional methods that require separate models, our framework flexibly unifies prediction tasks under various observation ($T_{obs}$) and prediction ($T_{pred}$) lengths. Extensive experiments on complex real-world datasets demonstrate that ShipTraj-R1 establishes a new state-of-the-art, significantly reducing both FDE and ADE compared to existing baselines.

We acknowledge that the current short-term prediction horizon (up to 20 s) primarily validates the model's capability for high-precision, immediate course-keeping. However, a complete long-term collision avoidance system requires a much longer forecast window. We plan to explore the performance of LLMs in long-term prediction, which holds greater value for practical applications.

**Acknowledgments.** This work is supported in part by grants from the National Natural Science Foundation of China (No.624B2113).

## References

1. Bae, I., Lee, J., Jeon, H.G.: Can language beat numerical regression? Language-based multimodal trajectory prediction. In: Proceedings of the IEEE/CVF Conference on Computer Vision and Pattern Recognition, pp. 753–766(2024)
2. Cao, S., et al.: Hunyuanimage 3.0 technical report. arXiv preprint arXiv:2509.23951 (2025)
3. Chib, P.S., Singh, P.: LG-Traj: LLM guided pedestrian trajectory prediction. In: Proceedings of the IEEE/CVF International Conference on Computer Vision, pp. 6802–6812 (2025)
4. Gan, L., Gao, Z., Zhang, X., Xu, W., Cai, Y., Song, L.: Maritime vessel trajectory prediction based on sequential long short-term memory and u-net architectures. Ocean Eng. **334**, 121598 (2025)
5. Grattafiori, A., et al.: The Llama 3 herd of models. arXiv preprint arXiv:2407.21783 (2024)
6. Guo, D., et al.: DeepSeek-R1 incentivizes reasoning in LLMs through reinforcement learning. Nature **645**(8081), 633–638 (2025)
7. Lan, Z., Liu, L., Fan, B., Lv, Y., Ren, Y., Cui, Z.: Traj-LLM: a new exploration for empowering trajectory prediction with pre-trained large language models. IEEE Trans. Intell. Veh. (2024)
8. Li, H., et al.: Bi-directional information fusion-driven deep network for ship trajectory prediction in intelligent transportation systems. Transp. Res. Part E: Logist. Transp. Rev. **192**, 103770 (2024)
9. Li, Y., Chen, B.Y., Liu, Q., Zhang, Y.: Geohash coding-powered deep learning network for vessel trajectory prediction using clustered ais data in maritime internet of things industries. Comput. Electr. Eng. **120**, 109611 (2024)
10. Li, Y., Zhan, Y., Liang, M., Zhang, Y., Liang, J.: UPTM-LLM: large language models-powered urban pedestrian travel modes recognition for intelligent transportation system. Appl. Soft Comput. 113999 (2025)
11. Li, Y., Yu, Q., Yang, Z.: Vessel trajectory prediction for enhanced maritime navigation safety: a novel hybrid methodology. J. Mar. Sci. Eng. **12**(8), 1351 (2024)
12. Liang, M., Su, J., Gao, R., Liu, R.W., Zhan, Y.: Estimation of vessel link-level travel time distribution: a directed network-driven approach. Ocean Eng. **313**, 119371 (2024)

13. Liu, H., Zhao, H., Wang, J., Yuan, S., Feng, W.: LSTM-GAN-AE: a promising approach for fault diagnosis in machine health monitoring. IEEE Trans. Instrum. Meas. **71**, 1–13 (2021)
14. Lu, B., Lin, R., Zou, H.: A novel CNN-LSTM method for ship trajectory prediction. In: 2021 IEEE 23rd International Conference on High Performance Computing & Communications; 7th International Conference on Data Science & Systems; 19th International Conference on Smart City; 7th Int Conf on Dependability in Sensor, Cloud & Big Data Systems & Application (HPCC/DSS/SmartCity/DependSys), pp. 2431–2436. IEEE (2021)
15. Peng, M., et al.: Lc-LLM: explainable lane-change intention and trajectory predictions with large language models. Commun. Transp. Res. **5**, 100170 (2025)
16. Shao, Z., et al.: DeepSeekMath: pushing the limits of mathematical reasoning in open language models. arXiv preprint arXiv:2402.03300 (2024)
17. Shen, H., et al.: VLM-R1: a stable and generalizable r1-style large vision-language model. arXiv preprint arXiv:2504.07615 (2025)
18. Team, Q.: Qwen2 technical report. arXiv preprint arXiv:2407.10671 (2024)
19. Wang, N.: An intelligent spatial collision risk based on the quaternion ship domain. J. Navig. **63**(4), 733–749 (2010)
20. Wang, S., Li, Y., Xing, H., Zhang, Z.: Vessel trajectory prediction based on spatio-temporal graph convolutional network for complex and crowded sea areas. Ocean Eng. **298**, 117232 (2024)
21. Yang, A., et al.: Baichuan 2: open large-scale language models. arXiv preprint arXiv:2309.10305 (2023)
22. Yang, A., et al.: Qwen3 technical report. arXiv preprint arXiv:2505.09388 (2025)
23. Yang, K., et al.: Trajectory-LLM: a language-based data generator for trajectory prediction in autonomous driving. In: The Thirteenth International Conference on Learning Representations (2025)
24. Yu, L., Zhang, W., Wang, J., Yu, Y.: SeqGAN: sequence generative adversarial nets with policy gradient. In: Proceedings of the AAAI Conference on Artificial Intelligence, vol. 31 (2017)
25. Zhan, Y., Xiong, Z., Yuan, Y.: SkyEyeGPT: unifying remote sensing vision-language tasks via instruction tuning with large language model. ISPRS J. Photogramm. Remote. Sens. **221**, 64–77 (2025)
26. Zhang, Y., Tu, P., Zhao, Z., Chen, X.Y.: Incorporating prior knowledge of collision risk into deep learning networks for ship trajectory prediction in the maritime internet of things industry. Eng. Appl. Artif. Intell. **146**, 110311 (2025)
27. Zhao, Z., Zhan, Y.: DiffSCTG: diffusion-based safety-critical trajectory generation for autonomous vehicles. In: 2025 IEEE 7th International Conference on Communications, Information System and Computer Engineering, pp. 405–414 (2025)

# Enhancing Graph Neural Networks with Adaptive Relation Learning via Variational Inference

Juyeob Lee, Yunho Jeong, Serin Kim, Dayeon Han, and Eunil Park(✉)

Sungkyunkwan University, Seoul, South Korea
juyblee@gmail.com, {0429dbsgh,serin0829,lunejjang}@g.skku.edu,
eunilpark@skku.edu

**Abstract.** Graph Neural Networks are highly effective in learning representations from graph-structured data. To better capture edge heterogeneity, prior approaches often partition edges using pre-defined rules or assign weights using attention mechanisms. However, these approaches suffer from critical limitations: pre-defined rules can be suboptimal, and deterministic methods fail to account for the inherent uncertainty in real-world graphs. To address these challenges, we propose the variational Relational Graph Convolutional Network (vRGCN), a novel framework that leverages Variational Inference to stochastically learn a distribution over latent relations directly from data. By employing a relation learner and a Gumbel-Softmax sampler, vRGCN adaptively infers the most effective relational structure for a given task without relying on any pre-defined information. Our experiments on graph and node classification benchmarks demonstrate that vRGCN consistently achieves superior performance against strong baselines. Furthermore, vRGCN is remarkably parameter-efficient, outperforming these baselines with significantly fewer parameters. Extensive ablation studies validate our design choices, confirming that stochastic sampling is crucial for robust exploration and that the ELBO objective is a vital regularizer for model generalization. vRGCN offers a flexible, powerful, and efficient paradigm for adaptive relation learning in graphs.

**Keywords:** Graph neural networks · Relation learning · Variational inference · Multi-relational graph

## 1 Introduction

Graph Neural Networks (GNNs) have emerged as a promising framework for learning representations from graph-structured data such as social networks [16, 22], user-item interactions [8,21], knowledge graphs [23,32], and 3-dimensional meshes [19,27]. By iteratively aggregating information from neighboring nodes, GNNs effectively capture structural information for various graph-oriented tasks, such as graph classification or node classification. However, many GNNs often


R. C.-W. Wong et al. (Eds.): PAKDD 2026, LNAI 16597, pp. 69–81, 2026.
https://doi.org/10.1007/978-981-92-1300-9_6

oversimplify the nature of edge relations, limiting their performance, particularly in graphs with heterogeneous interactions [20,25].

To handle edge heterogeneity, prior approaches fall into two main categories. The first relies on pre-defined relations or rules to partition edges [3,20]. Although effective in some cases, this approach requires explicit relations to be defined in advance, which may not always be available, accurate, or optimal for the downstream task [17,24]. Establishing these rules often demands significant domain knowledge and may not generalize across different graph structures. The second category uses deterministic mechanisms to assign importance scores to the edges [26,27]. However, these deterministic methods fail to account for the inherent uncertainty in real-world graphs [30].

Probabilistic frameworks have emerged to manage uncertainty in graph-structured data [4,9,28]. Although these methods, including Bayesian networks [31] and Variational Inference (VI) [15], have shown promise in handling noisy or incomplete graphs, existing work has focused primarily on inferring the graph structure or the node labels. To the best of our knowledge, there has been no attempt to explicitly learn a distribution over latent relation types directly from graph data without pre-defined rules.

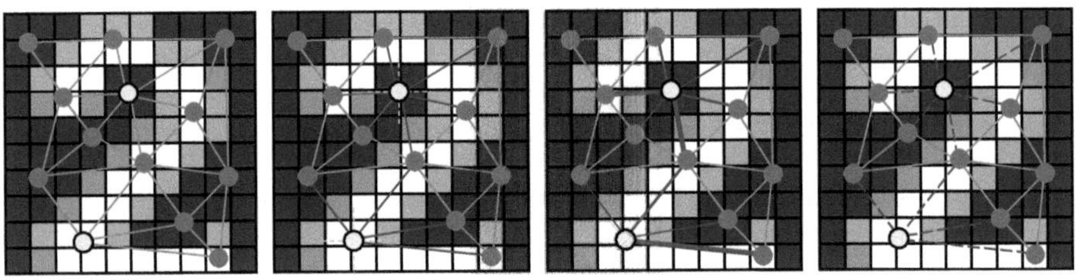

(a) Homogeneous (b) Pre-defined (c) Attention (d) vRGCN

**Fig. 1.** Illustrations of different edge differentiation methods on a superpixel graph. (a) A homogeneous GNN treats all edges uniformly, ignoring their functional roles. (b) Pre-defined rules (e.g., based on spatial coordinates) are rigid and often fail to capture the underlying structure. (c) Attention-based methods learn scalar importance scores but do not distinguish between different types of relations. (d) vRGCN learns a distribution over latent relations, adaptively identifying the functional roles of edges.

In this paper, we introduce the variational Relational Graph Convolutional Network (vRGCN), a novel framework that leverages VI to address these challenges. vRGCN models the diversity of interactions by stochastically learning a distribution over latent relations directly from the data. By employing a relation learner and a Gumbel-Softmax sampler, our framework adaptively infers the most effective relational structure for a given task without relying on any pre-defined information. This allows vRGCN to differentiate the functional roles of edges, capturing complex dependencies and providing more nuanced representations. Figure 1 illustrates this key difference from prior edge differentiation methods. We demonstrate the capability of vRGCN through comprehensive experiments on graph and node classification tasks, where vRGCN consistently exhibits enhanced performance and parameter efficiency. Extensive ablation studies validate our design choices, confirming the effectiveness of our key components. The implementation code is available[1]. Our contributions are as follows:

- We propose vRGCN, a novel framework that learns latent relations directly from data without pre-defined rules, making it adaptable to diverse graphs.
- We integrate VI to explicitly model the uncertainty in graph relations, allowing for more robust and flexible representations.
- We demonstrate that vRGCN achieves superior performance on multiple benchmarks compared to several baselines, while also being parameter-efficient.
- We conduct comprehensive ablation studies verifying that stochastic sampling and the ELBO are crucial for robust exploration and generalization.

## 2 Related Work

### 2.1 Multi-Relational Modeling in GNNs

Relational Graph Convolutional Network (RGCN) [20] learns distinct transformation matrices for each pre-defined relation type. This line of work has been extended by models such as CompGCN [25], which also learns relation representations to reduce over-parameterization. Although effective, the primary limitation of these models is their dependency on a pre-defined relational schema. In many real-world scenarios, such information is unavailable, incomplete, or the pre-defined relations may not be optimal for the downstream task [17,30].

On the other hand, there have been attempts to differentiate edges without relying on explicit relation types. The Graph Attention Network (GAT) [26] and its successor GATv2 [2] assign a scalar importance score to each edge, allowing the model to weigh neighbors differently. Other methods, such as EdgeConv [27], dynamically compute edge features based on local node features, which is particularly effective for spatial data such as point clouds. SelectionGCN [6] proposes partitioning edges into groups using heuristic rules (e.g., spatial direction) and applies distinct weights to each group.

However, these approaches are fundamentally deterministic. This deterministic nature prevents them from modeling the inherent uncertainty and ambiguity in graph relations. This reliance limits the flexibility and robustness of the models, motivating the exploration of probabilistic frameworks.

### 2.2 Probabilistic Approach in GNNs

To manage uncertainty in graph-structured data, various probabilistic approaches using Bayesian methods have emerged. Bayesian GCNN [31] extends the Bayesian framework to GCNs by jointly inferring the graph structure and

[1] https://github.com/jjuyeah/vRGCN.

node labels. Others employed Tempered MCMC to perform Bayesian inference, enabling robust uncertainty quantification [4], or applied Bayesian learning to dynamic spatio-temporal graphs [28]. However, a common thread in these methods is their reliance on sampling-based inference, such as Markov Chain Monte Carlo. Although accurate, these techniques are often computationally intensive and face significant scalability challenges, limiting their applicability.

To address these computational limitations, VI has been introduced as a more scalable and efficient alternative for probabilistic GNNs. VI reformulates the intractable posterior inference problem into an optimization problem, thereby avoiding the costly sampling process. Foundational works such as VGAE [9] incorporate VI to learn probabilistic node representations in an unsupervised manner, while later works expand this by using Dirichlet distributions as priors for tasks such as graph generation [11]. Other advancements, such as VGCN [15], use VI to handle incomplete or adversarially perturbed graphs by applying a prior distribution over the graph structure.

Despite these notable achievements, these probabilistic approaches predominantly focused on learning the graph topology. The problem of learning the semantic types or properties of the relations represented by the edges has been relatively underexplored. Our work addresses this specific gap by proposing a VI-based framework designed to learn a distribution over latent relations rather than the graph structure.

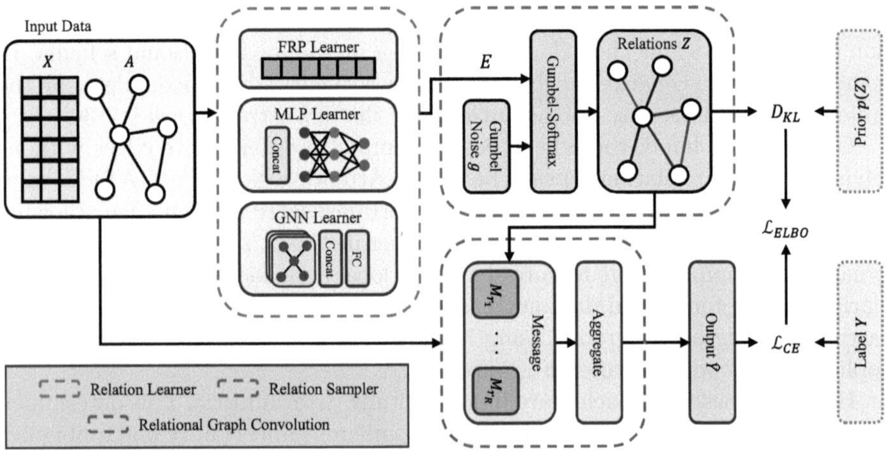


**Fig. 2.** An illustration of variational Relational Graph Convolutional Network (vRGCN) architecture.

## 3 Method

We describe the notations and introduce a variational Relational Graph Convolutional Network (vRGCN). vRGCN consists of three main modules: the relation learner, the relation sampler, and the relational graph convolution. Figure 2 illustrates the overall architecture of vRGCN.

### 3.1 Notations

Let $\mathcal{G} = (\mathcal{V}, \mathcal{E})$ represents an directed graph with set of $N$ nodes $\mathcal{V} = \{v_1, \ldots, v_N\}$ and set of $M$ edges $\mathcal{E}$. Nodes contain $d$-dimensional node features $X \in \mathbb{R}^{N \times d}$. Each edge belongs to one of $R$ relations in multi-relational graphs, where $\mathcal{R}$ is the set of relations. The relations of the edges are represented as $\{r_1, \ldots, r_M\}, \forall r \in \mathcal{R}$. Edges can be represented as an adjacency matrix $A \in \mathbb{R}^{n \times n}$, where $A_{ij} = 1$ if there is an edge from node $v_i$ to node $v_j$, otherwise 0. Labels can fully or partially exist per node $Y = \{y_1, \ldots, y_N\}$, or per graph $Y = \{y_1, \ldots, y_{|\mathcal{G}|}\}$.

### 3.2 Relation Learner

The role of the relation learner is to extract edge features $E \in \mathbb{R}^{N \times R}$ from the input graph. Motivated by [12], we introduce three architectures that explore different hypotheses about how relations are formed.

The full relation parameterization (FRP) learner operates on the premise that relations are independent of input data, assigning a unique set of learnable parameters for each edge as follows:

$$E = W_{FRP}, \tag{1}$$

where $W_{FRP} \in \mathbb{R}^{N \times R}$ is a parameter matrix of FRP learner.

The multi-layer perceptron (MLP) learner is based on the idea that relations are determined by the nodes forming the edge, computing features by applying an MLP to the concatenated features of the source and target nodes as follows:

$$E = \text{MLP}(X_{src} || X_{tgt}), \tag{2}$$

where $X_{src}$ and $X_{tgt}$ are the features of the source and target nodes. $(\cdot||\cdot)$ is a concatenation function.

The GNN learner posits that relations also depend on the neighborhood, first processing node features with a GNN, and then feeding the concatenated representations of node pairs into a final layer as follows:

$$H^{(t)} = \text{GNN}(H^{(t-1)}), \tag{3}$$

$$E = \text{FC}(H_{src}^{(T)} || H_{tgt}^{(T)}), \tag{4}$$

where $H^{(0)} = X$ is the node features, $H_{src}^{(T)}$ and $H_{tgt}^{(T)}$ are the final node embeddings of the source and target nodes, and FC is a fully-connected layer.

### 3.3 Relation Sampler

The relation of each edge is sampled with the relation sampler using the edge features $E$ from the relation learner. We assume that the relations are sampled from categorical distributions consisting of the Softmax values of the edge features. The relation sampler is formulated as follows:

$$p_{uv} = \frac{\exp(E_{uv})}{\sum_{k \in \mathcal{N}(u)} \exp(E_{uk})}, \quad \forall (u,v) \in \mathcal{E}, \tag{5}$$

$$r_{uv} \sim \text{Cat}\left(R, p_{uv}\right). \tag{6}$$

Although Eqs. 5 and 6 successfully sample the relations in the forward process, it has difficulty with backpropagation owing to the sampling procedure. Therefore, we adopt the Gumbel-Softmax function [7] for the gradients to flow to the relation learner. This function uses Gumbel noise $g$ to trick the computation into a sampling procedure, which is formulated as follows:

$$p_{uv} = \frac{\exp((E_{uv} + g_{uv})/\tau)}{\sum_{k \in \mathcal{N}(u)} \exp(E_{uk} + g_{uk})/\tau)}, \quad \forall (u,v) \in \mathcal{E}, \tag{7}$$

where $g = -\log(-\log(\epsilon))$, $\epsilon \sim \text{Uniform}(0,1)$, and $\tau$ is a temperature parameter that controls the smoothness of the approximation. We use a relation with maximum probability from $p_{uv}$ in Eq. 7 in the forward process. For the backward pass, we use a straight-through estimator for gradients to be passed directly through the continuous Gumbel-Softmax probabilities $p_{uv}$. This allows the relation learner to be effectively trained using gradient-based optimization, refining the edge features that guide the stochastic relation sampling process.

### 3.4 Relational Graph Convolution

The relational graph convolution generates the final node representations using the original node features, the adjacency matrix, and the latent relations $Z$ sampled by the relation sampler. We adopt the well-established message passing scheme of [6]. In this framework, messages are computed separately for each of the $R$ latent relations, and a distinct, learnable weight matrix $W_r^{(t)}$ is applied to the neighbors belonging to each relation $r$. The final node representation is produced by aggregating these relation-specific messages.

### 3.5 Objective Function

We optimize the proposed model by maximizing the ELBO $\mathcal{L}_{ELBO}$ instead of $\log p(\hat{Y}|\mathcal{G})$. Regarding the sampled relations as latent variables $Z$, the ELBO is formulated as follows:

$$\mathcal{L}_{ELBO} = \mathbb{E}_{q_\phi(Z|\mathcal{G})}[\log p(Y|Z,\mathcal{G})] - D_{KL}[q_\phi(Z|\mathcal{G})||p(Z)] \tag{8}$$

where $\mathbb{E}_{q_\phi(Z|\mathcal{G})}[\log p(Y|Z,\mathcal{G})]$ is the expected log-likelihood that is computed by the relational graph convolution. $D_{KL}[q_\phi(Z|\mathcal{G})||p(Z)]$ is the KL divergence between the approximate posterior, which is computed by the relation learner and relation sampler, and the prior. We assume the prior $p(\mathbf{R})$ as a Uniform distribution where the probabilities of edges belonging to each relation are $R^{-1}$. Maximizing the expected log-likelihood is typically achieved by minimizing cross-entropy in classification tasks and mean squared error in regression tasks. Therefore, maximizing the ELBO is reformulated as minimizing the following loss functions for each task:

$$\mathcal{L}_{classification} = \mathcal{L}_{CE}(\hat{Y}, Y) + D_{KL}[q_\phi(Z|\mathcal{G})||p(Z)], \tag{9}$$

**Table 1.** Statistics of the Euclidean benchmark datasets

| Dataset | # Graphs | # Nodes | # Edges | # Classes |
|---|---|---|---|---|
| MNIST Sup. [13] | 70,000 | 75 | 1,393.0 | 10 |
| CoMA [18] | 20,465 | 5,023 | 29,990 | 12 |
| FAUST [1] | 100 | 6,890 | 41,328 | 10 |

**Table 2.** Statistics of the non-Euclidean benchmark datasets

| Dataset | # Nodes | # Edges | # Relations | # Classes |
|---|---|---|---|---|
| AIFB [20] | 8,285 | 29,043 | 45 | 4 |
| MUTAG [20] | 23,644 | 74,227 | 23 | 2 |
| BGS [20] | 333,845 | 916,199 | 103 | 2 |
| AM [20] | 1,666,764 | 5,988,321 | 133 | 11 |

$$\mathcal{L}_{regression} = \mathcal{L}_{MSE}(\hat{Y}, Y) + D_{KL}[q_\phi(Z|\mathcal{G})||p(Z)], \tag{10}$$

where $\hat{Y}$ is the prediction of the model. $\mathcal{L}_{CE}$ and $\mathcal{L}_{MSE}$ are cross-entropy and mean squared error, respectively.

## 4 Experiment

### 4.1 Experiment Setup

**Datasets.** We evaluate vRGCN on graph and node classification tasks using seven benchmark datasets. These are categorized into 3 Euclidean graphs, used for graph classification, and 4 non-Euclidean graphs, used for node classification. Detailed statistics for the datasets are provided in Table 1 and 2, respectively.

**Baselines.** We compare the performance of vRGCN with four conventional GNNs that do not model the difference in edges: GCN [10], GraphSAGE [5], $k$-GNN [14], and GIN [29]. We further consider four GNNs that differentiate edges: GAT [26], GATv2 [2], EdgeConv [27], and SelectionGCN [6].

**Implementation Details.** We follow the dataset split strategies and hyperparameter settings of [6] for graph classification tasks and [20] for node classification tasks to ensure a fair comparison. For vRGCN, we conduct a grid search to find optimal values for the number of learner layers, the number of relations, and the ratio between cross-entropy and KL divergence in ELBO for vRGCN. All reported results are the average of five independent runs with random seeds.

**Table 3.** The accuracy (%) and standard deviation of the graph and node classification tasks. The best results are in **bold**, and the second are in *italic*.

| Method | MNIST Sup. | CoMA | FAUST | AIFB | MUTAG | BGS | AM |
|---|---|---|---|---|---|---|---|
| GCN [10] | 64.76 ± 1.42 | 40.51 ± 6.02 | 55.00 ± 6.32 | 90.00 ± 2.22 | 60.88 ± 5.47 | 60.69 ± 8.33 | 34.85 ± 0.00 |
| GraphSAGE [5] | 86.45 ± 0.40 | 45.39 ± 2.12 | 58.00 ± 4.00 | 82.78 ± 2.08 | 63.82 ± 4.22 | 57.24 ± 9.41 | 40.71 ± 1.80 |
| $k$-GNN [14] | 11.37 ± 0.03 | 55.79 ± 1.34 | *73.00* ± 7.48 | 91.11 ± 3.24 | 73.53 ± 2.94 | 62.76 ± 3.38 | *70.30* ± 4.77 |
| GIN [29] | 86.42 ± 0.22 | 66.89 ± 0.57 | 58.00 ± 10.30 | 87.78 ± 5.44 | *75.59* ± 2.73 | 65.52 ± 4.36 | 57.88 ± 9.03 |
| GAT [26] | 94.07 ± 0.36 | 15.74 ± 1.95 | 65.00 ± 5.48 | **94.44** ± 3.04 | 70.29 ± 3.28 | 62.76 ± 3.38 | 64.34 ± 6.91 |
| GATv2 [2] | *96.87* ± 0.12 | 11.75 ± 0.89 | 46.00 ± 13.56 | **94.44** ± 2.48 | 66.47 ± 5.98 | *66.21* ± 7.36 | 61.92 ± 3.83 |
| EdgeConv [27] | 95.34 ± 2.03 | 42.00 ± 9.01 | 28.00 ± 10.77 | 70.56 ± 4.84 | 62.65 ± 4.97 | 57.93 ± 3.38 | 63.03 ± 7.83 |
| SelectionGCN [6] | 95.84 ± 0.26 | *73.14* ± 2.12 | 57.00 ± 51.0 | – | – | – | – |
| vRGCN | **98.29** ± 0.07 | **90.88** ± 1.09 | **82.00** ± 2.45 | *93.33* ± 1.36 | **80.00** ± 2.73 | **75.17** ± 2.58 | **72.12** ± 2.22 |

## 4.2 Performance Comparison

**Graph Classification.** Table 3 presents the results of the graph classification task. vRGCN achieves state-of-the-art performance across all benchmark graphs. The superiority of our method is particularly pronounced on the complex FAUST dataset, where vRGCN surpasses the second-best model by a significant margin of +9.00%. This result stands in stark contrast to other approaches. Attention-based models and rule-based methods show notable performance degradation on complex datasets such as the CoMA and FAUST datasets, highlighting the limitations of deterministic or fixed-heuristic approaches. In contrast, vRGCN's ability to stochastically model a distribution over latent relations provides the flexibility and robustness needed to adapt to these diverse structures.

**Node Classification.** As presented in Table 3, vRGCN demonstrates strong generalizability, achieving state-of-the-art accuracy on three of the four datasets. Although attention-based models secure the best performance on AIFB, vRGCN significantly outperforms them on the other, more complex datasets. This suggests that deterministic attention, though effective on certain graphs, may be less universally applicable than our stochastic relation learning approach. Crucially, vRGCN achieves this without using any pre-defined relations, underscoring its ability to learn powerful and task-specific latent relations directly from data.

### 4.3 Ablation Study

We conduct ablation studies to evaluate the core components of vRGCN: the stochastic sampler, the VI framework, the three relation learner architectures, and the sensitivity to the number of latent relations $R$.

As shown in Table 4, removing the ELBO and using cross-entropy only for the objective function (*w/o ELBO*) causes a catastrophic performance collapse, particularly on the CoMA dataset (90.88% to 40.00%). This confirms that regularization of the VI framework is vital to prevent overfitting. Similarly, the deterministic *w/o Gumbel Sampling* variant shows a consistent performance drop, especially on the FAUST dataset (82.00% to 69.00%), demonstrating that stochastic exploration is crucial for finding robust solutions.

Table 5 shows that the optimal learner architecture is dataset-dependent. The MLP learner excels on Euclidean graphs by effectively capturing pairwise node features. Conversely, the GNN learner is superior for non-Euclidean graphs, successfully leveraging the surrounding neighborhood context, which is critical for semantic graphs. The performance of the FRP learner confirms that ignoring node features is insufficient despite its intuitiveness.

Figure 3 highlights the parameter efficiency of vRGCN. Performance on the MNIST Superpixels dataset peaks with as few as $R = 2$ relations, already surpassing the SelectionGCN baseline. On the MUTAG dataset, performance peaks at $R = 3$, outperforming the RGCN baseline, which uses far more pre-defined relations. This powerfully demonstrates that learning a small, adaptive set of relations is more effective than relying on larger, fixed ones, and establishes $R$ as a key, dataset-dependent hyperparameter.

**Table 4.** The accuracy (%) of the ablation studies on the Gumbel sampling and ELBO of the graph and node classification tasks.

| Method | MNIST Sup. | CoMA | FAUST | AIFB | MUTAG | BGS | AM |
|---|---|---|---|---|---|---|---|
| w/o Gumbel Sampling | 98.11 | 90.34 | 69.00 | 82.22 | 74.12 | 67.59 | 60.00 |
| w/o ELBO | 98.21 | 40.00 | 63.00 | 84.44 | 74.12 | 67.59 | 67.07 |
| vRGCN | **98.29** | **90.88** | **82.00** | **93.33** | **80.00** | **75.17** | **72.12** |

**Table 5.** The accuracy (%) of the ablation studies on the relation learner types of the graph and node classification tasks.

| Dataset | MNIST Sup. | CoMA | FAUST | AIFB | MUTAG | BGS | AM |
|---|---|---|---|---|---|---|---|
| FRP Learner | – | – | – | 92.22 | 79.12 | 74.48 | 67.68 |
| MLP Learner | **98.29** | **90.88** | **82.00** | **93.33** | 79.12 | **75.17** | 69.90 |
| GNN Learner | 96.35 | 87.74 | 81.00 | **93.33** | **80.00** | **75.17** | **72.12** |

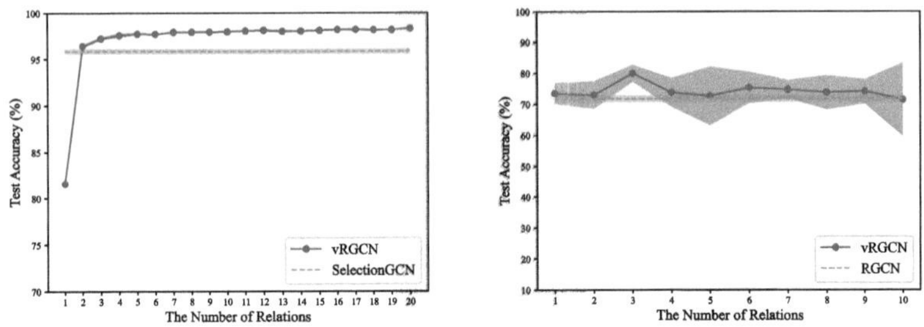


(a) MNIST Superpixels (b) MUTAG

**Fig. 3.** The accuracy (%) of the ablation studies on the number of relations on (a) MNIST Superpixels and (b) MUTAG. The blurred region denotes standard deviation.

## 4.4 Complexity Analysis

Although vRGCN introduces computational overhead from the relation learner ($O(M)$) and relational convolution ($O(R \cdot M)$), this is a worthwhile trade-off for superior performance and significant parameter efficiency. As shown in Table 6, vRGCN ($R = 2$) on the MNIST Superpixels dataset outperforms both GAT and SelectionGCN while using 3.5 times fewer parameters than GAT and 3 times fewer than SelectionGCN. This result confirms that the performance of vRGCN stems from learning a compact and highly effective set of relations rather than from sheer model size. This highlights a flexible trade-off of vRGCN. One can deploy a highly parameter-efficient model that still surpasses strong baselines, or scale it up ($R = 20$) with a comparable parameter count to achieve state-of-the-art results.

**Table 6.** The number of parameters and the accuracy (%) of models that differentiate edges on the MNIST Superpixels dataset. # denotes the number of.

| Method | # Parameters | Accuracy |
|---|---|---|
| GAT | 138,250 | 94.07 |
| SelectionGCN (R = 9) | 120,202 | 95.84 |
| vRGCN (R = 2) | **39,120** | 96.41 |
| vRGCN (R = 20) | 262,662 | **98.29** |

# 5 Conclusion

In this paper, we introduce vRGCN, a novel framework that addresses the fundamental limitations of existing GNNs in handling edge heterogeneity. By leveraging Variational Inference, vRGCN moves beyond rigid pre-defined rules and deterministic attention mechanisms to stochastically learn a distribution over latent relations directly from data. This approach enables our model to capture the inherent uncertainty in graph structures, leading to more robust and expressive representations. Our comprehensive experiments on various graph and node classification benchmarks confirm the superiority of vRGCN, which consistently outperforms the baseline models. The extensive ablation studies further validate our design choices, demonstrating that stochastic sampling is crucial for exploration and that the ELBO objective is a vital regularizer for generalization. Furthermore, we show that vRGCN is highly parameter-efficient, achieving state-of-the-art performance with significantly fewer parameters than established methods. Ultimately, vRGCN offers a flexible, powerful, and efficient paradigm for adaptive relation learning in graphs. Future work could extend this framework to dynamic or temporal graphs, explore more sophisticated prior distributions for latent relations, or investigate its applicability to other graph-based tasks such as link prediction and graph generation.

**Acknowledgments.** This research was supported by the MSIT (Ministry of Science and ICT), Korea, under the ICAN (ICT Challenge and Advanced Network of HRD; RS-2024-00436934), the Graduate School of Virtual Convergence (RS-2023-00254129), and the ITRC (Information Technology Research Center; RS-2024-00436936) support programs, supervised by the IITP (Institute for Information & Communications Technology Planning & Evaluation).

# References

1. Bogo, F., Romero, J., Loper, M., Black, M.J.: Faust: dataset and evaluation for 3d mesh registration. In: Proceedings of CVPR '14, pp. 3794–3801 (2014)
2. Brody, S., Alon, U., Yahav, E.: How attentive are graph attention networks? In: Proceedings of ICLR '22, pp. 1–26 (2022)

3. Busbridge, D., Sherburn, D., Cavallo, P., Hammerla, N.Y.: Relational graph attention networks (2019). https://arxiv.org/abs/1904.05811
4. Chandra, R., Bhagat, A., Maharana, M., Krivitsky, P.N.: Bayesian graph convolutional neural networks via tempered MCMC. IEEE Access **9**, 130353–130365 (2021)
5. Hamilton, W.L., Ying, R., Leskovec, J.: Inductive representation learning on large graphs. In: Proceedings of NeurIPS '17, pp. 1025–1035 (2017)
6. Hart, D., Morse, B.: Improving graph networks through selection-based convolution. In: Proceedings of WACV '24, pp. 1794–1804 (2024)
7. Jang, E., Gu, S., Poole, B.: Categorical reparameterization with Gumbel-Softmax. In: Proceedings of ICLR '17, pp. 1–12 (2017)
8. Kim, S., Park, E.: Stad-GCN: spatial-temporal attention-based dynamic graph convolutional network for retail market price prediction. Expert Syst. Appl. **255**, 124553 (2024)
9. Kipf, T.N., Welling, M.: Variational graph auto-encoders (2016). https://arxiv.org/abs/1611.07308
10. Kipf, T.N., Welling, M.: Semi-supervised classification with graph convolutional networks. In: Proceedings of ICLR '17, pp. 1–14 (2017)
11. Li, J., et al.: Dirichlet graph variational autoencoder. In: Proceedings of NeurIPS '20, pp. 5274–5283 (2020)
12. Liu, Y., Zheng, Y., Zhang, D., Chen, H., Peng, H., Pan, S.: Towards unsupervised deep graph structure learning. In: Proceedings of WWW '22, pp. 1392–1403 (2022)
13. Monti, F., Boscaini, D., Masci, J., Rodola, E., Svoboda, J., Bronstein, M.M.: Geometric deep learning on graphs and manifolds using mixture model cnns. In: Proceedings of CVPR '17, pp. 5115–5124 (2017)
14. Morris, C., et al.: Weisfeiler and leman go neural: higher-order graph neural networks. In: Proceedings of AAAI '19, vol. 33, pp. 4602–4609 (2019)
15. Pantelis, E., Bonilla, E.V., Tiao, L.C.: Variational inference for graph convolutional networks in the absence of graph data and adversarial settings. In: Proceedings of NeurIPS '20, pp. 18648–18660 (2020)
16. Park, J., Ahn, H., Kim, D., Park, E.: GNN-IR: examining graph neural networks for influencer recommendations in social media marketing. J. Retail. Consum. Serv. **78**, 103705 (2024)
17. Pujara, J., Augustine, E., Getoor, L.: Sparsity and noise: where knowledge graph embeddings fall short. In: Proceedings of EMNLP '17, pp. 1751–1756 (2017)
18. Ranjan, A., Bolkart, T., Sanyal, S., Black, M.J.: Generating 3D faces using convolutional mesh autoencoders. In: Proceedings of ECCV '18, pp. 704–720 (2018)
19. Sanchez-Gonzalez, A., Godwin, J., Pfaff, T., Ying, R., Leskovec, J., Battaglia, P.: Learning to simulate complex physics with graph networks. In: Proceedings of ICML '20, pp. 8459–8468 (2020)
20. Schlichtkrull, M., Kipf, T.N., Bloem, P., Van Den, Berg, R., Titov, I., Welling, M.: Modeling relational data with graph convolutional networks. In: Proceedings of ESWC '18, pp. 593–607 (2018)
21. Shchur, O., Mumme, M., Bojchevski, A., Günnemann, S.: Pitfalls of graph neural network evaluation (2019). https://arxiv.org/abs/1811.05868
22. Sun, T., Qian, Z., Dong, S., Li, P., Zhu, Q.: Rumor detection on social media with graph adversarial contrastive learning. In: Proceedings of WWW '22, pp. 2789–2797 (2022)
23. Sun, Z., et al.: Knowledge graph alignment network with gated multi-hop neighborhood aggregation. In: Proceedings of AAAI '20, vol. 34, pp. 222–229 (2020)

24. Tang, L., Liu, H.: Relational learning via latent social dimensions. In: Proceedings of KDD '09, pp. 817–826 (2009)
25. Vashishth, S., Sanyal, S., Nitin, V., Talukdar, P.: Composition-based multi-relational graph convolutional networks. In: Proceedings of ICLR '20, pp. 1–16 (2020)
26. Velickovic, P., Cucurull, G., Casanova, A., Romero, A., Lio, P., Bengio, Y.: Graph attention networks. In: Proceedings of ICLR '18, pp. 1–12 (2018)
27. Wang, Y., Sun, Y., Liu, Z., Sarma, S.E., Bronstein, M.M., Solomon, J.M.: Dynamic graph CNN for learning on point clouds. ACM Trans. Graph. **38**(5), 1–12 (2019)
28. Xia, J., Wang, S., Wang, X., Xia, M., Xie, K., Cao, J.: Multi-view bayesian spatio-temporal graph neural networks for reliable traffic flow prediction. Int. J. Mach. Learn. Cybern. **15**(1), 65–78 (2024)
29. Xu, K., Hu, W., Leskovec, J., Jegelka, S.: How powerful are graph neural networks? In: Proceedings of ICLR '19, pp. 1–17 (2019)
30. Yang, H., Lin, Z., Zhang, M.: Rethinking knowledge graph evaluation under the open-world assumption. In: Proceedings of NeurIPS '22, pp. 8374–8385 (2022)
31. Zhang, Y., Pal, S., Coates, M., Ustebay, D.: Bayesian graph convolutional neural networks for semi-supervised classification. In: Proceedings of AAAI '19, vol. 33, pp. 5829–5836 (2019)
32. Zhang, Y., Yao, Q.: Knowledge graph reasoning with relational digraph. In: Proceedings of WWW '22, pp. 912–924 (2022)

# EBVC: Electronic Bee Veterinarian – Beyond Monitoring and Onto Control

Mst Shamima Hossain[1(✉)], Meng-Chieh Lee[2], Christos Faloutsos[2], Boris Baer[1], Hyoseung Kim[1], and Vassilis Tsotras[1]

[1] University of California, Riverside, USA
{mhoss037,boris.baer,hyoseung,vtsotras}@ucr.edu
[2] Carnegie Mellon University, Pittsburgh, USA
{mengchil,christos}@cs.cmu.edu

**Abstract.** Given a fleet of beehives, how can we help with thermoregulation, for healthier bees and increased honey production? The objective is to develop a complete thermal control system, comprising hardware and software, that monitors and analyzes internal hive temperature in real-time and autonomously activates the heater or cooler as needed. We propose such a framework, '**EBVC**' (Electronic Bee Veterinarian), which is: (a) *White-Box:* entirely based on first principles, physics equations, and control theory; (b) *End-to-End:* does sensing, analysis, optimization, and control; (c) *Practical:* beekeepers can 'set it and forget it', as the underlying methods are all scalable and linear on the input size, have no parameters or hyperparameters to tune, and require no customization for specific hive type, geographical location, etc.. With reasonable assumptions that we list below, we estimate that **EBVC** will yield up to **$71** benefits per hive per year, thanks to increased honey production.

## 1 Introduction

How can we design a closed-loop, cost-effective hive core temperature regulation system that improves honey yield by reducing bees' energy expenditure?

**Motivation.** Honeybees are vital pollinators [26], supporting one-third of global food production [27] and a $500 billion global pollination industry [28]. Maintaining brood (core hive) temperature at 33–36 °C is critical [17], but each 1 °C temperature deviation forces bees to use energy on thermoregulation rather than honey production, reducing beekeepers' yields.

*Temperature Control Improves Honey Yield.* Beekeepers often assist thermoregulation with sugar feeding and heating pads for cold [22]. Studies show external temperature assistance directly boosts honey production: a 1 °C increase in June temperature correlates with 3.7 kg additional honey yield [18]; thermo-insulated hives produce about 35% more honey than conventional wooden ones [10].

M.-C. Lee—The work was done prior to joining Amazon.


R. C.-W. Wong et al. (Eds.): PAKDD 2026, LNAI 16597, pp. 82–95, 2026.
https://doi.org/10.1007/978-981-92-1300-9_7

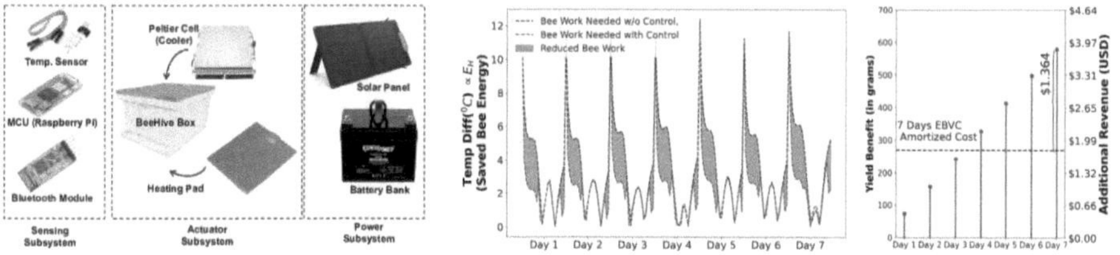


*(a) System Overview (Image Courtesy: Amazon.com)*

*(b)* ***EBVC*** *in action for one week*

**Fig. 1.** <u>`EBVC` **boosts honey yield**</u> (a) hardware components; (b) left: estimated savings in honeybee cooling effort (green areas) (b) right: resulting honey (left-y axis) and dollar savings (right-y axis). The black dashed line is the estimated amortized cost of `EBVC`. (Color figure online)

*Automatic Control is Technically Feasible with Current IoT Technology.* Off-the-shelf sensors and microcontroller units (MCUs) can continuously monitor hive temperatures, while inexpensive heating pads and Peltier coolers can be operated via simple on-off switching. These actuators can run for several hours per day using a solar-recharged battery pack (Sect. 4.1).

**Limitations of Prior Work.** Only `EBVC` matches all the specs (Table 1), namely both forecasting as well as control, tailored to beehives.

**Our Approach.** We propose `EBVC`, an end-to-end framework for automatic hive thermal regulation that integrates heating/cooling actuators with minimal in-hive sensing (Fig. 1). Our system consists of:

(1) $\texttt{EBVC}_{\texttt{Model}}$: a physics-informed thermal model to predict core temperatures using in-hive sensor and easily-obtainable meteorological data; extended to model wall-temperature dynamics needed for our control optimization.
(2) $\texttt{EBVC}_{Honey}$: a honey production model based on bee thermoregulation energy costs; and
(3) $\texttt{EBVC}_{Opt}$: a control optimization algorithm that governs actuators to maximize honey yield.

**Explainability is a 'must':** The reader may ask why not use deep-learning or foundation models (like Chronos [2,25] or tabPFN forecasting [12]) with reinforcement learning [29]. The answer is explainability: bee-keepers emphatically and repeatedly mentioned [7] that they must know why a model takes an action.

**Contributions.** The main contributions of `EBVC` are:

1. **White-Box:** `EBVC` is completely transparent, by design: based on standard physics (thermal diffusion), P-control, and white-box optimization methods.
2. **End-to-End:** `EBVC` is a complete hardware-software system that automatically handles sensing, forecasting, and control with no human intervention.

3. **Practical:** Our EBVC system is designed after feedback from entomologists and bee-keepers; and we made it scalable, with no parameters to tune, and applicable to all hive types, geographical locations, and bee genotypes.

**Reproducibility.** Our code is open-sourced at https://github.com/rtenlab/EBVC.

## 2 Related Work

**Table 1.** EBVC wins, fulfilling all the specifications. '?' means 'maybe, depending on implementation'.

| Property \ Method | Prior systems[9] | Trad. control systems [6] | EBV+ [14] | **EBVC (proposed)** |
|---|---|---|---|---|
| Monitoring | ✔ | | ✔ | ✔ |
| Forecasting | | | ✔ | ✔ |
| Controlling | | ✔ | | ✔ |
| White-Box | | ? | ✔ | ✔ |
| End-to-End | | | | ✔ |
| Practical | ? | ? | ✔ | ✔ |

Table 1 summarizes the features of each family of methods. Notice that only our proposed EBVC fulfills all the specs.

**Forecasting:** Forecasting methods, such as traditional ones like ARIMA and derivatives [3,4] up to very recent ones with foundation models (Chronos [2], tabPFN forecasting [12]) can not do actuator control (Table 1).

**Control:** There is also huge literature in control with Model Predictive Control (MPC) [32] and Reinforcement Learning (RL) being closely related (e.g., book by [29]). However, these methods cannot do precise forecasting, nor are they tailored towards the problem at hand.

**Sensor-Based Beehive Monitoring.** Several sensor-based tools, such as temperature [9], humidity, CO2 [24] , acoustic signals [19], IR sensors or cameras at entrance [16], have been proposed for hive monitoring, but lack capabilities for analysis or temperature control. Existing control systems [5,6,15] are domain-specific and require significant modification for bee hive adaptation.

**Thermal Model-Based Beehive Monitoring.** Recent approaches like EBV and EBV+ [13,14] model hive thermoregulation. But, they lack external control.

Other studies [21–23] modeled thermophysical process and microclimate dynamics in a Dadant hive, simulating optimal electric heating under varying winter conditions. However, the analysis is limited to specific season and hive design.

## 3 Dataset and Experimental Setup

**Hive Data.** We used the dataset from [14]: two months (August to September 2021: late summer) of hive temperature data per 20-minute interval from ten homogeneous Langstroth hives. In this dataset, five hives were treated by ice when the environmental temperature exceeded 35 °C; the remaining control hives received no treatment . For EBVC, we focus on the following temperature data:

1. Peripheral temperature (top wall) data from the top outer frame near the food and honey preservation area.
2. Core temperature data from developing brood preservation area (susceptible to heat stress): Bees try to maintain the core area temperature of $33 - 36$ °C.

Our aim is to regulate the top wall temperature to support optimal core temperature.

**CIMIS Data.** We used the hourly environmental temperature and solar radiation data (same date range as 2021 data) from the California Irrigation Management Information System (CIMIS) [1]. We selected CIMIS weather station 44, which matches our hive dataset location. We extracted "Air Temp ($^0$F)" and "Sol Rad (Ly/day)" from the database and converted them to °C and Wm$^{-2}$.

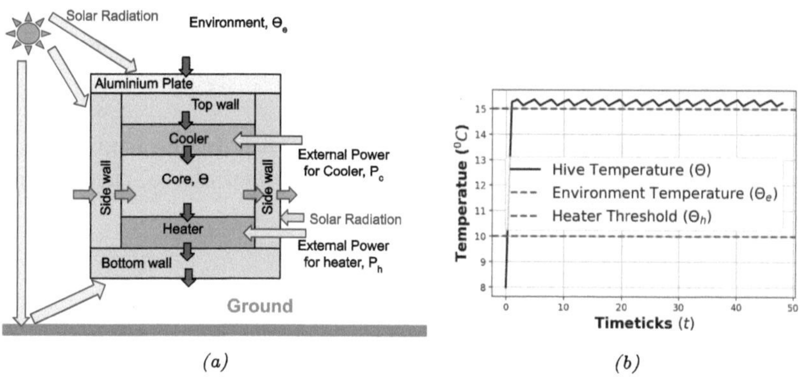


**Fig. 2.** (a) Simplified overview of our proposed system, EBVC. (b) Sanity check: Our fixed power on-off P-controller exhibits the expected 'saw-tooth' behavior (solid black line).

## 4 Proposed Framework: EBVC

Our EBVC consists of hardware and software. We describe each in the next subsection, and elaborate on the software parts in the rest of the subsections.

### 4.1 System Overview and Preliminaries

**Hardware.** Figure 2(a) gives a simplified overview the hardware of our system, with the following components:

1. *Actuation Subsystem.* A cooling plate mounted beneath the top wall and a heating plate placed above the bottom wall of the hive.
2. *Sensing Subsystem.* Temperature sensors at hive core area. This data feeds into our optimization algorithm, with publicly available meteorological data.
3. *Power Subsystem.* A solar panel with a rechargeable battery, whose limited daily energy budget requires careful scheduling of heater/cooler activation.
4. *Control Subsystem.* Figure 2(b) shows how the control subsystem works (heater example). It activates when hive temperature drops below (or exceeds, for cooling) a daily dynamic threshold, then toggles on and off to maintain it, resulting in a sawtooth temperature pattern.

**Software.** To control such hardware properly, we need:

1. *Thermal Model* for hive core temperature dynamics ($\texttt{EBVC}_{\texttt{Model}}$, Sect. 4.2)
2. *Honey Production Model* to estimate the amount of honey saved through thermal regulation ($\texttt{EBVC}_{Honey}$, Sect. 4.3) and
3. *Optimization Algorithm* to decide when to activate the heater/cooler, for the given energy budget ($\texttt{EBVC}_{Opt}$, Sect. 4.4)

**Device Cost Breakdown.** The estimated amortized cost is between \$62-\$90 per device per year (we omitted the details for lack of space). The price could go even lower, with mass production. Even as-is, the cost is below the break-even point (see Figs. 1(b) and 5b).

**Symbols and Definitions.** Table 2 gives such a list. In the next subsections we describe our three software compoments mentioned above: thermal model, honey production model, optimization.

**Table 2.** Symbols & Definitions

| Symbols | Definitions | Units |
|---|---|---|
| $\Theta_e(t), \Theta(t)$ | Environment, Hive core temperature | °C |
| $\Theta_{w_i}(t)$ | Hive wall temperature: $i = \{b(\text{bottom}), t(\text{top}), s(\text{side})\}$ | |
| $\theta_{ideal}$ | Ideal core temperature | |
| $S(t)$ | Solar radiation | $\text{Wm}^{-2}$ |
| $h$ | Hive health factor | |
| $r_h, r_c$ | *Bees' thermoregulation capacity:* heating, cooling | N/A |
| $P_h, P_c$ | *External source power:* heater, cooler | W |
| $\alpha_w$ | *Hive wall lumped heat absorption coefficient* | $^\circ\text{CJ}^{-1}\text{m}^{-2}$ |
| $k_w, k_a, k_h, k_c$ | *Lumped thermal conductivity:* wall, core, heater, cooler | $\text{s}^{-1}$ |
| $C_h, C_c$ | *Thermal Capacitance:* heater, cooler | $\text{J}^\circ\text{C}^{-1}$ |

### 4.2 Thermal Model: $\texttt{EBVC}_{\texttt{Model}}$

**Model 1** ($\texttt{EBVC}_{\texttt{Model}}$). *Given the environment temperature $\Theta_e(t)$, solar radiation $S(t)$, maximum thermoregulation capacity limit of bees $(r_h, r_c)$, hive health factor $h$ and controller power $(P_h, P_c)$, hive core temperature $\Theta(t)$ obeys:*

$$\frac{\partial\theta(t)}{\partial t} = 6[(k_w + k' + k_a)(\theta_e(t) - \theta(t)) + \alpha_w S(t)] - h\sigma(r_c, r_h; \theta(t)) - P(t) \quad (1)$$

$$\textit{where} \quad \sigma(r_c, r_h; \theta(t)) = \frac{r_c r_h (1 - e^{\theta(t)})}{r_h + r_c e^{-\theta(t)}} \quad \textit{and}$$

$$(P, k') = \begin{cases} (-\frac{P_c}{C_c}, k_c) & \text{if } \theta(t) > +\delta_{\delta_{\theta_s}} \,\&\, \textit{top wall} \\ (\frac{P_h}{C_h}, k_h) & \text{if } \theta(t) < -\delta_{\delta_{\theta_s}} \,\&\, \textit{bottom wall} \\ (0, 0) & \textit{otherwise} \end{cases}$$

For simplicity, the $(t)$ notation is omitted onwards. Next, we justify $\texttt{EBVC}_{\texttt{Model}}$, using concepts of heat transfer (physics) and P-controllers (control theory).

**Justification.** In $\texttt{EBVC}_{\texttt{Model}}$, we use relative temperatures (noted as $\theta_X = \Theta_X - \theta_{ideal}$ i.e., $X = e$ if environment temperature and so on) to ideal core temperature, $\theta_{ideal}$ as bees try to maintain $\theta_{ideal}$ within $33 - 36\,°\text{C}$.

*Step 1: Physics - Thermal Diffusion & Newton's Law of Cooling.* The first term in Eq. (1) model heat flow into an empty hive core without solar and bee activity based on heat transfer principles [8].

*Step 2: Physics - Solar Radiation Effect.* The fourth term in Eq. (1) represents heat gain from solar radiation.

*Step 3: Control Theory - Sigmoid P Controller.* The second-last term in Eq. (1) models bees' thermoregulation via a feedback loop [14] (if bees are left alone). Bee count limits cooling/heating capacity ($r_c$, $r_h$): fixed for similar structured hives and geography. The health factor ($0 < h \leq 1$) scales this response, $\sigma(r_c, r_h; \theta)$: higher $h$ indicates a stronger hive with less $\Theta$ fluctuation and vice-versa.

*Step 4: Control Theory - On-Off Controller.* The last term models a fixed-power on-off P-controller [20] for the installed heater and cooler. Since bees regulate $\Theta$ toward $\theta_{ideal}$, our controller mimics this behavior [13]: activating cooling when $\theta > \delta_{\theta_s}$ and heating when $\theta < \delta_{\theta_s}$, i.e., when no bees are present.

*Extension of* $\texttt{EBVC}_{\texttt{Model}}$. We also extend $\texttt{EBVC}_{\texttt{Model}}$ to model hive wall temperature dynamics, $\theta_{w_i}$, assuming no active thermoregulation effect from bees on $\theta_{w_i}$,

$$\frac{\partial\theta_{w_i}}{\partial t} = (k_w + k')(\theta_e - \theta_{w_i}) + k_a(\theta - \theta_{w_i}) + \alpha_w S - P \quad (2)$$

Here, $\theta_{w_i}$ denotes relative wall temperature for $i \in \{\text{top: t, side: s, bottom: b}\}$. With uniform wall properties, all walls have the same temperature unless a heater/cooler is installed. $k_w$ is structure-dependent, while $k_a$ and $\alpha_w = x_w k_w$ vary with internal hive conditions and sun exposure.

### 4.3 Honey-Production Model: $\texttt{EBVC}_{Honey}$

Here, we introduce $\texttt{EBVC}_{Honey}$, a model to estimate honey yield gains from external thermal regulation. Since real-world data do not provide simultaneous controlled and uncontrolled temperature measurements for the same hive, we rely on reconstructed temperature for both scenarios. Assuming accurate reconstruction, this allows for a fair comparison to estimate yield under external control.

**Model 2** ($\textit{EBVC}_{Honey}$). *Given we activate the heater or cooler to reduce bees' heating effort (when $\hat{\theta} > \hat{\theta}_w$ or $\hat{\theta} < \hat{\theta}_w$), the saved honey consumption is,*

$$H_H(t) = 3|(\hat{\theta}_w - \hat{\theta}) - (\frac{1}{6}(\sum_{n=1}^{5} \hat{\theta}_w + \hat{\theta}_{w_{b,adj}}) - \hat{\theta}_{adj})| \tag{3}$$

$$H_C(t) = 12|(\hat{\theta}_w - \hat{\theta}) - (\frac{1}{6}(\sum_{n=1}^{5} \hat{\theta}_w + \hat{\theta}_{w_{t,adj}}) - \hat{\theta}_{adj})| \tag{4}$$

We use $H_H(t)$ for saved honey consumption during heating and $H_C(t)$ for cooling. $\hat{\theta}_w$ and $\hat{\theta}$ are the reconstructed wall and core temperatures without external thermal regulation, while $\hat{\theta}_{w_{b,adj}}$, $\hat{\theta}_{w_{t,adj}}$ and $\hat{\theta}_{adj}$ represent the regulated case. Bees' heating effort is measured by the wall and core temperature gap; reducing this gap reflects energy (and honey) savings. During heating, we use the average wall temperature, since the heater needs to offset environmental effects on all walls. Based on expert opinion and experiments, bees consume 3g of honey to reduce the wall-core temperature gap by 1 °C. Thus, reducing this difference by 1 °C yields an equivalent honey saving. Similarly, the bees' cooling effort is proportional to the core and wall temperature difference, and each 1 °C reduction translates to an equivalent amount of honey preserved. Here, we assume reducing the wall-core temperature gap by 1 °C during cooling yields 12 g of honey saving. Detailed explanations are omitted due to space constraints. In the next section, we present our optimization objective function, $\texttt{EBVC}_{Opt}$ (alternatively, profit function) to maximize the profit in terms of the bees' saved energy and find out the corresponding controller parameters.

### 4.4 Control Optimization Objective: $\texttt{EBVC}_{Opt}$

We formulate our control optimization problem using a Model Predictive Control (MPC) approach. Given a daily electrical energy budget $B$ and our predictive thermal model $\texttt{EBVC}_{\texttt{Model}}$, we solve an optimization problem to schedule heater and cooler operations to maximize yield benefit, quantified as honey savings through reduced bee energy expenditure:

$$\max_{T_H, T_C} \sum_{i,j=1;i\neq j}^{I} (T_{H_i} E_{H_i} + T_{C_j} E_{C_j}), \quad \text{such that} \quad (T_{H_i} P_h + T_{C_j} P_c) \leq B \tag{5}$$

$$\text{where,} \quad E_H \propto |(\hat{\theta}_w - \hat{\theta}) - (\frac{1}{6}(\sum_{n=1}^{5} \hat{\theta}_w + \hat{\theta}_{w_{b,adj}}) - \hat{\theta}_{adj})| \tag{6}$$

$$E_C \propto e_C|(\hat{\theta}_w - \hat{\theta}) - (\frac{1}{6}(\sum_{n=1}^{5} \hat{\theta}_w + \hat{\theta}_{w_{t,adj}}) - \hat{\theta}_{adj})| \tag{7}$$

$$T_H \in \{\mathbf{0}_{\{\hat{\theta}_{adj}(i) < -\delta_{\theta_s}\}}, \mathbf{1}_{\{\hat{\theta}_{adj}(i) < -\delta_{\theta_s}\}}\} \tag{8}$$

$$T_C \in \{\mathbf{0}_{\{\hat{\theta}_{adj}(j) > +\delta_{\theta_s}\}}, \mathbf{1}_{\{\hat{\theta}_{adj}(j) > +\delta_{\theta_s}\}}\} \tag{9}$$

$$\hat{\theta}_{w_{b,adj}} \leq H_L - \theta_{ideal}; \hat{\theta}_{w_{t,adj}} \geq C_L - \theta_{ideal} \tag{10}$$

Here, $\hat{\theta}_w$ and $\hat{\theta}$ are the one-day ahead forecasted wall and core temperatures without external thermal regulation, while $\hat{\theta}_{w_{b,adj}}$ and $\hat{\theta}_{adj}$ represent the same for regulated case. $I$ denotes the optimization period (i.e., number of samples per day), and $e_C = \frac{r_h}{r_c} \approx 4$ captures the relative energy cost: bees expend approximately 4 times more energy for cooling than for heating (see Sect. 5), consistent with our earlier calculation that cooling requires roughly 4× more honey than heating. $E_H$ and $E_C$ represent the energy saved by bees when the heater and cooler are activated, respectively, and are directly proportional to the amount of honey saved through external control. $T_H$ and $T_C$ are binary activation schedules for the heater and cooler (1 = on, 0 = off). The controller operates under the following conditions: (i) the current time point yields the highest bee energy saving relative to others; and (ii) sufficient budget remains, e.g., if turning the heater on at time $i$ yields more bee energy savings than at $i+1$, and only one activation is affordable, time $i$ is chosen. To prevent thermal damage, $H_L$ and $C_L$ set safety limits on the minimum and maximum wall temperatures during heating and cooling.

Our goal is to find $T_H, T_C$ such that total electrical energy spent fits $B$, and maximizes bees' saved energy.

**Dynamic Optimization.** How can we dynamically schedule controller operation for maximum yield benefit? Algorithm 1 outlines $\texttt{EBVC}_{Opt}$ that takes past $p$ days of temperature sequence $(\Theta_{\mathcal{E}}, \Theta)_p$ and solar radiation data $(\mathcal{S}_{R,p})$ as input and returns optimized controller operation schedule $(T_C, T_H)$ as outputs.

**Algorithm 1:** $\texttt{EBVC}_{Opt}$

**Input:** (i) Past data, $\mathcal{D}_p = \{\Theta_{\mathcal{E}}, \Theta, \mathcal{S}_R\}_p$ (ii) $\{r_c, r_h\}$
**Output:** $\mathcal{T} = \{T_C, T_H\}$
Reconstruct $\Theta_p$ to find $h$ and $\theta_{ideal}$ for the last day
Forecast $\hat{\Theta}_{p+1}$ based on $h$ and $\theta_{ideal}$
Find $\mathcal{T}$ to optimize Eq. (5) for $\leq B$ // *e.g. greedy algo*

**One-Day Ahead Forecast.** Given, the past $p$ days of temperature sequences and the maximum thermoregulation ability $(r_c, r_h)$ (Details in Sect. 5), we first reconstruct $\Theta$ by minimizing reconstruction error and change in consecutive health factor $h$ values [14]. This gives us an estimate of daily $h$ and ideal core temperature $\theta_{ideal}$. Then we use the last day's $\{h, \theta_{ideal}\}$ to forecast the next day's $\Theta$ assuming the same thermoregulation parameters as the previous day.

**Find Optimal Operation Schedule for $B$.** Finally, we use the greedy algorithm to plan for an optimal controller operation schedule. We choose such a combination of operation times $T_H$ (Eq. (8)) and $T_C$ (Eq. (9)) within the given energy budget $B$ that maximizes Eq. (5).

## 5 Experiments

We conduct experiments to answer the following questions:

Q1. How accurate is $\texttt{EBVC}_{\texttt{Model}}$ in forecasting $\Theta$ and $\Theta_{w_t}$?
Q2. How accurately $\texttt{EBVC}_{Honey}$ captures yield benefit of the ice-pack experiment?
Q3. How well does $\texttt{EBVC}_{Opt}$ perform in maximizing yield for control hives?

**Experimental Setup.** All experiments are done on a Samsung Galaxy Book4 Pro 360 laptop (Intel i7, 16 GB RAM). We used hourly-averaged temperature data. To ensure field deployability, EBVC avoids interpolation and handles missing data directly. To estimate $\{r_c, r_h\} = \{124, 428\}$ for reconstruction and forecasting, we used Eq. (1) and bi-level optimization objective described in [14].

**Hive-Specific Parameter Estimation.** We use ten (non-treated) days of peripheral $\Theta_{w_t}$, core $\Theta$, and corresponding $(\Theta_e, S)$ to estimate $k_w, k_a, x_w$ using Eq. (2). We use the estimated $k_w$ from one hive (chosen randomly) to estimate $k_a$ and $x_w$ for other hives, e.g., control hive-1: $k_w = 0.29, x_w = 0.012, k_a = 0.46$ and treated hive-1: $k_w = 0.29, x_w = 0.0017, k_a = 0.41$. Parameter values for other hives are omitted due to space constraints.

### 5.1 Q1: Forecasting Capability

**Experimental Setup.** Given (i) seven days of data ($\Theta$, $\Theta_e$, $S, P$), we forecast hourly $\Theta$ for the next 14 days and (ii) per day ($\Theta_e, S, \Theta$) and fixed ($k_w, k_a, \alpha_w$), we forecast (reconstruct) daily $\Theta_{w_t}$ to be used for EBVC$_{Honey}$ and EBVC$_{Opt}$ later.

**Baselines.** For $\Theta$ forecast, we compared EBVC$_{\texttt{Model}}$ with state-of-the-art time series models: (i) ARX and seasonal ARX [3], (ii) Holt-Winters [11], (iii) DeepAR [30], and (iv) thermal-model based EBV+. For EBV+ and EBVC$_{\texttt{Model}}$, we estimate parameters ($h$, $\theta_{ideal}$) from the inputs, then use the final day's parameters along with $\Theta_e$, $S$, and $P = 0$ (for EBVC$_{\texttt{Model}}$) for forecasting $\Theta$.

We show the impact of introducing thermal conductance and solar radiation for $\Theta_{w_t}$ forecast. Baselines are: (i) EBV+ ($k_w = k_a = 1$ with no solar radiation $\alpha_w = 0$), (ii) $k_w = k_a = 1$ and location-specific $\alpha_W$, (iii) hive-specific $k_w, k_a$ with $\alpha_W = 0$, and (iv) EBVC$_{Opt}$ (hive-specific $k_w, k_a$ and location specific $\alpha_W$).

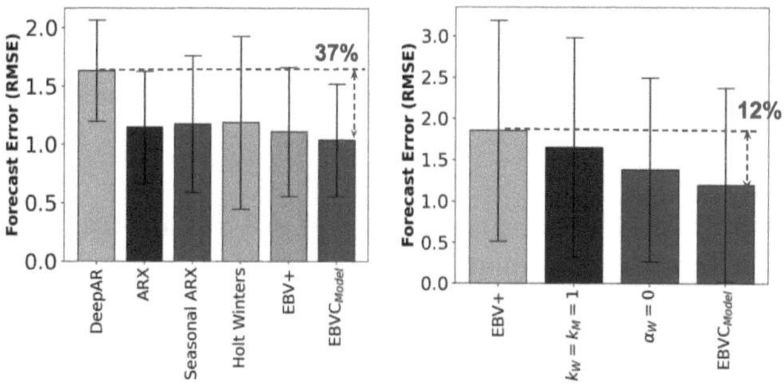


*(a) Core Temperature, $\Theta$* *(b) Top Wall Temperature, $\Theta_{w_t}$*

**Fig. 3.** <u>EBVC$_{\texttt{Model}}$ **wins**</u> against baseline in terms of core and wall temperature forecasting accuracy (up to 37% & 12% improvement). Error bars show 1 standard deviation.

**Accuracy.** Figure 3(a) shows the average core temperature forecasting RMSE for all methods. EBVC$_{\texttt{Model}}$ shows an average improvement of up to 37% over baselines in terms of accuracy and performs comparably to EBV+. However, EBVC$_{\texttt{Model}}$ offers greater robustness due to its inclusion of additional parameters such as thermal conductance, solar radiation effects, and external control. As shown in Fig. 3(b), EBVC$_{\texttt{Model}}$ improves top wall temperature reconstruction accuracy by 33% over EBV+; critical for the effectiveness of EBVC$_{Honey}$ and EBVC$_{Opt}$.

### 5.2 Q2: Capturing the Effect of Ice-Packs

**Baseline.** We validate EBVC$_{Honey}$ by comparing its estimated honey savings with the actual increase in stored honey measured between the start and end of the experiment. Honey storage was recorded in terms of the number of nectar frames, with inspections conducted biweekly on Aug 4, Aug 18, Sep 1, Sep 15, and Sep 29. For estimation, we assumed that one deep Langstroth nectar frame yields approximately 4.5lb ($\approx$ 2039g) of honey after dehydration and capping.

For $\texttt{EBVC}_{Honey}$, we used Eq. (4) to estimate the net daily honey savings (ice-treatment effect minus residual effect). According to beekeeper notes, ice packs were applied around 9 AM that melted throughout the day. Daily estimates are not available for all days. We therefore compute the average savings between each pair of inspections and multiply by the corresponding gap to approximate total honey savings. Note that ice treatment was done 15 days in 2-month period.

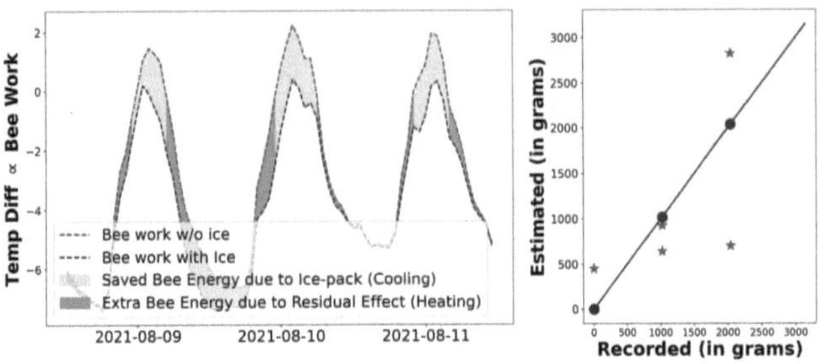


**Fig. 4.** $\texttt{EBVC}_{\texttt{Model}}$ **is accurate**: it estimates the yield benefit of ice-pack on $\Theta(t)$: saved honey estimate over 2 months (red stars) match closely with recorded ones (blue line). (Color figure online)

**Results.** Figure 4(a) illustrates the impact of a 3-day ice treatment. The blue and red dotted lines show estimated bee workload (temperature difference via Eq. (7)). The cyan shaded area highlights saved energy during the day due to ice cooling, while the orange shaded area shows the added burden at night due to residual cooling effects.

Figure 4(b) compares the observed increase in stored honey during the experiment (blue dots) with $\texttt{EBVC}_{Honey}$'s estimated increase (red stars). Each point represents one treated hive with ice-packs. The red points lie close to the reference line with slope 1, indicating strong agreement. As the values were recorded based on changes in the number of nectar frames (rather than direct weight), and $\texttt{EBVC}_{Honey}$ estimates honey in grams, minor discrepancies are expected.

### 5.3 Q3: Profit Optimization

**Baselines.** We compare $\texttt{EBVC}_{Opt}$ against the following baselines: (i) dynamic thresholds ($\Theta_{H_T}$, $\Theta_{L_T}$) without energy budget scheduling (FCFS: first-come, first-served); (ii) fixed thresholds ($\Theta_{H_T} = 36\,^\circ\mathrm{C}$, $\Theta_{L_T} = 33\,^\circ\mathrm{C}$) with adaptive budget scheduling; and (iii) fixed controller operation windows (9AM – 1PM and 12AM – 4AM) with dynamic thresholds. The fixed threshold and controller operation window are based on existing literature and expert recommendations.

**Results.** Figure 5b presents the yield benefit (in grams of honey) and the beekeeper revenue (in USD) for each control hive over the two-month experiment period. $\texttt{EBVC}_{Opt}$ consistently achieves the highest yield benefit across all hives, thanks to adaptive energy budget scheduling based on available resources. In

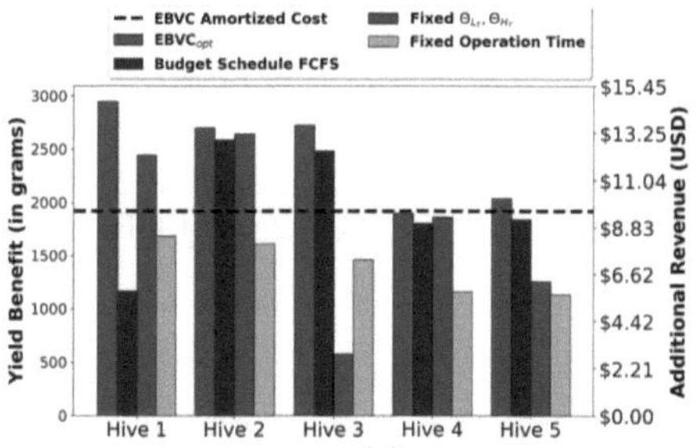


(a)

| Parameters | Values |
|---|---|
| $B$ | 200W |
| $P_H, P_C$ | 20W, 50W |
| $\frac{P_H}{C_H}, -\frac{P_C}{C_C}$ | 30 °C, −30°C |
| $k_c, k_h$ | $0.1 * k_w, k_w$ |
| $H_L, C_L$ | 45°C, 25°C |
| beekeeper's revenue | 2/lb(= 453g) of honey [31] |

*(b) Experimental Setup*

**Fig. 5.** EBVC$_{Opt}$ **gives the highest** yield benefit (in grams) and beekeeper's revenue (in USD) among all baselines; moreover, it is consistently profitable (at or above the black dashed 'break-even' line). Twin y-axes (left: grams; right $ revenue).

California (where the dataset was collected), bees typically produce honey for about six months each year, e.g., for control hive 1, beekeepers are projected to gain an annual economic benefit of ∼ $71 compared to no external control.

The black line in Fig. 5b represents the estimated amortized cost of operating our system, EBVC, over a two-month period. The amortized cost for the system is calculated considering $62 per device per year because the difference in the cost comes from battery size. For our experiments, we used the lowest assumed power requirement of 200 W. As shown, the additional revenue generated by EBVC consistently exceeds or matches this cost —resulting in a net annual revenue gain of up to $8.62 per hive. Bees produce honey only during part of the year, but they consume it year-round for thermoregulation. Since our system helps reduce this consumption, we amortize its cost over the full year, while calculating additional honey yield over the six-month production period.

## 6 Conclusion

The proposed EBVC method proposes a system to help hives maintain a healthy temperature, to improve honey production, honey-bee health, with all the additional benefits (pollination, agriculture). Our method has the following desirable characteristics that no other method enjoys:

- **White-Box:** it is based on well-understood components (thermal diffusion, 'P'-controllers)
- **End-to-End:** it is a complete, autonomous system, with software and hardware, without human intervention and parameter tuning.
- **Practical:** it is carefully designed to match the specs of end users (beekeepers): it is scalable, explainable, and general (applicable to any geographical location, any honey-bee subspecies). Moreover, we estimate that EBVC will provide significant savings (up to ≈$71 per year)

*Reproducibility.* Our code is open-sourced at https://github.com/rtenlab/EBVC.

**Acknowledgments.** This work is supported by the USDA/NIFA grants (2024-67022-43695 and 2024-67021-43696), UC MRPI award (# M21PR2306), and UCR Delfino Seed Fund.

## References

1. California irrigation management information system (CIMIS) (2025). https://cimis.water.ca.gov/Default.aspx
2. Ansari, A.F.: Chronos: learning the language of time series. Trans. Mach. Learn. Res. **2024** (2024)
3. Box, G.E., et al.: Time Series Analysis: Forecasting and Control. John Wiley & Sons, Hoboken (2015)
4. Brockwell, P.J., Davis, R.A.: Introduction to Time Series and Forecasting. Springer Texts in Statistics. Springer, Heidelberg (2016)
5. Cao, X., et al.: igrow: A smart agriculture solution to autonomous greenhouse control. In: AAAI, vol. 36, no. 11, pp. 11837–11845 (2022)
6. Chen, Y., et al.: Marlp: time-series forecasting control for agricultural managed aquifer recharge. In: KDD, KDD '24, New York, NY, USA, pp. 4862–4872 (2024)
7. CIBER, Center for Integrative Bee Research: Ciber bee health conference at UC riverside (2022). https://ciber.ucr.edu/bee-health-conference-2022
8. Crank, J.: The Mathematics of Diffusion. Oxford University Press, Cambridge (1975)
9. Davidson, P., et al.: Anomaly detection in beehives using deep recurrent autoencoders (2020). CoRR arxiv:2003.04576
10. Erdoğan, Y.: Comparison of colony performances of honeybee (apis mellifera l.) housed in hives made of different materials. Ital. J. Anim. Sci. (2019)
11. Hamilton, J.D.: Time Series Analysis, vol. 2. Princeton university press, Princeton (1994)
12. Hoo, S.B.: The tabular foundation model tabpfn outperforms specialized time series forecasting models based on simple features (2025). CoRR arxiv:2501.02945
13. Hossain, M.S.: EBV: electronic bee-veterinarian for principled mining and forecasting of honeybee time series. In: SDM, pp. 298–306. SIAM (2024)
14. Hossain, M.S., et al.: Principled mining, forecasting, and monitoring of honeybee time series with EBV+. ACM TKDD **19**(5) (2025)
15. Jeen, S., et al.: Low emission building control with zero-shot reinforcement learning. In: AAAI, vol. 37, no. 12, pp. 14259–14267 (2023)
16. Jiang, J.A., et al.: A wsn-based automatic monitoring system for the foraging behavior of honey bees and environmental factors of beehives. Comput. Electron. Agric. **123**, 304–318 (2016)
17. Kleinhenz, M., et al.: Hot bees in empty broodnest cells: heating from within. J. Exp. Biol. **206**, 4217–31 (2003)
18. Langowska, A., et al.: Long-term effect of temperature on honey yield and honeybee phenology. Int. J. Biometeorol. **61**(6), 1125–1132 (2017)
19. Murphy, F.E., et al.: An automatic, wireless audio recording node for analysis of beehives. In: ISSC, pp. 1–6 (2015)

20. Nise, N.S.: Control Systems Engineering. John Wiley & Sons, Hoboken (2019)
21. Oskin, S., et al.: Modeling the main physical processes in beehives. Biophysics **64**, 129–136 (2019)
22. Oskin, S., et al.: Modeling of thermophysical processes in electrically heated hives. Biophysics **65**, 331–337 (2020)
23. Oskin, S.: Modeling beehive microclimate at the end of wintering. Bio-physics **67**, 85–91 (2022)
24. Pandimurugan, V.: IoT based smart beekeeping monitoring system for bee- keepers in India, pp. 65–70. IEEE (2021)
25. Pineda-Arango, S.: Chronosx: adapting pretrained time series models with exogenous variables (2025). CoRR arxiv:2503:12107
26. Potts, S.G., et al.: Global pollinator declines: trends, impacts and drivers. Trends Ecol. Evol. **25**(6), 345–353 (2010)
27. Potts, S.G.: The assessment report on pollinators, pollination and food pro- duction: summary for policymakers. In: Secretariat of the Intergovernmental Science-Policy Platform on Biodiversity (2016)
28. Potts, S.G., et al.: Safeguarding pollinators and their values to human well-being. Nature **540**(2), 220–229 (2016)
29. Sutton, R.S., Barto, A.G.: Reinforcement Learning - An Introduction, 2nd edn. MIT Press, Cambridge (2018)
30. Turkmen, C., et al.: Easy and accurate forecasting with autogluon-timeseries (2022)
31. USDA: National honey report (2025). https://www.ams.usda.gov/mnreports/fvmhoney.pdf
32. Williams, G., et al.: Information theoretic MPC for model-based reinforcement learning. In: ICRA, pp. 1714–1721. IEEE Press (2017)

# Cross-Domain Aspect Sentiment Triplet Extraction Based on Generative Data Augmentation and Pseudo-Label Optimization

Zhou Zou[1], Jianxia Chen[2(✉)], Long Yang[1], Ninglong Ding[1], Liang Xiao[3], and Zhongwei Huang[1]

[1] School of Computer Science, Hubei University of Technology, Wuhan, China
{102311292,102401127,102311280,huangzhongwei}@hbut.edu.cn
[2] School of Computer Science, Hubei Provincial Key Laboratory of Green Intelligent Computing Power Network, Hubei University of Technology, Wuhan, China
chenjianxia@hbut.edu.cn
[3] School of Computer Science, Hubei Provincial Engineering Research Center for Digital and Intelligent Manufacturing Technologies and Applications, Hubei University of Technology, Wuhan, China
lx@mail.hbut.edu.cn

**Abstract.** Cross-domain aspect sentiment triplet extraction (CD-ASTE) aims to extract sentiment triplets from unlabeled target domains by transferring knowledge from labeled source domains. However, existing CD-ASTE methods still suffer from insufficient augmented samples and semantic misalignment in pseudo-labels. To address these issues, we propose a multi-stage CD-ASTE framework that integrates extraction and generation tasks. Specifically, we employ paraphrasing-based augmentation to generate diverse target-domain texts to alleviate data sparsity. In addition, we introduce a pseudo-label optimization mechanism to filter noisy pseudo-labels and reduce semantic errors, effectively mitigating error propagation. By combining multi-stage augmentation and pseudo-label filtering, our approach fully exploits unlabeled target-domain data and enhances cross-domain transferability. Experiments on six cross-domain settings show that our model achieves new state-of-the-art (SOTA) performance compared to strong baselines.

**Keywords:** Pseudo-label Optimization · Data Augmentation · Aspect Sentiment Triplet Extraction · Data Sparsity · Cross-domain

## 1 Introduction

Aspect Sentiment Triplet Extraction (ASTE) is a challenging subtask in aspect-based sentiment analysis (ABSA) that extracts sentiment triplets from user reviews [18]. As shown in Fig. 1, for the sentence *"The pizza was so delicious."*,


R. C.-W. Wong et al. (Eds.): PAKDD 2026, LNAI 16597, pp. 96–107, 2026.
https://doi.org/10.1007/978-981-92-1300-9_8

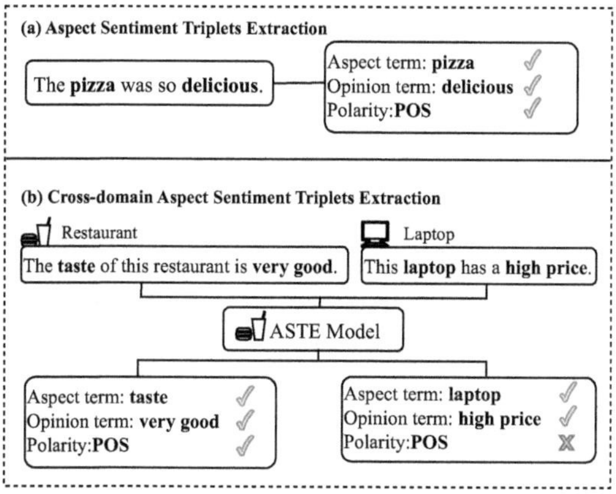


**Fig. 1.** An example of ASTE and CD-ASTE.

the triplet consists of aspect term *"pizza"*, opinion term *"delicious"*, and sentiment polarity *"positive"*. Compared to traditional sentiment classification, ASTE can more comprehensively understand semantics by obtaining fine-grained sentiment elements.

Existing ASTE models [13,17] mainly focus on single domains, where domain discrepancies and data sparsity limit cross-domain transfer. As shown in Fig. 1(b), domain-specific terms like *"taste"* (restaurant domain) and *"laptop"* (laptop domain) prevent models from generalizing across domains. Although pretrained language models (PLMs) such as T5 [12] and BART [5] achieve strong generalization in Seq2seq tasks, the scarcity of fine-grained annotations and limited sample diversity remain major obstacles in cross-domain ASTE.

To address these challenges, we leverage two key methods. First, following [3], we use pseudo-labeled target-domain samples as additional training resources. We further employ a paraphrasing model[1] to generate diverse samples and reduce data sparsity. Second, inspired by generative scoring methods [20], we design a pseudo-label optimizer (PLO) that treats pseudo-labels as conditional sequences and calculates likelihood-based match scores. Unlike confidence-based filtering [16], this generative approach provides fine-grained evaluation and effectively mitigates error propagation from noisy pseudo-labels.

In summary, we propose DAPO, a generative CD-ASTE framework with data augmentation and pseudo-label optimization, with the following contributions:

- We propose a pseudo-label optimization approach and construct label-distinguished datasets to train and evaluate it, demonstrating its effectiveness in CD-ASTE.
- We employ paraphrasing-based augmentation to enhance target-domain diversity and provide sufficient inputs for downstream tasks.
- Experimental results show that DAPO achieves SOTA performance compared to strong baselines.

[1] https://huggingface.co/humarin/chatgpt_paraphraser_on_T5_base.

# 2 Related Work

## 2.1 Aspect Sentiment Triplets Extraction

Presently, ASTE methods can be roughly divided into three categories. Grid-based approaches [13] formulate ASTE as a structured prediction problem over a 2D grid to detect token-pair interactions, but they cannot handle overlapping triplets. Span-based approaches [1,17,18] treat aspect and opinion terms as text spans, capturing long-range dependency and overlapping entities. Moreover, generative approaches linearize ASTE into text generation, [5] proposes a unified BART-based framework, while GAS [19] designs extraction-type and annotation-type formats to reduce error propagation. Inspired by GAS, we adopt a similar annotation idea but design a unified structured format $< polarity > a < opinion > o$, which preserves full triplets and reduces ambiguity for better pseudo-label filtering.

## 2.2 Cross-Domain Aspect Sentiment Triplets Extraction

Unlike ASTE, CD-ASTE additionally suffers from domain shift and data sparsity. Currently, many studies use pseudo-labeled data to improve cross-domain transfer. BGCA [3] adopts text-to-label and label-to-text generation to mitigate domain shift based on a bidirectional Sequence-to-Sequence (Seq2Seq) task. RSDA [15] performs pseudo-label refinement and data synthesis. Moreover, [1] uses dependency-based graph representations to learn domain-consistent syntactic features, aiming to enhance cross-domain robustness in ASTE. Motivated by BGCA and RSDA, we argue that pseudo-label quality remains critical and further design data augmentation and pseudo-label filtering to suppress error propagation.

## 2.3 Large Language Models

Recently, LLMs have demonstrated strong effectiveness in ABSA and cross-domain scenarios. They can serve as pretrained encoders, data generators, pseudo-labelers, and instruction followers. [6] uses LLM-generated outputs to mitigate domain gaps. GeDa [8] generates high-quality sentences from sentiment triplets, [21] iteratively enhances pseudo-labeled datasets, and [4] applies continual learning to reduce catastrophic forgetting. Additionally, CASE [2] introduces confidence-aware sentiment extraction into LLMs to enhance domain generalization.

# 3 Preliminaries

Formally, given a sentence $\mathbf{x} = \{x_1, x_2, ..., x_n\}$, the goal of ASTE is to extract a set of sentiment triplets $\mathbf{y} = \{(a_i, o_i, s_i)\}_{i=1}^{N}$, where $a_i \in \mathbf{x}$ is the aspect term, $o_i \in \mathbf{x}$ is opinion term, $s_i \in \{positive, negative, neutral\}$ is the sentiment polarity. In a cross-domain setting $S \rightarrow T$, we are given the labeled source

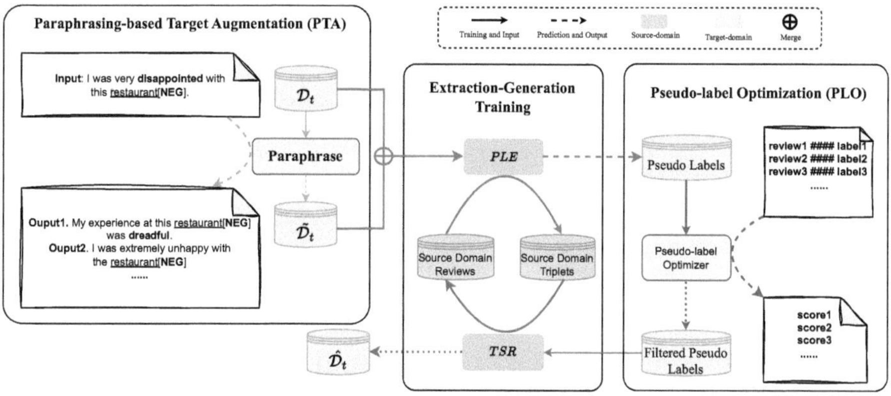


**Fig. 2.** Overview of DAPO framework.

domain $\mathcal{D}_s = \{(\mathbf{x}_i, \mathbf{y}_i)\}_{i=1}^{N_s}$ and unlabeled target-domain $\mathcal{D}_t = \{(\mathbf{x}_i\}_{i=1}^{N_t}$. Our purpose is to utilize the $\mathcal{D}_s$ and $\mathcal{D}_t$ to perform ASTE on test data in the target domain.

In this paper, the DAPO model is entirely based on the T5 encoder-decoder to model the ASTE task in a unified Seq2Seq manner. Specifically, we define a marker set $\mathcal{M} = \{< m_i >\}_{i=1}$ to mark sentiment elements, each triplet is formatted into $k$ continuous sequences: $< polarity >$ $a$ $< opinion >$ $o$, $< opinion >$ is an mark of the opinion term $o$. This format can effectively use generative language models, which can naturally handle sequence generation tasks and benefit from large-scale pretraining.

# 4 DAPO Model

To address the data scarcity and noisy pseudo-labels, our approach is proposed as a unified multi-stage framework including the following modules: Paraphrasing-based Target Augmentation (PTA), Pseudo-label Extraction (PLE), Pseudo-label Optimization (PLO), and Target Sentence Reconstruction (TSR). The entire DAPO procedure is shown in Fig. 2.

## 4.1 Paraphrasing-Based Target Augmentation

We perform paraphrasing augmentation on unlabeled target-domain data, thereby improving semantic diversity by generating additional samples. Given an unlabeled sentence $\mathbf{x}^T$, we use a paraphrasing model[2] to generate multiple paraphrases:

$$\tilde{\mathbf{x}}^T = x_1, x_2, ..., x_K = \mathcal{P}(\mathbf{x}^T) \tag{1}$$

Here, $K$ is the number of paraphrases per sentence. These paraphrases retain the original meaning but update aspect and opinion terms. Then, the augmented corpus $\tilde{\mathcal{D}}_t = \{\tilde{\mathbf{x}}_i^T\}_{i=1}^{N}$ is used for pseudo-label extraction in the next PLE module.

[2] https://huggingface.co/humarin/chatgpt_paraphraser_on_T5_base.

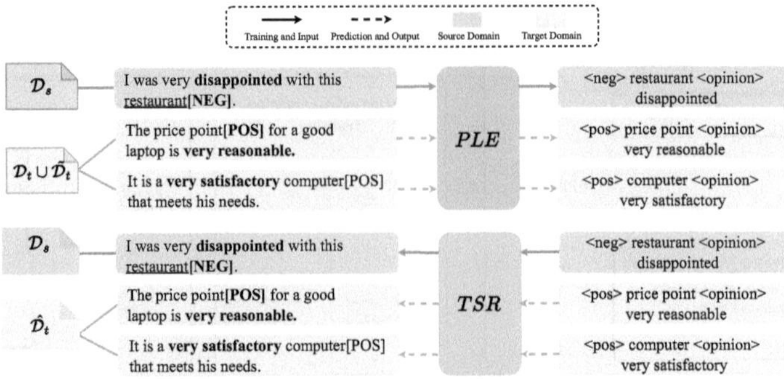


**Fig. 3.** Illustration of PLE and TSR.

### 4.2 Pseudo-Label Extraction

As shown in Fig. 3, the pseudo-label extraction (PLE) aims to extract pseudo sentiment labels for $\mathcal{D}_t$ and $\tilde{\mathcal{D}}_t$ to alleviate the label scarcity of the target domain. We utilize T5-base [12] as the backbone of PLE, then infer the target domain by fine-tuning it on the source domain $\mathcal{D}_s = \{(\mathbf{x}_i, \mathbf{y}_i)\}_{i=1}^{N_s}$. The likelihood loss function is as follows:

$$\mathcal{L} = -\sum_{i=-1}^{l} \log p(y_i \mid \mathbf{x}; y_{\leq i-1}) \tag{2}$$

where $l$ denotes the length of the label sequence. Given a sentence $\mathbf{x}^T \subset \mathcal{D}_t \cup \tilde{\mathcal{D}}_t$, pseudo-labels are represented as $\hat{\mathbf{y}}^T$, where each triplet also follows "$< polarity > a < opinion > o$". Moreover, to reduce error prediction, we stipulate that each prediction token $y_j$ can only come from $\mathcal{U} = \{x_i\}_{i=1}^{n} \cup \mathcal{M}$:

$$\hat{y}_i^T = \text{argmax}_{y_j \in \mathcal{U}} p(y_j \mid \mathbf{x}^T; \hat{y}_{\leq i-1}^T), \tag{3}$$

Ultimately, pseudo-labeled samples $(\mathbf{x}^T, \hat{\mathbf{y}}^T)$ inevitably contain noise from the source domain and need to be further filtered.

### 4.3 Pseudo-Label Optimization

For noisy labels, we propose a pseudo-label optimization (PLO) method to filter them out. Given a sentence $\mathbf{x}$ and pseudo-label set $\mathbf{y}$, PLO evaluates the match level between $\mathbf{x}$ and $\mathbf{y}$ by assigning a probability score $s(\mathbf{x}, \mathbf{y})$. The filtered pseudo-labels are then used to generate sentences. Meanwhile, we construct a label-distinguished dataset to train and evaluate PLO.

**Label-Distinguished Datasets.** To further enhance PLO, we construct a label-distinguished dataset (LDD). Each sample consists of a sentence and several candidate pseudo-labels. Specifically, we collect unlabeled reviews from Yelp

**Table 1.** Statistics of the LDD. P1-P4 indicate the position number of the correct labels.

| Datasets | Preferred Option | | | | |
|---|---|---|---|---|---|
| | P1 | P2 | P3 | P4 | Total |
| L16-AI | 1680 | 403 | 206 | 122 | 2411 |
| R14-AI | 699 | 749 | 294 | 295 | 2037 |
| R15-AI | 1449 | 345 | 142 | 84 | 2020 |
| R16-AI | 1736 | 274 | 107 | 59 | 2176 |

and Amazon [12], dividing them into single sentences. Then, we use the GAS [19] model to generate multiple pseudo-labels per sentence. Finally, we employ gpt-3.5-turbo[3] to annotate them. The annotation prompt[4] design follows the guidelines from SemEval2014 [11] and SemEval2016 [9]. The statistics of the LDD are shown in Table 1.

**Training.** Given sentence $\mathbf{x}$ and a set of candidate pseudo-label set $\mathcal{Y} = \{y_p, y_n, ..., y_n\}$ from LDD, we adopt a listwise loss function to encourage the model to assign higher probabilities to positive labels and lower probabilities to negative ones:

$$\mathcal{L}_{LIST} = -\log \frac{p(y_p \mid \mathbf{x})}{Z} \tag{4}$$

$$Z = p(y_p \mid \mathbf{x}) + \sum_{y_n} p(y_n \mid \mathbf{x}) \tag{5}$$

where $y_p$ and $y_n$ denote positive and negative labels, respectively; $Z$ is the normalization factor over all candidates in $\mathcal{Y}$. Moreover, we further incorporate labeled target-domain $\mathcal{D}_t^l$ as additional positive samples, to enhance the training of PLO and define the total loss as:

$$\mathcal{L} = \mathcal{L}_1 + \alpha\mathcal{L}_2 \tag{6}$$

$$\mathcal{L}_1 = \mathbb{E}_{(\mathbf{x},\mathcal{Y})\sim D_{LDD}}[\mathcal{L}_{LIST}(\mathbf{x}, \mathcal{Y})] \tag{7}$$

$$\mathcal{L}_2 = \mathbb{E}_{(\mathbf{x},y_p)\sim \mathcal{D}_{LDD}\cup\mathcal{D}_t^l}[-\log p(y_p|x)] \tag{8}$$

where $\mathcal{L}_1$ and $\mathcal{L}_2$ denote pseudo-label loss and augmented positive loss, respectively; $\alpha$ is a weighting hyperparameter, and $\mathcal{D}_t^l$ is the labeled target-domain samples.

[3] API: https://api.gpt.ge/v1/chat/completions.
[4] https://github.com/Jeoland/DAPO.

**Inference.** The PLO scores each label $\mathbf{y}$ for a given sentence $\mathbf{x}$ using a semantic matching score $s(\mathbf{x}, \mathbf{y})$.

$$s(\mathbf{x}, \mathbf{y}) \propto p(\mathbf{y} \mid \mathbf{x}) = \prod_{t=1}^{T} p(y_t \mid y_{<t}, \mathbf{x}) \tag{9}$$

where $p(\mathbf{y} \mid \mathbf{x})$ is the likelihood of generating sequence $\mathbf{y}$ conditioned on $\mathbf{x}$.

However, sentences with excessively high scores are often very simple, while those with too low scores are low-quality. Both are unsuitable for training. Therefore, we apply a dual-threshold $(\gamma_1, \gamma_2)$ and retain a label $\mathbf{y}$ only if its score falls within this interval:

$$\gamma_1 \leq s(\mathbf{x}, \mathbf{y}) \leq \gamma_2 \tag{10}$$

### 4.4 Target Sentence Reconstruction

We utilize the TSR model to reconstruct target-domain sentences from pseudo-labels to bridge the domain gap. TSR is implemented using a T5-base [12] model, structurally similar to PLE. Unlike PLE, TSR is trained on $\mathcal{D}_s$ using likelihood loss, with inputs and outputs reversed: $\mathbf{y} \rightarrow \mathbf{x}$.

$$\mathcal{L} = -\sum_{j=-1}^{l'} \log p(x_j \mid \mathbf{y}; x_{\leq j-1}) \tag{11}$$

where $l'$ denotes the length of the sentence sequence. After training, we ensure that the predicted token $\hat{x}_i^T$ exists in the model's vocabulary $\mathcal{V}$:

$$\hat{x}_i^T = \text{argmax}_{x_j \in \mathcal{V}} p(x_j \mid \hat{\mathbf{y}}^T; \hat{x}_{\leq i-1}^T), \tag{12}$$

Finally, $\mathcal{D}_s$ and $\hat{\mathcal{D}}_t = \{\hat{x}_i^T\}_{i=1}^n$ are jointly used to train the ASTE model for target-domain prediction.

## 5 Experiment

### 5.1 Settings

**Datasets.** Our benchmark dataset in the experiment includes two domains: Restaurant and Laptop, which we collectively refer to as the ASTE datasets. Restaurant14 (R14), Restaurant15 (R15), and Restaurant16 (R16) are from the SemEval 2014–2016 challenges [9–11]. Laptop16 (L16) [20] contains user reviews from the laptop domain. The statistics of ASTE datasets are shown in Table 2.

**Implementation.** We construct six cross-domain settings: $R14 \rightarrow L16$, $R15 \rightarrow L16$, $R16 \rightarrow L16$, $L16 \rightarrow R14$, $L16 \rightarrow R15$, and $L16 \rightarrow R16$. We use Micro-F1 as a metric, where a predicted triplet is considered correct only if all sentiment elements exactly match the gold labels.

**Table 2.** Statistics of ASTE datasets. #S and #T represent the number of sentences and triplets.

| Datasets | Train | | Dev | | Test | |
|---|---|---|---|---|---|---|
| | #S | #T | #S | #T | #S | #T |
| L16 | 1599 | 2310 | 197 | 257 | 460 | 658 |
| R14 | 1226 | 2338 | 310 | 577 | 492 | 994 |
| R15 | 642 | 1069 | 160 | 285 | 371 | 572 |
| R16 | 967 | 1645 | 105 | 166 | 361 | 585 |

**Baselines.** Due to the limited prior work on cross-domain ASTE, we select three available cross-domain methods: BGCA [3], SRC [14], and RSDA [15]. In addition, we include several strong in-domain ASTE models for comparison, including MiniConGTS [13], SpanASTE [18], SBN [1], SSFF [17], GAS [19], RoBMRC [7].

### 5.2 Comparison of Performance

We evaluate all in-domain baselines under the zero-shot setting. As shown in Table 3, DAPO achieves consistently superior performance on the CD-ASTE task. We summarize the key observations as follows:

Firstly, DAPO surpasses the strongest CD-ASTE models (SRC, BGCA, RSDA) by 4.70%, 2.84%, and 2.07% average F1, respectively, verifying that PTA and PLO effectively mitigate domain discrepancy and pseudo-label noise. Secondly, MiniConGTS achieves the best performance among in-domain baselines, with 1.91% and 1.84% F1 improvements over GAS and SSFF, showing that its tagging structure and contrastive token-level supervision can better exploit pre-trained features. Finally, cross-domain models still outperform strong in-domain baselines under domain shift, indicating that domain-specific models cannot generalize well to other domains.

### 5.3 Ablation Study

As shown in Table 4, we conduct ablation studies to verify each component of DAPO on all cross-domain settings. First, removing TSR yields the largest performance drop, confirming the importance of retrieving target-domain sentences for self-training. Furthermore, removing PLO and replacing it with confidence filtering also leads to noticeable degradation, illustrating the superiority of PLO in label filtering. Finally, removing PTA further degrades performance, showing that target-domain enhancement is also indispensable.

### 5.4 Analysis of PLO

To comprehensively evaluate the effectiveness and superiority of PLO, we analyze it from two perspectives: data and algorithm.

**Table 3.** Result of comparison experiment. All baselines are reproduced in our datasets. SpanASTE* denotes that it employs an LSTM encoder proposed in the paper [18]. RSDA* denotes that we supplement the missing *composition-based diversity enhancement* component. The best is bold, the second best is underlined.

| Methods | $S \rightarrow T$ | | | | | | Avg. |
|---|---|---|---|---|---|---|---|
| In-domain | R14→L16 | R15→L16 | R16→L16 | L16→R14 | L16→R15 | L16→R16 | |
| RoBMRC | 42.46 | 45.31 | 40.28 | 47.64 | 38.53 | 34.82 | 41.34 |
| SpanASTE* | 40.04 | 42.45 | 51.82 | 39.42 | 30.58 | 51.39 | 42.62 |
| SBN | 44.44 | 48.15 | 45.02 | 43.92 | 33.34 | 44.01 | 43.31 |
| GAS | 45.94 | 43.27 | 47.75 | 46.21 | 47.98 | 48.26 | 46.57 |
| SSFF | 46.62 | 42.99 | 40.56 | 48.58 | 48.64 | 52.43 | 46.64 |
| MiniConGTS | 43.80 | 39.62 | 42.09 | 60.05 | 47.96 | 57.38 | 48.48 |
| Cross-domain | R14→L16 | R15→L16 | R16→L16 | L16→R14 | L16→R15 | L16→R16 | Avg. |
| SRC | <u>62.33</u> | **64.91** | <u>67.01</u> | 55.61 | 52.18 | 60.33 | 60.40 |
| BGCA | 61.34 | 58.40 | 65.56 | <u>63.86</u> | <u>61.35</u> | 67.99 | 62.26 |
| RSDA* | 61.04 | 60.61 | 65.82 | 62.54 | 60.13 | <u>68.03</u> | <u>63.03</u> |
| DAPO | **63.76** | <u>62.16</u> | **67.61** | **64.14** | **63.31** | **69.61** | **65.10** |
| Δ(%) | ↑2.72 | ↑1.55 | ↑1.79 | ↑1.60 | ↑3.18 | ↑1.58 | ↑2.07 |

**Table 4.** Ablation Study. w/o denotes removing the corresponding module. * denotes replacing the PLO with the corresponding method.

| Method | Precision Avg. | Recall Avg. | F1 Avg. |
|---|---|---|---|
| **DAPO** | **64.54** | **65.66** | **65.10** |
| w/o PTA | 63.29 | 64.53 | 63.90 |
| *confidence | 63.26 | 62.81 | 63.03 |
| w/o PLO | 62.94 | 61.60 | 62.26 |
| w/o TSR | 61.46 | 61.72 | 61.59 |

*Data.* To further analyze PLO, we conduct self-training on PLE using three types of pseudo-labels: unfiltered, filtered, and randomly sampled unfiltered labels. As shown in Table 5, the filtered pseudo-labels from PLO achieve the best performance on average, despite not having the largest quantity. Moreover, randomly sampled labels yield worse performance, showing that pseudo-label volume also matters.

*Algorithm.* To investigate the influence of score ranges in PLO, we conduct a fine-grained analysis in different intervals. As shown in Fig. 4, each setting has an optimal score range that yields the highest F1. However, extremely high scores mean easy samples, while low scores introduce noise. Therefore, it is important to set thresholds in Eq. 10.

**Table 5.** Analysis of pseudo-label quantity and quality. Random denotes a random sampling of pseudo-labels. Unfiltered denotes fully unfiltered pseudo-labels.

| Type | $S \rightarrow T$ | | | | | | F1 Avg. |
|---|---|---|---|---|---|---|---|
| | R14→L16 | R15→L16 | R16→L16 | L16→R14 | L16→R15 | L16→R16 | |
| Random | 62.33 | 64.91 | 67.01 | 55.61 | 52.18 | 60.33 | 60.40 |
| Unfiltered | 61.34 | 58.40 | 65.56 | 63.86 | 61.35 | 67.99 | 62.26 |
| Filtered | 63.76 | 62.16 | 67.61 | 64.14 | 63.31 | 69.61 | 65.10 |

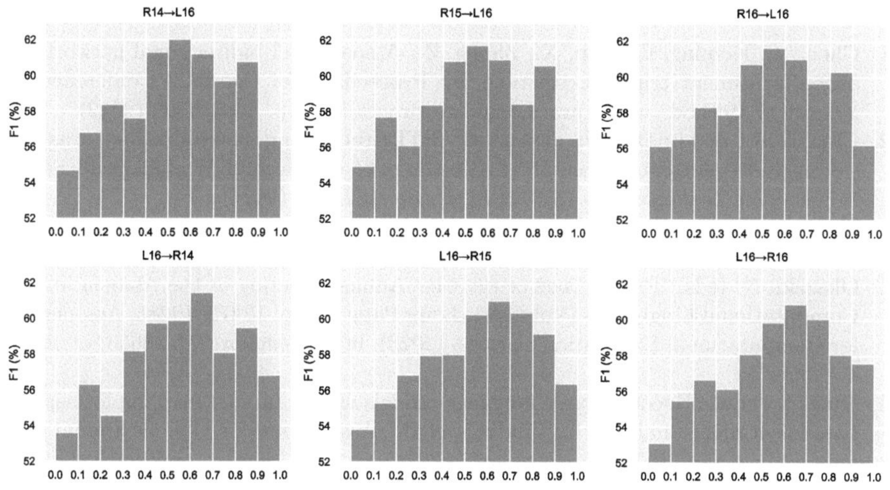


**Fig. 4.** Analysis of score range in PLO.

# 6 Conclusion

In this paper, we primarily solve two main challenges: data sparsity and noise pseudo-labels in the cross-domain ASTE task. First, we introduce a paraphrasing-based augmentation strategy to increase the semantic diversity of target-domain samples. Second, we design a pseudo-label optimization strategy to filter noisy labels. Besides, we comprehensively evaluate its effectiveness and advantages from both data and algorithmic perspectives. In future work, we plan to extend our analysis to additional domains and investigate how different annotation strategies for label-distinguished datasets influence the performance of PLO. Furthermore, enhancing the robustness and generalization ability of PLO remains an important direction for continued research.

**Acknowledgment.** This research was funded by the National Natural Science Foundation of China (grant number 62306106) and the Natural Science Foundation of Hubei Province (grant number 2023AFB377).

**Disclosure of Interests.** The authors declare that they have no known competing financial interests or personal relationships that could have appeared to influence the work reported in this paper.

**AI Declaration.** During the preparation of this work, the authors used GPT-3.5-turbo to annotate LLD. After using this tool, the authors reviewed and edited the content as needed and take full responsibility for the content of the publication.

## References

1. Chen, Y., Keming, C., Sun, X., Zhang, Z.: A span-level bidirectional network for aspect sentiment triplet extraction. In: Proceedings of the 2022 Conference on Empirical Methods in Natural Language Processing, pp. 4300–4309 (2022)
2. Chia, Y.K., et al.: Domain-expanded ASTE: rethinking generalization in aspect sentiment triplet extraction. In: Proceedings of the Second Workshop on Social Influence in Conversations (SICon 2024), pp. 152–165 (2024)
3. Deng, Y., Zhang, W., Pan, S.J., Bing, L.: Bidirectional generative framework for cross-domain aspect-based sentiment analysis. In: Rogers, A., Boyd-Graber, J., Okazaki, N. (eds.) Proceedings of the 61st Annual Meeting of the Association for Computational Linguistics (Volume 1: Long Papers), pp. 12272–12285. Association for Computational Linguistics, Toronto (2023). https://doi.org/10.18653/v1/2023.acl-long.686
4. Ding, X., et al.: Boosting large language models with continual learning for aspect-based sentiment analysis. In: Findings of the Association for Computational Linguistics: EMNLP 2024, pp. 4367–4377 (2024)
5. Lewis, M., et al.: Bart: denoising sequence-to-sequence pre-training for natural language generation, translation, and comprehension. In: Proceedings of the 58th Annual Meeting of the Association for Computational Linguistics (2020)
6. Li, S., Wang, Z., Zhou, G.: LLM enhanced cross-domain aspect-based sentiment analysis. J. Softw. **36**(2), 644–659 (2024)
7. Liu, S., Li, K., Li, Z.: A robustly optimized BMRC for aspect sentiment triplet extraction. In: Carpuat, M., de Marneffe, M.C., Meza Ruiz, I.V. (eds.) Proceedings of the 2022 Conference of the North American Chapter of the Association for Computational Linguistics: Human Language Technologies, pp. 272–278. Association for Computational Linguistics, Seattle (2022)
8. Mai, W., Zhang, Z., Chen, Y., Li, K., Xue, Y.: GeDa: improving training data with large language models for aspect sentiment triplet extraction. Knowl.-Based Syst. **301**, 112289 (2024). https://doi.org/10.1016/j.knosys.2024.112289
9. Pontiki, M., et al.: SemEval-2016 task 5: aspect based sentiment analysis. In: International Workshop on Semantic Evaluation, pp. 19–30 (2016)
10. Pontiki, M., Galanis, D., Papageorgiou, H., Manandhar, S., Androutsopoulos, I.: SemEval-2015 task 12: aspect based sentiment analysis. In: Proceedings of the 9th International Workshop on Semantic Evaluation (SemEval 2015) (2015)
11. Pontiki, M., Galanis, D., Pavlopoulos, J., Papageorgiou, H., Manandhar, S.: SemEval-2014 task 4: aspect based sentiment analysis. In: Proceedings of International Workshop on Semantic Evaluation at (2014)
12. Raffel, C., et al.: Exploring the limits of transfer learning with a unified text-to-text transformer. J. Mach. Learn. Res. **21**(140), 1–67 (2020)

13. Sun, Q., Yang, L., Ma, M., Ye, N., Gu, Q.: MiniConGTS: a near ultimate minimalist contrastive grid tagging scheme for aspect sentiment triplet extraction. arXiv preprint: arXiv:2406.11234 (2024)
14. Wang, G., Wang, B., Xu, F., Wang, R., Zhu, Z., Liu, P.: Domain-consistent syntactic representation for cross-domain aspect sentiment triplet extraction. Expert Syst. Appl. **256** (2024)
15. Wang, H., et al.: Refining and synthesis: a simple yet effective data augmentation framework for cross-domain aspect-based sentiment analysis. In: Ku, L.W., Martins, A., Srikumar, V. (eds.) Findings of the Association for Computational Linguistics: ACL 2024, pp. 10318–10329. Association for Computational Linguistics, Bangkok (2024)
16. Xie, Q., Dai, Z., Hovy, E., Luong, T., Le, Q.: Unsupervised data augmentation for consistency training. In: Advances in Neural Information Processing Systems, vol. 33, pp. 6256–6268 (2020)
17. Xu, G., Yang, Z., Xu, B., Luo, L., Lin, H.: Span-based syntactic feature fusion for aspect sentiment triplet extraction. Inf. Fusion **120**, 103078 (2025). https://doi.org/10.1016/j.inffus.2025.103078
18. Xu, L., Chia, Y.K., Bing, L.: Learning span-level interactions for aspect sentiment triplet extraction. In: Proceedings of the 59th Annual Meeting of the Association for Computational Linguistics and the 11th International Joint Conference on Natural Language Processing (Volume 1: Long Papers) (2021)
19. Zhang, W., Li, X., Deng, Y., Bing, L., Lam, W.: Towards generative aspect-based sentiment analysis. In: Proceedings of the 59th Annual Meeting of the Association for Computational Linguistics and the 11th International Joint Conference on Natural Language Processing (Volume 2: Short Papers) (2021). https://doi.org/10.18653/v1/2021.acl-short.64
20. Zhang, Y., Zeng, J., Hu, W., Wang, Z., Chen, S., Xu, R.: Self-training with pseudo-label scorer for aspect sentiment quad prediction. In: Ku, L.W., Martins, A., Srikumar, V. (eds.) Proceedings of the 62nd Annual Meeting of the Association for Computational Linguistics (Volume 1: Long Papers), pp. 11862–11875. Association for Computational Linguistics, Bangkok, Thailand (2024)
21. Zhong, Q., Li, H., Zhuang, L., Liu, J., Du, B.: Iterative data generation with large language models for aspect-based sentiment analysis (2024)

# TARP: An Effective and Efficient Route Prediction Method via Uncertainty-Aware Traffic Imputation and Divide-and-Conquer Inference

Xingyu Xian[1], Xiaowei Mao[1], Yang Jiang[1], Tingrui Wu[1], Shengnan Guo[1,2], Youfang Lin[1,2], and Huaiyu Wan[1,2(✉)]

[1] School of Computer Science and Technology, Beijing Jiaotong University, Beijing, China
{xingyuxian,maoxiaowei,jiangyang,25120430,guoshn,yflin}@bjtu.edu.cn
[2] Beijing Key Laboratory of Traffic Data Mining and Embodied Intelligence, Beijing, China
hywan@bjtu.edu.cn

**Abstract.** Route prediction aims to infer the most probable route between a given Origin and Destination under a specific departure Time (ODT), based on historical trajectories and the road network. However, it remains challenging due to two key limitations. First, existing route prediction models struggle to reliably leverage real-time traffic conditions, as the data are often highly missing and noisy, making it difficult to accurately capture dynamic road conditions. Second, the large model sizes and the autoregressive decoding of long sequences incur high inference latency, making efficiency difficult to maintain while preserving accuracy. In this paper, we propose a novel **T**raffic-**A**ware **R**oute **P**rediction model (TARP). To tackle the challenge of missing and noisy real-time traffic data, we propose a spatio-temporal uncertainty-aware traffic imputation module that initializes traffic conditions as Gaussian distributions and uses variance-guided attention to propagate uncertainty and emphasize reliable information. To improve inference efficiency, we propose a divide-and-conquer strategy with landmark guidance to shorten prediction horizons, and further introduce a linear-time **S**patial **B**ottleneck **A**ttention **M**echanism (SBAM) to significantly reduce latency without compromising accuracy. Experiments on two real-world datasets show that our model consistently outperforms state-of-the-art methods in both prediction quality and efficiency. Our code is available at https://github.com/XingYuXian/TARP_main.

**Keywords:** route prediction · uncertainty · imputation · divide-and-conquer · spatial bottleneck attention mechanism

## 1 Introduction

Route prediction aims to infer the most probable route given an ODT, based on historical trajectories. While many route prediction approaches have developed, existing approaches still face two critical challenges.


R. C.-W. Wong et al. (Eds.): PAKDD 2026, LNAI 16597, pp. 108–121, 2026.
https://doi.org/10.1007/978-981-92-1300-9_9

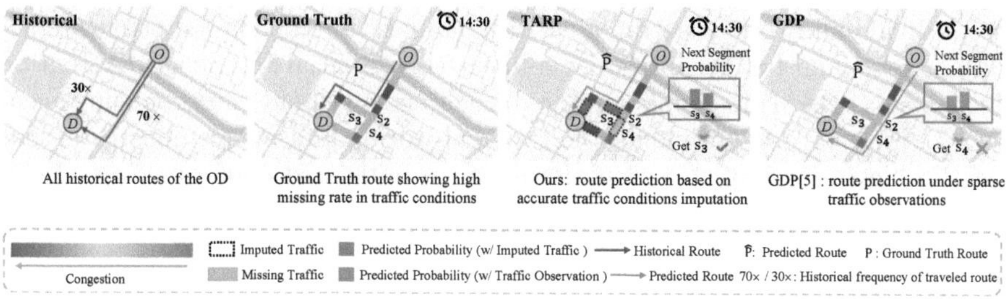


**Fig. 1.** Route Prediction Results Under Sparse Traffic Observations.

**Modeling Highly Missing Real-Time Traffic Conditions.** Existing route prediction methods [2,3,5,6,9,14,15,17] often leverage real-time traffic conditions to enhance prediction accuracy. However, these route prediction methods typically assume data completeness and do not model the missingness of traffic conditions, making it difficult to handle the high sparsity commonly observed in real-world scenarios. In fact, traffic conditions are frequently incomplete due to sensor malfunctions or unstable communications. For instance, as illustrated in Fig. 1, relying solely on historical route frequencies will favor the historically dominant route (e.g., choosing the route traveled 70% of the time), even if a less frequent route (e.g., traveled 30% of the time) is currently much faster. Although existing methods attempt to incorporate traffic conditions, they typically adopt coarse imputation strategies such as historical averages, which overlook real-time variations under high missingness. Consequently, unreliable traffic observations cannot overturn historical preferences, leading to suboptimal route decisions. Therefore, a key challenge is to effectively handle the sparsity and uncertainty of real-time traffic conditions to ensure reliable guidance for route prediction.

**Improving Efficiency While Preserving Accuracy.** In real-world navigation services, accurate route prediction must be delivered with extremely low latency to ensure timely and effective guidance for users. Search-based methods [2,3] are efficient but struggle to model complex routing behaviors, resulting in limited prediction accuracy. To overcome these limitations, deep learning-based approaches [5,6,18,19] leverage neural networks to learn transition patterns from data and autoregressively generate routes, significantly improving prediction performance. However, the sequential nature of autoregressive decoding makes the inference inherently slow and non-parallelizable, especially for long OD routes. Therefore, a challenge lies in accelerating inference while maintaining high prediction accuracy.

To address the above challenges, we propose a novel **T**raffic-**A**ware **R**oute **P**rediction model (TARP). The main contributions of this paper are summarized as follows: 1) We propose a spatio-temporal uncertainty-aware traffic imputation module that initializes traffic conditions as Gaussian distributions and employs variance-guided spatio-temporal attention to propagate uncertainty and emphasize reliable information for downstream route prediction. 2) We propose a divide-and-conquer strategy with landmark guidance to shorten prediction horizons, and further introduce a linear-time **S**patial **B**ottleneck **A**ttention **M**echanism (SBAM) to significantly reduce latency without compromising accuracy. 3) Extensive experiments on two real-world datasets demonstrate that TARP achieves state-of-the-art performance in both prediction quality and efficiency.

## 2 Related Work

Existing route prediction methods generally fall into two paradigms: Search-based methods and Deep learning-based methods. We summarize representative works in each category below.

**Search-Based Methods.** NMLR [3] employs Lipschitz regularized GCN embeddings to estimate transition probabilities and guide graph search. NASR [8] jointly learns observable and estimated edge costs using RNN and GNN modules, subsequently applying A* search for route generation. DRPK [2] derives destination-aware transition preferences statistically and predicts routes through greedy segment selection. While these methods leverage modern learning techniques to optimize cost functions, they fundamentally rely on classical deterministic graph search algorithms (e.g., A*, Dijkstra) to sequentially select segments based on cumulative cost minimization. This inherently myopic and cost-driven paradigm limits their expressive power, making it difficult to capture complex human routing behaviors and often resulting in inferior accuracy compared with Deep learning–based methods.

**Deep Learning–Based Methods.** GDP [5] adopts a graph-constrained diffusion generation model that leverages OD-conditioned attention and road-network feasibility constraints to synthesize realistic and valid routes. HOSER [18] employs a GAT- and GCN-based hierarchical semantic fusion architecture to learn routing preferences at multiple spatial granularities and autoregressively generates the next road segment. ProgRPGAN [7] utilizes a progressive GAN to refine route generation from global coarse structures to detailed road segments. DutyTTE [1] employs deep reinforcement learning to optimize alignment between predicted and ground-truth routes, mitigating step-by-step error accumulation. While these methods effectively capture human routing preferences through autoregressive predictions, they often rely on large models and lengthy sequential decoding, which results in high inference latency. Additionally, their use of coarse or incomplete real-time traffic data limits their robustness in dynamic environments.

## 3 Preliminary

**Route.** We represent a route $\mathcal{P} = [s_1, \ldots, s_N]$ as an ordered sequence of directed road segments. The road network is a directed graph $\mathcal{G} = (V, E)$ with intersections $V$ and directed links $E \subseteq V \times V$, where each link $(u, v)$ has a unique segment identifier $s = \phi(u, v)$. All GPS trajectories are map-matched to $\mathcal{G}$ to obtain such segment sequences.

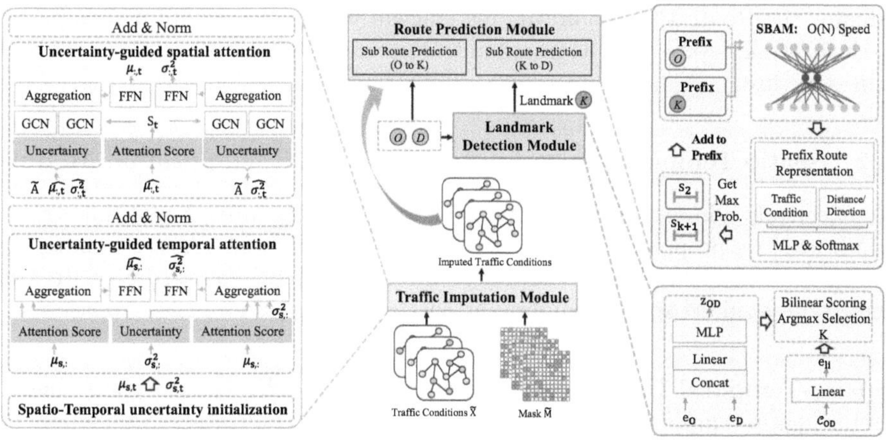


**Fig. 2.** Overview of the proposed TARP model.

**Traffic Condition.** We define the traffic condition as the average travel time of each road segment. For segment $s$ and time slice $t$, let $\mathcal{R}_{s,t}$ be the set of traversals on $s$ in $t$. For each traversal $r \in \mathcal{R}_{s,t}$ with entry and exit timestamps $T_{\text{entry}}(r)$ and $T_{\text{exit}}(r)$, the observed traffic condition is

$$x_{s,t} = \frac{1}{|\mathcal{R}_{s,t}|} \sum_{r \in \mathcal{R}_{s,t}} \big(T_{\text{exit}}(r) - T_{\text{entry}}(r)\big). \tag{1}$$

We collect $\{x_{s,t}\}$ into $\tilde{\mathbf{X}}$. We use a binary mask $\tilde{\mathbf{M}}$ where $\tilde{M}_{s,t} = 1$ iff $|\mathcal{R}_{s,t}| > 0$.

**Route Prediction Problem.** Given a historical trajectory dataset $\mathcal{D}$ on a road network $\mathcal{G}$ and an ODT query consisting of an origin segment $O$, a destination segment $D$, and a departure time $T$, $\mathcal{P}(O, D)$ means the set of all possible routes from $O$ to $D$, the goal is to give a route $\hat{P}$ that maximizes the conditional probability:

$$\hat{P} = \underset{P \in \mathcal{P}(O,D)}{\arg\max}\, Pr(P|O, D, T, \mathcal{G}, \mathcal{D}). \tag{2}$$

# 4 The TARP Model

## 4.1 Overview

As illustrated in Fig. 2, the proposed TARP model consists of three integrated components designed to address the high sparsity of traffic conditions and enable efficient route prediction. First, the Traffic Imputation Module reconstructs reliable traffic conditions from sparse observations by explicitly modeling uncertainty. Second, the Landmark Detection Module decomposes the complex global routing task into manageable sub-route prediction tasks by identifying landmark.

Finally, the Route Prediction Module efficiently predicts the optimal route for each sub-task, leveraging the imputed traffic conditions and the simplified problem structure.

### 4.2 Spatio-Temporal Uncertainty-Aware Traffic Imputation Module

To effectively address the challenge of modeling highly missing real-time traffic conditions, we propose a novel Spatio-Temporal Uncertainty-Aware Imputation module, which includes an uncertainty initialization block followed by a stack of uncertainty-guided spatio-temporal attention blocks.

**Spatio-Temporal Uncertainty Initialization Block.** To incorporate multi-factor spatio-temporal influences into uncertainty initialization, we model the latent traffic representation $\boldsymbol{H}_{s,t}$ for each road segment $s$ and time step $t$ as a Gaussian variable parameterized by a mean $\boldsymbol{\mu}_{s,t}$ and variance $\boldsymbol{\sigma}^2_{s,t}$, which respectively capture the expected traffic state and its uncertainty. The initialization of these parameters integrates multiple contextual embeddings to encode spatio-temporal heterogeneity. Specifically, a spatial embedding $\boldsymbol{SE} \in \mathbb{R}^{N\times 1\times d}$ assigns each road segment a learnable vector representing spatial characteristics, while a temporal embedding $\boldsymbol{TE} \in \mathbb{R}^{1\times T\times d}$ captures the temporal pattern of each time slice. To preserve relative temporal order, a sin–cos positional embedding $\boldsymbol{TPE} \in \mathbb{R}^{1\times T\times d}$ is added to the mean, where sine and cosine functions of different frequencies are used to encode temporal positions. The input traffic data $\tilde{\boldsymbol{X}}$ is first projected through a feed-forward layer, defined as $g(\tilde{\boldsymbol{X}}) = \tilde{\boldsymbol{X}}\boldsymbol{W}_E + \boldsymbol{b}_E$, while the missingness mask $\tilde{\boldsymbol{M}}$ is mapped by a variance embedding function $V_E(\tilde{\boldsymbol{M}})$ to initialize uncertainty.

The overall initialization can be expressed as:

$$\boldsymbol{\mu}^{(0)}_{s,t} = \boldsymbol{TE}+\boldsymbol{SE}+\boldsymbol{TPE}+g(\tilde{\boldsymbol{X}}), \quad \boldsymbol{\sigma}^{(0)2}_{s,t} = \mathrm{softplus}(\boldsymbol{TE}+\boldsymbol{SE}+V_E(\tilde{\boldsymbol{M}})), \quad (3)$$

where (0) indicates the initialization stage of the Gaussian parameters, corresponding to the mean and variance tensors over all road segments and time steps. This initialization jointly encodes spatial, temporal, and missingness information. The mean integrates deterministic inputs and contextual embeddings to represent expected traffic states, while the variance leverages the missingness mask to reflect uncertainty. The softplus function ensures a positive and learnable variance, enabling a stable probabilistic initialization before the attention-based imputation stage.

**Uncertainty-Guided Spatio-Temporal Attention Block.** To reduce the negative effect of unreliable traffic observations, we propose an Uncertainty-Guided Spatio-Temporal Attention block that emphasizes high-confidence information. The uncertainty-guided temporal attention block filters noisy time steps, while the uncertainty-guided spatial attention block attenuates unreliable road segments, enabling more robust spatio-temporal interaction modeling under sparse observations. Residual connections and layer normalization are applied between the two blocks to stabilize training and accelerate convergence.

In the temporal dimension, we aim to selectively aggregate information from high-confidence time steps to prevent noisy observations from distorting dependency modeling. Given the initialized latent states $(\boldsymbol{\mu}_{s,t}^{(0)}, \boldsymbol{\sigma}_{s,t}^{(0)2})$, we first compute an uncertainty weight to suppress unreliable representations. Layer superscripts are omitted for clarity:

$$\boldsymbol{\beta}_{s,:} = \exp\left(-\tau\, \boldsymbol{\sigma}_{s,:}^2\right), \qquad \tau > 0. \tag{4}$$

Clearly, the larger the variance, the smaller the attention weight. The scaled dot-product attention is performed on the mean states:

$$\boldsymbol{E} = \mathrm{Softmax}\left(\frac{\boldsymbol{Q}\boldsymbol{K}^\top}{\sqrt{d_k}}\right), \quad \boldsymbol{Q} = \boldsymbol{\mu}_{s,:}\boldsymbol{W}_Q,\ \boldsymbol{K} = \boldsymbol{\mu}_{s,:}\boldsymbol{W}_K, \tag{5}$$

The matrix $\boldsymbol{E}$ provides the semantic dependencies among time steps, capturing relationships based on feature similarity. This is then combined with the uncertainty weight $\boldsymbol{\beta}_{s,:}$ for the final uncertainty-aware aggregation:

$$\boldsymbol{z}_\mu = \boldsymbol{E}(\boldsymbol{\beta}_{s,:} \odot \boldsymbol{\mu}_{s,:}), \qquad \boldsymbol{z}_\sigma = (\boldsymbol{E} \odot \boldsymbol{E})(\boldsymbol{\beta}_{s,:} \odot \boldsymbol{\beta}_{s,:} \odot \boldsymbol{\sigma}_{s,:}^2). \tag{6}$$

Since weighted sums of independent Gaussian variables remain Gaussian, the latent distribution is consistently propagated while down-weighting unreliable time steps. The squared weights are applied to the variance $(\boldsymbol{z}_\sigma)$ to maintain mathematical consistency in the propagation of uncertainty. The aggregated temporal states $(\boldsymbol{z}_\mu, \boldsymbol{z}_\sigma)$ are then passed through separate linear transformations and non-linear activation functions for feature refinement and projection.

$$\hat{\boldsymbol{\mu}_{s,:}} = \mathrm{ReLU}(\boldsymbol{z}_\mu \boldsymbol{W}_T^\mu), \qquad \hat{\boldsymbol{\sigma}}_{s,:}^2 = \mathrm{Softplus}(\boldsymbol{z}_\sigma \boldsymbol{W}_T^\sigma). \tag{7}$$

Analogously, in the spatial dimension, we design a mechanism to selectively aggregate information from high-confidence neighbors. Using the temporal-refined latent states $(\hat{\boldsymbol{\mu}}, \hat{\boldsymbol{\sigma}}^2)$, we compute a spatial uncertainty weight:

$$\boldsymbol{\alpha}_{:,t} = \exp(-\nu \hat{\boldsymbol{\sigma}}^2), \quad \nu > 0, \tag{8}$$

which is used to suppress road segments with large uncertainty.

The static adjacency matrix $\tilde{\boldsymbol{A}}$ is insufficient as it fails to capture both the dynamically evolving spatial correlations and the latent, non-physical dependencies between road segments. Therefore, to capture both fixed topology and evolving dependencies, we utilize $\tilde{\boldsymbol{A}}$ alongside a learned dynamic correlation matrix $\boldsymbol{S}_t$:

$$\boldsymbol{S}_t = \mathrm{Softmax}\left(\frac{\hat{\boldsymbol{\mu}}_{:,t}\hat{\boldsymbol{\mu}}_{:,t}^\top}{\sqrt{d}}\right). \tag{9}$$

Both matrices are modulated by $\boldsymbol{\alpha}_{:,t}$ for uncertainty-aware spatial propagation:

$$\boldsymbol{c}_\mu = \tilde{\boldsymbol{A}}(\boldsymbol{\alpha}_{:,t} \odot \hat{\boldsymbol{\mu}_{:,t}}) + \boldsymbol{S}_t(\boldsymbol{\alpha}_{:,t} \odot \hat{\boldsymbol{\mu}_{:,t}}), \tag{10}$$

$$\boldsymbol{c}_{\sigma} = (\tilde{\boldsymbol{A}} \odot \tilde{\boldsymbol{A}})(\boldsymbol{\alpha}_{:,t}^2 \odot \hat{\boldsymbol{\sigma}}_{:,t}^2) + (\boldsymbol{S}_t \odot \boldsymbol{S}_t)(\boldsymbol{\alpha}_{:,t}^2 \odot \hat{\boldsymbol{\sigma}}_{:,t}^2). \tag{11}$$

Here, $\boldsymbol{S}_t$ complements $\tilde{\boldsymbol{A}}$ by capturing latent structural dependencies that vary over time. Similar to the temporal block, the aggregated spatial states $(\boldsymbol{c}_{\mu}, \boldsymbol{c}_{\sigma})$ are then processed by separate FFN layers for feature refinement, yielding the final output states:

$$\boldsymbol{\mu}_{:,t} = \text{ReLU}(\boldsymbol{c}_{\mu}\boldsymbol{W}_S^{\mu}), \qquad \boldsymbol{\sigma}_{:,t}^2 = \text{Softplus}(\boldsymbol{c}_{\sigma}\boldsymbol{W}_S^{\sigma}). \tag{12}$$

**Training Objective.** To learn both accurate reconstructions and calibrated predictive uncertainty, we adopt a hybrid loss based on the final outputs $(\boldsymbol{\mu}, \boldsymbol{\sigma}^2)$, defined over the observed entries $\Omega$:

$$\mathcal{L} = \frac{1}{|\Omega|} \sum_{(s,t)\in\Omega} \left[ \frac{(x_{s,t} - \mu_{s,t})^2}{2\sigma_{s,t}^2} + \frac{1}{2}\log \sigma_{s,t}^2 + \lambda |x_{s,t} - \mu_{s,t}| \right], \tag{13}$$

The loss combines the Gaussian Negative Log-Likelihood (NLL), which minimizes reconstruction error and ensures consistency with predicted uncertainty, with an $\ell_1$ regularization term $(\lambda|x_{s,t} - \mu_{s,t}|)$ for improved robustness under sparse supervision.

### 4.3 Traffic-Aware Divide-and-Conquer Route Inference

Deep learning-based route prediction models generally rely on autoregressive decoding, where each segment is generated based on the previously predicted ones, leading to high inference latency and error accumulation on long $O \rightarrow D$ routes. To overcome this limitation, we develop a traffic-aware divide-and-conquer strategy that detects an intermediate landmark $K$ to shorten the prediction horizon, and efficiently predicts the two sub-routes ($O \rightarrow K$ and $K \rightarrow D$) in parallel. By jointly leveraging reliable traffic conditions and reduced prediction dependency, our approach achieves faster and more stable route prediction.

**Landmark.** A landmark $K$ is an intermediate road segment used to decompose an $O \rightarrow D$ route into two shorter sub-routes. Selecting $K$ balances horizon reduction and prediction accuracy, so we construct candidates with two constraints: frequency provides a spatial prior under scarce exact OD matches, while fidelity ensures $K$ lies on the target route. Since routes with exact OD overlap account for only 13% of our dataset, we collect training trajectories whose origins and destinations fall within the 2-hop neighborhoods of $O$ and $D$, and retain the top-$k$ most frequent segments as $\mathcal{C}_{OD}$. For each training sample, we further intersect $\mathcal{C}_{OD}$ with its ground-truth route segments to construct supervision and avoid off-route candidates. The landmark detector is trained to assign high probability to candidates in the supervision set, and outputs the landmark as $K$ at inference. All frequency statistics are computed on training data only, avoiding test-time leakage.

**Landmark Detection Module.** While the dual constraints provide the statistical definition for the ideal landmark $K$, the Landmark Detection Module realizes this definition by learning to map the input ODT to the segment $K$ which best satisfies the defined constraints. The model predicts $K$ to serve as the decomposition anchor, and is trained using the ground-truth indicator labels $y_i$ over candidates.

We first encode the origin and destination into a joint embedding $\boldsymbol{h}_{OD}$ via a learnable OD encoder, where $\boldsymbol{e}_O, \boldsymbol{e}_D \in \mathbb{R}^{d_{\text{seg}}}$ are segment embeddings:

$$\boldsymbol{h}_{OD} = \text{Concat}(\boldsymbol{e}_O, \boldsymbol{e}_D)\boldsymbol{W}_{OD}, \tag{14}$$

and apply a two-layer transformation to capture nonlinear relation:

$$\boldsymbol{z}_{OD} = \text{ReLU}((\boldsymbol{h}_{OD}\boldsymbol{W}_1 + \boldsymbol{b}_1)\boldsymbol{W}_2 + \boldsymbol{b}_2)\,. \tag{15}$$

Following the frequency constraint defined above, we construct the candidate set $\mathcal{C}_{OD}$ by collecting frequent road segments from the corresponding *training* trajectories, yielding a reliable candidate pool. Each candidate $l_i \in \mathcal{C}_{OD}$ is represented by its segment embedding $\boldsymbol{e}_{l_i}$. We compute the compatibility $q_i$ between $\boldsymbol{z}_{OD}$ and each candidate $l_i$ and normalize the scores $p_i$:

$$q_i = \boldsymbol{z}_{OD}^{\top}\boldsymbol{W}_s\boldsymbol{e}_{l_i} + b_s, \quad p_i = \frac{\exp(q_i)}{\sum_{j\in\mathcal{C}_{OD}}\exp(q_j)}. \tag{16}$$

The landmark is selected as the highest-probability candidate, $K = \arg\max_i p_i$. For a training instance with ground-truth route $\mathcal{P}$, we set $y_i = 1$ if $l_i \in \mathcal{C}_{OD} \cap \mathcal{P}$ and 0 otherwise, and normalize $\tilde{y}_i = y_i / \sum_{j\in\mathcal{C}_{OD}} y_j$. The module is optimized using a cross-entropy loss between the predicted distribution $p_i$ and the normalized target $\tilde{y}_i$:

$$\mathcal{L}_{\text{landmark}} = -\sum_{i\in\mathcal{C}_{OD}} \tilde{y}_i \log p_i. \tag{17}$$

**Efficient Route Prediction Module.** The Route Prediction module performs core autoregressive prediction to generate the predicted route $\hat{\mathcal{P}} = [\hat{s}_1, \ldots, \hat{s}_N]$. At each step $n$, the predictor takes the prefix embeddings $\boldsymbol{H}_{\hat{\mathcal{P}}_n}$ together with geometric features and traffic conditions to infer the next segment $\hat{s}_{n+1}$.

Instead of computing full pairwise attention across all segments, SBAM introduces $r$ learnable spatial reference points ($r \ll N$) to aggregate and distribute global spatial information.

Let $\boldsymbol{H}_{\hat{\mathcal{P}}_n}^{(b-1)} \in \mathbb{R}^{n\times d_{\text{model}}}$ denote the prefix representation at the $(b-1)$-th SBAM block, and $\boldsymbol{R}^{(b)} \in \mathbb{R}^{r\times d_{\text{model}}}$ the reference set at the $b$-th block. SBAM consists of two Multi-Head Self-Attention (MHSA) layers:

$$\boldsymbol{R}' = \text{MHSA}(\boldsymbol{R}^{(b)}, \boldsymbol{H}_{\hat{\mathcal{P}}_n}^{(b-1)}, \boldsymbol{H}_{\hat{\mathcal{P}}_n}^{(b-1)}), \qquad \boldsymbol{H}_{\hat{\mathcal{P}}_n}^{(b)} = \text{MHSA}(\boldsymbol{H}_{\hat{\mathcal{P}}_n}^{(b-1)}, \boldsymbol{R}', \boldsymbol{R}'). \tag{18}$$

The first MHSA aggregates global spatial information into $\boldsymbol{R}'$, while the second distributes it back to each prefix element to obtain the refined contextual representation $\boldsymbol{H}_{\hat{\mathcal{P}}_n}^{(b)}$.

At step $n$, the contextual representation $\boldsymbol{h}_n$ is obtained from $\boldsymbol{H}_{\hat{\mathcal{P}}_n}$ and fused with geometric features $\boldsymbol{f}_{s'}^{\text{geo}}$ and traffic condition features $\boldsymbol{f}_{s'}^{\text{traf}}$ to compute the logits for a candidate segment $s'$:

$$z_{s'} = \text{Concat}\left(\boldsymbol{h}_n, \boldsymbol{f}_{s'}^{\text{geo}}, \boldsymbol{f}_{s'}^{\text{traf}}\right) \boldsymbol{W}_{\text{out}} + b. \tag{19}$$

where $\boldsymbol{f}_{s'}^{\text{traf}} = \mu_{s',t}$ is the imputed mean travel time of $s'$ at the departure-time slice $t$ of $T$. The probability over neighbors $\mathcal{N}(\hat{s}_n)$ is:

$$p_n(s') = \frac{\exp(z_{s'})}{\sum_{s^* \in \mathcal{N}(\hat{s}_n)} \exp(z_{s^*})}, \qquad \hat{s}_{n+1} = \arg \max_{s' \in \mathcal{N}(\hat{s}_n)} p_n(s'). \tag{20}$$

The prediction module is optimized using the negative log-likelihood of the ground-truth next segment:

$$\mathcal{L}_{\text{route}} = -\sum_{n=1}^{N-1} \log p_n(s_{n+1}). \tag{21}$$

**Route Merging and Final Output.** The complete predicted route is obtained by concatenating the two sub-routes inferred from the parallel prediction process:

$$\hat{\mathcal{P}}_{O \to D} = \hat{\mathcal{P}}_{O \to K} \oplus \hat{\mathcal{P}}_{K \to D}, \tag{22}$$

where $K$ is the detected landmark.

# 5 Experiments

## 5.1 Experimental Setup

**Datasets.** In our experiments, we use two real-world cab trajectory datasets collected from Didi Chuxing. The table lists the statistics of the datasets (Table 1).

**Table 1.** Dataset statistics.

| Dataset | City A | City B |
|---|---|---|
| Time span | 11.01–11.16, 2018 | 11.01–11.16, 2018 |
| Sampling rate | 3 s | 3 s |
| Number of trajectories | 1,613,355 | 1,951,585 |
| Average travel time | 810.18 s | 770.32 s |
| Area (width * height km$^2$) | 14.38 * 9.98 | 10.14 * 8.87 |
| Number of nodes | 4599 | 5259 |
| Number of edges | 6737 | 7784 |
| Traffic data missing rate | 46% | 49% |

**Baselines.** To evaluate our model, we compare it with representative route prediction baselines, including DA [16], NMLR [3], DRPK [2], Transformer [4], and GDP [5]. We further benchmark our sequence model against efficient neural architectures designed for long-range dependency modeling, including Performer [10], FlashAttention2 [11], RetNet [12], and Mamba [13], to assess computational efficiency and scalability. To examine the effectiveness of our uncertainty-aware traffic imputation module, we compare its performance against two baselines: Historical Average (HA), and an ablation where the uncertainty-aware spatio-temporal attention block is replaced by the standard MHSA mechanism. This comparison specifically evaluates the contribution of uncertainty guidance.

**Evaluation Metrics.** For route prediction, we use Longest Common Subsequence (LCS) and Dynamic Time Warping (DTW), where higher LCS and lower DTW indicate better trajectory similarity. For traffic imputation, we report MAE, RMSE, and MAPE. Computational efficiency is measured by OD/s, i.e., the number of OD queries processed per second for online decoding. All results are averaged over five runs.

**Evaluation Details.** We implement TARP in PyTorch and train it using Adam with a learning rate of 0.001, batch size of 16, and 25 epochs. The hidden size is 100 with 4 attention heads. All experiments are conducted on a single NVIDIA A40 GPU.

### 5.2 Experimental Results

**Table 2.** Performance comparison of our model and baselines in route prediction.

| Methods | City A | | | City B | | |
|---|---|---|---|---|---|---|
| | LCS (↑) | DTW (↓) | OD/s(↑) | LCS (↑) | DTW (↓) | OD/s(↑) |
| DA | 4.86 | 0.103 | **285** | 5.89 | 0.093 | **302** |
| NMLR | 7.92 | 0.112 | 24 | 8.65 | 0.102 | 27 |
| DRPK | 15.66 | 0.059 | 62 | 16.51 | 0.056 | 64 |
| Transformer | 17.55 | 0.031 | 22 | 17.66 | 0.029 | 23 |
| GDP | <u>17.59</u> | <u>0.029</u> | 7 | <u>17.71</u> | <u>0.028</u> | 7 |
| TARP | **18.81** | **0.022** | <u>69</u> | **18.95** | **0.017** | <u>70</u> |

**Overall Performance.** Table 2 shows that TARP not only achieves the highest accuracy across both cities but also maximizes efficiency improvements while maintaining strong predictive performance. Search-based methods (DA, NMLR,

DRPK) are fast but rely on predefined heuristics, making it difficult to capture complex human routing behaviors. Deep learning methods (Transformer, GDP) improve accuracy but incur high latency. By integrating traffic signals with a divide-and-conquer architecture, TARP achieves a superior balance of accuracy and real-time inference.

**Table 3.** Performance comparison of our model and baselines in imputation.

| Methods | City A | | | City B | | |
|---|---|---|---|---|---|---|
| | MAE (↓) | RMSE (↓) | MAPE(↓) | MAE (↓) | RMSE (↓) | MAPE(↓) |
| HA | 23.85 | 36.20 | 94.02 | 22.08 | 34.35 | 93.25 |
| MHSA | <u>19.56</u> | <u>28.66</u> | <u>87.77</u> | <u>18.08</u> | <u>26.77</u> | <u>85.10</u> |
| TARP | **13.05** | **25.66** | **56.88** | **12.56** | **24.86** | **56.85** |

**Table 4.** Impact of different traffic imputation strategies on route prediction.

| Methods | City A | | City B | |
|---|---|---|---|---|
| | LCS (↑) | DTW (↓) | LCS (↑) | DTW (↓) |
| HA | 18.40 | 0.025 | 18.33 | 0.022 |
| TARP | **18.81** | **0.022** | **18.95** | **0.017** |

**Traffic Imputation Quality and Its Impact on Route Prediction.** As shown in Table 3, TARP achieves the lowest MAE, RMSE, and MAPE in both cities, indicating more reliable traffic recovery under sparse observations. This improvement is further reflected in Table 4: using our imputed traffic yields higher LCS and lower DTW than the HA strategy. These results demonstrate that more accurate traffic imputation effectively enhances route prediction performance.

**Efficient Sequence Modeling Comparison.** Table 5 compares efficient sequence architectures for route prediction. Although FlashAttention2 and Performer improve efficiency, their approximations lead to accuracy drops. RetNet and Mamba perform well but show limited gains under short-horizon trajectories. In contrast, our SBAM module achieves the best accuracy and the fastest inference, demonstrating its suitability for route prediction.

**Table 5.** Comparison with efficient sequence modeling baselines.

| Methods | City A | | | City B | | |
|---|---|---|---|---|---|---|
| | LCS (↑) | DTW (↓) | OD/s(↑) | LCS (↑) | DTW (↓) | OD/s(↑) |
| FlashAttention2 | 16.17 | 0.046 | 52 | 16.48 | 0.032 | 53 |
| Performer | 16.36 | 0.041 | 30 | 16.27 | 0.043 | 25 |
| RetNet | 18.40 | 0.025 | 31 | 18.58 | 0.026 | 33 |
| Mamba | 18.56 | 0.023 | 25 | 18.80 | 0.022 | 26 |
| SBAM | **18.81** | **0.022** | **69** | **18.95** | **0.017** | **70** |

**Table 6.** Ablation Experiments.

| Methods | City A | | | City B | | |
|---|---|---|---|---|---|---|
| | LCS (↑) | DTW (↓) | OD/s(↑) | LCS (↑) | DTW (↓) | OD/s(↑) |
| TARP | 18.81 | 0.022 | 69 | 18.95 | 0.017 | 70 |
| TARP w/o L | 18.51 | 0.023 | 55 | 18.77 | 0.022 | 57 |
| TARP w/o T | 18.30 | 0.029 | 75 | 18.44 | 0.025 | 76 |

**Ablation Experiments.** Table 6 evaluates the contributions of the divide-and-conquer strategy (w/o L) and traffic condition modeling (w/o T). Removing the divide-and-conquer strategy lowers accuracy and slows inference due to longer-horizon route generation. Removing traffic condition modeling improves speed but causes a clear accuracy drop. These results demonstrate that both components are essential to achieving a strong balance between accuracy and efficiency.

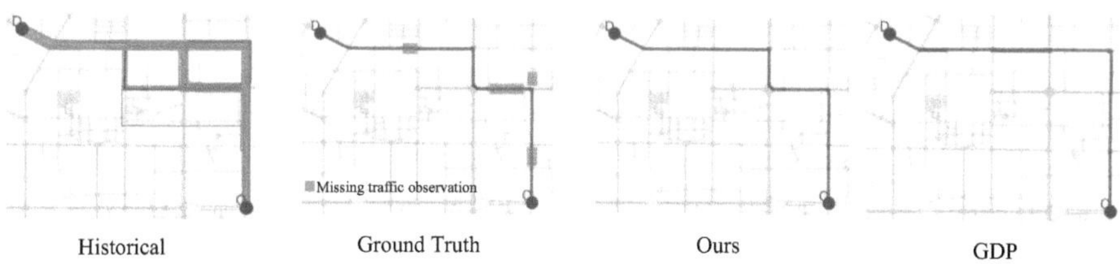


**Fig. 3.** Case study in City B under sparse traffic observations.

**Case Study.** As shown in Fig. 3, GDP predicts a geometrically reasonable route but overlooks traffic conditions. With traffic-aware modeling, TARP produces routes that better match real travel behavior and the ground truth. Moreover, even in road segments where traffic observations are missing, TARP can still infer the correct direction and continue along the ground-truth route, demonstrating strong robustness to incomplete data.

## 6 Conclusion

We propose the TARP model, effectively and efficiently predicting routes by uncertainty-aware traffic imputation and divide-and-conquer strategy. Extensive experiments on real-world datasets confirm its superiority over strong baselines.

## References


1. Mao, X., et al.: DutyTTE: deciphering uncertainty in origin-destination travel time estimation. In: AAAI, pp. 12390–12398 (2025)
2. Tian, W., Shi, J., Luo, S., Li, H., Xie, X., Zou, Y.: Effective and efficient route planning using historical trajectories on road networks. Proc. VLDB Endow. **16**(10), 2512–2524 (2023)
3. Jain, J., Bagadia, V., Manchanda, S., Ranu, S.: NeuroMLR: robust & reliable route recommendation on road networks. In: NeurIPS, pp. 22070–22082 (2021)
4. Vaswani, A., et al.:Attention is all you need. In: NeurIPS, pp. 5998–6008 (2017)
5. Shi, D., Tong, Y., Zhou, Z., Xu, K., Wang, Z., Ye, J.:Graph-constrained diffusion for end-to-end path planning. In: ICLR (2024)
6. Zhang, P., et al.: Route planning using divide-and-conquer: a GAT enhanced insertion transformer approach. Transp. Res. Part E: Logist. Transp. Rev. **176**, 103176 (2023)
7. Fu, T.-Y., Lee, W.-C.: ProgRPGAN: progressive GAN for route planning. In: Proc. ACM SIGKDD, pp. 393–403 (2021)
8. Wang, J., Wu, N., Zhao, W.X., Peng, F., Lin, X.: Empowering A* search algorithms with neural networks for personalized route recommendation. In: Proc. ACM SIGKDD, pp. 539–547 (2019)
9. Wang, J., Wu, N., Zhao, W.X.: Personalized route recommendation with neural network enhanced A* search algorithm. IEEE Trans. Knowl. Data Eng. **34**(12), 5910–5924 (2022)
10. Choromanski, K., et al.:Rethinking attention with performers (2020). arXiv:2009.14794
11. Dao, T.: FlashAttention-2: faster attention with better parallelism and work partitioning (2023). arXiv:2307.08691
12. Sun, Y., et al.: Retentive network: a successor to transformer for large language models (2023). arXiv:2307.08621
13. Gu, A., Dao, T.: Mamba: linear-time sequence modeling with selective state spaces. In: COLM (2024)
14. Mao, X., Ding, H., Lin, Y., et al.: Spatial-temporal feedback diffusion guidance for controlled traffic imputation (2026). arXiv:2601.04572
15. Mao, X., Cai, T., Peng, W., et al.: Estimated time of arrival prediction via modeling the spatial-temporal interactions between links and crosses. In: Proc. ACM SIGSPATIAL, pp. 658–661 (2021)
16. Dijkstra, E.W.: A note on two problems in connexion with graphs. Numer. Math. **1**, 269–271 (1959)

17. Mao, X., Wan, H., Wen, H., et al.: GMDNet: a graph-based mixture density network for estimating packages' multimodal travel time distribution. In: AAAI, pp. 4561–4568 (2023)
18. Cao, J., et al.: Holistic semantic representation for navigational trajectory generation (2025). arXiv:2501.02737
19. Mao, X., Wen, H., Zhang, H., et al.: DRL4Route: a deep reinforcement learning framework for pick-up and delivery route prediction. In: Proc. ACM SIGKDD, pp. 4628–4637 (2023)

# Boosting Black-Box Graph Reconstruction Attacks via Adjacency Relationship Recovery of Representative Nodes

Xinyu Jiang and Qiao Yan(✉)

Shenzhen University, Shenzhen, China
2310274015@email.szu.edu.cn, yanq@szu.edu.cn

**Abstract.** Graph Neural Networks (GNNs) have been widely applied across various fields due to their effective modeling of graph data. However, the training data used in GNNs poses privacy risks as it contains sensitive information. Recently, graph reconstruction attacks have been proposed to reconstruct the adjacency matrix of the training dataset in graph neural networks, receiving significant attention due to their strong attack performance. Existing studies have demonstrated graph reconstruction attacks in both white-box and black-box settings. To further advance research in black-box scenarios, we propose ASM-GRA, a black-box graph reconstruction attack that reconstructs the adjacency matrix by attacking a substitute model trained on the target model. Furthermore, we propose a node selection strategy to reconstruct the adjacency relationships of representative nodes across multiple categories, which effectively enhances the accuracy of the entire reconstructed matrix. Extensive experiments on several public datasets demonstrate the effectiveness of our method. For instance, on the Citeseer dataset, we achieve an attack AUC of 0.95 and an attack AP of 0.951. Our code is available at https://github.com/jiangxinyu118/ASM-GRA.

**Keywords:** Graph Neural Network · Graph Reconstruction Attack

## 1 Introduction

Graph neural networks (GNNs) have been widely applied in various domains such as social networks [3], recommendation systems [13], and molecular property prediction [16]. By adopting a message-passing mechanism to aggregate information from neighboring nodes [7], GNNs effectively update node representations, achieving outstanding performance in downstream tasks like node classification and graph classification. Typically, many trained GNNs are publicly released for downstream applications. However, such publicly available models may be exploited by malicious attackers to conduct privacy attacks, especially when the training data contain sensitive information from data providers.

In recent years, graph reconstruction attacks (GRA) have drawn increasing attention. These attacks aim to reconstruct private links within the target


R. C.-W. Wong et al. (Eds.): PAKDD 2026, LNAI 16597, pp. 122–134, 2026.
https://doi.org/10.1007/978-981-92-1300-9_10

model's training dataset [15]. This poses a serious privacy threat, as links may reveal sensitive relationship information or proprietary knowledge embedded in the dataset. For example, many trained GNN models are used for user relationship classification or advertisement recommendation. If an attacker obtains such a trained GNN and performs a graph reconstruction attack, it could reconstruct sensitive social connections among users.

Depending on the level of access to the target model, graph reconstruction attacks can be categorized into white-box attacks [8,15] and black-box attacks [2,5,10,14,17]. In the case of white-box attacks, the attacker has complete access to the parameters of the target model, including weights and gradients. In the case of black-box attacks, the attacker can only interact with the target model by querying its outputs, such as prediction probabilities or final classification results. In this work, we focus on black-box graph reconstruction attacks. Specifically, we consider the scenario where an attacker has access to a publicly released graph neural network and attempts to perform a graph reconstruction attack using background knowledge (e.g., node features) to infer the link structure of the training data. In this setting, there are two main challenges in achieving effective graph reconstruction under the black-box scenario. First, when gradient-based optimization methods are not directly applicable, how can the adversary extract effective information from the released target model to reconstruct the original training graph? Second, how to maintain the realism and quality of the reconstructed training graph while ensuring its similarity to the original graph? These challenges make exploring graph reconstruction attacks in black-box settings a pressing research problem.

To address the aforementioned challenges, we propose a novel black-box attack framework named ASM-GRA. Specifically, ASM-GRA first trains a substitute model that approximates the behavior of the target model using auxiliary information. Then, a gradient-based optimization attack is performed on the substitute model to reconstruct the graph structure. During this process, a node selection strategy is employed to reconstruct adjacency relations of representative nodes from multiple classes, thereby improving the overall reconstruction accuracy. Laplacian and sparsity regularization terms are introduced to ensure feature smoothness and structural sparsity of the reconstructed graph. Finally, the node features and the optimized adjacency matrix are fed into a graph autoencoder to obtain the final reconstructed adjacency matrix. We systematically evaluate ASM-GRA against baseline methods, and results show that ASM-GRA achieves more accurate graph reconstruction.

In summary, our contributions are as follows:

- We have identified two critical challenges in achieving black-box graph reconstruction attacks and proposed corresponding strategies for extracting effective information from the target model and ensuring the accuracy of reconstructed graphs.
- We propose a novel black-box graph reconstruction attack framework, ASM-GRA, which reconstructs the link relationships of private training data by attacking a substitute model that approximates the target model.

- Extensive experimental results across diverse datasets demonstrate that the proposed ASM-GRA effectively reconstructs link relationships in private training data. Furthermore, results from attacking different substitute model structures indicate that the framework is independent of the specific substitute model architecture, highlighting ASM-GRA's broad adaptability.

# 2 Preliminaries

## 2.1 Graph Structures

Given an undirected and unweighted graph $\mathbf{G} = (\mathbf{V}, \mathbf{E})$, where $\mathbf{V} = \{v_i\}_{i=1}^{n}$ represents the set of nodes with $|\mathbf{V}| = n$, and $\mathbf{E} \subseteq \{(u, v) \mid u, v \in \mathbf{V}\}$ denotes the set of edges, the graph can also be represented as $\mathbf{G} = (\mathbf{A}, \mathbf{X})$, where $\mathbf{A} \in \{0, 1\}^{n \times n}$ is the adjacency matrix with $A_{ij}$ indicating the connection of nodes $v_i$ and $v_j$, and $\mathbf{X} \in R^{n \times d}$ is the node feature matrix of dimension $d$. The label $y_i \in \mathbf{Y}$ denotes the category of node $v_i$.

## 2.2 Graph Reconstruction Attacks

**Attack Setting and Objectives.** In the black-box setting, the attacker can only query the target model $f_\theta$ and obtain either prediction vectors or labels. Because the attacker lacks access to $f_\theta$'s internal parameters, gradient-based direct attacks are infeasible. Therefore, we seek to train a substitute model $f_s$ that approximates $f_\theta$, enabling graph reconstruction attacks under the black-box setting. The GRA is defined as follows:

**Definition 1 (Graph Reconstruction Attack).** *Given a set of prior knowledge $S$ and a trained node classification model $f_\theta$ as the target model, the goal of a Graph Reconstruction Attack is to reconstruct the adjacency matrix $\mathbf{A}_O$ that was used to train the target model $f_\theta$. Formally:*

$$\mathbf{A}_R^* = \arg\max_{\mathbf{A}_R} P(\mathbf{A}_R \mid f_\theta, \mathbf{S}), \tag{1}$$

*where $\boldsymbol{A}_R$ is the reconstructed adjacency matrix, $P$ denotes the reconstruction method, and $\boldsymbol{S}$ signifies background knowledge. In our attack scenario, $\boldsymbol{S} = \{\boldsymbol{X}, \boldsymbol{Y}\}$.*

# 3 ASM-GRA: Graph Reconstruction Attack

## 3.1 Attack Overview

We propose a black-box graph reconstruction attack framework, ASM-GRA, which consists of four main components: (1) substitute model training, (2) projected gradient descent algorithm, (3) node selection strategy, and (4) graph autoencoder. Figure 1 illustrates the attack pipeline of ASM-GRA. In summary, ASM-GRA first trains a substitute model based on the target model, then attacks the substitute to obtain an initial reconstructed adjacency matrix, and finally feeds this matrix along with the node feature matrix $\mathbf{X}$ into a graph autoencoder to generate the final adjacency matrix $\mathbf{A}_R$.

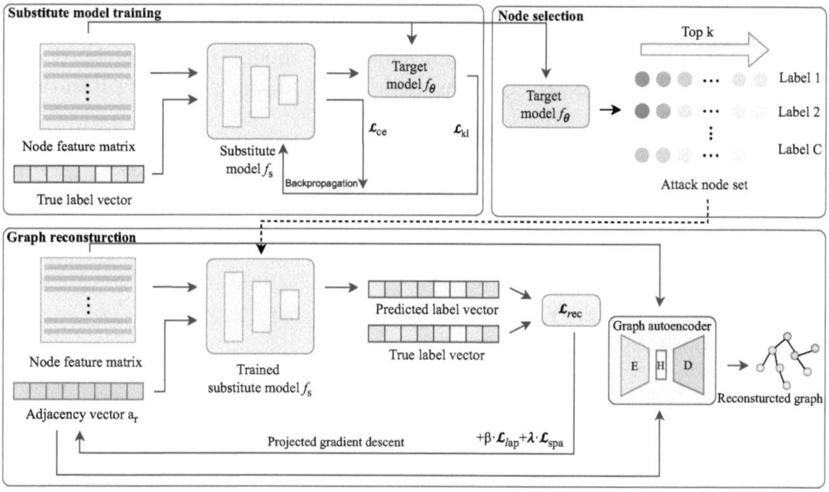


**Fig. 1.** Overview of the proposed ASM-GRA framework.

## 3.2 Substitute Model Training

When training the substitute model $f_s$, three key aspects must be considered:

**Substitute Model Structure:** In this work, since the target model performs node classification on graph-structured data, we adopt the Graph Attention Network (GAT) [12] as the structure of the substitute model. Moreover, as different substitute architectures may lead to varying attack effectiveness, the attacker can train multiple substitute models and select the one achieving the highest attack success rate.

**Generating the Training Dataset:** The training data required for GAT includes the node feature matrix $\mathbf{X}$ and edge index information (i.e., the indices of connected nodes). Given that the node feature matrix $\mathbf{X}$ is known, the key challenge lies in obtaining edge index information when the adjacency matrix $\mathbf{A}$ is unknown. In this work, we compute the cosine similarity between node feature vectors to derive the edge indices.

**Substitute Model Training Algorithm:** Algorithm 1 outlines the training procedure of the GAT-based substitute model. The detailed training process of the GAT substitute model is as follows.

Line 1 (Model Initialization): Initialize the GAT substitute model with a two-layer GATConv architecture.

Line 2 (Training Dataset Generation): To obtain the edge index information $I$ , we compute the cosine similarity between node feature vectors and compare it with a predefined threshold to infer edge existence.

Lines 3–9 (Model Training): Within the maximum number of iterations $E$, the substitute model $f_s$ is trained using a distillation loss function $\mathcal{L}_{distill}$, which

**Algorithm 1.** Training Algorithm for Substitute Model $f_s$

**Input:** Trained target model $f_\theta$ ; Initial substitute model $f_s$ ; Node label set $\mathbf{Y}$; Node feature matrix $\mathbf{X}$; Training node index $\mathrm{id}_{train}$ ; Training iterations $E$; Hyperparameters $\alpha$ , $m$

**Output:** Trained substitute model $f_s$

1: Initialize GAT model $f_s$
2: Compute edge index $I$ using node feature matrix $\mathbf{X}$
3: **for** *episode* $\leftarrow 0$ to $E-1$ **do**
4: $\quad \mathbf{z}_s \leftarrow f_s(\mathbf{X}, I, \mathrm{id}_{train})$
5: $\quad \mathbf{z}_t \leftarrow f_\theta(\mathbf{X}, \mathrm{id}_{train})$
6: $\quad \mathcal{L}_{distill} \leftarrow \alpha \cdot \mathcal{L}_{ce}(\mathbf{z}_s, \mathbf{Y}) + (1-\alpha) \cdot \mathcal{L}_{kl}(\mathbf{z}_s, \mathbf{z}_t, m)$
7: $\quad$ Backpropagate $\mathcal{L}_{distill}$
8: $\quad$ Update $\theta_{f_s}$
9: **end for**
10: **return:** Trained substitute model $f_s$

consists of the Kullback-Leibler (KL) divergence loss $\mathcal{L}_{kl}$ and the cross-entropy loss $\mathcal{L}_{ce}$.

$$\mathcal{L}_{kl} = \frac{1}{n}\sum_{i=1}^{n}\sum_{j=1}^{C}\frac{e^{z_{t,ij}/m}}{\sum_{k=1}^{C} e^{z_{t,ik}/m}} \log\left(\frac{\frac{e^{z_{t,ij}/m}}{\sum_{k=1}^{C} e^{z_{t,ik}/m}}}{\frac{e^{z_{s,ij}/m}}{\sum_{k=1}^{C} e^{z_{s,ik}/m}}}\right) \tag{2}$$

$$\mathcal{L}_{ce} = -\sum_{i=1}^{n}\sum_{j=1}^{C} y_{ij} \log\left(\frac{e^{z_{s,ij}}}{\sum_{k=1}^{C} e^{z_{s,ik}}}\right) \tag{3}$$

As shown in Eq. 2, $z_{t,ij}$ and $z_{s,ij}$ represent the values of the $j$-th component predicted by node $i$ in the target model $f_\theta$ and the substitute model $f_s$, respectively, where $C$ denotes the number of classes. $\mathcal{L}_{kl}$ facilitates smoother knowledge transfer from the target model to the substitute model by adjusting the temperature parameter $m$ in the temperature scaling technique. $\mathcal{L}_{ce}$ can be found in Eq. 3, where $y_{ij}$ represents the $j$-th component of the true label for node $i$.

$$\mathcal{L}_{distill} = \alpha \cdot \mathcal{L}_{\mathrm{ce}} + (1-\alpha) \cdot \mathcal{L}_{\mathrm{kl}} \tag{4}$$

Equation 4 summarizes the objective function for training the substitute model $f_s$, where the parameter $\alpha$ controls the contributions of $\mathcal{L}_{kl}$ and $\mathcal{L}_{ce}$. During the training, when the substitute model cannot fully learn the target model's behavior from $\mathcal{L}_{kl}$ due to architectural differences, $\mathcal{L}_{ce}$ helps the substitute model achieve better convergence in the node classification task and prevents large classification errors when the substitute model approaches the target model.

Line 10 (Model Saving): Save the trained parameters of the model $f_s$.

### 3.3 Projected Gradient Descent Algorithm

To facilitate gradient computation, we replace the matrix $\mathbf{A}_R$ with a vector $\mathbf{a}_r$ containing $N = n(n-1)/2$ elements, making $\mathbf{a}_r$ the variable to be optimized. The matrix $\mathbf{A}_R$ and the vector $\mathbf{a}_r$ can be converted into each other.

During the attack process, we reconstruct the adjacency matrix by minimizing the cross-entropy between the predicted labels $\hat{y}_i$ obtained from the substitute model $f_s$ and the true labels $y_i$ for node $i$. The attack objective function can be formalized as:

$$\mathcal{L}_{rec} = -\sum_{i=1}^{n}\sum_{j=1}^{C} y_{ij}\log(\hat{y}_{ij}). \tag{5}$$

Meanwhile, we employ two common regularization terms. The first is Laplacian regularization [1], defined as:

$$\mathcal{L}_{lap} = \mathrm{tr}(\mathbf{X}^T\mathbf{L}\mathbf{X}) = \sum_{i,j=1}^{n} A_{ij}\|\mathbf{x}_i - \mathbf{x}_j\|_2^2, \tag{6}$$

where $\mathbf{x}_i - \mathbf{x}_j$ represents the difference in feature vectors between nodes $i$ and $j$, $\mathbf{L} = \mathbf{D} - \mathbf{A}$ is the graph Laplacian matrix, with $\mathbf{D}$ being the degree matrix of $\mathbf{A}$. Laplacian regularization promotes the smoothness of node representations by constraining the feature differences between neighboring nodes [7].

Next, we apply graph sparsity regularization [6], as shown in Eq. 7:

$$\mathcal{L}_{spa} = \|\mathbf{a}_r\|_2^2 = \sum_{i=1}^{N} a_{r,i}^2, \tag{7}$$

where $\|\mathbf{a}_r\|_2^2$ is the square of the L2 norm of the vector $\mathbf{a}_r$, i.e., the sum of squares of all its elements. By enforcing the sparsity of $\mathbf{A}_R$, we ensure the plausibility and quality of the $\mathbf{A}_R$.

**Attack Objective.** Equation 8 summarizes the objective function for performing the graph reconstruction attack:

$$\mathcal{L} = \mathcal{L}_{rec} + \beta\mathcal{L}_{lap} + \lambda\mathcal{L}_{spa}, \tag{8}$$

where $\beta$ and $\lambda$ are the hyperparameters controlling the two regularizations.

Inspired by [14], the projection gradient descent (PGD) module is employed to address the optimization problem in Eq. 8. The update rule for $\mathbf{a}_r^{t+1}$ is given by [14]:

$$\mathbf{a}_r^{t+1} = P_{[0,1]}\left(\mathbf{a}_r^t - \eta_t g_t\right), \tag{9}$$

where $t$ denotes the iteration index of PGD, $\eta_t$ represents the learning rate, and $g_t$ is the gradients of the loss $\mathcal{L}$ in Eq. 8 with respect to $\mathbf{a}_r^t$. The projection operator is defined as [14]:

$$P_{[0,1]}[x] = \begin{cases} 0, & \text{if } x < 0 \\ 1, & \text{if } x > 1 \\ x, & \text{otherwise.} \end{cases} \tag{10}$$

### 3.4 Node Selection Strategy

In many graph neural networks, most nodes have low degrees, while a small number of nodes have high degrees. These highly connected representative nodes contribute a large proportion of the total edges in the graph. Based on this observation, our intuition is that recovering the connections of these representative nodes can facilitate reconstructing the overall graph structure.

To this end, we design a strategy that leverages the target model to select several representative nodes, referred to as attack nodes. Specifically, we feed the node feature matrix $\mathbf{X}$ into the target model $f_\theta$ to obtain the prediction vector $\hat{\mathbf{y}} = f_\theta(\mathbf{X})$. For each class, we then select the top-$k$ nodes with the highest prediction scores to form the attack node set $\mathbf{S}$. This set is used in Eq. (5) of Sect. 3.3. The node selection strategy can be formalized as follows:

$$\mathbf{S} = \bigcup_{j=1}^{C} \{i \mid \hat{y}_{ij} \in \text{top}_k(\hat{y}[:, j])\}\,. \tag{11}$$

Here, $\hat{y}_{ij}$ represents the predicted probability of node $i$ belonging to class $j$, and $\text{top}_k(\hat{y}[:, j])$ denotes the set of nodes with the top $k$ highest prediction scores for class $j$.

### 3.5 Generator: Graph Autoencoder

In ASM-GRA, we propose to use a Graph AutoEncoder (GAE) to generate the adjacency matrix, a structure that has been commonly adopted in prior works [10,15,17]. The GAE consists of two components: an encoder and a decoder, and it serves as a general framework that can be implemented with different structures. Specifically:

**Encoder:** The encoder adopts a GCN structure, which takes the node adjacency vector $\mathbf{a}_r$ and feature matrix $\mathbf{X}$ as inputs to obtain the latent representations of the nodes $\mathbf{H} \in R^{n \times k}$. This process can be expressed as:

$$\mathbf{H} = \text{GCN}(\mathbf{a}_r, \mathbf{X}), \tag{12}$$

**Decoder:** We adopt an inner-product decoder structure to reconstruct the adjacency matrix by calculating the inner product of $\mathbf{H}$, as shown in Eq. 13:

$$\mathbf{A} = \sigma(\mathbf{H}\mathbf{H}^T), \tag{13}$$

where $\sigma$ denotes the logistic sigmoid function. Intuitively, if the inner product of the embedding vectors of two nodes is larger, their positions in the embedding space are closer, indicating a higher likelihood of an edge between them. The output inner product $\mathbf{A}$ can be interpreted as a probability matrix, representing the likelihood of edge existence. By setting a threshold $M$, we obtain a binary reconstructed adjacency matrix $\mathbf{A}_R$.

# 4 Evaluation

## 4.1 Experimental Setup

**Datasets.** Our graph reconstruction attack is evaluated on five public datasets: Cora, Citeseer, Pubmed, AIDS, and Enzymes. These datasets are widely used as benchmark datasets for assessing GRA performance [5,10,14]. The detailed statistics of the evaluation datasets are presented in Table 1.

**Table 1.** Dataset Statistics.

| Dataset | Nodes | Edges | Classes | Features | Degree Dist. |
|---|---|---|---|---|---|
| Cora | 2,708 | 5,429 | 7 | 1,433 | Power-law |
| Citeseer | 3,327 | 4,732 | 6 | 3,703 | Sparse long-tail |
| Pubmed | 19,717 | 44,338 | 3 | 500 | Long-tail |
| AIDS | 31,385 | 64,780 | 38 | 4 | Low-degree |
| Enzymes | 19,580 | 74,564 | 3 | 18 | Concentrated |

**Baselines.** We compare a total of four baseline methods.

- *Attribute Similarity (Attr. Sim)*: A method estimates the probability of edge existence by measuring the cosine similarity between node attributes [5].
- *Embedding Reconstruction (ER)*: Reconstructing the adjacency matrix from node embeddings [2,10]. We take the node representations from the penultimate layer of the target graph neural network, then compute the cosine similarity between node representations in this layer to infer edges.
- *Gradient Estimation (GE)*: A black-box graph reconstruction attack using zeroth-order gradient estimation [14].
- *RL-GraphMI*: A black-box graph reconstruction method that employs deep reinforcement learning (DRL) to infer edges of the training graph [14].

**Metrics.** We adopt the following metrics: 1) **AUC** (Area Under the ROC Curve). AUC quantifies a model's ability to distinguish between different categories by calculating the area under the ROC curve. 2) **AP** (Average Precision). This metric estimates the proportion of correctly predicted edges among all predicted edges. Specifically, we randomly sample an equal number of node pairs with and without edges to compute AUC and AP. Generally, higher AUC and AP values indicate better attack performance.

**Implementation Details.** To evaluate the attack performance of ASM-GRA, we train three target models: a 2-layer GCN [7], a 2-layer GAT [12], and a 2-layer GraphSAGE [4]. During training, 80% of the nodes are used for the training set, and all GNN models are trained for 200 epochs with early stopping based on the highest validation accuracy. The validation set consists of 20% randomly

**Table 2.** Attack performance of ASM-GRA on different datasets. Higher AUC and AP values indicate better attack performance.

| Method | Cora | | Citeseer | | Pubmed | | AIDS | | Enzymes | |
|---|---|---|---|---|---|---|---|---|---|---|
| | AUC | AP | AUC | AP | AUC | AP | AUC | AP | AUC | AP |
| Attr. Sim | 0.798 | 0.804 | 0.885 | 0.899 | 0.797 | 0.796 | 0.781 | 0.799 | 0.609 | 0.630 |
| ER | 0.826 | 0.825 | 0.848 | 0.852 | 0.801 | 0.800 | 0.664 | 0.665 | 0.614 | 0.627 |
| GE | 0.833 | 0.842 | 0.842 | 0.857 | 0.807 | 0.793 | 0.782 | 0.793 | **0.659** | 0.643 |
| RL-GraphMI | 0.801 | 0.788 | 0.862 | 0.859 | 0.803 | 0.791 | 0.776 | 0.763 | 0.571 | 0.532 |
| **Ours** | **0.891** | **0.876** | **0.950** | **0.951** | **0.811** | **0.805** | **0.816** | **0.838** | 0.640 | **0.658** |

selected nodes. We also adopt GCN, GAT, and GraphSAGE as substitute model architectures and report results for both cases where the substitute model has the same or different architecture as the target model. During the attack, we set $\alpha = 0.5$, $\beta = 0.001$, $\lambda = 0.0001$, $\eta_t = 0.1$, $M = 0.5$, and the value of $k$ is determined based on different datasets. All experiments are conducted on a single A100 GPU with 80 GB of memory, and the reported results are averaged over five runs.

### 4.2 Reconstruction Results

**Comparison with Baselines.** The attack results are summarized in Table 2. We report the ASM-GRA performance using GCN as the target model and GAT as the substitute model. As shown in Table 2, ASM-GRA achieves the best performance on almost all datasets, demonstrating its effectiveness across datasets with different degree distributions. For example, ASM-GRA attains an average attack AUC of 0.95 and an average attack AP of 0.951 on the Citeseer dataset, outperforming the best baseline by at least 5% points. In addition, we observe that as the number of edges increases, the attack performance of all methods decreases, because denser graphs exhibit more complex structures that are harder to reconstruct. Nevertheless, on the AIDS dataset, our attack still achieves the best performance, reaching an average attack AUC of 0.816 and an average attack AP of 0.838.

**Performance on Various Target Models.** To evaluate the adaptability of ASM-GRA across different GNN models, we further adopt GAT and GraphSAGE as target models. Meanwhile, to examine the impact of different substitute models, we use GCN, GAT, and GraphSAGE as substitute models when attacking a given target model. Figure 2 presents the attack performance of ASM-GRA on the three target GNNs. For example, in Fig. 2 (a), where the GCN serves as the target model, the attack results obtained with the three substitute models are similar. Overall, although the attack performance varies slightly across substitute models, ASM-GRA consistently succeeds on all target models, indicating that it does not depend on a particular substitute model architecture.

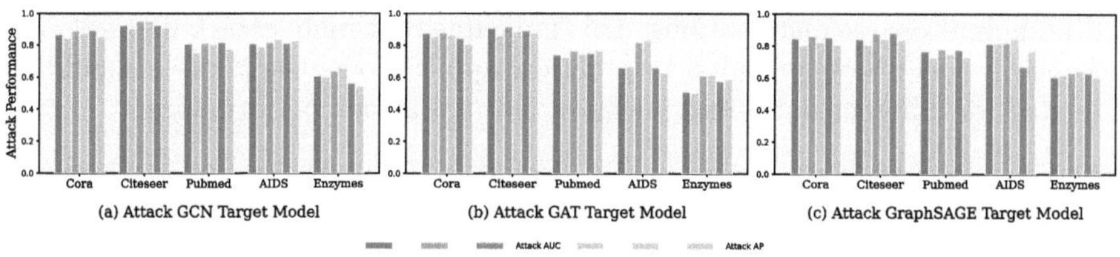


**Fig. 2.** Attack performance of ASM-GRA on different graph neural networks. Each dataset corresponds to 3 groups of results. From left to right, the 3 colors represent the average attack performance against the target model when GCN, GAT, and GraphSAGE are used as substitute models, respectively.

### 4.3 Ablation Study

**Node Selection Strategy.** To assess the impact of the Node Selection Strategy (NSS) module on the attack performance of ASM-GRA, we remove the NSS module and use randomly selected nodes as attack nodes while keeping all other attack procedures unchanged. Table 3 summarizes the ablation results. The results show that incorporating the NSS module enables ASM-GRA to achieve the best performance on all datasets. This is because the NSS module facilitates the reconstruction of representative nodes' adjacency relations across different classes, thereby improving the accuracy of the entire adjacency matrix.

**Table 3.** Ablation study of the Node Selection Strategy (NSS) module on 5 datasets.

| Dataset | Method | AUC | AP |
|---|---|---|---|
| Cora | ASM-GRA | **0.891** | **0.876** |
| | ASM-GRA w/o NSS | 0.865 | 0.834 |
| Citeseer | ASM-GRA | **0.950** | **0.951** |
| | ASM-GRA w/o NSS | 0.927 | 0.917 |
| AIDS | ASM-GRA | **0.816** | **0.838** |
| | ASM-GRA w/o NSS | 0.765 | 0.781 |
| Pubmed | ASM-GRA | **0.811** | **0.805** |
| | ASM-GRA w/o NSS | 0.785 | 0.794 |
| Enzymes | ASM-GRA | **0.640** | **0.658** |
| | ASM-GRA w/o NSS | 0.629 | 0.627 |

**The K Value.** We also investigate the impact of the parameter $k$ on the attack performance of ASM-GRA. Specifically, we vary $k$ from 20 to 120, and Fig. 3 illustrates how attack performance changes with $k$. As we have observed, the optimal $k$ value for ASM-GRA's best attack performance varies across datasets with different degree distributions. This indicates that more attack nodes do not necessarily yield better results. In practical applications, the $k$ value should be adjusted for different datasets to achieve optimal attack performance.

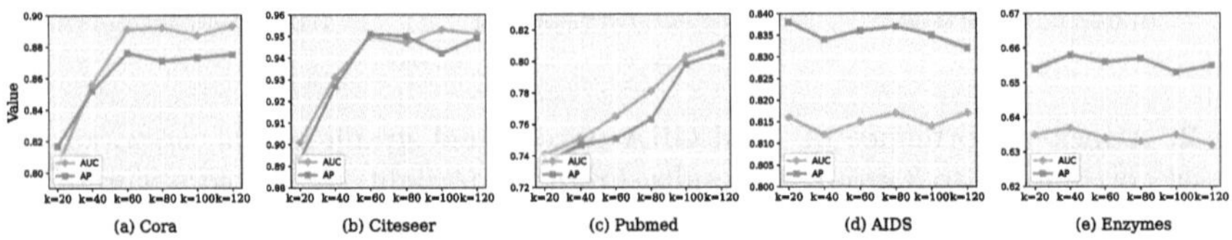


**Fig. 3.** The impact of the number of attack nodes $k$ on the performance of ASM-GRA.

## 5 Related Work

**Graph Reconstruction Attacks.** GRA was first proposed in [2] and has since been extensively studied. Existing attack methods can be categorized into white-box attacks [8,15] and black-box attacks [2,5,10,14,17]. In white-box attacks, the attacker has access to the target model's parameters and gradients, enabling graph reconstruction based on this information. For example, Zhang et al. [15] proposed an optimization-based white-box framework, GraphMI, which reconstructs the adjacency matrix by minimizing the cross-entropy loss between true and predicted node labels. In contrast, black-box attacks only allow the attacker to query the target model for outputs, making reconstruction much more challenging. To this end, this paper transforms black-box attacks into white-box attacks by attacking a substitute model approximating the target model, thereby enabling the reconstruction of the graph structure. This approach of training substitute models has played a significant role in multiple research domains, particularly in privacy attacks [9,11,18]. Given the effectiveness of substitute models in black-box scenarios, this paper aims to explore their application in graph reconstruction attacks, offering new insights for future research.

## 6 Conclusion

In this paper, we propose a black-box graph reconstruction attack framework, ASM-GRA, which reconstructs the adjacency matrix by attacking a substitute model. Extensive experimental results demonstrate that the proposed ASM-GRA effectively achieves reconstruction. Meanwhile, we also recognize that ASM-GRA can successfully reconstruct adjacency matrices only for datasets that possess node features, which constitutes a restrictive condition. Therefore, reconstructing adjacency matrices for datasets without node features will be left as future work.

**Acknowledgments.** This work is supported by the National Natural Science Foundation of China (61976142) and Shenzhen Science and Technology Plan Project (JCYJ20210324093609025).

## References

1. Belkin, M., Niyogi, P.: Laplacian eigenmaps and spectral techniques for embedding and clustering. In: Advances in Neural Information Processing Systems, vol. 14 (2001)
2. Duddu, V., Boutet, A., Shejwalkar, V.: Quantifying privacy leakage in graph embedding. In: MobiQuitous, pp. 76–85. ACM (2020)
3. Fan, W., et al.: Graph neural networks for social recommendation. In: The World Wide Web Conference, pp. 417–426 (2019)
4. Hamilton, W., Ying, Z., Leskovec, J.: Inductive representation learning on large graphs. In: Advances in Neural Information Processing Systems, vol. 30 (2017)
5. He, X., Jia, J., Backes, M., Gong, N.Z., Zhang, Y.: Stealing links from graph neural networks. In: 30th USENIX Security Symposium (USENIX Security 21), pp. 2669–2686 (2021)
6. Kalofolias, V.: How to learn a graph from smooth signals. In: Artificial Intelligence and Statistics, pp. 920–929. PMLR (2016)
7. Kipf, T.N., Welling, M.: Semi-supervised classification with graph convolutional networks. In: 5th International Conference on Learning Representations, ICLR 2017, Toulon, France, 24–26 April 2017, Conference Track Proceedings (2017)
8. Lin, M., Dai, E., Xu, J., Jia, J., Zhang, X., Wang, S.: Stealing training graphs from graph neural networks. In: KDD (1), pp. 777–788. ACM (2025)
9. Nguyen, B.N., Chandrasegaran, K., Abdollahzadeh, M., Cheung, N.M.M.: Label-only model inversion attacks via knowledge transfer. In: Advances in Neural Information Processing Systems, vol. 36, pp. 68895–68907 (2023)
10. Shen, Y., et al.: Finding MNEMON: reviving memories of node embeddings. In: CCS, pp. 2643–2657. ACM (2022)
11. Shokri, R., Stronati, M., Song, C., Shmatikov, V.: Membership inference attacks against machine learning models. In: 2017 IEEE Symposium on Security and Privacy (SP), pp. 3–18 (2017)
12. Velickovic, P., Cucurull, G., Casanova, A., Romero, A., Liò, P., Bengio, Y.: Graph attention networks. In: 6th International Conference on Learning Representations, ICLR 2018, Vancouver, BC, Canada, April 30 – May 3 2018, Conference Track Proceedings (2018)
13. Wu, S., Sun, F., Zhang, W., Xie, X., Cui, B.: Graph neural networks in recommender systems: a survey. ACM Comput. Surv. **55**(5), 1–37 (2022)
14. Zhang, Z., Liu, Q., Huang, Z., Wang, H., Lee, C.K., Chen, E.: Model inversion attacks against graph neural networks. IEEE Trans. Knowl. Data Eng. **35**(9), 8729–8741 (2022)
15. Zhang, Z., et al.: GraphMI: extracting private graph data from graph neural networks. In: Proceedings of the Thirtieth International Joint Conference on Artificial Intelligence, IJCAI 2021, Virtual Event/Montreal, Canada, 19–27 August 2021, pp. 3749–3755 (2021)
16. Zhang, Z., Liu, Q., Wang, H., Lu, C., Lee, C.K.: Motif-based graph self-supervised learning for molecular property prediction. In: Advances in Neural Information Processing Systems , vol. 34, pp. 15870–15882 (2021)

17. Zhang, Z., Chen, M., Backes, M., Shen, Y., Zhang, Y.: Inference attacks against graph neural networks. In: USENIX Security Symposium, pp. 4543–4560. USENIX Association (2022)
18. Zhao, X., Zhang, W., Xiao, X., Lim, B.: Exploiting explanations for model inversion attacks. In: Proceedings of the IEEE/CVF International Conference on Computer Vision, pp. 682–692 (2021)

# Unified Local and Global Structure Learning for Feature Selection

Xinjie Han[1,2], Cong Lei[3], Yuhao Jiang[1,2], Jiayang Su[4], Shuyu Li[1,2], and Guoqiu Wen[1,2(✉)]

[1] Key Laboratory of Education Blockchain and Intelligent Technology, Ministry of Education, Guangxi Normal University, Guilin 541004, China
{hanxinjie,jyhgxnu,lishuyu}@stu.gxnu.edu.cn, wenguoqiu2008@163.com
[2] Guangxi Key Laboratory of Multi-source Information Mining and Security, Guangxi Normal University, Guilin 541004, China
[3] The University of Sydney, Sydney, NSW 2006, Australia
clei9249@uni.sydney.edu.au
[4] Hainan University, Hainan, China

**Abstract.** Feature selection is an indispensable key technology for tackling the challenges of high-dimensional data. Although prior research indicates that simultaneously preserving local and global data structures provides complementary information, existing methods mostly perform simple linear superposition of the two, making it difficult to characterize their underlying correlations and interactions. This results in structural information being underutilized, thereby limiting the model's representational capacity and generalization performance. To overcome this limitation, we propose a unified feature selection model that collaboratively learns and aligns local and global structures within a shared low-dimensional subspace. The introduced alignment mechanism dynamically coordinates structural information, ensuring structural consistency during learning and enhancing representation quality. Furthermore, we develop an efficient iterative optimization algorithm and provide a rigorous theoretical proof of its convergence. Extensive experiments on multiple benchmark datasets demonstrate that our method significantly outperforms existing feature selection algorithms in various clustering tasks, which fully validates the model's effectiveness, robustness, and universality.

**Keywords:** Feature selection · Local and global structure alignment · Shared subspace · Clustering tasks

## 1 Introduction

As high-dimensional data continues to emerge across various domains [1–3], challenges including computational burden, model overfitting, and the "curse of dimensionality" have emerged accordingly [4]. Feature selection, as a key


R. C.-W. Wong et al. (Eds.): PAKDD 2026, LNAI 16597, pp. 135–148, 2026.
https://doi.org/10.1007/978-981-92-1300-9_11

dimensionality reduction technique, identifies and retains the most discriminative feature subsets to enhance model performance, efficiency, and interpretability. Existing methods often adhere to the "structure preservation" principle, leveraging either the local manifold structure [5] or global statistical properties [6] of the data to guide feature selection. Although local methods capture neighborhood relationships and global methods model overall directional variation, relying on either structure alone often leads to suboptimal feature selection. Recent studies attempt to integrate both within a unified framework [7,8], yet most simply combine them linearly without considering their intrinsic interaction. When local manifold geometry conflicts with global variance directions, such naive superposition may cause structural inconsistency, preventing the learned representation from faithfully preserving the original data structure.

To address these challenges, this paper proposes a robust feature selection model that unifies local and global structure perspectives (ULGFS). By introducing a shared low-dimensional subspace representation, both the local and global perspectives are forced to be represented and interact with each other through this subspace. Specifically, we no longer merely require samples to be close in the local subspace but further demand that their performance in the global feature selection space aligns with the local spatial structure. Consequently, the constraints on the local structure no longer function in isolation but are actively aligned with the global structural features, enabling the mutual refinement and reinforcement of the two types of information during the learning process. Overall, our main contributions can be summarized as follows:

- Explicitly incorporating a global preservation term into the objective function complements the local manifold preservation term, ensuring that the low-dimensional representation both captures the microscopic local geometry and reflects the macroscopic global distribution of the data.
- By introducing shared subspaces, local learning and global learning are forced to mutually align and constrain each other, dynamically reconciling differences between local and global perspectives to learn more discriminative and robust feature subsets.
- Our proposed method demonstrates superior performance over several selected comparison methods through clustering experiments conducted on seven distinct datasets.

## 2 Method

In many existing approaches, local and global structures are introduced as two independent regularization terms and optimized simultaneously. However, when the local manifold geometry and global variance direction fail to align completely, simply combining them may lead to inconsistent optimization objectives. Based on this observation, we aim to construct a unified framework through an innovative shared subspace alignment mechanism, enabling local and global structures not only to be preserved but also to interact during the learning process. Define the input high-dimensional data $\mathbf{X} \in \mathbb{R}^{d\times n}$ (where $d$ and $n$ denote the number of features and samples, respectively). The core variable $\mathbf{Z} \in \mathbb{R}^{k\times n}$ (where $k$ denotes the target subspace dimensionality) represents the low-dimensional subspace, $\mathbf{W} \in \mathbb{R}^{d\times k}$ denotes the subspace reconstruction projection matrix, $\mathbf{M} \in \mathbb{R}^{d\times k}$ is the feature selection projection matrix, and $\mathbf{S} \in \mathbb{R}^{n\times n}$ is the graph matrix.

### 2.1 Subspace Reconstruction

A fundamental premise of representation learning is that low-dimensional representation should retain sufficient information to reconstruct the original data [9]. By minimizing the reconstruction error between the original high-dimensional data $\mathbf{X}$ and its reconstructed representation $\mathbf{WZ}$ in the low-dimensional space, we force the subspace representation $\mathbf{Z}$ to preserve the most essential structural information in the data. This is mathematically defined as:

$$L_{\text{rec}} = \|\mathbf{X} - \mathbf{WZ}\|_F^2 = \sum_{i=1}^{n} \|\mathbf{x}_i - \mathbf{W}\mathbf{z}_i\|_2^2 \tag{1}$$

### 2.2 Joint Learning of Local and Global Structures

**Local Structure Preservation.** We construct the adjacency matrix $\mathbf{S}$ between samples through adaptive graph learning and impose constraints on local consistency [10] to obtain the following objective function

$$\sum_{i,j=1}^{n} s_{ij} \|\mathbf{z}_i - \mathbf{z}_j\|_2^2 \tag{2}$$

where $\mathbf{z}_i$ and $\mathbf{z}_j$ represent the low-dimensional representations of samples $\mathbf{x}_i$ and $\mathbf{x}_j$, respectively. This constraint forces the low-dimensional representations $\mathbf{z}_i$ and $\mathbf{z}_j$ corresponding to larger elements $s_{ij}$ in the relationship matrix $\mathbf{S}$ to be closer in Euclidean space. At the same time, to obtain a reasonable neighborhood structure, we impose constraints such as non-negativity, the exclusion of self-connections, and row normalization on the graph matrix. This endows it with probabilistic interpretability and avoids trivial solutions. Furthermore, introducing a graph regularization term effectively promotes the sparsity and stability of the neighborhood structure. Through the aforementioned constraints and regularization, the following local preservation term is derived:

$$\begin{aligned} \mathcal{L}_{\text{local}} &= \sum_{i,j=1}^{n} s_{ij} \|\mathbf{z}_i - \mathbf{z}_j\|_2^2 + \beta \sum_{i=1}^{n} \|\mathbf{s}_i\|_2^2 \\ \text{s.t.} \quad & \mathbf{s}_i^\top \mathbf{1} = 1, \quad s_{ii} = 0, \quad s_{ij} \geq 0, \quad \forall i, j. \end{aligned} \tag{3}$$

**Global Structure Preservation.** Inspired by the principles of Principal Component Analysis (PCA) [11], we introduce a global constraint term into our model. This term maximizes the variance of the projected data to preserve the most significant global variability information, thereby maintaining the overall representational capacity of the data.

Let the projection basis matrix $\mathbf{M}$ satisfy the orthogonality constraint $\mathbf{M}^\top\mathbf{M} = \mathbf{I}$. Its reconstruction operator can be expressed as $\mathbf{M}\mathbf{M}^\top$. This operator corresponds to the orthogonal projection that reconstructs data $\mathbf{X}$ back to the original space after projecting it onto the subspace spanned by $\mathbf{M}$. We aim to preserve the primary information of the data in the low-dimensional space by minimizing the reconstruction error $\left\|\mathbf{X} - \mathbf{M}\mathbf{M}^\top\mathbf{X}\right\|_F^2$. Based on the properties of matrix trace operations, the following derivation follows:

Let $\mathbf{P} = \mathbf{M}\mathbf{M}^\top$ with $\mathbf{M}^\top\mathbf{M} = \mathbf{I}$, then we get

$$\begin{aligned} \|\mathbf{X} - \mathbf{M}\mathbf{M}^\top\mathbf{X}\|_F^2 &= \|(\mathbf{I} - \mathbf{P})\mathbf{X}\|_F^2 \\ &= \operatorname{tr}\left(\mathbf{X}^\top(\mathbf{I} - \mathbf{P})^\top(\mathbf{I} - \mathbf{P})\mathbf{X}\right) \\ &= \operatorname{tr}\left(\mathbf{X}^\top(\mathbf{I} - \mathbf{P})\mathbf{X}\right) \\ &= \operatorname{tr}(\mathbf{X}^\top\mathbf{X}) - \operatorname{tr}(\mathbf{X}^\top\mathbf{P}\mathbf{X}) \\ &= \operatorname{tr}(\mathbf{X}^\top\mathbf{X}) - \operatorname{tr}(\mathbf{M}^\top\mathbf{X}\mathbf{X}^\top\mathbf{M}) \end{aligned} \tag{4}$$

Since $\operatorname{tr}(\mathbf{X}^\top\mathbf{X})$ is constant with respect to $\mathbf{M}$, we only retain $-\operatorname{tr}(\mathbf{M}^\top\mathbf{X}\mathbf{X}^\top\mathbf{M})$ in the objective function. To perform feature selection rather than simple dimensionality reduction, we further impose an $\ell_{2,1}$-norm regularization on $\mathbf{M}$, which encourages row sparsity and automatically eliminates irrelevant features. Thus, the global preservation term is formulated as:

$$\mathcal{L}_{\text{global}} = -\operatorname{tr}\left(\mathbf{M}^\top\mathbf{X}\mathbf{X}^\top\mathbf{M}\right) + \delta\,\|\mathbf{M}\|_{2,1} \tag{5}$$

### 2.3 Coordination Mechanism via Shared Subspace

Traditional methods often optimize local and global structures independently, which may result in inconsistent feature representations. To overcome this issue, we introduce a shared low-dimensional subspace in which both structures are jointly modeled, enabling effective interaction instead of separate optimization.

Equation (2) enforces similar samples $(\mathbf{x}_i, \mathbf{x}_j)$ to remain close in the low-dimensional subspace $\mathbf{Z}$, but this only preserves the internal structure of $\mathbf{Z}$ without ensuring consistency with the global feature selection space. As a result, neighborhood relationships in $\mathbf{Z}$ may not align with those in the selected feature space. To bridge this gap, we redefine the local term by evaluating neighborhood consistency with respect to the global projection. Specifically, we replace $\mathbf{z}_j$ with $\mathbf{M}\mathbf{x}_j$ and formulate the local preservation term as $\sum_{i,j=1}^{n} s_{ij} \left\|\mathbf{z}_i - \mathbf{M}\mathbf{x}_j\right\|_2^2$, thereby encouraging alignment between local manifold structure and global discriminative direction. The final objective function is then given by:

$$\begin{aligned} \min_{\mathbf{S},\mathbf{W},\mathbf{Z},\mathbf{M}} \quad & \sum_{i=1}^{n} \|\mathbf{x}_i - \mathbf{W}\mathbf{z}_i\|_2^2 + \alpha \sum_{i,j=1}^{n} s_{ij} \|\mathbf{z}_i - \mathbf{M}\mathbf{x}_j\|_2^2 \\ & + \beta \sum_{i=1}^{n} \|\mathbf{s}_i\|_2^2 - \gamma\, \mathrm{tr}(\mathbf{M}^\top \mathbf{X}\mathbf{X}^\top \mathbf{M}) + \delta \|\mathbf{M}\|_{2,1} \\ \text{s.t.} \quad & \mathbf{W}^\top\mathbf{W} = \mathbf{I},\ \mathbf{M}^\top\mathbf{M} = \mathbf{I},\ s_i^\top \mathbf{1} = 1,\ s_{ii} = 0,\ s_{ij} \geq 0,\ \forall i,j \end{aligned} \tag{6}$$

## 2.4 Robust Constraints

Given that the objective function employs quadratic loss, which is sensitive to outliers, we introduce a robust loss function to replace the quadratic loss in both the reconstruction term and the locality preservation term in equation (6), thereby enhancing the model's robustness and ultimately leading to the following optimized model:

$$\begin{aligned} \min_{\mathbf{S},\mathbf{W},\mathbf{Z},\mathbf{M}} \quad & \sum_{i=1}^{n} O_1\left(\|\mathbf{x}_i - \mathbf{W}\mathbf{z}_i\|_2^2\right) + \alpha \sum_{i,j=1}^{n} s_{ij}\, O_2\left(\|\mathbf{z}_i - \mathbf{M}\mathbf{x}_j\|_2^2\right) \\ & + \beta \sum_{i=1}^{n} \|\mathbf{s}_i\|_2^2 - \gamma\, \mathrm{tr}(\mathbf{M}^\top \mathbf{X}\mathbf{X}^\top \mathbf{M}) + \delta \|\mathbf{M}\|_{2,1} \\ \text{s.t.} \quad & \mathbf{W}^\top\mathbf{W} = \mathbf{I},\ \mathbf{M}^\top\mathbf{M} = \mathbf{I},\ s_i^\top \mathbf{1} = 1,\ s_{ii} = 0,\ s_{ij} \geq 0,\ \forall i,j \end{aligned} \tag{7}$$

where $O_1$ and $O_2$ are robust functions applied to the reconstruction residual and local structure residual [12], respectively. To ensure the feasibility and stability of the optimization process, $O(\mathbf{t})$ should be defined as a continuous even function that is non-decreasing with respect to $|t|$, satisfying $O(\mathbf{t}) \geq 0$ and $O(0) = 0$. Under these conditions, according to semi-quadratic (HQ) optimization theory [13], it follows that the problem $\min_{\mathbf{t}} \sum_{i=1}^{n} O(t_i)$ can be transformed into $\min_{\mathbf{t},\mathbf{p}} \sum_{i=1}^{n} (p_i t_i^2 + o(p_i))$, where $\mathbf{p}$ is the auxiliary variable, and $o(\mathbf{p})$ denotes the corresponding conjugate function of $O(\mathbf{t})$. Motivated by the optimization framework from prior work [14], we employ the following robust operator to replace both $O_1(x)$ and $O_2(x)$, that is,

$$O(x) = \frac{x^2}{2\,(1 + x^2)} \tag{8}$$

Substituting Eq. (8) into Eq. (7) and performing an equivalent rewrite using HQ theory, the objective function can be transformed into:

$$\begin{aligned} \min_{\mathbf{S},\mathbf{W},\mathbf{Z},\mathbf{M}} \quad & \sum_{i=1}^{n} b_i \|\mathbf{x}_i - \mathbf{W}\mathbf{z}_i\|_2^2 + \alpha \sum_{i,j=1}^{n} s_{ij} c_{ij} \|\mathbf{z}_i - \mathbf{M}\mathbf{x}_j\|_2^2 \\ & + o(\mathbf{b}) + o(\mathbf{C}) + \beta \sum_{i=1}^{n} \|\mathbf{s}_i\|_2^2 \\ & - \gamma\ \mathrm{tr}(\mathbf{M}^\top \mathbf{X}\mathbf{X}^\top \mathbf{M}) + \delta \|\mathbf{M}\|_{2,1} \\ \text{s.t.} \quad & \mathbf{W}^\top\mathbf{W} = \mathbf{I},\ \mathbf{M}^\top\mathbf{M} = \mathbf{I},\ \mathbf{s}_i^\top \mathbf{1} = 1,\ s_{ii} = 0,\ s_{ij} \geq 0,\ \forall i,j \end{aligned} \tag{9}$$

Here, $\mathbf{b} \in \mathbb{R}^{n\times 1}$ and $\mathbf{C} \in \mathbb{R}^{n\times n}$ respectively denote the auxiliary weight vector and auxiliary weight matrix introduced by the HQ transformation. According to the IRLS optimization strategy, during the alternating update process, the weights are adaptively adjusted based on the current residuals. The update form is:

$$b_i = \frac{1}{\sqrt{a_1 + \|\mathbf{x}_i - \mathbf{W}\mathbf{z}_i\|_2^2}}, \quad c_{ij} = \frac{1}{\sqrt{a_2 + \|\mathbf{z}_i - \mathbf{M}\mathbf{x}_j\|_2^2}} \tag{10}$$

Among these, $a_1, a_2 > 0$ is a numerical stability parameter used to prevent denominator division by zero and smooth weight changes. When residuals are large, the corresponding weights automatically decrease, thereby reducing the impact of outlier samples on the overall optimization process. The terms $o(\mathbf{b})$ and $o(\mathbf{C})$ are conjugate terms introduced by the HQ transformation. Through the above adaptive reweighting mechanism, the model significantly enhances robustness to outliers while preserving structural information.

### 2.5 Optimization Algorithms and Convergence Properties

We employ an alternating optimization strategy [15], iteratively updating variables in groups while providing a rigorous convergence proof. The variables in the objective function can be naturally grouped into five sets: **b**, **C**, **S**, **Z**, **W**, **M**. The algorithm fixes four of these sets and sequentially updates the remaining set. Thus, we formally express the optimization problem of objective function (8) as $J(\Theta^{t+1}) \leq J(\Theta^t)$, where $\Theta$ represents the set of all variables, $\Theta^t$ denotes the variables at the iteration as $\mathbf{b}^t, \mathbf{C}^t, \mathbf{S}^t, \mathbf{Z}^t, \mathbf{W}^t, \mathbf{M}^t$, based on Nikolova [13], and by determining $\mathbf{b}^{t+1}, \mathbf{C}^{t+1}, \mathbf{S}^{t+1}, \mathbf{Z}^{t+1}, \mathbf{W}^{t+1}, \mathbf{M}^{t+1}$, we have:

$$\begin{aligned}
J(\Theta^{t+1}) &\leq J(\mathbf{b}^t \mid \mathbf{C}^{t+1}, \mathbf{S}^{t+1}, \mathbf{Z}^{t+1}, \mathbf{W}^{t+1}, \mathbf{M}^{t+1}) \\
&\leq J(\mathbf{C}^t \mid \mathbf{S}^{t+1}, \mathbf{Z}^{t+1}, \mathbf{W}^{t+1}, \mathbf{M}^{t+1}, \mathbf{b}^t) \\
&\leq J(\mathbf{S}^t \mid \mathbf{Z}^{t+1}, \mathbf{W}^{t+1}, \mathbf{M}^{t+1}, \mathbf{b}^t, \mathbf{C}^t) \\
&\leq J(\mathbf{Z}^t \mid \mathbf{W}^{t+1}, \mathbf{M}^{t+1}, \mathbf{b}^t, \mathbf{C}^t, \mathbf{S}^t) \\
&\leq J(\mathbf{W}^t \mid \mathbf{M}^{t+1}, \mathbf{b}^t, \mathbf{C}^t, \mathbf{S}^t, \mathbf{Z}^t) \\
&\leq J(\mathbf{M}^t \mid \mathbf{b}^t, \mathbf{C}^t, \mathbf{S}^t, \mathbf{Z}^t, \mathbf{W}^t) \\
&= J(\Theta^t)
\end{aligned} \tag{11}$$

**Algorithm 1.** Optimization of ULGFS

**Require:** $\mathbf{X} \in \mathbb{R}^{d\times n}$, parameters $\alpha, \beta, \gamma, \delta$, target dimension $k$
**Ensure:** $\mathbf{W} \in \mathbb{R}^{d\times k}$, $\mathbf{M} \in \mathbb{R}^{d\times k}$, $\mathbf{Z} \in \mathbb{R}^{k\times n}$, $\mathbf{S} \in \mathbb{R}^{n\times n}$
**repeat**
  **Update b and C by equation (10)**
  **Update similarity matrix S:**
  For each sample $i$, compute $\mathbf{s}_i$ via normalized exponential form:
  $s_{ij} = \frac{\exp\left(-\frac{\alpha}{\beta}(d_{ij}-\eta_i)\right)}{\sum_{j=1}^{n}\exp\left(-\frac{\alpha}{\beta}(d_{ij}-\eta_i)\right)}$
  **Construct Laplacian matrix:** $\mathbf{L} = \mathrm{diag}(\mathbf{S} \odot \mathbf{C}) - (\mathbf{S} \odot \mathbf{C})$
  **Update Z:**
  Closed-form solution: $\mathbf{Z} = \mathbf{W}^\top \mathbf{X}\mathbf{B}\,(\mathbf{B} + \alpha\mathbf{L})^{-1}$
  **Update W:**
  Top-$k$ eigenvectors of $\frac{1}{2}\left(\mathbf{X}\mathbf{B}\,(\mathbf{B} + \alpha\mathbf{L})^{-1}\,\mathbf{B}\mathbf{X}^\top + \mathrm{sym.}\right)$
  **Update M:**
  $\mathbf{M} = \mathrm{eig}_k\left(\frac{1}{2}\left(\mathbf{X}\mathbf{B}\,(\mathbf{B} + \alpha\mathbf{L})^{-1}\,\mathbf{B}\mathbf{X}^\top + \gamma\,\mathbf{X}\mathbf{X}^\top\right) - \delta\,\mathbf{D}_\ell\right)$ where $\mathbf{D}_\ell$ denotes the diagonal weighting induced by iterative $\ell_{2,1}$ reweighting; enforces row sparsity
**until** Relative change of objective $< \varepsilon$

For clarity, we summarize the optimization process in Algorithm 1. Since each iteration solves one subproblem optimally while fixing the others, and each subproblem ensures a non-increasing objective value, the overall sequence $J(\Theta^t)$ is monotonically non-increasing and bounded below. By the Block Coordinate Descent theory, the algorithm therefore converges to a local optimum. Furthermore, the primary computational overhead per iteration stems from the singular value decomposition of the $d \times d$ matrix, resulting in a time complexity of approximately $O(d^3)$.

# 3 Experiments

## 3.1 Dataset and Comparison Methods

To comprehensively evaluate the method's performance, we select seven publicly available datasets from diverse domains shown in Table 1.

**Table 1.** The information of used datasets

| Dataset | Samples | Features | Classes | Source |
|---|---|---|---|---|
| COIL20 | 1440 | 1024 | 20 | http://www.cad.zju.edu.cn/home/dengcai/Data/data.html |
| ORL | 400 | 1024 | 40 | https://jundongl.github.io/scikit-feature/datasets.html |
| JAFEE | 213 | 256 | 10 | https://arxiv.org/abs/2403.10910 [16] |
| ATT40 | 400 | 1024 | 40 | https://doi.org/10.1109/TIP.2015.2479560 [17] |
| binalpha | 1404 | 320 | 36 | https://doi.org/10.1016/j.knosys.2021.106847 [18] |
| lung_small | 73 | 325 | 7 | https://jundongl.github.io/scikit-feature/datasets.html |
| lung | 203 | 3312 | 5 | https://jundongl.github.io/scikit-feature/datasets.html |

At the same time, we select five representative existing algorithms for comparative experiments: Laplace Scores (LS) [19] selects features based on local geometric structures of the data. Adaptive Unsupervised Feature Selection with robust graph regularization (AUFS) [20] conducts feature selection by minimizing a reconstructionregularization objective with an additional robust graph term. High-order Similarity Learning (HSL) [21] jointly learns projection matrices along with first- and higher-order similarity structures. The Robust Adaptive Flexible Graph method (RAFG) [22] integrates the $\ell_{2,1}$-norm into elastic regression and learns adaptive $\ell_1$-norm mappings to capture the clustering structure. The $\ell_{2,p}$-norm-based feature reconstruction method NFRFS [23] employs a flexible norm to model reconstruction geometry and adjusts the parameter $p$ to improve adaptability.

### 3.2 Experimental Setup

We employed grid search to tune hyperparameters and selected the optimal parameter combination for testing. Specifically, parameters $\alpha$, $\beta$, $\gamma$ and $\delta$ are searched within the range from $1 \times 10^{-3}$ to $1 \times 10^3$, while stability parameters $a_1$ and $a_2$ are selected from the interval $1 \times 10^{-4}$ to $1 \times 10^{-1}$. For the selected feature point set, we applied the K-means clustering algorithm to group the data. Given K-means' sensitivity to initialization, each experiment was repeated five times. To comprehensively evaluate clustering performance, we employed clustering accuracy (ACC) and normalized mutual information (NMI) as metrics, presenting final results as the average across all trials.

### 3.3 Experimental Analysis

Tables 2 and 3 report the ACC and NMI performance of all methods across seven benchmark datasets. Our method achieves the best results on all datasets, demonstrating strong generalization ability under varying data structures and noise conditions.

**Table 2.** Best ACC across different datasets for various methods (mean ± std (%))

| Dataset | LS | AUFS | HSL | RAFG | NFRFS | Ours |
|---|---|---|---|---|---|---|
| COIL20 | 53.89 ± 3.34 | 70.26 ± 1.82 | 71.05 ± 3.32 | 62.14 ± 2.40 | 69.94 ± 2.25 | 71.35 ± 2.17 |
| ORL | 40.09 ± 2.20 | 53.70 ± 1.92 | 54.89 ± 2.24 | 55.42 ± 2.40 | 58.66 ± 2.37 | 60.20 ± 0.94 |
| JAFFE | 66.38 ± 6.15 | 80.80 ± 3.52 | 80.00 ± 4.49 | 81.62 ± 2.97 | 83.43 ± 1.90 | 83.85 ± 0.64 |
| ATT40 | 49.16 ± 2.80 | 58.66 ± 2.75 | 55.18 ± 2.65 | 55.39 ± 3.05 | 59.31 ± 2.37 | 61.40 ± 0.94 |
| binalpha | 39.56 ± 1.51 | 43.82 ± 2.12 | 43.31 ± 1.72 | 45.47 ± 1.65 | 45.08 ± 1.36 | 48.38 ± 1.41 |
| lung_small | 67.74 ± 5.56 | 82.26 ± 5.47 | 78.15 ± 5.29 | 79.18 ± 5.49 | 83.97 ± 5.55 | 87.94 ± 1.34 |
| lung | 57.91 ± 7.48 | 79.68 ± 4.52 | 79.85 ± 3.94 | 77.64 ± 3.29 | 84.04 ± 4.14 | 91.63 ± 0.00 |

**Table 3.** Best NMI across different datasets for various methods (mean ± std (%))

| Dataset | LS | AUFS | HSL | RAFG | NFRFS | Ours |
|---|---|---|---|---|---|---|
| COIL20 | 70.53 ± 1.45 | 79.76 ± 1.02 | 80.91 ± 1.45 | 76.82 ± 0.86 | 80.04 ± 1.04 | 80.79 ± 0.59 |
| ORL | 63.93 ± 1.46 | 74.76 ± 1.20 | 75.59 ± 1.59 | 76.23 ± 1.63 | 78.19 ± 1.05 | 78.84 ± 0.43 |
| JAFFE | 71.27 ± 3.52 | 82.32 ± 2.04 | 83.26 ± 2.65 | 83.05 ± 2.63 | 84.67 ± 1.50 | 84.86 ± 0.72 |
| ATT40 | 72.23 ± 1.39 | 78.13 ± 0.95 | 76.13 ± 0.61 | 76.20 ± 0.68 | 78.53 ± 1.31 | 79.20 ± 0.44 |
| binalpha | 55.85 ± 1.24 | 59.05 ± 0.85 | 59.19 ± 0.51 | 60.64 ± 0.76 | 59.68 ± 0.93 | 60.77 ± 0.39 |
| lung_small | 64.70 ± 4.42 | 76.61 ± 4.23 | 74.30 ± 3.39 | 73.32 ± 3.01 | 78.36 ± 4.33 | 82.13 ± 1.97 |
| lung | 47.04 ± 3.14 | 64.74 ± 1.30 | 64.36 ± 1.89 | 65.87 ± 1.34 | 66.78 ± 3.16 | 74.94 ± 0.00 |

Compared with methods that also combine local and global information, such as RAFG and NFRFS, our alignment-based framework achieves more substantial improvements. For instance, on ATT40, it outperforms RAFG and NFRFS by 6.01% and 2.09% in ACC, and by 3% and 0.67% in NMI, respectively. These results suggest that simple integration is insufficient, while our alignment mechanism effectively alleviates structural inconsistency.

Across face datasets (ORL, JAFFE, ATT40), high-dimensional biological datasets (lung, lung_small), and object-image datasets (COIL20, binalpha), our method consistently achieves superior performance, demonstrating the robustness and broad applicability of the proposed collaborative learning mechanism.

### 3.4 Ablation Study

We conduct comprehensive ablation experiments to evaluate the impact of global structure preservation, local structure modeling, shared subspace alignment, and robust functions on model performance. Specifically, in addition to removing the global constraints, removing the local constraints, and eliminating the robustness term, we design two non-shared alternatives: one that constructs local structure solely in the latent representation space $\|\mathbf{z}_i - \mathbf{z}_j\|$, and another that constructs local structure solely in the feature selection space $\|\mathbf{Mx}_i - \mathbf{Mx}_j\|$, neither of which is coupled through a shared subspace. As shown in Table 4 and 5, the full model consistently achieves the best overall performance across all datasets, validating the importance of synergistic optimization of components for enhancing clustering effectiveness.

**Table 4.** ACC of our method across different modules (mean ± std (%))

| Dataset | Full | w/o Global | w/o Local | Latent Local | Feature Local | w/o Robust |
|---|---|---|---|---|---|---|
| COIL20 | 71.35 ± 2.17 | 69.50 ± 3.75 | 68.69 ± 2.22 | 68.29 ± 2.83 | 68.89 ± 2.00 | 67.89 ± 1.52 |
| ORL | 60.20 ± 0.94 | 54.75 ± 1.64 | 56.95 ± 1.30 | 57.75 ± 2.53 | 57.65 ± 2.61 | 56.95 ± 1.30 |
| JAFFE | 83.85 ± 0.64 | 81.22 ± 1.81 | 81.03 ± 1.53 | 81.69 ± 1.57 | 82.16 ± 1.15 | 82.19 ± 1.05 |
| ATT40 | 61.40 ± 0.94 | 60.25 ± 1.67 | 60.25 ± 1.67 | 58.75 ± 1.35 | 58.65 ± 1.39 | 59.70 ± 1.16 |
| binalpha | 48.38 ± 1.41 | 46.61 ± 1.13 | 47.09 ± 1.96 | 43.23 ± 1.00 | 40.94 ± 1.03 | 45.30 ± 0.99 |
| lung_small | 87.94 ± 1.34 | 81.37 ± 6.04 | 87.12 ± 1.10 | 64.93 ± 6.52 | 67.12 ± 4.33 | 78.90 ± 4.47 |
| lung | 91.63 ± 0.00 | 90.64 ± 0.00 | 62.36 ± 1.06 | 65.32 ± 2.10 | 62.27 ± 3.03 | 68.97 ± 0.00 |

**Table 5.** NMI of our method across different modules (mean ± std (%))

| Dataset | Full | w/o Global | w/o Local | Latent Local | Feature Local | w/o Robust |
|---|---|---|---|---|---|---|
| COIL20 | 80.79 ± 0.59 | 80.07 ± 1.08 | 78.97 ± 0.69 | 78.60 ± 1.10 | 79.07 ± 0.61 | 79.12 ± 0.43 |
| ORL | 78.84 ± 0.43 | 76.79 ± 0.70 | 76.20 ± 0.29 | 76.56 ± 0.77 | 76.62 ± 0.73 | 77.14 ± 0.80 |
| JAFFE | 84.86 ± 0.72 | 81.22 ± 1.81 | 81.83 ± 1.41 | 82.43 ± 1.44 | 83.16 ± 1.24 | 83.90 ± 1.10 |
| ATT40 | 79.20 ± 0.44 | 78.08 ± 1.02 | 78.08 ± 1.02 | 78.05 ± 0.93 | 77.95 ± 0.97 | 78.39 ± 0.58 |
| binalpha | 60.77 ± 0.30 | 60.41 ± 0.61 | 60.11 ± 0.73 | 59.12 ± 0.59 | 56.52 ± 0.14 | 59.17 ± 0.57 |
| lung_small | 82.13 ± 1.97 | 75.09 ± 3.61 | 80.03 ± 1.63 | 63.05 ± 3.91 | 60.35 ± 4.76 | 72.42 ± 2.91 |
| lung | 74.94 ± 0.00 | 72.99 ± 0.00 | 60.53 ± 0.07 | 61.76 ± 0.93 | 59.25 ± 2.39 | 60.72 ± 0.00 |

### 3.5 Parameter Sensitivity Analysis

To systematically assess the influence of the four hyperparameters in our model, we group them into two pairs based on their functional roles in the objective function. $\alpha$ and $\gamma$ control local alignment and global preservation, respectively, jointly determining the balance between the micro-level neighborhood structure and thr macro-level discriminative directions; $\beta$ and $\delta$ regulate the sparsity of the graph structure and the row-wise sparsity of the projection matrix, respectively, jointly affecting the stability and robustness of feature selection. In the experiments, we vary each parameter's value on a logarithmic scale from $1 \times 10^{-3}$ to $1 \times 10^{3}$ while fixing other parameters to their optimal values obtained on the validation set.

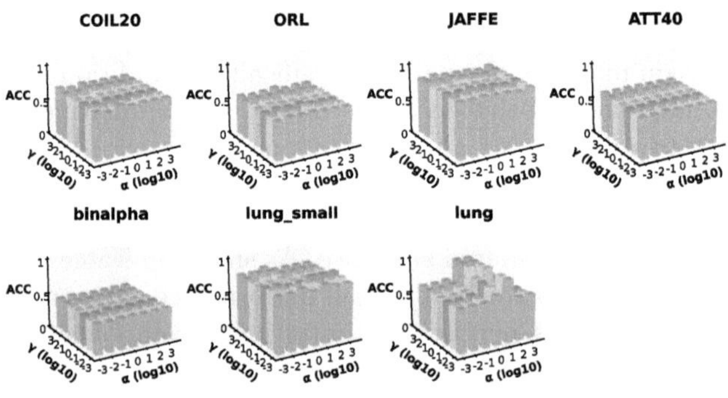


**Fig. 1.** The results of ULGFS's ACC on different $\alpha$ and $\gamma$.

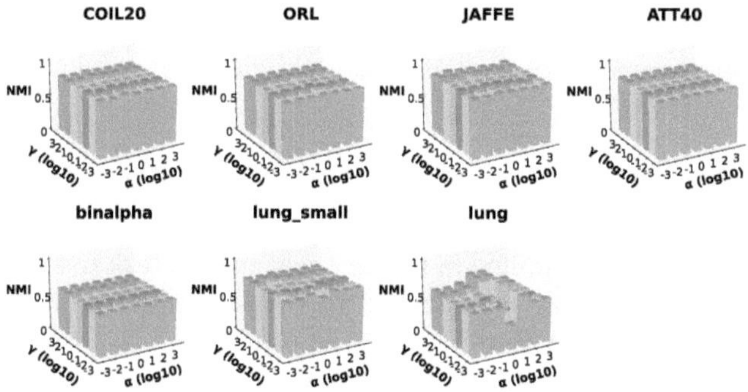


**Fig. 2.** The results of ULGFS's NMI on different $\alpha$ and $\gamma$.

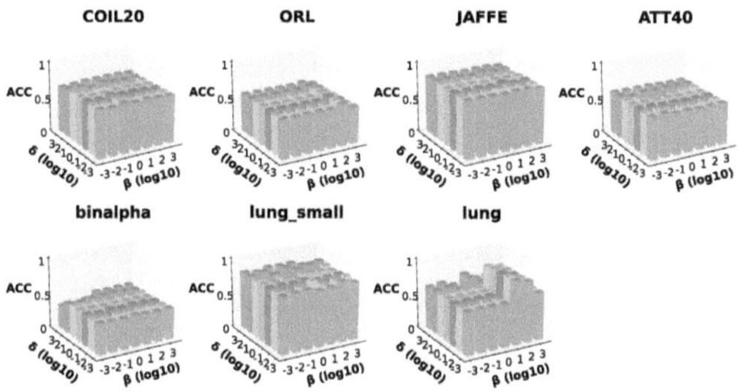


**Fig. 3.** The results of ULGFS's ACC on different $\beta$ and $\delta$.

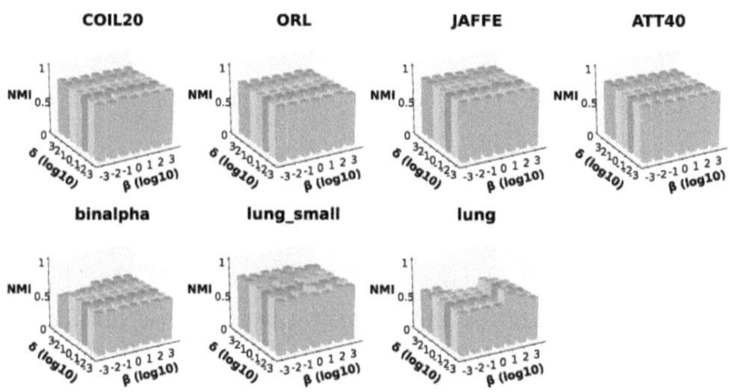


**Fig. 4.** The results of ULGFS's NMI on different $\beta$ and $\delta$.

Figures 1 and 2 depict the variations in clustering performance metrics (ACC and NMI) across seven datasets with respect to different values of $\alpha$ and $\gamma$, respectively, while the performance variations with respect to $\beta$ and $\delta$ are shown in Figs. 3 and 4. The results indicate that, except for minor fluctuations in $\alpha$ and $\beta$ over a narrow range on the lung dataset, our model exhibits relatively stable performance across a broad parameter range. This demonstrates the method's insensitivity to parameter perturbations, highlighting its robust stability and tunability.

### 3.6 Applications on ORL

Feature selection is widely applied in the face recognition task to extract the most informative pixels. We apply our method to the ORL dataset and visualize 10 randomly selected images. Figure 5 shows the top 10% to 40% selected features, in which the selected pixels are marked in red. As the selection ratio increases, ULGFS consistently identifies key facial regions, including the contour, nose, eyes, brow ridge, and mouth—demonstrating both its ability to capture the most representative features and the interpretability of the learned selection.

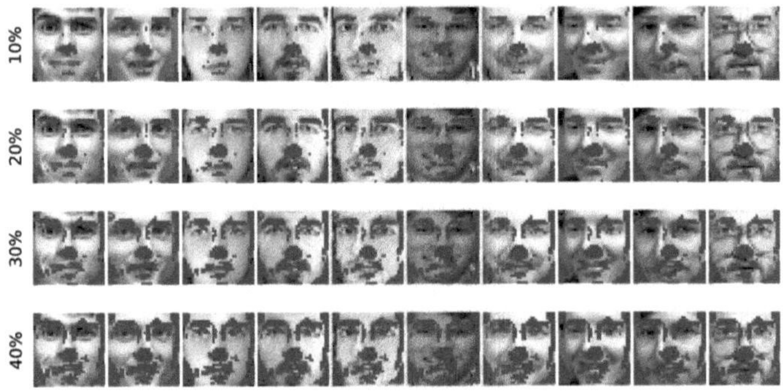


**Fig. 5.** Our method demonstrates its effectiveness on ORL.

## 4 Conclusion

This paper proposes a novel feature selection method that achieves structural unity and consistency by dynamically aligning local and global structural information within a shared subspace, thereby obtaining a more robust and discriminative feature subset. In future work, we will further evaluate and extend the proposed framework on more complex and large-scale datasets to explore its scalability and generalization ability.

**Acknowledgments.** The Guangxi Natural Science Foundation (2023GXNSFBA 026010), the Key Lab of Education Blockchain and Intelligent Technology, Ministry of Education (EBME24-09), and the Joint Cultivation Projects of the National Natural Science Foundation of China for Guangxi Normal University (2025PY039).

## References


1. LeCun, Y., Bengio, Y., Hinton, G.: Deep learning. Nature **521**(7553), 436–444 (2015)
2. Golub, T.R., et al.: Molecular classification of cancer: class discovery and class prediction by gene expression monitoring. Science **286**(5439), 531–537 (1999)
3. McInnes, L., Healy, J., Melville, J.: Umap: Uniform manifold approximation and projection for dimension reduction. Statistics **1050**, 18 (2020)
4. Gorban, A.N., Makarov, V.A., Tyukin, I.Y.: High-dimensional brain in a high-dimensional world: blessing of dimensionality. Entropy **22**(1), 82 (2020)
5. Zhao, Z., Liu, H.: Spectral feature selection for supervised and unsupervised learning. In: Proceedings of the 24th International Conference on Machine Learning, pp. 1151–1157 (2007)
6. Mitra, P., Murthy, C.A., Pal, S.K.: Unsupervised feature selection using feature similarity. IEEE Trans. Pattern Anal. Mach. Intell. **24**(3), 301–312 (2002)
7. Cai, D., Zhang, C., He, X.: Unsupervised feature selection for multi-cluster data. In: Proceedings of the 16th ACM SIGKDD International Conference on Knowledge Discovery and Data Mining, pp. 333–342 (2010)
8. Zhao, Z., Wang, L., Liu, H.: Efficient spectral feature selection with minimum redundancy. In: Proceedings of the AAAI Conference on Artificial Intelligence, vol. 24, pp. 673–678 (2010)
9. Van Der Maaten, L., Postma, E.O., Van Den Herik, H.J., et al.: Dimensionality reduction: a comparative review. J. Mach. Learn. Res. **10**(66–71), 1–41 (2009)
10. Nie, F., Wang, X., Huang, H.: Clustering and projected clustering with adaptive neighbors. In: Proceedings of the 20th ACM SIGKDD International Conference on Knowledge Discovery and Data Mining, pp. 977–986 (2014)
11. Abdi, H., Williams, L.J.: Principal component analysis. Wiley Interdisc. Rev. Comput. Stat. **2**(4), 433–459 (2010)
12. Huber, P.J.: Robust statistics. In: International Encyclopedia of Statistical Science, pp. 1248–1251. Springer, Cham (2011)
13. Nikolova, M., Ng, M.K.: Analysis of half-quadratic minimization methods for signal and image recovery. SIAM J. Sci. Comput. **27**(3), 937–966 (2005)
14. Zhu, X., Lei, C., Yu, H., Li, Y., Gan, J., Zhang, S.: Robust graph dimensionality reduction. In: Proceedings of the Twenty-Seventh International Joint Conference on Artificial Intelligence, pp. 3257–3263 (2018)
15. Lee, D., Seung, H.S.: Algorithms for non-negative matrix factorization. Adv. Neural. Inf. Process. Syst. **13**, 556–562 (2000)
16. Wang, Z., Min, W.: Graph regularized nmf with $\ell_{2,0}$-norm for unsupervised feature learning. arXiv preprint arXiv:2403.10910 (2024)
17. Li, Z., Tang, J.: Unsupervised feature selection via nonnegative spectral analysis and redundancy control. IEEE Trans. Image Process. **24**(12), 5343–5355 (2015)
18. Wang, S., Chen, J., Guo, W., Liu, G.: Structured learning for unsupervised feature selection with high-order matrix factorization. Expert Syst. Appl. **140**, 112878 (2020)
19. He, X., Cai, D., Niyogi, P.: Laplacian score for feature selection. Adv. Neural. Inf. Process. Syst. **18**, 507–514 (2005)
20. Cao, Z., Xie, X., Sun, F.: Adaptive unsupervised feature selection with robust graph regularization. Int. J. Mach. Learn. Cybern. **15**(2), 341–354 (2024)
21. Mi, Y., Chen, H., Luo, C., Horng, S.J., Li, T.: Unsupervised feature selection with high-order similarity learning. Knowl.-Based Syst. **285**, 111317 (2024)

22. Jiang, K., Cao, T., Zhu, L., Sun, Q.: Adaptive and flexible $\ell_1$-norm graph embedding for unsupervised feature selection. Appl. Intell. **54**(22), 11732–11751 (2024)
23. Liu, W., Ning, Q., Liu, G., Wang, H., Zhu, Y., Zhong, M.: Unsupervised feature selection algorithm based on $\ell_{2,p}$-norm feature reconstruction. PLoS ONE **20**(3), e0318431 (2025)

# When SpikeKANFormer Meets Angular Contrastive Learning

Liangjin Liu, Siwen Deng, Zhengzhong Zhu, and Pei Zhou(✉)

College of Computer Science, Sichuan University, Chengdu, China
{liuliangjin,dengsiwen,2024323040010}@stu.scu.edu.cn, zhoupei@scu.edu.cn

**Abstract.** In recent years, Spiking Transformer has emerged as a promising architecture for neuromorphic computing, leveraging MLPs as self-attention computation units to achieve competitive performance. However, this approach faces two critical limitations: the inherent constraints of MLPs in modeling highly nonlinear spatio-temporal dynamics in spiking data, and the lack of robust supervised representation learning mechanisms for spike-based paradigms. To address these challenges, we propose SpikeKANFormer, a novel architecture that replaces traditional MLPs with Kolmogorov-Arnold Networks (KANs). KANs enhance the models capacity to approximate complex spike-timing dependencies, thanks to their theoretically grounded function composition and adaptive structure, while also improving interpretability. Furthermore, we introduce Angular Contrastive Learning Loss (ACL) to tackle the representation learning bottleneck. By exploiting the temporal dynamics of Spiking Neural Networks (SNNs), ACL constructs positive and negative sample pairs across different time steps and optimizes their angular separability in feature space. This approach not only mitigates the reliance on labeled data but also strengthens feature discrimination for downstream tasks. Extensive experiments demonstrate that SpikeKANFormer, combined with Angular Contrastive Learning Loss, achieves state-of-the-art performance across multiple neuromorphic vision and text benchmarks.

**Keywords:** KAN Networks · Deep Learning · Contrastive learning

## 1 Introduction

Spiking Neural Networks (SNNs), as the third-generation neural network [22], are promising due to their low power consumption and biological plausibility [26]. SNNs have adopted advanced architectures from ANNs, such as ResNet-style SNNs [12], Spiking Recurrent Neural Networks [21], and Spiking Graph Neural Networks [19], achieving significant performance gains. Similarly, Transformers [28], initially for NLP, are now widely used in computer vision [1,3]. The self-attention mechanism, a key component of Transformers, is also a feature of human biological systems [4,30], making the integration of these two bio-inspired mechanisms compelling for advanced deep learning.


R. C.-W. Wong et al. (Eds.): PAKDD 2026, LNAI 16597, pp. 149–161, 2026.
https://doi.org/10.1007/978-981-92-1300-9_12

Previous research has implemented self-attention in SNNs using MLPs. However, MLPs struggle to capture nonlinear features and their fully-connected nature creates computational overhead for sparse spike-based processing, a drawback pronounced by the binary nature of spikes. Furthermore, the sparsity of spikes leads to insufficiently dense feature representations, limiting the model's ability to extract high-level semantic information and causing SpikeTransformers' representation learning capacity to fall short of ANNs. This gap necessitates new learning paradigms.

This work addresses these challenges by improving both network architecture and the objective function. We propose SpikeKANFormer, which replaces standard MLPs with Kolmogorov-Arnold Networks (KANs) [20,34]. KANs use learnable nonlinear activation functions to effectively learn complex function mappings, improving model performance, interpretability, and reducing training costs. For the objective function, we introduce Angular Contrastive Learning (ACL), a novel contrastive learning method. While contrastive learning is a powerful representation learning method, its application in SNNs is underexplored [23]. Traditional softmax-based losses like NT-Xent [5] often fail to achieve optimal discriminative performance. Moreover, SNNs generate different features across time steps. Relying solely on data augmentation for constructing positive/negative pairs is insufficient. We introduce supervision into the time domain, creating pairs from the same sample at different time steps to learn temporal correlations. To further enhance discriminative ability, ACL projects features into an angular space and maximizes the decision boundary. We then extend ACL to a Siamese Angular Contrastive Learning (SACL) framework, which uses data augmentation to strengthen both temporal and class constraints, achieving robust performance even at very low inference time steps. Our contributions are as follows:

(1) We design a KAN-based Spike-KAN self-attention (SKSA) that replaces spike QKV operations with masking and sparse addition, greatly reducing energy consumption.

(2) We introduce contrastive learning at each time step and propose a temporal-domain supervised framework that strengthens temporal modeling and alleviates severe accuracy drops under ultra-low-latency inference.

(3) We further extend Siamese Angular Contrastive Learning to jointly learn augmentation invariance, temporal and intra-class correlations, markedly improving performance in low-latency settings.

## 2 Related Work

### 2.1 Contrastive Learning

Contrastive learning has emerged as a prominent unsupervised learning paradigm, illustrating its advantages and maturity in representation learning [29,38]. It pre-specifies positive and negative sample pairs, reduces the distance between the representations of positive sample pairs using the contrastive loss function, and increases the distance between the representations of positive and negative sample pairs. Recently, researchers have incorporated label information into contrastive learning, enhancing the definition of positive and negative samples. This improvement enables contrastive learning to play a role in supervised learning paradigms [38], thereby further enhancing the model's representational power. Unlike these studies, we use sample features from different time steps to construct positive and negative sample pairs, thereby introducing the unique temporal information of spiking neural networks. Additionally, we have augmented the angular margin at the decision boundary to further enhance the model's discriminative ability.

### 2.2 Kolmogorov-Arnold Network (KAN)

The Kolmogorov-Arnold Network (KAN) [20] presents a promising alternative to the Multi-Layer Perceptron (MLP), which is inspired by the universal approximation theorem [6]. In contrast, KAN is based on the Kolmogorov-Arnold representation theorem [15]. Considering the training speed, in this paper, we replace the MLP with a variant of the KAN network, FasterKAN [27], which trains significantly faster than the original KAN. FasterKAN uses Reflectional Switch Activation Functions (RSWAFs). RSWAFs are activation functions that are easy to compute because of their uniform grid structure. The RSWAF function is shown as follows:

$$\phi_{RSWAF}(r) = 1 - \left(\tanh\left(\frac{r}{h}\right)\right)^2 \tag{1}$$

Then, the RSWAF network with $N$ centers:

$$RSWAF(x) = \sum_{i=1}^{N} w_i \phi_{RSWAF}(r_i) = \sum_{i=1}^{N} w_i \left(1 - \left(\tanh\left(\frac{\|x - c_i\|}{h}\right)\right)^2\right) \tag{2}$$

where $r = \|x - c\|$ is the distance between an input vector $x$ and a center $c$, and $h$ controls the width of the Gaussian function, which is also used for controlling the width of the RSWAF.

### 2.3 Bio-Inspired Spiking Neural Networks

Bio-inspired Spiking Neural Networks(SNNs) can greatly benefit from advanced deep learning techniques and insights from neuroscience [17,35]. Many biological mechanisms offer valuable guidance for modeling [14] and learning rules [33] of SNNs. Some research indicates that integrating SNNs with brain mechanisms, such as long short-term memory [24] and attention mechanisms [36,37], results in improved effectiveness. Furthermore, SNNs have significantly enhanced their task accuracy by incorporating deep learning techniques, including innovative network architectures [10,13] and normalization [8], while preserving their inherent spike-driven advantages. Our goal is to integrate KAN with Transformer architectures. One approach involves discretizing Transformers into spike form through neuron equivalence [31], i.e., ANN2SNN. However, this method necessitates a longer simulation time step, which in turn increases energy consumption.

To address this, we used direct training where the first SNN layer is the spike encoding layer, utilizing surrogate gradient training [32].

## 3 Overall Architecture

As shown in Fig. 1, our model consists of the SpikeKANFormer, followed by the Siamese Angular Contrastive Learning module.

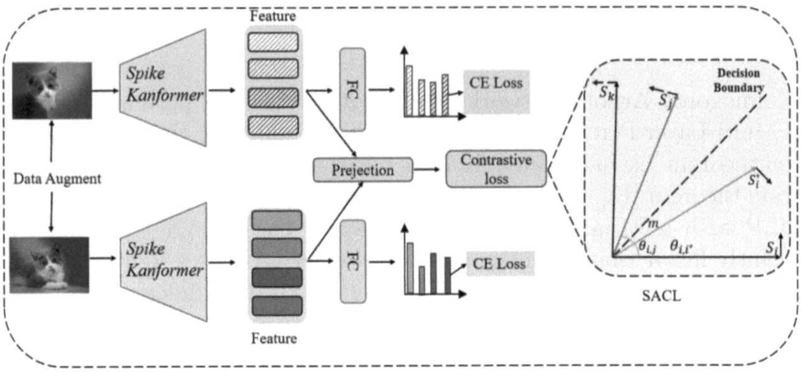


**Fig. 1.** The overall structure of SACL. Our proposed Spike-Angular Contrastive Learning: we perform data augmentation on samples and feed them into an encoder with a pulse transformer, injecting contrastive supervision in the temporal domain to model the correlation between representations at different time steps, and then optimizing the representations in the angular space.

### 3.1 SpikeKANFormer

Followed by an overview of the SpikeKANFormer and its components. Figure 2 illustrates the overall structure of the SpikeKANFormer, following the design outlined in [36]. SpikeKANFormer is an architecture based on Spiking Neural Networks (SNN), consisting primarily of the input and convolution module, the SpikeKANFormer encoder, and the output classification head. Initially, the input 2-dimensional image sequence is processed through five convolutional layers to produce flattened spike patch sequences, with the final convolutional layer generating relative positional embeddings to retain positional information. The extracted features are then passed into the core SpikeKANFormer encoder, which is composed of multiple repeated layers (with a total of L layers). Each layer includes a Spiking KAN Self-Attention (SKSA) module, which combines three KANs followed by batch normalization, and a self-attention mechanism, with mask and addition operations, which work together to enhance feature selectivity and processing efficiency. After passing through multiple encoding layers, the features are fed into a classification head (Head) for classification. Notably, since layer normalization (LN) is not suitable for SNNs, batch normalization (BN) is used instead to stabilize the training process, improving both the model's stability and efficiency during training.

$$S_0 = \mathrm{SN}(U_0) \tag{3}$$
$$U'_l = \mathrm{SKSA}(S_{l-1}) + U_{l-1}, \quad l = 1, \ldots, L \tag{4}$$
$$S'_l = \mathrm{SN}(U'_l), \quad l = 1, \ldots, L \tag{5}$$
$$S_l = \mathrm{SN}(\mathrm{KAN}(S'_l) + U'_l), \quad l = 1, \ldots, L \tag{6}$$
$$Y = \mathrm{CH}(\mathrm{GAP}(S_L)) \tag{7}$$

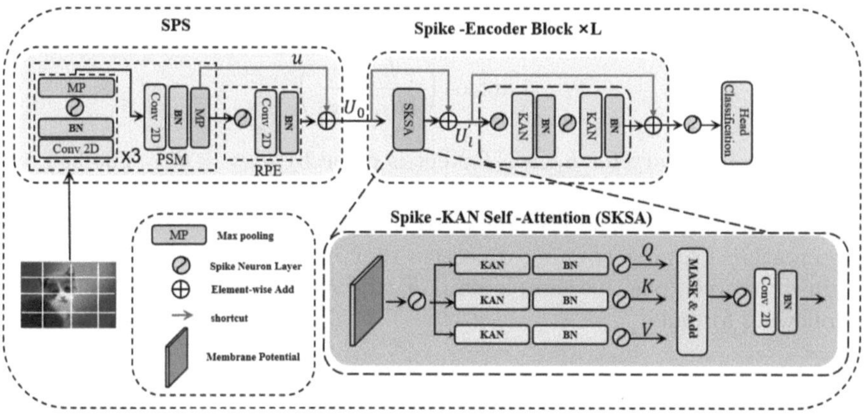


**Fig. 2.** The overall structure of SpikeKANFormer. Overview of Spike KANTransformer (SpikeKANformer): it consists of a spiking patch splitting module (SPS), a SpikKANformer encoder, and a KAN classification head. Our empirical results show that layer normalization (LN) is not applicable to spiking neural networks (SNNs), so we use batch normalization (BN) instead.

where $U_0$, $U'_l$, and $S'_l$ denote the membrane potential and the spike tensor output produced by SKSA at the $l$-th layer, respectively; $\mathrm{SN}(\cdot)$ denotes the spike neuron layer, and $\mathrm{CH}(\cdot)$ denotes the classification layer. By gradually replacing the KAN and QKV mapping matrices, the model benefits from KANs parameter compression and nonlinear representation at each step. After replacing the MLP, the reduction in parameter count and increase in computational efficiency provides a solid foundation for further replacing the QKV mapping matrices. When the QKV mapping matrices are also replaced with KAN, the overall computational complexity is further reduced, and parameter efficiency is further improved. This synergistic optimization ensures that the model achieves optimal performance at each stage. A global average-pooling (GAP) is utilized on the feature from the SpikeKANFormer encoder and outputs the $D$-dimension feature, which is sent to the KAN layer classification head (CH) to output the prediction $\mathbf{Y}$.

### 3.2 Angular Contrastive Learning

Given a set of images $D = \{s_i\}_{i=1}^{N}$, we generate their representations using SpikeKANFormer. For each sample $s_i$, combined with the augmented sample, we generate two different representations $h_i$ and $h_i^k$. These two representations with the same semantics constitute a positive pair, while the negative pairs are derived from the representations of different classes. After obtaining positive and negative sample pairs, we input them into the training objective for model fine-tuning. The most commonly utilized training objective is the NT-Xent loss [5,11], which has been extensively applied in image representation learning. Let us denote the angular variable $\theta_{i,j}$ as follows:

$$\theta_{i,j} = \arccos\left(\frac{h_i^\top h_j}{\|h_i\|\|h_j\|}\right) \tag{8}$$

The decision boundary for $h_i$ in NT-Xent is given by $\theta_{i,i^+} = \theta_{i,j}$. The lack of a decision margin can cause the model to make incorrect decisions due to small disturbances near the decision boundary. To address this, we propose a new image representation learning training objective that introduces an additive angular margin $m$ between positive sample pairs $h_i$ and $h_i^k$. This can be represented by the following formula:

$$L_{\text{ACL}} = -\log \frac{\exp(\cos(\theta_{i,i^+} + m)/\tau)}{\exp(\cos(\theta_{i,i^+} + m)/\tau) + \sum_{j\neq i} \exp(\cos(\theta_{i,j})/\tau)} \tag{9}$$

From this, it is evident that the decision boundary for $h_i$ is given by $\theta_{i,i^+} + m = \theta_{i,j}$. While this method narrows down representations with similar semantics, it also expands the differences between representations with distinct semantics, further pushing $h_i$ towards a region that reduces $\theta_{i,i^+}$ while increasing $\theta_{i,j}$. Besides, the decision boundary leaves an extra margin $m$ to boundary $\theta_{i,i^+} = \theta_{i,j}$, which significantly enhances the robustness and noise resistance of the model during the inference process .To extend ACL to handle representations from different time steps and samples, we made the following modifications to the formula:

$$L_{\text{SACL}} = -\frac{1}{T} \sum_{i\in I} \sum_{t=1}^{T} \sum_{t'\neq t} \log \frac{\exp(\cos(\theta_{i,i^*}^{t} + m)/\tau)}{\exp(\cos(\theta_{i,i^*}^{t} + m)/\tau) + \sum_{j\neq i, t''\neq t} \exp(\cos(\theta_{i,j}^{t,t''})/\tau)} \tag{10}$$

where $T$, $m$, and $\tau$ are the total time steps, additive angular margin, and temperature, respectively.

## 4 Objective Function

When training with positive and negative samples that are exclusively derived from different temporal information, the overall loss function is defined as follows:

$$L_{\text{SACL}} = L_{\text{ACLaug1}} + L_{\text{ACLaug2}} \tag{11}$$

$$L_{\text{Total}} = L_{\text{SACL}} + L_{\text{CEaug1}} + L_{\text{CEaug2}} \tag{12}$$

where $L_{\text{CE}}$ is Cross entropy loss.

## 5 Experiments

We evaluate our method on multiple text and image datasets. The image datasets include static datasets such as ImageNet [7] and CIFAR-10/100 [16], as well as neuromorphic datasets like CIFAR10-DVS [18] and DVS128 Gesture [2]. For a fair comparison, we followed the experimental setup in [40], using a $224 \times 224$ input and a batch size of 128 or 256.

### 5.1 Comparisons with Other Methods

We compared our method with previous state-of-the-art (SOTA) methods, and the results are shown in Table 1 and 2. It can be seen that SpikeKANFormer outperforms previous methods across all datasets. Additionally, we observe a gradual performance improvement after incorporating Angular Contrastive Learning (ACL) and Siamese Angular Contrastive Learning (SACL). This improvement is primarily attributed to ACL's ability to learn temporal information between different representations and SACL's simultaneous learning of temporal information along with label-supervised information.

### 5.2 Fine-Grained Image Classification

In this section, we use SpikeKANFormer as the base model to evaluate the performance of our proposed SACL framework across five fine-grained image classification tasks. Table 3 presents that the SACL framework significantly outperforms the baseline across all datasets. Specifically, the improvement for the CUB dataset is 6.8% (58.29% vs. 51.49%), for the Car dataset, it reaches 7.0% (85.21% vs. 78.21%), and for the Aircraft dataset, the enhancement is 5.48% (70.85% vs. 65.37%). Moreover, the SACL framework demonstrates robust performance improvements across four datasets.

### 5.3 Ablation Studies

**Influence of Time Steps.** We investigated the impact of time steps $T$ on our ACL and SACL methods using CIFAR-10/100 datasets. As shown in Fig. 5aand 5b, both methods consistently outperformed the baseline, underscoring our framework's effectiveness in modeling temporal correlations. Our

**Table 1.** Experimental Results on four datasets.

| Methods | Spike-driven | CIFAR10-DVS | | DVS128Gesture | | CIFAR-10 | | CIFAR-100 | |
|---|---|---|---|---|---|---|---|---|---|
| | | T | Acc | T | Acc | T | Acc | T | Acc |
| tdBN | | 10 | 67.8 | 40 | 96.9 | 6 | 93.2 | - | - |
| PLIF | ✓ | 20 | 74.8 | 20 | 97.6 | 8 | 93.5 | - | - |
| Dspike | | 10 | 75.4 | - | - | 6 | 94.3 | 6 | 74.2 |
| DSR | ✓ | 10 | 77.3 | - | - | 20 | 95.4 | 20 | 78.5 |
| spikeformer | ✓ | 16 | 80.9 | 16 | 98.3 | 4 | 95.5 | 4 | 78.2 |
| DIET-SNNI | | - | - | - | - | 5 | 92.7 | 5 | 69.67 |
| ANN(ResNet19) | | - | - | - | - | - | 94.97 | 1 | 75.4 |
| ANN(Transformer4-384) | | - | - | - | - | - | 96.7 | 1 | 81.02 |
| Spike drivenTransformer | ✓ | 16 | 80 | 16 | 99.3 | 4 | 95.6 | 4 | 78.4 |
| SpikeKANFormer | ✓ | 16 | **80.55** | 16 | 99.14 | 4 | **95.81** | 4 | **78.73** |
| SpikeKANFormer+ACL | ✓ | 16 | **81.25** | 16 | 99.42 | 4 | **96.34** | 4 | **79.62** |
| SpikeKANFormer+SACL | ✓ | 16 | **81.98** | 16 | **99.56** | 4 | **96.40** | 4 | **80.25** |

**Table 2.** Experimental Results on ImageNet.

| Methods | Architecture | Spike-driven | Param (M) | Power (mJ) | Time Step | Acc |
|---|---|---|---|---|---|---|
| Hybrid training [25] | ResNet-34 | ✓ | 21.79 | - | 250 | 61.48 |
| TET [9] | SEW-ResNet-34 | | 21.79 | - | 4 | 68.00 |
| Spiking ResNet [12] | ResNet-50 | ✓ | 25.56 | 70.93 | 350 | 72.75 |
| tdBN [39] | Spiking-ResNet-34 | | 21.79 | 6.39 | 6 | 63.72 |
| SEW ResNet [10] | | | | | | |
| SEW-ResNet-34 | SEW-ResNet-34 | | 21.79 | 4.04 | 4 | 67.04 |
| SEW-ResNet-50 | SEW-ResNet-50 | | 25.56 | 4.89 | 4 | 67.78 |
| SEW-ResNet-101 | SEW-ResNet-101 | | 44.55 | 8.91 | 4 | 68.76 |
| SEW-ResNet-152 | SEW-ResNet-152 | | 60.19 | 12.89 | 4 | 69.26 |
| MS ResNet [13] | | | | | | |
| MS-ResNet-18 | MS-ResNet-18 | ✓ | 11.69 | 4.29 | 4 | 63.10 |
| MS-ResNet-34 | MS-ResNet-34 | ✓ | 21.80 | 5.11 | 4 | 69.42 |
| MS-ResNet-104 | MS-ResNet-104 | ✓ | 77.28 | 10.19 | 4 | 76.02 |
| Att MS ResNet [37] | | | | | | |
| Att-MS-ResNet-18 | Att-MS-ResNet-18 | ✓ | 11.87 | 0.48 | 1 | 63.97 |
| Att-MS-ResNet-34 | Att-MS-ResNet-34 | ✓ | 22.12 | 0.57 | 1 | 69.15 |
| Att-MS-ResNet-104 | Att-MS-ResNet-104 | ✓ | 78.37 | 7.30 | 4 | 77.08 |
| ResNet | Res-CNN-104 | | 77.28 | 54.21 | 1 | 76.87 |
| Transformer | Transformer-8-512 | | 29.68 | 41.77 | 1 | **80.80** |
| SpikeTransformer | Transformer-8-512 | ✓ | 29.68 | 11.57 | 4 | 73.38 |
| Spike drivenformer | Transformer-8-512 | ✓ | 29.68 | 4.50 | 4 | 74.57 |
| SpikeKANFormer | KANFormer-8-512 | ✓ | 28.54 | 3.56 | 1 | **74.96** |
| SpikeKANFormer(ACL) | KANFormer-8-512 | ✓ | 28.76 | 3.89 | 4 | **75.12** |
| SpikeKANFormer(SACL) | KANFormer-8-512 | ✓ | 28.76 | 3.89 | 4 | **75.84** |

**Table 3.** Fine-Grained Image Classification. Numbers following datasets indicate category count.

| Dataset | $\mathbf{CUB}_{200}$ | $\mathbf{Car}_{196}$ | $\mathbf{Aircraft}_{102}$ | $\mathbf{Flowers}_{102}$ | $\mathbf{Dogs}_{120}$ |
|---|---|---|---|---|---|
| BaseLine | 51.49 | 78.21 | 65.37 | 90.42 | 59.42 |
| ACL | 53.51(+2.02) | 81.86(+3.65) | 67.43(+2.06) | 92.17(+1.75) | 60.12(+0.7) |
| SACL | **58.29**(+6.8) | **85.21**(+7.0) | **70.85**(+5.48) | **94.86**(+4.44) | **62.54**(+3.12) |

SACL method, leveraging data augmentation invariance and temporal correlation, achieved peak accuracy. At a time step of 4, it reached 80.25% on CIFAR-100 and 96.40% on CIFAR-10. Performance stabilizes or slightly diminishes beyond this optimal point, suggesting a trade-off: while increasing the time step initially boosts generalization, an excessively large step can lead to overfitting or reduced computational efficiency.

**Influence of Angular Margin.** The angular margin $m$ in the SACL loss directly affects the discriminative ability. To investigate the impact of $m$, we conducted an experiment where $m$ was varied from $0°$ to $20°$ in steps of $2°$. We adjusted the hyperparameters based on accuracy. As shown in Fig. 5c, we can observe that the performance decreases when $m = 10$. This aligns with our intuition, as a small $m$ may have a negligible effect, while a large $m$ could negatively impact the modeling of positive relationships.

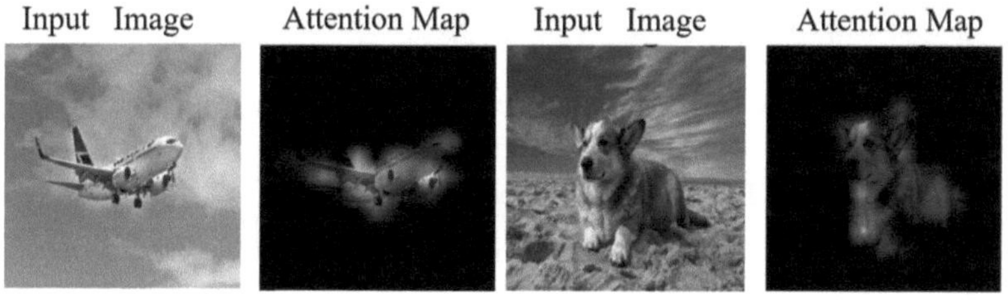


**Fig. 3.** Attention map examples of SKSA.

## 6 Visualization

Figure 3 illustrates an example of the attention map for the last encoder block at timestamp 8-512 during the fourth step. Through SKSA (Spiking KAN Self-Attention), we effectively identify image regions closely related to the semantic content of the classification task, assigning higher attention weights to these regions. At the same time, SKSA suppresses the attention weights of irrelevant image areas to near zero, rendering them black. As shown in Fig. 4, we visualized

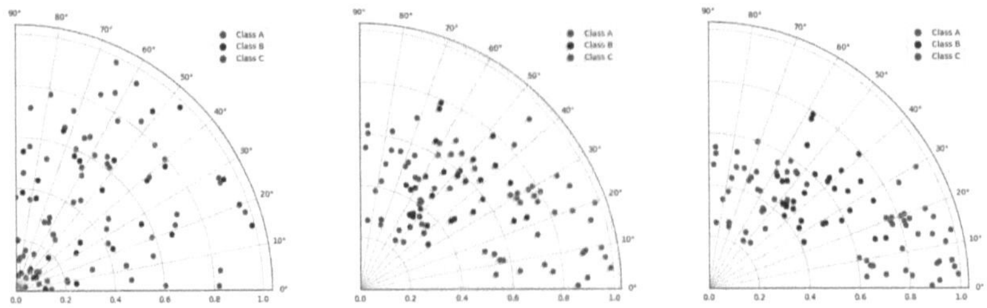


(a) Original (b) Supcon (c) SACL

**Fig. 4.** Feature visualization in Angular Space on CIFAR-10.

the original samples, the label-supervised contrastive learning method Supcon, and SACL in the representation space on the CIFAR-10 datasets. Both SACL and Supcon are able to distinguish the samples, but our method, by incorporating temporal information and maximizing the decision boundary in the angular space, demonstrates better discriminative ability compared to Supcon.

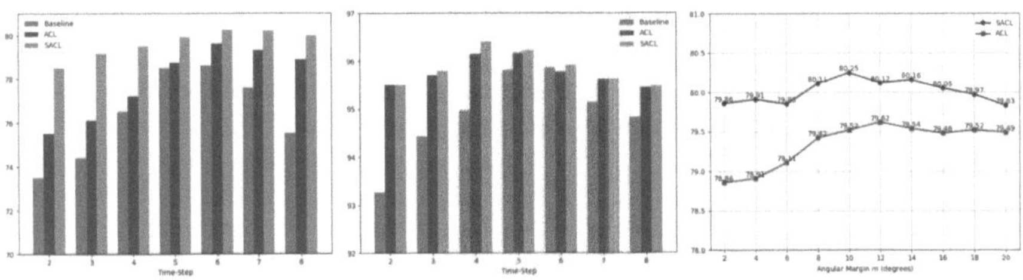


(a) CIFAR-10 (b) CIFAR-100 (c) Angular Margin $m$

**Fig. 5.** Ablation experiment results.

## 7 Conclusion

We propose a SpikeKANFormer network that combines the low power consumption of SNNs with the excellent accuracy of transformers, while enhancing the model's nonlinearity and interpretability. Furthermore, we observe that Spiking Neural Networks inherently exhibit temporal relationships. To leverage this, we introduce Angular Contrastive Learning and the Angular Contrastive Loss to model pairwise relationships with improved discriminative ability by maximizing the decision boundary in angular space. By expanding positive and negative pairs, we further propose SACL. Experimental results demonstrate substantial improvements in model performance with SpikeKANFormer and Angular Contrastive Learning, outperforming the latest state-of-the-art methods. In the future, we plan to explore the application of this architecture in other domains, such as natural language processing and graph classification.

## References


1. Alexey, D.: An image is worth 16x16 words: Transformers for image recognition at scale. arXiv preprint arXiv: 2010.11929 (2020)
2. Amir, A., et al.: A low power, fully event-based gesture recognition system. In: Proceedings of the IEEE Conference on Computer Vision and Pattern Recognition, pp. 7243–7252 (2017)
3. Carion, N., Massa, F., Synnaeve, G., Usunier, N., Kirillov, A., Zagoruyko, S.: End-to-End Object Detection with Transformers. In: Vedaldi, A., Bischof, H., Brox, T., Frahm, J.-M. (eds.) ECCV 2020. LNCS, vol. 12346, pp. 213–229. Springer, Cham (2020). https://doi.org/10.1007/978-3-030-58452-8_13
4. Caucheteux, C., King, J.R.: Brains and algorithms partially converge in natural language processing. Commun. Biol. **5**(1), 134 (2022)
5. Chen, T., Kornblith, S., Norouzi, M., Hinton, G.: A simple framework for contrastive learning of visual representations. In: International Conference on Machine Learning, pp. 1597–1607. PMLR (2020)
6. Cybenko, G.: Approximation by superpositions of a sigmoidal function. Math. Control Signals Syst. **2**(4), 303–314 (1989)
7. Deng, J., Dong, W., Socher, R., Li, L.J., Li, K., Fei-Fei, L.: Imagenet: A large-scale hierarchical image database. In: 2009 IEEE Conference on Computer Vision and Pattern Recognition, pp. 248–255. IEEE (2009)
8. Deng, L., et al.: Comprehensive SNN compression using ADMM optimization and activity regularization. IEEE Trans. Neural Netw. Learn. Syst. **34**(6), 2791–2805 (2021)
9. Deng, S., Li, Y., Zhang, S., Gu, S.: Temporal efficient training of spiking neural network via gradient re-weighting, arXiv preprint arXiv:2202.11946 (2022)
10. Fang, W., Yu, Z., Chen, Y., Huang, T., Masquelier, T., Tian, Y.: Deep residual learning in spiking neural networks. Adv. Neural. Inf. Process. Syst. **34**, 21056–21069 (2021)
11. Gao, T., Yao, X., Chen, D.: Simcse: Simple contrastive learning of sentence embeddings, arXiv preprint arXiv:2104.08821 (2021)
12. Hu, Y., Tang, H., Pan, G.: Spiking deep residual networks. IEEE Trans. Neural Netw. Learn. Syst. **34**(8), 5200–5205 (2021)
13. Hu, Y., Deng, L., Wu, Y., Yao, M., Li, G.: Advancing spiking neural networks toward deep residual learning. IEEE Trans. Neural Netw. Learn. Syst. (2024)
14. Izhikevich, E.M.: Simple model of spiking neurons. IEEE Trans. Neural Netw. **14**(6), 1569–1572 (2003)
15. Kolmogorov, A.N.: On the representation of continuous functions of several variables by superpositions of continuous functions of a smaller number of variables, American Mathematical Society (1961)
16. Krizhevsky, A., Hinton, G.: Learning multiple layers of features from tiny images (2009)
17. Li, G., et al.: Brain-inspired computing: A systematic survey and future trends. In: Proceedings of the IEEE (2024)
18. Li, H., Liu, H., Ji, X., Li, G., Shi, L.: Cifar10-dvs: an event-stream dataset for object classification. Front. Neurosci. **11**, 309 (2017)

19. Li, J., et al.: A graph is worth 1-bit spikes: When graph contrastive learning meets spiking neural networks, arXiv preprint arXiv:2305.19306 (2023)
20. Liu, Z., Wang, Y., Vaidya, S., Ruehle, F., Halverson, J., Soljačić, M., Hou, T.Y., Tegmark, M.: Kan: Kolmogorov-arnold networks, arXiv preprint arXiv:2404.19756 (2024)
21. Lotfi Rezaabad, A., Vishwanath, S.: Long short-term memory spiking networks and their applications. In: International Conference on Neuromorphic Systems 2020, pp. 1–9 (2020)
22. Maass, W.: Networks of spiking neurons: the third generation of neural network models. Neural Netw. **10**(9), 1659–1671 (1997)
23. Qiu, H., et al.: Temporal contrastive learning for spiking neural networks. In: International Conference on Artificial Neural Networks, pp. 422–436. Springer, Cham (2024)
24. Rao, A., Plank, P., Wild, A., Maass, W.: A long short-term memory for ai applications in spike-based neuromorphic hardware. Nat. Mach. Intell. **4**(5), 467–479 (2022)
25. Rathi, N., Srinivasan, G., Panda, P., Roy, K.: Enabling deep spiking neural networks with hybrid conversion and spike timing dependent backpropagation, arXiv preprint arXiv:2005.01807
26. Roy, K., Jaiswal, A., Panda, P.: Towards spike-based machine intelligence with neuromorphic computing. Nature **575**(7784), 607–617 (2019)
27. Ta, H.T., Thai, D.Q., Rahman, A.B.S., Sidorov, G., Gelbukh, A.: Fc-kan: Function combinations in kolmogorov-arnold networks, arXiv preprint arXiv:2409.01763 (2024)
28. Vaswani, A.: Attention is all you need. In: Advances in Neural Information Processing Systems (2017)
29. Wang, B., Tian, Z., Ye, A., Wen, F., Du, S., Gao, Y.: Generative variational-contrastive learning for self-supervised point cloud representation. IEEE Trans. Pattern Anal. Mach. Intell. (2024)
30. Whittington, J.C., Warren, J., Behrens, T.E.: Relating transformers to models and neural representations of the hippocampal formation. In: International Conference on Learning Representations
31. Wu, J., et al.: Progressive tandem learning for pattern recognition with deep spiking neural networks. IEEE Trans. Pattern Anal. Mach. Intell. **44**(11), 7824–7840 (2021)
32. Wu, Y., Deng, L., Li, G., Zhu, J., Xie, Y., Shi, L.: Direct training for spiking neural networks: Faster, larger, better. In: Proceedings of the AAAI Conference on Artificial Intelligence, vol. 33, pp. 1311–1318 (2019)
33. Wu, Y., et al.: Brain-inspired global-local learning incorporated with neuromorphic computing. Nat. Commun. **13**(1), 65 (2022)
34. Xu, J., et al.: Fourierkan-gcf: Fourier kolmogorov-arnold network-an effective and efficient feature transformation for graph collaborative filtering, arXiv preprint arXiv:2406.01034 (2024)
35. Yadav, S., Chaudhary, S., Kumar, R., Nkomozepi, P.: Consciousness driven spike timing dependent plasticity. Expert Syst. Appl. **269**, 126490 (2025)
36. Yao, M., et al.: Spike-driven transformer. In: Advances in Neural Information Processing Systems, p. 36 (2024)
37. Yao, M., et al.: Attention spiking neural networks. IEEE Trans. Pattern Anal. Mach. Intell. **45**(8), 9393–9410 (2023)
38. Zhang, L., Chen, X., Zhang, J., Dong, R., Ma, K.: Contrastive deep supervision. In: European Conference on Computer Vision, pp. 1–19. Springer, Cham (2022). https://doi.org/10.1007/978-3-031-19809-0_1

39. Zheng, H., Wu, Y., Deng, L., Hu, Y., Li, G.: Going deeper with directly-trained larger spiking neural networks. In: Proceedings of the AAAI Conference on Artificial Intelligence, vol. 35, pp. 11062–11070 (2021)
40. Zhou, Z., et al.: Spikformer: When spiking neural network meets transformer, arXiv preprint arXiv:2209.15425 (2022)

# NANSDE-Net: A Neural SDE Framework for Generating Time Series with Memory

Hiromu Ozai[1(✉)] and Kei Nakagawa[2]

[1] Hiroshima University, Hiroshima, Japan
d251239@hiroshima-u.ac.jp

[2] Graduate School of Business, Osaka Metropolitan University, Osaka, Japan

**Abstract.** Modeling time series with long- or short-memory characteristics is a fundamental challenge in many scientific and engineering domains. While fractional Brownian motion has been widely used as a noise source to capture such memory effects, its incompatibility with Itô calculus limits its applicability in neural stochastic differential equation (SDE) frameworks. In this paper, we propose a novel class of noise, termed Neural Network-kernel ARMA-type noise (NA-noise), which is an Itô-process-based alternative capable of capturing both long- and short-memory behaviors. The kernel function defining the noise structure is parameterized via neural networks and decomposed into a product form to preserve the Markov property. Based on this noise process, we develop NANSDE-Net, a generative model that extends Neural SDEs by incorporating NA-noise. We prove the theoretical existence and uniqueness of the solution under mild conditions and derive an efficient backpropagation scheme for training. Empirical results on both synthetic and real-world datasets demonstrate that NANSDE-Net matches or outperforms existing models, including fractional SDE-Net, in reproducing long- and short-memory features of the data, while maintaining computational tractability within the Itô calculus framework.

**Keywords:** Neural SDE · NANSDE-Net · ARMA-type noise · long-memory · short-memory

## 1 Introduction

Time series generation using deep learning has become an active area of research, particularly in modeling complex real-world dynamics. This is especially true in fields such as physics and finance, where it is important to model continuous-time dynamics in a natural way. One promising approach is the Neural Ordinary Differential Equation (Neural ODE) model [4], which combines differential equations with deep learning. Neural ODE views ResNet [9] as the limit of infinitely deep networks and models time evolution as a continuous process using ODEs. A new family of Neural ODEs was proposed to model various dynamic systems, allowing for flexible representation of complex temporal patterns [6,12].


R. C.-W. Wong et al. (Eds.): PAKDD 2026, LNAI 16597, pp. 162–174, 2026.
https://doi.org/10.1007/978-981-92-1300-9_13

However, real-world time series often include random fluctuations due to measurement noise or external shocks. To address this, the Neural Stochastic Differential Equation (Neural SDE) model has been proposed [10,11]. It extends Neural ODE by introducing randomness using Brownian motion (Bm), allowing the model to generate stochastic trajectories. Building on this framework, several variants of Neural SDEs have been developed to model various stochastic systems [15,16]. Still, a limitation of Neural SDE is that it assumes the Bm as the noise process. The Bm is a Gaussian process with independent and stationary increments, but many real-world systems show long- (resp. short-)memory or non-Markovian behavior[1], which Bm cannot model well. For example, in financial markets or natural phenomena, past events often influence future behavior for a long (resp. short) time [5,7,14].

To handle this, researchers have used fractional Brownian motion (fBm) [8, 13], which can represent long- (resp. short-)memory when the Hurst index $H >$ (resp. $<$)0.5. However, fBm is not an Itô process, so standard stochastic calculus methods cannot be applied. As a result, it is challenging to perform gradient-based optimization or derive closed-form representations for learning and inference.

To overcome this issue, [2,3] proposed a class of AR($\infty$)-type Gaussian processes, also known as AutoRegressive and Moving Average (ARMA)-type noise. These processes have stationary increments and memory, and they are also Itô processes. This means they can model long- (resp. short-)memory and still be analyzed using stochastic calculus. However, the main difficulty is that their kernel function $\ell(s,u)$, which determines the structure of the process, is hard to express explicitly in general.

In this paper, we propose a new type of ARMA-type noise called Neural Network (NN)-kernel ARMA-type noise (NA-noise), where the kernel function $\ell(s,u)$ is approximated by neural networks. Specifically, we express the NN kernel as a product $\ell(s,u) = \ell_1(s)\ell_2(u)$ and use NNs to learn $\ell_1$ and $\ell_2$. This allows us to keep the process Markovian and make training easier.

We then build a new stochastic differential equation model driven by this NA-noise, which we call *NANSDE (Neural Arma-type Noise SDE)*, and implement a generative model called *NANSDE-Net*. Our NANSDE is written in the Itô form, so it guarantees existence and uniqueness of the solution and can be solved numerically using methods like Euler–Maruyama.

From a theoretical perspective, we review known results for explicit kernel forms (Proposition 2) and extend them to NN kernels. We show that NANSDE-Net can reproduce both long- and short-memory behavior, similar to or better than fSDE-Net, while still being compatible with Itô calculus. In experiments, NANSDE-Net performs well on both $H > 0.5$ and $H < 0.5$ cases, unlike traditional models that struggle with one or the other.

[1] Long-memory refers to persistent statistical dependence across long time lags, in contrast to short-memory where correlations decay quickly. Non-Markovian behavior implies that the future evolution of a system depends not only on its current state but also on its entire history.

Our construction guarantees Itô form and a Markov augmentation that facilitate both simulation and training; it does not, in general, enforce stationary increments of the learned noise nor guarantee a prescribed asymptotic decay of autocorrelations associated with rigorous long-memory theory. Designing neural parameterizations that simultaneously ensure Gaussianity, stationary increments, and targeted long-memory asymptotics remains an open direction. Likewise, broad comparisons with state-of-the-art sequence generators and architecture search are outside our present scope. The present work should therefore be read as an Itô-compatible, proof-of-concept alternative to fBm-based approaches, clarifying when and how memory can be introduced into Neural SDEs without abandoning standard stochastic calculus.

## 2 Preliminaries

This section introduces the noise process employed in this study. In the works by [2] and [3], a class of Gaussian processes was proposed that possesses stationary increments and memory, while also being semimartingales. fBm is commonly used as a Gaussian process with stationary increments. However, since fBm is not a semimartingale, standard tools of stochastic calculus cannot be applied, making it analytically intractable in many cases. To address this limitation, [2] and [3] introduced a new class of noise processes specifically designed to overcome such difficulties. Building upon this framework, the present study adopts a modified version of the noise process proposed in those works.

### 2.1 Gaussian Processes with Stationary Increments and the Semimartingale Property

We assume that the process $Z(\cdot)$ is a continuous process with stationary increments such that $Z(0) = 0$, and satisfies one of the following continuous-time AR($\infty$)-type equations:

$$\frac{dZ}{dt}(t) + \int_{-\infty}^{t} a(t-s)\frac{dZ}{dt}(s)ds = \frac{dW}{dt}(t). \tag{1}$$

(see equation (2.10) and (2.17) in [2] for their precise formulation), where $\{W(t)\}_{t\in\mathbb{R}}$ is a one-dimensional standard Bm defined on $(\Omega, \mathcal{F}, P)$ satisfying $W(0) = 0$, and $dZ/dt$ and $dW/dt$ are the derivatives of $Z(\cdot)$ and $W(\cdot)$ respectively in the random distribution sense. The kernel $a(\cdot)$ is a nonnegative decreasing function with some adequate conditions to be specified in [2]. The case $a(\cdot) = 0$ yields the usual white noise.

As is clear from the definition, $\{Z(t)\}_{t\in\mathbb{R}}$ does not have independent increments. In other words, it is a stochastic process with memory. Under appropriate conditions related to $a(\cdot)$ (see [2,3]), $Z(\cdot)$ has the following MA($\infty$)-type representation. In the proposition below, a short range memory type kernel is used for $a(\cdot)$.

**Proposition 1 ([2,3]).** *For real parameter $q \in \mathbb{R}$ and positive parameter $p \in \mathbb{R}_{>0}$ that satisfy $p > q$, determine $a : (0, \infty) \to \mathbb{R}$ in (1) as follows:*

$$a(t) := qe^{-(p-q)t}. \tag{2}$$

*Then, $\{Z(t)\}_{t\in\mathbb{R}}$ in (1) has the following MA($\infty$)-type representation:*

$$Z(t) = W(t) - \int_0^t \left\{ \int_{-\infty}^s c(s-u)dW(u) \right\} ds, \quad t \in \mathbb{R} \tag{3}$$

*where $c : (0, \infty) \to \mathbb{R}$ is expressed in the following form:*

$$c(t) = qe^{-pt} \tag{4}$$

Furthermore, by transforming it according to the filtering theory, we obtain the representation of the Ito process of $\{Z(t)\}_{t\geq 0}$ shown below.

$$Z(t) = \hat{W}(t) - \int_0^t \left\{ \int_0^s \ell(s,u)d\hat{W}(u) \right\} ds, \quad t \geq 0. \tag{5}$$

where $\ell(s,u)$ is a deterministic function and $\{\hat{W}(t)\}_{t\geq 0}$ denotes the innovation process defined below, which is a one-dimensional Brownian motion.

$$\hat{W}(t) = Z(t) + \int_0^t E\left[ \int_{-\infty}^s c(s-u)dW(u) \middle| \mathcal{F}_s \right] ds, \quad t \geq 0, \tag{6}$$

To use this $\{Z(t)\}_{t\geq 0}$ as noise, we need to find the specific form of the Volterra kernel $\ell(s,u)$. The explicit form of this kernel was made available for the first time thanks to the results of [2] and [3]. The explicit form of $\ell(s,u)$ can be obtained from the following proposition.

**Proposition 2 ( [2,3]).** *We take $c(\cdot)$ in the form of equation* (4). *Then $\ell(s,u)$ in (5) has the following closed-form representation:*

$$\begin{aligned} \ell(s,u) &= e^{-ps}\ell(u), \qquad s > u > 0, \\ \ell(u) &:= qe^{pu}\left\{1 - \frac{2q(p-q)}{(2p-q)^2e^{2(p-q)u} - q^2}\right\}, \quad u > 0. \end{aligned} \tag{7}$$

By explicitly determining the shape of $\ell$ in this way, the Itô process representation of $\{Z(t)\}_{t\geq 0}$ can be specifically determined, making it possible to treat $\{Z(t)\}_{t\geq 0}$ as driving noise within the scope of Itô calculus. We refer to this type of noise as ARMA-type noise. Furthermore, although $\{Z(t)\}_{t\geq 0}$ itself is not a Markov process, it can be embedded as a component of a Markov process because $\ell(s,u)$ can be decomposed into functions of $s$ and $u$. That is, if $K(\cdot)$ is defined as follows:

$$K(t) := \int_0^t \ell(s)d\hat{W}(s), \quad t \geq 0, \tag{8}$$

$(Z(t), K(t))^\top$ is described as a solution to the following Markov-type SDE [1]:

$$\begin{cases} dZ(t) = -e^{-pt}K(t)dt + d\hat{W}(t), \\ dK(t) = \ell(t)d\hat{W}(t). \end{cases} \tag{9}$$

This makes it possible to use numerical calculation methods, such as the Euler–Maruyama method.

### 2.2 ARMA-Type Noise with the Kernel Approximated by a Neural Network

The kernel in equation (2) has been shown to represent a short-memory structure (see [2]). To reproduce long-memory data, it is necessary to employ a kernel that exhibits long range memory characteristics. [2] introduces such a kernel in the form $a(t) = \frac{p}{(t+1)^{p+1}}, t > 0$. However, an explicit expression for the corresponding $\ell(s, u)$ in this case has not yet been obtained, which remains an open problem. Therefore, in this paper, we adopt a method that approximates the kernel using neural networks in order to capture long-memory behavior. To ensure the Markov property, we adopt the form $\ell(s, u) = \ell_1(s)\ell_2(u)$. The noise process used in this study takes the following form.

$$\begin{cases} dZ(t) = -\ell_1(t)K(t)dt + dW(t), \\ dK(t) = \ell_2(t)dW(t). \end{cases} \tag{10}$$

where $\ell_1$ and $\ell_2$ are deterministic functions represented by neural networks and $W$ is a one-dimensional Wiener process. While $Z$ is indeed a Gaussian process, its increments are not guaranteed to be stationary. We refer to this type of noise as NN-kernel ARMA-Type noise (abbreviated as NA-noise).

### 2.3 SDE Driven by NN-Kernel ARMA-Type Noise

To implement NN-kernel ARMA-Type Noise SDE-Net (abbreviated as NANSDE-Net), we study SDEs where the standard Bm is replaced by NA-noise. We consider SDEs driven by NA-noise (10) of the form

$$\begin{cases} X_t = X_0 + \int_0^t b(s, X_s)ds + \int_0^t \sigma(s, X_s)dZ(s) \\ \quad = X_0 + \int_0^t \{b(s, X_s) - \ell_1(s)\sigma(s, X_s)K(s)\}ds \\ \qquad + \int_0^t \sigma(s, X_s)dW(s), \\ K_t = \int_0^t \ell_2(s)dW(s), \end{cases} \tag{11}$$

where $W(t)$ is standard Bm. Note that when $\ell_2(u)$ is 0, NANSDE-Net becomes SDE-Net. Therefore, NANSDE-Net can be regarded as an extension of SDE-Net. Furthermore, (11) is an Itô process, the results of conventional stochastic calculus can be applied to determine the existence and uniqueness of solutions. Furthermore, the Euler–Maruyama method can be used for numerical calculations.

## 3 Generative Modeling of Time Series Using NANSDE-Net

In [8], to compare the ability of different noise processes to reproduce long-memory effects, the authors deliberately limit the expressive power of the drift and diffusion terms. Specifically, time $t$ is not used as an input to the functions $b$ and $\sigma$. To conduct experiments under the same setting as in [8], in the following, we assume that the time evolution of a stochastic process $X = \{X_t\}_{t\geq 0}$ is described by the following NANSDE

$$\begin{cases} dX_t = \{b(X_t,\theta) - \ell_1(t,\theta)\sigma(X_t,\theta)K(t,\theta)\}dt \\ \qquad\qquad + \sigma(X_t,\theta)dW_t, \\ dK(t,\theta) = \ell_2(t,\theta)dW_t. \end{cases} \tag{12}$$

where the drift term $b$, the diffusion term $\sigma$, the kernel $\ell_1$ and $\ell_2$ are parameterized by NN. Hereafter, we optimize $b$, $\sigma$, $\ell_1$ and $\ell_2$ with respect to the NN-parameters $\theta$.

### 3.1 Network Architecture

To estimate drift, volatility and kernel functions of (12) by neural networks, we use $L$-layer multi-layer perceptron (MLP) as network architecture[2]:

$$h_i^\ell = \sum_{j=1}^{N_{\ell-1}} w_{ij}^{\ell-1} x_j^{\ell-1} + b_i^{\ell-1}, \quad x_i^\ell = \varphi(h_i^\ell)$$

for each $\ell = 1,\ldots,L$ and each $i = 1,\ldots,N_\ell$. In this case $x^0$ and $x^L$ be input and output data and NN function is such that $x^L = f_\theta(x^0)$ where $\theta = \{w_{ij}^\ell, b_i^\ell\}_{i,j,\ell}$. In the sequel, let $\Theta$ be the space of NN parameters on which $\theta$ takes values. Then existence and uniqueness of NANSDE, and precision assurance of numerical solutions are established as follows.

**Theorem 1.** *Assume that the functions $b(x,\theta)$, $\sigma(x,\theta)$, $\ell_1(t,\theta)$, and $\ell_2(t,\theta)$ are parameterized by MLP with* tanh *activation functions and finite network parameters. Then $b$ and $\sigma$ are globally Lipschitz continuous in $x$ and satisfy the linear growth condition. Moreover, $\ell_1$ and $\ell_2$ are bounded functions in $t$. Hence, the augmented system* (12) *admits a unique strong solution. Furthermore, the Euler–Maruyama scheme converges in the mean-square sense.*

[2] We use merely MLP for simplicity. We note that we can expect the improvement of fitting by optimizing network architecture.

**Algorithm 1.** Optimization of NANSDE-Net generator

**Input:** $\{X_0, \ldots, X_T\}$ - a realized path of a stochastic process until time horizon $T$, sample size $M$, learning rate $\eta$, number of optimization steps $k$

**Output:** $(\theta_b, \theta_\sigma, \theta_{\ell_1}, \theta_{\ell_2})$ - the optimized parameter for NANSDE-Net generator

**while** not converged **do**

**for** $k$ steps **do**

Let $\{\hat{X}_0^{(i)}, \ldots \hat{X}_T^{(i)}\}_{i=1}^M$ be the realization of generated paths of size $M$ and let $p_t(\theta, \cdot)$ be the probability density function of log-difference process, which is estimated from generated values $\{\hat{r}_t^{(i)}\}_{i=1}^M$ where $\hat{r}_t^{(i)} = \log(\hat{X}_{t+1}^{(i)}/\hat{X}_t^{(i)})$.

Compute the gradient of loss function

$$\mathcal{L}(\theta) = -\frac{1}{T}\sum_{t=0}^{T} \log p_\theta(t, r_t).$$

Descent parameters: $\theta \leftarrow \theta - \eta\nabla_\theta\mathcal{L}(\theta)$.

**end for**

**end while**

*Proof.* Since the activation function tanh is smooth with bounded derivatives, it is globally Lipschitz continuous. Under the assumption of finite network parameters, a multilayer perceptron is a composition of affine mappings and Lipschitz functions, hence remains globally Lipschitz. Consequently, $b$ and $\sigma$ satisfy global Lipschitz continuity and linear growth conditions. The boundedness of tanh also implies boundedness of $\ell_1$ and $\ell_2$. Defining the augmented state $Y_t = (X_t, K_t)^\top$, the system (12) becomes a finite-dimensional SDE with globally Lipschitz coefficients. The existence and uniqueness of a strong solution follow from classical SDE theory, and the convergence of the Euler–Maruyama scheme follows from standard results.

### 3.2 Algorithm

The generation procedure of NANSDE-Net is summarized in Algorithm 1. This algorithm follows the same procedure as Algorithm in Section V of [8]. For further details, please refer to [8].

## 4 Experiments

We evaluate our models on synthetic and real time series, and generate sample paths from each trained model. For fair comparison, we follow the dataset choices and settings of [8]. Readers can consult [8] for additional dataset details.

**Synthetic Data.** We simulate fractional Brownian motion (**fBm**) with Hurst indices $H \in \{0.2, 0.3\}$ to represent short-memory (anti-persistent) dynamics. Each path is discretized over $[0, 1]$ into 1000 equal steps ($\Delta t = 10^{-3}$).

**Table 1.** Numerical results for each performance metric and generative model for fBm and other types of data from the real world. Bold indicates the best performance.

| Data | Method | Hurst Index | Marginal Distribution | ACF | Weighted ACF | $R^2$ Score |
|---|---|---|---|---|---|---|
| fBm(H=0.2) | Original | 0.2 (True value) | - | - | - | - |
| | RNN | 0.464 ± 0.111 | **0.615 ± 0.024** | **1.215** | **0.633** | -1.086 ± 0.180 |
| | SDE | 0.463 ± 0.132 | 1.058 ± 0.024 | 1.220 | 0.639 | -0.121 ± 0.044 |
| | fSDE | 0.581 ± 0.144 | 1.518 ± 0.013 | 1.220 | 0.638 | **-0.017 ± 0.018** |
| | NANSDE | **0.453 ± 0.133** | 1.094 ± 0.070 | 1.276 | 0.727 | -0.109 ± 0.049 |
| fBm(H=0.3) | Original | 0.3 (True value) | - | - | - | - |
| | RNN | 0.463 ± 0.121 | **0.475 ± 0.024** | **1.217** | **0.666** | -1.139 ± 0.168 |
| | SDE | 0.489 ± 0.157 | 0.567 ± 0.199 | 1.995 | 1.279 | -0.513 ± 0.237 |
| | fSDE | 0.581 ± 0.130 | 1.186 ± 0.069 | 1.429 | 0.845 | **-0.069 ± 0.037** |
| | NANSDE | **0.422 ± 0.177** | 0.561 ± 0.191 | 1.995 | 1.415 | -0.716 ± 0.320 |
| SPX | Original | 0.614 | - | - | - | - |
| | RNN | 0.471 ± 0.097 | 0.704 ± 0.011 | 2.717 | **1.340** | -6.376 ± 0.605 |
| | SDE | 0.510 ± 0.117 | 0.204 ± 0.045 | 2.762 | 1.430 | -2.362 ± 0.946 |
| | fSDE | **0.591 ± 0.112** | 0.248 ± 0.043 | **2.716** | 1.343 | **-1.170 ± 0.293** |
| | NANSDE | 0.472 ± 0.115 | **0.202 ± 0.058** | 2.804 | 1.404 | -1.481 ± 0.575 |
| TPX | Original | 0.396 | - | - | - | - |
| | RNN | 0.455 ± 0.098 | 0.327 ± 0.012 | **1.889** | **0.829** | -1.524 ± 0.183 |
| | SDE | 0.451 ± 0.098 | 0.223 ± 0.024 | 1.900 | 0.842 | -3.471 ± 0.832 |
| | fSDE | 0.590 ± 0.137 | 0.261 ± 0.018 | 1.894 | 0.833 | **-1.053 ± 0.249** |
| | NANSDE | **0.396 ± 0.139** | **0.152 ± 0.038** | 2.018 | 0.972 | -2.371 ± 0.653 |
| SX5E | Original | 0.328 | - | - | - | - |
| | RNN | 0.468 ± 0.102 | 0.744 ± 0.011 | **2.479** | **1.269** | -2.940 ± 0.277 |
| | SDE | 0.464 ± 0.121 | **0.277 ± 0.042** | 2.609 | 1.343 | -5.410 ± 1.291 |
| | fSDE | 0.575 ± 0.130 | 0.277 ± 0.128 | 2.893 | 1.514 | -1.616 ± 0.435 |
| | NANSDE | **0.449 ± 0.130** | 0.351 ± 0.168 | 3.018 | 1.717 | **-1.606 ± 0.582** |
| NileMin | Original | 0.973 | - | - | - | - |
| | RNN | 0.477 ± 0.121 | 0.554 ± 0.040 | **1.482** | **0.948** | -4.900 ± 0.702 |
| | SDE | 0.455 ± 0.145 | 1.211 ± 0.102 | 1.527 | 0.997 | -29.771 ± 8.093 |
| | fSDE | 0.646 ± 0.151 | **0.233 ± 0.051** | 1.511 | 0.967 | **-1.543 ± 0.410** |
| | NANSDE | **0.952 ± 0.182** | 0.435 ± 0.192 | 2.938 | 2.463 | -2.837 ± 3.714 |
| ethernetTraffic | Original | 0.750 | - | - | - | - |
| | RNN | 0.442 ± 0.100 | 1.400 ± 0.009 | **1.646** | **1.124** | -9.297 ± 0.398 |
| | SDE | 0.668 ± 0.138 | 0.980 ± 0.043 | 3.902 | 2.562 | -0.753 ± 0.320 |
| | fSDE | **0.761 ± 0.134** | 0.931 ± 0.218 | 3.485 | 2.577 | -0.880 ± 0.508 |
| | NANSDE | 0.677 ± 0.135 | **0.801 ± 0.037** | 3.224 | 1.827 | **-0.739 ± 0.148** |
| NhemiTemp | Original | 1.066 | - | - | - | - |
| | RNN | 0.463 ± 0.120 | 0.652 ± 0.023 | **2.069** | **1.343** | -6.973 ± 0.614 |
| | SDE | 0.603 ± 0.075 | **0.172 ± 0.023** | 2.092 | 1.348 | -2.290 ± 0.296 |
| | fSDE | 0.599 ± 0.161 | 0.965 ± 0.187 | 2.575 | 1.890 | -17.534 ± 6.263 |
| | NANSDE | **1.025 ± 0.235** | 0.455 ± 0.239 | 4.804 | 3.258 | **-1.677 ± 2.245** |

**Real Data.** We evaluate on six real-world time series that cover both long- and short-memory behaviors: **SPX** (S&P 500), **TPX** (TOPIX), and **SX5E** (Euro

Stoxx 50) daily closing prices; **NileMin** (annual minimum Nile levels at the Roda gauge, 622–1281); **ethernetTraffic** (LAN traffic at Bellcore); and **NhemiTemp** (monthly Northern Hemisphere temperature anomalies, 1854–1989, CRU/UEA). The equity index series are obtained from Bloomberg. The other datasets are from the CRAN package `longmemo` (https://cran.r-project.org/web/packages/longmemo/index.html), which also provides detailed descriptions.

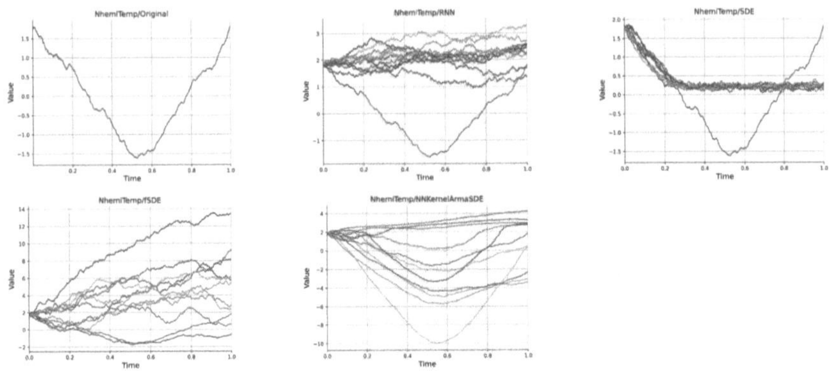


**Fig. 1.** Comparison of real and generated paths for the NhemiTemp dataset (long-memory regime). The top-left panel shows the original NhemiTemp time series. Synthetic paths generated after calibration by RNN, SDE-Net, fSDE-Net (with H>1/2), and NANSDE-Net are shown in the remaining panels. The NANSDE-Net model successfully reproduces the characteristics of long-memory in NhemiTemp data.

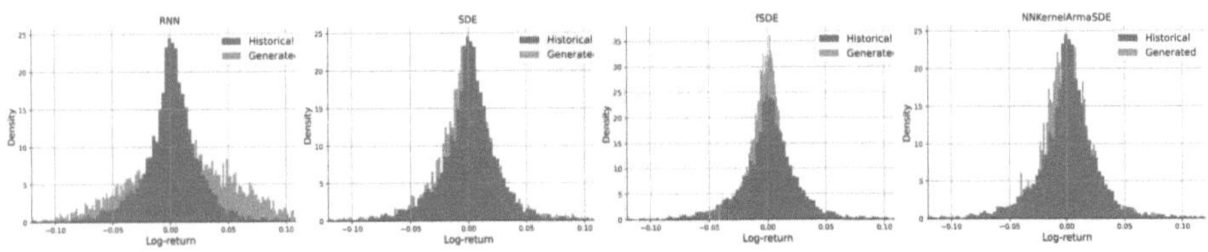


**Fig. 2.** Comparison of the histogram of the log return process for both synthetic and actual S&P 500 index data. The synthetic time series are generated after calibration by RNN, SDE-Net, fSDE-Net, and NANSDE-Net, from left to right, respectively.

### 4.1 Experimental Setup

In the following, we quantitatively evaluate the performance of the NANSDE-Net using a specified criterion. For comparison, existing time series generators—including recurrent neural networks (RNN), SDE-Net, and fSDE-Net with $H > 1/2$ are used in place of NANSDE-Net. For the network architecture, we employ a two-layer MLP with 20 hidden units for SDE-Net, fSDE-Net, and NANSDE-Net, while a vanilla RNN with 40 hidden units is used for the RNN generator.

### 4.2 Performance Metrics

We prioritize the **Hurst index** as the primary measure of memory fidelity, reflecting our goal of capturing long- and short-memory behavior.

As secondary checks, we report:

**(i) Marginal distribution (total variation).** Let $\{B_k\}_{k=1}^{K}$ be fixed bins. Define empirical frequencies

$$p_k = \frac{1}{T}\sum_{t=1}^{T} \mathbf{1}\{r_t \in B_k\}, \qquad \hat{p}_k = \frac{1}{MT}\sum_{i=1}^{M}\sum_{t=1}^{T} \mathbf{1}\{\hat{r}_t^{(i)} \in B_k\}.$$

The discrepancy is

$$\mathrm{TV} = \frac{1}{2}\sum_{k=1}^{K} |p_k - \hat{p}_k| \in [0,1] \quad (0 = \text{identical},\ 1 = \text{maximally different}).$$

**(ii) Time dependence (ACF scores).** Write $r_{0:T} = (r_1, \ldots, r_T)$ and let $\gamma(\tau) = \mathrm{Corr}(|r_t|, |r_{t+\tau}|)$ for lags $\tau = 1, \ldots, S$. Set $C(r_{0:T}) = (\gamma(1), \ldots, \gamma(S)) \in [-1,1]^S$ and define the score

$$\mathrm{ACF} = \left\| C(r_{0:T}) - \frac{1}{M}\sum_{i=1}^{M} C(\hat{r}_{0:T}^{(i)}) \right\|_2,$$

where $\|\cdot\|_2$ is the Euclidean norm. A weighted variant emphasizes long lags:

$$\mathrm{ACF}_w = \left\| C(r_{0:T}) \circ w - \frac{1}{M}\sum_{i=1}^{M} C(\hat{r}_{0:T}^{(i)}) \circ w \right\|_2,$$

with Hadamard product $\circ$ and weights $w_\tau = 2\tau/(S+1)$ (unit-mean).

**(iii) Predictive accuracy ($R^2$).** Using an 80%/20% train/test split, let $\tilde{r}_t$ be one-step-ahead predictions on the test set $\mathcal{T}_{\text{test}}$ and $\bar{r}_{\text{test}}$ its mean. Report

$$R^2 = 1 - \frac{\sum_{t\in\mathcal{T}_{\text{test}}}(r_t - \tilde{r}_t)^2}{\sum_{t\in\mathcal{T}_{\text{test}}}(r_t - \bar{r}_{\text{test}})^2}.$$

For fair comparison, we follow the metric definitions and settings in [8].

### 4.3 Results

Table 1 shows results for the above performance metrics after 1000 iteration steps with 200 early stops where Adam with learning rate 0.004 is used as the optimizer. Examining the results for data with high Hurst indices, such as NileMin, ethernetTraffic, NBSdiff and NhemiTemp, it can be seen from the table that NANSDE-Net achieves results that are equivalent to or better than those of fSDE-Net. This suggests that NANSDE-Net can reproduce noise exhibiting long-memory with $H > 1/2$. As shown in Fig. 1, for NhemiTemp paths with long-memory, NANSDE-Net performs as well as or better than fSDE-Net. Moreover, unlike SDE-Net, it more accurately captures the characteristic features of the paths. This indicates that NA-noise, unlike Brownian motion, possesses the ability to capture long-memory characteristics. Moreover, for datasets such as SPX, NileMin, and Ethernet, NANSDE-Net generates superior marginal distributions compared to SDE-Net. However, from an overall perspective, focusing on performance metrics related to marginal distributions and autocorrelation functions (ACF), it cannot be conclusively stated that NANSDE-Net clearly outperforms the other methods Fig. 2. Furthermore, looking at the results for data with small Hurst indices such as fBm, TPX and SX5E, RNN shows good results overall, but NANSDE-Net shows better results than other neural SDE methods such as SDE-Net and fSDE-Net. This demonstrates the superiority of NA-noise over other types of noise in data with $H < 1/2$. This shows that NANSDE-Net can reproduce not only $H > 1/2$ data but also $H < 1/2$ data. However, for data with $H < 1/2$, SDE-Net also achieves sufficiently good results, and thus NANSDE-Net cannot be said to perform significantly better.

## 5 Conclusion

Our contributions are as follows:

- We introduce NANSDE-Net, an extension of SDE-Net that incorporates NN-kernel ARMA-type noise. The proposed approach is based on the theoretical results presented in [2] and [3].
- We prove that the solution of NANSDE exists uniquely under standard Lipschitz and growth conditions, and derive a backpropagation formula for efficient gradient-based training.
- The experimental results demonstrate the ability of NANSDE-Net to accurately reproduce the estimate of the Hurst index and the distribution characteristics of the original time series, particularly in the context of long-memory data. Our code is available at https://github.com/ozhr/ArmaNoise-SDE-Net.

**Further Study**

Since our primary focus is on investigating the role of the noise process in capturing memory characteristics, we deliberately adopt a simple neural network architecture, without attempting to optimize the network design or improve training efficiency. Accordingly, the experimental performance could potentially be enhanced by employing more advanced network architectures or by designing more effective loss functions.

As discussed in Sect. 2, kernels of the form $a(t) = \frac{p}{(t+1)^{p+1}}$ satisfy the conditions for long-memory. However, the explicit expression of the corresponding function $\ell(s,u)$ remains unknown. Identifying this function constitutes an important direction for future research. In the present study, we instead approximate $\ell(s,u)$ by decomposing it as $\ell(s,u) = \ell_1(s)\ell_2(u)$ and learning each component via neural networks. Nevertheless, this decomposition does not ensure that the resulting process $Z(t)$ has stationary increments. Therefore, developing neural architectures that can enforce the stationary increment property remains a critical challenge. If such stationarity is guaranteed, it becomes possible to construct a noise process that is Gaussian, possesses stationary increments, and exhibits memory effects—thus offering a viable alternative to fBm.

Moreover, we observed that NANSDE-Net did not perform well on datasets characterized by very low Hurst indices. Addressing this issue requires a more comprehensive investigation into the general relationship between ARMA-type noise kernels $a(\cdot)$ and the Hurst index. The ultimate goal of this line of research is to identify kernel structures that enable the generation of sample paths, even for time series with low regularity.

A comprehensive analysis across varying network capacities and alternative kernel parameterizations remains an important direction for future work. In particular, investigating more expressive kernel forms and evaluating their impact on approximation power and training stability would provide deeper insight into the trade-off between Markovian tractability and functional flexibility. Furthermore, a systematic runtime comparison—including training time, inference speed, and memory usage—against diffusion-based models, transformer-based generators, and lightweight alternatives such as wavelet-based approaches would further clarify scalability and computational overhead in practical settings.

**Broader Impact**

Fractional Brownian motion (fBm) has traditionally been employed as the primary noise source with memory for modeling time series data exhibiting long- and short-memory behavior across various domains. However, the findings of this study reveal that alternative noise processes can also capture such memory characteristics. This raises a new question as to whether fBm is indeed the most suitable choice for modeling long- and short-memory phenomena. In particular, by appropriately designing the kernel function $\ell(s, u)$ to satisfy stationarity conditions, we demonstrate the existence of stationary-increment Gaussian processes with memory that are distinct from fBm.

**Acknowledgments.** This work was supported by JST BOOST, Japan Grant Number JPMJBS2424.

**Disclosure of Interests.** The authors have no competing interests to declare that are relevant to the content of this article.

## References

1. Inoue, A., Moriuchi, S., Nakamura, Y.: A vasicek-type short rate model with memory effect. Stoch. Anal. Appl. **33**, 1068–1082 (2015)
2. Anh, V., Inoue, A.: Financial markets with memory I: dynamic models. Stoch. Anal. Appl. **23**, 275–300 (2005)
3. Anh, V., Inoue, A.: Financial markets with memory II: innovation processes and expected utility maximization. Stoch. Anal. Appl. **23**, 301–328 (2005)

4. Chen, R.T., Rubanova, Y., Bettencourt, J., Duvenaud, D.: Neural ordinary differential equations. In: Advances in Neural Information Processing Systems, pp. 6572–6583 (2018)
5. Cont, R.: Empirical properties of asset returns: stylized facts and statistical issues. Quantit. Finance **1**(2), 223 (2001)
6. Dang, T., Dimitriadis, A., Wu, J., Sethu, V., Ambikairajah, E.: Constrained dynamical neural ode for time series modelling: a case study on continuous emotion prediction. In: ICASSP 2023–2023 IEEE International Conference on Acoustics, Speech and Signal Processing (ICASSP), pp. 1–5. IEEE (2023)
7. Gatheral, J., Jaisson, T., Rosenbaum, M.: Volatility is rough. In: Commodities, pp. 659–690. Chapman and Hall/CRC (2022)
8. Hayashi, K., Nakagawa, K.: Fractional SDE-net: generation of time series data with long-term memory. In: Proceedings of IEEE International Conference on Data Science and Advanced Analytics. IEEE (2022)
9. He, K., Zhang, X., Ren, S., Sun, J.: Deep residual learning for image recognition. In: Proceedings of the IEEE Conference on Computer Vision and Pattern Recognition, pp. 770–778 (2016)
10. Kidger, P., Foster, J., Li, X., Oberhauser, H., Lyons, T.: Neural SDEs as infinite-dimensional gans (2021). arXiv:2102.03657 arXiv preprint
11. Kong, L., Sun, J., Zhang, C.: SDE-net: equipping deep neural networks with uncertainty estimates. In: International Conference on Machine Learning, pp. 5405–5415. PMLR (2020)
12. Luo, Z., Kamata, S.I., Sun, Z., Zhou, W.: Deep neural networks with flexible complexity while training based on neural ordinary differential equations. In: ICASSP 2021-2021 IEEE International Conference on Acoustics, Speech and Signal Processing (ICASSP), pp. 1690–1694. IEEE (2021)
13. Nakagawa, K., Hayashi, K.: LF-net: generating fractional time-series with latent fractional-net. In: 2024 International Joint Conference on Neural Networks (IJCNN), pp. 1–8. IEEE (2024)
14. Strogatz, S.H.: Nonlinear dynamics and chaos: with applications to physics, biology, chemistry, and engineering. CRC Press (2018)
15. Sun, Z., El-Laham, Y., Vyetrenko, S.: Neural stochastic differential equations with change points: a generative adversarial approach. In: ICASSP 2024-2024 IEEE International Conference on Acoustics, Speech and Signal Processing (ICASSP), pp. 6965–6969. IEEE (2024)
16. Wu, S., Shi, Z.: Itôwave: Itô stochastic differential equation is all you need for wave generation. In: ICASSP 2022-2022 IEEE International Conference on Acoustics, Speech and Signal Processing (ICASSP), pp. 8422–8426. IEEE (2022)

# Dual-Modality Latent Gloss Alignment for Gloss-Free Sign Language Translation

Yunlong Yu[1], Ming Ji[1], Zhao Wang[2(✉)], and Ziqian Lu[3]

[1] College of Information Science and Electronic Engineering, Zhejiang University, Zhejiang, China
{yuyunlong,ming.ji}@zju.edu.cn
[2] Ningbo Global Innovation Center, Zhejiang University, Zhejiang, China
zhao_wang@zju.edu.cn
[3] Zhejiang Sci-Tech University, Zhejiang, China
ziqianlu@zstu.edu.cn

**Abstract.** Sign Language Translation (SLT) remains a challenging task due to the inherent cross-modal complexity of processing visual-gestural language. In this paper, we propose **GluNet**, a novel **gloss-free** framework that learns latent gloss representations through fine-grained video-text alignment, eliminating the dependency on gloss annotations. Our method constructs a joint cross-modal embedding space by aggregating temporally consistent segments from sign language videos and semantically aligned subword token units from textual translations. These aligned representations form latent visual and textual glosses that inherently capture the structural semantics of both modalities. Extensive experiments on two public benchmarks demonstrate that GluNet achieves state-of-the-art performance, surpassing all gloss-free methods by a significant margin, including a +2.57% BLEU-4 improvement on the Phoenix-2014T dataset compared to the second-best gloss-free competitor. Notably, GluNet even outperforms the leading some gloss-based methods on key metrics such as BLEU-4 and ROUGE, establishing a new benchmark for SLT. The codes are available at here.

**Keywords:** Sign language translation · Cross-modal alignment · Multimodal learning

## 1 Introduction

Sign languages, the primary communication mode for deaf and hard-of-hearing people, use hand, body, and facial cues, and Sign Language Translation (SLT) aims to translate continuous sign videos into spoken or written language to bridge the gap between signers and non-signers [5,9,12,26].

Despite recent progress, SLT remains difficult because it must jointly model fine-grained visualspatial cues and complex linguistic structure, while large-scale paired signtext datasets are still scarce. Earlier work inserted gloss annotations as a bridge between video and text, but these labels are costly and hard to scale. Recent gloss-free methods exploit self-supervised learning [28] or LLM-assisted


R. C.-W. Wong et al. (Eds.): PAKDD 2026, LNAI 16597, pp. 175–186, 2026.
https://doi.org/10.1007/978-981-92-1300-9_14

weak supervision [8,21], yet the granularity gap between unstructured video and hierarchical language still hurts fine-grained semantics and leaves a performance gap to gloss-based systems.

Inspired by the hierarchical nature of spoken language, we argue that SLT benefits from latent intermediate units even without explicit gloss labels. To bridge this granularity gap between visual and textual modalities, we propose **GluNet**, a gloss-free framework that learns temporally coherent video units aligned with latent linguistic substructures. GluNet projects both sign videos and sentences into a shared cross-modal latent space. Within this space:

- Video frames are grouped into temporally coherent segments via similarity-based aggregation, forming pseudo-visual glosses.
- Sentences are decomposed into semantically meaningful subword clusters, forming pseudo-textual glosses.

These latent gloss units act as structural proxies, "gluing" the two modalities without any gloss annotations.

We validate GluNet on PHOENIX-2014T [1] and CSL-Daily [29], it outperforms prior gloss-free methods and even surpasses several gloss-based models, all without gloss supervision.

## 2 Related Work

Existing SLT methods primarily follow two paradigms: gloss-based and gloss-free approaches.

Gloss-based models use explicit gloss annotations and CTC loss to align visual features with gloss sequences [2,4]. SignBERT+ [12], SLTUNet [24], and MSKA [10] fine-tune encoders so frames map to gloss tokens. These systems are effective but require costly labels that do not scale.

Gloss-free SLT reduces annotation cost via (1) end-to-end models that map frames directly to sentences with pretrained encoders [14,17] or contrastive objectives [25] (GFSLT-VLP [28], SignCL [22], and GASLT [23]), or (2) pseudo-gloss methods that add a lightweight intermediate layer (PGG-SLT [11], LOST-in-SLT [13], Sign2GPT [21]). Recent work also feeds visual tokens into frozen LLMs [8] or adds adapters [21], yet these designs are domain-specific and heavy.

Our method falls into the gloss-free paradigm but does not rely on LLMs or multi-stage pipelines. GluNet induces latent gloss-like representations for both visual and textual modalities within the network itself. It performs semantic clustering at the frame and subword levels and fuses them through dual InfoNCE losses, forming an **end-to-end**, **single-stage architecture** that requires **no external gloss generator** or **prompt engineering**.

## 3 Method

As illustrated in Fig. 1, our architecture comprises three components: (1) a visual encoder that maps sign videos to latent representations, (2) a text encoder that

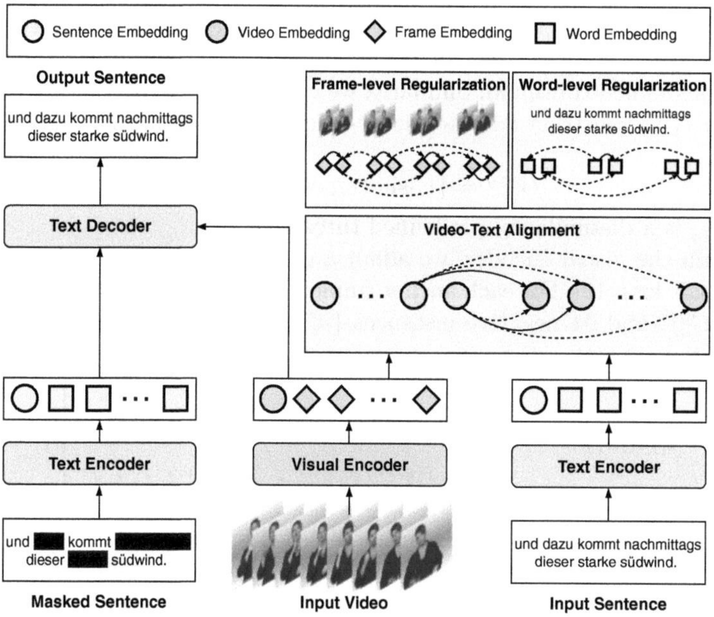


**Fig. 1.** Illustration of our GluNet framework. GluNet trains the SLT model by (1) performing masked sentence reconstruction for context learning, (2) direct sign language translation, and (3) visual-text alignment.

embeds sentences into the same latent space, and (3) a text decoder that reconstructs target sentences from these representations. The framework is trained end-to-end through a dual optimization strategy: cross-modal contrastive learning aligns video and text embeddings in the shared latent space, while the decoder is simultaneously optimized for SLT using the aligned representations.

Given an input sign video $V = \{v_t\}_{t=1}^{R}$ comprising $R$ frames and its paired spoken language transcription $S = \{s_u\}_{u=1}^{U}$ with $U$ words, the visual encoder $f_\theta$ and text encoder $g_\phi$ are used to extract the corresponding features in the latent representation space:

$$\begin{aligned}\mathbf{Z}^v &= f_\theta(V) \in \mathbb{R}^{R\times d},\\ \mathbf{Z}^s &= g_\phi(S) \in \mathbb{R}^{U\times d},\end{aligned} \tag{1}$$

where $\mathbf{Z}^v = [\mathbf{z}_1^v, ..., \mathbf{z}_R^v]$ and $\mathbf{Z}^s = [\mathbf{z}_1^s, ..., \mathbf{z}_U^s]$ are aligned embeddings for contrastive learning. $d$ denotes the dimenstionality of the latent space.

### 3.1 Latent Visual Gloss Learning

For each frame $v_t$ in video $V$, we identify its visually similar frames to construct a set termed $v_t$-centered visual positive gloss, defined as:

$$G_{v_t}^{pos} = \{v_i | \mathrm{sim}(v_i, v_t) > T_{pos}^{v}\},$$

where $v_i$ is a frame in video $V$, sim$(\cdot,\cdot)$ denotes the similarity metric (we use cosine similarity in this work) in the latent feature representation space, and $T^v_{pos}$ is a predefined threshold. Similarly, we could identify the visually dissimilar frames for $v_t$ to construct a set termed $v_t$-centered visual negative gloss, defined as:

$$G^{neg}_{v_t} = \{v_i | \text{sim}(v_i, v_t) < T^v_{neg}\},$$

where $T^v_{neg}$ is a dissimilarity predefined threshold.

To train the visual encoder, we adopt a contrastive learning objective using the InfoNCE loss [19]. For each anchor frame $v_t$, we sample one positive instance $v^+_t$ from $G^{pos}_{v_t}$ and $M$ negative instances $\{v^-_t\}^M_{i=1}$ from $G^{neg}_{v_t}$. The loss function is formulated as:

$$\mathcal{L}_{\text{INCE}}(f) = \mathbb{E}\left[\ln \frac{-e^{f(v_t)^T f(v^+_t)/\tau}}{e^{f(v_t)^T f(v^+_t)/\tau} + \sum_{i=1}^{M} e^{f(v_t)^T f(v^-_t)/\tau}}\right], \tag{2}$$

where $\tau$ is a temperature hyperparameter scaling the similarity distribution. This objective maximizes agreement between $v_t$ and its positives while pushing negative pairs apart in the embedding space. Through these explicit contrastive relationships, the model learns to cluster frames sharing the same latent gloss in the latent representation space.

### 3.2 Latent Text Gloss Learning

For each word $w_t$ in the sentence $S$, we construct two word sets centered at $w_u$, a text positive gloss and a text negative gloss. The positive gloss $G^{pos}_{w_u}$ groups words semantically similar to $w_u$, defined as:

$$G^{pos}_{w_u} = \{w_i | \text{sim}(w_i, w_u) > T^w_{pos}\},$$

where sim$(\cdot,\cdot)$ measures semantic similarity, and $T^w_{pos}$ is a predefined similarity threshold. Conversely, the negative gloss $G^{neg}_{w_u}$ contains words semantically dissimilar to $w_u$:

$$G^{neg}_{w_u} = \{w_i | \text{sim}(w_i, w_u) < T^w_{neg}\},$$

where $T^w_{neg}$ is a dissimilarity threshold. These sets impose structural constraints to cluster related words and separate unrelated ones in the latent space.

Similar to the visual framework, we train the text encoder $g$ using the InfoNCE loss. For each anchor word $w_u$, we sample one positive word $w^+_u$ from $G^{pos}_{w_u}$ and $K$ negative words $\{w^-_t\}^K_{i=1}$. The loss is defined as:

$$\mathcal{L}_{\text{INCE}}(g) = \mathbb{E}\left[\ln \frac{-e^{g(w_t)^T f(w^+_t)/\tau}}{e^{g(w_t)^T g(w^+_t)/\tau} + \sum_{i=1}^{K} e^{g(w_t)^T g(w^-_t)/\tau}}\right], \tag{3}$$

where $\tau$ is a temperature hyperparameter. This objective strengthens the discriminative power of text embeddings by maximizing agreement between $w_u$ and its positives while repelling negatives.

### 3.3 Proxy Tasks During Training

During training, our framework employs three proxy objectives to jointly optimize the visual encoder, the text encoder, and the text decoder.

**Visual-Text Alignment.** Given a batch of $N$ (video, text) pairs $\{(V_i, S_i)\}_{i=1}^{N}$, we adopt a CLIP-inspired contrastive objective [16,28] to distinguish true pairings from $N^2 - N$ non-matching candidates. Specifically, we jointly train a video encoder $f$ and text encoder $g$ to align matching pairs in a shared embedding space while repelling mismatched ones.

For each pair $(V_i, S_i)$, we compute cosine similarity scores:

$$s_{i,j} = \frac{f(V_i)^T g(S_j)}{\tau \|f(V_i)\| \|g(S_j)\|},$$

where $\tau$ is a learnable temperature parameter. These scores form an $N \times N$ similarity matrix $\mathbf{S}$. The model is optimized to maximize $s_{i,i}$ for aligned pairs and minimize $s_{i,j} (i \neq j)$ for mismatched pairs via a symmetric cross-entropy loss:

$$\mathcal{L}_{\text{Align}} = \frac{1}{2}(\mathcal{L}_{v2t} + \mathcal{L}_{t2v}) \quad = -\frac{1}{2N} \left( \sum_{i=1}^{N} \log \frac{e^{s_{i,i}}}{\sum_{j=1}^{N} e^{s_{i,j}}} + \sum_{j=1}^{N} \log \frac{e^{s_{j,j}}}{\sum_{i=1}^{N} e^{s_{i,j}}} \right), \tag{4}$$

This objective encourages the videos and their corresponding text descriptions to be mutually retrievable in the latent space, while non-matching pairs are distinctly separated.

**Masked Sentence Reconstruction.** Following [28], we incorporate a masked language reconstruction task to train both text encoder $g$ and text decoder $h$. For each input sentence $S_i$, we randomly mask a subset of tokens to generate a corrupted version $\widetilde{S}_i$. The decoder $h$ reconstructs the original $S_i$ conditioned on the text encoder $g$'s embeddings of $\widetilde{S}_i$. The reconstruction loss is defined as:

$$\mathcal{L}_{\text{Recon}} = -\frac{1}{N} \sum_{i=1}^{N} \sum_{w \in S_i} \log P\left(w \mid g(\widetilde{S}_i), h\right), \tag{5}$$

where $P(w \mid \cdot)$ is the decoder's predicted probability distribution over the vocabulary for token $w$. This objective trains $h$ to recover masked tokens by leveraging bidirectional context and semantic features from $g$, initializing it for the sign language translation task.

**Sign Language Translation.** Given an input sign video $V$, the model generates the corresponding target sentence $S = (s_1, \ldots, s_U)$ through three stages.

First, the visual encoder $f$ processes the video frames $v_{1:R}$ to extract spatiotemporal features:

$$r_{1:E} = f(v_{1:R}),$$

where $E = R/4$ denotes temporal downsampling ratio.

The second stage is autoregressive decoding. The text decoder $h$ generates tokens sequentially, conditioned on visual features $r_{1:E}$ and previous predictions $s_{1:u-1}$:

$$z_u = h(s_{1:u-1}, r_{1:E}),$$

with $s_0$ initialized as the beginning-of-sequence token $\langle$BOS$\rangle$. Generation terminates when $\langle$EOS$\rangle$ is predicted.

The final stage is probability prediction. The output logits $z_u$ are projected to vocabulary probabilities via:

$$p(s_u|\cdot) = \mathrm{Softmax}(\mathbf{W}z_u + b),$$

where $\mathbf{W}$ and $b$ are learnable parameters. The translation loss measures the negative log-likelihood of the ground-truth sentence:

$$\mathcal{L}_{\mathrm{SLT}} = -\sum_{u=1}^{U} \log p(s_u|s_{<u}, V). \tag{6}$$

### 3.4 Objective Function

The total training objective combines proxy task losses with latent space regularization through intra-modal contrastive learning:

$$\mathcal{L}_{\mathrm{Total}} = \mathcal{L}_{\mathrm{Align}} + \mathcal{L}_{\mathrm{Recon}} + \mathcal{L}_{\mathrm{SLT}} \quad + \alpha\mathcal{L}_{\mathrm{INCE}}(f) + \beta\mathcal{L}_{\mathrm{INCE}}(g), \tag{7}$$

where $\alpha$ and $\beta$ are two hyperparameters. $\mathcal{L}_{\mathrm{INCE}}(f)$ and $\mathcal{L}_{\mathrm{INCE}}(g)$ are regularizations on frame-level feature embeddings and word-level feature embeddings, respectively.

## 4 Experiments

### 4.1 Implementation Details

Following [23], ResNet18 pre-trained on ImageNet serves as our backbone. For temporal modeling, we employ convolutional layers (kernel size: 5, stride: 1) and max pooling layers (kernel size: 2, stride: 2). The Transformer encoder-decoder architecture consists of 3 layers with 8 attention heads, a hidden dimension of 1,024, and a feed-forward dimension of 4,096.

We train with batch size 4, resizing frames to 256 $\times$ 256 and cropping to 224 $\times$ 224 (random crops for training, center crops for testing). Optimization uses SGD with an initial learning rate of 0.01 and cosine decay to $10^{-5}$ for 80 epochs. At inference, we apply beam search with beam size 5 and length penalty 1.0.

**Table 1.** Comparison results (%) on PHOENIX14T and CSL-Daily datasets. The best results are marked in **bold**.

| Method | PHOENIX14T | | | | | | CSL-Daily | | | | | |
|---|---|---|---|---|---|---|---|---|---|---|---|---|
| | Dev | | | Test | | | Dev | | | Test | | |
| | R | B1 | B4 | R | B1 | B4 | R | B1 | B4 | R | B1 | B4 |
| **Gloss-based** | | | | | | | | | | | | |
| ConSLT [7] | 47.74 | - | 21.11 | 47.69 | - | 21.59 | 41.46 | - | 14.80 | 40.98 | - | 14.53 |
| STN-SLT [20] | - | 49.12 | 23.23 | - | 48.61 | 23.65 | - | - | - | - | - | - |
| STMC-T [30] | 48.24 | 47.60 | 24.09 | 46.65 | 46.98 | 23.65 | - | - | - | - | - | - |
| MMTLB [4] | 53.10 | 53.95 | 27.61 | 52.65 | 53.97 | 28.39 | 53.38 | 53.81 | 24.42 | 53.25 | 53.31 | 23.92 |
| TS-SLT [5] | 54.08 | 54.32 | 28.66 | 53.48 | 54.90 | 28.95 | 55.10 | 55.21 | 25.76 | 55.72 | 55.44 | 25.79 |
| SLTUNET [24] | 52.23 | - | 27.87 | 52.11 | 52.92 | 28.47 | 53.58 | - | 23.99 | 54.08 | 54.98 | 25.01 |
| CV-SLT [27] | 54.43 | 55.09 | 29.10 | 54.33 | 54.88 | 29.27 | 56.36 | 58.05 | 28.24 | 57.06 | 58.29 | 28.94 |
| MSKA-SLR [10] | 52.67 | 54.09 | 27.63 | 53.54 | 54.79 | 29.03 | 53.53 | 55.95 | 25.16 | 54.04 | 56.37 | 25.52 |
| SCOPE [15] | 67.09 | 61.80 | 32.83 | 60.06 | 61.74 | 32.84 | 60.18 | 60.37 | 31.74 | 60.68 | 60.48 | 32.08 |
| **Gloss-free** | | | | | | | | | | | | |
| GASLT [23] | - | - | - | 39.86 | 39.07 | 15.74 | - | - | - | 20.35 | 19.90 | 4.07 |
| GFSLT-VLP [28] | 43.72 | 44.08 | 22.12 | 42.49 | 43.71 | 21.44 | 36.70 | 39.20 | 11.07 | 36.44 | 39.37 | 11.00 |
| SignLLM [8] | 47.23 | 46.88 | 25.25 | 44.49 | 45.21 | 23.40 | 39.18 | 42.45 | 12.23 | 39.91 | 39.55 | 15.75 |
| SignCL [22] | - | - | - | 49.04 | 49.76 | 22.74 | - | - | - | 48.92 | 47.47 | 16.16 |
| VAP [14] | 51.47 | 52.78 | 26.62 | 51.28 | 53.07 | 26.16 | 51.19 | 53.31 | 23.84 | 51.09 | 52.98 | 23.65 |
| Sign2GPT [21] | - | - | - | 48.90 | 49.54 | 22.52 | - | - | - | 42.36 | 41.75 | 15.40 |
| $C^2RL$ [6] | - | - | - | 50.96 | 52.81 | 26.75 | - | - | - | 48.21 | 49.32 | 21.61 |
| PGG-SLT [11] | 52.85 | 53.84 | 27.53 | 52.56 | 54.02 | 27.32 | - | - | - | - | - | - |
| GluNet (Ours) | **61.58** | **54.93** | **28.66** | **61.81** | **54.25** | **29.89** | **57.43** | **54.32** | **24.16** | **58.16** | **53.87** | **24.05** |

## 4.2 Datasets and Evaluation Metrics

Following the prior work, we evaluate on two popular SLT datasets: Phoenix-2014T [1] and CSL-Daily [29], using their development (dev) and test splits for evaluation.

**Phoenix-2014T** is a German sign language dataset with 7,096/519/642 train/dev/test samples and 2,887 unique signs. **CSL-Daily** is a large-scale Chinese sign language dataset with 18,401/1,077/1,176 train/dev/test samples and a vocabulary of 2,343 signs. We adopt the established evaluation protocol from prior work [8,22,28], using BLEU and ROUGE-L to assess translation quality. Specifically, BLEU-n measures the precision of n-gram matches between predictions and ground-truth translations. We report results for BLEU-1 and BLEU-4 (denoted as B1 and B4). ROUGE-L computes an F1 score based on the longest common subsequence (LCS) between generated and reference texts, capturing semantic coherence.

## 4.3 Main Results

**Results on PHOENIX-2014T.** Table 1 compares GluNet with both gloss-based and gloss-free methods. GluNet is highly competitive on both dev and test splits and clearly outperforms recent *pseudo-gloss* approaches. Compared with Sign2GPT [21], GluNet improves BLEU-4 and ROUGE on the test set by up to **+7.37%** and **+12.91%** points, respectively. It also surpasses PGG-SLT [11], with gains of **+2.57%** BLEU-4 and **+9.25%** ROUGE, showing the

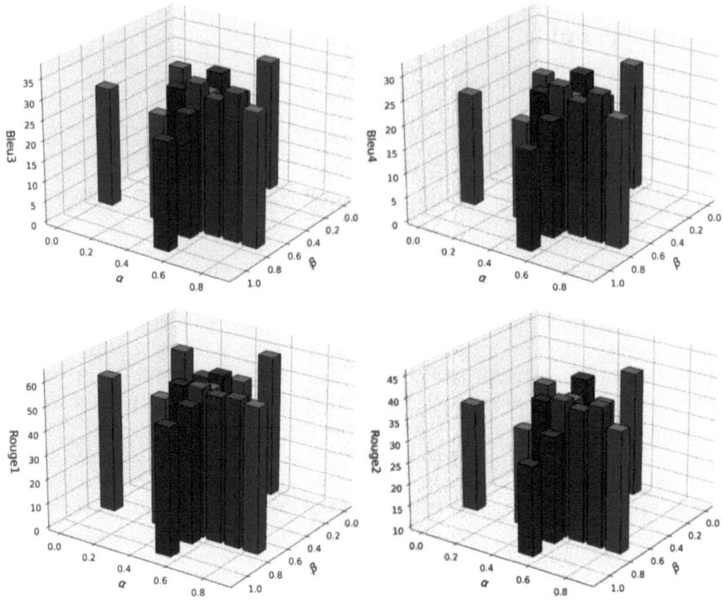


**Fig. 2.** Impacts of different hyperparameters on four metrics for Phoenix-2014T dataset.

benefit of our dual-modality latent gloss alignment over unimodal pseudo-gloss generation and post-hoc reordering.

**Results on CSL-Daily.** On CSL-Daily, GluNet again achieves the best results among gloss-free methods (24.05% BLEU-4, 58.16% ROUGE on the test set), outperforming GFSLT-VLP [28] (22.66%, 48.29%) and VAP [14] (23.65%, 51.09%). Its performance is comparable to or better than several gloss-based systems, indicating that latent gloss modeling can close much of the gap without explicit gloss supervision.

### 4.4 Ablation Studies

**Table 2.** Ablation results (%) evaluating component contributions of $\mathcal{L}_{\mathrm{INCE}}(f)$ and $\mathcal{L}_{\mathrm{INCE}}(g)$ on the Phoenix-2014T dataset.

| $\mathcal{L}_{\mathrm{INCE}}(f)$ | $\mathcal{L}_{\mathrm{INCE}}(g)$ | Dev | | | | | Test | | | | |
|---|---|---|---|---|---|---|---|---|---|---|---|
| | | B1 | B2 | B3 | B4 | ROUGE | B1 | B2 | B3 | B4 | ROUGE |
| ✗ | ✗ | 44.64 | 31.67 | 23.65 | 18.15 | 50.81 | 45.89 | 33.47 | 25.27 | 19.60 | 52.19 |
| ✓ | ✗ | 50.26 | 38.20 | 30.38 | 24.85 | 57.54 | 51.30 | 39.81 | 32.00 | 26.32 | 58.46 |
| ✗ | ✓ | 47.54 | 34.93 | 28.97 | 21.19 | 54.03 | 47.80 | 37.95 | 28.57 | 24.30 | 54.96 |
| ✓ | ✓ | 54.93 | 43.05 | 34.80 | 28.66 | 61.58 | 54.25 | 43.61 | 35.85 | 29.89 | 61.81 |

**Impacts of Regularization.** The ablation experiments presented in Table 2 offer insights into the individual and combined contributions of the $\mathcal{L}_{\text{INCE}}(f)$ and $\mathcal{L}_{\text{INCE}}(g)$ components. When used together, the full model achieves the highest performance across all evaluation metrics, notably reaching a Test-ROUGE of 61.81%, an improvement of +6.85% over $\mathcal{L}_{\text{INCE}}(g)$ alone, and a Test-B4 of 29.89%, representing a 10.29% gain over the baseline. $\mathcal{L}_{\text{INCE}}(f)$ emerges as the primary contributor to local sequence alignment, achieving a standalone Test-B4 of 26.32% and Test-ROUGE of 58.46%. While $\mathcal{L}_{\text{INCE}}(g)$ alone yields moderate improvements (Test-B4 = 24.30%, ROUGE = 54.96%), its integration with $\mathcal{L}_{\text{INCE}}(f)$ leads to synergistic gains, particularly in long-sequence modeling and enhanced contextual coherence (a 2.77% ROUGE increase). A potential data anomaly in the $\mathcal{L}_{\text{INCE}}(g)$-only configuration may indicate instability in capturing mid-range dependencies without $\mathcal{L}_{\text{INCE}}(f)$. Importantly, the combined model exhibits minimal dev-test degradation, underscoring its strong generalization capability. These findings highlight the complementary nature of $\mathcal{L}_{\text{INCE}}(f)$ and $\mathcal{L}_{\text{INCE}}(g)$.

**Results of Different Hyperparameters.** Figure 2 presents a comparative analysis of four metrics on the Phoenix-2014T dataset across different hyperparameter combinations ($\alpha$ and $\beta$ ranging 0.0∼1.0). The 3D bar charts reveal that: (1) all metrics exhibit parabolic trends, peaking at mid-range values with performance plateaus in $\alpha, \beta \in [0.4, 0.8]$ regions. (2) Extreme values ($\alpha/\beta$ =0.0 or 1.0) consistently underperform by 8∼10% relative to optima, indicating the necessity of balanced modality weighting.

### 4.5 Further Analysis

**T-SNE Visualization.** Figure 3 presents a comparative t-SNE analysis of two video samples (a-b) from the Phoenix-2014T dataset, contrasting feature distributions with/without our regularizations. The visualization reveals three key improvements: (1) Regularized features form tighter clusters, (2) Clearer separation emerges between dissimilar frames, and (3) Embeddings become more discriminative, with scattered points turning into well-formed clusters. These geometric transformations quantitatively validate our method's ability to enhance sign language feature discriminability.

**Qualitative Analysis.** Figure 4 presents four cases from the Phoenix-2014T test set, including both short and long sentences. Comparing predictions with references yields several observations: (1) For short phrases like "es bleibt winding" (It will remain windy), the model achieves nearly perfect translation result. (2) For longer weather reports, it preserves the main content (e.g., temperature ranges of 22–32°) while exhibiting some lexical errors ("gewitterschauer" for "nochitter"). (3) The visual alignment between signer gestures and textual outputs confirms the model's ability to maintain temporal synchronization, particularly evident in the consistent positioning of temporal markers ("am Abend", "morgen"). (4) Recurrent patterns in prediction errors (e.g., repeated words like "wechselwechsel") suggest specific decoding challenges in SLT.

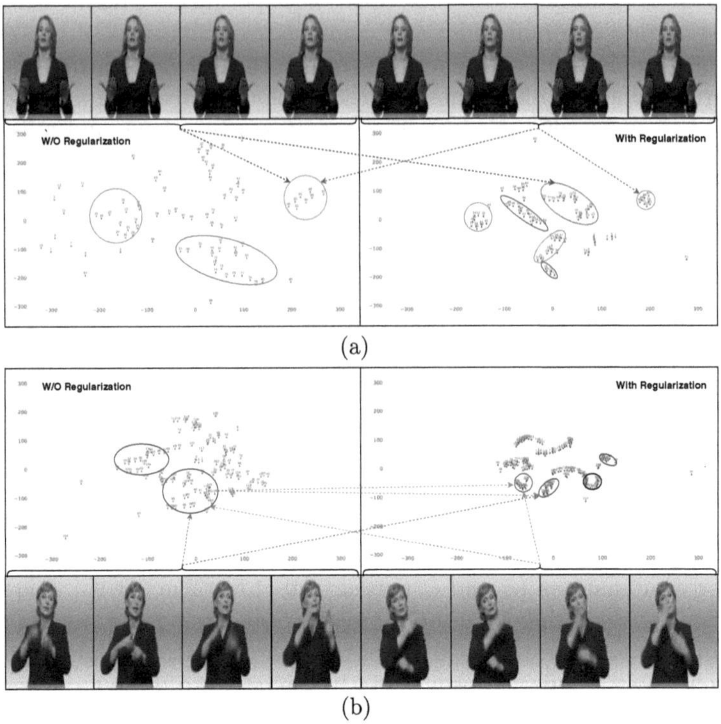


**Fig. 3.** T-SNE visualization of two video samples from Phoenix-2014T dataset using the models with and without our regularizations.

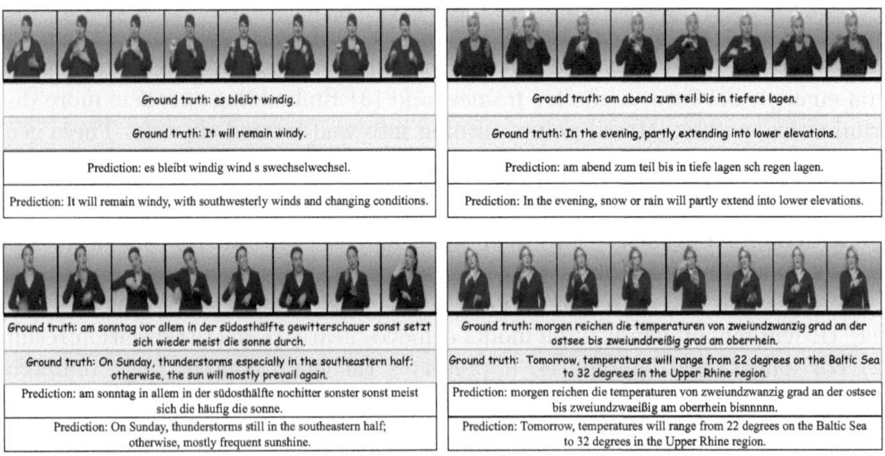


**Fig. 4.** Qualitative translation results on the Phoenix-2014T test set, evaluating both short and long sentences.

## 5 Conclusion

In this paper, we presented a novel gloss-free framework **GluNet** for SLT by learning latent gloss representations. By constructing a joint cross-modal embedding space, it aggregates temporally consistent visual segments with semantically aligned textual subwords, effectively capturing structural semantics without gloss supervision. Extensive experiments on Phoenix-2014T and CSL-Daily show that GluNet achieves state-of-the-art performance, surpassing all existing gloss-free methods by significant margins and even outperforming several gloss-based approaches on key metrics. Ablation studies validate the synergy of our proposed regularizations, while qualitative analyses confirm GluNet's ability to maintain temporal alignment and semantic coherence in translation.

**Acknowledgment.** This research was supported in part by the Key R&D Program of Zhejiang Province 2025C01075, National Natural Science Foundation of China under Grant 62576313 and 62501532, and CASIC Guangxin Intelligent Technology.

## References

1. Camgoz, N.C., Hadfield, S., Koller, O., Ney, H., Bowden, R.: Neural sign language translation. In: CVPR, pp. 7784–7793 (2018)
2. Camgoz, N.C., Koller, O., Hadfield, S., Bowden, R.: Multi-channel transformers for multi-articulatory sign language translation. In: ECCV Workshops, pp. 301–319 (2020)
3. Camgoz, N.C., Koller, O., Hadfield, S., Bowden, R.: Sign language transformers: Joint end-to-end sign language recognition and translation. In: CVPR, pp. 10023–10033 (2020)
4. Chen, Y., Wei, F., Sun, X., Wu, Z., Lin, S.: A simple multi-modality transfer learning baseline for sign language translation. In: CVPR, pp. 5120–5130 (2022)
5. Chen, Y., Zuo, R., Wei, F., Wu, Y., Liu, S., Mak, B.: Two-stream network for sign language recognition and translation. In: NeurIPS, pp. 17043–17056 (2022)
6. Chen, Z., et al.: C2rl: Content and context representation learning for gloss-free sign language translation and retrieval. In: IEEE Transactions on Circuits and Systems for Video Technology (2025)
7. Fu, B., et al.: A token-level contrastive framework for sign language translation. In: ICASSP 2023-2023 IEEE International Conference on Acoustics, Speech and Signal Processing, pp. 1–5 (2023)
8. Gong, J., Foo, L.G., He, Y., Rahmani, H., Liu, J.: LLMs are good sign language translators. In: CVPR, pp. 18362–18372 (2024)
9. Guan, M., Wang, Y., Ma, G., Liu, J., Sun, M.: Multi-stream keypoint attention network for sign language recognition and translation, arXiv:2405.05672 (2024)
10. Guan, M., Wang, Y., Ma, G., Liu, J., Sun, M.: Mska: Multi-stream keypoint attention network for sign language recognition and translation. Pattern Recogn. **165**, 111602 (2025)
11. Guo, J., Li, P., Cohn, T.: Bridging sign and spoken languages: Pseudo gloss generation for sign language translation. arXiv preprint arXiv:2505.15438 (2025)

12. Hu, H., Zhao, W., Zhou, W., Li, H.: Signbert+: Hand-model-aware self-supervised pre-training for sign language understanding. IEEE Trans. Pattern Anal. Mach. Intell. **45**(9), 11221–11239 (2023)
13. Jang, Y., Raajesh, H., Momeni, L., Varol, G., Zisserman, A.: Lost in translation. In: Sign language translation with contextual cues, arXiv preprint arXiv:2501.09754 (2025)
14. Jiao, P., Min, Y., Chen, X.: Visual alignment pre-training for sign language translation, pp. 349–367. ECCV (2024)
15. Liu, Y., Zhang, W., Ren, S., Huang, C., Yu, J., Xu, L.: Scope: Sign language contextual processing with embedding from LLMs, pp. 5739–5747. AAAI (2025)
16. Radford, A., et al.: Learning transferable visual models from natural language supervision. In: ICML, pp. 8748–8763 (2021)
17. Shi, B., Brentari, D., Shakhnarovich, G., Livescu, K.: Open-domain sign language translation learned from online video. In: EMNLP, pp. 6365–6379 (2022)
18. Sincan, O.M., Camgoz, N.C., Bowden, R.: Using an llm to turn sign spottings into spoken language sentences. arXiv preprint arXiv:2403.10434 (2024)
19. Sohn, K.: Improved deep metric learning with multi-class n-pair loss objective. NeurIPS (2016)
20. Voskou, A., Panousis, K.P., Kosmopoulos, D., Metaxas, D.N., Chatzis, S.: Stochastic transformer networks with linear competing units: Application to end-to-end sl translation. In: ICCV, pp. 11946–11955 (2021)
21. Wong, R.C., Camgöz, N.C., Bowden, R.: Sign2gpt: leveraging large language models for gloss-free sign language translation. In: ICLR. (2024)
22. Ye, J., Wang, X., Jiao, W., Liang, J., Xiong, H.: Improving gloss-free sign language translation by reducing representation density. In: NeurIPS (2024)
23. Yin, A., Zhong, T., Tang, L., Jin, W., Jin, T., Zhao, Z.: Gloss attention for gloss-free sign language translation. In: CVPR, pp. 2551–2562 (2023)
24. Zhang, B., Müller, M., Sennrich, R.: Sltunet: A simple unified model for sign language translation. In: ICLR (2024)
25. Zhang, B., Tanzer, G., Firat, O.: Scaling sign language translation. In: NeurIPS (2025)
26. Zhang, R., Hu, C., Yu, P., Chen, Y.: Improving multilingual sign language translation with automatically clustered language family information. In: Proceedings of the 31st International Conference on Computational Linguistics, pp. 3579–3588 (2025)
27. Zhao, R., Zhang, L., Fu, B., Hu, C., Su, J., Chen, Y.: Conditional variational autoencoder for sign language translation with cross-modal alignment, pp. 19643–19651. AAAI (2024)
28. Zhou, B., et al.: Gloss-free sign language translation: Improving from visual-language pretraining. In: ICCV, pp. 20871–20881 (2023)
29. Zhou, H., Zhou, W., Qi, W., Pu, J., Li, H.: Improving sign language translation with monolingual data by sign back-translation. In: CVPR, pp. 1316–1325 (2021)
30. Zhou, H., Zhou, W., Zhou, Y., Li, H.: Spatial-temporal multi-cue network for sign language recognition and translation. IEEE Trans. Multimedia **24**, 768–779 (2021)

# Subject Information Extraction for Novelty Detection with Domain Shifts

Yangyang Qu, Dazhi Fu, and Jicong Fan(✉)

School of Data Science, The Chinese University of Hong Kong, Shenzhen, Guangdong 518172, China

quyang9517@gmail.com, fudazhiaka@gmail.com, fanjicong@cuhk.edu.cn

**Abstract.** Unsupervised novelty detection (UND) is vital for applications such as medical diagnosis and cybersecurity. A prevalent assumption in current UND methods is that the normal data for training and testing are drawn from the same domain. This assumption is challenged in practice by domain shift, where a disparity exists between the training and testing domains (e.g., due to different data acquisition pipelines or sites). Domain shift often leads to the incorrect classification of normal data as novel by existing methods. A typical situation is that the normal testing data and normal training data describe the same underlying subject semantics, yet they differ in background/domain conditions. To address this problem, we introduce a novel method that separates subject information from background variation, encapsulating the domain information to enhance detection performance under domain shifts. The proposed method minimizes mutual information between the representations of the subject and background while modeling the background variation using a deep Gaussian mixture model; novelty detection is conducted solely on the subject representations, thus reducing sensitivity to domain variations. Experiments on Multi-background MNIST and Kurcuma demonstrate that our model generalizes effectively to unseen domains and outperforms strong baselines under domain shifts.

## 1 Introduction

Novelty detection [27] aims to identify samples that deviate from the training distribution, and is widely used in finance, healthcare, and security. It is closely related to outlier detection [10,12], anomaly detection [7,9], fault detection [13], and one-class classification [15]; in particular, unsupervised anomaly detection can be viewed as a special case of novelty detection where (almost) all training data are normal. In the past decades, numerous novelty detection methods [4,5,21,31,33,36,37,42] have been proposed. Classical approaches such as kernel density estimation (KDE) [28] estimate data density and treat low-density samples as novel. Recent deep methods include ALAD [39] (GAN-based reconstruction), DeepSVDD [31] (hypersphere enclosure in latent space), MO-GAAL [22] (adversarially generating informative outliers), PLAD [4] (perturbation learning), and DIF [38] (deep isolation forest [21]). More recently, Dai and Fan (2025)


R. C.-W. Wong et al. (Eds.): PAKDD 2026, LNAI 16597, pp. 187–198, 2026.
https://doi.org/10.1007/978-981-92-1300-9_15

[9] introduced AutoUAD, a framework for model selection and hyperparameter tuning in unsupervised anomaly detection, and Fu and Fan (2026) [10] proposed UniOD, a universal model designed for outlier detection across different domains.

In many deployments, normal test data may differ from training data due to acquisition-domain changes (e.g., different scanners/sites or protocols in medical imaging, different cameras/lighting in industrial inspection), while the underlying subject semantics remain unchanged. In this paper, we refer to subject information as the task-relevant semantic factor that should remain stable across domains, and background/domain information as nuisance variations induced by acquisition conditions. Our goal is to detect novelty in terms of subject information rather than background variations.

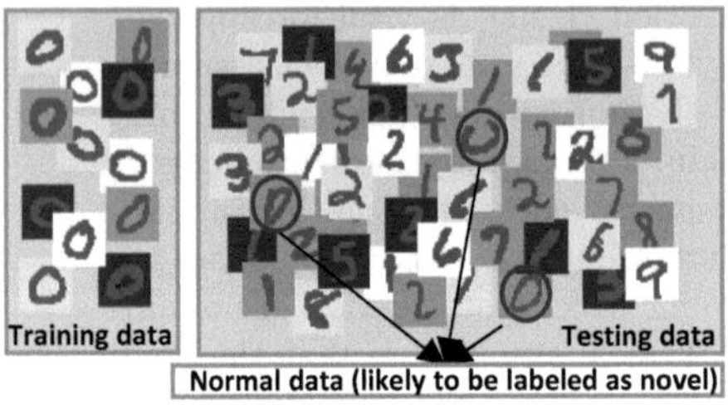


**Fig. 1.** An illustration of the novelty detection task. The training data (left) consists of images of the digit '0′ presented in three backgrounds. The testing data (right) includes images of multiple digits '0 − 9′ in seen backgrounds and entirely unseen backgrounds. Although the '0′ digits in the test set are normal, some of them are likely to be labelled as novel due to the shift in background.

A key limitation of many existing unsupervised novelty detection (UND) methods is that they implicitly assume that normal training data and normal testing data come from the same domain. In practice, normal testing data may come from different domains (domain shift), causing models to misclassify normal samples as novel, especially when training and test data describe the same subject but differ in background conditions (Fig. 1) [35].

Recent work has started to benchmark and address novelty detection under distribution shift (e.g., GNL) [6]. Meanwhile, disentanglement is often encouraged by enforcing statistical independence. In this work, we adopt mutual information minimization with a CLUB-style estimator [8] to reduce the dependence between subject and background content.

Related ideas such as multi-source augmentation, representation factorization, and removing nuisance information also appear beyond our setting (e.g., cross-modal augmentation [20], multi-source modelling [14], two-space decomposition [29], unlearning [17], and concept erasure in diffusion/flow models [11,23]). Beyond images, related directions include video anomaly detection [24,25] and vision-language/embodied factorization [40,41].

To contextualize our evaluation, we also include representative domain generalization techniques such as ERM [34] and IRM [2] as transferable baselines by pairing their learned representations with a KDE detector. Unlike these supervised DG methods that typically rely on task labels (and often domain labels), our approach targets unsupervised novelty detection and only assumes the number of background domains in the training set, reducing annotation burden.

To address background (domain) shifts between training and normal test data in UND, we propose Subject-Novelty Detection (SND). SND disentangles subject information from background features in the training data and performs novelty detection solely in the subject representation space, improving robustness to unseen backgrounds. Our main contributions are:

- We propose SND, which isolates subject information from background variations to enable robust novelty detection under significant domain shifts.
- SND requires only the number of domains in the training data (without per-sample domain labels), making it practical for low-annotation settings.
- SND outperforms strong novelty detection and domain-shift baselines in the experiments.

# 2 Unsupervised Subject Novelty Detection

## 2.1 Problem Formulation

To be precise, suppose we have a training dataset consisting of $N$ images, denoted as $\mathcal{D} = \{\mathbf{x}_1, \mathbf{x}_2, \ldots, \mathbf{x}_N\}$, in which each $\mathbf{x}_i \in \mathbb{R}^{C\times H\times W}$ has a background $b_i$ chosen from a set of $K$ different backgrounds $\mathcal{B} = \{B_1, B_2, \ldots, B_K\}$ and all or most of the $N$ samples are normal. Notably, although the total number of backgrounds $K$ is known, the specific background type of each image is unknown. This setting is practical since data or images collected often come from different backgrounds (or domains more generally), and labeling the backgrounds is costly. We consider a test set $\mathcal{D}' = \{\mathbf{x}'_1, \mathbf{x}'_2, \ldots, \mathbf{x}'_M\}$, where the background $b_i{}'$ of each $\mathbf{x}'_i$ is chosen from a larger set $\tilde{\mathcal{B}} = \{B_1, B_2, \ldots, B_K, B_{K+1}, \ldots, B_{K+K'}\}$. Note that $B_{K+1}, \ldots, B_{K+K'}$ are actually new backgrounds different from $B_1, B_2, \ldots, B_K$ and $K'$ is unknown. Our goal is to learn a model from $\mathcal{D}$ to determine whether a new sample from $\mathcal{D}'$ is a novel sample in terms of the subject information rather than the background information. This is a nontrivial task because the domain (or distribution) of normal data changed.

An example of the task is shown in Fig. 1, where the training set contains images of digit $'0'$ with 3 different colored backgrounds (white, yellow, blue), and the testing set contains images of digits $'0-9'$ with an additional unseen background (green). To evaluate performance under background shifts, the background of digit $'0'$ (normal testing data) is set to a completely unseen green background, while the backgrounds of digits $'1-9'$ (testing novel data) may appear in both seen and unseen backgrounds. Our aim is to identify digits $'1-9'$ as novel samples which contain different subject information, while treating digits $'0'$ as normal samples which differ from training data only in background information.

Classical ND tasks only consider the distribution difference between the training data and the testing novel data. Our tasks consider not only the distribution difference mentioned before but also the background (domain) shift between training data and testing normal data, which leads to a distribution difference between them. Therefore, unsupervised subject novelty detection is more challenging, as classical ND methods often mislabel normal samples with new backgrounds as novel, leading to high false positive rates.

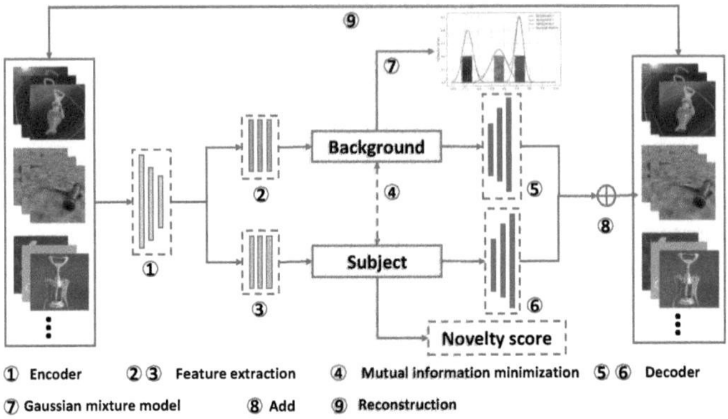


**Fig. 2.** An overview of the proposed SND model.

## 2.2 Proposed Model

We aim to address the challenge of isolating subject information from varying backgrounds for improved novelty detection. One key point is to learn representations that separate subject and background features in an unsupervised manner, allowing the model to detect novel subject information despite shifts in background domains. The overall process of our method is illustrated in Fig. 2.

The process begins with a feature extraction network $G_{\theta_f} : \mathbb{R}^{C\times H\times W} \to \mathbb{R}^d$ with parameters $\theta_f$, which processes the input image $\mathbf{x}$ and generates a feature representation $\mathbf{z}_f$, i.e.,

$$\mathbf{z}_f = G_{\theta_f}(\mathbf{x}). \tag{1}$$

This representation is then decomposed into two distinct components, a subject feature vector $\mathbf{z}_s$ and a background feature vector $\mathbf{z}_b$, through two neural networks $F_{\theta_s} : \mathbb{R}^d \to \mathbb{R}^d$ and $F_{\theta_b} : \mathbb{R}^d \to \mathbb{R}^d$, i.e.,

$$\mathbf{z}_s = F_{\theta_s}(\mathbf{z}_f), \quad \mathbf{z}_b = F_{\theta_b}(\mathbf{z}_f). \tag{2}$$

It is nontrivial to guarantee that $\mathbf{z}_s$ and $\mathbf{z}_b$ exclusively represent the subject and background information, respectively. Following the insights of CLUB [8], we propose to minimize the mutual information $I(\mathbf{z}_s; \mathbf{z}_b)$ between $\mathbf{z}_s$ and $\mathbf{z}_b$, which encourages the two parts to be statistically independent. The mutual information is estimated using a neural network $\xi_{\theta_m}$ based on the following formulation:

$$\hat{I}_{\mathrm{MI}}(\mathbf{z}_s; \mathbf{z}_b) = \frac{1}{N}\sum_{i=1}^{N}\left[\log\left(\xi_{\theta_m}(\mathbf{z}_b^{(i)} \mid \mathbf{z}_s^{(i)})\right) - \frac{1}{N}\sum_{j=1}^{N}\log\left(\xi_{\theta_m}(\mathbf{z}_b^{(j)} \mid \mathbf{z}_s^{(i)})\right)\right]. \tag{3}$$

In practice, we parameterize $\xi_{\theta_m}(\mathbf{z}_b \mid \mathbf{z}_s)$ as a Gaussian distribution and adopt an alternating training scheme: we first update $\theta_m$ by maximizing $\hat{I}_{\text{MI}}$ with encoders fixed, and then update the encoder parameters $G_{\theta_f}, F_{\theta_s}, F_{\theta_b}$ by minimizing $\hat{I}_{\text{MI}}$ while detaching the gradients through $\theta_m$.

It should be noted that making $\mathbf{z}_s$ and $\mathbf{z}_b$ independent alone cannot ensure that $\mathbf{z}_s$ is composed of subject information and $\mathbf{z}_b$ is composed of background information; they may swap roles. To address this, we assume that the number of background types is $K$ and that $K$ is different from the number of potential clusters in the subject information. Under this assumption, we can distinguish between subject and background information. Specifically, inspired by [46], we use a deep Gaussian mixture model (GMM) with $K$ components to model $\mathbf{z}_b$. A neural network $S_{\theta_g} : \mathbb{R}^d \to \mathbb{R}^K$ projects $\mathbf{z}_b^{(i)}$ to soft membership predictions $\hat{\gamma}^i \in \mathbb{R}^K$ for each mixture component:

$$\hat{\gamma}^i = \text{softmax}\big(S_{\theta_g}(\mathbf{z}_b^{(i)})\big). \tag{4}$$

This modeling encourages $\mathbf{z}_b$ to capture $K$ clusters, making it different from the subject information. Denoting $\hat{\pi}_k$ as the weight of the $k$-th Gaussian component and $\hat{\boldsymbol{\mu}}_k \in \mathbb{R}^d$ as the mean, we estimate the mixture parameters from a mini-batch $\{\mathbf{z}_b^{(i)}\}_{i=t_1}^{t_L}$ of size $L$ as

$$\begin{aligned}
\hat{\pi}_k &= \frac{1}{L}\sum_{i=t_1}^{t_L} \hat{\gamma}_k^i, \quad \hat{\boldsymbol{\mu}}_k = \frac{\sum_{i=t_1}^{t_L} \hat{\gamma}_k^i \mathbf{z}_b^{(i)}}{\sum_{i=t_1}^{t_L} \hat{\gamma}_k^i}, \\
\hat{\boldsymbol{\sigma}}_k^2 &= \frac{\sum_{i=t_1}^{t_L} \hat{\gamma}_k^i \big(\mathbf{z}_b^{(i)} - \hat{\boldsymbol{\mu}}_k\big) \odot \big(\mathbf{z}_b^{(i)} - \hat{\boldsymbol{\mu}}_k\big)}{\sum_{i=t_1}^{t_L} \hat{\gamma}_k^i}, \\
\mathbf{s}_k &= \log\big(\hat{\boldsymbol{\sigma}}_k^2 + \varepsilon\big), \quad \hat{\boldsymbol{\Sigma}}_k = \text{diag}\big(\exp(\mathbf{s}_k)\big).
\end{aligned} \tag{5}$$

where $\odot$ denotes element-wise multiplication, and $\log(\cdot)$ and $\exp(\cdot)$ are applied element-wise. $\varepsilon > 0$ is a small constant for numerical stability.

Given a randomly sampled batch of data $\{\mathbf{x}_i\}_{i=t_1}^{t_L} \subseteq \mathcal{D}$ and their background feature vectors $\{\mathbf{z}_b^{(i)}\}_{i=t_1}^{t_L}$, we define the following background energy function:

$$E(\mathbf{z}_b^{(i)}) = -\log\left(\sum_{k=1}^{K} \hat{\pi}_k (2\pi)^{-d/2} \big|\hat{\boldsymbol{\Sigma}}_k\big|^{-1/2} \exp\Big(-\tfrac{1}{2}(\mathbf{z}_b^{(i)} - \hat{\boldsymbol{\mu}}_k)^\top \hat{\boldsymbol{\Sigma}}_k^{-1} (\mathbf{z}_b^{(i)} - \hat{\boldsymbol{\mu}}_k)\Big)\right). \tag{6}$$

The identification of $\mathbf{z}_b$ as the background latent variable together with its enforced independence from $\mathbf{z}_s$ encourages $\mathbf{z}_s$ to capture the subject information.

Nevertheless, we still need to ensure that $\mathbf{z}_s$ and $\mathbf{z}_b$ preserve the original information of the input $\mathbf{x}$. This is achieved by reconstructing $\mathbf{x}$ from the two latents. Specifically, we feed $\mathbf{z}_s$ and $\mathbf{z}_b$ into two decoders $H_{\theta'_s} : \mathbb{R}^d \to \mathbb{R}^{C\times H\times W}$ and $H_{\theta'_b} : \mathbb{R}^d \to \mathbb{R}^{C\times H\times W}$, i.e.:

$$\mathbf{x}_s = H_{\theta'_s}(\mathbf{z}_s), \quad \mathbf{x}_b = H_{\theta'_b}(\mathbf{z}_b), \tag{7}$$

and combine their outputs as the reconstruction of $\mathbf{x}$:

$$\hat{\mathbf{x}} = \mathbf{x}_s + \mathbf{x}_b. \tag{8}$$

By isolating subject and background information in this way, our method focuses on detecting novelty in the subject information, even when there is significant variation in the background. This feature decomposition and reconstruction mechanism ensures robustness to background changes and facilitates accurate novelty detection. Note that the mixture parameters $\{\hat{\pi}_k, \hat{\boldsymbol{\mu}}_k, \hat{\boldsymbol{\Sigma}}_k\}_{k=1}^K$ are estimated from each mini-batch using the soft responsibilities $\hat{\gamma}_k^i$ in (5), and gradients are back-propagated through both the responsibilities and the batch statistics during training.

### 2.3 Training and Evaluation

Here, we summarize the entire process of the proposed method. Due to the mutual information estimation and GMM parts, we can ensure that both $\mathbf{z}_s$ and $\mathbf{z}_b$ contain necessary information about the subject and background, respectively, without worrying that one has learned most of the information while the other has not learned anything. The loss function $L_{\text{rec}}(\mathbf{x}, \hat{\mathbf{x}})$ represents the reconstruction error between the original image $\mathbf{x}$ and the reconstructed output image $\hat{\mathbf{x}}$, which is expressed as

$$L_{\text{rec}}(\mathbf{x}, \hat{\mathbf{x}}) = \|\mathbf{x} - \hat{\mathbf{x}}\|_2^2 . \tag{9}$$

We calculate the weighted sum of all the loss terms to obtain the total loss for the proposed model:

$$L_{\text{total}} = L_{\text{rec}}(\mathbf{x}, \hat{\mathbf{x}}) + \omega_1 E(\mathbf{z}_b) + \omega_2 \hat{I}_{\text{MI}}(\mathbf{z}_s; \mathbf{z}_b), \tag{10}$$

where $\omega_1$ and $\omega_2$ are non-negative hyperparameters, and the parameters to learn are $\{\theta_f, \theta_s, \theta_b, \theta_m, \theta_g, \theta'_s, \theta'_b\}$.

After our model is well-trained, we can use KDE [28], a simple yet effective method, to perform novelty detection. Specifically, we denote the subject feature vectors of the training set as $\mathcal{D}_s = \{\mathbf{z}_s^{(1)}, \mathbf{z}_s^{(2)}, \ldots, \mathbf{z}_s^{(N)}\} = \{F_{\theta_s}(G_{\theta_f}(\mathbf{x})) : \mathbf{x} \in \mathcal{D}\}$. Given a test sample $\mathbf{x}_{\text{new}}$, its subject feature vector is $\mathbf{z}_s^{\text{new}} = F_{\theta_s}(G_{\theta_f}(\mathbf{x}_{\text{new}}))$. The novelty score (NS) of $\mathbf{x}_{\text{new}}$ is given by the negative density of $\mathbf{z}_s^{\text{new}}$, i.e.,

$$\text{NS}(\mathbf{x}_{\text{new}}) = -\hat{p}(\mathbf{z}_s^{\text{new}}) = -\frac{1}{N(2\pi h^2)^{d/2}} \sum_{i=1}^{N} \exp\left(-\frac{\|\mathbf{z}_s^{\text{new}} - \mathbf{z}_s^{(i)}\|^2}{2\,h^2}\right) \tag{11}$$

where $h$ is the bandwidth parameter controlling the smoothness of the estimated density, and $d$ is the dimensionality of $\mathbf{z}_s$. A higher novelty score $\text{NS}(\mathbf{x}_{\text{new}})$ indicates that the subject of $\mathbf{x}_{\text{new}}$ has a lower likelihood of belonging to the distribution of subjects in the training data $\mathcal{D}$.

In general, the proposed method learns comprehensive subject features $\mathbf{z}_s$ and background features $\mathbf{z}_b$ using our objective function defined in (10), which includes the weighted sum of reconstruction loss, energy of $\mathbf{z}_b$, and mutual information between $\mathbf{z}_s$ and $\mathbf{z}_b$. For novelty detection, KDE fitted on the training set is applied to the subject feature $\mathbf{z}_s$ of the test sample, and the novelty score for a test sample is determined using (11).

# 3 Experiments

In this section, we benchmark our method and various baselines using numerical experiments on several challenging and widely used datasets. We adopt a leave-one-class-out protocol: for each run, we select 9 out of 10 classes as normal for training and treat the remaining class as novel for testing; we report per-class results and the average across all 10 splits. For testing, images from a different unseen background are used.

## 3.1 Datasets, Compared Methods, and Evaluation Metrics

We evaluated the proposed method on two challenging datasets: Multi-background MNIST and Kurcuma [30]. To address the limitations in variability in the original MNIST datasets [16], we introduced domain shifts by altering background colors. For the Multi-background MNIST dataset, the model was trained using 'blue', 'yellow', and 'white' backgrounds and tested on a previously unseen 'green' background. These setups evaluated the model's generalization to unseen domains. Additionally, we used Kurcuma, a benchmark for kitchen-utensil recognition in home-assistance robotics that comprises seven corpora/domains with diverse appearance shifts (e.g., synthetic, clipart, and real-image settings) [30].

We conducted an extensive performance evaluation by comparing our model against a wide range of recent state-of-the-art novelty detection methods. We include both classical and deep novelty detection baselines; all methods are implemented using PyOD defaults when applicable. The compared methods include AnoGAN [32], DeepSVDD [31], XGBOD [43], ALAD [39], INNE [3], MO-GAAL [22], COPOD [18], ROD [1], SUOD [44], and ECOD [19]. The hyperparameters for the methods were set according to the default settings provided by the PyOD [45]. Furthermore, we evaluated our approach against GNL, a recently proposed method for novelty detection across domain transformations [6]. We also compared our method with two key domain adaptation techniques, ERM and IRM [2], both followed by a KDE step for novelty detection. This allowed us to evaluate our model's effectiveness in handling domain shifts and identifying novel data in unseen environments.

We employed two common metrics to evaluate the performance of novelty detection: (i) Area Under the Receiver Operating Characteristic curve (AUROC), which can be interpreted as the probability that a positive sample has a higher discriminative score than a negative sample; and (ii) Area Under the Precision-Recall curve (AUPRC), an ideal metric for adjusting extreme differences between positive and negative base rates.

**Table 1.** Average AUROCs (%) in novelty detection on Multi-background MNIST. In each case, the best result is marked in bold.

| Method | 0 | 1 | 2 | 3 | 4 | 5 | 6 | 7 | 8 | 9 | Average |
|---|---|---|---|---|---|---|---|---|---|---|---|
| COPOD | 62.82 | 70.33 | 63.77 | 64.41 | 65.80 | 64.33 | 64.44 | 66.05 | 63.81 | 65.95 | 65.17 |
| SUOD | 64.52 | 67.42 | 65.79 | 67.72 | 70.17 | 69.37 | 65.46 | 67.24 | 65.10 | 67.70 | 67.05 |
| MO_GAAL | 61.41 | 72.20 | 69.40 | 77.73 | 65.74 | 58.00 | 71.48 | 71.23 | 73.13 | 73.63 | 69.40 |
| DeepSVDD | 63.92 | 58.84 | **72.46** | 56.69 | 48.24 | **88.63** | 66.82 | 76.00 | 62.02 | 62.95 | 65.66 |
| ALAD | 27.97 | 29.07 | 7.58 | 17.22 | 8.85 | 25.06 | 13.84 | 9.77 | 22.37 | 16.93 | 17.87 |
| ECOD | 57.29 | 61.46 | 60.37 | 60.41 | 62.13 | 61.19 | 60.22 | 61.29 | 59.67 | 61.53 | 60.56 |
| INNE | 61.23 | 57.83 | 63.72 | 63.15 | 61.80 | 66.64 | 62.72 | 65.09 | 58.84 | 61.37 | 62.24 |
| AnoGAN | 4.86 | 0.36 | 32.28 | 52.98 | 52.43 | 43.97 | 9.63 | 33.07 | 22.85 | 36.38 | 28.88 |
| ERM | 36.42 | 95.36 | 42.00 | 42.25 | 38.25 | 40.29 | 51.65 | 51.96 | 48.00 | 40.54 | 48.67 |
| IRM | 35.65 | 96.32 | 40.41 | 37.54 | 38.47 | 37.09 | 47.72 | 63.56 | 47.82 | 41.29 | 48.59 |
| GNL | 61.47 | 93.07 | 50.04 | 82.73 | 63.20 | 54.30 | 60.23 | 68.56 | 60.58 | 56.51 | 65.07 |
| SND | **85.74** | **97.68** | 71.35 | **84.40** | **75.55** | 74.59 | **90.39** | **85.09** | **80.24** | **74.08** | **82.27** |

**Table 2.** Average AUROC (%) on Kurcuma across seven target domains. Columns 0–8 correspond to the held-out novel class (9 categories in total).

| Method | 0 | 1 | 2 | 3 | 4 | 5 | 6 | 7 | 8 | Average |
|---|---|---|---|---|---|---|---|---|---|---|
| ALAD | 47.60 | 46.97 | 48.78 | 55.30 | 47.37 | 49.41 | 48.87 | 50.12 | 50.27 | 49.41 |
| COPOD | 45.51 | 43.85 | 55.37 | 50.76 | 50.99 | 48.32 | 54.48 | 47.18 | 51.75 | 49.80 |
| DeepSVDD | 48.79 | 46.12 | 49.64 | 50.87 | 51.27 | 48.31 | 52.12 | 50.55 | 50.22 | 49.77 |
| ECOD | 45.97 | 43.95 | 55.23 | 50.63 | 49.10 | 49.31 | 54.56 | 47.37 | 49.52 | 49.51 |
| INNE | 47.09 | 41.52 | 58.66 | 53.47 | 45.48 | 42.89 | 59.00 | 47.60 | 52.87 | 49.84 |
| AnoGAN | 47.39 | 52.70 | 47.23 | 50.48 | 57.89 | 47.59 | 49.42 | 46.73 | 48.82 | 49.81 |
| ERM | 55.37 | 45.96 | 47.05 | 50.42 | 51.17 | 48.36 | 47.08 | 55.49 | 50.18 | 50.12 |
| IRM | 53.71 | 47.71 | 47.23 | 50.19 | 51.90 | 48.76 | 50.25 | 53.45 | 43.56 | 49.64 |
| GNL | 49.29 | 41.46 | **77.32** | 65.75 | 48.81 | **62.32** | **71.97** | 61.11 | **71.67** | 61.08 |
| SND | **72.70** | **71.56** | 74.13 | **65.85** | **78.67** | 59.98 | 64.18 | **71.72** | 70.85 | **69.96** |

### 3.2 Results and Discussion

In this section, we evaluate and analyze the performance of our method compared to recent novelty detection baseline methods across the mentioned datasets. SND consistently demonstrates superior performance in the extensive experiments.

In Table 1, we compare the performance of various methods on novelty detection using the Multi-background MNIST dataset, focusing on AUROC scores. Our proposed method, SND, achieves the highest average AUROC of 82.27%, outperforming baseline methods like COPOD (65.17%) and SUOD (67.05%). ERM and IRM, two domain adaptation techniques followed by KDE for novelty detection, perform significantly lower with averages of 48.67% and 48.59%, respectively. Notably, SND excels in digits such as 0 (85.74%) and 1 (97.68%), demonstrating superior generalization across different digits. Due to the space limitations, we will not report the results of AUPRC, though our method still outperformed the baselines in terms of AUPRC.

**Table 3.** Average AUPRC (%) on Kurcuma across seven target domains. Columns 0–8 correspond to the held-out novel class (9 categories in total).

| Method | 0 | 1 | 2 | 3 | 4 | 5 | 6 | 7 | 8 | Average |
|---|---|---|---|---|---|---|---|---|---|---|
| ALAD | 91.82 | 94.30 | 90.33 | 80.77 | 95.29 | 82.79 | 78.94 | 93.48 | 93.08 | 88.98 |
| COPOD | 91.31 | 94.45 | 91.74 | 79.00 | 94.96 | 82.72 | 81.30 | 93.22 | 93.24 | 89.10 |
| DeepSVDD | 92.14 | 94.15 | 90.22 | 79.04 | 95.52 | 83.07 | 81.11 | 93.72 | 92.96 | 89.10 |
| ECOD | 91.55 | 94.26 | 91.52 | 79.04 | 94.92 | 83.30 | 80.90 | 93.18 | 93.19 | 89.10 |
| INNE | 91.80 | 93.80 | 92.77 | 80.09 | 94.74 | 79.96 | 83.45 | 93.29 | 93.90 | 89.31 |
| AnoGAN | 92.36 | 95.32 | 89.04 | 79.28 | 96.49 | 82.30 | 80.20 | 93.05 | 93.29 | 89.04 |
| ERM | 93.41 | 94.52 | 89.41 | 79.38 | 95.86 | 82.08 | 77.95 | 94.62 | 92.93 | 88.91 |
| IRM | 93.45 | 94.79 | 88.99 | 79.95 | 95.81 | 83.22 | 77.84 | 94.42 | 92.96 | 89.05 |
| GNL | 91.88 | 93.97 | **96.47** | 85.65 | 95.08 | **88.77** | **89.14** | 95.23 | **96.52** | 92.52 |
| SND | **96.00** | **97.67** | 96.43 | **87.08** | **98.36** | 87.55 | 88.73 | **96.82** | 96.38 | **93.89** |

In Table 2, we show the AUROC results on the Kurcuma dataset, which includes seven distinct scenarios: SYNTHETIC, AKUD, CLIPART, EKUD, EKUD-M1, EKUD-M2, and EKUD-M3. Each scenario corresponds to a specific category: 0 for bottle opener, 1 for can opener, 2 for fork, 3 for knife, 4 for pizza cutter, 5 for spatula, 6 for spoon, 7 for tongs, and 8 for whisk. SND achieves the best average AUROC of 69.96%, outperforming other methods across most categories, including 0 (72.70%) and 4 (78.67%). GNL shows strong results in category 2 (77.32%) but falls short overall with an average of 61.08%. ERM and IRM trail behind with averages of 50.12% and 49.64%. Table 3 further shows that SND achieves the best average AUPRC.

Additionally, we use t-SNE [26] to qualitatively evaluate SND's feature representation under domain shift. As shown in Figs. 3a, 3b, and 3c, the visualizations confirm the model's ability to isolate subject features by forming distinct clusters for different domains, effectively separating subject from background and original image features.

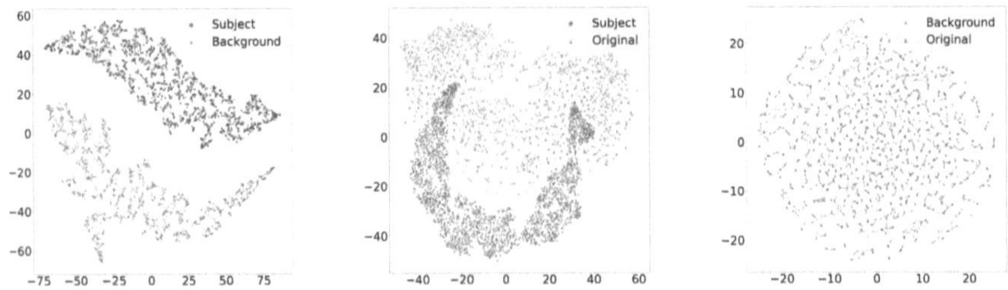


(a) Subject vs. Background (b) Subject vs. Original (c) Background vs. Original

**Fig. 3.** t-SNE visualization of the learned features: (a) Subject vs. Background, (b) Subject vs. Original, and (c) Background vs. Original. We chose the class of "0" in the Multi-background MNIST to provide the visualization result.

# 4 Conclusion

This paper proposed Subject-Novelty Detection (SND), which disentangles subject and background information and performs novelty detection using subject features only. By reducing the mutual information between subject and background variables and modeling background variation with a deep GMM, SND improves robustness when normal test data come from unseen domains.

**Acknowledgements.** The work was partially supported by the General Program of the Natural Science Foundation of Guangdong Province under Grant No.2024A1515011771, the Shenzhen Stability Science Program 2023, and the National Natural Science Foundation of China under Grant No.62376236.

**Disclosure of Interests.** The authors have no competing interests.

# References

1. Almardeny, Y., Boujnah, N., Cleary, F.: A novel outlier detection method for multivariate data. IEEE Trans. Knowl. Data Eng. **34**(9), 4052–4062 (2020)
2. Arjovsky, M., Bottou, L., Gulrajani, I., Lopez-Paz, D.: Invariant risk minimization (2019). arXiv:1907.02893 arXiv preprint
3. Bandaragoda, T.R., et al.: Isolation-based anomaly detection using nearest-neighbor ensembles. Comput. Intell. **34**(4), 968–998 (2018)
4. Cai, J., Fan, J.: Perturbation learning based anomaly detection. Adv. Neural. Inf. Process. Syst. **35**, 14317–14330 (2022)
5. Cai, J., Zhang, Y., Fan, J.: Self-discriminative modeling for anomalous graph detection. In: Forty-second International Conference on Machine Learning (2025)
6. Cao, T., Zhu, J., Pang, G.: Anomaly detection under distribution shift. In: Proceedings of the IEEE/CVF International Conference on Computer Vision, pp. 6511–6523 (2023)
7. Chandola, V., Banerjee, A., Kumar, V.: Anomaly detection: a survey. ACM Comput. Surv. (CSUR) **41**(3), 1–58 (2009)

8. Cheng, P., et al.: Club: a contrastive log-ratio upper bound of mutual information. In: International conference on machine learning, pp. 1779–1788 (2020) PMLR
9. Dai, W., Fan, J.: AutoUAD: hyper-parameter optimization for unsupervised anomaly detection. In: The Thirteenth International Conference on Learning Representations (2025)
10. Fu, D., Fan, J.: UniOD: a universal model for outlier detection across diverse domains. In: The Fourteenth International Conference on Learning Representations (2026)
11. Gao, D., et al.: Eraseanything: enabling concept erasure in rectified flow transformers. arXiv preprint arXiv:2412.20413 (2024)
12. Hodge, V., Austin, J.: A survey of outlier detection methodologies. Artif. Intell. Rev. **22**, 85–126 (2004)
13. Isermann, R.: Process fault detection based on modeling and estimation methods—a survey. Automatica **20**(4), 387–404 (1984)
14. Ke, Z., et al.: Early warning of cryptocurrency reversal risks via multi-source data. Finance Res. Lett. 107890 (2025)
15. Khan, S.S., Madden, M.G.: One-class classification: taxonomy of study and review of techniques. Knowl. Eng. Rev. **29**(3), 345–374 (2014)
16. LeCun, Y., Bottou, L., Bengio, Y., Haffner, P.: Gradient-based learning applied to document recognition. Proc. IEEE **86**(11), 2278–2324 (1998)
17. Li, Y., et al.: Multi-objective unlearning in recommender systems via preference guided pareto exploration. IEEE Trans. Serv. Comput. (2025)
18. Li, Z., Zhao, Y., Botta, N., Ionescu, C., Hu, X.: COPOD: Copula-Based Outlier Detection. In: 2020 IEEE International Conference on Data Mining, pp. 1118–1123
19. Li, Z., et al.: ECOD: unsupervised outlier detection using empirical cumulative distribution functions. IEEE Trans. Knowl. Data Eng. **35**(12), 12181–12193 (2022)
20. Li, Z., Ke, Z.: Cross-modal augmentation for low-resource language understanding and generation. In: Proceedings of the 1st Workshop on Multimodal Augmented Generation via Multimodal Retrieval (MAGMaR 2025), pp. 90–99 (2025)
21. Liu, F.T., Ting, K.M., Zhou, Z.H.: Isolation forest. In: eighth IEEE international conference on data mining, pp. 413–422. IEEE (2008)
22. Liu, Y., et al.: Generative adversarial active learning for unsupervised outlier detection. IEEE Trans. Knowl. Data Eng. **32**(8), 1517–1528 (2019)
23. Lu, S., Wang, Z., Li, L., Liu, Y., Kong, A.W.K.: Mace: mass concept erasure in diffusion models. In: Proceedings of the IEEE/CVF Conference on Computer Vision and Pattern Recognition, pp. 6430–6440 (2024)
24. Lyu, J., et al.: Vadmamba: exploring state space models for fast video anomaly detection. In: 2025 IEEE International Conference on Multimedia and Expo (ICME), pp. 1–6 (2025)
25. Lyu, J., et al.: Forward consistency learning with gated context aggregation for video anomaly detection (2026)
26. Van der Maaten, L., Hinton, G.: Visualizing data using t-SNE. J. Mach. Learn. Res. **9**(11) (2008)
27. Markou, M., Singh, S.: Novelty detection: a review–part 1: statistical approaches. Signal Process. **83**(12), 2481–2497 (2003)
28. Parzen, E.: On estimation of a probability density function and mode. Ann. Math. Stat. **33**(3), 1065–1076 (1962)
29. Qiu, S., Wang, H., Zhang, Y., Ke, Z., Li, Z.: Convex optimization of Markov decision processes based on z transform: a theoretical framework for two-space decomposition and linear programming reconstruction. Mathematics **13**(11), 1765 (2025)

30. Rosello, A., Valero-Mas, J.J., Gallego, A.J., Saez-Perez, J., Calvo-Zaragoza, J.: Kurcuma: a kitchen utensil recognition collection for unsupervised domain adaptation. Pattern Anal. Appl. **26**(4), 1557–1569 (2023)
31. Ruff, L., et al.: Deep one-class classification. In: International Conference on Machine Learning, pp. 4393–4402. PMLR (2018)
32. Schlegl, T., Seeböck, P., Waldstein, S.M., Schmidt-Erfurth, U., Langs, G.: Unsupervised anomaly detection with generative adversarial networks to guide marker discovery. In: International Conference on Information Processing in Medical Imaging, pp. 146–157. Springer (2017)
33. Schölkopf, B., Williamson, R.C., Smola, A., Shawe-Taylor, J., Platt, J.: Support vector method for novelty detection. Adv. Neural. Inf. Process. Syst. **12** (1999)
34. Vapnik, V.: Principles of risk minimization for learning theory. Adv. Neural. Inf. Process. Syst. **4** (1991)
35. Wu, X., Lu, J., Fang, Z., Zhang, G.: Meta OOD learning for continuously adaptive OOD detection. In: Proceedings of the IEEE/CVF International Conference on Computer Vision, pp. 19353–19364 (2023)
36. Xiao, F., Fan, J.: Unsupervised anomaly detection in the presence of missing values. Adv. Neural. Inf. Process. Syst. **37**, 138130–138162 (2024)
37. Xiao, F., Wang, Y., Joe Qin, S., Fan, J.: Semi-Supervised anomaly detection using restricted distribution transformation. IEEE Trans. Neural Netw. Learn. Syst. **36**(10), 17966–17977 (2025)
38. Xu, H., Pang, G., Wang, Y., Wang, Y.: Deep isolation forest for anomaly detection. IEEE Trans. Knowl. Data Eng. **35**(12), 12591–12604 (2023)
39. Zenati, H., Romain, M., Foo, C.S., Lecouat, B., Chandrasekhar, V.: Adversarially learned anomaly detection. In: 2018 IEEE International Conference on Data Mining (ICDM), pp. 727–736. IEEE (2018)
40. Zeng, S., et al.: Futuresightdrive: thinking visually with spatio-temporal cot for autonomous driving (2025). arXiv:2505.17685 arXiv preprint
41. Zeng, S., et al.: JanusVLN: decoupling semantics and spatiality with dual implicit memory for vision-language navigation (2025). arXiv:2509.22548 arXiv preprint
42. Zhang, Y., Sun, Y., Cai, J., Fan, J.: Deep orthogonal hypersphere compression for anomaly detection. In: The Twelfth International Conference on Learning Representations (2024)
43. Zhao, Y., Hryniewicki, M.K.: XGBOD: improving supervised outlier detection with unsupervised representation learning. In: 2018 International Joint Conference on Neural Networks (IJCNN), pp. 1–8. IEEE (2018)
44. Zhao, Y., et al.: SUOD: accelerating large-scale unsupervised heterogeneous outlier detection. Proc. Mach. Learn. Syst. **3**, 463–478 (2021)
45. Zhao, Y., Nasrullah, Z., Li, Z.: PyOD: a python toolbox for scalable outlier detection. J. Mach. Learn. Res. **20**(96), 1–7 (2019)
46. Zong, B., et al.: Deep autoencoding gaussian mixture model for unsupervised anomaly detection. In: International Conference on Learning Representations (2018)

# A VMD-Prior-Guided Adaptive Learning Method for Multi-step Industrial Time Series Forecasting

Zhenya Chen[1,2,3], Yameng Zhang[1,2,3], Ming Yang[1,2,3], Xuguo Jiao[3,4(✉)], Xiaoming Wu[1,2,3], and Xin Wang[1,2,3]

[1] Key Laboratory of Computing Power Network and Information Security, Ministry of Education, Shandong Computer Science Center, Qilu University of Technology (Shandong Academy of Sciences), Jinan 250014, China

[2] Shandong Provincial Key Laboratory of Industrial Network and Information System Security, Shandong Fundamental Research Center for Computer Science, Jinan 250014, China

[3] Quan Cheng Laboratory, Jinan 250014, China

[4] School of Information and Control Engineering, Qingdao University of Technology, Qingdao 266520, China

xuguojiao@gmail.com

**Abstract.** Multi-step industrial time series forecasting is crucial for production scheduling and predictive maintenance. However, industrial data often contain noise, nonstationarity, and multiscale patterns, which make accurate prediction more difficult. Although signal decomposition is an effective approach to address this issue, its practical application faces a dilemma: offline decomposition may cause data leakage and violate deployment constraints, while online rolling decomposition is computationally expensive and prone to boundary effects, reducing predictive accuracy. To address these challenges, this paper proposes a Variational Mode Decomposition (VMD) prior-guided adaptive prediction network (VPALNet). The model uses VMD to construct modal anchors as physical supervision signals. Combined with frequency alignment loss and time-frequency orthogonality constraints, it guides the network to learn implicit mode decomposition. Based on this, the model first uses a shared encoder to integrate window-level statistical features. Then, multiple prediction heads capture the dynamic characteristics of different modes. Finally, a gating network adaptively fuses the predictions from all modes. This approach enables efficient, reliable industrial time series forecasting without online decomposition. Experiments on three real-world industrial datasets across multiple forecasting horizons show that VPALNet achieves superior forecasting accuracy in most cases, reducing MSE by up to 22.6% compared to state-of-the-art baselines.

**Keywords:** Multi-step forecasting · Variational Mode Decomposition · Modal prior · Adaptive fusion · Frequency alignment


R. C.-W. Wong et al. (Eds.): PAKDD 2026, LNAI 16597, pp. 199–211, 2026.
https://doi.org/10.1007/978-981-92-1300-9_16

# 1 Introduction

Multi-step industrial time series forecasting plays a key role in modern intelligent systems. It is widely used in many areas, including power load scheduling [15], energy optimization [19], and traffic flow management [20]. Achieving high forecasting accuracy is crucial for improving energy efficiency and ensuring operational safety. However, due to complex working conditions and external disturbances, real-world industrial data often exhibit strong nonstationarity and multiscale characteristics. These complex patterns pose significant challenges for reliable multi-step forecasting [12]. Furthermore, prediction errors tend to accumulate across horizons, hindering the capture of long-term dependencies and leading to degraded stability in extended forecasts.

In recent years, numerous approaches have been applied to multi-step industrial time series forecasting. Early statistical models, such as autoregressive moving-average (ARMA) [9] and its extensions (e.g., autoregressive integrated moving-average (ARIMA) and seasonal ARIMA (SARIMA)), offer high computational efficiency and strong interpretability, and can address non-stationarity through differencing operations. However, these linear models still have limitations in modeling the nonlinear dynamics, multiscale characteristics, and complex temporal dependencies commonly observed in industrial signals. With the rapid advancement of deep learning, neural networkbased models—including long short-term memory (LSTM) networks [6], convolutional neural networks (CNNs) [13], and multilayer perceptrons (MLPs) [2] have demonstrated strong capability in capturing temporal dependencies and nonlinear mappings.

However, existing models still face challenges in capturing long-term dependencies and non-stationary characteristics in multi-step forecasting. To address this, hybrid methods combining signal decomposition with deep learning have been developed to separate multimodal features and better capture complex temporal patterns. Common decomposition techniques include seasonal decomposition [1], empirical mode decomposition (EMD) [3], wavelet transform (WT) [10], and variational mode decomposition (VMD) [17]. Among them, VMD shows superior adaptability in isolating stable and independent modal components in the frequency domain, making it well-suited for handling industrial signals characterized by operational fluctuations and noise interference. For example, Ma et al. [8] integrated VMD with change-point detection for load forecasting, effectively distinguishing different operating states. Similarly, Li et al. [7] combined VMD with RNNs and attention mechanisms for wind speed prediction, significantly improving the robustness of industrial forecasting models. However, decomposition-based methods have limitations: pre-splitting decomposition can cause future information leakage, while online rolling decomposition, although avoiding leakage, introduces boundary effects that distort modal components and ultimately degrade prediction accuracy.

This dilemma largely stems from the fact that existing approaches treat decomposition as a preprocessing step for prediction rather than as an intrinsic capability that can be learned by the model itself. To systematically investigate this issue, this study proposes the following research question (RQ): Can modal prior information extracted from offline decomposition guide neural networks to implicitly learn the decomposition structure, thereby effectively eliminating data leakage and boundary effects while maintaining predictive accuracy?

To answer the above research question, this paper proposes a Variational Mode Decomposition (VMD)- prior-guided adaptive learning network (VPALNet) for multi-step industrial time series forecasting. During training, VMD is applied to extended sequences for offline decomposition to extract the spectral centroid and energy ratio of each mode as modal anchors, guiding the model to learn implicit mode decomposition via a shared encoder and multiple prediction heads. Window-level statistical features enhance dynamic perception, while frequency alignment and timeâĂŞfrequency orthogonality constraints ensure efficient prediction without online decomposition. Finally, a gating network adaptively fuses modal outputs to generate the final forecast. Experimental results on multiple industrial datasets demonstrate that, under identical experimental settings, VPALNet outperforms representative baseline forecasting methods in multi-step prediction tasks. Furthermore, by eliminating the need for online decomposition, the proposed model achieves higher computational efficiency during inference.

The main contributions of this paper can be summarized as follows:

- We propose a VMD prior-guided learning framework that reformulates signal decomposition as an implicit representation learning process, eliminating the reliance on online decomposition while mitigating data leakage and boundary effects.
- We introduce a modal learning scheme with frequency alignment and orthogonality constraints, enhancing spectral consistency, modal complementarity, and structured representation of decomposed components.
- We develop an adaptive multi-head forecasting architecture with gate mechanism fusion to enable structured integration of modal predictions for accurate and efficient multi-step forecasting.

## 2 Problem Statement

In multi-step industrial time series forecasting, decomposition-based methods are widely used to decompose non-stationary signals into multiple intrinsic modal components, enabling multi-scale feature extraction and improving prediction accuracy. However, similar to many decomposition algorithms, VMD is also prone to data leakage and boundary effects. For example, in VMD, existing studies typically decompose the entire sequence before splitting it into training and test sets, which introduces data leakage. In real-world online forecasting, however, future data are unavailable, and the model can only decompose local sequences within a sliding window. Because the window is relatively short, almost every data point is close to a boundary. As a result, boundary effects can contaminate the decomposition results across the entire window, leading to severe distortion of the decomposed components.

As illustrated in Fig. 1, there exist significant differences between global and local VMD decompositions. Figure 1(a) shows the original power load series and its selected local windows. Figure 1(b)-(d) compares three modal components under both decompositions, while Fig. 1(e)-(g) provides a magnified view, offering a clearer illustration of the mode distortions caused by boundary effects. Figure 1(h)-(j) further verifies the frequency centroid shift and energy redistribution caused by local decomposition. The results indicate that, compared with global decomposition, local decomposition introduces instabilities such as spectral drift, energy perturbations, and mode distortion, making it difficult to maintain consistent modal characteristics during training and inference, thereby reducing the model's generalization and forecasting performance. Although iterative online decomposition can preserve global consistency, it is computationally expensive and unsuitable for real-time forecasting. Therefore, VMD's reliance on full-sequence decomposition fundamentally limits its applicability to real-time forecasting tasks.

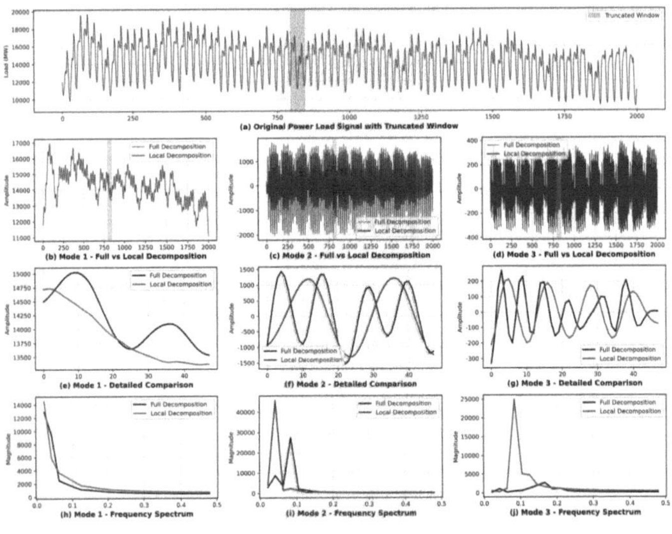


**Fig. 1.** Comparison of full-sequence and local VMD decomposition.

To address this issue, this paper proposes a Variational Mode Decomposition Prior-Guided Adaptive Learning Network (VPALNet). During training, modal information is extracted offline via VMD and used as physical prior supervision to guide the network's learning of implicit mode decomposition, enabling efficient multi-step forecasting without online VMD.

## 3 Methodology

This section elaborates on the proposed VPALNet model. As illustrated in Fig. 2, VPALNet is an end-to-end forecasting framework whose core innovation is to embed physical prior knowledge from VMD into the neural network, thereby guiding the model to implicitly learn signal decomposition capabilities. This approach effectively circumvents the data leakage and online application bottlenecks inherent in the traditional "decomposition-prediction-reconstruction" paradigm, aiming to comprehensively enhance the accuracy and stability of multi-step forecasting for univariate non-stationary time series. VPALNet primarily consists of three core components: (1) Offline VMD Decomposition and Modal Anchor Construction; (2) VPALNet Network Architecture; and (3) Composite Loss Function. The details of each essential module are explained in the following subsections.

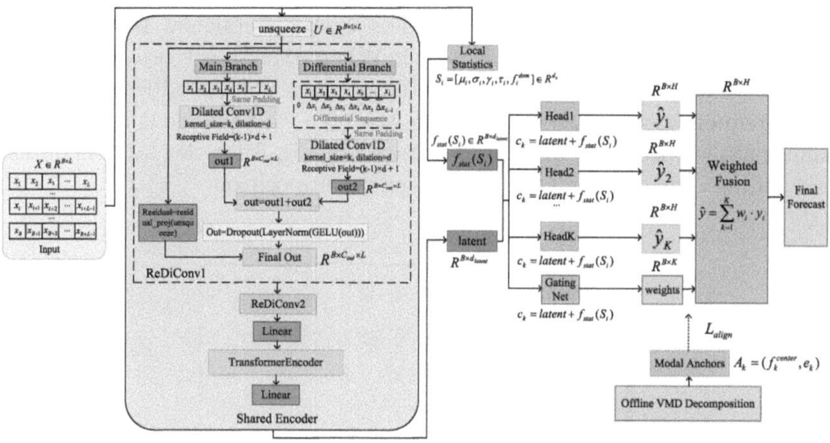

**Fig. 2.** Overall architecture of the proposed VPALNet model.

### 3.1 Offline VMD Decomposition and Modal Anchor Construction

**Data Preprocessing.** This work focuses on the multi-step forecasting task for univariate time series. Given historical observations $\{x_1, x_2, ..., x_L\}$, the model aims to predict its future $H$ steps $\{x_{L+1}, x_{L+2}, ..., x_{L+H}\}$.

First, data cleaning and missing value imputation are performed on the raw sequence to obtain a continuous series $\{f_t\}_{t=1}^{N}$. Subsequently, the data is partitioned into training set $D_{train} = f(1 : T_{train})$, validation set $D_{val} = f(T_{train} - L + 1 : T_{train} + T_{val})$, and test set $D_{test} = f(T_{train} + T_{val} - L + 1 : N)$. To maintain temporal continuity, overlapping windows of length $L$ (the input sequence length) are used during partitioning. The series is then standardized to obtain $f'_t$.

**Offline VMD and Modal Anchor Construction.** To embed physical priors, the training and validation sets are concatenated into an extended sequence

$s(t) = f'(1 : T_{train} + T_{val})$, which is decomposed offline using VMD. VMD adaptively separates $s(t)$ into $K$ band-limited intrinsic mode functions (IMFs) $\{u_k(t)\}_{k=1}^{K}$, each concentrated around a center frequency $\omega_k$. The decomposition is obtained by solving:

$$\min_{\{u_k\},\{\omega_k\}} \left\{ \sum_{k=1}^{K} \left\| \partial_t \left[ \left( \delta(t) + \frac{j}{\pi t} \right) * u_k(t) \right] e^{-j\omega_k t} \right\|_2^2 \right\}, \quad \sum_{k=1}^{K} u_k = s(t) \tag{1}$$

The VMD hyperparameters are set as follows: penalty factor $\alpha = 2000$, and convergence tolerance $\epsilon = 10^{-7}$. To determine the optimal number of modes $K$, we propose a criterion based on the variation of the center frequency: the difference of the last mode's center frequency is computed as $\Delta\omega_{last}(K) = \|\omega_{last}(K) - \omega_{last}(K-1)\|$. When $\Delta\omega_{last}(K)$ converges close to zero, it indicates that adding the $K$-th mode does not introduce any new significant frequency component. Therefore, the $(K-1)$-th mode is considered the optimal number of modes $K^*$ (as illustrated in Fig. 3 for the ISONE dataset), thus avoiding over-decomposition or under-decomposition.

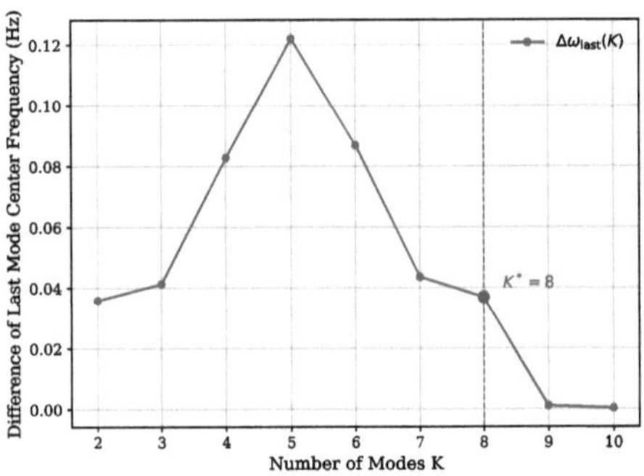


**Fig. 3.** Criterion for selecting the optimal number of VMD modes.

Modal components aligned with the original training set timeframe are extracted from the VMD results to mitigate boundary effects. Subsequently, a two-dimensional **Modal Anchor** $A_k$ is constructed for each mode $k$:

$$A_k = (f_k^{center}, e_k) \tag{2}$$

Here, the Spectral Centroid $f_k^{center}$ describes the dominant frequency component of the mode, and the Energy Ratio $e_k$ reflects the relative contribution of the mode to the original signal:

$$f_k^{center} = \frac{\sum_v v \cdot |\hat{u}_k(v)|^2}{\sum_v |\hat{u}_k(v)|^2}, \quad e_k = \frac{\|u_k\|_2^2}{\sum_{j=1}^{K} \|u_j\|_2^2}, \tag{3}$$

where $\hat{u}_k(v)$ is the Fourier transform of $u_k(t)$, $v$ is the normalized frequency index, and $\|u_k\|_2^2$ denotes the total energy of the $k$-th mode. The set of modal anchors $\mathcal{A} = \{A_1, A_2, ..., A_K\}$ serves as a fixed physical supervision signal to guide the neural network during training.

### 3.2 VPALNet Architecture

VPALNet is an end-to-end neural network for multi-step forecasting. A shared encoder first extracts latent representations from the input sequence to capture global context. These features are then fused with local statistical features to drive $K$ parallel prediction heads, where $K$ is determined by offline VMD decomposition. Each head models the dynamics of a specific mode. Finally, a gated fusion network adaptively integrates all head outputs to produce the final $H$-step forecast. The framework includes three main components: a Shared Encoder, a Multi-Head Prediction Module, and a Gated Fusion Network. Given a normalized input $X_i \in \mathbb{R}^L$ and its local statistics $S_i \in \mathbb{R}^{d_s}$, the model outputs $\hat{\mathbf{y}} \in \mathbb{R}^H$.

**Shared Encoder.** The shared encoder is responsible for distilling an information-rich latent representation from the input sequence. It consists of a novel *Residual Differential Convolutional Unit (ReDiConv)* followed by a compact Transformer Encoder.

The ReDiConv unit is designed to capture both local patterns and instantaneous change trends simultaneously. Given an input $X \in \mathbb{R}^{B \times L}$, two parallel branches process the sequence as follows.

The *main branch* applies 1D dilated convolution to extract macro patterns and periodicity over an expanded receptive field:

$$\mathbf{y}^{(main)} = \text{Conv1D}_{dilated}(X; \mathbf{W}_{main}, \mathbf{b}_{main}, d), \tag{4}$$

where $d$ denotes the dilation factor, and $\mathbf{W}_{main}$, $\mathbf{b}_{main}$ are learnable parameters.

The *differential branch* emphasizes short-term fluctuations by computing first-order differences and convolving them in the same manner:

$$\begin{aligned} \Delta X &= [0, X_{:,2} - X_{:,1}, \ldots, X_{:,L} - X_{:,L-1}], \\ \mathbf{y}^{(diff)} &= \text{Conv1D}_{dilated}(\Delta X; \mathbf{W}_{diff}, \mathbf{b}_{diff}, d_{diff}), \end{aligned} \tag{5}$$

where $d_{diff}$, $\mathbf{W}_{diff}$, and $\mathbf{b}_{diff}$ correspond to the dilation factor and parameters of the differential branch.

The outputs of both branches are fused through element-wise summation, followed by GELU activation and Layer Normalization:

$$\mathbf{y} = \text{GELU}(\text{LayerNorm}(\mathbf{y}^{(main)} + \mathbf{y}^{(diff)})). \tag{6}$$

A residual shortcut preserves the original information, producing the ReDiConv output:

$$Z = \text{Dropout}(\mathbf{y}) + \text{Proj}(X), \tag{7}$$

where Proj(·) is a $1 \times 1$ convolution for potential channel alignment. Two ReDi-Conv layers are stacked for multi-scale feature extraction.

Finally, a lightweight Transformer encoder captures global dependencies, and its last hidden state is projected into a latent variable **latent** $\in \mathbb{R}^{B \times d_{latent}}$ shared across prediction heads.

**Multi-head Prediction and Gated Fusion.** The prediction module consists of $K$ parallel heads $\{f_k\}_{k=1}^{K}$, each corresponding to a VMD-derived mode. For each head, the input is conditioned by local statistics through a mapping function $f_{stat}$:

$$\mathbf{c}_k = \mathbf{latent} + f_{stat}(S_i), \tag{8}$$

enabling adaptive alignment with short-term dynamics. Each prediction head $f_k$ is a sub-network composed of an LSTM layer followed by fully connected layers, which predicts its mode-specific output $\hat{\mathbf{y}}_k \in \mathbb{R}^{B \times H}$. As the prediction head, LSTM effectively models sequential dependencies and long-term temporal correlations, improving the accuracy of multi-step industrial load forecasts.

**Gated Fusion Network.** A lightweight gating network is employed to dynamically assess the importance of each modal prediction and adaptively fuse them:

$$\mathbf{g} = \text{Softmax}(\mathbf{W}_g \mathbf{c}_k + \mathbf{b}_g), \quad \mathbf{g} \in \mathbb{R}^{B \times K}, \tag{9}$$

$$\hat{\mathbf{y}} = \sum_{k=1}^{K} g_k \odot \hat{\mathbf{y}}_k, \quad \hat{\mathbf{y}} \in \mathbb{R}^{B \times H}. \tag{10}$$

where $\odot$ denotes element-wise multiplication (with broadcasting), $g_k \in \mathbb{R}^{B}$ is the fusion weight vector for the $k$-th mode, and $\hat{\mathbf{y}}_k \in \mathbb{R}^{B \times H}$ is the predicted output of the $k$-th mode. This mechanism adaptively weights each mode's contribution, enhancing both robustness in non-stationary environments.

### 3.3 Composite Loss Function

To simultaneously optimize prediction accuracy and enforce the learning of VMD-based physical priors, a composite loss function with three components is designed for end-to-end joint training:

$$\mathcal{L} = \mathcal{L}_{pred} + \lambda_{ortho}\mathcal{L}_{ortho} + \lambda_{align}\mathcal{L}_{align} \tag{11}$$

**Prediction Loss $\mathcal{L}_{pred}$.** The Mean Squared Error (MSE) is used to measure the deviation between the final fused prediction $\hat{\mathbf{y}}$ and the ground truth $\mathbf{Y}$:

$$\mathcal{L}_{pred} = \frac{1}{H}\sum_{j=1}^{H}(\hat{\mathbf{y}}_j - \mathbf{Y}_j)^2 \tag{12}$$

**Orthogonality Constraint Loss $\mathcal{L}_{ortho}$.** This term jointly enforces temporal and spectral independence among different predicted modes to reduce redundancy and enhance mode separability. For each sample $i$ and mode $k$, the demeaned modal sequence is defined as $\tilde{\mathbf{y}}_i^k = \hat{\mathbf{y}}_i^k - \frac{1}{H}\sum_{t=1}^{H}\hat{\mathbf{y}}_i^k(t)$. Its frequency-domain representation is obtained using the Real Fast Fourier Transform (RFFT): $F_i^k = \mathrm{RFFT}(\tilde{\mathbf{y}}_i^k)$. The orthogonality constraint combines both the time-domain and frequency-domain decorrelation terms as follows:

$$\mathcal{L}_{ortho} = \frac{1}{B}\sum_{i=1}^{B}\sum_{k=1}^{K}\sum_{\substack{k'=1\\k'\neq k}}^{K}\left(\left|\left\langle \tilde{\mathbf{y}}_i^k, \tilde{\mathbf{y}}_i^{k'}\right\rangle\right| + \left|\left\langle F_i^k, F_i^{k'}\right\rangle\right|\right) \tag{13}$$

where $\langle\cdot,\cdot\rangle$ denotes the inner product, $B$ is the batch size, and $K$ is the number of modes. This constraint encourages the model to learn orthogonal representations across both time and frequency domains, leading to physically meaningful and stable decompositions.

**Frequency Alignment Loss$\mathcal{L}_{align}$.** This is the key component for injecting VMD-based physical priors. For each predicted mode, the spectral centroid $f_k^{pred}$ is computed and matched to the pre-computed VMD modal anchors $\{f_m^{center}\}$ using the Hungarian algorithm to obtain the optimal one-to-one correspondence $\pi^*$. The loss minimizes the post-matching spectral discrepancy:

$$\mathcal{L}_{align} = \frac{1}{B}\sum_{i=1}^{B}\min_{\pi\in\Pi}\sum_{k=1}^{K}\left|f_k^{pred} - f_{\pi(k)}^{center}\right| \tag{14}$$

where $\pi$ is a bijective mapping and $\Pi$ denotes the set of all $K!$ possible permutations. This term ensures that the learned modal frequency structure remains consistent with the offline VMD decomposition results.

In summary, the proposed composite loss function incorporates classical signal processing theory as prior knowledge, enabling VPALNet to achieve stronger generalization without sacrificing prediction accuracy. The hyperparameters $\lambda_{ortho}$ and $\lambda_{align}$ balance the contributions of the respective loss terms.

# 4 Experiments

## 4.1 Datasets and Baseline Models

This study selects three publicly available datasets to demonstrate the generalization capability of the proposed model. The datasets include the New England Independent System Operator (ISO-NE), the Malaysian power system [4], and the Global Energy Forecasting Competition 2014 (GEFCom2014). All datasets are recorded at an hourly resolution and are split into training, validation, and test subsets with ratios of 70%, 20%, and 10%, respectively.

To benchmark the performance of VPALNet, we compare it with five state-of-the-art models: AMD [5], TimeMixer [14], iTransformer [11], DLinear [16], and TiDE [18]. All baseline models are trained and evaluated using the same data splits, input sequence lengths, and preprocessing strategies to ensure a fair comparison. Experiments run on NVIDIA RTX 3090 with PyTorch. We use Adam optimizer (lr=$5 \times 10^{-4}$, batch=128, epochs=100), input length $L = 48$, and forecast horizons $H \in \{12, 96, 192, 336\}$. Early stopping and standardized preprocessing are applied uniformly.

## 4.2 Results Analysis

Table 1 shows the performance comparison of the proposed VPALNet with five advanced baseline models across multiple datasets and forecasting horizons, with the best results bolded in black and the second-best results underlined.

**Table 1.** Forecasting performance comparison on multiple datasets.

| Dataset | Horizon | VPALNet (ours) | | | AMD | | | TimeMixer | | | iTransformer | | | DLinear | | | TiDE | | |
|---|---|---|---|---|---|---|---|---|---|---|---|---|---|---|---|---|---|---|---|
| | | MSE | MAE | $R^2$ | MSE | MAE | $R^2$ | MSE | MAE | $R^2$ | MSE | MAE | $R^2$ | MSE | MAE | $R^2$ | MSE | MAE | $R^2$ |
| **ISO-NW** | 12 | **0.0248** | **0.1096** | **0.9744** | 0.0301 | 0.1188 | 0.9690 | 0.0292 | 0.1217 | 0.9699 | 0.0302 | 0.1219 | 0.9689 | 0.0737 | 0.1895 | 0.9241 | 0.0857 | 0.2031 | 0.9117 |
| | 96 | **0.1460** | **0.2700** | **0.8496** | 0.1649 | 0.2880 | 0.8301 | 0.1782 | 0.3020 | 0.8164 | 0.1689 | 0.2927 | 0.8259 | 0.2274 | 0.3482 | 0.7657 | 0.2591 | 0.3686 | 0.7331 |
| | 192 | **0.1618** | **0.2914** | **0.8332** | 0.1665 | 0.2888 | 0.8284 | 0.1905 | 0.3084 | 0.8037 | 0.1802 | 0.3075 | 0.8143 | 0.2100 | 0.3335 | 0.7837 | 0.2155 | 0.3358 | 0.7780 |
| | 336 | **0.1767** | **0.3057** | **0.8180** | 0.1866 | 0.3020 | 0.8077 | 0.1935 | 0.3178 | 0.8006 | 0.2109 | 0.3272 | 0.7827 | 0.2373 | 0.3547 | 0.7554 | 0.2457 | 0.3611 | 0.7468 |
| **Malaysia** | 12 | **0.0606** | **0.1382** | **0.9359** | 0.0752 | 0.1456 | 0.9204 | 0.0624 | 0.1435 | 0.9339 | 0.0682 | 0.1427 | 0.9278 | 0.1433 | 0.2499 | 0.8483 | 0.1783 | 0.2559 | 0.8112 |
| | 96 | 0.1482 | 0.2762 | 0.8451 | 0.1548 | 0.2276 | 0.8361 | 0.1498 | 0.2444 | 0.8414 | **0.1438** | **0.2300** | **0.8477** | 0.2102 | 0.3526 | 0.7773 | 0.2450 | 0.3585 | 0.7405 |
| | 192 | **0.0997** | **0.1979** | **0.8944** | 0.1288 | 0.2106 | 0.8636 | 0.1318 | 0.2329 | 0.8604 | 0.1283 | 0.2190 | 0.8642 | 0.1590 | 0.2889 | 0.8316 | 0.1712 | 0.2913 | 0.8187 |
| | 336 | **0.1163** | **0.2181** | **0.8768** | 0.1405 | 0.2234 | 0.8512 | 0.1479 | 0.2504 | 0.8434 | 0.1396 | 0.2348 | 0.8521 | 0.1649 | 0.2952 | 0.8254 | 0.1789 | 0.3006 | 0.8105 |
| **GEFCom2014** | 12 | **0.0238** | **0.1128** | **0.9751** | 0.0252 | 0.1167 | 0.9737 | 0.0254 | 0.1183 | 0.9735 | 0.0249 | 0.1170 | 0.9740 | 0.0629 | 0.1796 | 0.9343 | 0.0728 | 0.1921 | 0.9239 |
| | 96 | **0.1012** | **0.2371** | **0.8942** | 0.1292 | 0.2595 | 0.8650 | 0.1357 | 0.2666 | 0.8582 | 0.1290 | 0.2623 | 0.8652 | 0.1819 | 0.3149 | 0.8098 | 0.2129 | 0.3422 | 0.7774 |
| | 192 | **0.1291** | **0.2625** | **0.8651** | 0.1307 | 0.2587 | 0.8633 | 0.1509 | 0.2860 | 0.8423 | 0.1360 | 0.2704 | 0.8578 | 0.1590 | 0.2953 | 0.8338 | 0.1679 | 0.3067 | 0.8245 |
| | 336 | 0.1375 | 0.2732 | 0.8562 | **0.1360** | **0.2648** | **0.8579** | 0.1504 | 0.2837 | 0.8428 | 0.1458 | 0.2814 | 0.8476 | 0.1727 | 0.3089 | 0.8195 | 0.1814 | 0.3203 | 0.8104 |

The results show that the proposed VPALNet consistently achieves the best overall performance across all experimental settings, outperforming other methods in terms of MSE, MAE, and $R^2$. These results indicate the effectiveness of the proposed framework. Specifically, on the ISO-NE dataset, VPALNet exhibits strong forecasting capability. For the 12-step horizon, VPALNet achieves an MSE of 0.0248, which is about 15.1% lower than the second-best model, TimeMixer (0.0292). As the horizon extends to 336 steps, the performance of all models decreases reasonably; however, VPALNet shows the highest robustness to error accumulation, maintaining an MSE of 0.1767, which is 5.3% lower than AMD (0.1866). This indicates that the VMD priorâĂŞguided implicit modal decomposition effectively captures stable temporal patterns and mitigates long-horizon error propagation. On the Malaysia and GEFCom2014 datasets, VPALNet also maintains stable, superior performance across various forecasting horizons, further demonstrating its strong generalization across different load scenarios and data distributions.

To further investigate the effectiveness of modal learning in VPALNet, Fig. 4 shows the correlation heatmap of the output sequences from different prediction heads. The correlation coefficients among the eight prediction heads are generally close to zero. Most coefficients have absolute values below 0.05. This indicates that the prediction heads have successfully learned highly independent temporal patterns. Notably, the maximum off-diagonal correlation coefficient is only −0.04, observed between Mode6. The minimum coefficient is −0.03, found between Mode2 and Mode6. This very low correlation level validates the effectiveness of the time-frequency orthogonality constraints. It ensures that different prediction heads capture distinct temporal features.

Figure 5 presents a detailed comparison of 12-step forecasting results on the ISO-NE dataset (time steps 85–135). The actual load values (black line with circle markers) show typical power consumption patterns. Among the compared methods, VPALNet (blue line) shows the closest alignment with the ground truth. It accurately captures both overall trends and local variations.

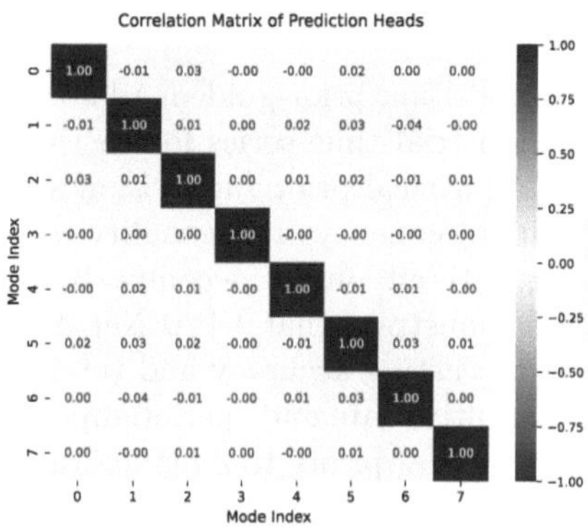


Fig. 4. Correlation Matrix.

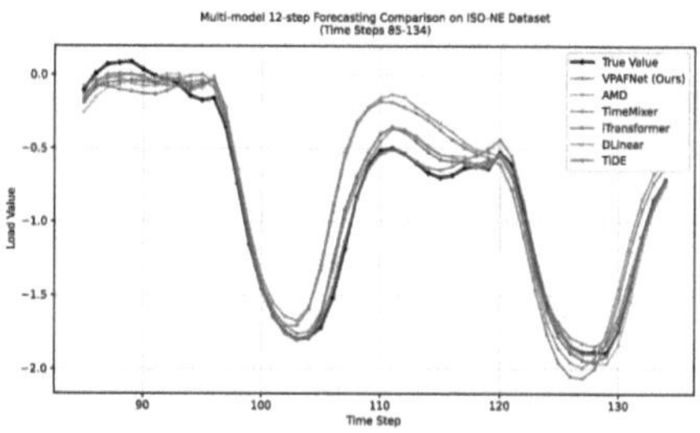


Fig. 5. Visual comparison of 12-step forecasting results on ISO-NE dataset (time steps 85–135).

### 4.3 Ablation Study

To validate the effectiveness of each component in VPALNet, we conduct a systematic ablation study on the ISO-NE dataset. Table 2 shows the results. The VMD modal anchor mechanism plays a critical role. Removing the time-frequency orthogonality loss causes significant performance degradation (MSE increases to 0.0514). Removing the frequency alignment loss also shows a clear performance drop (MSE=0.0283). This demonstrates that VMD anchors ensure frequency alignment with physical priors. They also promote modal diversity through orthogonality constraints.

Removing the gating fusion network results in a noticeable performance decrease (MSE=0.0280). This proves the value of adaptive weighting in multimodal fusion. Removing local statistical features also negatively affects performance (MSE = 0.0284). This confirms their contribution to local dynamic perception. All components positively enhance model performance. The VMD physical prior is most important. It effectively guides the network to learn physically meaningful modal decomposition.

**Table 2.** Ablation study results on ISO-NE dataset (Horizon=12)

| Model Variant | MSE | MAE | $R^2$ |
|---|---|---|---|
| **VPALNet (Full)** | **0.0248** | **0.1096** | **0.9744** |
| w/o Frequency Alignment | 0.0283 | 0.1180 | 0.9708 |
| w/o T-F Orthogonality | 0.0514 | 0.1643 | 0.9470 |
| w/o Gating Fusion | 0.0280 | 0.1162 | 0.9712 |
| w/o Local Features | 0.0284 | 0.1192 | 0.9708 |

## 5 Conclusion

This paper proposes a Variational Mode Decomposition prior-guided Adaptive Learning Network (VPALNet) for multi-step industrial time series forecasting. The network introduces VMD modal anchors as physical prior supervision signals. It employs frequency alignment loss and time-frequency orthogonality constraints. This enables implicit modal learning without online decomposition. Experiments across three real-world datasets demonstrate that VPALNet outperforms state-of-the-art methods in terms of forecasting accuracy and robustness. Future work will extend the framework to multivariate and spatiotemporal forecasting tasks. We will also explore adaptive modal priors to enhance their general applicability.

**Acknowledgmen.** This work was supported in part by the Key R&D Program of Shandong Province, China under Grant 2023CXGC011101, in part by the Research Project of Quancheng Laboratory, China under Grant QCL20250201 and Grant QCL20250302, in part by the Pilot Project for the Integration of Science, Education and Industry of Qilu University of Technology (Shandong Academy of Sciences) under Grant 2025ZDZX01, and in part by the Shandong Provincial Natural Science Foundation of China under Grant ZR2025MS1100.

**Disclosure of Interests.** The authors have no competing interests to declare that are relevant to the content of this article.

## References

1. Bandara, K., Hyndman, R.J., Bergmeir, C.: MSTL: a seasonal-trend decomposition algorithm for time series with multiple seasonal patterns. Int. J. Oper. Res. **52**(1), 79–98 (2025)
2. Challu, C., Olivares, K.G., Oreshkin, B.N., Ramirez, F.G., Canseco, M.M., Dubrawski, A.: NHITS: neural hierarchical interpolation for time series forecasting. In: Proceedings of the AAAI Conference on Artificial Intelligence, vol. 37, pp. 6989–6997 (2023)
3. Cheng, G., Wang, X., He, Y.: Remaining useful life and state of health prediction for lithium batteries based on empirical mode decomposition and a long and short memory neural network. Energy **232**, 121022 (2021)

4. Hasanat, S.M., Ullah, I., Aurangzeb, K., Rizwan, M., Alhussein, M., Anwar, M.S.: Hybrid long short-term memory and bidirectional multichannel network cascaded with split convolution for short-term load forecasting. Eng. Appl. Artif. Intell. **147**, 110268 (2025)
5. Hu, Y., Liu, P., Zhu, P., Cheng, D., Dai, T.: Adaptive multi-scale decomposition framework for time series forecasting. In: Proceedings of the AAAI Conference on Artificial Intelligence, vol. 39, pp. 17359–17367 (2025)
6. Kim, D., Seomun, G., Lee, Y., Cho, H., Chin, K., Kim, M.H.: Forecasting building energy demand and on-site power generation for residential buildings using long and short-term memory method with transfer learning. Appl. Energy **368**, 123500 (2024)
7. Li, S., et al.: Medium-term offshore wind speed multi-step forecasting based on VMD and GRU-MATNET model. Ocean Eng. **325**, 120737 (2025)
8. Ma, K., Nie, X., Yang, J., Zha, L., Li, G., Li, H.: A power load forecasting method in port based on VMD-ICSS-hybrid neural network. Appl. Energy **377**, 124246 (2025)
9. Ouyang, Y., Yin, H.: Multi-step time series forecasting with an ensemble of varied length mixture models. Int. J. Neural Syst. **28**(04), 1750053 (2018)
10. Parvini, N., Abdollahi, M., Seifollahi, S., Ahmadian, D.: Forecasting bitcoin returns with long short-term memory networks and wavelet decomposition: a comparison of several market determinants. Appl. Soft Comput. **121**, 108707 (2022)
11. Pei, J., Liu, N., Shi, J., Ding, Y.: Tackling the duck curve in renewable power system: a multi-task learning model with itransformer for net-load forecasting. Energy Convers. Manage. **326**, 119442 (2025)
12. Tian, G., Zhang, C., Shi, Y., Li, X.: Multiwavenet: a long time series forecasting framework based on multi-scale analysis and multi-channel feature fusion. Expert Syst. Appl. **251**, 124088 (2024)
13. Wang, H., Peng, J., Huang, F., Wang, J., Chen, J., Xiao, Y.: MICN: multi-scale local and global context modeling for long-term series forecasting. In: The Eleventh International Conference on Learning Representations (2023)
14. Wang, S., et al.: Timemixer: decomposable multiscale mixing for time series forecasting (2024). arXiv:2405.14616 arXiv preprint
15. Yang, F., Fu, X., Yang, Q., Chu, Z.: Decomposition strategy and attention-based long short-term memory network for multi-step ultra-short-term agricultural power load forecasting. Expert Syst. Appl. **238**, 122226 (2024)
16. Zeng, A., Chen, M., Zhang, L., Xu, Q.: Are transformers effective for time series forecasting? In: Proceedings of the AAAI Conference on Artificial Intelligence. vol. 37, pp. 11121–11128 (2023)
17. Zhang, Y., Zhao, Y., Kong, C., Chen, B.: A new prediction method based on VMD-PRBF-ARMA-e model considering wind speed characteristic. Energy Convers. Manage. **203**, 112254 (2020)
18. Zheng, H., Lu, Y., Sun, Z., Panneerselvam, J., Sun, X., Liu, L.: Energy optimisation in cloud datacentres with mc-tide: mixed channel time-series dense encoder for workload forecasting. Appl. Energy **374**, 123903 (2024)
19. Zhou, X., Zhai, N., Li, S., Shi, H.: Time series prediction method of industrial process with limited data based on transfer learning. IEEE Trans. Industr. Inf. **19**(5), 6872–6882 (2022)
20. Zou, G., Lai, Z., Wang, T., Liu, Z., Li, Y.: MT-STNET: a novel multi-task spatiotemporal network for highway traffic flow prediction. IEEE Trans. Intell. Transp. Syst. **25**(7), 8221–8236 (2024)

# SGTRec: Integrating Spectral Encoding with Graph Neural Networks and Transformers for Recommendation

Sichan Oh[1], Byungmoon Heo[1], Namjun Lee[1], Seonah Kim[1], Sejong Yoon[2], and Jaekwang Kim[1](✉)

[1] Sungkyunkwan University, Suwon, Republic of Korea
{ohsichan,ssiass,namejun12000,rlatjsdk2025,linux}@skku.edu
[2] The College of New Jersey, Ewing, NJ, USA
yoons@tcnj.edu

**Abstract.** Recommender systems are essential for mitigating information overload in e-commerce, streaming, and social platforms. Graph neural network (GNN) recommenders exploit user–item interaction topology but can suffer from oversmoothing and local bias as depth increases. Transformer-based models capture long-range dependencies with self-attention, yet may under-utilize graph structure and become costly at scale. We propose SGTRec, a hybrid framework that combines GNN message passing with Transformer attention and injects structural context via Spectral Positional Encodings (SPE). SPEs are derived from Laplacian eigenvectors through the Graph Fourier Transform and integrate low- and high-frequency components in a unified space. To scale spectral updates across layers, we approximate spectral filtering using Chebyshev polynomials, injecting locality without repeated eigendecomposition while preserving spectral characteristics for stable learning on large interaction graphs. Experiments on multiple public benchmarks show that SGTRec consistently improves accuracy and robustness over strong baselines.

**Keywords:** Recommender Systems · Spectral Encoding · Chebyshev Filtering

## 1 Introduction

Recommender systems play a pivotal role in alleviating information overload across various domains, including e-commerce, streaming services, and social platforms [14,21,22]. Traditional recommendation methods have primarily relied on Collaborative Filtering (CF), which generates recommendations by calculating user-user or item-item similarities based on historical interaction data [11,26,27]. However, these approaches suffer from inherent limitations such as data sparsity and the cold-start problem, which restrict their performance and scalability in large-scale real-world environments [12,15,25].


R. C.-W. Wong et al. (Eds.): PAKDD 2026, LNAI 16597, pp. 212–223, 2026.
https://doi.org/10.1007/978-981-92-1300-9_17

To overcome these limitations, Graph-based Recommender Systems have been proposed. This approach models users and items as nodes and interactions as edges in a graph, enabling the capture of not only direct relationships but also high-order connectivity [5,9,33]. By leveraging the graph structure, information can propagate between neighboring nodes, allowing the system to effectively learn complex relationships that go beyond simple similarity. Within this paradigm, Graph Neural Networks (GNNs) have significantly improved recommendation performance through message-passing mechanisms that aggregate information from neighbors. However, as GNN layers become deeper, the oversmoothing phenomenon [19] occurs, making node embeddings indistinguishable [1,18]. Moreover, GNNs are inherently biased toward local structures, limiting their ability to capture long-range dependencies across the graph [9,33,38].

To mitigate these shortcomings, Transformer architectures [30] have recently been introduced to recommender systems [11,29]. Transformers leverage the self-attention mechanism to assign weights to important interactions regardless of graph distance, thereby effectively modeling global dependencies. Nevertheless, Transformer based models often struggle to sufficiently encode fine-grained structural information inherent in graphs and incur substantial computational costs [23]. Consequently, recent research has explored hybrid models that combine the structural encoding ability of GNNs with the global contextual learning capability of Transformers [36,40]. Additionally, complementary learning paradigms such as Self-Supervised Learning and Contrastive Learning have been investigated to mitigate data sparsity and enhance representation robustness [20,35]. A key component of such hybrid GNN-Transformer models is Positional Encoding (PE) [24,31,40], which provides structural context to Transformer layers. Among various PE strategies, Graph Fourier Transform (GFT) based spectral encodings have been shown to be particularly effective [4,16,34]. GFT decomposes signals into frequency components via eigendecomposition of the graph Laplacian, where low-frequency components capture community level smooth patterns, while high-frequency components capture sharp variations such as niche preferences and community boundaries. Integrating both is essential for modeling global trends and fine grained variations.

It is important to note that while the overall graph structure remains fixed, the embeddings are continuously updated through GNN and Transformer layers [3,13]. Capturing the structural context for these updated embeddings is crucial. However, recomputing eigenvectors of the graph Laplacian each time embeddings are updated is computationally infeasible for large scale graphs. Accordingly, we retain the spectral PE for global structure while additionally using Chebyshev polynomial filtering [7,8] as a local PE: it enriches newly updated embeddings with neighborhood-level information and avoids repeated eigendecomposition by approximating spectral filtering.

In this study, we propose **SGTRec**, a hybrid framework that integrates GFT-based spectral positional encodings with Chebyshev polynomial filtering into a GNN-Transformer architecture. Specifically, (1) we design a dual-frequency spectral PE that jointly encodes low frequency and high frequency components, enabling the model to capture global trends and sharp local variations within a unified representation space. (2) We propose a block-wise structural bias update that efficiently approximates spectral filtering via Chebyshev polynomial approximation, which injects and updates $K$-hop neighborhood information for subsequent blocks without repeated eigendecomposition. (3) We develop a GNN-Transformer design that combines global self-attention with graph message passing through adaptive multi-hop aggregation to alleviate local bias and improve the robustness of representation learning.

## 2 Related Work

**Recommendation with Graph Neural Networks.** Graph Neural Networks have become a central paradigm for graph-based collaborative filtering by propagating signals on the user-item interaction graph and refining node representations via neighborhood aggregation. Early autoencoder-style methods such as GCMC [2] framed recommendation as graph-aware reconstruction. At industrial scale, PinSage [38] introduced importance-based neighbor sampling and random-walk proximity to preserve locality. To explicitly exploit high-order relations, NGCF [33] stacked multiple message-passing layers with feature transforms, whereas LightGCN [9] distilled the core effect of linear propagation with layer-wise fusion, removing nonlinearities to improve stability. Generalized collaborative filtering variants such as GCCF [6] revisited normalization and weighting strategies, and hypergraph extensions such as HyRec [32] and MHCN [39] modeled group-wise or multi-channel relations to capture semantics beyond pairwise edges. Self-supervised graph learning methods such as SGL [35] and SLRec [37] further improved robustness under sparsity via contrastive objectives on perturbed graph views. Despite these advances, deep GNN stacks may suffer from oversmoothing and remain biased toward local neighborhoods, limiting their ability to capture long-range dependencies and fine-grained structural signals.

**Recommendation with Transformers.** Transformer-based recommenders approach the interaction graph from a complementary angle to message passing, using self-attention to aggregate signals from distant nodes and relax locality constraints. Structure-aware variants inject topological bias into attention so that graph cues are preserved: SHT [36] leverages hypergraph context to model many-to-many relations, and Gformer [17] directly encodes graph topology to strengthen global context under sparse interactions. More recently, TransGNN [40] interleaves graph message passing with Transformer attention within a unified block. Such hybrids enlarge the effective receptive field while preserving locality, mitigating oversmoothing and improving long-range dependency modeling, and they have demonstrated strong performance on public graph recommendation benchmarks.

## 3 Method

This section introduces SGTRec as a single pipeline. We first provide an overview (Fig. 1). We then describe Spectral Positional Encoding, the SGTRec Block, Adaptive Multi-hop Aggregation, and the training loss.

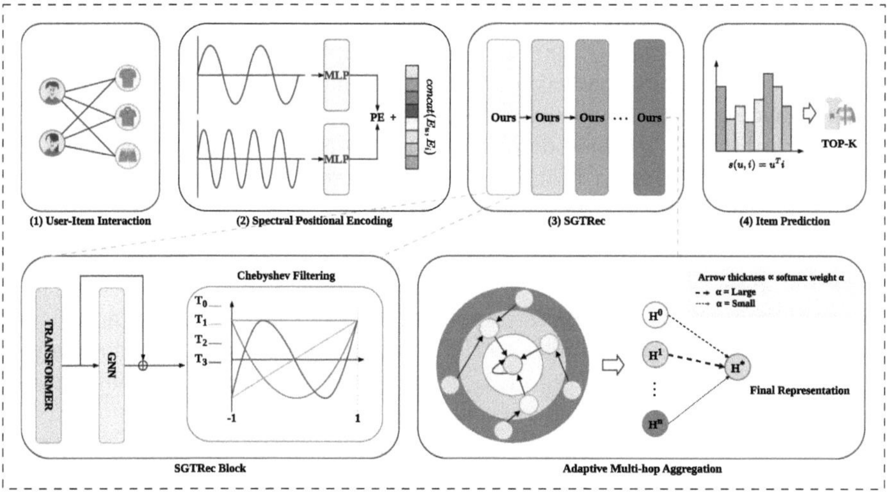


**Fig. 1.** Overview of the SGTRec framework.

### 3.1 User-Item Interaction Graph

Let $\mathcal{U}$ and $\mathcal{I}$ denote the user and item sets with $m = |\mathcal{U}|$ and $n = |\mathcal{I}|$, and let $N = m + n$. We model implicit feedback with a user-item matrix $R \in \mathbb{R}^{m \times n}$, where $R_{ui} = 1$ if user $u$ interacted with item $i$ and 0 otherwise. We then build a bipartite graph whose vertices are $\mathcal{U} \cup \mathcal{I}$ and whose edges $(u, i)$ exist exactly when $R_{ui} = 1$. From this bipartite structure, we construct an adjacency matrix $A \in \mathbb{R}^{N \times N}$ with nonzero entries for observed interactions, and a diagonal degree matrix $D$ storing node degrees.

To stabilize message passing and spectral operations, we use the symmetric normalized adjacency $\hat{A} = D^{-\frac{1}{2}} A D^{-\frac{1}{2}}$ and the associated normalized Laplacian $L = I - \hat{A}$. Intuitively, $\hat{A}$ controls diffusion on the user-item graph, while $L$ is the core operator for spectral positional encodings. Both are computed once from the fixed interaction graph and reused across the model.

### 3.2 Spectral Positional Encoding

We compute the normalized Laplacian and its eigendecomposition $L = U \Lambda U^\top$, where eigenvalues are sorted as $\lambda_1 \leq \lambda_2 \leq \cdots \leq \lambda_N$. We define the low- and high-frequency index sets as $\mathcal{I}_{\text{low}} = \{1, \ldots, d_{\text{low}}\}$ and $\mathcal{I}_{\text{high}} = \{N - d_{\text{high}} + 1, \ldots, N\}$, and select the corresponding eigenvector matrices as $U_{\text{low}} = U[:, \mathcal{I}_{\text{low}}] \in \mathbb{R}^{N \times d_{\text{low}}}$ and $U_{\text{high}} = U[:, \mathcal{I}_{\text{high}}] \in \mathbb{R}^{N \times d_{\text{high}}}$. We project them with a multi layer perceptron (MLP) and define the initial spectral positional encoding as

$$E_{\text{low}} = \text{MLP}(U_{\text{low}}), \quad E_{\text{high}} = \text{MLP}(U_{\text{high}}), \quad E^{(0)} = E_{\text{low}} + \beta^2 E_{\text{high}}, \tag{1}$$

where $\beta$ is a learnable scalar that controls the contribution of the high-frequency component, and we use $\beta^2$ to ensure a non-negative scaling factor.

**Chebyshev Polynomial Approximation.** Because embeddings are updated at every depth, repeatedly recomputing the Laplacian eigendecomposition to obtain spectral information becomes computationally inefficient and impractical for large-scale graphs. Following the standard Chebyshev polynomial approach for spectral filtering without decomposition, we constrain the filter to a truncated polynomial form so that it can be evaluated using only sparse matrix multiplications. We set $T_0(x) = 1$ and $T_1(x) = x$, and define

$$T_k(x) = 2x\,T_{k-1}(x) - T_{k-2}(x), \qquad \forall\, k \geq 2. \tag{2}$$

To apply them stably, the operator spectrum should lie in $[-1, 1]$. In our case, we apply Chebyshev polynomials to the symmetrically normalized adjacency $\hat{A}$, whose spectrum is bounded in $[-1, 1]$, so no additional rescaling is required. We then define a $K$-th order Chebyshev filter as a matrix polynomial

$$g_\Theta(\hat{A}) \;=\; \sum_{k=0}^{K} \theta_k\, T_k(\hat{A}), \tag{3}$$

where $K$ denotes the maximum polynomial order, and $\theta_k$ are learnable scalar coefficients that weight the contribution of each Chebyshev basis term. Here $T_k(\cdot)$ denotes the $k$-th Chebyshev polynomial applied to the graph operator as a matrix polynomial, so $T_k(\hat{A}) \in \mathbb{R}^{N \times N}$ for $\hat{A} \in \mathbb{R}^{N \times N}$. We apply the filter at block $\ell$ to obtain the per-block locality signal

$$\Delta^{(\ell)} \;=\; g_\Theta(\hat{A})\, H^{(\ell)} \;=\; \sum_{k=0}^{K} \theta_k\, T_k(\hat{A})\, H^{(\ell)}, \tag{4}$$

Each term $T_k(\hat{A})H^{(\ell)}$ is computed recursively using a small number of sparse matrix multiplications, which injects up to $K$-hop structural information without repeated eigendecomposition. Finally, we update the positional encoding by

$$E^{(\ell+1)} \;=\; E^{(\ell)} \;+\; \gamma\, \Delta^{(\ell)}, \tag{5}$$

where $\gamma$ is a learnable residual gate that controls the strength of the injected locality signal across depths.

### 3.3 SGTRec Block

At depths $\ell = 0, \ldots, L-1$, the model maintains node features $H^{(\ell)} \in \mathbb{R}^{N \times d}$ that are produced by the $\ell$-th block, and it initializes $H^{(0)}$ by concatenating learnable user embeddings and item embeddings. The first $m$ rows correspond to users and the last $n$ rows to items, so we write $H^{(\ell)} = [\, H^{(\ell)}_{\text{usr}};\, H^{(\ell)}_{\text{itm}}\,]$. A positional encoding $E^{(\ell)} \in \mathbb{R}^{N \times d}$ of the same shape is maintained and added residually at each depth to anchor structure, while the normalized adjacency $\hat{A} = D^{-\frac{1}{2}} A D^{-\frac{1}{2}}$ is computed once from the training graph and shared across all blocks. Each block takes $\big(H^{(\ell)}, E^{(\ell)}, \hat{A}\big)$ and outputs updated representations $\big(H^{(\ell+1)}, E^{(\ell+1)}\big)$ as follows.

**Residual PE Injection.** We anchor features to structure by a residual addition:

$$\widetilde{H}^{(\ell)} = H^{(\ell)} + E^{(\ell)}. \tag{6}$$

**Transformer Layer.** We partition the residually biased features into the user half and the item half. Each partition is passed through its dedicated Transformer encoder, and a skip connection from the corresponding input is added so the output preserves the original information. Denote the resulting user and item features by $\bar{H}^{(\ell)}_{\text{usr}}$ and $\bar{H}^{(\ell)}_{\text{itm}}$. Finally, we reassemble them in the original order to obtain the stacked representation $\breve{H}^{(\ell)}$.

**GNN Message Passing.** Local diffusion is performed by one-step propagation:

$$H^{(\ell+1)} = \hat{A}\,\breve{H}^{(\ell)}. \tag{7}$$

**Chebyshev-Based PE.** To adapt the positional signal to the updated representations without redoing eigendecomposition, we update the positional encoding as in Eq. 5. This preserves a global spectral anchor while injecting $K$-hop locality tied to the current state.

### 3.4 Adaptive Multi-hop Aggregation

Deeper propagation expands the receptive field but risks over-smoothing, while shallow features preserve locality yet miss long-range cues. To aggregate these complementary signals in a data-adaptive manner, we keep the intermediate representations $\{H^{(0)}, \dots, H^{(L)}\}$ produced by each block and learn hop importances with a temperature-stable simplex parameterization:

$$w \in \mathbb{R}^{L+1}, \qquad \alpha = \text{softmax}(w), \qquad \sum_{\ell=0}^{L} \alpha_\ell = 1. \tag{8}$$

The final representation is a convex fusion of multi-hop states,

$$H^* = \sum_{\ell=0}^{L} \alpha_\ell\, H^{(\ell)}. \tag{9}$$

This design yields an adaptive aggregation of shallow and deep receptive fields: (i) the softmax constraint prevents uncontrolled scaling and makes the mixture identifiable; (ii) gradients propagate through all hops so the model can raise or suppress specific depths according to supervision; and (iii) the fusion can be interpreted as a low-degree polynomial response over the propagation operator, selecting an effective receptive-field length from data. In practice, the hop weights are learned end-to-end jointly with all parameters and the fused $H^*$ is passed to the scoring head.

### 3.5 Training Loss

We adopt pairwise Bayesian Personalized Ranking (BPR) with inner-product scores $\hat{y}_{ui} = \langle P_u^*, Q_i^* \rangle$:

$$\mathcal{L}_{\text{BPR}} = -\,\mathbb{E}_{(u,i^+,i^-)} \log \sigma\big(\langle P_u^*, Q_{i+}^* \rangle - \langle P_u^*, Q_{i-}^* \rangle\big). \tag{10}$$

Here, $P_u^*$ is the $u$-th user vector; $Q_{i+}^*$ and $Q_{i-}^*$ are the positive and negative sampled item vectors, and $\sigma$ is the sigmoid function. Negatives are randomly sampled online from the entire item set. We optimize with Adam and rank items by top-$K$ scores at inference after masking training interactions.

## 4 Experiments

### 4.1 Experimental Settings

**Datasets.** We evaluate on three benchmarks spanning different domains and sparsity levels: *Yelp*, *Tmall*, and *Gowalla*. Following prior work [10,36], we binarize ratings/reviews into interactions, sort user logs chronologically, and filter users/items with insufficient activity.

**Evaluation.** We follow the all-rank evaluation protocol [9,35]. For each user, we score all candidate items, mask the user's training interactions, and report Recall@{20, 40} and NDCG@{20, 40}.

**Hyperparameters.** We set the embedding dimension to $d$=128 and use $L$=5 SGTRec blocks with 4 attention heads. For spectral positional encoding, we use 64-dimensional low- and high-frequency bands. We set the Chebyshev polynomial order to $K$=2. We perform a grid search over $L \in \{4, 5, 6\}$ and $K \in \{2, 3\}$. All models are optimized with Adam (learning rate $10^{-3}$) and a batch size of 4096.

**Table 1.** Performance comparison with baseline models. Best results are marked in bold, second best results underlined.

| Methods | Yelp | | | | Tmall | | | | Gowalla | | | |
|---|---|---|---|---|---|---|---|---|---|---|---|---|
| | R@20 | N@20 | R@40 | N@40 | R@20 | N@20 | R@40 | N@40 | R@20 | N@20 | R@40 | N@40 |
| AutoR [28] | 0.0348 | 0.0236 | 0.0515 | 0.0358 | 0.0235 | 0.0120 | 0.0405 | 0.0103 | 0.1298 | 0.0963 | 0.1359 | 0.1178 |
| GCMC [2] | 0.0462 | 0.0379 | 0.0836 | 0.0443 | 0.0292 | 0.0147 | 0.0511 | 0.0200 | 0.1395 | 0.1060 | 0.1783 | 0.1204 |
| PinSage [38] | 0.0471 | 0.0393 | 0.0895 | 0.0417 | 0.0373 | 0.0221 | 0.0594 | 0.0310 | 0.1380 | 0.1035 | 0.1852 | 0.1196 |
| NGCF [33] | 0.0579 | 0.0477 | 0.1131 | 0.0693 | 0.0402 | 0.0294 | 0.0613 | 0.0405 | 0.1570 | 0.1102 | 0.2270 | 0.1327 |
| LightGCN [9] | 0.0653 | 0.0532 | 0.1203 | 0.0828 | 0.0525 | 0.0372 | 0.0867 | 0.0508 | 0.1820 | 0.1270 | 0.2531 | <u>0.1547</u> |
| GCCF [6] | 0.0462 | 0.0398 | 0.0930 | 0.0704 | 0.0327 | 0.0257 | 0.0619 | 0.0402 | 0.1577 | 0.1150 | 0.2348 | 0.1285 |
| HyRec [32] | 0.0467 | 0.0355 | 0.0891 | 0.0628 | 0.0333 | 0.0262 | 0.0677 | 0.0428 | 0.1649 | 0.1162 | 0.2333 | 0.1452 |
| MHCN [39] | 0.0646 | 0.0410 | 0.1226 | 0.0810 | 0.0405 | 0.0337 | 0.0783 | 0.0473 | 0.1710 | 0.1256 | 0.2347 | 0.1510 |
| SLRec [37] | 0.0639 | 0.0358 | 0.1221 | 0.0806 | 0.0394 | 0.0354 | 0.0782 | 0.0413 | 0.1656 | 0.1213 | 0.2331 | 0.1473 |
| SGL [35] | 0.0675 | 0.0555 | 0.1269 | 0.0871 | 0.0436 | 0.0363 | 0.0802 | 0.0505 | 0.1709 | 0.1259 | 0.2502 | 0.1529 |
| SHT [36] | 0.0651 | 0.0546 | 0.1091 | 0.0709 | 0.0387 | 0.0262 | 0.0645 | 0.0352 | 0.1232 | 0.0881 | 0.1804 | 0.0731 |
| Gformer [17] | <u>0.0906</u> | <u>0.0781</u> | <u>0.1509</u> | <u>0.1007</u> | <u>0.0609</u> | <u>0.0431</u> | 0.0938 | 0.0545 | <u>0.1892</u> | 0.1168 | <u>0.2656</u> | 0.1380 |
| TransGNN [40] | 0.0817 | 0.0734 | 0.1506 | 0.0999 | 0.0608 | 0.0422 | <u>0.0989</u> | <u>0.0555</u> | 0.1887 | **0.1318** | 0.2640 | **0.1602** |
| SGTRec(Ours) | **0.0956** | **0.0821** | **0.1519** | **0.1015** | **0.0781** | **0.0551** | **0.1226** | **0.0708** | **0.2107** | <u>0.1279</u> | **0.2921** | 0.1484 |

### 4.2 Overall Performance Comparison

Across all three benchmarks and both ranking metrics, SGTRec consistently surpasses representative baselines from the CF, GNN, SSL-based hypergraph, and Transformer families (Table 1). Relative to classic CF methods, the performance gap is substantial, reflecting the benefit of modeling higher-order connectivity on user-item graphs. Moreover, compared with strong GNN baselines, SGTRec further improves ranking quality, suggesting that simply stacking message-passing layers is insufficient without incorporating global context and frequency aware signals.

Hypergraph and SSL-enhanced models alleviate sparsity and often narrow the gap with vanilla GNNs, yet SGTRec remains ahead, especially on datasets with complex communities or skewed popularity distributions. In such cases, our low-/high-frequency spectral PE together with block-wise Chebyshev-based PE updates captures both smooth trends and sharp boundaries. Compared with Transformer baselines that cover structure-aware attention and hybrid designs, SGTRec yields more stable gains across datasets, suggesting that our SGTRec block and adaptive hop fusion better balance locality preservation with expansion of the global receptive field.

### 4.3 Ablation Study

To quantify the contribution of each module in SGTRec, we perform controlled ablations under identical training/evaluation settings on Yelp. We retrain variants by toggling four primary switches and one architectural order switch, and report results in Table 2a: *w/o Spectral PE*, *w/o Chebyshev*, *w/o Hop-weight*, *w/o Update PE*, and *switch Layer*. The full model keeps all modules enabled and uses the **Transformer** → **GNN** order.

- **Spectral PE.** Removing eigenvector-based anchors (*w/o Spectral PE*) weakens the global coordinate frame for attention, degrading long-range alignment across the bipartite graph.
- **Chebyshev refinement.** Without the per-block local update (*w/o Chebyshev*), PE cannot adapt to evolving embeddings, leading to either overfitting to the initial spectrum or oversmoothing that erases high-frequency cues.
- **Multi-hop fusion.** Disabling learnable hop weighting (*w/o Hop-weight*) collapses the representation to a single effective depth, hurting users/items with heterogeneous locality needs.
- **Frozen PE.** Keeping PE fixed across blocks (*w/o Update PE*) makes structural signals stale relative to the feature space, weakening gradient guidance.
- **Block order.** The *switch Layer* variant underperforms the default order. Applying **Transformer** first aggregates global dependencies; the subsequent **GNN** diffusion then consolidates these signals locally, reducing oversmoothing. Reversing the order mixes neighbors too early, making later attention less discriminative.

**Table 2.** (a) Ablation studies on Yelp. (b) Complexity

(a)

| Variant | Recall@20 | NDCG@20 |
|---|---|---|
| w/o Spectral PE | 0.0912 | 0.0775 |
| w/o Chebyshev | 0.0921 | 0.0783 |
| w/o Hop-weight | 0.0373 | 0.0327 |
| w/o Update PE | 0.0912 | 0.0779 |
| switch Layer | 0.0911 | 0.0779 |
| SGTRec (Ours) | **0.0956** | **0.0821** |

(b)

| | SHT | GFormer | SGTRec |
|---|---|---|---|
| # Params | 1.75M | 1.74M | 7.38M |
| # GPU Mem | 2.33 | 11.00 | 2.97 |
| # FLOPs | 385.50 | 28,626.47 | 64,361.43 |
| Infer. Cost (s) | 12.59 | 17.25 | **5.15** |
| Recall@20 | 0.0651 | 0.0906 | **0.0956** |

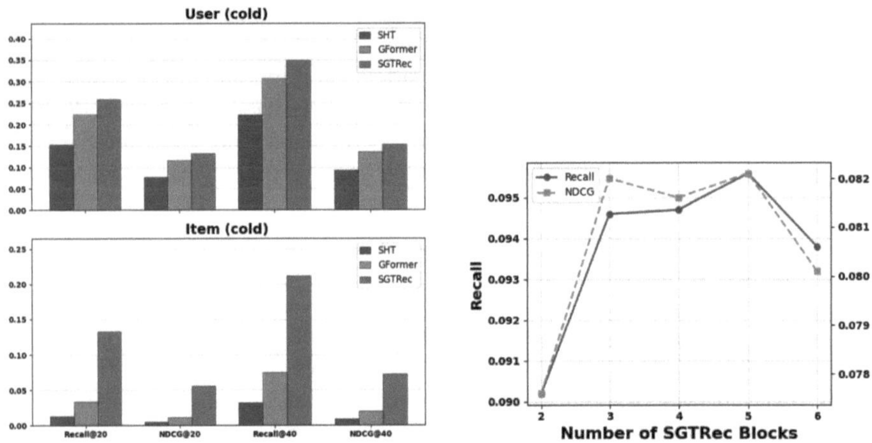


(a) (b)

**Fig. 2.** (a) Cold-start results on Gowalla. (b) Depth Sensitivity.

*Complexity.* (Table 2b). Although SGTRec uses more parameters, it is both faster and more memory-efficient than strong Transformer baselines. This stems from one-step propagation and Chebyshev filtering over sparse operators, and it also achieves the highest accuracy, yielding a favorable accuracy–efficiency trade-off.

*Cold-start Analysis.* (Figure 2a). We evaluate with *I-ALL-COLD*: include only test cases whose positives are all cold, and compute Recall/NDCG via full-catalog ranking with training interactions masked. On Gowalla, SGTRec leads on both user- and item-cold, with the largest margin on item that spectral PEs and per-block Chebyshev refinement transfer structure to sparse or unseen nodes.

*Depth Sensitivity.* Figure 2b shows that increasing the number of blocks improves performance up to a moderate depth, after which returns diminish. Unlike conventional GNNs that often degrade beyond 3 layers due to oversmoothing,

SGTRec sustains deeper stacks by preserving high-frequency cues with spectral PE and refreshing locality via Chebyshev updates.

## 5 Conclusion

SGTRec is a hybrid GNN–Transformer recommender that anchors attention with Laplacian-eigenvector spectral positional encodings and refreshes them via Chebyshev polynomial refinement to inject $K$-hop locality without re-eigendecomposition. Using a hybrid block and an adaptive multi-hop fusion, it balances global context with local propagation. Across public benchmarks, SGTRec consistently improves Recall and NDCG, while ablations verify the contribution of each component.

**Acknowledgments.** This research was supported by the Basic Science Research Program of the National Research Foundation (NRF) funded by the Korean government (MSIT) (No. IITP-2026-RS-2024-00346737); and by three additional grants from the Institute for Information & Communications Technology Planning & Evaluation (IITP) funded by the Ministry of Science and ICT (MSIT), Korea, through the Global Scholars Invitation Program (No. IITP-2026-RS-2024-00459638), the Graduate School of Metaverse Convergence at Sungkyunkwan University (No. IITP-2026-RS-2023-00254129), and the ICT Challenge and Advanced Network of HRD (ICAN) support program (No. IITP-2026-RS-2023-00259497).

**Disclosure of Interests.** The authors declare no competing interests.

## References

1. Alon, U., Yahav, E.: On the bottleneck of graph neural networks and its practical implications, arXiv preprint arXiv:2006.05205 (2020)
2. Berg, R.v.d., Kipf, T.N., Welling, M.: Graph convolutional matrix completion. arXiv preprint arXiv:1706.02263 (2017)
3. Bo, D., Shi, C., Wang, L., Liao, R.: Specformer: Spectral graph neural networks meet transformers, arXiv preprint arXiv:2303.01028
4. Bruna, J., Zaremba, W., Szlam, A., LeCun, Y.: Spectral networks and locally connected networks on graphs. arXiv preprint arXiv:1312.6203 (2013)
5. Cai, X., Huang, C., Xia, L., Ren, X.: Lightgcl: Simple yet effective graph contrastive learning for recommendation, arXiv preprint arXiv:2302.08191 (2023)
6. Chen, L., Wu, L., Hong, R., Zhang, K., Wang, M.: Revisiting graph based collaborative filtering: A linear residual graph convolutional network approach. In: Proceedings of the AAAI Conference on Artificial Intelligence, vol. 34, pp. 27–34 (2020)
7. Defferrard, M., Bresson, X., Vandergheynst, P.: Convolutional neural networks on graphs with fast localized spectral filtering. Adv. Neural. Inf. Process. Syst. **29** (2016)
8. He, M., Wei, Z., Wen, J.R.: Convolutional neural networks on graphs with chebyshev approximation, revisited. Adv. Neural. Inf. Process. Syst. **35**, 7264–7276 (2022)

9. He, X., Deng, K., Wang, X., Li, Y., Zhang, Y., Wang, M.: Lightgcn: Simplifying and powering graph convolution network for recommendation. In: Proceedings of the 43rd International ACM SIGIR Conference on Research and Development in Information Retrieval, pp. 639–648 (2020)
10. Huang, C., et al.: Knowledge-aware coupled graph neural network for social recommendation. In: Proceedings of the AAAI Conference on Artificial Intelligence, vol. 35, pp. 4115–4122 (2021)
11. Kang, W.C., McAuley, J.: Self-attentive sequential recommendation. In: 2018 IEEE International Conference on Data Mining (ICDM), pp. 197–206. IEEE (2018)
12. Kim, S., Kang, H., Choi, S., Kim, D., Yang, M., Park, C.: Large language models meet collaborative filtering: An efficient all-round llm-based recommender system. In: Proceedings of the 30th ACM SIGKDD Conference on Knowledge Discovery and Data Mining, pp. 1395–1406 (2024)
13. Kipf, T.N., Welling, M.: Semi-Supervised Classification with Graph Convolutional Networks, pp. 1680–1689. ICLR (2017)
14. Ko, H., Lee, S., Park, Y., Choi, A.: A survey of recommendation systems: recommendation models, techniques, and application fields. Electronics **11**(1), 141 (2022)
15. Koren, Y., Bell, R., Volinsky, C.: Matrix factorization techniques for recommender systems. Computer **42**(8), 30–37 (2009)
16. Kreuzer, D., Beaini, D., Hamilton, W., Létourneau, V., Tossou, P.: Rethinking graph transformers with spectral attention. Adv. Neural. Inf. Process. Syst. **34**, 21618–21629 (2021)
17. Li, C., Xia, L., Ren, X., Ye, Y., Xu, Y., Huang, C.: Graph transformer for recommendation. In: Proceedings of the 46th International ACM SIGIR Conference on Research and Development in Information Retrieval, pp. 1680–1689 (2023)
18. Li, G., Muller, M., Thabet, A., Ghanem, B.: Deepgcns: Can GCNs go as deep as CNNs? In: Proceedings of the IEEE/CVF International Conference on Computer Vision, pp. 9267–9276 (2019)
19. Li, Q., Han, Z., Wu, X.M.: Deeper insights into graph convolutional networks for semi-supervised learning. In: Proceedings of the AAAI Conference on Artificial Intelligence, vol. 32 (2018)
20. Liu, Y., Xia, L., Huang, C.: Selfgnn: Self-supervised graph neural networks for sequential recommendation. In: Proceedings of the 47th International ACM SIGIR Conference on Research and Development in Information Retrieval, pp. 1609–1618 (2024)
21. Lu, J., Wu, D., Mao, M., Wang, W., Zhang, G.: Recommender system application developments: a survey. Decis. Support Syst. **74**, 12–32 (2015)
22. Mahajan, K.C., Porobo Dharwadker, A., Gupta, S., Schumitsch, B.: Videorecsys 2023: First workshop on large-scale video recommender systems. In: Proceedings of the 17th ACM Conference on Recommender System, pp. 1269–1271 (2023)
23. Min, E., et al.: Transformer for graphs: An overview from architecture perspective. arXiv preprint arXiv:2202.08455 (2022)
24. Park, W., Chang, W., Lee, D., Kim, J., Hwang, S.W.: Grpe: Relative positional encoding for graph transformer, arXiv preprint arXiv:2201.12787 (2022)
25. Rendle, S.: Factorization machines. In: 2010 IEEE International Conference on Data Mining, pp. 995–1000. IEEE (2010)
26. Rendle, S., Freudenthaler, C., Schmidt-Thieme, L.: Factorizing personalized markov chains for next-basket recommendation. In: Proceedings of the 19th International Conference on World Wide Web, pp. 811–820 (2010)

27. Sarwar, B., Karypis, G., Konstan, J., Riedl, J.: Item-based collaborative filtering recommendation algorithms. In: Proceedings of the 10th International Conference on World Wide Web, pp. 285–295 (2001)
28. Sedhain, S., Menon, A.K., Sanner, S., Xie, L.: Autorec: Autoencoders meet collaborative filtering. In: Proceedings of the 24th International Conference on World Wide Web, pp. 111–112 (2015)
29. Sun, F., Liu, J., Wu, J., Pei, C., Lin, X., Ou, W., Jiang, P.: Bert4rec: Sequential recommendation with bidirectional encoder representations from transformer. In: Proceedings of the 28th ACM International Conference on Information and Knowledge Management, pp. 1441–1450 (2019)
30. Vaswani, A., et al.: Attention is all you need. Adv. Neural Inf. Process. Syst. **30** (2017)
31. Wang, H., Yin, H., Zhang, M., Li, P.: Equivariant and stable positional encoding for more powerful graph neural networks, arXiv preprint arXiv:2203.00199 (2022)
32. Wang, J., Ding, K., Hong, L., Liu, H., Caverlee, J.: Next-item recommendation with sequential hypergraphs. In: Proceedings of the 43rd International ACM SIGIR Conference on Research and Development in Information Retrieval, pp. 1101–1110 (2020)
33. Wang, X., He, X., Wang, M., Feng, F., Chua, T.S.: Neural graph collaborative filtering. In: Proceedings of the 42nd International ACM SIGIR Conference on Research and Development in Information Retrieval, pp. 165–174 (2019)
34. Wang, X., Zhang, M.: How powerful are spectral graph neural networks. In: International Conference on Machine Learning, pp. 23341–23362. PMLR (2022)
35. Wu, J., et al.: Self-supervised graph learning for recommendation. In: Proceedings of the 44th International ACM SIGIR Conference on Research and Development in Information Retrieval, pp. 726–735 (2021)
36. Xia, L., Huang, C., Zhang, C.: Self-supervised hypergraph transformer for recommender systems. In: Proceedings of the 28th ACM SIGKDD Conference on Knowledge Discovery and Data Mining, pp. 2100–2109 (2022)
37. Yao, T., et al.: Self-supervised learning for large-scale item recommendations. In: Proceedings of the 30th ACM International Conference on Information & Knowledge Management, pp. 4321–4330 (2021)
38. Ying, R., He, R., Chen, K., Eksombatchai, P., Hamilton, W.L., Leskovec, J.: Graph convolutional neural networks for web-scale recommender systems. In: Proceedings of the 24th ACM SIGKDD International Conference on Knowledge Discovery & Data Mining. pp. 974–983 (2018)
39. Yu, J., Yin, H., Li, J., Wang, Q., Hung, N.Q.V., Zhang, X.: Self-supervised multi-channel hypergraph convolutional network for social recommendation. In: Proceedings of the Web Conference 2021, pp. 413–424 (2021)
40. Zhang, P., Yan, Y., Zhang, X., Li, C., Wang, S., Huang, F., Kim, S.: Transgnn: Harnessing the collaborative power of transformers and graph neural networks for recommender systems. In: Proceedings of the 47th International ACM SIGIR Conference on Research and Development in Information Retrieval, pp. 1285–1295 (2024)

# PAED: Physics-Anchored and Emotion Disentangled Memory Diffusion for Talking Face Generation

Xingchu Zhang[1], Chuwen Zhang[2(✉)], Rui Mao[3], and Zehao Deng[4]

[1] Tianjin University, Tianjin, China
zhangxingchu@tju.edu.cn
[2] Stony Brook University, Stony Brook, NY, USA
chuwen.zhang@stonybrook.edu
[3] Central South University, Changsha, China
8212220705@csu.edu.cn
[4] Soochow University, Suzhou, China
2327406010@stu.suda.edu.cn

**Abstract.** The ability to generate emotionally expressive audio-based videos has seen significant advancements with end-to-end diffusion models, which have greatly improved video quality. However, these models still struggle with expression inconsistencies and unnatural transitions, particularly in long-duration sequences and complex emotional shifts. To address these limitations, we propose Physics-Anchored and Emotion Disentangled Memory Diffusion (PAED), a hybrid framework that enhances pre-trained memory architectures through two novel modules: (1) a Physics-Guided AU Momentum Regulator, which models facial dynamics as damped harmonic systems to enforce biomechanically valid AU transitions through inertial constraints, and (2) a Cross-Modal Emotion Disentangler, which employs contrastive momentum queues to align audio-visual emotion representations while minimizing mutual information between emotion and articulation features. Experiments show that our model reduces expression inconsistency compared to previous models, offering a more emotionally nuanced user experience. Access to the source code can be found at: https://github.com/YummyBamboo/PAED_MEMO.

**Keywords:** Diffusion model · Talking head · Facial expression · Action unit · Emotion disentangler

## 1 Introduction

Audio-driven video generation has shown great potential in areas such as digital human interaction and movie production, ensuring more natural interactive experiences by synchronizing audio signals with facial animations. In


R. C.-W. Wong et al. (Eds.): PAKDD 2026, LNAI 16597, pp. 224–236, 2026.
https://doi.org/10.1007/978-981-92-1300-9_18

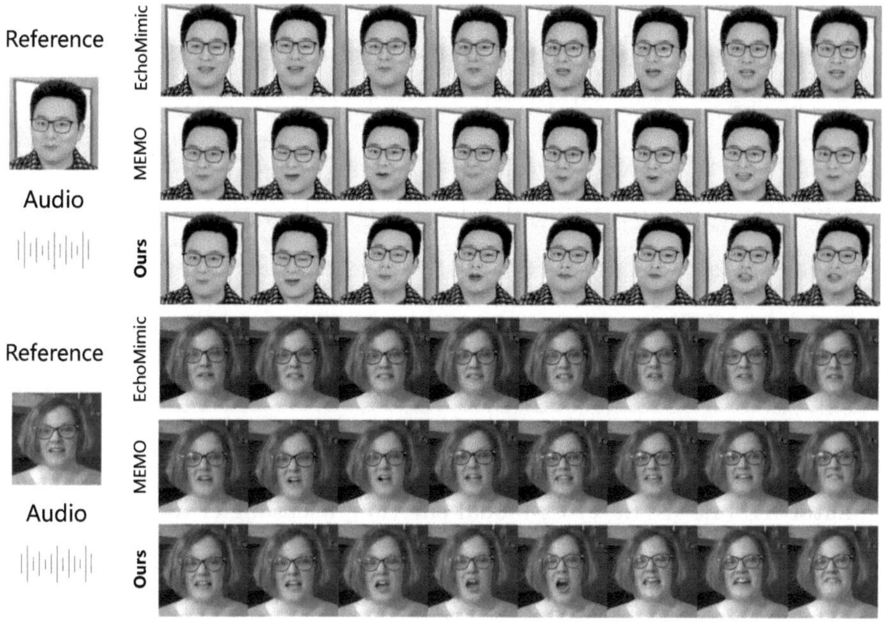


**Fig. 1.** Our PAED generates talking videos with improved emotional consistency and motion smoothness.

recent years, progress has been made in audio-driven video generation, especially in improving its quality. However, realizing long-term emotional consistency remains a challenge (Fig. 1).

Prior studies [1–6] have achieved notable progress in lip synchronization and identity consistency, yet remain limited by insufficient facial details and constrained emotional controllability. The introduction of diffusion models [10,11] has subsequently improved overall synthesis quality. However, these models still face challenges in preserving fine-grained emotional consistency across long duration. Expanding temporal inference windows persistently accumulate frame-level errors, and interframe inconsistencies remain unresolved. Despite advancements of the MEMO model [13] in these areas, it struggles to ensure stable emotion over extended durations. Additionally, current methods often entangle speech articulation with emotional expressions, limiting diversity and realism of facial expressions.

To address these limitations, we propose the Physically Anchored and Emotionally Disentangled Memory Diffusion (PAED) framework. The PAED enhances emotional expression and long-term coherence through two pivotal modules: (1) the Physical-guided Action Unit (AU) Momentum Regulator and (2) the Cross-modal Emotion Disentangler Module. To prevent unnatural motion transitions in long-term video synthesis, PAED introduces a physical-guided AU momentum regulator, which models facial dynamics as a damped harmonic oscillator system. Additionally, PAED incorporates a cross-modal emotion dis-

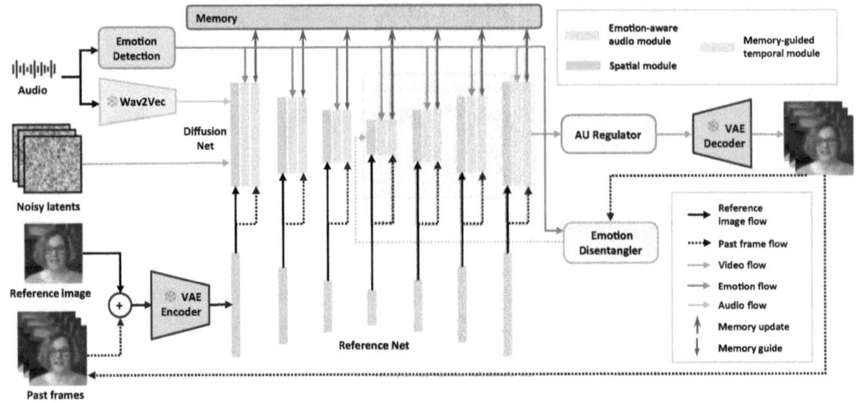


**Fig. 2.** PAED framework architecture extends conventional memory-based diffusion approaches through two innovative modules: the **physics-guided AU momentum regulator module** and the **cross-modal emotion disentangler module**. Through the synergistic operation of the two modules, PAED addresses limitations in existing approaches by ensuring smooth emotional transitions and consistent facial animations

entangler, which aligns audio and video emotion representations via contrastive momentum queues. It minimizes mutual information between audio features and facial expressions, thereby diversifying natural emotional outputs. In summary, our contributions include:

- We propose a novel Physics-Anchored and Emotion Disentangled Memory Diffusion (PAED) framework to improve emotional consistency in long-duration audio-driven video generation.
- We introduce the Physics-Guided AU Momentum Regulator to enforce biomechanically valid action unit transitions.
- We introduce the Cross-Modal Emotion Disentanglement technique to align the emotional representations of audio and video while minimizing their mutual information with articulation feature.

## 2 Related Work

Emotional consistency is crucial for audio-driven video generation, where most methods are based on GANs, VAEs, and diffusion models.

GAN-based models [1,2] excel in synthesizing high-fidelity talking faces. For example, Wav2Lip [1] achieves accurate lip synchronization, and Few-Shot Talking Heads [3] do a better job of identity consistency. However, due to unstable training, GANs can cause a loss of facial expression details, struggling with limited emotional control. VAE-based models such as EVP [4], PC-AVS [5] and MakeItTalk [6] also attempt to enhance coherence or personalized expressions, but their reliance on latent space representations results in blurred facial details and inadequate expressiveness.

Diffusion-based models, such as EMO [8], Hallo [9], DiffTalk [10], and DAE-Talker [11], have demonstrated superior performance by gradually denoising images, leading to greater emotional consistency and end-to-end synthesis. However, a persistent challenge lies in maintaining long-term emotional consistency and preventing lip distortion in extended videos. Current approaches, typically relying on short temporal windows for autoregressive inference, suffer from error accumulation and quality degradation over time. While Loopy [12] expands the frame window, it still struggles to achieve global temporal coherence. Frame-level generation often leads to inconsistent facial expressions, and audio feature fusion based on cross-attention lacks precision, resulting in unstable audio-expression alignment. EmoSpeaker [7] and Diff2Lip [14], respectively, attempt to address emotional consistency and lip sync, but both fall short in stabilizing facial expressions over longer durations. The MEMO [13] then integrates a memory-guided temporal module and an emotion-aware audio module, improving long-term identity consistency and emotional alignment. However, it still faces challenges in maintaining fine-grained emotional consistency.

Overall, while current diffusion models enhance emotional expression in video generation, they still struggle with long-term stability and fine-grained consistency. We introduce the PAED to improve them through physics-based dynamics and cross-modal disentanglement, which is a promising direction.

## 3 Methods

As illustrated in Fig. 2, PAED is a diffusion framework for generating audio-driven videos with enhanced emotional coherence. Our architecture extends conventional memory-based diffusion approaches through two innovative modules: the **physics-guided AU momentum regulator module** and the **cross-modal emotion disentangler module**.

In recent research, Stable Diffusion 3 (SD3) [17] refines typical diffusion model objective by introducing rectified flow loss, which modifies the conventional Denoising Diffusion Probabilistic Model (DDPM) training procedure. The new loss function is expressed as:

$$\mathcal{L} = \mathbb{E}_{z_t, c, \epsilon\ \mathcal{N}(0,1), t}[\lambda(t)||\epsilon - \epsilon_\theta(z_t, t, c)||_2^2], \tag{1}$$

where

$$\lambda(t) = \frac{1}{(1-t)^2}, \tag{2}$$

and the variable $z_t$ is obtained by reparameterizing,

$$z_t = (1-t)z_0 + t\epsilon, \tag{3}$$

where $z_0 = \mathcal{E}(I)$, and $\mathcal{E}$ represents the pre-trained encoder. $\mathcal{L}$ represents the flow loss, $t$ denotes the timestep, and $z$ is the noisy latent feature. The term $\epsilon$ signifies the added noise, while $\epsilon_\theta$ is the noise prediction made by the U-net network. $c$ stands for the conditioning information.

This strategy not only yields greater stability during training but also improves inference efficiency. Motivated by these benefits, we incorporate the rectified flow loss [18] from SD3 into our own training process as well.

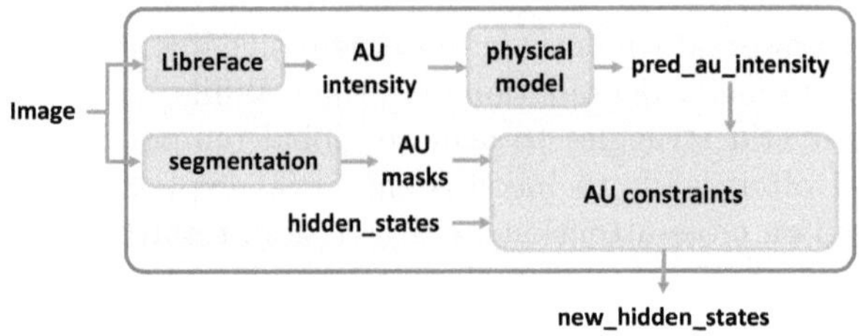


**Fig. 3.** Physics-Guided AU Momentum Regulator Module

## 3.1 Physics-Guided AU Momentum Regulator Module

Facial muscle activation triggers specific facial actions, associating muscles with different action units. The Facial Action Coding System (FACS) [19] comprehensively depicts human facial changes and identifies 46 types of facial actions related to movement. FACS dissects facial expressions into action units by identifying anatomical features, enabling a detailed understanding of facial movements and their association with emotional expressions. Each action unit is carefully designed to encapsulate facial activities associated with emotional experiences. By combining various AUs, a range of different facial expressions can be defined. The temporal continuity of AU intensity trajectories is a key determinant of the nature of expressions, as biological facial movements inherently exhibit inertia and viscoelastic tissue resistance (Fig. 3).

Current diffusion models face challenges in maintaining continuity due to inherent architectural constraints. Their frame-wise denoising paradigm induces discontinuous AU trajectories that manifest as expression jitter – rapid, unnatural fluctuations between successive frames – by ignoring the inertial properties of facial musculature. This discontinuity is exacerbated by the absence of explicit biomechanical constraints, which allow physically implausible AU transitions, thus violating the fundamental principles of facial tissue dynamics.

Inspired by damped oscillation phenomenon in biomechanical systems, an inertial-elastic-damping model is employed to simulate the transformation of action units. The parameters of this model are derived through learning. Moreover, our physical formula effectively addresses the discontinuity problem by enforcing second-order motion continuity:

$$m\frac{d^2\mathbf{x}(\mathbf{t})}{dt^2} + c\frac{d\mathbf{x}(\mathbf{t})}{dt} + k\mathbf{x}(\mathbf{t}) = \mathbf{F}(t), \quad (4)$$

where $m$ represents mass, $c$ is the damping coefficient, and $k$ is the resilience constant. Furthermore, $x(t)$ denotes the AU intensity over time, and $F(t)$ signifies the external driving force.

The general solution to this equation can be obtained as follows:

$$\mathbf{x}(\mathbf{t}) = Ae^{-\zeta\omega_n t}cos(\omega_d t + \phi), \tag{5}$$

where $A$ and $\phi$ are the constant (determined by the initial conditions, reflect the initial amplitude and phase of the AU action), $\zeta = \frac{c}{2\sqrt{mk}}$ is the damping ratio, $\omega_n = \sqrt{\frac{k}{m}}$ is the natural frequency (determines the oscillation speed of the system), $\omega_d = \omega_n\sqrt{1-\zeta^2}$ is the damping oscillation frequency.

If $\zeta < 1$, the system is underdamped so the expression transitions smoothly with a slight rebound effect. If $\zeta = 1$, the system is critically damped so the system is stable at maximum speed and suitable for fast facial movements. If $\zeta > 1$, the system is overdamped so the facial movements change slowly and smoothly, suitable for slow motion.

**Inertial Constraints.** Inertial constraints are used to ensure that dynamic changes of AU movements conform to the requirements of continuity and nature, and to avoid sudden changes in facial expressions that do not conform to physical laws. During training, inertial constraints can be embedded in the loss function:

$$\mathcal{L}_{\text{i}} = \lambda_v \cdot \sum_t (v_{t+1} - v_t)^2 + \lambda_a \cdot \sum_t (a_{t+1} - a_t)^2), \tag{6}$$

where $v_t = \frac{dx}{dt}$ indicates the change rate of the AU action unit, $a_t = \frac{d^2x}{dt^2}$ represents the acceleration of the AU action unit, $\lambda_v, \lambda_a$ is a hyper parameter that controls the smoothness of velocity and acceleration.

This loss term acts as a soft constraint, ensuring that predicted AU trajectories adhere to biomechanical principles while preserving the flexibility to capture diverse emotional expressions.

## 3.2 Cross-Modal Emotion Disentangler Module

We leverage the emotion2vec model [15], which captures audio-based emotional representations, and compare representations with visual emotion recognition output from the DeepFace [16], which processes facial expressions to recognize emotions.

**Emotion Extraction from Audio.** The emotion2vec is a generalized speech emotion representation model that is pre-trained on unlabeled emotional data through self-supervised online distillation. It incorporates both sentence-level and frame-level losses during pre-training process. The emotion2vec performs speech emotion recognition solely by training a linear layer and maps emotional information into a continuous vector space through transfer learning and emotion embedding models. This approach enables the model to effectively capture and recognize emotional states. For each audio frame, we extract the emotion vector, denoted as $E_{audio}$, which represents the probability distribution over predefined set of emotional classes. Formally, the emotion vector can be given by:

$$\mathbf{E}_{audio} = [p_1, p_2, ..., p_C] \in \mathbb{R}^C, \tag{7}$$

where $C$ is the number of emotion classes, $p_i$ represents the confidence of the audio in the $i$-th emotion class.

**Emotion Extraction from Visual Input.** The DeepFace is a deep learning-based face recognition method that extracts features from facial images using a multi-layer perceptron (MLP) and classifies these features through fully connected layers to achieve face recognition.

In this module, the DeepFace model is used to extract emotion-related features from visual input by analyzing facial expressions. For each frame of the input video, the DeepFace outputs a dictation that includes both emotional categories and corresponding confidence values. The resulting vector represents the emotion confidence distribution for the visual modality

$$\mathbf{E}_{visual} = [q_1, q_2, ..., q_C] \in \mathbb{R}^C, \tag{8}$$

where $q_i$ represents the confidence of the visual input in the $i$-th emotion class.

**Cross-Modal Consistency Constraint.** To ensure emotional consistency between audio and visual modalities, we introduce a consistency loss that measures the discrepancy between emotional confidence scores from the two modalities. Loss function is based on the Kullback-Leibler divergence, which quantifies the difference between two probability distributions. The consistency loss is computed as follows.

$$\mathcal{L}_{con} = D_{KL}(\mathbf{E}_{audio}||\mathbf{E}_{visual}) = \sum_{i=1}^{C} p_i log(\frac{p_i}{q_i}), \tag{9}$$

where $D_{KL}$ is the Kullback-Leibler divergence, which measures how much the two distributions diverge.

Controlling this consistency loss within a certain threshold ensures that the emotional states predicted by audio and visual modalities align closely, promoting cross-modal coherence in emotion recognition.

# 4 Experiments

## 4.1 Implementation Setups

**Dataset.** We conduct experiments on multiple datasets to cover a wide range of speaking styles and emotional expressions. The VoxCeleb2 dataset provides high-quality videos including eight categories of emotions. We split the Voxceleb2 dataset into training set, validation set and test set in a 7:1:2 ratio based on identity. However, the VoxCeleb2 dataset may exhibit certain demographic biases. Specifically, it may not comprehensively represent the full spectrum of the population in terms of race and gender. To ensure comprehensive coverage of diverse speaking scenarios, we additionally incorporate YouTube sourced datasets into our test dataset. For uniformity, the videos utilized in our experiments were resized to 512 × 512. We fine-tuned our model for 72 h on 4 NVIDIA A100 GPUs.

**Evaluation Metrics.** Our evaluation framework employs three principal metrics to comprehensively assess the performance of video generation. Fréchet Inception Distance (FID) [20] quantifies visual fidelity by measuring feature-space similarity between randomly sampled frames from generated and reference videos, with lower scores indicating better photorealism. Fréchet Video Distance (FVD) [21] extends this temporal analysis by evaluating 256-frame video clips through a 3D CNN architecture, capturing both spatial quality and motion coherence. For emotion expressiveness, we compute AU Score, measuring the cosine similarity between AU intensity histograms of generated and ground truth sequences.

**Baselines.** Our evaluation benchmarks against three state-of-the-art approaches using their official implementations and pretrained checkpoints. Hallo2 [22] represents GAN-based video synthesis with emotion conditioning, EchoMimic [23] employs neural radiance fields for disentangled emotion control, and Memo implements transformer-based cross-modal alignment.

### 4.2 Quantitative Results

**Performance on Test Datasets.** We quantitatively summarizes performance of PAED versus the baselines. As shown in Table 1, PAED demonstrates exceptional emotion expressiveness with 3.3% improvement in AU Score. This performance pattern suggests that the physics-based AU momentum regulator effectively enhances emotional articulation. In the generation process, our emphasis was on intensifying the attention paid to the AU area. Through adjustment of coefficients in physical model, we deliberately enhanced effect of facial expressions. This approach might lead to a minor decline in FID and FVD scores. However, we consider this a justifiable compromise when aiming to improve expressiveness of facial expressions. Notably, PAED reduces expression jitter compared to Memo through biomechanical constraints.

**Human Evaluation.** To better evaluate the quality of generated videos, we recruited 30 participants to evaluate the generated outcomes across motion smoothness, expression-audio alignment, audio-lip synchronization and identity consistency. Each person evaluated three videos and gave a score for each video. We calculate the average of those scores. As shown in Fig. 4, our method achieves higher win rate in overall quality, further demonstrating its effectiveness.

**Table 1.** Quantitative results of video quality and AU synchronization on test datasets. PAED outperforms existing talking video baselines.

| Methods | FID Score | FVD Score | Au Score |
|---|---|---|---|
| Hallo2 | 45.6 | 364.0 | 0.82 |
| EchoMimic | 47.8 | 351.2 | 0.86 |
| Memo | 32.8 | 294.9 | 0.91 |
| **PAED** | **34.0** | **310.4** | **0.94** |

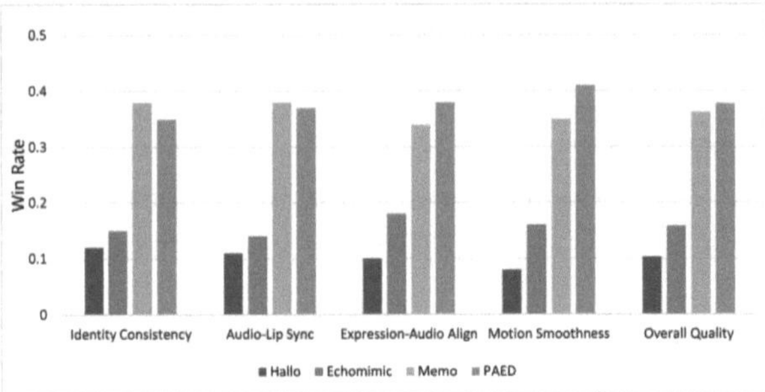


**Fig. 4.** Human preference comparisons between PAED and baseline methods, where users evaluate and select the best method according to each metric.

### 4.3 Qualitative Analysis

**Improved Emotional Expression.** Figure 5 illustrates sample frames from generated videos. Compared with MEMO, videos generated by PAED exhibit characteristic intensification phenomena aligned with human facial movement patterns during key emotional frames: 1) The lip region demonstrates more precise articulatory organ dynamics; 2) Periocular muscle groups manifest coordinated contraction patterns that correlate with emotional intensity levels. These biomechanically plausible fine-grained motion patterns validate the effectiveness of proposed muscular momentum constraint mechanism in coordinating dynamic equilibrium across multiple facial regions.

### 4.4 Ablation Studies

To gain a deeper understanding of individual contributions of each module within PAED, we conduct comprehensive ablation experiments by systematically removing or modifying key components.

**Effects of the Physics-Guided AU Momentum Regulator.** We substitute the physics-guided AU momentum regulator with a baseline AU consistency loss that does not incorporate the physical dynamics of facial motion. As a result, generated videos exhibit a marked increase in expression jitter. Results in Table 2 underscore the effectiveness of physics-guided AU momentum regulator in serval

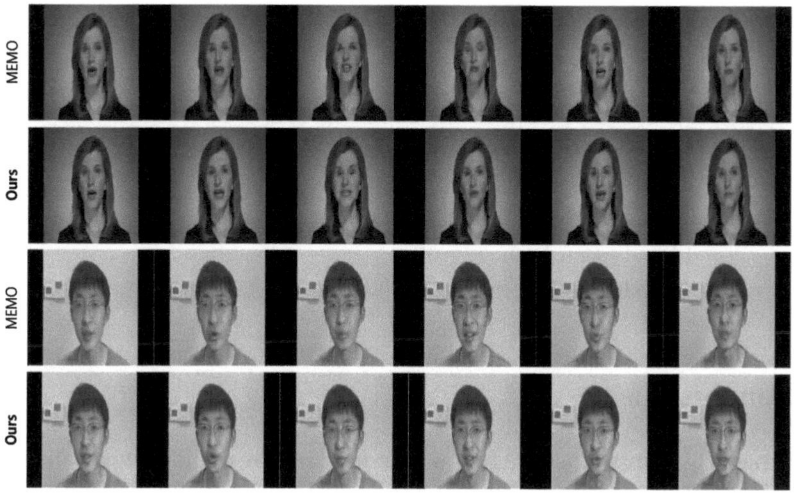


**Fig. 5.** PAED can generate talking face videos featuring more expressive emotions compared with baseline.

dimensions. Qualitatively, the absence of momentum constraints leads to abrupt transitions in facial expressions, as the model is unable to enforce smooth, inertia-based motion continuity. This confirms that physics-based approach is crucial for preserving natural facial dynamics over time.

**Table 2.** User preference proportions for the ablation study on the AU (physics-guided AU momentum regulator).

| | With AU (%) | Without AU (%) |
|---|---|---|
| Motion Smoothness | **79** | 21 |
| Expression-Audio Align | **73** | 27 |
| Audio-Lip Sync | **65** | 35 |
| Identity Consistency | **62** | 38 |
| Overall Quality | **68** | 32 |

**Effects of the Cross-Modal Emotion Disentangler.** We eliminate the cross-modal emotion disentangler module and directly fuse audio features with visual cues without explicit disentanglement. This change disrupts the fine-grained alignment between audio-based emotion predictions and visual expressions. The KL divergence between predicted audio and visual emotion distributions increases, indicating a poorer match. Results in Table 3 underscore the effectiveness of cross-modal emotion disentangler in mitigating emotion leakage and ensuring consistent emotion alignment.

**Table 3.** User preference proportions for the ablation study on the ED (cross-modal Emotion Disentangler).

| | With ED (%) | Without ED (%) |
|---|---|---|
| Motion Smoothness | **60** | 40 |
| Expression-Audio Align | **79** | 21 |
| Audio-Lip Sync | **72** | 28 |
| Identity Consistency | **69** | 31 |
| Overall Quality | **76** | 24 |

## 5 Conclusions

In this paper, we introduce PAED, a new method for audio-driven talking face generation. By integrating a Physics-Guided AU Momentum Regulator with a Cross-Modal Emotion Disentangler, our approach effectively addresses persistent challenge of long-term emotional inconsistency. Overall, PAED offers a promising direction for advancing the field of emotionally expressive talking face generation.

However, we acknowledge that the damping resonator used is a linear approximation of facial biomechanics, which may not fully capture all the nuances of facial movements. It also lacks a formal guarantee of complete separation between emotions and speech. Future work may focus on enhancing the model's ability to generate diverse emotional expressions to improve the reliability of emotion recognition.

**Acknowledgements.** Xingchu Zhang, Chuwen Zhang, Rui Mao, and Zehao Deng contributed equally to this work and should be considered co-first authors. Correspondence and requests for materials should be addressed to email chuwen.zhang@stonybrook.edu.

## References

1. Prajwal, K.R., Mukhopadhyay, R., Namboodiri, V.P., Jawahar, C.V.: A lip sync expert is all you need for speech to lip generation in the wild. In: Proceedings of the 28th ACM International Conference on Multimedia, Seattle, WA, USA, 12–16 October 2020, pp. 484–492 (2020)
2. Thies, J., Elgharib, M., Tewari, A., Theobalt, C., Nießner, M.: Neural voice puppetry: audio-driven facial reenactment. In: Vedaldi, A., Bischof, H., Brox, T., Frahm, J.-M. (eds.) ECCV 2020. LNCS, vol. 12361, pp. 716–731. Springer, Cham (2020). https://doi.org/10.1007/978-3-030-58517-4_42
3. Zakharov, E., Shysheya, A., Burkov, E., Lempitsky, V.: Few-shot adversarial learning of realistic neural talking head models. In: Proceedings of the IEEE/CVF International Conference on Computer Vision, Seoul, Republic of Korea, 27 October–2 November 2019, pp. 9459–9468 (2019)

4. Ji, X., et al.: Audio-driven emotional video portraits. In: Proceedings of the IEEE/CVF Conference on Computer Vision and Pattern Recognition (CVPR), Nashville, TN, USA, 20–25 June 2021, pp. 14080–14089 (2021)
5. Zhou, H., Sun, Y., Wu, W., Loy, C.C., Wang, X., Liu, Z.: Pose-controllable talking face generation by implicitly modularized audio-visual representation. In: Proceedings of the IEEE/CVF Conference on Computer Vision and Pattern Recognition, Nashville, TN, USA, 20–25 June 2021, pp. 4176–4186 (2021)
6. Zhou, Y., Han, X., Shechtman, E., Echevarria, J., Kalogerakis, E., Li, D.: Makelttalk: speaker-aware talking-head animation. ACM Trans. Graph. **39**, 221:1–221:15 (2020)
7. Feng, G., et al.: EmoSpeaker: One-shot Fine-grained Emotion-Controlled Talking Face Generation. arXiv arXiv:2402.01422 (2024)
8. Tian, L., Wang, Q., Zhang, B., Bo, L.: EMO: emote portrait alive generating expressive portrait videos with audio2video diffusion model under weak conditions. In: Leonardis, A., Ricci, E., Roth, S., Russakovsky, O., Sattler, T., Varol, G. (eds.) Computer Vision - ECCV 2024, pp. 244–260. Springer, Cham (2024)
9. Xu, M., et al.: Hallo: hierarchical audio-driven visual synthesis for portrait image animation. arXiv arXiv:2406.08801 (2024)
10. Shen, S., et al.: DiffTalk: crafting diffusion models for generalized audio-driven portraits animation. In: Proceedings of the IEEE/CVF Conference on Computer Vision and Pattern Recognition (CVPR), Vancouver, BC, Canada, 17–24 June 2023, pp. 1982–1991 (2023)
11. Du, C., et al.: DAE-talker: high fidelity speech-driven talking face generation with diffusion autoencoder. In: Proceedings of the 31st ACM International Conference on Multimedia, Ottawa, ON, Canada, 29 October–3 November 2023, pp. 4281–4289 (2023)
12. Jiang, J., Liang, C., Yang, J., Lin, G., Zhong, T., Zheng, Y.: Loopy: taming audio-driven portrait avatar with long-term motion dependency. In: Proceedings of the Thirteenth International Conference on Learning Representations, Singapore, pp. 24–28 (2025)
13. Zheng, L., et al.: MEMO: Memory-Guided Diffusion for Expressive Talking Video Generation. arXiv arXiv:2412.04448 (2024)
14. Mukhopadhyay, S., Suri, S., Gadde, R.T., Shrivastava, A.: Diff2lip: audio conditioned diffusion models for lip-synchronization. In: Proceedings of the IEEE/CVF Winter Conference on Applications of Computer Vision, Waikoloa, HI, USA, 3–8 January 2024, pp. 5292–5302 (2024)
15. Ma, Z., et al.: emotion2vec: self-supervised pre-training for speech emotion representation. In: Ku, L.W., Martins, A., Srikumar, V. (eds.) Findings of the Association for Computational Linguistics: ACL 2024, Bangkok, Thailand, pp. 15747–15760. Association for Computational Linguistics (2024)
16. Taigman, Y., Yang, M., Ranzato, M.A., Wolf, L.: DeepFace: closing the gap to human-level performance in face verification. In: Proceedings of the 2014 IEEE Conference on Computer Vision and Pattern Recognition, Columbus, OH, USA, 23–28 June 2014, pp. 1701–1708 (2014)
17. Esser, P., et al.: Scaling rectified flow transformers for high-resolution image synthesis. In: Proceedings of the 41st International Conference on Machine Learning, Vienna, Austria, 21–27 July 2024, pp. 12606–12633 (2024)
18. Liu, X., Gong, C., Liu, Q.: Flow straight and fast: learning to generate and transfer data with rectified flow. In: Proceedings of the NeurIPS 2022 Workshop on Score-Based Methods, New Orleans, LA, USA, 28 November–9 December 2022 (2022)

19. Ekman, P., Friesen, W.V.: Facial action coding system (FACS): a technique for the measurement of facial actions. Riv. Psichiatr. **47**, 38–126 (1978)
20. Heusel, M., Ramsauer, H., Unterthiner, T., Nessler, B., Hochreiter, S.: GANs trained by a two time-scale update rule converge to a local nash equilibrium. In: Proceedings of the 31st International Conference on Neural Information Processing Systems, Long Beach, CA, USA, 4–9 December 2017, pp. 6629–6640 (2017)
21. Unterthiner, T., Van Steenkiste, S., Kurach, K., Marinier, R., Michalski, M., Gelly, S.: Towards accurate generative models of video: a new metric & challenges. arXiv arXiv:1812.01717 (2018)
22. Cui, J., et al.: Hallo2: long-duration and high-resolution audio-driven portrait image animation. In: Proceedings of the Thirteenth International Conference on Learning Representations, Singapore, 24–28 April 2025 (2025)
23. Chen, Z., Cao, J., Chen, Z., Li, Y., Ma, C.: Echomimic: lifelike audio-driven portrait animations through editable landmark conditions. arXiv arXiv:2407.08136 (2024)

# CeeDet: A Class-Incremental Learning Method with Early-Exit Mechanism for Malicious Traffic Detection in IIoT

Jiakun Sun[1], Yabo Wang[1], Ruizhi Xiao[1], Weilong Li[1], and Shuyuan Jin[1,2(✉)]

[1] School of Computer Science and Engineering, Sun Yat-sen University, Guangzhou, China
{sunjk3,wangyb58,xiaorz3,liwl23}@mail2.sysu.edu.cn,
jinshuyuan@mail.sysu.edu.cn

[2] Guangdong Province Key Laboratory of Information Security Technology, Guangzhou, China

**Abstract.** The Industrial Internet of Things is increasingly threatened by continuously evolving cyberattacks, where new attack types emerge and older patterns persist. Existing detection methods often fail to adapt to this non-stationary data due to catastrophic forgetting and computational inefficiency. This paper introduces CeeDet, a class-incremental learning method designed for adaptive malicious traffic detection in IIoT networks. CeeDet integrates a hierarchical Mamba-based temporal feature extractor with an early-exit inference mechanism, allowing for dynamic adjustment of the inference path to achieve significant time savings. A Knowledge Distillation along Temporal dimension with soft-DTW strategy facilitates knowledge retention across evolving attacks without replaying previous data. Evaluations on the CIC-IIoT-2025 and NF-ToN-IoT-v2 benchmarks show that CeeDet achieves up to 87% average accuracy and forgets less than 12% of previously learned knowledge. Moreover, the early-exit mechanism provides an average 27.2% inference time reduction within the imbalanced dataset, demonstrating the method's performance in continuously evolving IIoT scenarios.

**Keywords:** Class-incremental learning · Early-exit mechanism · Mamba · Malicious traffic detection

## 1 Introduction

With the rapid evolution of the Industrial Internet of Things (IIoT), industries continuously deploy an increasing number of heterogeneous devices that generate massive network traffic. While enhanced connectivity drives the advancement of the smart industry, it also introduces unprecedented challenges to securing IIoT networks. The enormous volume of IIoT traffic demands substantial computational resources to enable effective intrusion detection. In addition, attackers continuously evolve techniques, producing more sophisticated threats. Existing


R. C.-W. Wong et al. (Eds.): PAKDD 2026, LNAI 16597, pp. 237–249, 2026.
https://doi.org/10.1007/978-981-92-1300-9_19

methods struggle to identify new attacks, particularly those that leverage encryption or stealthy traffic obfuscation. Furthermore, method updating is often hindered by the unavailability of historical training data due to industrial confidentiality, privacy protection regulations, and strict data isolation policies [12].

To address the aforementioned challenges, we propose CeeDet, a Class incremental learning method with early-exit mechanism for efficient malicious traffic detection in IIoT environments. CeeDet utilizes the Mamba architecture [7] to extract deep hierarchical features for precise malicious traffic discrimination, while multiple early-exit units dynamically generate inference paths for each sample, thereby reducing global inference latency. Furthermore, a $DT^2W$-based class-incremental learning strategy [15] enables CeeDet to learn new attack categories while alleviating the forgetting of previously learned ones, thus maintaining adaptability in evolving IIoT environments. The main contributions of this paper are summarized as follows:

1. We propose CeeDet, a novel method for efficient malicious traffic detection in IIoT environments that employs a Mamba-based hierarchical feature extractor to enhance traffic representation. It outperforms the best method by 8.97%/15.99% (NF-ToN-IoT-v2/CIC-IIoT-2025) in accuracy.
2. We integrate a $DT^2W$-based class incremental learning strategy to detect newly emerging attack classes without full retraining, thereby achieving an average forgetting rate of 8.69%/11.42% on the two datasets.
3. We construct an imbalanced dataset based on real network data distributions and integrate multiple early-exit modules into CeeDet, enabling benign traffic to exit earlier and thus significantly reducing overall inference time.

## 2 Related Work

**Malicious Traffic Detection in IIoT.** With the rapid proliferation of connected devices, the IIoT has become an attractive target for cyberattacks. Existing intrusion detection approaches typically rely on handcrafted features combined with classical machine learning classifiers [1,4]. Although these methods achieve satisfactory performance in controlled scenarios, they require extensive manual feature engineering and often fail to capture the complex temporal dependencies inherent in IIoT traffic. In recent studies, deep learning based methods have been introduced to automatically learn spatio-temporal representations from raw traffic data to improve detection performance [9,13]. Despite these advances, most methods fail to account for the data distributions in real IIoT networks, limiting the practical deployment of deep learning methods.

**Class-Incremental Learning.** Class-incremental learning (CIL) provides a promising solution for adaptive network intrusion detection by enabling models to learn new attack classes sequentially without revisiting prior training data. Typical approaches include Fine-Tuning, which updates model parameters for new data but suffers from catastrophic forgetting [8,14]; Learning without Forgetting, which constrains output consistency between old and new models [5];

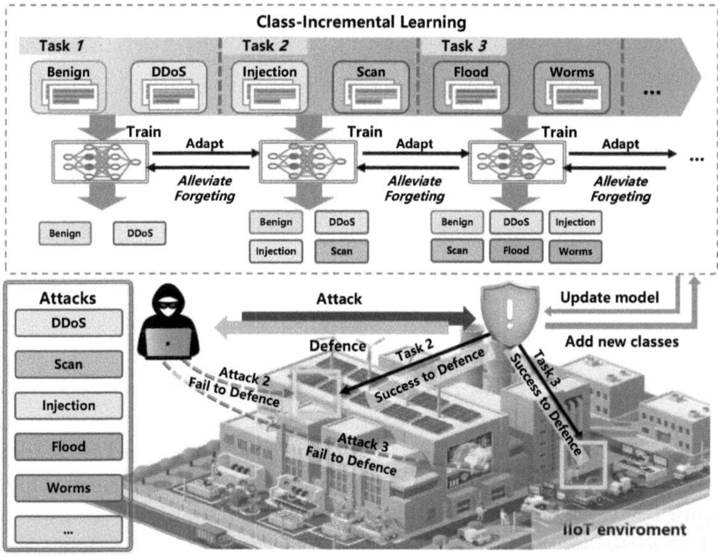


**Fig. 1.** Illustration of the CIL scenario for IIoT malicious traffic detection.

and Memory Replay, which reuses stored exemplars to retain knowledge of previous classes [2]. Although recent work, such as HFIN [15], takes an initial step toward addressing the ever-evolving network intrusion detection problem, CIL in IIoT environments still faces significant challenges due to the need to maintain temporal feature consistency, balance inter-class representations, and adapt effectively to continuously evolving threat landscapes.

## 3 Problem Definition

In the context of the IIoT, network traffic evolves continuously as new devices, communication protocols, and attack patterns emerge in production networks. As illustrated in Fig. 1, the malicious traffic detection task in such ever-evolving IIoT environments can be formulated as a CIL process. A detection model $f^t$ is trained over a sequence of tasks $\mathcal{T} = \{\mathcal{T}^t\}_{t=1}^{T}$, where each task $\mathcal{T}^t = (D^t, C^t)$ corresponds to traffic data collected within a specific time window. Each task contains a distinct class set $C^t = \{c_1^t, c_2^t, \ldots, c_{K_t}^t\}$ representing newly discovered attack categories (*e.g.*, DDoS, Injection, Scanning). For privacy and security reasons, the raw training data used in previous tasks cannot be accessed during subsequent tasks, rendering replay-based incremental learning methods infeasible in this scenario [12]. Classes across tasks are non-overlapping, *i.e.*, $C^i \cap C^j = \emptyset$ for $i \neq j$. The training data for task $t$ is denoted as $D^t = \{(\mathbf{x}_i^t, y_i^t)\}_{i=1}^{N_t}$, where each sample $\mathbf{x}_i^t \in \mathcal{X}$ represents a sequence of packets within a network flow, and the corresponding label $y_i^t \in C^t$ specifies the flow-level class.

The goal of class-incremental malicious traffic detection is to learn a mapping:

$$f : \mathcal{X} \rightarrow C^1 \cup C^2 \cup \cdots \cup C^t, \tag{1}$$

which can correctly classify both previously learned and newly emerging attack classes without accessing historical data, maintaining stability in old tasks and plasticity for new attacks.

# 4 CeeDet

## 4.1 Overall Workflow

Figure 2 shows the overall workflow of CeeDet. Raw packet-capture (PCAP) traces are first segmented into flows and transformed into compact multivariate representations. These embedded flows are then processed by a hierarchical Mamba-based feature extractor that models multi-scale temporal dependencies and incorporates a layer-wise early-exit mechanism to achieve dynamic inference. During class-incremental learning, the DT$^2$W strategy aligns feature distributions between new and old tasks via prototype-based matching, enabling stable knowledge transfer without replaying historical data.

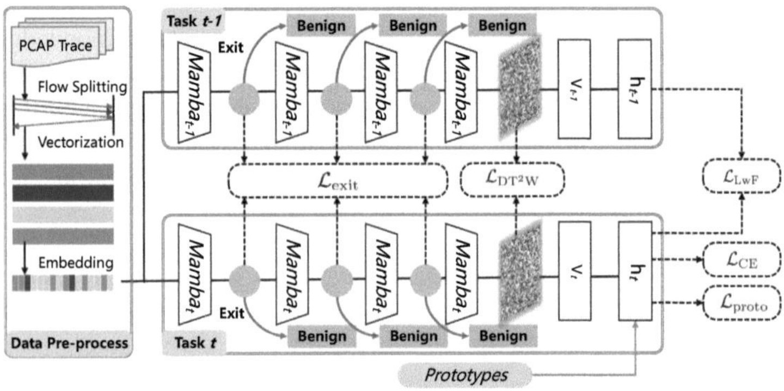


**Fig. 2.** Overall workflow of CeeDet.

## 4.2 Data Pre-process

In CeeDet, PCAP files are first parsed into network flows according to the five-tuple ($\text{IP}_{src}$, $\text{IP}_{dst}$, $\text{port}_{src}$, $\text{port}_{dst}$, and protocol). Since most contemporary attacks are encapsulated using encryption techniques (e.g., TLS or VPN), the payload content usually provides little discriminative information. Therefore, CeeDet focuses on the structural and statistical patterns within packet headers.

For each flow, the sequence of packet headers is concatenated in chronological order to form a one-dimensional temporal sequence. Each packet header is directly represented in hexadecimal format and converted into a fixed-length embedding vector, preserving low-level protocol information without hand-crafted feature extraction. To maintain uniform dimensionality, flows shorter than a predefined length $L$ are zero padded, while longer flows are truncated. The resulting input tensor is denoted as $\mathbf{X} \in \mathbb{R}^{L \times D}$, where $L$ represents the standardized sequence length corresponding to the number of packets preserved for each flow, and $D$ denotes the dimensionality of the packet-header embedding. $L = 128$ and $D = 64$ in our setting.

This one-dimensional representation seamlessly matches the Mamba-based feature extractor in CeeDet, which excels at modeling long-range temporal dependencies in sequential data. By leveraging chronological header sequences while avoiding the exposure of encrypted payloads, CeeDet achieves an efficient and privacy-preserving input format that captures key temporal behaviors of network flows for incremental malicious traffic detection.

### 4.3 Mamba-Based Dynamic Neural Network

The embedded flow $\mathbf{X}$ is encoded by a stacked Mamba architecture that learns hierarchical temporal dependencies across packet sequences. Each Mamba block $m_k$ produces an intermediate representation $\mathbf{h}_k$ [7], while deeper layers capture progressively higher-level semantics within the network flow.

To enable dynamic inference, a lightweight multi-layer perceptron (MLP) head is attached to each intermediate Mamba block. Each MLP consists of two fully connected layers with ReLU activation, followed by a softmax output layer that estimates the class probabilities. Each head produces an auxiliary prediction $\hat{p}^{(k)}$ and contributes to an early-exit classification loss defined as

$$\mathcal{L}_{\text{exit}} = \frac{1}{K} \sum_{k=1}^{K} \Big( - \frac{1}{N} \sum_{i=1}^{N} \sum_{c \in C^1} \mathbb{I}(y_i = c) \log \hat{p}_{i,c}^{(k)} \Big), \tag{2}$$

where $K$ is the total number of early-exit units. During the initial task ($t = 1$), CeeDet is trained in a fully supervised manner using a cross-entropy classification objective. Given a ground-truth label $y \in C^1$ and the prediction $\hat{p}$ from the final classifier attached to the deepest Mamba layer, the primary loss function is defined as

$$\mathcal{L}_{\text{CE}} = -\frac{1}{N} \sum_{i=1}^{N} \sum_{c \in C^1} \mathbb{I}(y_i = c) \log \hat{p}_{i,c}, \tag{3}$$

where $\hat{p}_{i,c}$ denotes the predicted probability of class $c$ at the final output layer, and $\mathbb{I}(\cdot)$ is the indicator function. The total training objective for the initial task becomes

$$\mathcal{L}_{\text{init}} = \mathcal{L}_{\text{CE}} + \lambda_{\text{exit}} \mathcal{L}_{\text{exit}}, \tag{4}$$

with $\lambda_{\text{exit}}$ controlling the contribution of early-exit supervision.

At inference time, each MLP head evaluates the confidence of its prediction (e.g., $\max_c \hat{p}_{i,c}^{(k)} > \tau$). If the confidence exceeds a predefined threshold $\tau$, the inference process terminates early; otherwise, the representation is propagated to deeper Mamba layers for further refinement.

This joint training strategy ensures that shallow Mamba blocks learn discriminative low-level patterns sufficient for recognizing most benign flows, while deeper layers specialize in modeling complex temporal dependencies associated with sophisticated attack behaviors, thus achieving an optimal trade-off between detection accuracy and inference efficiency.

### 4.4 CIL with Ever-Evolving Attacks

To continuously adapt to evolving malicious attacks, CeeDet adopts a CIL paradigm guided by the $\text{DT}^2\text{W}$ strategy [15]. At the beginning of each incremental step $t$, a frozen copy of the model trained on previous tasks $\{1,\ldots,t-1\}$, denoted as $f^{t-1}$, serves as the teacher, while the current model $f^t$ is the student to be optimized. The student incrementally learns new attack classes in task $t$ while retaining historical knowledge without accessing earlier data.

After finishing the initial training $T_1$, the representations from the deepest Mamba layer are aggregated into class prototypes that capture the centroids of each category. Formally, for class $c \in C^t$, its prototype is

$$\mathbf{p}_c^t = \frac{1}{|S_c^t|} \sum_{\mathbf{x}_i^t \in S_c^t} f^t(\mathbf{x}_i^t), \tag{5}$$

where $S_c^t$ is the sample set of class $c$, and $f^t(\cdot)$ denotes the feature extractor. These prototypes serve as semantic anchors for feature alignment and reduce classifier bias between old and new classes. For each incremental task, CeeDet jointly optimizes a composite objective consisting of five components:

$$\mathcal{L}_{\text{inc}} = \mathcal{L}_{\text{CE}} + \lambda_1 \mathcal{L}_{\text{DT}^2\text{W}} + \lambda_2 \mathcal{L}_{\text{LwF}} + \lambda_3 \mathcal{L}_{\text{proto}} + \lambda_4 \mathcal{L}_{\text{exit}}. \tag{6}$$

To retain temporal knowledge of the feature extractor, a Soft-DTWbased distillation loss $\mathcal{L}_{\text{DT}^2\text{W}}$ aligns the temporal feature maps of the student $F^t$ with those generated by its teacher $F^{t-1}$:

$$\mathcal{L}_{\text{DT}^2\text{W}} = \text{dtw}_\gamma\big(F^t(\mathbf{x}), F^{t-1}(\mathbf{x})\big). \tag{7}$$

The differentiable Soft-DTW metric [3] computes the shape-aligned discrepancy between two temporal sequences. To preserve global decision boundaries, a Learning-without-Forgetting loss ($\mathcal{L}_{\text{LwF}}$) is applied [10]:

$$\mathcal{L}_{\text{LwF}} = -\sum_{i=1}^{|C^{t-1}|} y_{t-1}^{\prime(i)} \log \hat{y}_{t-1}^{\prime(i)}, \tag{8}$$

where temperature-scaled targets ensure that the teacher's soft class distribution guides the student progressively. A prototype-based regularization $\mathcal{L}_{\text{proto}}$

encourages inter-class compactness and inter-class separability in the learned feature space:

$$\mathcal{L}_{\text{proto}} = \frac{1}{N} \sum_{i=1}^{N} \left\| f^t(\mathbf{x}_i^t) - \mathbf{p}_{y_i^t}^t \right\|_2^2, \tag{9}$$

and augmented pseudo-features $\tilde{f}_c = \mathbf{p}_c^{t-1} + r\mathbf{z}$ $(\mathbf{z} \sim \mathcal{N}(0,1))$ are reused to further stabilize the classifier.

The calculations of $\mathcal{L}_{\text{CE}}$ and $\mathcal{L}_{\text{exit}}$ are consistent with Eq. 3 and Eq. 2, and thus are not repeated here. Overall, this unified objective enables CeeDet to incrementally learn new attack classes while maintaining the temporal consistency of representations, preserving old knowledge, and achieving adaptive, low-latency inference in dynamic IIoT environments.

# 5 Experimental Setup

## 5.1 Datasets and Task Configuration

This study employs two public datasets to evaluate CeeDet: the CIC-IIoT-2025 and the NF-ToN-IoT-v2. CIC-IIoT-2025 contains 50 types of malicious attacks, which were collected from a realistic five-layer IIoT testbed integrating approximately 40 heterogeneous devices [6]. NF-ToN-IoT-v2 [11] covers 9 distinct attack types, which are generated from publicly available PCAP files of the original ToN-IoT dataset.

To simulate a CIL intrusion detection scenario, both datasets are divided into multiple sequential tasks, where each new task contains unseen attack classes. An example of detailed task configuration for the NF-ToN-IoT-v2 is shown in Table 1, illustrating the progressive introduction of new attack classes at each incremental step. Specifically, in the setting with $T = 9$, the CeeDet starts with Benign and Scanning traffic and subsequently adds with *step=1* until all nine attack types are encountered. For the settings with $T = 5$ and $T = 3$, the numbers of incremental *steps* are 2 and 4, respectively. The CIC-IIoT-2025 follows a similar partitioning strategy. At $t = 1$, the model is initialized with benign traffic, along with 7 representative attack classes. The dataset is further organized under configurations with $T \in \{9, 6, 5\}$, corresponding respectively to *step* = 5, 8, and 10. Detailed class mappings are omitted here for brevity, as the incremental partitioning strictly follows the same principle as that applied to NF-ToN-IoT-v2.

**Table 1.** Class-Incremental Task Configuration for NF-ToN-IoT-v2

| Classid ID | | 1 | 2 | 3 | 4 | 5 | 6 | 7 | 8 | 9 | 10 |
|---|---|---|---|---|---|---|---|---|---|---|---|
| Class | | Benign | Scanning | XSS | DDoS | Password | Dos | Injection | Backdoor | MITM | Ransomware |
| Incremental Task | $T = 9$ | $t = 1$ | | $t = 2$ | $t = 3$ | $t = 4$ | $t = 5$ | $t = 6$ | $t = 7$ | $t = 8$ | $t = 9$ |
| | $T = 5$ | $t = 1$ | | $t = 2$ | | $t = 3$ | | $t = 4$ | | $t = 5$ | |
| | $T = 3$ | $t = 1$ | | $t = 2$ | | | | $t = 3$ | | | |

### 5.2 Baselines

We compare CeeDet with representative methods spanning both exemplar-free continual learning and pre-training-based malicious traffic detection methods. Specifically, four classical exemplar-free continual learning algorithms are included: eiGenvalues ExplorAtion Regularization (GEAR) [8], Synaptic Intelligence (SI) [14], Memory-Aware Synapses (MAS) [2], and Split Learning Without Forgetting (SLwF) [5]. In addition, two pre-training-based traffic detection methods, ET-BERT [9] and NETMAMBA [13], are adopted as strong non-incremental baselines. During fine-tuning, we allow the pre-trained methods to replay a small subset of samples (5%) for fairness. Furthermore, the lower bound ("Naive", trained sequentially without any regularization) and the upper bound ("Offline", jointly trained on all tasks with full data access) are reported to provide comprehensive reference points.

### 5.3 Evaluation Metrics

To measure both adaptation and knowledge retention, we employ two widely adopted metrics in CIL research. Average accuracy after learning the $i$-th task is defined as:

$$\mathrm{ACC}_i = \frac{1}{i}\sum_{j=1}^{i} a_{i,j}, \tag{10}$$

where $a_{i,j}$ denotes the performance of the model on the $j$-th task after learning the $i$-th task, and $i$ represents the current incremental step. Average forgetting quantifies performance deterioration on earlier tasks:

$$\mathrm{FGT}_i = \frac{1}{i-1}\sum_{j=1}^{i-1} \big(\max_{l<i} a_{l,j} - a_{i,j}\big). \tag{11}$$

The term $\max_{l<i} a_{l,j}$ denotes the highest performance achieved on task $j$ among all previous learning stages, where $l$ serves as an index variable iterating over all previous training steps ($1 \le l < i$). Hence, $\max_{l<i} a_{l,j} - a_{i,j}$ quantifies the performance degradation of task $j$ after step $i$, and $\mathrm{FGT}_i$ computes the average forgetting across all previously learned tasks.

### 5.4 Implementation Details

We implement CeeDet using PyTorch on an NVIDIA RTX 3090 GPU. The model is trained using the Adam optimizer with an initial learning rate of $1e^{-3}$, which decays by a factor of 0.1 every 10 epochs. The weight for the early-exit supervision is set to $\lambda_{exit} = 0.5$. For the composite incremental objective in Eq. 6, we balance the plasticity and stability trade-off by setting the weights as follows: $\lambda_1 = 2.0$ , $\lambda_2 = 1.0$, $\lambda_3 = 0.1$ , and $\lambda_4 = 0.5$.

# 6 Experimental Results and Analysis

## 6.1 CIL Performance of Baselines

First, as summarized in Fig. 3, the experimental results reveal CeeDet's dominance in CIL performance under various task configurations. It exhibits the most gradual accuracy decline as new tasks arrive, retaining more than 83% of its initial performance even after the final task. This demonstrates CeeDet's strong ability to preserve previously learned knowledge while adapting to new attack categories. The performance gap between CeeDet and other approaches widens as the number of tasks increases, highlighting its robustness against catastrophic forgetting. Second, when each incremental step introduces a larger number of new classes (i.e., smaller $T$ values), the overall accuracy decreases more rapidly. For example, the average accuracy drop from $T = 9$ to $T = 5$ exceeds 10% on the NF-ToN-IoT-v2 dataset and nearly 15% on the CIC-IIoT-2025 dataset. This indicates that larger distributional shifts between tasks intensify forgetting, as the model faces greater variation in decision boundaries and feature distributions. Third, compared with NF-ToN-IoT-v2, the CIC-IIoT-2025 dataset exhibits more severe accuracy degradation after each task. This is mainly attributed to its larger number of attack categories and stronger feature heterogeneity, which increase inter-class overlap and confusion between old and new decision boundaries. In addition, ET-BERT and NETMAMBA achieve higher accuracy on the first task due to their large-scale pre-training, which provides richer and more generalizable representations. However, during class-incremental

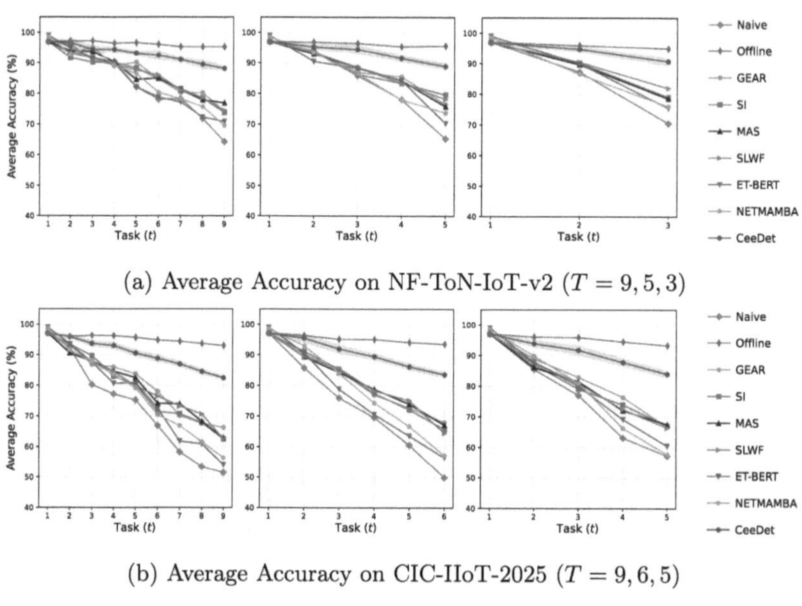


(a) Average Accuracy on NF-ToN-IoT-v2 ($T = 9, 5, 3$)

(b) Average Accuracy on CIC-IIoT-2025 ($T = 9, 6, 5$)

**Fig. 3.** Average accuracy comparison across sequential tasks.

learning, both show rapid accuracy decreases of more than 20% because they lack replayed samples and cannot retain knowledge of earlier classes.

The final average ACC and the average FGT after learning all incremental tasks are summarized in Table 2. CeeDet achieves the highest overall performance on both datasets, improving the final average accuracy by 8.97% over SLwF on NF-ToN-IoT-v2 and 15.99% over SI on CIC-IIoT-2025. In terms of forgetting, CeeDet also shows the most stable performance, with its average forgetting reduced by 25.8% and 28.71% when compared to the Naive method on the two datasets. Such consistent advantages indicate that CeeDet effectively preserves previously acquired knowledge while integrating new tasks, achieving a better balance between plasticity and stability.

### 6.2 Efficiency Analysis

The number of attacks targeting IIoT has been increasing in recent years; however, the proportion of benign traffic remains relatively small. To better simulate realistic network behavior, we constructed an imbalanced dataset with a benign-to-malicious ratio of 7:3. This setting reflects the practical scenario where benign flows dominate, allowing for a more accurate evaluation of the proposed model's detection efficiency and robustness under imperfect traffic distributions.

**Table 2.** Overall performance of different methods

| Method | NF-ToN-IoT-v2 | | CIC-IIoT-2025 | |
|---|---|---|---|---|
| | ACC (%) | FGT (%) | ACC (%) | FGT (%) |
| Naive (lower bound) | 63.02 (±1.80) | 33.87 (±1.80) | 56.12 (±1.52) | 40.13 (±1.52) |
| SLwF | 78.04 (±2.10) | 18.11 (±2.10) | 61.08 (±1.86) | 24.83 (±1.86) |
| GEAR | 76.01 (±2.37) | 20.20 (±2.37) | 64.05 (±2.20) | 21.66 (±2.20) |
| SI | 78.03 (±2.24) | 18.08 (±2.24) | 67.04 (±2.11) | 28.47 (±2.11) |
| MAS | 75.06 (±2.58) | 21.24 (±2.58) | 62.03 (±2.39) | 23.80 (±2.39) |
| ET-BERT | 70.08 (±3.46) | 34.30 (±3.46) | 55.13 (±3.24) | 40.47 (±3.24) |
| NETMAMBA | 70.04 (±4.24) | 32.29 (±4.24) | 57.02 (±3.81) | 42.49 (±3.81) |
| **CeeDet** | **87.01 (± 3.06)** | **8.69 (± 3.06)** | **83.03 (± 2.94)** | **11.42 (± 2.94)** |
| Offline (upper bound) | 95.29 | – | 93.72 | – |

Figure 4a illustrates the relationship between the confidence threshold $\tau$ and two metrics: the accuracy and the inference time savings achieved by shallow exits. Figure 4b presents the distribution of exited samples across different layers under various confidence thresholds. The inference time saving is computed relative to the case of $\tau = 1$, which corresponds to the condition where all samples are forced to pass through the final layer, effectively disabling shallow exits. As the threshold increases, fewer samples terminate inference at earlier layers, resulting in higher overall accuracy but reduced time savings. Conversely, lower thresholds allow more samples to exit earlier, effectively reducing inference time but slightly decreasing accuracy. This trade-off reflects the model's ability to balance efficiency and performance in dynamic inference scenarios.

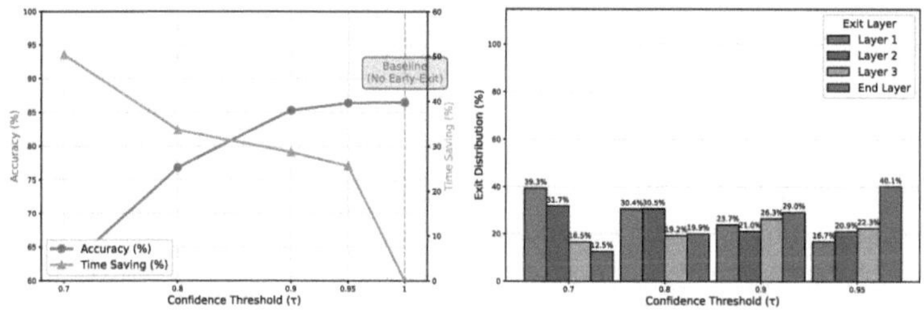


(a) Effect of performance. (b) Effect of exit distribution.

**Fig. 4.** Effect of threshold on performance and exit distributions.

## 6.3 Ablation and Layer Depth Analysis

Figure 5a reports the impact of removing key components from the proposed CeeDet, including the $DT^2W$ alignment, prototype learning, and LwF regularization modules. It can be observed that each component contributes positively to model stability and overall accuracy. In particular, the $DT^2W$ strategy plays a critical role in sequential knowledge preservation, while the prototype constraint further enhances inter-task discrimination. Figure 5b shows the effect of varying stacked Mamba block numbers. As the layer depth increases, accuracy improves steadily until four layers, whereas further deepening brings marginal gains but noticeably reduces inference efficiency. This indicates that an appropriate depth provides a balance between detection performance and time savings.

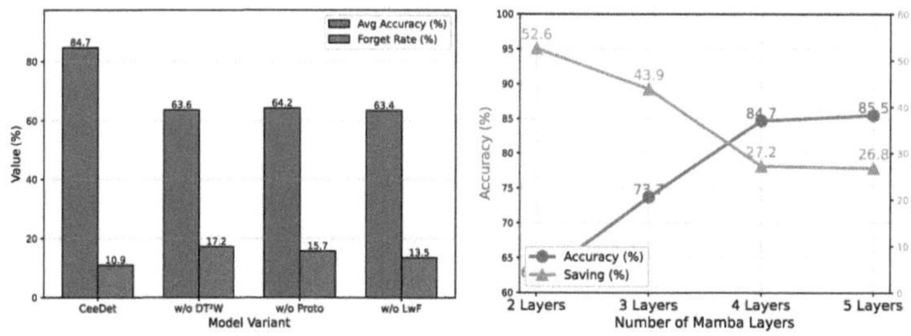


(a) Ablation Results (b) Effect of Mamba Layers.

**Fig. 5.** Ablation and Mamba layer depth analysis.

## 7 Conclusion

This work presents CeeDet, an efficient class-incremental learning method tailored for evolving malicious traffic detection in IIoT networks. By combining hierarchical temporal modeling through the Mamba architecture with an early-exit mechanism, CeeDet adaptively balances inference depth and decision accuracy, achieving substantial time savings during detection. The $DT^2W$ strategy and prototype-based regularization effectively mitigate catastrophic forgetting, enabling the model to preserve acquired knowledge while learning new attack patterns. Comprehensive experiments on benchmark IIoT datasets demonstrate that CeeDet consistently outperforms existing exemplar-free CIL methods in both accuracy and stability, while achieving significant improvements in inference efficiency. Future research will explore extending CeeDet to distributed and federated incremental learning methods to further enhance scalability and multi-domain adaptability in industrial cyber defense applications.

## References

1. Alhowaide, A., Alsmadi, I., Tang, J.: PCA, random-forest and Pearson correlation for dimensionality reduction in IoT IDS. In: 2020 IEEE International IOT, Electronics and Mechatronics Conference (IEMTRONICS), pp. 1–6 (2020)
2. Aljundi, R., Babiloni, F., Elhoseiny, M., Rohrbach, M., Tuytelaars, T.: Memory aware synapses: learning what (not) to forget. In: Ferrari, V., Hebert, M., Sminchisescu, C., Weiss, Y. (eds.) ECCV 2018, Part III. LNCS, vol. 11207, pp. 144–161. Springer, Cham (2018). https://doi.org/10.1007/978-3-030-01219-9_9
3. Cuturi, M., Blondel, M.: Soft-DTW: a differentiable loss function for time-series. In: International Conference on Machine Learning, pp. 894–903. PMLR (2017)
4. Davahli, A., Shamsi, M., Abaei, G.: A lightweight anomaly detection model using SVM for WSNs in IoT through a hybrid feature selection algorithm based on GA and GWO. J. Comput. Secur. **7**(1) (2020)
5. Feng, X., Jia, R., Luo, C., Leung, V.C.M., Xu, W.: SLwF: a split learning without forgetting framework for internet of things. IEEE Internet Things J. **12**(9), 12008–12020 (2025). https://doi.org/10.1109/JIOT.2024.3519732
6. Firouzi, A., Dadkhah, S., Maret, S.A., Ghorbani, A.A.: DataSense: a real-time sensor-based benchmark dataset for attack analysis in IIoT with multi-objective feature selection. Electronics **14**(20) (2025). https://www.mdpi.com/2079-9292/14/20/4095
7. Gu, A., Dao, T.: Mamba: linear-time sequence modeling with selective state spaces. CoRR abs/2312.00752 (2023). https://doi.org/10.48550/arXiv.2312.00752
8. Kong, Y., Liu, L., Chen, H., Kacprzyk, J., Tao, D.: Overcoming catastrophic forgetting in continual learning by exploring eigenvalues of hessian matrix. IEEE Trans. Neural Netw. Learn. Syst. **35**(11), 16196–16210 (2024). https://doi.org/10.1109/TNNLS.2023.3292359
9. Lin, X., Xiong, G., Gou, G., Li, Z., Shi, J., Yu, J.: ET-BERT: a contextualized datagram representation with pre-training transformers for encrypted traffic classification. In: WWW 2022: The ACM Web Conference 2022, Lyon, France, 25–29 April 2022, pp. 633–642. ACM (2022). https://doi.org/10.1145/3485447.3512217

10. Rebuffi, S.A., Kolesnikov, A., Sperl, G., Lampert, C.H.: ICARL: incremental classifier and representation learning. In: Proceedings of the IEEE conference on Computer Vision and Pattern Recognition, pp. 2001–2010 (2017)
11. Sarhan, M., Layeghy, S., Portmann, M.: Towards a standard feature set for network intrusion detection system datasets. Mob. Netw. Appl. **27**(1), 357–370 (2022). https://doi.org/10.1007/s11036-021-01843-0
12. Sun, D., et al.: A comprehensive survey on collaborative data-access enablers in the IIoT. ACM Comput. Surv. **56**(2), 37 (2024)
13. Wang, T., Xie, X., Wang, W., Wang, C., Zhao, Y., Cui, Y.: NetMamba: efficient network traffic classification via pre-training unidirectional mamba. In: 32nd IEEE International Conference on Network Protocols, ICNP 2024, Charleroi, Belgium, 28–31 October 2024, pp. 1–11. IEEE (2024). https://doi.org/10.1109/ICNP61940.2024.10858569
14. Zenke, F., Poole, B., Ganguli, S.: Continual learning through synaptic intelligence. In: Proceedings of the 34th International Conference on Machine Learning, ICML 2017, Sydney, NSW, Australia, 6–11 August 2017, vol. 70, pp. 3987–3995. PMLR (2017). http://proceedings.mlr.press/v70/zenke17a.html
15. ZhongZheng, Q., Hu, M., Jiang, X., Suganthan, P.N., Ramasamy, S.: Class-incremental learning on multivariate time series via shape-aligned temporal distillation. In: IEEE International Conference on Acoustics, Speech and Signal Processing ICASSP 2023, Rhodes Island, Greece, 4–10 June 2023, pp. 1–5. IEEE (2023). https://doi.org/10.1109/ICASSP49357.2023.10094960

# MUNet: A Multi-outcome Uplift Network for Modeling Multi-dimensional Treatment Effects

Ludan Zhang(✉)

Shanghai, China
1264425526@qq.com

**Abstract.** Uplift modeling estimates individual-level causal effects, yet most existing methods address only a single outcome, overlooking the fact that real-world interventions often influence multiple correlated outcomes. Independent modeling fails to exploit shared causal structure and may produce inconsistent estimates. We propose MUNet, a unified multi-outcome framework for multi-dimensional uplift estimation. MUNet learns shared causal representations while enforcing additive consistency between category-level and total uplifts. The extended MUNet-X incorporates content-based gating and outcome-wise uncertainty weighting to model sample-dependent outcome relations and handle heterogeneous noise. Experiments on controlled synthetic data and the Dunnhumby Complete Journey dataset show that MUNet consistently outperforms representative single-outcome uplift methods, achieving higher ranking quality and stronger structural coherence. Repeated-split evaluation further indicates reduced variance for MUNet-X, demonstrating improved generalization under observational noise. These results highlight the value of structured multi-task modeling for realistic uplift applications.

**Keywords:** Uplift Modeling · Multi-Outcome Learning · Structural Consistency

## 1 Introduction

Uplift modeling estimates the **Individual Treatment Effect** (ITE) and is widely applied in marketing, recommendation, and healthcare. Classical meta-learners [9] and tree-based uplift methods [2,17] remain standard approaches, while deep models such as TARNet, CFRNet [18], and DragonNet [19] improve representation balance. However, existing methods predominantly assume a single outcome.

In practice, interventions often influence multiple correlated outcomes (e.g., category-level and total spending) that exhibit additive or hierarchical structure. Naively modeling outcomes independently—or via parameter sharing alone—fails to enforce structural coherence, leading to inconsistent uplift estimates


R. C.-W. Wong et al. (Eds.): PAKDD 2026, LNAI 16597, pp. 250–262, 2026.
https://doi.org/10.1007/978-981-92-1300-9_20

across levels. Existing multi-task uplift methods treat tasks as separate KPIs [22], and multi-treatment models [12,17,20,21,25] focus on treatment multiplicity rather than structured outcomes.

We propose **MUNet**, a Multi-Outcome Uplift Network for structured multi-outcome modeling. **MUNet-Base** introduces a structural consistency constraint —the core contribution of this work—to enforce additive coherence between component and total uplifts. **MUNet-X** further incorporates content-based gating and uncertainty weighting [8,11] to enhance robustness under heterogeneous noise.

Our contributions are summarized as follows:

1. We formalize the **structured multi-outcome uplift** problem, where treatment-sensitive outcomes exhibit additive or hierarchical relations.
2. We introduce a **structural consistency** constraint to enforce coherence between aggregate and component uplifts.
3. We integrate a shared multi-task backbone with optional robustness extensions (content-based gating and uncertainty weighting).
4. Experiments demonstrate that structural consistency—rather than parameter sharing alone—drives consistent gains in uplift ranking and cross-outcome coherence.

## 2 Related Work

*Uplift Modeling and ITE Estimation.* Uplift modeling estimates individual treatment effects (ITE) under the potential outcome framework. Classical meta-learners (S-, T-, X-, DR-Learner) [9] and deep causal models such as TARNet, CFRNet [18], and DragonNet [19] primarily assume a single outcome and do not exploit dependencies among multiple treatment-sensitive outputs.

*Multi-task Uplift Modeling.* Multi-task extensions share representations across objectives [4,5,16], and mixture-of-experts models address multi-treatment or multi-output prediction [11]. While these approaches enable parameter sharing, they typically treat tasks as independent and do not enforce structural coherence among outcomes.

*Multi-output Causal Learning.* Recent work studies vector-valued or multi-output causal estimation using Gaussian Processes [1] and deep multi-task models [7]. These methods leverage cross-outcome correlations but focus on regression accuracy rather than uplift-based ranking, and they do not impose additive or hierarchical constraints across outcomes.

*Gap.* Existing approaches either assume a single outcome or treat multiple outcomes as loosely related tasks. None explicitly model structured outcome relations—such as category-level effects aggregating into a total uplift—while enforcing causal consistency across them. This motivates our structured multi-outcome uplift formulation with explicit consistency constraints.

# 3 Methodology

## 3.1 Problem Definition

Given a dataset

$$\mathcal{D} = \{(x_i, t_i, y_i^{(1)}, \ldots, y_i^{(K)}, y_i^{total})\}_{i=1}^{N},$$

where $x_i \in \mathbb{R}^d$ denotes user features, $t_i \in \{0, 1\}$ is the treatment indicator, and $y_i^{(k)}$ represents the $k$-th outcome (e.g., category spending) with total outcome $y_i^{total}$. Each user only observes one potential outcome $Y_i = Y_i(T_i)$.

Following the standard definition of individualized treatment effects [9,14], the individual uplift for outcome $k$ is defined as:

$$U_k(x) = \mathbb{E}[Y_k \mid T = 1, X = x] - \mathbb{E}[Y_k \mid T = 0, X = x], \quad (1)$$

and for the total outcome:

$$U_{total}(x) = \mathbb{E}[Y_{total} \mid T = 1, X = x] - \mathbb{E}[Y_{total} \mid T = 0, X = x]. \quad (2)$$

Our objective is to estimate the multi-task uplift vector:

$$U(x) = [U_1(x), U_2(x), \ldots, U_K(x), U_{\text{total}}(x)].$$

where the sub-outcomes and total outcome may satisfy additive or hierarchical consistency.

## 3.2 Base Model: MUNet-Base

**Motivation.** Conventional uplift models treat each outcome independently, ignoring shared causal structures and additive relationships. MUNet-Base addresses this by learning a shared causal representation while enforcing consistency among related outcomes.

**Shared Encoder.** A shared MLP encoder transforms the input features $X$ into a latent representation:

$$H = \text{MLP}_\theta(X). \quad (3)$$

This shared representation $H$ captures underlying causal factors and serves as a common basis for all outcome-specific decoders.

**Task-specific Decoders.** Each outcome has two decoder heads for control and treatment predictions:

$$\hat{Y}_0^{(k)} = D_k^0(H), \qquad \hat{Y}_1^{(k)} = D_k^1(H), \qquad \hat{U}_k = \hat{Y}_1^{(k)} - \hat{Y}_0^{(k)}, \quad k = 1, \ldots, K.$$

The total uplift $\hat{U}_{total}$ is computed in the same manner.

**Uplift Loss.** For each outcome $k$, we adopt a standard potential-outcome-based loss to supervise uplift prediction. Specifically, given treatment indicator $T \in \{0,1\}$ and observed outcome $Y^{(k)}$, the per-task uplift loss is defined as

$$\mathcal{L}_{uplift}^{(k)} = \frac{1}{N}\sum_{i=1}^{N}\Big[\mathbb{I}(T_i=1)\big(\hat{Y}_1^{(k)}(x_i)-Y_i^{(k)}\big)^2+\mathbb{I}(T_i=0)\big(\hat{Y}_0^{(k)}(x_i)-Y_i^{(k)}\big)^2\Big]. \quad (4)$$

**Structural Consistency.** To ensure additive coherence between total and sub-outcomes, a soft consistency constraint is applied:

$$\mathcal{L}_{cons} = \frac{1}{N}\sum_{i=1}^{N}\|\hat{U}_{total}(x_i) - \sum_{k=1}^{K}\hat{U}_k(x_i)\|_2^2. \quad (5)$$

**Objective Function.** MUNet jointly minimizes per-task uplift losses, the total-outcome loss, and the structural consistency penalty:

$$\mathcal{L}_{Base} = \sum_{k=1}^{K}\mathcal{L}_{uplift}^{(k)} + \mathcal{L}_{uplift}^{(total)} + \lambda\,\mathcal{L}_{cons}, \quad (6)$$

where $\mathcal{L}_{uplift}^{(k)}$ denotes the per-outcome uplift loss defined above.

*Why soft consistency instead of hard summation?* Hard additivity could be enforced by defining $\hat{U}_{total} = \sum_{k=1}^{K}\hat{U}_k$ by construction. We adopt a soft constraint for three reasons: (i) real-world outcomes often exhibit heterogeneous noise and zero inflation, and hard coupling may amplify errors from sparse components; (ii) a dedicated total head can capture global uplift signals not fully recoverable from category-level predictions; (iii) the penalty serves as an inductive bias toward coherence without imposing a rigid architectural assumption.

### 3.3 Enhanced Model: MUNet-X

**Motivation.** While MUNet-Base captures shared structure, it assumes fixed inter-outcome relations. Built upon the same MLP-based shared encoder, MUNet-X introduces content-based gating to model sample-dependent outcome coupling and incorporates outcome-wise uncertainty weighting to handle heterogeneous noise across outcomes.

**Content-Based Gating (MMoE) Decoders.** vFollowing the multi-gate mixture-of-experts design [11], each outcome adopts a content-based gating network to combine a shared set of $M$ expert transformations $\{f_m(\cdot)\}_{m=1}^{M}$. Given the shared representation $H$, outcome $k$ generates sample-specific mixture weights

$$\alpha_k = \text{softmax}\big(W_k^{(\text{gate})}H\big),$$

where the gating parameters $W_k^{(\text{gate})}$ are learnable. The resulting weights $\alpha_{k,m}$ vary across both samples and outcomes, forming a multi-gated mixture-of-experts (MMoE) structure. The outcome-specific representation is then obtained as

$$H_k = \sum_{m=1}^{M} \alpha_{k,m} f_m(H).$$

Following the decoder design in MUNet-Base, the representation $H_k$ serves as the shared input to two task-specific decoder heads for control and treatment predictions:

$$\hat{Y}_0^{(k)} = D_k^0(H_k), \quad \hat{Y}_1^{(k)} = D_k^1(H_k), \quad \hat{U}_k = \hat{Y}_1^{(k)} - \hat{Y}_0^{(k)}, \quad k = 1, \ldots, K.$$

**Uncertainty-Weighted Extension.** To adaptively balance heterogeneous outcome noise, MUNet-X learns outcome-specific uncertainties $\sigma_k$ following [8]:

$$\mathcal{L}_X = \sum_{k=1}^{K} \frac{1}{2\sigma_k^2} \mathcal{L}_{uplift}^{(k)} + \frac{1}{2\sigma_{total}^2} \mathcal{L}_{uplift}^{(total)} + \log \sigma_k + \log \sigma_{total} + \lambda \mathcal{L}_{cons}.$$

### 3.4 Training Objective and Bias Adjustment

In observational settings, treatment assignment may be non-random, introducing confounding bias. To mitigate this, we apply inverse propensity weighting (IPW) [15] in the prediction losses: a separate propensity model estimates $e(x_i) = P(T_i = 1 \mid X_i)$, and each sample is reweighted by

$$w_i = \frac{T_i}{e(x_i)} + \frac{1 - T_i}{1 - e(x_i)}.$$

We apply clipping at 0.01 and 0.99 to ensure numerical stability. These weights adjust for selection bias in outcome prediction while the structural consistency term remains unweighted, providing a simple yet effective bias correction without modifying the model architecture.

## 4 Experiments

We conduct comprehensive experiments to evaluate the proposed MUNet framework on both synthetic and real-world datasets. Our goals are to assess its predictive performance, structural consistency, and robustness compared with representative uplift modeling baselines. Unless otherwise specified, all models share comparable capacity and are trained under identical data splits and evaluation protocols.

### 4.1 Dataset

Existing uplift datasets such as Hillstrom or Criteo focus on single-outcome, binary-treatment settings, which do not capture the hierarchical and correlated structures central to our study. To evaluate multi-outcome causal uplift with structural consistency, we follow prior work on uplift evaluation [14,24] and employ a two-stage design: a controlled synthetic dataset for methodological validation and the real-world Dunnhumby Complete Journey dataset for robustness analysis.

**Synthetic Dataset.** We construct a synthetic multi-outcome uplift dataset that balances *interpretability*, *controllability*, and *behavioral realism*. The design introduces nonlinear feature effects, heterogeneous noise, inter-task dependencies, and zero inflation.

*Base Outcomes.* Each instance $X_i \in \mathbb{R}^{32}$ generates latent base outcomes

$$g_k(X_i) = \beta_k^\top X_i + \sum_j \gamma_{k,j} \tanh(\alpha x_{i,j}) + \varepsilon_{i,k}^{(g)},$$

combining linear and smooth nonlinear effects with heteroscedastic noise.

*Treatment assignment.* Treatments follow a logistic exposure model

$$T_i \sim \text{Bernoulli}(\text{sigmoid}(\beta_e^\top X_i)),$$

yielding heterogeneous but overlapping propensities.

*Uplift generation.* Uplifts $\tau_i \in \mathbb{R}^K$ combine global synergy, nonlinear perturbation, individual coupling, shared latent factors, and competition:

$$\begin{aligned}\tau_i = {} & Ag(X_i) + \xi_A \,\text{MLP}(g(X_i)) + \xi_B \, A^{(i)} g(X_i) + BZ_i + \eta_i \\ & - \lambda_{\text{comp}} \, \overline{\tau}_{i,\mathcal{N}(k)}^{(0)} + \epsilon_{i,k}^{(\text{cross})}.\end{aligned}$$

This structure induces realistic cross-task relations, including moderate substitution ($-0.3 \leq \rho \leq 0.5$).

*Potential and observed outcomes.* Potential outcomes are

$$Y_k^*(0) = g_k(X_i) + b_k, \qquad Y_k^*(1) = g_k(X_i) + \tau_{i,k} + b_k.$$

Observed outcomes apply a GaussianâĂŞLaplace noise mixture and a Bernoulli gate ($\approx 60\%$ zeros) to model zero inflation and purchase sparsity.

*Summary.* This synthetic design yields synergistic yet partially substitutable multi-outcome uplifts with controllable structure and noise, forming a realistic benchmark with 100,000 samples for evaluating multi-task uplift models.

**Dunnhumby Dataset.** We evaluate on the *Dunnhumby – The Complete Journey* retail dataset [6], which includes 2,500 households observed over 102 weeks and more than 2.5M transactions, together with demographic and promotional information.

*Temporal Setup and Treatment.* To ensure causal ordering, the timeline is divided into an observation window (Weeks 33–80), a short buffer (Weeks 81–85), and an outcome window (Weeks 86–101). A household is considered treated ($T_i = 1$) if it redeemed at least one coupon during the observation window. After filtering low-activity households, 801 remain, including 237 treated.

*Features and Outcome Grouping.* Features are aggregated from transaction logs (spending, frequency, basket value, promotion ratio, and category distribution) and combined with demographic attributes. Product categories are merged into three macro-groups: Grocery & Essentials (50.8%), Perishables & Ready-to-eat (25.2%), and Non-food & Lifestyle (24.0%). Each record contains covariates $X_i$, treatment $T_i$, and both total and per-category spending in the outcome window, maintaining clear temporal separation.

### 4.2 Evaluation Metrics

We evaluate ranking quality, individual-level accuracy, and structural coherence using scale-independent metrics.

*Ranking Metrics.* The **Qini coefficient** [14] measures uplift ranking performance via the normalized area between the model's cumulative uplift curve and a random baseline:

$$\text{Qini} = \frac{1}{n}\sum_{k=1}^{n}\left(\frac{k\,u(k)}{n\,u(n)} - \frac{k}{n}\right).$$

We also report **Uplift@20** and **Spearman** rank correlation.

*Individual-Level Accuracy (synthetic only).* With access to ground-truth effects $\tau_i$, we compute **PEHE** [18]:

$$\text{PEHE} = \frac{\frac{1}{N}\sum_i(\hat{\tau}_i - \tau_i)^2}{\text{Var}(\tau)}.$$

We additionally report **Kendall** correlation for local ordering agreement.

*Structural Consistency.* To measure cross-task coherence, we use the normalized **Consistency** score:

$$\text{Consistency} = 1 - \frac{\|\hat{U}_{\text{total}} - \sum_{k=1}^{K}\hat{U}_k\|_2}{\|\hat{U}_{\text{total}}\|_2 + \sum_{k=1}^{K}\|\hat{U}_k\|_2}.$$

Higher values indicate closer alignment between total and component uplifts.

### 4.3 Experimental Setup

*Baselines.* We compare **MUNet** with representative uplift modeling approaches, including both tree-based meta-learners and neural causal architectures:

- **S-Learner:** trains a single LightGBM model with treatment $T$ appended as a feature [9].
- **T-Learner:** fits separate models for treated and control groups, estimating uplift as the difference of predicted outcomes [9].
- **X-Learner:** refines T-Learner by imputing counterfactuals and learning second-stage treatment effect models [9].
- **TARNet** [18]: learns shared representations with treatment-specific heads.
- **CFRNet** [18]: extends TARNet with an IPM regularizer for balanced representation learning.
- **DragonNet** [19]: jointly predicts outcomes and propensity scores with targeted regularization.

We additionally include a naive multi-task baseline (TARNet-MTL), which shares a common encoder across outcomes without enforcing structural consistency. This enables a fair comparison to isolate the effect of the proposed constraint. All baselines share identical inputs, data partitions, and optimization settings for fairness.

*Training and Evaluation Protocol.* All models share identical preprocessing, data splits, and evaluation metrics to ensure a fair comparison. Neural architectures are trained with Adam (learning rate $10^{-3}$, batch size 128, weight decay $3 \times 10^{-4}$) for 100 epochs. Hidden dimensions, network depth, and dropout rates are coarsely tuned such that all neural baselines have comparable parameter counts. LightGBM-based meta-learners use 500 trees with maximum depth 5, learning rate 0.05, and minimum child weight 10. All metrics—except for the structural consistency score—are reported as outcome-weighted averages, where outcome weights correspond to each outcome's proportion of the total observed magnitude.

For evaluation, the synthetic dataset is assessed over 10 independently generated splits, capturing variability in latent factors, task correlations, and zero inflation. The Dunnhumby dataset is evaluated using 15 temporal splits, each repeated with 5 random seeds (75 runs in total) to account for sampling variability in real-world observational data. We report mean $\pm$ standard deviation across all repetitions.

*Ablation Studies.* We conduct a cumulative ablation study in which modules are added sequentially to a TARNet-style baseline (no shared encoder). This ladder isolates the marginal contribution of each component. The incremental steps are:

- **Shared Encoder**: introduces cross-task representation sharing across all outcomes.
- **Consistency Constraint** (MUNet-Base): enforces additive and hierarchical coherence between individual outcomes and the total outcome.

- **Content-Based Gating**: an MMoE-style mechanism that enables sample-dependent inter-task coupling by mixing shared experts using learned gating weights.
- **Outcome-Wise Uncertainty Weighting** (MUNet-X): adaptively balances prediction losses across outcomes based on learned noise levels.

All other settings (including IPW weighting) are kept identical across variants. This staged design allows us to quantify the incremental gains in uplift accuracy, ranking quality, and structural consistency contributed by each module.

### 4.4 Results

*Metric Usage.* For the synthetic dataset, we report **Qini**, **PEHE**, **Kendall**, and **Consistency**, leveraging access to ground-truth potential outcomes. For the Dunnhumby dataset—where true effects are unobservable—we evaluate **Qini**, **Uplift@20**, **Spearman**, and **Consistency**. All metrics are reported as mean $\pm$ standard deviation across repeated runs.

**Baselines vs. MUNet**

*Synthetic Dataset.* On the synthetic benchmark (Table 1), **MUNet-X** achieves the best Qini, Kendall, and Consistency. Compared with the naive multi-task baseline (TARNet-MTL), which already shares representations across outcomes, MUNet still yields consistent improvements, indicating that gains are not solely due to parameter sharing but are primarily driven by the structural consistency constraint. Relative to DragonNet, Qini improves from 0.650 to 0.692. X-Learner attains the lowest PEHE, reflecting the typical bias–variance trade-off.

*Dunnhumby Dataset.* On the real-world dataset (Table 2), MUNet-Base and MUNet-X outperform both independent neural baselines and TARNet-MTL. Qini increases from 0.156 (DragonNet) and 0.159 (TARNet-MTL) to 0.187, while Spearman rises from 0.540 to 0.612. Consistency reaches 0.959, the highest among all models, demonstrating the benefit of explicit structural regularization under noisy conditions.

**Ablation Studies**

*Synthetic Dataset.* Table 3 shows that the shared encoder provides moderate gains over independent modeling, while the structural consistency constraint produces the dominant improvement, particularly in cross-outcome coherence ($0.702 \rightarrow 0.842$). Additional components in MUNet-X further enhance ranking correlation.

**Table 1.** Baselines vs. MUNet on the Synthetic Dataset.

| Model | Qini | PEHE | Kendall | Consistency |
|---|---|---|---|---|
| S-Learner | 0.598 ± 0.289 | 0.926 ± 0.014 | 0.284 ± 0.013 | 0.796 ± 0.004 |
| T-Learner | 0.446 ± 0.166 | 0.956 ± 0.013 | 0.284 ± 0.005 | 0.716 ± 0.010 |
| X-Learner | 0.227 ± 0.102 | **0.836** ± **0.005** | 0.393 ± 0.007 | 0.631 ± 0.006 |
| TARNet | 0.587 ± 0.236 | 1.027 ± 0.018 | 0.283 ± 0.014 | 0.614 ± 0.037 |
| CFRNet | 0.532 ± 0.236 | 1.021 ± 0.017 | 0.289 ± 0.016 | 0.620 ± 0.029 |
| DragonNet | 0.650 ± 0.329 | 1.091 ± 0.016 | 0.236 ± 0.009 | 0.565 ± 0.028 |
| TARNet-MTL | 0.612 ± 0.322 | 0.984 ± 0.012 | 0.296 ± 0.015 | 0.702 ± 0.024 |
| MUNet-Base | 0.667 ± 0.313 | 0.841 ± 0.003 | 0.309 ± 0.017 | 0.842 ± 0.008 |
| MUNet-X | **0.692** ± **0.300** | 0.922 ± 0.006 | **0.400** ± **0.006** | **0.881** ± **0.006** |

**Table 2.** Baselines vs. MUNet on the Dunnhumby dataset.

| Model | Qini | Uplift@20 | Spearman | Consistency |
|---|---|---|---|---|
| S-Learner | 0.089 ± 0.015 | 1.364 ± 0.417 | 0.431 ± 0.187 | 0.846 ± 0.048 |
| T-Learner | 0.131 ± 0.107 | 1.265 ± 0.161 | 0.548 ± 0.197 | 0.732 ± 0.023 |
| X-Learner | 0.088 ± 0.089 | 1.268 ± 0.368 | 0.279 ± 0.121 | 0.871 ± 0.051 |
| TARNet | 0.147 ± 0.033 | 1.755 ± 0.441 | 0.537 ± 0.194 | 0.895 ± 0.018 |
| CFRNet | 0.150 ± 0.032 | 1.801 ± 0.495 | 0.535 ± 0.167 | 0.939 ± 0.022 |
| DragonNet | 0.156 ± 0.035 | 1.925 ± 0.485 | 0.540 ± 0.190 | 0.954 ± 0.030 |
| TARNet-MTL | 0.159 ± 0.039 | 1.874 ± 0.427 | 0.552 ± 0.188 | 0.946 ± 0.021 |
| MUNet-Base | 0.178 ± 0.031 | 1.949 ± 0.486 | 0.576 ± 0.231 | 0.958 ± 0.019 |
| MUNet-X | **0.187** ± **0.030** | **2.042** ± **0.553** | **0.612** ± **0.180** | **0.959** ± **0.021** |

**Table 3.** Ablation on the Synthetic dataset.

| Model Variant | Qini | PEHE | Kendall | Consistency |
|---|---|---|---|---|
| Independent models (K+1 nets) | 0.587 | 1.027 | 0.283 | 0.614 |
| TARNet-MTL (shared encoder) | 0.612 | 0.984 | 0.296 | 0.702 |
| + Consistency constraint (MUNet-Base) | 0.639 | 0.841 | 0.309 | 0.842 |
| + Content-based gating | 0.668 | 0.890 | 0.356 | 0.862 |
| + Uncertainty weighting (MUNet-X) | **0.692** | 0.922 | **0.400** | **0.881** |

*Dunnhumby Dataset.* A similar pattern holds in real data (Table 4). Parameter sharing improves alignment, but structural consistency drives the main performance jump. MUNet-X achieves the strongest overall results across ranking and coherence metrics.

**Sensitivity Analysis.** We further observed that MUNet's gains remain stable across variations in zero inflation (50–70%), uplift correlation ($\rho \in [-0.2, 0.5]$),

**Table 4.** Ablation on the Dunnhumby dataset.

| Model Variant | Qini | Uplift@20 | Spearman | Consistency |
|---|---|---|---|---|
| Independent models (K+1 nets) | 0.147 | 1.755 | 0.537 | 0.895 |
| TARNet-MTL (shared encoder) | 0.159 | 1.874 | 0.552 | 0.946 |
| + Consistency constraint (MUNet-Base) | 0.178 | 1.949 | 0.576 | 0.958 |
| + Content-based gating | 0.181 | 1.982 | 0.588 | 0.953 |
| + Uncertainty weighting (MUNet-X) | **0.187** | **2.042** | **0.612** | **0.959** |

and noise levels. While extreme substitution reduces absolute performance, MUNet-X consistently outperforms all baselines, indicating robustness to structural shifts.

## 5 Conclusion and Future Work

This paper proposed **MUNet**, a unified framework for *structured* multi-outcome uplift modeling that enforces coherent treatment-effect estimates across multiple correlated outcomes. By centering on a structural consistency regularization, supported by shared representations and optional robustness extensions, MUNet provides a principled approach to multi-dimensional causal estimation and ranking-based decision making in real-world intervention systems.

Experiments on both synthetic and real-world retail datasets demonstrate that **structural consistency regularization** is the primary driver of the performance gains beyond naive multi-task parameter sharing: it improves uplift ranking quality while simultaneously increasing cross-outcome coherence between component-level and aggregate effects. The shared encoder serves as a stable representation backbone that reduces redundancy across outcomes and enhances generalization. On top of this base, content-based gating helps capture sample-dependent cross-outcome relations, while uncertainty weighting improves robustness under heterogeneous noise and zero-inflated outcomes. Together, these components yield consistent improvements in uplift ranking and structurally coherent multi-outcome predictions.

Future work includes extending MUNet to **multi-armed** or **continuous** treatments [22], integrating **generative counterfactual** models such as CEVAE [10] and GANITE [23] to enhance counterfactual calibration, and generalizing the consistency principle to richer structured outcome relations (e.g., deeper hierarchies or graph-structured outcomes) via structured causal modeling [13] and multi-output causal learning methods [3]. Another promising direction is to couple uplift estimation with **bandit** or **reinforcement learning** to enable end-to-end policy optimization under budget and constraint-aware decision objectives.

Overall, MUNet offers a solid foundation for structured multi-outcome uplift modeling, and suggests that explicitly encoding cross-outcome causal coherence can be an effective inductive bias for reliable uplift-based decision systems.

## References

1. Alaa, A.M., Van Der Schaar, M.: Bayesian inference of individualized treatment effects using multi-task gaussian processes. In: Proceedings of the Advances in neural information processing systems, p. 30 (2017)
2. Athey, S., Imbens, G.: Recursive partitioning for heterogeneous causal effects. Proc. Natl. Acad. Sci. **113**(27), 7353–7360 (2016)
3. Caron, A., Baio, G., Manolopoulou, I.: Counterfactual learning with multioutput deep kernels (2022). arXiv preprint arXiv:2211.11119
4. Caruana, R.: Multitask learning. Mach. learn. **28**(1), 41–75 (1997)
5. Crawshaw, M.: Multi-task learning with deep neural networks: a survey (2020). arXiv preprint arXiv:2009.09796
6. Dunnhumby: the complete journey retail dataset. https://www.dunnhumby.com/source-files/. Accessed 01 Jan 2025
7. Jiang, Z., et al.: Estimating causal effects using a multi-task deep ensemble. In: Proceedings of the International Conference on Machine Learning, pp. 15023–15040. PMLR (2023)
8. Kendall, A., Gal, Y., Cipolla, R.: Multi-task learning using uncertainty to weigh losses for scene geometry and semantics. In: Proceedings of the IEEE Conference on Computer Vision and Pattern Recognition, pp. 7482–7491 (2018)
9. Künzel, S.R., Sekhon, J.S., Bickel, P.J., Yu, B.: Metalearners for estimating heterogeneous treatment effects using machine learning. Proc. Natl. Acad. Sci. **116**(10), 4156–4165 (2019)
10. Louizos, C., et al.: Causal effect inference with deep latent-variable models. In: Proceedings of the Advances in Neural Information Processing Systems, p. 30 (2017)
11. Ma, J., et al.: Modeling task relationships in multi-task learning with multi-gate mixture-of-experts. In: Proceedings of the 24th ACM SIGKDD International Conference on Knowledge Discovery & Data Mining, pp. 1930–1939 (2018)
12. Olaya, D., Coussement, K., Verbeke, W.: A survey and benchmarking study of multitreatment uplift modeling. Data Min. Knowl. Disc. **34**(2), 273–308 (2020). https://doi.org/10.1007/s10618-019-00670-y
13. Pearl, J.: Causality. Cambridge University Press (2009)
14. Radcliffe, N.: Using control groups to target on predicted lift: building and assessing uplift model. Direct Mark. Analytics J. 14–21 (2007)
15. Rosenbaum, P.R., Rubin, D.B.: The central role of the propensity score in observational studies for causal effects. Biometrika **70**(1), 41–55 (1983)
16. Ruder, S.: An overview of multi-task learning in. In: Proceedings of the Deep Neural Networks (2017) arXiv preprint arXiv:1706.05098
17. Rzepakowski, P., Jaroszewicz, S.: Decision trees for uplift modeling with single and multiple treatments. Knowl. Inf. Syst. **32**(2), 303–327 (2012)
18. Shalit, U., Johansson, F.D., Sontag, D.: Estimating individual treatment effect: generalization bounds and algorithms. In: Proceedings of the International Conference on Machine Learning, pp. 3076–3085. PMLR (2017)
19. Shi, C., Blei, D., Veitch, V.: Adapting neural networks for the estimation of treatment effects. Adv. Neural. Inf. Process. Syst. **32** (2019)
20. Sun, Z., Chen, X.: M 3 tn: multi-gate mixture-of-experts based multi-valued treatment network for uplift modeling. In: ICASSP 2024–2024 IEEE International Conference on Acoustics, Speech and Signal Processing (ICASSP), pp. 5065–5069. IEEE (2024)

21. Sverdrup, E., Wu, H., Athey, S., Wager, S.: Qini curves for multi-armed treatment rules. J. Comput. Graph. Stat. **34**(3), 948–960 (2025)
22. Wei, Y., Qiu, Z., Li, Y., Sun, Y., Li, X.: Multi-treatment multi-task uplift modeling for enhancing user growth (2024). arXiv preprint arXiv:2408.12803
23. Yoon, J., Jordon, J., Van Der Schaar, M.: Ganite: estimation of individualized treatment effects using generative adversarial nets. In: Proceedings of the International Conference on Learning Representations (2018)
24. Zhang, W., Li, J., Liu, L.: A unified survey of treatment effect heterogeneity modelling and uplift modelling. ACM Comput. Surv. (CSUR) **54**(8), 1–36 (2021)
25. Zhao, Y., Fang, X., Simchi-Levi, D.: Uplift modeling with multiple treatments and general response types. In: Proceedings of the 2017 SIAM International Conference on Data Mining, pp. 588–596. SIAM (2017)

# ProvGrapher: A Top-K Approximate Tuple-Level Data Provenance Framework via Graph Neural Network

Dehua Chen, Zhifen Yang(✉), and Ziteng He

School of Computer Science and Technology, Donghua University, Shanghai, China
yangzhifen@mail.dhu.edu.cn

**Abstract.** With the explosive growth of data scale, traditional annotation-based tuple-level data provenance methods face escalating challenges in storage overhead and inference efficiency. Recent vectorized approximate approaches for non-aggregate queries embed tuples into continuous spaces and infer lineage via vector similarity, improving scalability at the expense of some accuracy. However, these methods rely solely on tuple text to construct static representations, overlooking relational structures and query semantics, which limits provenance accuracy and generalization. To address this, we propose ProvGrapher, a graph-structured modeling approach for Top-K approximate tuple-level data provenance. ProvGrapher constructs data provenance graphs and integrates a Gated Multi-scale Multi-head Graph Attention Encoder (MSMH-GAT) with re-ranking optimization, which effectively capturing both table structures and query-driven data flow semantics to learn high-quality tuple embeddings. Experiments on the Olist and Movielens datasets, covering both single-table and multi-table SQL query scenarios, demonstrate that ProvGrapher consistently outperforms existing methods in terms of provenance coverage and ranking quality, while maintaining robustness with increasing data scale. Ablation studies further validate the effectiveness of key components including graph compression, gated multi-scale multi-head attention, and re-ranking. ProvGrapher provides a practical and scalable solution for approximate tuple-level data provenance in complex environments.

**Keywords:** Approximate Tuple-Level Data Provenance · Graph Neural Network · Representation Learning · Top-K Re-ranking

## 1 Introduction

Data provenance is a crucial technique for ensuring data quality and trustworthiness by tracing the origins and transformation paths of target data. Its core value lies in providing critical support for effective root cause localization when anomalous data occurs. To describe and organize data dependencies, provenance research has established a conceptual framework centered on three core semantics: why-provenance and where-provenance [1], which respectively identify contributing source data and their original locations, and how-provenance, which describes transformation processes. The


R. C.-W. Wong et al. (Eds.): PAKDD 2026, LNAI 16597, pp. 263–275, 2026.
https://doi.org/10.1007/978-981-92-1300-9_21

W3 model [2] integrates these three aspects into a consistent representation, while later standards such as PROV [3], PROV-JSON [4], and BDMP [5] extend it to support interoperability across diverse systems and application domains. Within relational data environments, provenance tasks are concretely manifested as tracing the origins and transformation paths of results generated by SQL queries. By granularity, provenance can be categorized into three levels: table-level, column-level and tuple-level (corresponding to why-provenance). Table-level provenance aims to identify the set of source tables contributing to a query result; column-level provenance focuses on the mapping relationships and transformation paths between columns; and tuple-level provenance further identifies the specific set of source tuples that contribute to each result tuple.

In this paper, we focus on tuple-level data provenance, the finest-grained form, which enhances data explainability but incurs high overhead. Existing methods to this task can be broadly categorized into annotation-based and annotation-free methods. Annotation-based methods rely on explicit annotation mechanisms, introducing substantial metadata during data processing to construct tuple-level lineage paths [6–10]. To overcome the reliance on explicit annotations, some efforts [11–13] infer source tuples by constructing inverse queries or using schema mappings, but their applicability is limited, particularly for non-reversible or complex queries where reliable inverse reasoning is infeasible. In contrast, recent work [14] has made preliminary progress in non-aggregate scenarios by text embeddings and vectorized similarity to infer potential lineage relationships approximately, offering valuable insights for annotation-free data provenance. However, it still has two main limitations: (1) it overlooks structural information inherent in relational data, which is essential for accurate semantic modeling of tuples; (2) it relies on static embeddings that cannot adapt to dynamic query semantics. These limitations lead to significantly reduced provenance accuracy, especially in complex queries involving cross-table joins.

To address the limitations of existing methods in non-aggregate query scenarios, this paper proposes ProvGrapher, a Top-K approximate tuple-level data provenance framework that formulates provenance as a graph-based representation learning task. The main contributions are as follows:

1. We propose semantics-enhanced data provenance graphs that integrate table structures and data flow semantics within query execution, combined with a graph compression strategy to improve tuple embedding expressiveness and efficiency.
2. We design a unified graph representation learning architecture that integrates a Gated Multi-scale Multi-head Graph Attention Encoder (MSMH-GAT) for capturing multi-granularity semantic dependencies, and a jointly optimized attention-based re-ranking network for refining Top-K approximate provenance results.
3. We conduct comprehensive experiments and demonstrate that our method outperforms baselines on approximate tuple-level data provenance for non-aggregate queries, particularly in cross-table query scenarios and large-scale data environments.

## 2 Related Work

In annotation-based tuple-level data provenance, source data and transformation steps are explicitly labeled. These annotations are stored and propagated together with the data, enabling direct access to provenance information in subsequent queries. Among existing works, provenance semiring approaches [6–8] are representative, using relational algebra to capture fine-grained lineage and ensure consistent results for equivalent queries. ProvSQL [9] is an open-source PostgreSQL extension built on provenance semiring theory, supporting provenance semantics as defined by the W3 model to efficiently trace the origins of SQL query results. [10] proposed PET (Provenance Explains Transformation), introducing origin graphs and origin paths for multi-granularity provenance. Although accurate and interpretable, annotation-based methods incur high storage overhead, as annotations can grow with query depth and even surpass the original data size.

Annotation-free tuple-level data provenance methods do not precompute and store provenance information as annotations. Instead, they infer data lineage retrospectively using limited system metadata, such as SQL statements and table schemas. [11] employed schema mappings to model transformations between input and output schemas, reconstructing provenance by reversing these mappings. [12, 13] constructed inverse queries to trace result tuples back to contributing source tuples and diagnose unexpected outputs, supporting aggregation and multi-relational queries, but remaining ineffective for nested queries. [14] proposed an approximate tuple-level provenance method for non-aggregate queries based on word embeddings. It represents tuples as vectors and estimates potential lineage by measuring similarity, thereby avoiding explicit tracking of all data flow paths and significantly reducing storage costs. However, the use of static embeddings limits its ability to capture table structures and query semantics, resulting in reduced accuracy.

# 3 Problem Formulation

## 3.1 Tuple-Level Data Provenance

This work focuses on provenance analysis for non-aggregate queries within the SELECT-FROM-WHERE query structure, supporting operations such as JOIN, WHERE, SELECT, and ORDER BY. Given a SQL query $Q$ executed on a set of tables $T = \{T_1, T_2, \ldots, T_m\}$, the execution produces a set of result tuples $R = Q(T)$. For any result tuple $r \in R$, its data provenance, denoted as $Prov(r)$, is defined as follows:

$$Prov(r, T_i) = \{(t, T_i) | t \in T_i \wedge t \xrightarrow{Q} r\} \tag{1}$$

$$Prov(r) = \bigcup_{T_i \in S} Prov(r, T_i) \tag{2}$$

where $S \subseteq T$ denotes the subset of source tables that actually contribute column values to the result tuples, $t \xrightarrow{Q} r$ denotes that a source tuple $t$ contributes column values to the result tuple $r$ via query $Q$.

It should be emphasized that the result set $R$ has been deduplicated, as tuples with identical values can be regarded as representing the same semantic unit. Consequently, all such duplicate result tuples are merged into a single entity $r$, and the corresponding source tuples from $T_i$ are aggregated into a label set $Prov(r, T_i)$, thereby establishing a one-to-many provenance relationship.

### 3.2 K-Approximate Tuple-Level Data Provenance

For any result tuple $r \in R$, its Top-K approximate (K-approximate) tuple-level data provenance, denoted as $ApproxProv(r, K)$, is defined as follows:

$$ApproxProv(r, T_i, K) = \{(t, T_i) | t \in argTopK_{x \in T_i} Sim(\phi(x), \phi(r))\} \quad (3)$$

$$ApproxProv(r, K) = \bigcup_{T_i \in S} ApproxProv(r, T_i, K) \quad (4)$$

where $\phi$ is the tuple embedding function that maps tuples into the $\mathbb{R}^d$ vector space. $Sim(\cdot)$ is the similarity function measuring semantic relevance between tuples in the embedding space. $argTopK_{x \in T_i} Sim(\phi(x), \phi(r))$ Returns the Top-K tuples $x \in T_i$ ranked in descending order of the similarity score. It is worth noting that $ApproxProv(r, T_i, K)$ is defined as a set for simplicity and deduplicated, like $Prov(r, T_i)$, although it is originally derived from an ordered list of Top-K tuples. In subsequent evaluation, the ranking information is explicitly incorporated into the metrics MRR@K and MAP@K to assess provenance quality.

## 4 Methodology

The overall architecture of our ProvGrapher model is shown in Fig. 1. Unlike approaches that directly vectorize entire tuples or simply average column embeddings, we construct a provenance graph that integrates table structures and data flow semantics. Based on this graph, we design a graph neural network to obtain more discriminative tuple embeddings, thereby enhancing provenance accuracy and generalization. During training, we introduce a re-ranking network for joint optimization to further improve the ranking quality of Top-K approximate provenance tuples.

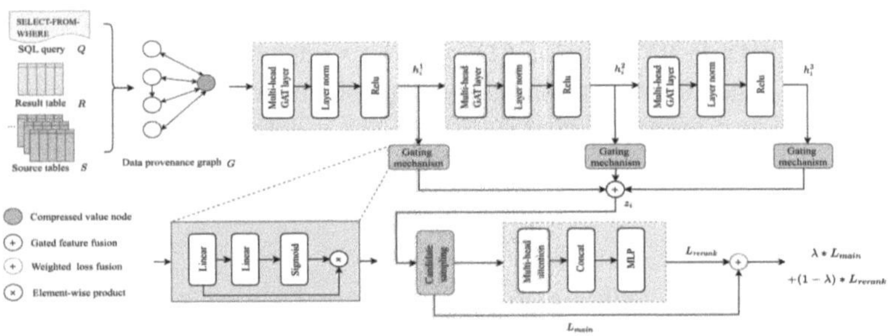


**Fig. 1.** Overall architecture of the ProvGrapher model.

### 4.1 Data Provenance Graph Construction

We propose a directed heterogeneous data provenance graph model that integrates table structures and data flow semantics. Given a SQL query $Q$, the corresponding data provenance graph $G = (V, E)$ is constructed, where nodes represent columns, values, and tuples, and edges model data flow and structural associations (see Table 1).

**Table 1.** Node and edge types in the data provenance graph.

| Element Type | Symbol | Description |
|---|---|---|
| Node | $C^G$ | a column node |
| | $V^G$ | a value node |
| | $T^G$ | a tuple node |
| Edge | $C_s{}^G \rightarrow C_t{}^G$ | a source-to-target column dependency |
| | $C^G \leftrightarrow V^G$ | an association between a column and a value |
| | $T^G \leftrightarrow V^G$ | an association between a tuple and a value |

In this work, we develop a SQL parser based on ANTLR4 [15] to automatically extract the set of contributing source tables $S$, as well as the set of column pairs $\{(C_s{}^G, C_t{}^G)\}$ that capture data flow between source and target columns. Together with the result set $R$, we instantiate nodes of $C^G$, $V^G$, $T^G$ and their corresponding edges to generate the data provenance graph $G$. To enhance semantic aggregation, we propose a value node compression strategy that merges identical values across multiple columns into shared nodes, strengthening semantic links along $T^G \leftrightarrow V^G \leftrightarrow C^G$ paths. This design enables tuple nodes to better capture the semantic context inherent in relational data during representation learning, particularly enhancing cross-table semantic modeling in multi-table query scenarios. Moreover, it reduces graph size, thereby enhancing training efficiency and scalability.

## 4.2 Gated MSMH-GAT

We propose a Multi-Scale Multi-Head Graph Attention Encoder (MSMH-GAT), which adopts a stacked GAT [16] architecture. Multiple layers are applied in series to progressively expand the receptive field and extract multi-scale semantic features at different structural depths. Within each scale, multi-head attention is employed to enhance representation diversity and improve training stability. Given a graph $G = (V, E)$, for any node $v_i \in V$, at layer $l \in \{1,2,\ldots,L\}$, the GAT employs $K$ attention heads. The node representation at layer $l$, denoted as $h_i{}^l$, is computed as follows:

$$\alpha_{ij}{}^{(l,k)} = \frac{exp(LeakyReLU(a^{(l,k)^T}[W^{(l,k)}h_i{}^{(l-1)} \| W^{(l,k)}h_j{}^{(l-1)}]))}{\sum_{o \in N(i)} exp(LeakyReLU(a^{(l,k)^T}[W^{(l,k)}h_i{}^{(l-1)} \| W^{(l,k)}h_o{}^{(l-1)}]))} \tag{5}$$

$$h_i{}^{(l,k)} = \sigma(\sum\nolimits_{j \in N(i)} \alpha_{ij}{}^{(l,k)} W^{(l,k)} h_j{}^{(l-1)}) \tag{6}$$

$$h_i{}^l = ReLU(LayerNorm\left(\|_{k=1}^K h_i{}^{(l,k)}\right)) \tag{7}$$

where $N(i)$ denotes the neighborhood of $v_i$, $W^{(l,k)}$ is a learnable weight matrix for the $k$-th attention head at layer $l$, and $a^{(l,k)}$ is the corresponding attention vector. LayerNorm and ReLU are applied to enhance non-linear representation capacity and improve training stability. $\cdot^T$ denotes the transpose operation and $\|$ denotes concatenation.

Multi-scale features capture hierarchical information ranging from local structures to global semantics. To adaptively fuse multi-scale features, we design a dynamic aggregation mechanism based on gating weights. For each node $v_i \in V$, the fusion of its multi-scale features $\{{h_i}^1, \ldots, {h_i}^L\}$ is computed as follows:

$$\hat{h_i}^l = {W_p}^l h_i^l + {b_p}^l \tag{8}$$

$$g_i^l = \sigma({W_g}^l \hat{h_i}^l + {b_g}^l) \tag{9}$$

$$z_i = \sum\nolimits_{l=1}^{L} g_i^l \odot \hat{h}_i^l \tag{10}$$

where $\hat{h_i}^l$ denotes $l$-th scale feature projected into a unified representation space, and ${g_i}^l \in (0,1)^d$ is a gating vector generated by the gating network, indicating the importance of the $l$-th scale feature. The operator $\odot$ denotes element-wise multiplication, and $z_i$ is the final fused representation of node $v_i$.

### 4.3 Re-ranking Network

First, for each source table $T_i$ involved with the result tuple $r$, a set of candidate source tuples is constructed, consisting of $M$ positive samples and $N$ negative samples. The positive sample set and negative sample set are constructed sequentially as follows:

$$\mathcal{N}(r, T_i)^+ = \{\, t^+ | (t^+, T_i) \in Prov(r, T_i)\} \tag{11}$$

$$\mathcal{N}(r, T_i)^- = argTopN_{x \in Neg} Sim(z_r, z_x) \tag{12}$$

where $Neg = \{t^- | (t^-, T_i) \notin Prov(r, T_i) \wedge t^- \in T_i\}$ represents the set of all non-contributing tuples, while $\mathcal{N}(r, T_i)^-$ is the set of $N$ hard negative samples selected via Top-N similarity retrieval. Correspondingly, the embedding vectors of the candidate source tuples constitute the set $\{{z_t}^{(i)}\}_{i=1}^{M+N}$.

Then, we employ the standard multi-head attention mechanism [17] to compute the interaction representation between the result tuple and the candidate source tuples. The attention output is an enhanced representation of the result tuple:

$$o_r = MultiHead(z_r, \{{z_t}^{(i)}\}_{i=1}^{M+N}, \{{z_t}^{(i)}\}_{i=1}^{M+N}) \tag{13}$$

Subsequently, $o_r \in \mathbb{R}^d$ is concatenated with each candidate source tuple embedding and passed through a multilayer perceptron (MLP) to obtain the corresponding candidate scores:

$$s^{(i)} = W_2 \cdot ReLU(W_1[o_r \| {z_t}^{(i)}] + b_1) + b_2 \tag{14}$$

where $W_1, W_2, b_1, b_2$ are learnable parameters, and $\|$ denotes vector concatenation.

### 4.4 Joint Loss

The overall training objective jointly optimizes the encoder and the re-ranking module by combining a main encoding loss $L_{main}$ and an auxiliary re-ranking loss $L_{rerank}$, represented as follows:

$$L = \lambda \cdot L_{main} + (1 - \lambda) \cdot L_{rerank} \tag{15}$$

where $\lambda$ is the weighting coefficient between the two loss terms.

For the encoder, we design an adaptive-margin Top-N hinge loss, which extends the classical hinge loss with Top-N hard negative mining and a dynamic margin mechanism to enhance the model's ability to separate positives from hard negatives and adaptively adjust the margin constraint. Given a given query $Q$ with result set $R = Q(T)$, the encoder's loss $L_{main}(Q)$ is represented as follows:

$$L_{main}(Q, r, T_i) = \frac{1}{|\mathcal{N}^+||\mathcal{N}^-|} \sum\nolimits_{t^+ \in \mathcal{N}^+} \sum\nolimits_{t^- \in \mathcal{N}^-} max(0, \delta - Sim(z_r, z_{t^+}) + Sim(z_r, z_{t^-})) \tag{16}$$

$$L_{main}(Q) = \frac{1}{|R|} \sum\nolimits_{r \in R} \frac{1}{|S|} \sum\nolimits_{T_i \in S} L_{main}(Q, r, T_i) \tag{17}$$

where $\delta$ is a learnable parameter, $\mathcal{N}^+ = \mathcal{N}(r, T_i)^+$ and $\mathcal{N}^- = \mathcal{N}(r, T_i)^-$ are constructed as in Eqs. 11 and 12, respectively.

For the re-ranking module, we introduce a listwise re-ranking loss as an auxiliary supervision signal. The re-ranking network outputs a score vector $s = [s^{(1)}, s^{(2)}, ..., s^{(M+N)}]$, which is transformed into a predicted ranking distribution $\tilde{y} = softmax(s)$. The loss minimizes the KL divergence between this predicted distribution and the ground-truth distribution $y$, as follows:

$$L_{rerank} = KL(y||\tilde{y}) \tag{18}$$

## 5 Experiments

### 5.1 Datasets

In this work, we construct two relational databases based on publicly available datasets: Movielens [18] and Olist [19], denoted as the M and O datasets, respectively (see Table 2). To evaluate the stability and generalization under different data scales, we divide each dataset into three subsets of increasing size: I, II, and III, corresponding to 1/4, 1/2, and the full dataset, denoted as MI-MIII and OI-OIII. For each dataset, we uniformly construct 50 single-table and 50 multi-table SQL queries. We use S and M to denote the single-table and multi-table query scenarios, respectively. Considering the number of columns in each dataset and GPU memory constraints of our experimental environment, multi-table queries on the O dataset are limited to two-table joins to ensure stable evaluation under controlled graph complexity. In contrast, 50% of multi-table queries on the M dataset involve three-table joins, introducing higher structural complexity for evaluating model generalization.

**Table 2.** Details of the datasets used in this work.

| Dataset | Table | #columns | #tuples |
|---|---|---|---|
| Olist | customers | 5 | 10000 |
| | sellers | 4 | 3095 |
| | orders | 8 | 10000 |
| | order_items | 7 | 22590 |
| Movielens | movies | 3 | 3883 |
| | users | 5 | 6040 |
| | ratings | 4 | 40000 |

### 5.2 Evaluation Metrics

For evaluation, ground-truth provenance labels are obtained by rewriting SQL queries with annotation fields that propagate source tuple identifiers during execution. Since DISTINCT is not enforced in non-aggregate queries, we therefore deduplicate result tuples by value and aggregate their associated source tuples, forming a one-to-many multi-label evaluation setting consistent with Eq. 3. We evaluate K-approximate tuple-level provenance using four standard Top-K metrics: Hit@K, Recall@K, MRR@K, and MAP@K. Each metric is computed according to its standard definition, averaged over source tables and result tuples within each query, and then averaged across all queries to assess overall model performance.

### 5.3 Baselines and Implementations

We set up six baseline models for comparison. First, based on the two vectorization strategies proposed in [14]: Tuple-based (TV) and Column-based (CV), combined with the two current mainstream text embedding models: BGE-base-en-v1.5 [20] and Nomic-embed-text-v1.5 [21], four static vectorization baselines are constructed: BGE_TV, NOMIC_TV, BGE_CV, and NOMIC_CV. Among them, the TV strategy encodes entire tuples as a whole, while the CV strategy encodes each column independently and then applies mean pooling. In addition, we include two classic GNN methods: GCN [22] and Graph Transformer Network (denoted as GTN for brevity) [23], as comparative baselines for graph representation learning.

Although the provenance graph is theoretically heterogeneous, it is simplified into a homogeneous graph during training and inference to reduce computational cost. To preserve type-distinguishing information, different node types are initialized using type-specific strategies: column nodes with column name text embeddings, value nodes with value text embeddings, and tuple nodes as the average of their value node vectors. All node embeddings are generated using Nomic-embed-text-v1.5. Models are trained using Adam with early stopping (max 300 epochs). To reduce computational overhead during training, Top-N hard negative mining randomly samples 1000 negative candidates and selects the 20 most similar as hard negatives. The re-ranking network also uses 20 hard negatives.

### 5.4 Results and Analysis

We evaluated K-approximate tuple-level data provenance on the OII and MII data subsets, covering both single-table and multi-table queries, for $K \in \{1, 5, 10, 20\}$. Due to space limitations, Table 3 reports only Recall@K and MAP@K, reflecting provenance coverage and ranking quality respectively.

**Table 3.** Performance comparison on OII-S, MII-S, OII-M and MII-M.

| Dataset | Method | Recall@1 | MAP@1 | Recall@5 | MAP@5 | Recall@10 | MAP@10 | Recall@20 | MAP@20 |
|---|---|---|---|---|---|---|---|---|---|
| OII-S | BGE_TV | 0.170 | 0.170 | 0.241 | 0.201 | 0.268 | 0.205 | 0.305 | 0.208 |
| | NOMIC_TV | 0.184 | 0.184 | 0.261 | 0.218 | 0.297 | 0.223 | 0.342 | 0.227 |
| | BGE_CV | 0.618 | 0.618 | 0.815 | 0.720 | 0.852 | 0.726 | 0.883 | 0.728 |
| | NOMIC_CV | 0.522 | 0.522 | 0.706 | 0.616 | 0.748 | 0.623 | 0.784 | 0.626 |
| | GCN | 0.813 | 0.813 | 0.955 | 0.893 | 0.971 | 0.895 | 0.981 | 0.896 |
| | GTN | 0.808 | 0.808 | 0.934 | 0.882 | 0.957 | 0.886 | 0.973 | 0.887 |
| | **ProvGrapher** | **0.913** | **0.913** | **0.996** | **0.973** | **0.998** | **0.973** | **0.999** | **0.974** |
| MII-S | BGE_TV | 0.284 | 0.284 | 0.407 | 0.330 | 0.455 | 0.337 | 0.503 | 0.340 |
| | NOMIC_TV | 0.374 | 0.374 | 0.452 | 0.405 | 0.490 | 0.410 | 0.517 | 0.412 |
| | BGE_CV | 0.657 | 0.657 | 0.847 | 0.737 | 0.900 | 0.744 | 0.940 | 0.747 |
| | NOMIC_CV | 0.649 | 0.649 | 0.834 | 0.724 | 0.874 | 0.729 | 0.911 | 0.732 |
| | GCN | 0.837 | 0.837 | 0.940 | 0.879 | 0.960 | 0.882 | 0.974 | 0.883 |
| | GTN | 0.831 | 0.831 | 0.947 | 0.879 | 0.968 | 0.882 | 0.981 | 0.883 |
| | **ProvGrapher** | **0.914** | **0.914** | **0.975** | **0.939** | **0.984** | **0.941** | **0.992** | **0.941** |
| OII-M | BGE_TV | 0.176 | 0.176 | 0.247 | 0.202 | 0.274 | 0.205 | 0.304 | 0.207 |
| | NOMIC_TV | 0.172 | 0.172 | 0.253 | 0.201 | 0.289 | 0.206 | 0.330 | 0.209 |
| | BGE_CV | 0.343 | 0.343 | 0.542 | 0.418 | 0.623 | 0.429 | 0.695 | 0.435 |
| | NOMIC_CV | 0.255 | 0.255 | 0.458 | 0.333 | 0.541 | 0.344 | 0.624 | 0.350 |
| | GCN | 0.677 | 0.677 | 0.793 | 0.724 | 0.824 | 0.728 | 0.849 | 0.730 |
| | GTN | 0.735 | 0.735 | 0.851 | 0.784 | 0.886 | 0.789 | 0.913 | 0.791 |
| | **ProvGrapher** | **0.874** | **0.874** | **0.964** | **0.916** | **0.980** | **0.919** | **0.990** | **0.920** |
| MII-M | BGE_TV | 0.192 | 0.192 | 0.205 | 0.197 | 0.215 | 0.198 | 0.228 | 0.199 |
| | NOMIC_TV | 0.196 | 0.196 | 0.214 | 0.202 | 0.224 | 0.204 | 0.238 | 0.205 |
| | BGE_CV | 0.425 | 0.425 | 0.578 | 0.485 | 0.651 | 0.496 | 0.715 | 0.501 |
| | NOMIC_CV | 0.475 | 0.475 | 0.645 | 0.542 | 0.710 | 0.553 | 0.765 | 0.557 |
| | GCN | 0.781 | 0.781 | 0.854 | 0.814 | 0.883 | 0.820 | 0.900 | 0.822 |
| | GTN | 0.779 | 0.779 | 0.853 | 0.812 | 0.882 | 0.818 | 0.903 | 0.821 |
| | **ProvGrapher** | **0.830** | **0.830** | **0.910** | **0.864** | **0.936** | **0.869** | **0.958** | **0.872** |

**The Single-Table Query Scenario.** TV baselines generally perform poorly because they ignore structural relationships and semantic distinctions across columns, which leads to substantial information loss. In contrast, CV baselines perform better because they preserve some degree of column-level semantics. All graph-based models outperform both TV and CV on all metrics, with our proposed model ProvGrapher achieving the best performance. Specifically, on all data subsets, it achieves Recall@10 above 0.980 and MAP@10 above 0.940, demonstrating superior coverage and ranking capabilities. Notably, it shows clear advantages at K = 1, achieving approximately 9%–13% relative improvement on all metrics over GCN and GTN.

**The Multi-table Query Scenario.** In this setting, the task becomes more challenging due to cross-table joins and dependencies. The static vectorization strategies (TV, CV) show limited effectiveness, struggling to capture cross-table semantics. And the graph-based baselines (GCN, GTN) exhibit clear bottlenecks when modeling complex dependency chains. In contrast, ProvGrapher maintains top performance on the OII and MII data subsets, achieving Recall@10 of 0.980 and 0.936, and MAP@10 of 0.919 and 0.869, respectively. The slight performance drop on MII-M is expected due to three-table joins, compared with two-table joins on OII-M.

**Model Performance under Data Scaling.** Figure 2 illustrates the performance trends of each model as the dataset scales from I to III. CV baselines exhibit significant performance degradation, particularly in multi-table query scenarios. For example, the performance of BGE_CV drops by more than 0.1 on all metrics from OI-M to OIII-M. In contrast, TV models remain consistently poor. Graph-based methods exhibit strong robustness, with ProvGrapher maintaining performance fluctuations within a narrow range of 0.03–0.07 on all metrics. Even under the most complex setting, MIII-M, it achieves Recall@10 of 0.919, MAP@10 of 0.852, and Recall@1(=MAP@1) of 0.815, all significantly outperforming baselines. These results suggest that our method offers strong robustness to noise and structural complexity caused by data scaling, making it well-suited for high-quality provenance in complex environments.

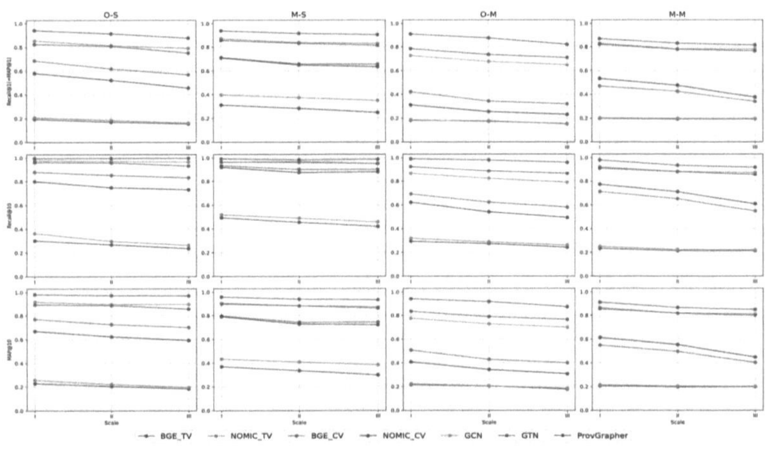


**Fig. 2.** Model performance across different data scales for single- and multi-table query settings.

### 5.5 Ablation Studies

To comprehensively evaluate the contribution of each key module to model performance, we conducted ablation experiments. Specifically, we progressively removed three core components: the graph compression mechanism (w/o Compression), the gated MSMH attention mechanism (w/o Gated MSMH), and the re-ranking network(w/o Re-ranking). The main results are summarized in Table 4.

As shown in Fig. 3, the graph compression rate of each data configuration ranges from 39.9% to 59.1%, significantly lowering storage and computational costs. In terms of performance, on both OII-S and OII-M, the model with graph compression significantly outperforms the one without compression across all four metrics ($K \in \{1, 10\}$). The improvement is particularly notable at $K = 1$, with all metrics increasing by approximately 0.2. These results show that graph compression not only alleviates computational overhead but also enhances overall provenance performance.

**Table 4.** Performance comparison of ablation experiments on OII-S and OII-M.

| Dataset | Method | Hit@1 | Recall@1 | Hit @10 | Recall@10 | MRR@10 | MAP@10 |
|---|---|---|---|---|---|---|---|
| OII-S | **ProvGrapher** | 0.955 | 0.913 | 0.999 | 0.998 | 0.974 | 0.973 |
| | w/o Compression | 0.754 | 0.716 | 0.874 | 0.872 | 0.794 | 0.792 |
| | w/o Gated MSMH | 0.854 | 0.815 | 0.972 | 0.971 | 0.899 | 0.899 |
| | w/o Re-ranking | 0.874 | 0.834 | 0.977 | 0.976 | 0.914 | 0.913 |
| OII-M | **ProvGrapher** | 0.886 | 0.874 | 0.981 | 0.980 | 0.920 | 0.919 |
| | w/o Compression | 0.667 | 0.659 | 0.777 | 0.774 | 0.701 | 0.699 |
| | w/o Gated MSMH | 0.780 | 0.770 | 0.921 | 0.919 | 0.827 | 0.826 |
| | w/o Re-ranking | 0.809 | 0.801 | 0.968 | 0.966 | 0.864 | 0.864 |

Note: Hit@1 = MRR@1 and Recall@1 = MAP@1. Thus, only Hit@1 and Recall@1 are shown

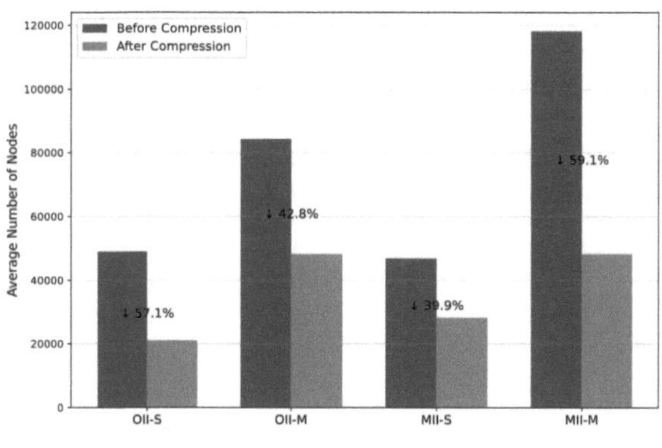


**Fig. 3.** Reduction in graph size through compression mechanism.

To verify the effectiveness of the MSMH-GAT and gated fusion mechanism, we designed a comparative experiment to remove the module and replace it with a single-scale attention layer, while keeping other structures unchanged. On OII-S, Hit@10 and Recall@10 (coverage-oriented metrics) drop by approximately 0.03, while on complex OII-M, the decline increases to around 0.06. More importantly, MAP@10 and MRR@10 (ranking-oriented metrics) drop by around 0.1 on both data configurations. Moreover, for $K = 1$, all metrics consistently show a decline of approximately 0.1, further indicating the model's weakened ability to accurately identify the most relevant tuples. These results demonstrate that the MSMH architecture and the gated mechanism jointly enhance the semantic expressiveness of node representations and play a crucial role in improving ranking quality and handling complex queries.

After removing the re-ranking module during training, we observe a notable decline on ranking-oriented metrics. On OII-S, MRR@10 and MAP@10 drop by nearly 0.06, while Hit@10 and Recall@10 show only minor fluctuations (around 0.02). Similar trends are observed on OII-M. These results highlight that the re-ranking network plays a crucial role in improving the encoder's ability to rank candidate source tuples.

## 6 Conclusion

In this paper, we propose ProvGrapher, a novel approximate tuple-level data provenance model, which constructs data provenance graphs and integrates the Gated MSMH-GAT module with the re-ranking optimization module. ProvGrapher effectively captures both table structural semantics and query-driven data flow semantics, generating highly discriminative tuple embeddings to support similarity-based Top-K approximate provenance. To enhance efficiency, ProvGrapher adopts a node-level graph compression strategy during the graph construction phase to significantly reduce computational and storage overhead. Experiments on both Olist and Movielens datasets, covering single-table and multi-table K-approximate provenance tasks, demonstrate that ProvGrapher significantly outperforms existing baselines on multiple metrics, with especially strong advantages in complex multi-table scenarios. Overall, ProvGrapher provides an effective solution for approximate tuple-level provenance in non-aggregate query scenarios. In the future, we plan to further explore approximate provenance mechanisms that support aggregate semantics, as well as improve graph modeling and inference strategies.

**Acknowledgments.** This work was supported in part by the Fundamental Research Funds for the Central Universities, and the DHU Distinguished Young Professor Program.

## References

1. Buneman, P., Khanna, S., Tan, W.-C.: Why and where: a characterization of data provenance. In: Database Theory-ICDT 2001, pp. 316–330. Springer (2001)
2. Cheney, J., Chiticariu, L., Tan, W.-C.: Provenance in databases: why, how, and where. Found. Trends Databases **1**(4), 379–474 (2009)

3. Missier, P., Belhajjame, K., Cheney, J.: The W3C PROV family of specifications for modelling provenance metadata. In: Proceedings of the 16th International Conference on Extending Database Technology, pp. 773–776. ACM (2013)
4. Niu, X., et al.: Interoperability for provenance-aware databases using PROV and JSON. In: Proceedings of the 7th USENIX Conference on Theory and Practice of Provenance. USENIX Association (2015)
5. Gao, Y., Chen, X., Du, X.: A big data provenance model for data security supervision based on PROV-DM model. IEEE Access **8**, 38742–38752 (2020)
6. Green, T.J., Karvounarakis, G., Tannen, V.: Provenance semirings. In: Proceedings of the Twenty-Sixth ACM SIGMOD-SIGACT-SIGART Symposium on Principles of Database Systems, pp. 31–40. ACM (2007). https://doi.org/10.1145/1265530.1265535
7. Amsterdamer, Y., Deutch, D., Tannen, V.: Provenance for aggregate queries. In: Proceedings of the Thirtieth ACM SIGMOD-SIGACT-SIGART Symposium on Principles of Database Systems, pp. 153–164. ACM (2011). https://doi.org/10.1145/1989284.1989302
8. Senellart, P.: Provenance and probabilities in relational databases. SIGMOD Record **46**(4), 5–15 (2018). https://doi.org/10.1145/3186549.3186551
9. Sen, A., Maniu, S., Senellart, P.: ProvSQL: a general system for keeping track of the provenance and probability of data. arXiv preprint arXiv:2504.12058 (2025)
10. Tang, D., Zhao, R., Lin, Y., Zhang, T., Zhang, P.: Modeling the data provenance of relational databases supporting full-featured SQL and procedural languages. Appl. Sci. **13**(1), 64 (2022). https://doi.org/10.3390/app13010064
11. Alexe, B., Chiticariu, L., Tan, W.-C.: SPIDER: a schema mapping debugger. In: Proceedings of the 32nd International Conference on Very Large Data Bases, pp. 1179–1182. VLDB Endowment (2006)
12. Srivastava, D., Velegrakis, Y.: Intensional associations between data and metadata. In: Proceedings of the 2007 ACM SIGMOD International Conference on Management of Data, pp. 401–412. ACM (2007). https://doi.org/10.1145/1247480.1247526
13. Uddin, M.S., Alexandrov, D.V., Rahman, A.: Query inversion to find data provenance. In: Computer Science & Information Technology 11th International Conference on Security and its Applications, pp. 17–31. AIRCC (2018). https://doi.org/10.5121/csit.2018.80102
14. Leybovich, M., Shmueli, O.: ML based lineage in databases. arXiv preprint arXiv:2109.06339 (2021)
15. Parr, T., Fisher, K.: LL(*): the foundation of the ANTLR parser generator. SIGPLAN Notices **46**(6), 425–436 (2011). https://doi.org/10.1145/2345156.1993548
16. Veličković, P., Cucurull, G., Casanova, A., Romero, A., Liò, P., Bengio, Y.: Graph attention networks. arXiv preprint arXiv:1710.10903 (2018)
17. Vaswani, A., et al.: Attention is all you need. arXiv preprint arXiv:1706.03762 (2023)
18. Harper, F.M., Konstan, J.A.: The MovieLens datasets: history and context. ACM Trans. Interact. Intell. Syst. **5**(4), 1–19 (2015)
19. Sionek, A.: Brazilian E-Commerce Public Dataset by Olist. Kaggle (2018)
20. Xiao, S., et al.: C-pack: packed resources for general Chinese embeddings. arXiv preprint arXiv: 2309.07597 (2024)
21. Nussbaum, Z., Morris, J.X., Duderstadt, B., Mulyar, A.: Nomic embed: training a reproducible long context text embedder. arXiv preprint arXiv:2402.01613 (2025)
22. Kipf, T.N., Welling, M.: Semi-supervised classification with graph convolutional networks. arXiv preprint arXiv:1609.02907 (2017)
23. Shi, Y., Huang, Z., Feng, S., Zhong, H., Wang, W., Sun, Y.: Masked label prediction: unified message passing model for semi-supervised classification. In: Proceedings of the Thirtieth International Joint Conference on Artificial Intelligence, pp. 1548–1554. International Joint Conferences on Artificial Intelligence Organization (2021)

# SDMC-Mamba: Steel Surface Defect Detection Based on DDIM Enhanced Mamba with Multi-module Collaboration

Yujie Li[1], Haozhe Zhang[1(✉)], and Ziming Wang[2]

[1] School of Artificial Intelligence, Guilin University of Electronic Technology, Guilin, China
haozhezhang@mails.guet.edu.cn
[2] Department of Intelligence and Information Engineering, Taiyuan University, Taiyuan, China

**Abstract.** Steel surface defects directly impact industrial product quality and production safety, making high-precision detection critical for high-quality steel production. Existing methods have clear limitations: CNNs, Transformers, and traditional Mamba struggle to effectively extract corresponding features and coordinate global-local feature extraction. To address these challenges, we propose SDMC-Mamba—a DDIM-enhanced Mamba with multi-module collaboration for high-precision steel surface defect detection. We design the Collaborative Vector Selective Scan (CVSS) module, which improves SparX-Mamba's VSS block by fusing convolution and attention mechanisms to enhance collaborative extraction of local features and global information. Second, we propose the Hybrid Enhancement Branch (HEB), adopting a CNN-VSS parallel structure to assist the main branch in supplementing edge details and suppressing overfitting. Furthermore, we use DDIM for data augmentation, expanding the original NEU-DET dataset into a balanced 3600-image dataset. Experiments on the augmented dataset show SDMC-Mamba achieves 82.4% mAP@50 (2.9% higher than the state-of-the-art) and 0.464 mAP@95 (0.09 higher than the state-of-the-art), outperforming mainstream advanced models such as YOLO, RT-DETR, and SparX-Mamba.

**Keywords:** Denoising Diffusion Implicit Models · Mamba · Hybrid Enhancement Branch · object detection · VSS

## 1 Introduction

With industrial manufacturing moving toward high quality and precision, steel—an indispensable raw material—finds wide application in shipbuilding, construction, aerospace, and industrial production lines. It can also be processed into secondary products with extended performance. However, high temperatures


R. C.-W. Wong et al. (Eds.): PAKDD 2026, LNAI 16597, pp. 276–287, 2026.
https://doi.org/10.1007/978-981-92-1300-9_22

during steel processing often cause unpredictable surface cracks, seriously affecting steel quality and product safety [15], and threatening production and personnel safety. Thus, high-precision steel surface defect detection has become essential for high-quality steel development. For downstream enterprises, it enables quick defect localization and timely replacement of unqualified materials to avoid losses, while simplifying detection processes, reducing costs, and improving production efficiency. For example, microcracks from high-temperature welding are hard to identify with the naked eye, and traditional ultrasonic testing has limited resolution leading to missed detections—undetected cracks in hull sections can cause structural issues, requiring dismantling and re-welding with direct economic losses and construction delays.

Owing to the complex steel production and processing environment, steel surfaces may have multi-scale defects such as cracks, inclusions, patches, pits, scale curling, and scratches [19]. These defects not only sharply increase manual visual inspection costs but also pose major safety risks to on-site inspectors. Minor defects are hard to detect manually, leading to misdetection and missed detection that compromise production quality.

With the advancement of CNN [9], its robust local feature extraction enables CNN-based methods to capture subtle differences and minor steel surface defects. Among single-stage methods, the YOLO [17] series leads research: YOLOv10 [22] introduced dual-label assignment for NMS-free inference, reducing latency via lightweight components while preserving accuracy for real-time inspection; YOLOv11 [8] integrated a Transformer backbone and dynamic head for continuous defect capture and NMS-free training, improving precision and deployment. Other single-stage methods include CenterNet [31] and SSD [10]. For two-stage methods, Fast R-CNN [5], Mask R-CNN [7], Yang et al. [27] embedded dual attention and FPN in ResNet to adapt to multi-scale defects, enhancing feature distinguishability; Ren et al. [18] extended Faster R-CNN with multi-branch multi-task learning for simultaneous defect coordinate, category, and severity output. However, CNN-based methods lack sufficient global and long-range semantic extraction, leading to suboptimal performance for blurred and minor defects.

Transformer architecture [21] mitigates traditional CNN's weak global information extraction and Vision Transformer [3] revitalizes computer vision. Transformer-based models like DETR [30] and RT-DETR [13] address DETR's small target omission. Chen et al. [1] proposed a global-local feature fusion scheme retaining global modeling while enhancing tiny defect local texture extraction to boost small-target detection. For lightweight deployment TinyViT [26] adopts semi-dynamic pruning and knowledge distillation reducing parameters while maintaining accuracy and lowering hardware costs ERF-NAS [14] leverages zero-shot neural architecture search with efficient receptive fields to automatically construct steel-adapted lightweight models balancing precision and deployment efficiency. RT-DETR optimizes multi-scale feature processing its Efficient Hybrid Encoder decouples scale interaction and cross-scale fusion strengthening adaptation to multi-size defects and improving overall performance. Mr. DETR [28] introduces multi-path training treating detection Transformer as a multi-task framework learning one-to-one and one-to-many prediction dynamically adjusting attention to prioritize defect edges and textures enhancing complex shape localization precision. However Transformer's inherent computational complexity and poor detailed feature capture limit its ability to distinguish subtle differences.

Mamba [6] addresses CNN's insufficient global modeling Transformer's weak local capture and high complexity. Mamba-based methods excel at image scanning and detailed extraction with linear complexity and efficient global modeling simpler than Transformer. Targeting complex rolling textures multi-scale defects and real-time industrial needs they offer diversified pathways Memory Mamba [24] adopts coarse-fine memory enhancement to tackle sample imbalance and pseudo-label noise achieving strong robustness. Pyramid-Mamba [23] uses pyramid scanning and context-aware modules for hierarchical multi-scale processing enhancing tiny defect detection with faster inference than Transformer. Sparx-Mamba combines Dynamic Multi-Channel Attention with SparX and optimizes skip connections providing new detection ideas. Samba [32] fuses spatial proximity scanning and dual attention optimizing defect continuity and multi-scale alignment under complex backgrounds to balance accuracy and real-time performance. Overall via memory enhancement multi-scale modeling and high-resolution output these methods offset CNN's long-range dependency flaws avoid Transformer's high computation adapt to hot and cold rolling welding and other steel scenarios and deliver high-precision efficient industrial quality inspection solutions.

The core work of this study is illustrated in Fig. 1. To acquire clearer features, integrate local-global feature information, and improve steel surface defect detection accuracy, we adopt Denoising Diffusion Implicit Models (DDIM) [20] for data augmentation—expanding the dataset to generate more distinct feature data. To enhance feature capture capability, we propose the SDMC-Mamba model based on SparX-Mamba [11], integrating multi-type convolutions and attention to strengthen the VSS structure's local feature capture and global modeling capabilities. We further introduce a HEB to provide auxiliary output for the main branch. Benefiting from these improvements, SDMC-Mamba effectively enhances steel defect detection quality. Evaluations on the NEU-DET [29] dataset validate the model's effectiveness in classical steel defect detection tasks. The main contributions of our work are as follows:

1. We propose a novel SDMC-Mamba model for steel surface defect detection. Based on DDIM enhanced Mamba with multi-module collaboration, it enhances global-local feature collaborative extraction via the improved CVSS module, supplements edge details and suppresses overfitting with a CNN-VSS parallel HEB, and adopts DDIM for data augmentation to ultimately boost detection accuracy significantly.
2. We propose a CVSS module into the VSS block of SparX-Mamba, and adopts a collaborative approach combining multiple types of convolution and atten-

tion mechanisms to enhance the VSS block's ability to utilize local features and global features.
3. We propose the Hybrid Enhancement Branch (HEB), which can capture detailed texture differences in defect images, and assists the output of the model's main branch by means of linear weights.
4. The proposed approach effectively extracts feature information from the NEU-DET dataset. While retaining strong feature capture capability, it enables better identification of defect features. Experimental results on the NEU-DET dataset show that our model outperforms existing advanced methods in key metrics such as mAP@50 and mAP@95, demonstrating excellent detection performance.

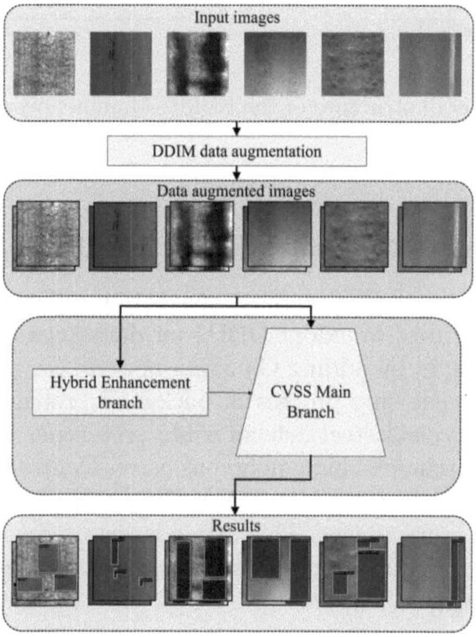


**Fig. 1.** The main task flow of our research.

# 2 Theory

## 2.1 Overview

To address low accuracy and high computational complexity, we introduce DDIM for data augmentation and design the SDMC-Mamba model (network structure in Fig. 2). Configured with cross-layer connection step S = 2 and parallel feature channels M = 2, SparX constructs two input-interleaved alternating layers: ganglion layers (DPE → DMCA → VSS module) and ordinary layers (DPE

→ VSS module). The feedforward network enables effective cross-layer feature interaction modeling while preserving sparsity per the above connection rules.

SDMC-Mamba integrates CVSS in the main branch, combining multi-type convolutions and attention to enhance VSS's multi-scale feature capture. It also introduces a HEB (CNN-VSS fusion) to assist main branch output, ensuring integration of CNN-derived local features and VSS-acquired global features.

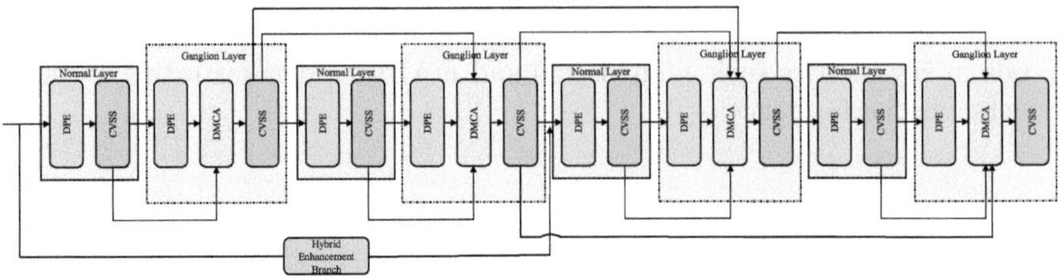


**Fig. 2.** Overall structure of the SDMC-Mamba network model.

## 2.2 DDIM Augmentation for Steel Surface Defect Detection

To address industrial steel data scarcity, acquisition difficulties, and multi-scale ambiguous defect features, we adopt DDIM for data augmentation. DDIM generates new defect images by adding Gaussian noise in forward propagation and learning the inverse denoising process in backward propagation. These images accurately replicate typical steel defects while preserving inter-type subtle differences critical for distinguishing ambiguous cases. Controlled sampling parameters avoid extra noise, ensuring data authenticity and enhancing representation of small defects and blurry edges. This expands the original NEU-DET dataset to match real-world detection complexity, laying a solid foundation for training models with robust generalization.

## 2.3 Collaborative Vector Selective Scan (CVSS)

To address traditional VSS's weak local capture and incomplete fine-grained extraction, we propose the CVSS (Convolution-VSS Synergy Module) by integrating VSS with multi-type convolutions and attention, boosting local-global synergy (Fig. 4). Input first undergoes VSS for global context and long-range semantics, then enters DWConv [2]—a channel-wise lightweight convolution extracting local details with reduced complexity. Extracted features split into three branches: a second DWConv refines local details; Star block [12] expands receptive field for multi-scale context; FFA [16] enhances long-range dependencies via attention. The latter two parallel branches' fusion is regulated by a sigmoid weight to balance contributions, with residual connection to VSS output mitigating gradient vanishing and preserving global features. Intermediate

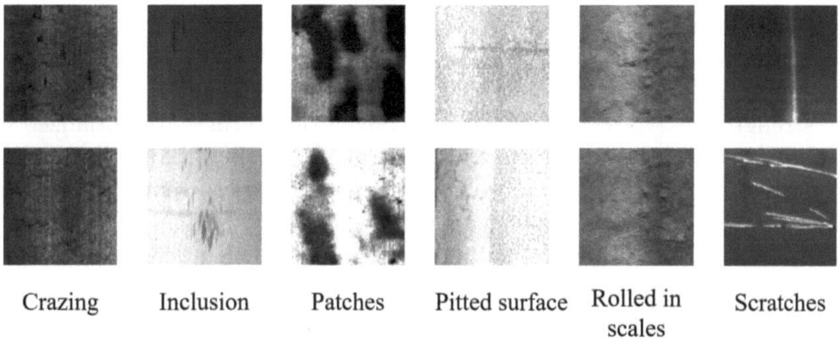


**Fig. 3.** Various defect images after data augmentation.

output feeds into CBAM [25] dual attention: channel attention (global average pooling + fully connected layers) highlights critical channels; spatial attention focuses on defect regions to boost saliency. Finally, CBAM-processed output fuses with the second DWConv result, combining refined local and attention-enhanced global features for comprehensive transmission. Integrating convolution and attention, CVSS strengthens local extraction and enables detailed feature capture, improving the model's representation capability.

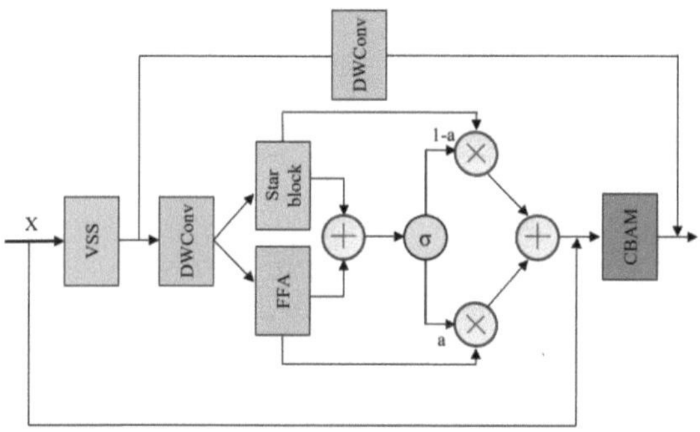


**Fig. 4.** Structure of the CVSS block.

## 2.4 Hybrid Enhancement Branch (HEB)

To enhance the local feature utilization and supplement details, we propose the HEB as an auxiliary component see Fig. 5. Adopting a CNN-VSS parallel structure for global-local feature integration, it supplements the main branch's details via a Branch Weight Scheduler—linearly increasing weights from 0 to 0.5 in the first 5 epochs to avoid noise, then stabilizing at 0.5–0.8. A 3D enhanced regularization method integrating spatial Dropout, channel noise and L1 sparsity suppresses overfitting, while the branch receives high-semantic features from the main branch to generate a target mask, realizing bidirectional semantic guidance-edge focusing interaction and ensuring edge-target alignment.

To strengthen CNN's edge extraction, convolution kernels are initialized with the Sobel operator. Input features pass through two $3 \times 3$ convolution layers (each with Batch Normalization and GELU activation) to refine coarse edges, with a residual connection supplementing original information. Meanwhile, the branch's VSS performs parallel global modeling; its output undergoes channel normalization and fuses with CNN-derived local edges. A residual connection between initial input and module output ensures effective feature transmission. Combining VSS's global modeling and CNN's local capture advantages, the HEB improves defect edge extraction and enhances the main branch's complex detail capture capability.

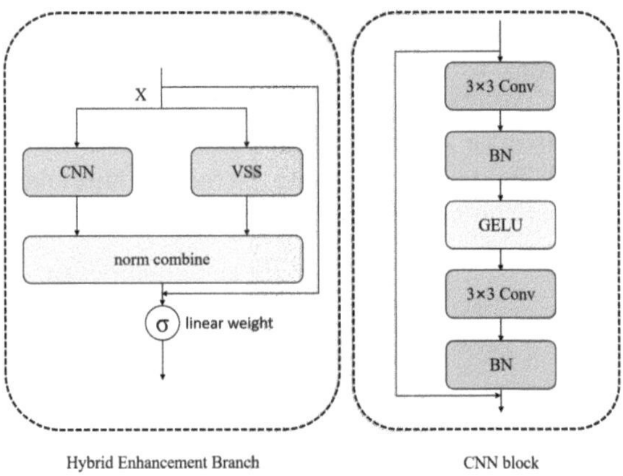


**Fig. 5.** Structure of the HEB and CNN block.

# 3 Discussion

In this section, we evaluate the performance of SDMC-Mamba in the task of steel surface defect detection. We first introduce the benchmark dataset, implementation details, and evaluation metrics. Then, we compare our method with current classic methods and state-of-the-art methods, and conduct ablation studies to verify the advanced design of the proposed architecture.

## 3.1 Experimental Setup

**Dataset:** We conducted experiments on Northeastern University's NEU-DET dataset, designed to alleviate steel surface defect detection data scarcity and advance deep learning-based automated detection. It includes 6 typical defects:

Crazing, Inclusion, Patches, Pitted Surface, Rolled-in Scales, and Scratches (as shown in Fig. 3), with 1800 original grayscale images (300 per category, mostly $224 \times 224$ pixels). After DDIM data augmentation, we added 1800 images, resulting in a total of 3600 images (600 per category) to ensure sufficient data for effective model training.

**Implementation Details:** The SDMC-Mamba model is implemented based on PyTorch and trained on an Nvidia GeForce RTX 4090 GPU with 24GB memory. No pre-trained model is used for this model; instead, it is randomly initialized and trained from scratch. We set the input image size of the dataset to $224 \times 224$ pixels, the initial learning rate to 0.0001, the batch size to 2, and the number of training epochs to 12.

**Evaluation Metrics:** To accurately evaluate the model's detection performance, we use mAP@50 (Mean Average Precision at IoU = 0.5) [4] and mAP@95 to quantitatively analyze the model's detection accuracy. Given that $P_i(R)$ denotes the precision-recall curve of the $i$-th category, the integral is calculated under the condition of IoU $\geq 0.5$. mAP@50 is calculated as:

$$\text{mAP@50} = \frac{1}{N}\sum_{i=1}^{N}(\int_{0}^{1} P_i(R)\, dR) \tag{1}$$

where $N$ represents the total number of categories. mAP@50 integrates the Average Precision (AP) values of all categories, reflecting the model's overall detection capability when IoU $\geq 0.5$.

$$\text{mAP@95} = \frac{1}{C}\sum_{i=1}^{C} \text{AP}_{95,i} \tag{2}$$

Here, $C$ denotes the total number of target categories in the dataset, and $\text{AP}_{95,i}$ stands for the "strict Average Precision" of the $i$-th category—i.e., the Average Precision of this category under the threshold of IoU $\geq 0.95$. mAP@95 is calculated as:

### 3.2 Performance of the Model on the Task

We compared SDMC-Mamba with several classical detection methods and state-of-the-art detection methods. Due to the stochastic nature of model results, each experimental data point was obtained by repeating the experiment multiple times and taking the average. To ensure the validity of the results, we conducted multiple comparative experiments on the DDIM-augmented NEU-DET dataset and obtained consistent results.

**Comparsion Results.** We compared SDMC-Mamba with several classical detection methods as well as state-of-the-art detection approaches. Given the inherent randomness in model outputs, each experimental result was derived by averaging data from multiple repeated trials. To validate the reliability of the findings, we conducted repeated experiments on the NEU-DET dataset augmented via DDIM, consistently yielding comparable outcomes.

Table 1 demonstrates that our model achieves superior performance in steel surface defect detection, outperforming state-of-the-art methods. Compared to SparX-Mamba, our mAP@50 increases from 73.5% to 82.4% (12.1% improvement) and mAP@95 rises from 0.366 to 0.464 (26.7% gain), which is attributed to the integration of complementary convolution and attention mechanisms enhancing the Mamba model's feature discriminability. Against the YOLO series (v3–v11), our mAP@50 reaches 82.4%, surpassing the series' highest value (70.4%) by 17%, while mAP@95 (0.464) outperforms the best variant (0.39) by 18.9%. Relative to the latest Transformer-based RT-DETR, our mAP@50 exceeds its 79.5% by 3.6% to 82.4%, and mAP@95 (0.464) outperforms its 0.455 by 1.9%. Compared with DE_RetinaNet, our mAP@50 climbs from 78.3% to 82.4%, representing a 5.2% advancement.

**Table 1.** different method experimental results on the NEU-DET dataset

| Methods | Crazing | Inclusion | Patches | Pitted-surface | Rolled-in-scales | Scratches | mAP@50 | mAP@95 |
|---|---|---|---|---|---|---|---|---|
| SparX-Mamba | 0.51 | 0.842 | 0.869 | 0.847 | 0.595 | 0.748 | 0.735 | 0.366 |
| YOLOv3 | 0.388 | 0.846 | 0.827 | 0.862 | 0.499 | 0.803 | 0.704 | 0.39 |
| YOLOv5 | 0.411 | 0.784 | 0.873 | 0.841 | 0.56 | 0.715 | 0.697 | 0.37 |
| YOLOv8 | 0.364 | 0.785 | 0.867 | 0.817 | 0.595 | 0.766 | 0.696 | 0.375 |
| YOLOv9 | 0.366 | 0.814 | 0.867 | 0.828 | 0.48 | 0.77 | 0.687 | 0.377 |
| YOLOv11 | 0.396 | 0.775 | 0.854 | 0.849 | 0.556 | 0.747 | 0.696 | 0.374 |
| RT-DETR | 0.562 | 0.864 | 0.901 | 0.907 | 0.68 | 0.874 | 0.795 | 0.455 |
| DE_RetinaNet | 0.558 | 0.819 | **0.947** | 0.892 | 0.702 | 0.777 | 0.783 | - |
| SDMC-Mamba | **0.646** | **0.864** | 0.869 | **0.947** | **0.743** | **0.877** | **0.824** | **0.464** |

**Visualization Results.** As illustrated in Fig. 6, the proposed model exhibits favorable performance in the detection of various types of defects. Notably, it demonstrates excellent capability even for small-sized defects—these defects, which typically have limited pixel coverage and insufficient visual feature information in the figure , can still be stably identified by the proposed model.

For densely distributed defects, the model accurately distinguishes individual boundaries and enables precise localization and detection, effectively alleviating adjacent defect interference that causes misjudgment or missed detection in dense scenarios. These results fully confirm the proposed model's excellent overall performance in meeting diverse defect inspection needs.

**Ablation Experiments.** As shown in Table 2, after the dataset was augmented by DDIM, mAP@50 increased from 73.5% to 76.9%, while mAP@95 rose from 0.366 to 0.396. This improvement in metrics can be attributed to the fact that DDIM, as a data augmentation method, effectively enhances the feature clarity

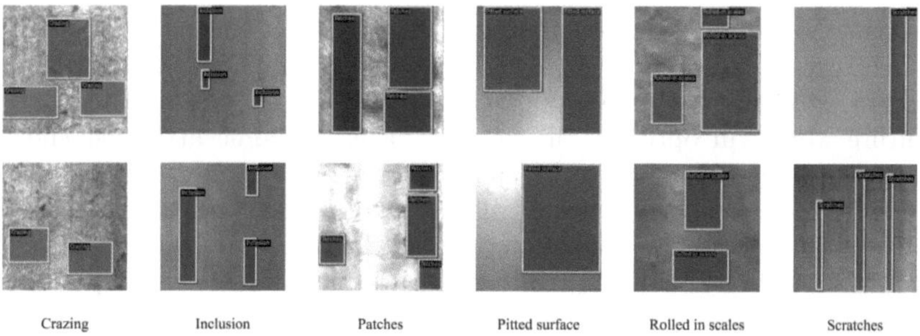


**Fig. 6.** Detection Results of Different Defects.

of the original dataset; meanwhile, data augmentation provides more feature maps, enabling the model to learn adequately.

**Table 2.** Ablation experiment results

| Methods | DDIM | CVSS | CNN | Crazing | Inclusion | Patches | Pitted | Rolled-in | Scratches | mAP@50 |
|---|---|---|---|---|---|---|---|---|---|---|
| Baseline | × | × | × | 0.51 | 0.842 | 0.869 | 0.847 | 0.595 | 0.748 | 0.735 |
| Baseline+DDIM | ✓ | × | × | 0.546 | 0.803 | 0.87 | 0.875 | 0.703 | 0.817 | 0.769 |
| Baseline+DDIM+CVSS | ✓ | ✓ | × | 0.671 | 0.846 | **0.882** | 0.927 | 0.699 | 0.894 | 0.820 |
| Baseline+DDIM+CVSS+HEB | ✓ | ✓ | ✓ | **0.646** | **0.864** | 0.869 | **0.947** | **0.743** | 0.877 | **0.824** |

Adding the CVSS module to the DDIM-augmented dataset raised model accuracy from 76.9% to 82.0% and mAP@95 from 0.396 to 0.471. Improved metrics stem from CVSS fusing multi-type convolutions and attention to better extract local features, integrate them with global features, and realize comprehensive detailed feature capture, enhancing the model's representation capability. Further adding the HEB to the DDIM-augmented dataset with CVSS increased mAP@50 from 82.0% to 82.4% while mAP@95 slightly decreased to 0.464. Overall improvement comes from the HEB integrating CNN-VSS advantages and optimizing training rates to enhance the main branch's complex detailed feature extraction. Ablation experiments confirm that DDIM augmentation combined with CVSS and HEB effectively improves mAP@50 and mAP@95. These components fully utilize feature information to complement the original model's global features, notably boosting detection accuracy and stability.

## 4 Conclusion

This paper proposes SDMC-Mamba, a steel surface defect detection network with three core designs: CVSS fuses multi-convolutions and attention for local-global feature synergy, HEB uses dynamic weight scheduling for complex feature extraction, and DDIM augmentation expands NEU-DET to mitigate data scarcity. Experiments on augmented NEU-DET demonstrate its superiority over traditional models, verifying accuracy and robustness. It offers a practical industrial steel inspection solution and lays the groundwork for edge device adaptation and complex on-site scenarios.

Future work will optimize the network using dynamic attention and contrastive learning, construct real-scene datasets for complex defects, develop lightweight models via knowledge distillation and channel pruning for edge adaptation, expand defect and production coverage, and integrate with on-site equipment and traceability platforms to form a "detection-adjustment-improvement" loop, enhancing industrial efficiency and safety.

**Acknowledgments.** This work is supported by the Regional Undergraduate Training Program for Innovation and Entrepreneurship (S202510595348).

## References

1. Chen, J., Huang, H., Li, Z.: Feature enhancement and metric optimization for defect detection on steel surfaces. Laser Optoelectron. Progress 1–16 (2024)
2. Chollet, F.: Xception: deep learning with depthwise separable convolutions. In: Proceedings of the IEEE Conference on Computer Vision and Pattern Recognition, pp. 1251–1258 (2017)
3. Dosovitskiy, A.: An image is worth 16x16 words: transformers for image recognition at scale. arXiv preprint arXiv:2010.11929 (2020)
4. Everingham, M., Eslami, S.A., Van Gool, L., Williams, C.K., Winn, J., Zisserman, A.: The pascal visual object classes challenge: a retrospective. Int. J. Comput. Vision **111**(1), 98–136 (2015)
5. Girshick, R.B.: Fast R-CNN. CoRR abs/1504.08083 (2015)
6. Gu, A., Dao, T.: Mamba: linear-time sequence modeling with selective state spaces. In: First Conference on Language Modeling (2024)
7. He, K., Gkioxari, G., Dollár, P., Girshick, R.: Mask R-CNN. In: Proceedings of the IEEE International Conference on Computer Vision, pp. 2961–2969 (2017)
8. Khanam, R., Hussain, M.: Yolov11: an overview of the key architectural enhancements. arXiv preprint arXiv:2410.17725 (2024)
9. LeCun, Y., Bottou, L., Bengio, Y., Haffner, P.: Gradient-based learning applied to document recognition. Proc. IEEE **86**(11), 2278–2324 (2002)
10. Liu, W., et al.: SSD: single shot multibox detector. In: European Conference on Computer Vision, pp. 21–37. Springer (2016)
11. Lou, M., Fu, Y., Yu, Y.: Sparx: a sparse cross-layer connection mechanism for hierarchical vision mamba and transformer networks. In: Proceedings of the AAAI Conference on Artificial Intelligence, vol. 39, pp. 19104–19114 (2025)
12. Lu, Z., et al.: Star-yolo: a lightweight and efficient model for weed detection in cotton fields using advanced yolov8 improvements. Comput. Electron. Agric. **235**, 110306 (2025)
13. Lv, W., Zhao, Y., Chang, Q., Huang, K., Wang, G., Liu, Y.: Rt-detrv2: improved baseline with bag-of-freebies for real-time detection transformer. arXiv preprint arXiv:2407.17140 (2024)
14. Otmani, K.E., Mateo-Gabín, A., Rubio, G., Ferrer, E.: Accelerating high order discontinuous galerkin solvers through a clustering-based viscous/turbulent-inviscid domain decomposition. Eng. Comput. **41**(2), 949–964 (2025)

15. Peng, Z., Mei, T., Zheng, J., Yuan, Y., Wang, L.: Cracking behavior and high-temperature thermoplastic analysis of 09crcusb steel billets. Metals **13**(6), 1058 (2023)
16. Qin, X., Wang, Z., Bai, Y., Xie, X., Jia, H.: FFA-net: feature fusion attention network for single image dehazing. arXiv preprint arXiv:1911.07559 (2019)
17. Redmon, J., Divvala, S., Girshick, R., Farhadi, A.: You only look once: unified, real-time object detection. In: Proceedings of the IEEE Conference on Computer Vision and Pattern Recognition, pp. 779–788 (2016)
18. Ren, S., He, K., Girshick, R., Sun, J.: Faster R-CNN: towards real-time object detection with region proposal networks. IEEE Trans. Pattern Anal. Mach. Intell. **39**(6), 1137–1149 (2016)
19. Singhka, D.K.H., Neogi, N., Mohanta, D.: Surface defect classification of steel strip based on machine vision. In: International Conference on Computing and Communication Technologies, pp. 1–5. IEEE (2014)
20. Song, J., Meng, C., Ermon, S.: Denoising diffusion implicit models. arXiv preprint arXiv:2010.02502 (2020)
21. Vaswani, A., et al.: Attention is all you need. In: Advances in Neural Information Processing Systems, vol. 30 (2017)
22. Wang, A., Chen, H., Liu, L., Chen, K., Lin, Z., Han, J., et al.: Yolov10: real-time end-to-end object detection. Adv. Neural. Inf. Process. Syst. **37**, 107984–108011 (2024)
23. Wang, L., Li, D., Dong, S., Meng, X., Zhang, X., Hong, D.: Pyramidmamba: rethinking pyramid feature fusion with selective space state model for semantic segmentation of remote sensing imagery. arXiv preprint arXiv:2406.10828 (2024)
24. Wang, Q., Hu, H., Zhou, Y.: Memorymamba: memory-augmented state space model for defect recognition. arXiv preprint arXiv:2405.03673 (2024)
25. Woo, S., Park, J., Lee, J.Y., Kweon, I.S.: CBAM: convolutional block attention module. arXiv preprint arXiv:1807.06521 (2018)
26. Wu, K., et al.: Tinyvit: fast pretraining distillation for small vision transformers. In: European Conference on Computer Vision, pp. 68–85. Springer (2022)
27. Yang, L., Bai, S., Huang, H., Kong, S.: An automatic surface defect detection method with residual attention network. In: CAAI International Conference on Artificial Intelligence, pp. 194–205. Springer (2022)
28. Zhang, C.B., Zhong, Y., Han, K.: Mr. DETR++: instructive multi-route training for detection transformers. arXiv preprint arXiv:2412.10028 (2024)
29. Zhang, Y., et al.: Development of a cross-scale weighted feature fusion network for hot-rolled steel surface defect detection. Eng. Appl. Artif. Intell. **117**, 105628 (2023)
30. Zhao, Y., et al.: DETRs beat YOLOs on real-time object detection. In: Proceedings of the IEEE/CVF Conference on Computer Vision and Pattern Recognition, pp. 16965–16974 (2024)
31. Zhou, X., Wang, D., Krähenbühl, P.: Objects as points. arXiv preprint arXiv:1904.07850 (2019)
32. Zhu, Q., et al.: Samba: semantic segmentation of remotely sensed images with state space model. Heliyon **10**(19) (2024)

# TabTokWak: Token(less)-Value Watermarking for Tabular Foundational Models

Jeroen M. Galjaard[1(✉)], Chaoyi Zhu[1], Robert Birke[2], Pin-Yu Chen[3], Cornelis Bos[1,4], and Lydia Y. Chen[1,5]

[1] Delft University of Technology, Delft, The Netherlands
{J.M.Galjaard,C.Zhu-2,C.Bos,Y.Chen-10}@tudelft.nl,
kees.bos@tatasteeleurope.com, lydiaychen@ieee.org
[2] University of Turin, Turin, Italy
robert.birke@unito.it
[3] IBM Research, New York, USA
Pin-Yu.Chen@ibm.com
[4] Tata Steel, IJmuiden, The Netherlands
[5] Université de Neuchâtel, Neuchâtel, Switzerland

**Abstract.** Shareable tables generated by Foundational Models (FM) are increasingly adopted in industry and scientific applications. With the advent of such data, governance of it becomes an ever more pressing matter. Towards this, watermarking is shown to be a promising approach. While existing watermarking approaches successfully embed watermarks in natural language, they fall short for tabular data due to the susceptibility of token-based representations to post-processing and the arbitrary column ordering. We propose a novel watermarking scheme for tabular data, **Tab**ular **To**ken(less) **Wa**termar**k** (TabTokWak), which embeds a watermarking signal on FM-generated tabular value *and* its tokenized representation. To be invariant to table generation order, our token watermark uses a seeded pattern to derive its watermark signal. In our experiments on five real-world datasets and a representative tabular FM, we demonstrate that TabTokWak enables effective data governance with high detection efficiency and minimal impact on data quality. Therewith, empirically demonstrating TabTokWak's detectability and its robustness with a favorable quality trade-off. Code online: https://gitlab.ewi.tudelft.nl/dmls/publications/tabtokwak.

## 1 Introduction

Tabular data is a rich source of information, yet privacy requirements [2,14], data quality issues, e.g., missing values [13], and class imbalance [5] often hinder its use in practice. To overcome such limitations, Foundational Models (FMs) like GPT-3 [17] have been proposed to generate synthetic tables [4,16]. Such generative tabular models train FMs to synthesize tabular data using a natural-language-like representation of tabular records [4,16]. Compared to tabular generative adversarial networks [6,18], such tabular FMs overcome the challenge of data encoding for heterogeneous columns while effectively leveraging contextual information [4,20].


R. C.-W. Wong et al. (Eds.): PAKDD 2026, LNAI 16597, pp. 288–300, 2026.
https://doi.org/10.1007/978-981-92-1300-9_23

The increasing use of FMs necessitates regulating the data generated by such models [3]. Watermarking model output is an essential and reliable strategy to mitigate the risks of misuse and manipulation [19]. To mitigate high computational costs and quality degradation, sampling-phase watermarks are increasingly adopted. Towards this, soft-red/green watermarks [11] embed an invisible signal by favoring green-listed tokens during sampling. Although green-red watermarking can be applied to any FM, it exhibits two significant challenges when adapted to tabular FMs. First, tabular data exhibit arbitrary column ordering, allowing for flexible generation by FMs. However, because the prior art leverages context windows built on *preceding* tokens, the detector benefits from knowing the generation order. Second, tables contain substantially more numerical data compared to natural language. To limit vocabulary size, FMs [4,16] render numerical values as numeric strings of "simpler" digit tokens. However, this approach is susceptible to (even small) distortions that have limited *semantic* impact, thereby limiting detection efficiency. For example, right-padding decimals with trailing 0s do not affect their value but does change their corresponding token sequence.

In this paper, we propose **Tab**ular **To**ken(less) **Wa**terma**rk** (`TabTokWak`). `TabTokWak` addresses the challenge of column permutation, using *a pattern* wherein we leverage the column names to determine red/green-listed tokens while simultaneously sampling that column's value from the whole prompt. Jointly, it uses a value-watermark that hides a key-derived signal directly into the numerical values *themselves*. Doing so by limiting numerical values during decoding to allowed "buckets" in each column's domain. As such, the detection resists column permutations, while the larger prompt context maintains the semantic coherence of generated values. In short, our contributions are:

**Column invariant token watermark.** `TabTokWak` makes the watermark column-order agnostic while preserving semantic context to improve data coherence.

**Token-less value-watermark.** In which selected columns, the interpretation in its numerical domain is watermarked.

**Robust against post-editing attacks.** We demonstrate that `TabTokWak` is unaffected by column permutation and achieves strong robustness against attacks based on missingness and numerical representation.

## 2 Background and Related Work

**Tabular Foundational Models (FMs)** generate high-fidelity data relying on token-based models, such as Large Tabular Models (LTM), which can be finetuned on datasets using discretized encodings of tabular rows. To achieve this, rows are represented as structured natural language. For example, a tabular row consisting of tuples of columns and values $R_i = \{(C_j, v_{j,i})\}_{j=1}^{|C|}$, gets represented into a sentence-like structure [4]; "$\texttt{column}_j$ $\texttt{is}$ $\texttt{value}_j, \ldots$". It is important to note that feature permutation is commonly employed to facilitate arbitrary conditioning [4,16]. Conditioning is done via pre-filled 'prompts' containing subject-predicate-value phrases for the relevant columns. Since conditioning can involve different columns, this leads to arbitrary order generation during data synthesis. To generate data, the model owner iteratively samples from the next token likelihood estimates $\hat{p}_k^{(t+1)}$ for each token $k$ in the vocabulary $V$. The previous tokens then serve as an input prompt to the model. The likelihoods are obtained by applying softmax to the model's next-token activations $l_{(\cdot)}^{t+1}$.

**Prior art on Watermarks** for FMs typically embeds a detectable signal in the output text during generation. This is done either by changing the sampling process [1, 10, 12] or modifying the pre-activation logits of the model [8, 11, 19].

The deterministic exponential scheme (EXP) [1] elects tokens that maximize a pseudo-random function on previous token(s). Calculating the sum of this function on tokens in watermarked text gives an anomalously large value enabling detection via thresholding. Since EXP is deterministic given a context and seed, it degrades in quality whenever repeated tokens occur. GumbelSoft [10] alleviates this issue by sampling from the EXP distribution conjoined with the unaltered pre-activation logits. In contrast, the other type of watermark pseudo-randomly partitions the model vocabulary into Red/Green sets [11], and boosts the logits of favored tokens. Detection uses the same vocabulary split and performs a statistical test on the ratio of favored tokens in the output.

**Soft Red/Green Watermarks** splits the model vocabulary $V$ into two disjoint sets: favored (green) and normal (red) tokens [11]. The split is determined by $L$ preceding tokens as context, a secret key $\kappa_p$ as seed, and a controllable *favored-ratio* $\gamma$. As such, the favored terms are $\gamma\,|V|$. The next-token sampling is biased by adding a constant $\delta$ to the logits of favored tokens,

$$\hat{p}_v^{(t)} = \frac{\exp(l_v^{(t)} + \mathbb{1}\,[k \in F]\,\delta)}{\sum_{i \in V} \exp(l_i^{(t)} + \mathbb{1}\,[i \in F]\,\delta)}, \; \forall v \in V. \tag{1}$$

where $\mathbb{1}\,[\cdot]$ is the indicator function which yields 1 when its clause holds, else 0.

A non-watermark text is expected to contain $\sim \gamma \cdot T$ favored tokens. Similarly, if the text is watermarked with a key different from $\kappa_p$. Only text watermarked with the correct key can increase the observed green token count $O$. Hence, a watermark can be detected when the score $z = (O - \gamma T)\,/\sqrt{T\gamma(1-\gamma)}$ exceeds a pre-determined $z$-threshold [11] where $O$ is the observed green token count. Following, we obtain the likelihood of the text being watermarked using the inverse cumulative density function of the normal distribution, $\mathcal{P} = 1 - \Phi_{\mathcal{N}}^{-1}(z)$.

## 3 Tabular Token(less) Watermark (TabTokWak)

Current token-based watermarking approaches rely on sequence-based embedding and discrete perturbation of tokens. As a result, they struggle with tabular data because it is sensitive to reordering and continuous numerical perturbations. We motivate the difficulty of watermarking tabular data with a small example. Consider a generated column-value value $v_k = 29.75$ for a row's column $C_j$. Through tokenization, the tabular FM represents numeric values as a sequence of digit tokens [4, 7, 16, 20]. For simplicity, we consider single-digit tokens, yielding: '`[2][9][.][7][5]`'. Even small perturbations that have little to no effect on the numeric value, e.g., rounding $v_k = 30$ or changing the precision to $v_k = 29.7500$, lead to significantly different tokenizations.

Prior art methods are: i) inherently sensitive to ordering due to the use of preceding tokens to construct a context to derive the watermark signal; and, ii) suffer from reduced

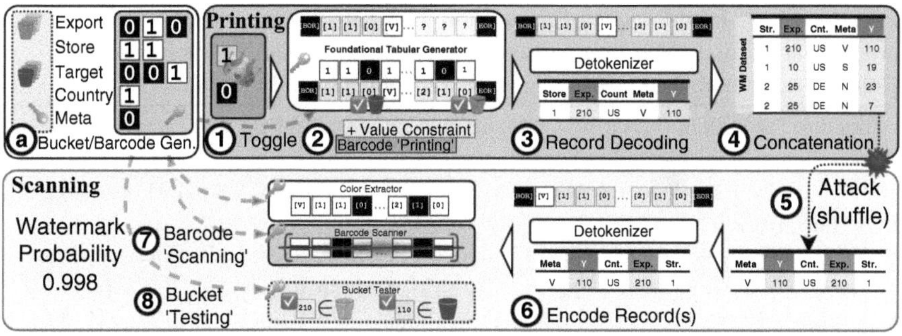


**Fig. 1.** Tabular Data Synthesis and detection with TabTokWak; using BarCode printing and value-watermark with a private key. During the generation of synthetic tabular rows, each record is individually watermarked with a private key $\kappa_p$ on a per-token basis (steps ①– ③). A secret is embedded using preferential sampling according to a key-derived BarCode pattern. Selected numerical columns have their values restricted to fall in allowed 'buckets'. The BarCode-colors {1, 0} indicate which token color to prefer during sampling ②.

variability of generated data due to the bias towards a subset of favored, green-listed tokens. Limitation i) imposes a dependency on the generation order, which may not hold during watermark detection. Requiring keeping track of column order during generation. Consequently, a considerable portion of the watermark signal may be altered as context can overlap with values from previous 'context' tokens. As a result, without access to generation order, rows generated with a context length of $L$ may lose up to $L$ tokens *per-cell*. As a result, it severely degrades the embedded watermark signal. Limitation ii) follows from the need to favor a subset of tokens. For text generation, this is often not a problem due to a large vocabulary size and the presence of synonyms likely spread across both favored and normal token sets. Instead, numerical values are represented by a small vocabulary where each token has a unique semantic meaning, e.g., single digits. Favoring a subset of such tokens yields numerical values with overly frequent digits, thereby biasing the distribution of represented values and degrading data quality.

Here, we aim to design a suitable watermarking method, TabTokWak, that addresses these limitations. At the heart of TabTokWak is the use of a BarCode and value watermark. The BarCode is a pseudo-random binary string associated with columns and the tokens for its corresponding cell value. Joined by the value watermark, specifically designed for numerical values and imposing specific patterns onto numerical ranges, termed buckets.

### 3.1 Watermark Embedding

In Fig. 1, we show the general watermarking process under TabTokWak using a private key $\kappa_p$. A key-emoticon indicates that the generator uses the private key. TabTokWak watermarks data on two levels, tokens and numerical values, both in an order-invariant way. Before generation proceeds (Step ⓐ), a balanced BarCode is derived leveraging column names and $\kappa_p$. Afterwards, we determine a random color toggle $h$ for a record

**Pseudocode 1:** Pseudo-code of generation with `TabTokWak`, consisting of the order-invariant and value-aware watermark during the model's sampling phase. Colors used to highlight corresponding components from Figure 1.

1 **function** **TabTokWak** (bias $\delta > 0$, columns names $C_j \in \mathcal{C}$, Tabular FM $f_\theta$, private key $\kappa_p$, empty/partial tokenized row $R$)
2 h $\sim$ Uniform{0, 1} ▷ **Step** ①: Random color toggle
3 **for** `Column` $C_j \in \mathcal{C}$ **do**
4 $R \leftarrow R \oplus$ Tokenize($C_j \oplus P$) ▷ Start phrase with subject-predicate
5 $B_j \leftarrow$ `GenerateBarcode`($\kappa_p, C_j$) ▷ **Step** ⓐ: Compute BarCode for $C_j$
6 $\mathcal{I}_j \leftarrow$ `GenerateAlowIntervals`($\kappa_p, C_j$) ▷ **Step** ⓐ: Compute $C_j$'s buckets.
7 $m \leftarrow 0$
8 **while** `Generating column value` $S_j$ **do**
9 f $\leftarrow B_j[m] +$ h mod 2 ▷ **Step** ② Color with $C_j$, index $m$, toggle $h$
10 $F \leftarrow$ `FavorList`($V, \kappa_p,$ f) ▷ Compute favored tokens
11 $s_m^{(j)} \sim f_\theta(R)$ under $F$, constrained by $\mathcal{I}_j$ ▷ Sample biased distribution Eq. 1
12 $R \leftarrow R \oplus s_m^{(j)}$
13 $m \leftarrow m + 1$
14 **return** `Decode`($R$) ▷ ③ Detokenize value to table format.

(Step ①). Afterwards, we auto-regressively generate values for each column using the name-is-value structures. Any previously generated columns serve as input prompt. To watermark the value, we generate a BarCode seeded with the private key $\kappa_p$ and column name $C_j$ plus the allowed intervals in which the final decoded value may fall (Step ②). Then, during next-token prediction, the BarCode is printed on the column's value by first filtering tokens that lead to values outside an allowed bucket, then boosting tokens in a favored list governed by the BarCode. Afterwards, the sequence of name-is-value tuples are converted to a tabular format (Step ③) and appended as a new row (Step ④).

**Token Watermarking** divides the model vocabulary $V$, i.e., tokens, into two subsets: $V_0$ and $V_1$. At its core, `TabTokWak` uses a pseudo-random sequence of 0s (0) and 1s (1), i.e., the 'BarCode'. In contrast to methods akin KGW, which bias *strictly* the prefer list. During generation, the barcode chooses the sub-vocabulary to boost, $V_0$ or $V_1$, based on the index of next value token. As each barcode is associated with a column, the process ignores column ordering and reduces biasing on column values.

`TabTokWak` starts by drawing a random color toggle h $\sim$ Uniform{0, 1} (line 2). When the toggle is 1 the two sub-vocabularies are swapped. The toggle $h$ adds row-level randomness, preventing bias toward certain tokens. Our watermark evaluation handles this choice automatically with two-sided detection, since $h$ only determines which side of the score the signal shows up on (see Subsect. 3.2 for details).

Next, `TabTokWak` auto-regressively generates the tokens representing the values for each column $C_j \in \mathcal{C}$ (line 3). For each column, `TabTokWak` creates the BarCode $B_j$ seeded with the column name $C_j$ and the secret key $\kappa_p$ that identifies the user. Using the column name makes the barcode independent on the generation order during watermark detection. Then, the column's value tokens are generated one-by-one (line 8).

TabTokWak applies BarCode $B_j$ using the *relative* column-token index $m$ of the next predicted value token, and the random color toggle $h$ to compute $\boxed{f}$ selecting which sub-vocabulary to boost (line 9) when sampling the next token.

**Value Watermarking** adds another watermark signal *directly* into synthesized numerical data, by restricting the model's numerical output directly. Instead of splitting the token vocabulary, for each selected numerical column $C_j$ we split its domain into sub-ranges, i.e., *buckets*. Then, dividing the buckets evenly into *allow* and *deny* types. Again, using a pseudo-random function seeded with private key $\kappa_p$ and column-name $C_j$ (line 6). During generation, we steer the process so that the final decoded column value ends up in an *allow* bucket. Hence, $\kappa_p$-watermarked data has all generated values falling into allowed buckets. For numerical columns to be value-watermarked, they must contain at least 100 distinct values. Then, using the original column's numerical range $R_j = \max(C_j) - \min(C_j)$, we derive the bucket count using $\#buckets(D_j) = 10^{\lfloor \log_{10} D_j \rfloor}\left(1 + \mathbb{1}\left[\frac{D_j}{10^{\lfloor \log_{10} D_j \rfloor}} \geq 5\right]\right)$.

Having determined the sub-vocabulary $F$ and allowed buckets $\mathcal{I}_j$, TabTokWak proceeds with generating the next numerical value token. First, TabTokWak removes any candidate that cannot lead to a value in an allowed bucket by speculatively decoding. Then, it positively biases the likelihoods of remaining tokens from $F$ following the barcode favor list. Finally, sampling the next token (line 11).

### 3.2 Watermark Detection

Similar to embedding, watermark detection is performed at both the token and value levels. In Fig. 1, after tokenizing each row (Step ⑥), we compute its support for the watermark's presence with two score-functions $\phi_{token}$, testing for the BarCode (Step ⑦), and $\phi_{value}$ testing for the allowed buckets (Step ⑧).

**Token watermarks** generally leverage a single 'favored' list, requiring only to test for the ratio of occurrences of favored tokens. However, TabTokWak randomly swaps the favored sub-vocabularies across records. Hence, the information about which sub-vocabulary was used is lost at the single-row level. To circumvent this issue in watermark detection, we propose a two-sided Z-score based on a score function $\phi_{token}$. This function works as a barcode scanner, measuring how well the record matches the expected BarCode pattern. In this context, we formulate the following hypotheses.

**Null Hypothesis** ($\mathcal{H}_0$): The observed tabular record without $\kappa_p$'s influence.
**Alternative Hypothesis** ($\mathcal{H}_1$): The record was generated under the key $\kappa_p$.

During detection, TabTokWak extracts the tokenized column values $S_j \in \mathcal{S}$ from row $R$ and divides their tokens $s_m^{(j)}$ into two disjoint sets based on the color predicted by the barcode $B_j[m] + h$. While the column name $C_j$ and relative token position $m$ are readily available, TabTokWak is unaware of the used toggle $h$. Therefore, without loss of generality, we assume $h = 0$. The two sets of tokens can be written as,

$$\mathcal{W} = \left\{ s_m^{(j)} \mid \forall s_m^{(j)} \in \mathcal{S},\ B_j[m] = \boxed{1} \right\} \qquad \mathcal{B} = \left\{ s_m^{(j)} \mid \forall s_m^{(j)} \in \mathcal{S},\ B_j[m] = \boxed{0} \right\},$$

where $\mathcal{W}$ contains all white tokens $s_m^{(j)}$, i.e., $B_j[m] = \boxed{1}$, and $\mathcal{B}$ contains all other expected black tokens. With $\mathcal{W}$ and $\mathcal{B}$, we define the scoring function as follows:

$$\phi_{token}(\mathcal{W}, \mathcal{B}, V, \kappa_p) = \sum_{s \in \mathcal{W}} (-1)^{\mathbb{1}[s \in \texttt{FavorList}(V, \kappa_p, \blacksquare)]} + \sum_{s \in \mathcal{B}} (-1)^{\mathbb{1}[s \in \texttt{FavorList}(V, \kappa_p, \boxed{1})]}.$$

Intuitively, $\phi_{token}$ measures how well the signal aligns with the expected watermark, by counting the tokens that *match* the expected color. In case a row $R$ is watermarked and the assumption on $h$ is correct, all terms are 1 and $\phi_{token}$ adds to the size of $|S|$. If the row is not watermarked, each $s_m^{(j)}$ is with probability $1/2$ correct; with on average half of the terms being 1 and the other half being $-1$. Hence, its resulting score $\phi_{token} \approx 0$. Finally, in case $R$ is $\kappa_p$-watermarked, but the assumption on $h$ is false, it means that all terms are exactly the opposite, i.e., all are $-1$ and $\phi_{token} = -|S|$. Thus allowing $\phi_{token}$ as evidence in a two-sided z-test to compute a p-value $\mathcal{P}^{token}$.

**Value Watermark** tests for column values falling in 'allow' buckets. Towards this end, we define a score function $\phi_{value}$ that counts the occurrences of column values falling inside their corresponding allow buckets over a set of rows $\mathcal{R}$ using a key $\kappa$:

$$\phi_{value}(\mathcal{R}, \kappa) = \sum_{R \in \mathcal{R}} \sum_{(c_j, v_{j,i}) \in R} \mathbb{1}\left[\exists b \in \texttt{AllowBuckets}(C_j, \kappa) \wedge v_{j,i} \in b\right].$$

Then, TabTokWak performs $T$ trials using random watermark keys $\kappa_t$ to compute an empirical p-value. Doing so by comparing the scores from random keys against the score from the correct private key $\phi_{value}(\mathcal{R}, \kappa_p)$, counting the wins of the key $\kappa_p$, i.e.,

$$\mathcal{P}^{\text{value}}(\mathcal{R}, \kappa_p) = \frac{1}{T+1}\left(1 + \sum_{t=1}^{T} \mathbb{1}\left[\phi_{value}(\mathcal{R}, \kappa_t) \leq \phi_{value}(\mathcal{R}, \kappa_p)\right]\right).$$

This empirical p-value reflects the likelihood that the numerical values in a record were watermarked. Following, we utilize Fisher's method [9] to derive the joint evidence from the separately obtained p-values of $\mathcal{P}^{\text{BarCode}}$ and $\mathcal{P}^{\text{value}}$. Lastly, the obtained statistic is used to evaluate whether $\mathcal{H}_0$ can be rejected, i.e., the data were watermarked.

## 4 Evaluation

In this section, we evaluate the effectiveness of TabTokWak on watermarking tables generated by a Tabular FM. We assess degradation in data quality and detectability with and without post-editing attacks. We compare TabTokWak against two watermarking schemes: KGW and GumbelSoft, representing both red/green-listed and EXP-style watermarks. Both sporting their author recommended hyper-parameters.

**Experimental Setting.** We evaluate TabTokWak and baseline watermarks on five real-world datasets: Adult, Diabetes, King, Loan, and Magic. We provide statistics of these datasets, and the corresponding downstream objective in Table 1. The backbone tabular FM is REaLTabFormer [16], serving as a representative example of an autoregressive

**Table 1.** Statistical properties of used dataset. #Records, #Cat., #Num indicate counts of; rows, and categorical and numerical columns, respectively.

| Name | Domain | #Records | #Cat. | # Num. | Objective |
|---|---|---|---|---|---|
| Adult | Social Science | 48,842 | 9 | 6 | Classification |
| Diabetes | Health-care | 678 | 1 | 8 | Classification |
| King | Commerce | 21,000 | 13 | 7 | Regression |
| Loan | Finance | 5000 | 7 | 6 | Classification |
| Magic | Physics | 19,019 | 1 | 10 | Classification |

generative tabular model, with the author-provided hyperparameters and stopping condition. Herein, categorical values are represented as a discrete distribution. Numerical values are represented as a discretized distribution consisting of a fixed-integer representation, although our method generalizes to an arbitrary FM tabular model.

**Metrics.** We investigate the quality of synthetic data using two groups of metrics: statistical resemblance to real data and downstream machine learning efficiency (MLE). We compute statistical scores with `synthcity` [15]. These are split into *Shape* and *Trend* based on the distributions of the data columns and pairwise column-feature correlations. Downstream MLE is evaluated using ML models. To evaluate watermark detectability, we compute $p$-values for each watermarking algorithm's statistical score.

**Baselines.** We compare with GumbelSoft [10] and KGW [11]. We make all watermarking strategies 'aware' of permissible token values for cell values, so that these watermarks operate on tabular data. We compare against KGW ($\gamma = 0.25, \delta = 2.0$) and GumbelSoft ($\tau = 0.3$). Moreover, we fix the context length of all methods to $K = 1$ for a fair comparison. As the watermark signals with larger context (KGW and GumbelSoft) are significantly distorted under various attacks [12]. Moreover, a shorter context, i.e., using 'self-hashing' with $K = 0$, results in a significant degradation in data quality.

**Data Quality** as shown in Table 2, presents the average quality metrics of Watermarked models. Alongside, we report the true-positive rate (TPR) at a fixed false-positive rate (FPR = 0.001) for base detection. In all cases, higher is better. Additionally, we provide quality results for the FM base model *without* watermarking, indicated by '-'. Therewith, allowing for the evaluation of each watermark's impact on data quality. In this table, the best-performing results are in **boldface**, and the second-best are <u>underlined</u>.

Overall, we observe that watermarks degrade data quality. Compared to their non-watermarked counterparts, both statistical and downstream utility metrics are reduced. This degradation, expressed as a percentage relative to the non-watermarked ('-') results, can be attributed to the bias introduced by each watermark. Notably, GumbelSoft shows a pronounced drop across all scores. With degradations in average metrics, shape by 37%, the trends dropping by 33%, and MLE decreasing by 22%, across the five datasets. Furthermore, we observe that the qualitative metrics of our proposed method surpass those of KGW, except for magic's trend statistic. This improvement is due to

**Table 2.** Quality of (non-)/watermarked ('-') data generated and detectability under varying record counts, generated by REaLTabFormer [16] . Reporting (higher better ↑) of statistical (Shape, Trend) and down-stream utility (MLE), of (non-/)watermarked ('-') data by pre-trained model, showing mean score ± standard deviation.

| | | Quality (↑) | | | TPR@FPR=1e-3 (↑) | | | | |
|---|---|---|---|---|---|---|---|---|---|
| | Watermark | Shape | Trend | MLE | 1 | 5 | 10 | 50 | 100 |
| adult | - | 0.970±0.000 | 0.948±0.003 | 0.914±0.026 | - | - | - | - | - |
| | GUMBEL | 0.630±0.068 | 0.400±0.114 | 0.750±0.137 | **0.752** | **1.00** | **1.00** | **1.00** | **1.00** |
| | KGW | 0.835±0.021 | 0.766±0.039 | 0.849±0.077 | 0.208 | 0.918 | 0.998 | **1.00** | **1.00** |
| | TabTokWak | **0.911**±0.016 | **0.835**±0.027 | **0.854**±0.075 | 0.583 | 0.924 | 0.890 | 0.967 | 0.994 |
| diabetes | - | 0.960±0.006 | 0.947±0.005 | 0.845±0.033 | - | - | - | - | - |
| | GUMBEL | 0.619±0.117 | 0.779±0.035 | 0.726±0.075 | 0.562 | **1.00** | **1.00** | **1.00** | **1.00** |
| | KGW | 0.842±0.028 | 0.915±0.007 | 0.775±0.078 | 0.272 | 0.737 | 0.917 | **1.00** | **1.00** |
| | TabTokWak | **0.894**±0.031 | **0.919**±0.011 | **0.781**±0.070 | **0.602** | 0.958 | **1.00** | **1.00** | **1.00** |
| king | - | 0.976±0.001 | 0.914±0.015 | 0.833±0.146 | - | - | - | - | - |
| | GUMBEL | 0.667±0.065 | 0.700±0.022 | 0.423±0.395 | 0.582 | **0.987** | **1.00** | **1.00** | **1.00** |
| | KGW | 0.831±0.061 | 0.838±0.012 | 0.671±0.176 | 0.090 | 0.716 | 0.974 | **1.00** | **1.00** |
| | TabTokWak | **0.915**±0.027 | **0.909**±0.023 | **0.679**±0.223 | **0.709** | 0.688 | 0.869 | **1.00** | **1.00** |
| loan | - | 0.939±0.002 | 0.883±0.001 | 0.959±0.025 | - | - | - | - | - |
| | GUMBEL | 0.612±0.028 | 0.479±0.042 | 0.854±0.078 | 0.571 | **1.00** | **1.00** | **1.00** | **1.00** |
| | KGW | 0.815±0.007 | **0.726**±0.004 | 0.910±0.067 | 0.207 | 0.808 | 0.990 | **1.00** | **1.00** |
| | TabTokWak | **0.840**±0.024 | **0.726**±0.026 | **0.921**±0.050 | **0.818** | 0.981 | 0.971 | **1.00** | **1.00** |
| magic | - | 0.974±0.000 | 0.957±0.009 | 0.845±0.037 | - | - | - | - | - |
| | GUMBEL | 0.682±0.032 | 0.802±0.003 | 0.732±0.099 | **1.00** | **1.00** | **1.00** | **1.00** | **1.00** |
| | KGW | 0.870±0.003 | **0.926**±0.010 | 0.806±0.030 | 0.930 | **1.00** | **1.00** | **1.00** | **1.00** |
| | TabTokWak | **0.907**±0.032 | 0.920±0.013 | **0.818**±0.035 | 0.997 | **1.00** | **1.00** | **1.00** | **1.00** |

KGW's biasing strategy, which is sensitive to low-entropy sequences, thereby reducing variability during sampling. In contrast, our watermarking approach permits a more relaxed strategy, due to its dual-color nature. Across all datasets, our TabTokWak consistently provides the highest MLE scores.

**Detectability** we consider in terms of the positive rate with a fixed false-positive rate ($10^{-3}$. Here, we consider the best-case scenario in which the generation order is known. In general, we observe that GumbelSoft achieves superior detectability, requiring at most 10 samples to reach a TPR of 1. However, this detectability comes at the cost of the highest quality penalty. Second comes KGW, requiring at most 50 samples to remain detectable. Although KGW's quality penalty is lower than Gumbel's, it remains higher than that of TabTokWak. While TabTokWak shows strong detection performance on longer tokenized records, its performance is lower on shorter records. Requiring 100+ records on adult to achieve a true positive rate of 1. However, its detectability under

**Table 3.** Attack robustness of watermarks in term of TPR@FPR=1e-3 on multi-record detection. Cell deletion represents a random deletion of a fraction of cell values.

| | | Shuffle Attack | | | Deletion = 0.05 | | | Deletion = 0.1 | | | Deletion = 0.2 | | |
|---|---|---|---|---|---|---|---|---|---|---|---|---|---|
| | Scheme | 5 | 10 | 100 | 5 | 10 | 100 | 5 | 10 | 100 | 5 | 10 | 100 |
| adult | GUMBEL | **0.951** | **1.00** | **1.00** | **0.971** | **1.00** | **1.00** | **0.915** | **0.995** | **1.00** | **0.754** | **0.929** | **1.00** |
| | KGW | 0.410 | 0.791 | **1.00** | 0.848 | 0.990 | **1.00** | 0.767 | 0.970 | **1.00** | 0.607 | **0.883** | **1.00** |
| | TabTokWak | 0.925 | 0.899 | 0.995 | 0.880 | 0.847 | 0.976 | 0.821 | 0.782 | 0.932 | 0.680 | 0.647 | 0.702 |
| diabetes | GUMBEL | 0.904 | 0.999 | **1.00** | **0.998** | **1.00** | **1.00** | **0.986** | **1.00** | **1.00** | **0.926** | **0.997** | **1.00** |
| | KGW | 0.524 | 0.756 | **1.00** | 0.667 | 0.842 | **1.00** | 0.607 | 0.773 | **1.00** | 0.517 | 0.661 | 0.995 |
| | TabTokWak | **0.959** | **1.00** | **1.00** | 0.898 | 0.996 | **1.00** | 0.832 | 0.985 | **1.00** | 0.692 | 0.930 | **1.00** |
| king | GUMBEL | **0.946** | 0.999 | **1.00** | **0.971** | **1.00** | **1.00** | 0.474 | 0.807 | **1.00** | 0.360 | 0.638 | **1.00** |
| | KGW | 0.568 | 0.802 | **1.00** | 0.650 | 0.947 | **1.00** | 0.572 | 0.912 | **1.00** | 0.424 | 0.798 | **1.00** |
| | TabTokWak | 0.945 | **0.995** | **1.00** | 0.912 | 0.988 | **1.00** | 0.866 | 0.970 | **1.00** | **0.758** | **0.905** | **1.00** |
| loan | GUMBEL | 0.606 | 0.942 | **1.00** | **0.999** | **1.00** | **1.00** | **0.994** | **1.00** | **1.00** | **0.952** | **1.00** | **1.00** |
| | KGW | 0.451 | 0.846 | **1.00** | 0.685 | 0.952 | **1.00** | 0.560 | 0.879 | **1.00** | 0.323 | 0.611 | **1.00** |
| | TabTokWak | **0.985** | **0.976** | **1.00** | 0.965 | 0.946 | **1.00** | 0.937 | 0.913 | **0.999** | 0.852 | 0.818 | 0.962 |
| magic | GUMBEL | **1.00** | **1.00** | **1.00** | **1.00** | **1.00** | **1.00** | **1.00** | **1.00** | **1.00** | **1.00** | **1.00** | **1.00** |
| | KGW | **1.00** | **1.00** | **1.00** | **1.00** | **1.00** | **1.00** | **0.999** | **1.00** | **1.00** | 0.996 | **1.00** | **1.00** |
| | TabTokWak | **1.00** | **1.00** | **1.00** | **1.00** | **1.00** | **1.00** | **1.00** | **1.00** | **1.00** | 0.997 | **1.00** | **1.00** |

1–5 samples provides similar detection, remaining close with more samples. Herein, showing the trade-off that the bi-color detection imposes on the detection results.

**Robustness Against Post-editing Attacks.** In addition to the previous best-case, we also evaluate the method's resistance to post-processing attacks. We consider three attacks on columns with floating-point data: column reordering (Shuffling), random cell deletion (Deletion), and rounding (Rounding). *Shuffling* randomly permutes the columns of each record, emulating detection under an unknown data generation order. *Deletion* randomly sets a fraction $d$ to a null value, following a Missing Completely at Random strategy. *Rounding* emulates post-processing by reducing the numerical precision by $n$ decimal places. Herein, we reduce the numerical precision of floating-point cell values by $n \in \{1, 2, 3\}$ decimal places and padding them with $n$ trailing zeros.

In Table 3, we present the detectability results and an overview of the detection efficiency of the re-ordering and cell-deletion attacks. Herein, reporting the TPR@FPR=1e-3 on 5, 10, and 100 samples. Considering the column shuffle attack, TabTokWak's detection remains unaffected. However, both KGW and GumbelSoft's performance degrades, particularly at low sample counts (<50). While GumbelSoft shows a more nuanced reduction in its TPR, KGW's TPR degrades below 0.5. With a lesser difference under the wider magic dataset. As magic's numerical values contain more decimal places. As a result, the fraction of tokens with invalidated context is diminished due to each cell's tokenized length, i.e., $|\mathcal{C}| << |S|$.

**Table 4.** Watermark's TPR@FPR=1e-3 following rounding attacks with multi-record detection. Rounding attack truncates floating-point values by $n$.

| | | Rounding = 1 | | | Rounding = 2 | | | Rounding = 3 | | |
|---|---|---|---|---|---|---|---|---|---|---|
| | Samples | 5 | 10 | 100 | 5 | 10 | 100 | 5 | 10 | 100 |
| diabetes | GUMBEL | **0.985** | **1.00** | **1.00** | **0.983** | **1.00** | **1.00** | **0.906** | **0.999** | **1.00** |
| | KGW | 0.624 | 0.723 | **1.00** | 0.600 | 0.707 | **1.00** | 0.600 | 0.706 | **1.00** |
| | TabTokWak | 0.793 | 0.974 | **1.00** | 0.619 | 0.879 | **1.00** | 0.667 | 0.800 | 0.984 |
| king | GUMBEL | **0.943** | **0.999** | **1.00** | **0.963** | **1.00** | **1.00** | **0.933** | **1.00** | **1.00** |
| | KGW | 0.536 | 0.864 | **1.00** | 0.425 | 0.754 | **1.00** | 0.490 | 0.799 | **1.00** |
| | TabTokWak | 0.853 | 0.955 | **1.00** | 0.805 | 0.925 | **1.00** | 0.822 | 0.937 | **1.00** |
| magic | GUMBEL | **1.00** | **1.00** | **1.00** | **0.990** | **1.00** | **1.00** | **0.487** | **0.940** | **1.00** |
| | KGW | 0.992 | **1.00** | **1.00** | 0.926 | 0.999 | **1.00** | 0.445 | 0.734 | **1.00** |
| | TabTokWak | 0.999 | **1.00** | **1.00** | 0.965 | 0.998 | **1.00** | 0.579 | 0.811 | **1.00** |

As the deletion attack strength increases, a negative trend is observed across methods. At a deletion strength of 20%, all but GumbelSoft require 100(+) samples to attain perfect TPR@FPR=1e-3. Again, we observe that especially with fewer samples (5–10), KGW and GumbelSoft are more sensitive to perturbations. Due to the construction of their context window in their sampling scheme, they suffer from changes in representation after deletion. Meanwhile, `TabTokWak` consistently shows only moderate degradation across datasets, except for adult. In this latter case, requiring 100+ samples to attain a TPR of 1, due to the high number of categoricals and a single value-watermarked column. Showing that in extreme cases, the trade-off from the dual coloring scheme cannot be won back by the additional value of the watermark signal.

Lastly, we evaluate rounding attacks limited to datasets with floating-point values: magic, king, and diabetes. We report the detectability in Table 4. As with the preceding attacks, for KGW and GumbelSoft, a decrease in TPR is observed for 5–10 samples. Meanwhile, `TabTokWak` shows stable detectability, attaining a TPR $\gtrsim 0.9$ with at most 10 samples with moderate attack strength 1–2. However, at higher attack strengths, the value-watermark's values increasingly fall outside their allowed buckets, reducing detectability. Additionally, the number of perturbed tokens increases significantly, reducing the barcode signal, in turn lowering the evidence for `TabTokWak`'s joint watermark detection.

## 5 Conclusion

In this paper, we identify two significant challenges when using soft red/green lists and sampling-based watermarking in auto-regressive tabular models, namely, column permutation and its impact on data quality. We propose Tabular Token(less) Watermarking (`TabTokWak`), a dual watermarking strategy for tabular FM. `TabTokWak` combines token and value watermarks that are column order invariant through column-based seeding. We demonstrate our method's performance using REaLTabFormer, an auto-regressive tabular model, on five real-world datasets. Therewith, providing a balanced trade-off between improved data quality and a minor reduction in detectability relative to the baselines. TabTokWak achieves reliable detection performance, even under post-processing attacks.

**Acknowledgments.** This research is part of the Priv-GSyn project, 200021E_229204 of Swiss National Science Foundation and the DEPMAT project, P20-22/N21022, of the research programme Perspectief, which is partly financed by the Dutch Research Council (NWO). This research was partly funded by the Spoke "FutureHPC & BigData" of the ICSC-Centro Nazionale di Ricerca in "High Performance Computing, Big Data and Quantum Computing", funded by the European Union - NextGenerationEU, and by the DYMAN project funded by the European Union - European Innovation Council under G.A. n. 101161930.

## References

1. Aaronson, S.: My AI Safety Lecture for UT Effective Altruism (2023)
2. Abdelhameed, S.A., Moussa, S.M., Khalifa, M.E.: Privacy-preserving tabular data publishing: a comprehensive evaluation from web to cloud. Comput. Secur. **72**, 74–95 (2018). https://doi.org/10.1016/J.COSE.2017.09.002
3. Bergman, A.S., et al.: Guiding the release of safer E2E conversational AI through value sensitive design. In: SIGdial, pp. 39–52 (2022). https://doi.org/10.18653/V1/2022.SIGDIAL-1.4
4. Borisov, V., Seßler, K., Leemann, T., Pawelczyk, M., Kasneci, G.: Language models are realistic tabular data generators. In: ICLR (2023)
5. Chawla, N.V., Bowyer, K.W., Hall, L.O., Kegelmeyer, W.P.: SMOTE: synthetic minority over-sampling technique. JAIR **16**, 321–357 (2002). https://doi.org/10.1613/JAIR.953
6. Choi, E., Biswal, S., Malin, B.A., Duke, J., Stewart, W.F., Sun, J.: Generating multi-label discrete patient records using generative adversarial networks. In: MLHC, vol. 68, pp. 286–305 (2017)
7. Cromp, S., et al.: Tabby: a language model architecture for tabular and structured data synthesis. TMLR **2026** (2026). https://openreview.net/forum?id=b9FPVnb0Bn
8. Fernandez, P., Chaffin, A., Tit, K., Chappelier, V., Furon, T.: Three bricks to consolidate watermarks for large language models. In: WIFS, pp. 1–6 (2023). https://doi.org/10.1109/WIFS58808.2023.10374576
9. Fisher, R.A.: On the interpretation of $\chi^2$ from contingency tables, and the calculation of p. J. Roy. Stat. Soc. **85**(1), 87–94 (1922). https://doi.org/10.1111/j.2397-2335.1922.tb00768.x
10. Fu, J., Zhao, X., Yang, R., Zhang, Y., Chen, J., Xiao, Y.: GumbelSoft: diversified language model watermarking via the GumbelMax-trick. In: ACL (2024). https://doi.org/10.18653/v1/2024.acl-long.315
11. Kirchenbauer, J., Geiping, J., Wen, Y., Katz, J., Miers, I., Goldstein, T.: A watermark for large language models. In: ICML, vol. 202, pp. 17061–17084 (2023). https://proceedings.mlr.press/v202/kirchenbauer23a.html
12. Kuditipudi, R., Thickstun, J., Hashimoto, T., Liang, P.: Robust distortion-free watermarks for language models. TMLR **2024** (2024). https://openreview.net/forum?id=FpaCL1MO2C
13. Lin, W., Tsai, C.: Missing value imputation: a review and analysis of the literature (2006-2017). Artif. Intell. Rev. **53**, 1487–1509 (2020). https://doi.org/10.1007/S10462-019-09709-4

14. Patki, N., Wedge, R., Veeramachaneni, K.: The synthetic data vault. In: DSAA, pp. 399–410 (2016). https://doi.org/10.1109/DSAA.2016.49
15. Qian, Z., Davis, R., van der Schaar, M.: Synthcity: a benchmark framework for diverse use cases of tabular synthetic data. In: NeurIPS (2023)
16. Solatorio, A.V., Dupriez, O.: Realtabformer: generating realistic relational and tabular data using transformers (2023). https://doi.org/10.48550/ARXIV.2302.02041
17. Team, O.: Language models are few-shot learners. In: NeurIPS (2020)
18. Xu, L., Skoularidou, M., Cuesta-Infante, A., Veeramachaneni, K.: Modeling tabular data using conditional GAN. In: NeurIPS, pp. 7333–7343 (2019)
19. Yoo, K., Ahn, W., Kwak, N.: Advancing beyond identification: multi-bit watermark for large language models. In: NAACL, pp. 4031–4055 (2024). https://doi.org/10.18653/V1/2024.NAACL-LONG.224
20. Zhang, T., Wang, S., Yan, S., Jian, L., Liu, Q.: Generative table pre-training empowers models for tabular prediction. In: EMNLP, pp. 14836–14854. ACL (2023). https://doi.org/10.18653/V1/2023.EMNLP-MAIN.917

# DENI: A Density-Enhanced Hybrid Sampling Framework with Neighborhood Information for Noisy Imbalanced Classification

Tian Tan, Yuhong Xu(✉), and Peijie Huang

College of Mathematics and Informatics, South China Agricultural University, Guangzhou, China
tantian@stu.scau.edu.cn,{xuyuhong,pjhuang}@scau.edu.cn

**Abstract.** The class-imbalance problem is a prevalent challenge in real-world applications. Label noise aggravates the complexity of classification. Conventional classifiers tend to learn a bias toward the majority class, which results in poor performance on the minority class. For noisy imbalanced data, the resampling method is one of the most competitive methods among the existing techniques. However, it is susceptible to improper sampling strategies. To this end, this paper proposes a density-enhanced hybrid sampling framework with neighborhood information (DENI). DENI first introduces the global density gap of the dataset to distinguish the noisy samples, which can reduce the impact of noise on classification. Second, a density-enhanced oversampling process with neighborhood information is designed, which enables the potential regions of the minority class to be appropriately expanded without generating the scattered samples or compressing the minority-class regions. Finally, random rebalance is implemented to obtain a balanced dataset. Extensive experiments on diverse benchmark noisy imbalanced datasets demonstrate the effectiveness and robustness of the proposed DENI.

**Keywords:** Imbalanced classification · Label noise · Hybrid sampling

## 1 Introduction

Classification models serve as a cornerstone of machine learning with critical applications across numerous domains. However, their effectiveness often relies on an idealized assumption that the training data exhibits a relatively balanced distribution across classes. In real-world scenarios, this assumption is frequently violated, leading to the prevalent challenge of class imbalance, where one class (i.e. the majority class) contains a significantly larger number of instances than the other (i.e. the minority class). This imbalance causes standard classification models, which are typically optimized for overall accuracy, to develop a strong predictive bias towards the majority class. Consequently, they often fail to recognize the minority-class examples, which are usually of greater interest and


R. C.-W. Wong et al. (Eds.): PAKDD 2026, LNAI 16597, pp. 301–313, 2026.
https://doi.org/10.1007/978-981-92-1300-9_24

value, such as fraudulent transactions [1], rare diseases [21], or network intrusion detection [13]. Therefore, enhancing the ability to identify the minority class has become a crucial and challenging research area in machine learning.

To address class imbalance, existing approaches can be broadly categorized into three groups: data-level resampling techniques, algorithm-level cost-sensitive learning, and ensemble methods [20]. Cost-sensitive learning modifies algorithm itself by assigning a higher misclassification cost to the minority class, thereby forcing the model to pay more attention to it [9]. Ensemble methods combine the strengths of multiple learners to enhance the overall robustness [19]. Resampling methods have been widely adopted as mainstream solutions due to their simplicity, efficiency, and independence from specific models [6].

In this study, we focus on the data-level resampling methods, which can be divided into three categories: undersampling, oversampling and hybrid sampling [16]. Oversampling balances datasets by generating synthetic samples for the minority class, such as SMOTE [3]. Undersampling balances datasets by reducing the number of majority-class instances, such as Random Undersampling (RUS) and CC [11]. Hybrid sampling techniques combine the above two strategies, aiming to achieve better results, such as SMOTE-TomekLinks [2]. However, oversampling techniques tend to generate redundant and noisy samples, which leads to overfitting, while undersampling techniques may remove critical samples that lead to underfitting. Although hybrid sampling can help mitigate the above issues, there is still the problem of how to properly generate high-quality minority samples and remove majority samples without information loss.

This paper proposes a density-enhanced hybrid sampling framework with neighborhood information (DENI) to address the above problems. The core idea is to resample datasets with the combination of global density distribution and local information, which enables the synthetic samples to fit the local structure and to conform to the global distribution. In this way, DENI achieves reasonable expansion for the minority-class regions, which can provide clearer decision boundaries for classifying noisy imbalanced data.

The contributions of this paper are summarized as follows: **(1)** DENI enhances the robustness of classification models to noise by introducing a global density-based noise filter, which accurately distinguishes noises without the need to determine local scope manually. **(2)** DENI expands potential regions of the minority class by density-enhanced oversampling with neighborhood information, which neither generates scattered samples nor compresses the minority-class regions. **(3)** Extensive experiments are conducted over diverse noisy imbalanced datasets to prove the effectiveness and robustness of the proposed method.

The rest of this paper is organized as follows: Sect. 2 briefly introduces the related work. Section 3 describes the proposed DENI framework. Section 4 presents the experimental studies. Section 5 concludes the paper.

## 2 Related Work

In past decades, a number of resampling methods have been proposed to handle imbalanced classification from different perspectives. Categorized by sampling strategies, these methods can be divided into three typical types: distance-based methods, density-based methods and noise-robust methods.

**Distance-based methods** rely on the local information provided by spatial distance between samples to guide samples processing. SMOTE [3] generates new samples by linear interpolation between minority samples and their nearest neighbors. SMOTE only generates samples in the safe regions, which compresses the boundary of the minority class. To address this issue, Borderline-SMOTE [7] and ADASYN [8] are proposed, which emphasize specific local regions while sampling. However, these methods focus only on the local regions, which tend to distort the boundaries between classes.

**Density-based methods** generate or remove samples based on the whole distribution of the dataset. PDFOS [5] oversamples the minority-class samples based on the probability density distribution. KernelADASYN [17] reweights the probability density of the minority class to generate more samples on class boundaries. However, it is easy for these methods to generate scattered samples, which blur the decision boundaries.

**Noise-robust methods** integrate noise detection mechanism into sampling process. SMOTE-TomekLinks [2] detects noise based on Tomek Links. SMOTE-IPF [14] reduces noise by an iterative noise filter. RSMOTE [4] and WRND [10] distinguish noise by local neighbors. NGOS [16] differentiates noise by a local region radius. IGBS [18] removes noisy samples while undersampling granular balls. Most of the existing methods need to determine a proper local scope for noise defining, which are sensitive to corresponding hyperparameters.

## 3 Proposed Method

The principle of the proposed DENI framework can be divided into three steps as follows: **(1)** Global Density-Based Noise Filtration, **(2)** Density-Enhanced Oversampling with Neighborhood Information, and **(3)** Random Rebalance. A schematic diagram of DENI is depicted in Fig. 1. The detailed process of each component is described in the following subsections.

### 3.1 Global Density-Based Noise Filtration

Given a data set $D = \{\mathbf{x}_i, y_i\}_{i=1}^{N} = D_{maj} \cup D_{min}$, where $\mathbf{x}_i \in \mathbb{R}^d, y_i \in \{0, 1\}$, $N$ is the size of $D$ and $d$ denotes the dimensions. $D_{maj} = \{\mathbf{x}_i, y_i = 0\}_{i=1}^{N_{maj}}$ is the majority and $D_{min} = \{\mathbf{x}_i, y_i = 1\}_{i=1}^{N_{min}}$ is the minority. For a specific set $\mathcal{D}$, $h_{\mathcal{D}}$ is the optimal bandwidth of the kernel function for $\mathcal{D}$ obtained by Scott's rule [15], where $\sigma_{\mathcal{D}}$ denotes the standard deviation of $\mathcal{D}$.

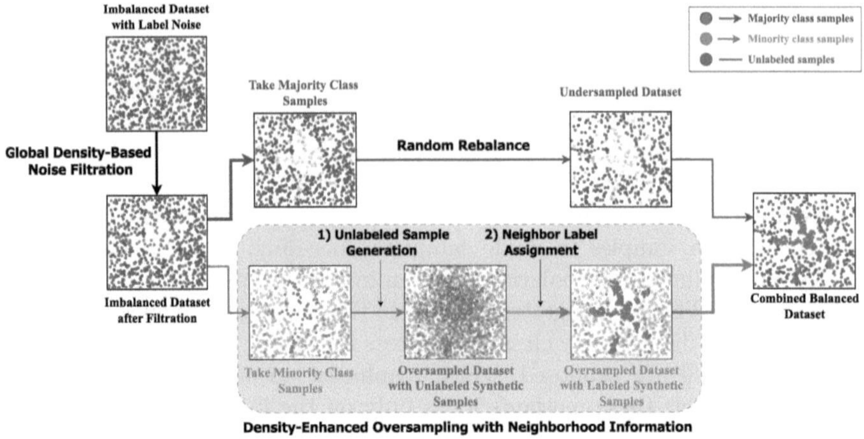


**Fig. 1.** Schematic diagram of the proposed DENI framework.

**Definition 1 (Global Majority Density, $p_{maj}(\mathbf{x})$).** *The probability density function estimated based on entire majority-class samples. For any point $\mathbf{x} \in \mathbb{R}^d$, its global majority density is defined as follows. Here, $h_{D_{maj}} = \sigma_{D_{maj}} \cdot N_{maj}{}^{-\frac{1}{d+4}}$.*

$$p_{maj}(\mathbf{x}) = \frac{1}{N_{maj}(h_{D_{maj}}\sqrt{2\pi})^d} \sum_{i=1}^{N_{maj}} \exp\left(-\frac{\|\mathbf{x} - \mathbf{x}_i^{maj}\|^2}{2h_{D_{maj}}^2}\right), \mathbf{x}_i^{maj} \in D_{maj} \quad (1)$$

**Definition 2 (Global Minority Density, $p_{min}(\mathbf{x})$).** *The probability density function estimated based on entire minority-class samples. For any point $\mathbf{x} \in \mathbb{R}^d$, its global minority density is defined as follows. Here, $h_{D_{min}} = \sigma_{D_{min}} \cdot N_{min}{}^{-\frac{1}{d+4}}$.*

$$p_{min}(\mathbf{x}) = \frac{1}{N_{min}(h_{D_{min}}\sqrt{2\pi})^d} \sum_{i=1}^{N_{min}} \exp\left(-\frac{\|\mathbf{x} - \mathbf{x}_i^{min}\|^2}{2h_{D_{min}}^2}\right), \mathbf{x}_i^{min} \in D_{min} \quad (2)$$

**Definition 3 (Density Gap, $DG(\mathbf{x})$).** *The density gap at any point $\mathbf{x} \in \mathbb{R}^d$ is defined as the arithmetic difference between the normalized global minority density and global majority density at the point, which is formulated by:*

$$p'_{min}(\mathbf{x}) = \frac{p_{min}(\mathbf{x})}{\sum_{\mathbf{x}_i \in D} p_{min}(\mathbf{x}_i)}, p'_{maj}(\mathbf{x}) = \frac{p_{maj}(\mathbf{x})}{\sum_{\mathbf{x}_i \in D} p_{maj}(\mathbf{x}_i)} \quad (3)$$

$$DG(\mathbf{x}) = p'_{min}(\mathbf{x}) - p'_{maj}(\mathbf{x}) \quad (4)$$

**Definition 4 (Majority Class Noise, $D_{maj}^{noise}$).** *A majority class sample $\mathbf{x}_i^{maj}$ is identified as noise if it lies in the region where global minority density exceeds global majority density. Thus the majority class noise is defined as*

$$D_{maj}^{noise} = \{\mathbf{x} | \mathbf{x}_i \in D_{maj} \ \& \ DG(\mathbf{x}_i) > 0\} \quad (5)$$

As discussed, local structure-based noise differentiation is unreliable due to the difficulty of determining the optimal local scope. A larger scope loses useful samples, while a smaller one retains most noise. To address this, a global density-based approach for noise reduction is proposed. Due to the rarity of minority-class samples, only majority-class noises are filtered. Referring to Definition 1 and Definition 2, we obtain the global density of majority and minority classes. Afterwards, the data space of each class can be clearly divided according to density gap in Definition 3. Isolated noisy samples lie in regions dominated by the density of heterogeneous class, enabling easy identification and filtering of majority-class noises according to Definition 4. Our method aligns better with original distribution of the dataset, eliminating the need to determine local scopes.

### 3.2 Density-Enhanced Oversampling with Neighborhood Information

SMOTE-based methods tend to generate safe samples, but they weaken the capacity of classifiers to capture the characteristics of minority class. The oversampling methods that rely solely on specific density distribution may generate scattered samples, which hinder the classifier from building clear decision boundaries. In DENI, we design a novel density-enhanced method to obtain information-rich minority-class samples with the aid of labeled neighboring samples, which neither retains scattered samples nor compresses the minority-class regions.

**Unlabeled Sample Generation.** To expand the data space of the minority, we construct a Gaussian distribution model for sample generation. For any minority instance $\mathbf{x}_i^{min}$, we define a Gaussian distribution $\mathcal{N}(0, h_{D_{min}}^2 I_d)$. Based on the Gaussian distribution, a new unlabeled sample is generated by Eq. (6).

$$\mathbf{x}_{new} = \mathbf{x}_i^{min} + \mathcal{N}(0, h_{D_{min}}^2 I_d) \tag{6}$$

The number of unlabeled samples to be generated is determined by:

$$G = \lceil N_{maj} \cdot \alpha \rceil - N_{min} \tag{7}$$

where $\alpha$ is a parameter that specifies the balance level.

**Neighbor Label Assignment.** In the previous stage, we have generated enough candidate unlabeled samples based on minority instances. However, not all of them can effectively augment the minority class, because the label information they absorb from neighbors varies significantly. The basic idea of this stage is to assign a label to each unlabeled sample, where the assigned labels contain information from their neighbors' labels. Subsequently, we keep only the samples that are assigned to the minority class.

For a data set $D = \{\mathbf{x}_i, y_i\}_{i=1}^{N}, y_i \in \{1, 2, \cdots, C\}$, we first build a label matrix $\mathbf{Y}_D \in \mathbb{R}^{N \times C}$, whose entries are defined as in Eq. (8).

$$\mathbf{Y}_{ic} = \begin{cases} 1 & \text{if } y_i = c \\ 0 & \text{otherwise} \end{cases} \tag{8}$$

Next, for each unlabeled sample $\mathbf{x}_u$, we find its $k$-nearest neighbors $NN_k(\mathbf{x}_u) = \{\mathbf{z}_1, \mathbf{z}_2, \cdots, \mathbf{z}_k\}$ in data set $D$ based on the Euclidean distance and compute a weight vector $\mathbf{w}_{\mathbf{x}_u} \in \mathbb{R}^{1 \times k}$ for each of the neighbors, as in Eq. (9).

$$\mathbf{w}_j = \frac{\exp(-\|\mathbf{x}_u - \mathbf{z}_j\|^2)}{\sum_{l=1}^{k} \exp(-\|\mathbf{x}_u - \mathbf{z}_l\|^2)}, \quad j = 1, 2, \cdots, k \tag{9}$$

After that, labels are assigned to them based on the spatial and label information of their neighbors:

$$\mathbf{P}(y = c|\mathbf{x}_u) = \mathbf{w}_{\mathbf{x}_u} \cdot \mathbf{Y}_{NN_k(\mathbf{x}_u)} \tag{10}$$

$$\hat{y}_{\mathbf{x}_u} = \arg\max_{c} \mathbf{P}(y = c|\mathbf{x}_u) \tag{11}$$

where $\mathbf{Y}_{NN_k(\mathbf{x}_u)} \in \mathbb{R}^{k \times C}$ is the label matrix built from the $k$-nearest neighbors of $\mathbf{x}_u$, and $\hat{y}_{\mathbf{x}_u}$ is the label assigned to the unlabeled instance $\mathbf{x}_u$. One can see that, in Eq. (10), weight vector $\mathbf{w}_{\mathbf{x}_u}$ contains the spatial information from the neighbors, and label matrix $\mathbf{Y}_{NN_k(\mathbf{x}_u)}$ contains the label information from the neighbors. To obtain information-rich minority samples, we only retain those who are labeled as the minority class according to Eq. (11).

### 3.3 Random Rebalance

To obtain a balanced dataset, a random rebalance process is implemented. In this stage, we undersample the new majority class samples randomly to rebalance the dataset after the oversampling stage, which means that the number of the majority samples will ultimately equal to the number of the minority samples.

The proposed DENI is described in Algorithm 1.

**Algorithm 1:** The DENI Framework

**Input**: Imbalanced dataset $D = D_{maj} \cup D_{min}$, $k$, $\alpha$
**Output**: Balanced dataset $S$
**Step 1. Noise Filtration**
Initialize $D^{clear} = \emptyset, U = \emptyset, S = \emptyset$;
Calculate density $p_{maj}$ and $p_{min}$ via Eq. (1) and Eq. (2);
Get $p'_{maj}$ and $p'_{min}$ via Eq. (3);
Calculate the density gap $DG(\mathbf{x})$ via Eq. (4);
Obtain $D^{noise}_{maj}$ by Eq. (5);
$D^{clear} \leftarrow D - D^{noise}_{maj}$;
**Step 2. Oversampling**
Calculate the number of unlabeled samples $G$ via Eq. (7);
Randomly select $G$ minority samples $D^{select}_{min}$ with replacement;
**for** $\mathbf{x}_i$ *in* $D^{select}_{min}$ **do**
  Generate a new sample $\mathbf{x}_{new}$ by Eq. (6);
  $U \leftarrow U \cup \mathbf{x}_{new}$;
**end**
**for** $\mathbf{x}_u$ *in* $U$ **do**
  Find $k$ nearest neighbors $NN_k(\mathbf{x}_u)$;
  Get label matrix $\mathbf{Y}_{NN_k(\mathbf{x}_u)}$ via Eq. (8);
  Get weight vector $\mathbf{w}_{\mathbf{x}_u}$ via Eq. (9);
  Assign label for $\mathbf{x}_u$ via Eq. (10) and Eq. (11);
**end**
$S \leftarrow S \cup U$;
**Step 3. Random Rebalance**
$S \leftarrow S \cup D_{clear}$;
Divide $S$ into new majority class $S^{new}_{maj}$ and new minority class $S^{new}_{min}$;
**while** *len*$(S^{new}_{maj}) \neq$ *len*$(S^{new}_{min})$ **do**
  Select a sample $\mathbf{s}_i \in S^{new}_{maj}$ randomly;
  $S \leftarrow S - \mathbf{s}_i$;
**end**
**return** $S$;

### 3.4 Computational Complexity Analysis

Given the data set $D = D_{maj} \cup D_{min}$, the feature dimension of $D$ is $d$. The time complexity of DENI primarily stems from three core components. In step 1, the probability density of the majority and the minority is calculated by kernel density estimation. Since we assume that the dimensions are independent of each other, the time complexity of this stage is $O(d \times N)$. The time complexity for computing $DG(\mathbf{x})$ and filtering majority-class noises can be expressed as $2O(N)+O(N_{maj})$. In step 2, the time complexity for unlabeled sample generation is $O(d \times G)$. The time complexity is $O(d \times G \times \log N)$ for the $k$-nearest neighbor search and $2O(k \times G) + O(G)$ for label assignment per unlabeled sample. In step 3, the time complexity of random rebalance is $O(N_{maj})$. In summary, the time complexity of DENI is $O(d \times N) + O(d \times G \times \log N) \approx O(dN \log N)$.

## 4 Experimental Study

### 4.1 Basic Settings

**Datasets.** To verify the effectiveness of our proposed framework, 20 imbalanced datasets from KEEL[1] and UCI[2] repository are adopted. The datasets vary in scale and complexity, with sample sizes ranging from small to large, and imbalance ratios (IR) ranging from low to high. More details are presented in Table 1.

[1] https://sci2s.ugr.es/keel/imbalanced.php.
[2] https://archive.ics.uci.edu/datasets.

**Table 1.** Detailed description of the experimental datasets.

| ID | Name | Size | $\#d$ | #Samples | IR | ID | Name | Size | $\#d$ | #Samples | IR |
|---|---|---|---|---|---|---|---|---|---|---|---|
| D1 | abalone20vs8910 | 1916 | 8 | {+26,−1890} | 72.69 | D11 | ionosphere | 351 | 34 | {+126,−225} | 1.79 |
| D2 | bank | 4521 | 16 | {+521,−4000} | 7.68 | D12 | mammography | 11183 | 6 | {+260,−10923} | 42.01 |
| D3 | cargood | 1728 | 6 | {+69,−1659} | 24.04 | D13 | pima | 768 | 8 | {+268,−500} | 1.87 |
| D4 | cleveland0vs4 | 173 | 13 | {+13,−160} | 12.31 | D14 | yeast0256vs3789 | 1004 | 8 | {+99,−905} | 9.14 |
| D5 | ecoli2 | 336 | 7 | {+52,−284} | 5.46 | D15 | yeast02579vs368 | 1004 | 8 | {+99,−905} | 9.14 |
| D6 | ecoli3 | 336 | 7 | {+35,−301} | 8.60 | D16 | yeast1458vs7 | 693 | 8 | {+30,−663} | 22.10 |
| D7 | ecoli4 | 336 | 7 | {+20,−316} | 15.80 | D17 | yeast1 | 1484 | 8 | {+429,−1055} | 2.46 |
| D8 | flareF | 1066 | 11 | {+43,−1023} | 23.79 | D18 | yeast3 | 1484 | 8 | {+163,−1321} | 8.10 |
| D9 | glass0 | 214 | 9 | {+70,−144} | 2.06 | D19 | yeast4 | 1484 | 8 | {+51,−1433} | 28.10 |
| D10 | haberman | 306 | 3 | {+81,−225} | 2.78 | D20 | yeast6 | 1484 | 8 | {+35,−1449} | 41.40 |

**Baselines.** To demonstrate the superiority of our method, we compare our method with 9 other resampling methods, including SMOTE (SMO) [3], Borderline-SMOTE (BSM) [7], PDFOS [5], RSMOTE (RSM) [4], NGOS [16], CC [11], IGBS [18], SMOTE-IPF (SIPF) [14] and SMOTE-TomekLinks (STL) [2].

**Evaluation Metrics.** As accuracy is a misleading metric for imbalanced classification, we apply F1-measure, G-mean and AUC [12] as the evaluation metrics of our experiments which are commonly used in imbalance learning.

**Base Classifiers.** To validate the robustness of our method, we employ 5 different classifiers in our experiments, including Decision Tree (DT), Naive Bayes (NB), Random Forest (RF), LightGBM and XGBoost. To ensure the correctness of the results, we use 10-fold stratified cross validation with 5 repetitions. The number of estimators in RF, LightGBM and XGBoost is set to 50.

### 4.2 Main Results

**Comparing with Baseline Resampling Algorithms.** This section compares the proposed DENI with 9 other resampling methods listed in Sect. 4.1. Due to space constraints, we only provide the AUC for each dataset. In Table 2, we can see that DENI achieves the best performance with the highest average score, the best average rank and the most datasets to win. It demonstrates that our DENI is effective in generating and preserving high-quality samples.

**Noise Robustness Study.** To verify the robustness of our method under noisy conditions, we compared the 10 algorithms under different noise rates. In the training set, given the noise rate $nr \in \{0\%, 10\%, 20\%, 30\%\}$, we flip $nr\%$ of the minority samples to heterogeneous label randomly and an equal number for the majority samples. Figure 2 shows the average scores of the 10 algorithms under varying noise rates. As can be seen, our proposed DENI outperforms the other methods on F1-measure, G-mean and AUC under different noise rates by a significant margin. In addition, as the noise rate rises, our method declines more slowly, which shows that our method is more robust to noise.

**Comparison on Different Base Classifiers.** To verify the general applicability of DENI, we conducted further experiments on different base classifiers. Figure 3 and 4 show the average rank of all the methods using different base classifiers. Limited by the space, we only show the results at 0% *nr* and 30% *nr*. It can be observed that DENI achieves competitive performance across different base classifiers regardless of individual classifiers or ensemble classifiers. Besides, as the noise rate rises, the average rank of DENI improves correspondingly, which demonstrates that DENI is more effective in enhancing the performance of classifiers compared with other resampling methods.

**Table 2.** AUC results on 20 datasets obtained by DT, noise rates = 0%.

| Dataset | SMO | BSM | PDFOS | RSM | NGOS | CC | IGBS | SIPF | STL | DENI |
|---|---|---|---|---|---|---|---|---|---|---|
| D1 | 0.7360 | 0.7147 | 0.7026 | 0.7004 | 0.7275 | 0.7752 | 0.6332 | 0.7946 | 0.7365 | **0.8269** |
| D2 | 0.6915 | 0.6812 | 0.6723 | 0.7039 | 0.6771 | 0.6457 | 0.6690 | 0.6831 | 0.6793 | **0.7125** |
| D3 | 0.8741 | 0.8806 | 0.9193 | 0.8918 | 0.9399 | 0.8779 | 0.9511 | 0.8889 | 0.8741 | **0.9604** |
| D4 | 0.8063 | 0.8594 | 0.7375 | 0.6969 | 0.8094 | 0.6781 | 0.6125 | 0.8375 | 0.8063 | **0.8781** |
| D5 | 0.8703 | 0.8673 | 0.8732 | 0.8525 | 0.8572 | 0.8042 | 0.8419 | 0.8887 | 0.8738 | **0.8892** |
| D6 | 0.7977 | 0.7368 | 0.8194 | 0.7619 | 0.8060 | 0.7627 | 0.7826 | 0.7293 | 0.8052 | **0.8885** |
| D7 | 0.8686 | 0.8937 | 0.9155 | 0.8390 | 0.8890 | 0.8697 | 0.9015 | 0.8671 | 0.8686 | **0.9219** |
| D8 | 0.6626 | 0.6519 | 0.6568 | 0.6568 | 0.6562 | 0.6125 | 0.6107 | 0.5418 | 0.6842 | **0.7644** |
| D9 | 0.7883 | 0.7960 | 0.7948 | 0.7950 | 0.7733 | 0.7255 | 0.7667 | 0.7914 | 0.7636 | **0.8033** |
| D10 | 0.5771 | 0.5685 | 0.5810 | 0.5251 | 0.5606 | 0.4931 | 0.5808 | 0.5636 | 0.5575 | **0.6152** |
| D11 | 0.8686 | 0.8477 | 0.8700 | 0.8378 | 0.8679 | 0.8882 | 0.8678 | 0.8703 | 0.8625 | **0.8991** |
| D12 | 0.8228 | 0.7763 | 0.7821 | 0.7768 | 0.8155 | 0.6461 | 0.7893 | 0.8535 | 0.8399 | **0.8582** |
| D13 | 0.6719 | 0.6622 | 0.6738 | 0.6805 | 0.6738 | 0.6479 | 0.6776 | 0.6780 | 0.6832 | **0.7098** |
| D14 | 0.7758 | 0.7352 | 0.7141 | 0.7447 | 0.7347 | 0.6352 | 0.7019 | 0.7219 | 0.7808 | **0.7985** |
| D15 | 0.8718 | 0.8495 | 0.8529 | 0.8717 | 0.8634 | 0.7777 | 0.8457 | 0.8480 | 0.8746 | **0.8834** |
| D16 | 0.5655 | 0.5434 | 0.5213 | 0.5624 | 0.5691 | 0.5219 | 0.5464 | 0.5934 | 0.5503 | **0.6542** |
| D17 | 0.6617 | 0.6631 | 0.6509 | 0.6507 | 0.6698 | 0.6437 | 0.6456 | 0.6557 | 0.6551 | **0.6807** |
| D18 | 0.8192 | 0.8371 | 0.8090 | 0.8546 | 0.8449 | **0.9091** | 0.8314 | 0.8351 | 0.8328 | 0.9001 |
| D19 | 0.7210 | 0.6185 | 0.6797 | 0.6192 | 0.6607 | 0.7263 | 0.6331 | 0.6996 | 0.6976 | **0.7547** |
| D20 | 0.7574 | 0.7094 | 0.7445 | 0.7341 | 0.7435 | 0.7214 | 0.7244 | 0.7598 | 0.7085 | **0.8758** |
| Avg. Rank | 4.90 | 6.40 | 5.65 | 6.35 | 5.25 | 7.65 | 7.20 | 5.00 | 5.40 | **1.05** |
| Avg. Score | 0.7604 | 0.7446 | 0.7485 | 0.7378 | 0.7570 | 0.7181 | 0.7307 | 0.7551 | 0.7567 | **0.8138** |
| Wins | 0/20 | 0/20 | 0/20 | 0/20 | 0/20 | 1/20 | 0/20 | 0/20 | 0/20 | **19/20** |

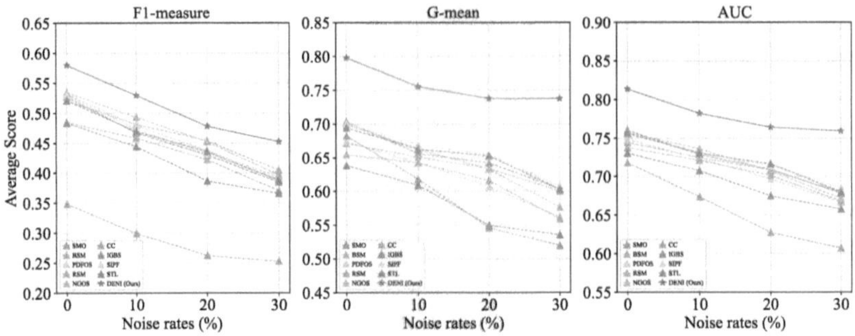


**Fig. 2.** The average results of 10 algorithms under varying noise rates.

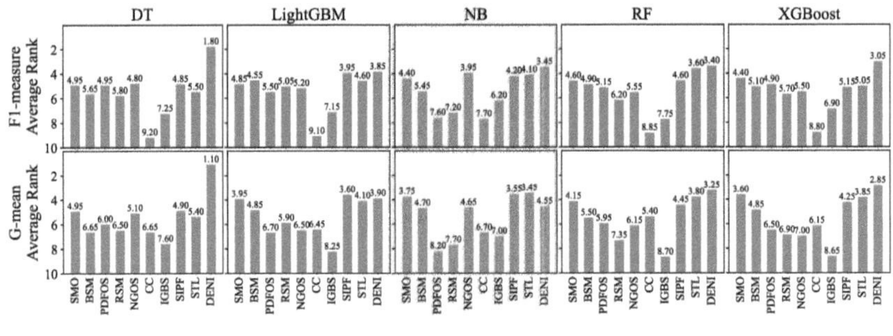


**Fig. 3.** Mean ranking of 10 resamplers for different classifiers and metrics at 0% $nr$.

### 4.3 Ablation Study

To verify that our framework is well designed, we conducted an ablation experiment by removing the components of DENI. DENI is composed of 3 components: Global Density-Based Noise Filtration (**component** $A$), Density-Enhanced Oversampling with Neighborhood Information (**component** $B$) and Random Rebalance (**component** $C$) as discussed in Sect. 3. Table 3 shows the results across all metrics at 0% $nr$ and 30% $nr$. We can see that the removal of component $A$ impairs the performance of the classifier, especially at a high noise rate. It demonstrates that component $A$ is indispensable for handling imbalanced noisy data. With the removal of component $B$, a significant decrease of the average score is observed, which shows that component $B$ is a critical component to obtain information-rich samples. Without component $C$, the average score decreases accordingly. This is most noticeable in G-mean and AUC because the final dataset remains imbalanced, which makes classifiers biased to the majority class. Therefore, component $C$ is vital for dealing with imbalanced datasets.

### 4.4 Parameters Analysis

To investigate the potential impact of the parameters and to choose the optimal parameters, we carried out a parametric experiment on the datasets. DENI has

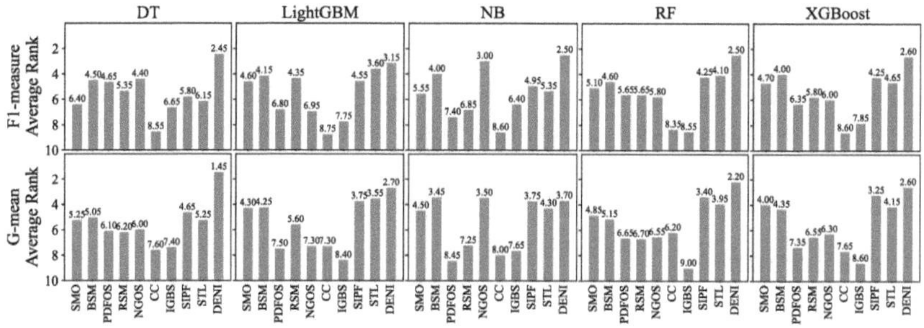

**Fig. 4.** Mean ranking of 10 resamplers for different classifiers and metrics at 30% $nr$.

**Table 3.** Ablation experiments.

| Algorithm | $nr = 0\%$ | | | $nr = 30\%$ | | |
|---|---|---|---|---|---|---|
| | F1-measure | G-mean | AUC | F1-measure | G-mean | AUC |
| DENI without $A$ | 0.5539 | 0.7169 | 0.7662 | 0.4031 | 0.6397 | 0.6976 |
| DENI without $B$ | 0.4368 | 0.7843 | 0.7913 | 0.3255 | 0.7044 | 0.7183 |
| DENI without $C$ | 0.5618 | 0.7550 | 0.7946 | 0.4521 | 0.6927 | 0.7377 |
| DENI | **0.5789** | **0.7891** | **0.8138** | **0.4527** | **0.7377** | **0.7593** |

two parameters: the balance level $\alpha$ in the stage of unlabeled sample generation and the number of nearest neighbors $k$ at the stage of neighbor label assignment. Figure 5 shows the results on F1-measure and G-mean. We can see that as parameter $k$ increases from 1 to 15, the average scores of the 20 datasets in all metrics exhibit only minor fluctuations. For each single dataset, the fluctuation is also limited to a specific range. It is reasonable to choose $k$ between 5 and 11, because an extreme value of $k$ will adversely affects the performance.

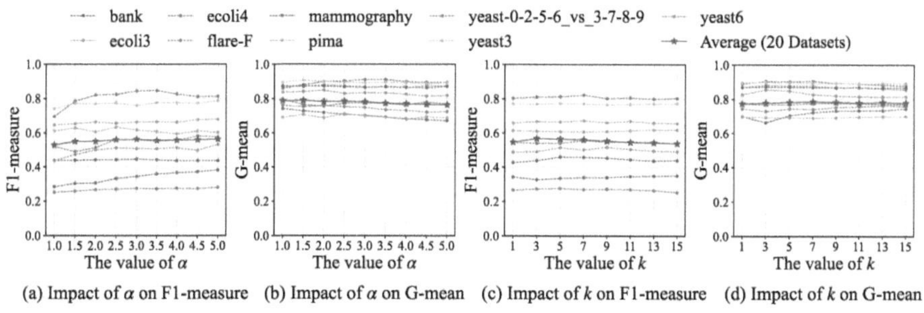

**Fig. 5.** The impact of $k$ and $\alpha$ on F1-measure and G-mean.

The average scores of the 20 datasets exhibit minor fluctuations in G-mean as parameter $\alpha$ varies. However, we can see that the F1-measure score increases and then plateaus as $\alpha$ increases from 1.0 to 5.0. This is because a smaller $\alpha$ leads to fewer unlabeled samples, resulting in insufficient supplementation of minority class samples. Considering that a larger $\alpha$ will not bring a significant improvement but increases computation time, we suggest choosing $\alpha$ to be larger than 1.0 to prevent insufficient supplementation of minority samples.

## 5 Conclusion

In this paper, we proposed a novel density-enhanced hybrid sampling framework with neighborhood information to address the class-imbalance problem with label noise. Our framework focuses on leveraging global density distribution and local neighborhood information to generate robust synthetic samples while effectively mitigating the impact of label noise. To verify the superiority and robustness of the proposed DENI, extensive experiments are implemented on five classifiers and 20 imbalanced datasets with varying noise rates. The experimental results show that our method significantly outperforms other state-of-the-art methods across all evaluation metrics, underscoring its potential for more complex scenarios of real-world applications.

**Acknowledgments.** This work was supported by the National Natural Science Foundation of China (62306119) and the Science and Technology Projects in Guangzhou (2025A04J3436).

## References

1. Anyfantis, D., Karagiannopoulos, M., Kotsiantis, S., Pintelas, P.: Robustness of learning techniques in handling class noise in imbalanced datasets. In: IFIP International Conference on Artificial Intelligence Applications and Innovations, pp. 21–28. Springer (2007)
2. Batista, G.E., Bazzan, A.L., Monard, M.C., et al.: Balancing training data for automated annotation of keywords: a case study. Wob **3**, 10–18 (2003)
3. Chawla, N.V., Bowyer, K.W., Hall, L.O., Kegelmeyer, W.P.: SMOTE: synthetic minority over-sampling technique. J. Artif. Intell. Res. **16**, 321–357 (2002)
4. Chen, B., Xia, S., Chen, Z., Wang, B., Wang, G.: RSMOTE: a self-adaptive robust smote for imbalanced problems with label noise. Inf. Sci. **553**, 397–428 (2021)
5. Gao, M., Hong, X., Chen, S., Harris, C.J., Khalaf, E.: PDFOS: PDF estimation based over-sampling for imbalanced two-class problems. Neurocomputing **138**, 248–259 (2014)
6. Haixiang, G., Yijing, L., Shang, J., Mingyun, G., Yuanyue, H., Bing, G.: Learning from class-imbalanced data: review of methods and applications. Expert Syst. Appl. **73**, 220–239 (2017)
7. Han, H., Wang, W.Y., Mao, B.H.: Borderline-SMOTE: a new over-sampling method in imbalanced data sets learning. In: International Conference on Intelligent Computing, pp. 878–887. Springer (2005)

8. He, H., Bai, Y., Garcia, E.A., Li, S.: ADASYN: adaptive synthetic sampling approach for imbalanced learning. In: 2008 IEEE International Joint Conference on Neural Networks (IEEE World Congress on Computational Intelligence), pp. 1322–1328. IEEE (2008)
9. Lee, H.K., Kim, S.B.: An overlap-sensitive margin classifier for imbalanced and overlapping data. Expert Syst. Appl. **98**, 72–83 (2018)
10. Li, M., Zhou, H., Liu, Q., Gong, X., Wang, G.: WRND: a weighted oversampling framework with relative neighborhood density for imbalanced noisy classification. Expert Syst. Appl. **241**, 122593 (2024)
11. Lin, W.C., Tsai, C.F., Hu, Y.H., Jhang, J.S.: Clustering-based undersampling in class-imbalanced data. Inf. Sci. **409**, 17–26 (2017)
12. López, V., Fernández, A., García, S., Palade, V., Herrera, F.: An insight into classification with imbalanced data: empirical results and current trends on using data intrinsic characteristics. Inf. Sci. **250**, 113–141 (2013)
13. Rodriguez, D., Herraiz, I., Harrison, R., Dolado, J., Riquelme, J.C.: Preliminary comparison of techniques for dealing with imbalance in software defect prediction. In: Proceedings of the 18th International Conference on Evaluation and Assessment in Software Engineering, pp. 1–10 (2014)
14. Sáez, J.A., Luengo, J., Stefanowski, J., Herrera, F.: SMOTE-IPF: addressing the noisy and borderline examples problem in imbalanced classification by a resampling method with filtering. Inf. Sci. **291**, 184–203 (2015)
15. Scott, D.W.: Multivariate Density Estimation: Theory, Practice, and Visualization. Wiley (2015)
16. Shao, X., Yan, Y.: Noise-robust gaussian distribution based imbalanced oversampling. In: International Conference on Algorithms and Architectures for Parallel Processing, pp. 221–234. Springer (2023)
17. Tang, B., He, H.: KernelADASYN: kernel based adaptive synthetic data generation for imbalanced learning. In: 2015 IEEE Congress on Evolutionary Computation (CEC), pp. 664–671. IEEE (2015)
18. Xia, S., Zheng, S., Wang, G., Gao, X., Wang, B.: Granular ball sampling for noisy label classification or imbalanced classification. IEEE Trans. Neural Netw. Learn. Syst. **34**(4), 2144–2155 (2021)
19. Xu, Y., Yu, Z., Chen, C.P.: Classifier ensemble based on multiview optimization for high-dimensional imbalanced data classification. IEEE Trans. Neural Netw. Learn. Syst. **35**(1), 870–883 (2022)
20. Xu, Y., Yu, Z., Chen, C.P.: Improved contraction-expansion subspace ensemble for high-dimensional imbalanced data classification. IEEE Trans. Knowl. Data Eng. **36**(10), 5194–5205 (2024)
21. Zhu, T., Liu, X., Zhu, E.: Oversampling with reliably expanding minority class regions for imbalanced data learning. IEEE Trans. Knowl. Data Eng. **35**(6), 6167–6181 (2022)

# Probabilistic-Gap-Driven Relabeling Method for Positive-Negative-Unlabeled Learning with Label Selection Bias

Yuzhe Han and Peng Zhang(✉)

School of Mathematical Sciences, Zhejiang University, Zhejiang, People's Republic of China
pengz@zju.edu.cn

**Abstract.** Label selection bias is common across different domains. In chronic disease management, many records lack gold standard tests which provide a definite label, but whether a record is labeled often depends on the patient's health status. Our method introduces a probabilistic-gap–driven relabeling method that assigns pseudo-labels to unlabeled samples using probabilistic gaps rather than commonly used feature similarity. This algorithm employs a model-free design, which requires no parameter tuning. We integrate the relabeled data into a pseudo-likelihood framework with KMM-based reweighting method, mitigating selection bias and distributional shifts. Experiments on synthetic and real world medical datasets demonstrated that our approach effectively leverages mitigates selection bias and improves classification performance.

**Keywords:** Semi-Supervised · Positive-Unlabeled Learning · Selection Bias · Probabilistic Gap

## 1 Introduction

Label selection bias is a common challenge in medical research, where high-quality labeled data are limited and most available data remain unlabeled. In chronic diseases such as hypertension, a definitive diagnosis depends on a gold standard test, for example standardized blood pressure measurements. Patients who undergo a gold standard test are assigned clear diagnostic labels, whereas those without such testing remain unlabeled. Importantly, the administration of a gold standard test is rarely random and is often driven by suspicious symptoms or higher clinical risk. As a result, labeled patients tend to have a higher pre-test probability of disease than untested individuals. Although some asymptomatic individuals receive testing during routine examinations, many patients without confirmatory evaluation remain undiagnosed. This non-random labeling mechanism introduces selection bias and limits the representativeness and generalizability of predictive models.


R. C.-W. Wong et al. (Eds.): PAKDD 2026, LNAI 16597, pp. 314–326, 2026.
https://doi.org/10.1007/978-981-92-1300-9_25

Semi-supervised learning (SSL) and positive-unlabeled (PU) learning provide solutions to this problem. However, selection bias violates the assumption that labeled and unlabeled samples originate from the same distribution, often resulting in degraded performance. PU learning aims to learn a classifier from labeled positive and mixed unlabeled samples to identify all positive instances [10,14,17], but does not utilize information from labeled negative samples. Existing SSL and PU methods therefore do not explicitly address label selection bias. In practice, labeled data tend to represent more distinguishable cases, leading to a larger probabilistic gap between positive and negative classes, we propose a semi-supervised relabeling framework for binary classification under the Positive-Negative-Unlabeled (PNU) setting, where the dataset contains biased labeled positives, biased labeled negatives, and unlabeled samples. Unlike traditional methods that rely on distance-based measures (such as spatial or feature similarity), our approach focuses on the probabilistic gap, effectively mitigating bias caused by mismatches between the labeled and full data distribution. Our contributions can be summarized as follows.

1. We proposed a probability-gap–driven relabeling method for PNU learning, ensuring consistency with the Bayesian optimal classifier.
2. We integrated the relabeled data into a pseudo-likelihood framework with KMM-based reweighting, mitigating selection bias and distributional shifts.
3. On a real-world chronic disease diagnosis dataset, the proposed approach achieved an average accuracy improvement of 4.69%, where the combination of ensemble-based relabeling and neural network classification delivered the best performance.

## 2 Related Work

**Semi-Supervised Learning (SSL):** SSL has attracted significant attention in numerous practical tasks due to its ability to bridge the gap between labeled and unlabeled data distributions. Early SSL approaches primarily focused on self-training [1,8], where a model makes predictions on unlabeled data and uses these predictions as pseudo-labels for training. Representative algorithms include Freematch [13], SoftMatch [5] and so on. This approach was subsequently extended through generative models, which enhance learning effectiveness by learning the data distribution and inferring labels for unlabeled data. Additionally, graph-based methods have gained widespread adoption, including label propagation and GNNs [9,12]. These methods assume that labeled and unlabeled data are drawn from the same underlying distribution $P(X, Y)$. To relax this assumption, the focus shifts to scenarios in which the source and target data are drawn from different distributions [16].

**Positive-Unlabeled Learning:** Selection bias in labeled data has been extensively explored in PU learning [10,11,14,15,17]. In PU learning, the labeled data are strictly positive, which typically exhibit clear positive characteristics. The original PU learning problem followed the SCAR assumption [2], which assumes that the probability of being labeled is independent of the data's characteristics and remains as a constant. Furthermore, most studies focus on "selected at random" (SAR) to consider the selection bias, suggesting that the probability of labeling depends on the observed characteristics. In the non-parametric framework, the concept of a propensity score has been utilized, representing the instance-specific probability of an example being labeled [2,3,6]. To address selection bias in the labeling process, methods have been proposed to learn a scoring function that preserves probability order across all samples [7]. Furthermore, the log odds has been explored, with Bayesian optimal classifiers used to automatically label groups of positive and negative examples.

# 3 Methodology

## 3.1 Probabilistic-Gap-Driven Relabeling Algorithm

Consider a binary classification task in the setting of a PNU dataset:

$$\{(\boldsymbol{x}_1, y_1, s_1), \ldots, (\boldsymbol{x}_n, y_n, s_n)\},$$

where each sample $(\boldsymbol{x}_i, y_i, s_i)$ is drawn independently and identically from an unknown distribution $P(X, Y, S)$. $\boldsymbol{x}_i \in \mathbb{R}^d$ is the feature vector, $y_i \in \{+1, -1\}$ is the unobserved true label, and $s_i \in \{+1, -1, 0\}$ is the observed label. In this framework, $+1$ denotes a positive label, $-1$ denotes a negative label, and 0 indicates an unlabeled sample. The dataset is divided into three subsets: labeled positives $(S = +1, Y = +1)$, labeled negatives $(S = -1, Y = -1)$, and unlabeled $(S = 0)$. For each subset, we denote the manually assigned label as $\tilde{Y}$. For unlabeled data, $\tilde{Y}$ can be either $\tilde{Y}_+ = +1$ or $\tilde{Y}_- = -1$. In the labeled set, instances with $S = +1$ are assigned $\tilde{Y}_+ = \tilde{Y}_- = +1$, and instances with $S = -1$ are assigned $\tilde{Y}_+ = \tilde{Y}_- = -1$. Let $\rho(X)$ denotes the misclassification rate and defines $\rho_-(X)$, $\rho_+(X)$ as follows:

$$\rho_-(X) = P(\tilde{Y} = +1 \mid Y = -1, X), \quad \rho_+(X) = P(\tilde{Y} = -1 \mid Y = +1, X).$$

We analyze misclassification rates under two scenarios: one where unlabeled data is assigned a positive label $(\tilde{Y}_+)$ and another where it is assigned a negative label $(\tilde{Y}_-)$. When $\tilde{Y}_+$ is assigned, we have $\rho_+(X) = 0$ and $\rho_-(X) > 0$; when $\tilde{Y}_-$ is assigned, $\rho_-(X) = 0$ and $\rho_+(X) > 0$. The probabilistic gap $\Delta P(X)$, $\Delta\tilde{P}_+(X)$ and $\Delta\tilde{P}_-(X)$ is defined as $\Delta P(X) = P(Y = +1 \mid X) - P(Y = -1 \mid X)$ and $\Delta\tilde{P}_+(X) = P(\tilde{Y}_+ = +1 \mid X) - P(\tilde{Y}_+ = -1 \mid X), \Delta\tilde{P}_-(X) = P(\tilde{Y}_- = +1 \mid X) - P(\tilde{Y}_- = -1 \mid X)$.

Under both conditions, the following theorems hold and the proofs are presented in Appendix A:

**Theorem 1.** *When all unlabeled data are assigned the positive label ($\tilde{Y}_+$), then*

1. *Observed probabilistic gap $\Delta\tilde{P}_+(X)$ is not smaller than probabilistic gap $\Delta P(X)$; i.e., for any $X$,*

$$\Delta\tilde{P}_+(X) \geq \Delta P(X).$$

2. *Observed probabilistic gap $\Delta\tilde{P}_+(X)$ has the same ordering of probabilistic gap $\Delta P(X)$; i.e., for any $X_1$ and $X_2$,*

$$\Delta\tilde{P}_+(X_1) = \Delta\tilde{P}_+(X_2) \iff \Delta P(X_1) = \Delta P(X_2),$$

$$\Delta\tilde{P}_+(X_1) \geq \Delta\tilde{P}_+(X_2) \iff \Delta P(X_1) \geq \Delta P(X_2).$$

This implies that, even if the initial labeling contains errors, the ranking information remains consistent, enabling reliable identification of high-confidence samples for relabeling.

**Theorem 2.** *When all unlabeled data are assigned the negative label ($\tilde{Y}_-$), then*

1. *Observed probabilistic gap $\Delta\tilde{P}_-(X)$ is not larger than probabilistic gap $\Delta P(X)$; i.e., for any $X$,*

$$\Delta\tilde{P}_-(X) \leq \Delta P(X).$$

2. *Observed probabilistic gap $\Delta\tilde{P}_-(X)$ has the same ordering of probabilistic gap $\Delta P(X)$; i.e., for any $X_1$ and $X_2$,*

$$\Delta\tilde{P}_-(X_1) = \Delta\tilde{P}_-(X_2) \iff \Delta P(X_1) = \Delta P(X_2),$$

$$\Delta\tilde{P}_-(X_1) \leq \Delta\tilde{P}_-(X_2) \iff \Delta P(X_1) \leq \Delta P(X_2).$$

**Theorem 3.** *When all unlabeled data are assigned the positive label ($\tilde{Y}_+$), let*

$$l_+ = \rho_-(X)\big|_{\Delta P(X)=0} \in (0,1).$$

*The unlabeled instances in the area $\Delta\tilde{P}_+(X) \in [l_+, 1]$ can be labeled as positive points; in the area $\Delta\tilde{P}_+(X) \in [-1, l_+)$ can be labeled as negative.*

This theorem allows for automatic assignment of pseudo-labels to unlabeled data, effectively expanding the labeled dataset while controlling the mislabeling risk, thereby facilitating more robust model training.

**Theorem 4.** *When all unlabeled data are assigned the negative label ($\tilde{Y}_-$), we define*

$$l_- = -\rho_+(X)\big|_{\Delta P(X)=0} \in (-1, 0).$$

*The unlabeled instances in the area $\Delta\tilde{P}_-(X) \in [-1, l_-]$ can be labeled as negative points; in the area $\Delta\tilde{P}_-(X) \in (l_-, 1]$ can be labeled as positive.*

By combining Theorem 3 and Theorem 4, we obtain the following corollary:

**Corollary 1.** *The unlabeled instances in the area $\Delta\tilde{P}_-(X) \in [-1, l_-]$ can be labeled as negative points. Meanwhile, the unlabeled instances in the area $\Delta\tilde{P}_+(X) \in [l_+, 1]$ can be labeled as positive.*

The new fully labeled dataset is denoted as $\mathcal{X}'$, and pseudo-label as $\tilde{s}_i'$.

$$\mathcal{X}' = \{(\boldsymbol{x}_i', \tilde{s}_i'),\ i = 1, \ldots, N'\}.$$

The probabilistic gap measures the dominance between posterior probabilities of the positive and negative classes, where a larger absolute gap indicates higher prediction confidence under the Bayes optimal classifier. Unlike feature-similarity-based pseudo-labeling, this approach operates directly in the posterior space and remains consistent with the Bayes decision rule.

### 3.2 Classifier with Relabeled Dataset

This section focused on constructing a classification model using the relabeled data $\mathcal{X}'$, in which $\tilde{S}'$ represents the pseudo-label, $X'$ is the feature vector, and $\Delta_i$ is a binary indicator of whether the true label $y_i$ is observed ($\Delta_i = 1$) or unobserved ($\Delta_i = 0$). Using Bayes' theorem, the likelihood function for the observed label $\tilde{S}'$ given the feature vector $X'$ is expressed as:

$$f_{\tilde{S}'|X'}(\tilde{S}' \mid X') = \sum_{Y} f_{\tilde{S}'|Y,X'}(\tilde{S}' \mid Y, X') f_{Y|X'}(Y \mid X'). \tag{1}$$

The likelihood function of $\tilde{S}'$ given $X'$ is:

$$\prod_{i=1}^{N'} f(y_i \mid \boldsymbol{x}_i')^{\Delta_i} \times \left\{ f(\tilde{s}_i' \mid \boldsymbol{x}_i') \right\}^{1-\Delta_i}. \tag{2}$$

And the log-likelihood of Eq. (2) is:

$$\sum_{i=1}^{N'} \Delta_i \left\{ \Psi_1(y_i, \boldsymbol{x}_i') \right\} + \sum_{i=1}^{N'} (1 - \Delta_i) \mathbb{E} \left\{ \Psi_1(y_i, \boldsymbol{x}_i') + \Psi_2(\tilde{s}_i', y_i, \boldsymbol{x}_i') \;\middle|\; \boldsymbol{x}_i' \right\} = 0, \tag{3}$$

where $\Psi_1 = \left(\frac{\partial}{\partial\beta}\log f_{Y|X}\right)^T$ and $\Psi_2 = \left(\frac{\partial}{\partial\gamma}\log f_{S|Y,X}\right)^T$. $\Psi_1$ and $\Psi_2$ can be separated completely. Since $\Psi_2$ is redundant and unrelated to the final prediction, the estimation equation is simplified as:

$$\sum_{i=1}^{N'} \Delta_i\left\{\Psi_1(y_i, \boldsymbol{x}_i')\right\} + \sum_{i=1}^{N'}(1-\Delta_i)\mathbb{E}\left\{\Psi_1(y_i, \boldsymbol{x}_i') \,\middle|\, \boldsymbol{x}_i'\right\} = 0. \tag{4}$$

We demonstrated the effectiveness of the fully labeled dataset $\mathcal{X}'$ in a double logistic $P(Y = 1 \mid X) = \sigma(\beta_0 + \beta_1^\top X), P(S = 1 \mid Y, X) = \sigma(\gamma_0 + \gamma_1 Y + \gamma_2^\top X)$, where the parameters $\beta = (\beta_0, \beta_1)$ and $\gamma = (\gamma_0, \gamma_1, \gamma_2)$. $\beta_1$ and $\gamma_2$ are $d$-dimensional weight vectors. For estimating the parameters, we employed the EM algorithm.

E-STEP(Expectation Step): For the unlabeled data with $\Delta_i = 0$, we calculated the expectation $\hat{E}[\Psi_1(y_i, \boldsymbol{x}_i', \beta) \mid \boldsymbol{x}_i']$:

$$\hat{E}\left[\Psi_1(y_i, \boldsymbol{x}_i', \beta) \mid \boldsymbol{x}_i'\right] = \begin{pmatrix} \hat{E}[y_i \mid \boldsymbol{x}_i'] - p_X \\ \left(\hat{E}[y_i \mid \boldsymbol{x}_i'] - p_X\right)\boldsymbol{x}_i' \end{pmatrix}. \tag{5}$$

$\hat{E}[y_i \mid \boldsymbol{x}_i']$ was estimated by $p(y_i \mid \boldsymbol{x}_i')$ using the labeled set.

M-STEP(Maximization Step): Update $\beta$ with the estimating equation:

$$\sum_{i=1}^{N'} \left[\Delta_i \Psi_1(y_i, \boldsymbol{x}_i', \beta) + (1-\Delta_i)\hat{E}[\Psi_1 \mid \boldsymbol{x}_i']\right] = 0. \tag{6}$$

We substituted the pseudo-label $\tilde{S}'$ derived in Sect. 3.1 into Eq. (6) rather than estimating $\hat{E}[\Psi_1(y_i, \boldsymbol{x}_i', \beta) \mid \boldsymbol{x}_i']$ from the labeled set.

$$\hat{E}\left[\Psi_1(y_i, \boldsymbol{x}_i', \beta) \mid \boldsymbol{x}_i'\right] = \begin{pmatrix} \tilde{s}_i' - p_X \\ (\tilde{s}_i' - p_X)\boldsymbol{x}_i' \end{pmatrix}. \tag{7}$$

The replacement approach was theoretically justified because we leveraged probability gaps through pseudo-labels to enhance labeling accuracy. This avoids the assumption that the labeled set has the same distribution as the unlabeled dataset while eliminating the need to estimate $E\left[\Psi_1(y_i, \boldsymbol{x}_i', \beta) \mid \boldsymbol{x}_i'\right]$ on the labeled set. However, the empirical risk must be adjusted because only newly labeled dataset $\hat{S}$ is used, which satisfies a distribution $\hat{D}$. To correct for this distribution bias, the probability density ratio must be estimated: $\frac{P_D(X,Y)}{P_{\hat{D}}(X,Y)} = \frac{P_D(X)P(Y|X)}{P_{\hat{D}}(X)P(Y|X)} = \frac{P_D(X)}{P_{\hat{D}}(X)}$. The likelihood function was also adjusted as:

$$L_{\hat{D}}(\beta,\gamma) = \prod_{i=1}^{N'}\left\{\frac{P_D(\boldsymbol{x}_i', y_i)}{P_{\hat{D}}(\boldsymbol{x}_i', y_i)} \cdot f(y_i \mid \boldsymbol{x}_i', \beta)\right\}^{\Delta_i} \cdot \left\{\frac{P_D(\boldsymbol{x}_i', y_i)}{P_{\hat{D}}(\boldsymbol{x}_i', y_i)} \cdot f(\tilde{s}_i' \mid \boldsymbol{x}_i', \beta, \gamma)\right\}^{1-\Delta_i}, \tag{8}$$

To estimate the density ratio, we used the kernel mean matching (KMM) method, which minimized the distribution discrepancy in reproducing kernel hilbert space [4].

Constraints are imposed as $w_i \geq 0$ and $\frac{1}{n}\sum_{i=1}^{n} w_i = 1$. The optimal solution $\boldsymbol{w}^*$ is obtained by solving:

$$\boldsymbol{w}^* = \arg\min_{\boldsymbol{w}} \frac{1}{2}\boldsymbol{w}^\top \boldsymbol{K}\boldsymbol{w} - \boldsymbol{\kappa}^\top \boldsymbol{w} + \frac{\lambda}{2}\|\boldsymbol{w}\|_2^2, \quad \text{s.t.} \quad \mathbf{1}^\top \boldsymbol{w} = n, \quad w_i \geq 0,$$

where $\boldsymbol{K}_{ij} = k(\boldsymbol{x}_i^s, \boldsymbol{x}_j^s)$ and $\boldsymbol{\kappa}_i = \frac{n}{m}\sum_{j=1}^{m} k(\boldsymbol{x}_i^s, \boldsymbol{x}_j^t)$. The solution $\boldsymbol{w}^*$ represents the density ratio $\frac{P_D(\boldsymbol{x}_i')}{P_{\hat{D}}(\boldsymbol{x}_i')}$ for each sample $i$. As proven in Theorem 5 of [6], even in the presence of distribution shift, the framework that combines bayesian optimal relabeling with KMM guarantees that the risk of the learned classifier converges to the risk of the original distribution's Bayesian optimal classifier. This theorem provides a rigorous theoretical foundation for our method.

**Table 1.** Summary of Different $\rho$ Settings and Corresponding Parameters

| $\rho$ Setting | Formulation | Parameters |
|---|---|---|
| inverse | $\rho(\lvert \Delta P\rvert) = \frac{\alpha}{\alpha+\lvert \Delta P\rvert} \quad (\alpha > 0)$ | $\alpha = 1.0, 5.0, 8.0$ |
| linear | $\rho(\lvert \Delta P\rvert) = \alpha(1 - 0.5 * \lvert \Delta P\rvert) \quad (\alpha \in (0,1])$ | $\alpha = 0.3, 0.8, 1.0$ |
| exponential | $\rho(\lvert \Delta P\rvert) = \beta e^{-\alpha\lvert \Delta P\rvert} \quad (\alpha > 0, \beta \in (0,1])$ | $(\alpha, \beta) = (3.0, 0.6)$ |
| exponential mixture (EM) | $\rho(\lvert \Delta P\rvert) = e^{-\alpha\lvert \Delta P\rvert} + \beta\lvert \Delta P\rvert(1 - \lvert \Delta P\rvert) \quad (\alpha > 0, \beta \geq 0)$ | $(\alpha, \beta) = (0.4, 0.1), (0.2, 0.05), (0.7, 0)$ |
| power-series-based (PSB) | $\rho(\lvert \Delta P\rvert) = 1 - \alpha\lvert \Delta P\rvert^\beta \quad (\alpha \in (0, 0.5], \beta \geq 1)$ | $(\alpha, \beta) = (0.3, 2), (0.4, 1), (0.15, 4)$ |

# 4 Experiments

## 4.1 Simulation on Synthetic Datasets

The performance of the relabeling algorithm was evaluated under various settings. $N = 2000$ instances $(X_1, X_2)$ were uniformly sampled from the square domain $[-1,1]^2$. We set the true model as a logistic model, $P(Y = 1 \mid X) = \text{logit}(\beta_0 + \beta_1 x_1 + \beta_2 x_2)$, and true coefficient values were set as $(\beta_0, \beta_1, \beta_2) = (0, 3, -3)$. The settings for the unlabeled probability $\rho$ and corresponding parameters were summarized in Table 1. The probability-gap-driven relabeling method, combined with random forest for label prediction and logistic regression for classification, consistently outperformed the labeled-set-only classifier. As shown in Fig. 1 (a)–(c) and Table 2 (Default Scenario), improvements ranged from 0.11% to 2.25% (average 1.11%), with further gains (0.33% to 5.75%, average 2.51%) when the density ratio approach was incorporated. However, as the labeled set size decreased, both the number of relabeled points and their accuracy declined.

In another experiment with a two-dimensional binary classification dataset, where positive and negative instances were sampled from overlapping triangular regions, we tested the bias-correction capability of our method. The dataset

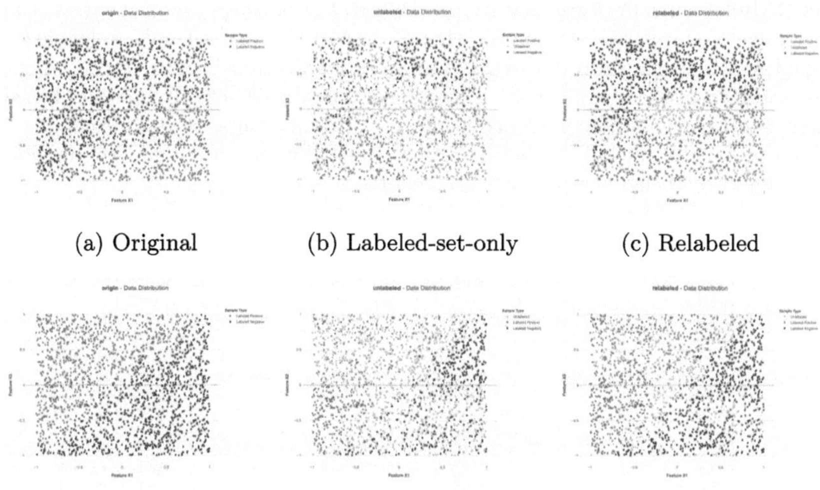

(a) Original (b) Labeled-set-only (c) Relabeled

(d) Original (e) Labeled-set-only (f) Relabeled

**Fig. 1.** (a) to (c) show the original, labeled-set-only, and relabeled data using the semi-supervised relabeling algorithm under the inverse $\rho$ setting with $\alpha = 1.0$. (d) to (f) display the same, but for overlap and biased scenarios under the same setting.

included 1,000 samples per class, with 23.45% in the overlapping region. A distributional bias was intentionally introduced by adding excessive points near $(-1, -1)$ and $(1, 1)$ in the labeled set, leading to significant estimation bias (Fig. (e)). The relabeling method effectively corrected this bias, aligning the dataset with the original distribution (Fig. (f)). Results (Table 2, Overlap and Biased Scenario) show the relabeled classifier outperformed the biased classifier, improving accuracy by 0.34%–2.65% (average 1.83%). The density ratio approach further increased accuracy by 0.14%–4.15% (average 2.74%).

### 4.2 Experiments on Real World Datasets

We evaluated the algorithm on a real-world dataset comprising 2.88 million patient records and 1,170 clinical features, targeting five chronic diseases including stroke, diabetes, hypertension, coronary heart disease, and chronic kidney disease. Features with substantial missingness were excluded, and feature selection using eXtreme Gradient Boosting (XGBoost) retained the top 23 features (e.g., age, homocysteine, D-dimer, HbA1c, and LDL cholesterol) for model training. Patients were categorized as definitively diagnosed ($S = \pm 1$) or undefined ($S = 0$). The data were split into training, validation, and test sets at a 6:2:2 ratio, and evaluation was conducted using only annotated cases in the validation and test sets.

**Table 2.** Relabeling Performance Across Different $\rho$ Settings and Scenarios. Classifier accuracy is compared across the original dataset, labeled-set-only, relabeled, and relabeled with density ratio datasets. Values are highlighted in blue or red when the corresponding method significantly outperforms the labeled-set-only method based on a paired t-test at the 95% confidence level. A ✓ indicates superior performance of the relabeled or density-refined relabeled setting. Relabeling accuracy and the number of unlabeled and relabeled samples are also reported.

| Scenario | $\rho$ Setting | Original | Labeled-set-only | Relabeled | Relabel with density ratio | Relabel Acc. | #Unlabel | #Relabel |
|---|---|---|---|---|---|---|---|---|
| | inverse $\alpha = 1.0$ | 89.51% (0.92) | 87.43% (2.13) | 88.02% (1.04) | 88.72% (0.89) ✓ | 99.32% (1.21) | 1012 | 738 |
| | inverse $\alpha = 5.0$ | 89.69% (0.98) | 84.09% (2.37) | 85.52% (0.99) | 87.32% (1.16) ✓ | 97.37% (1.74) | 1607 | 723 |
| | inverse $\alpha = 8.0$ | 89.65% (0.87) | 82.21% (2.23) | 84.11% (0.91) | 86.55% (1.12) ✓ | 96.67% (1.85) | 1708 | 240 |
| | linear $\alpha = 0.3$ | 89.67% (0.93) | 89.00% (2.10) | 89.35% (0.75) ✓ | 89.33% (0.64) | 98.81% (0.98) | 300 | 253 |
| | linear $\alpha = 0.8$ | 89.64% (0.85) | 87.96% (2.25) | 88.17% (0.99) | 89.16% (1.01) ✓ | 99.07% (0.94) | 832 | 753 |
| | linear $\alpha = 1.0$ | 89.52% (0.88) | 87.14% (2.43) | 88.81% (0.67) | 88.95% (0.83) ✓ | 97.77% (1.36) | 1050 | 898 |
| Default | exponential $(\alpha, \beta) = (3.0, 0.6)$ | 89.37% (0.68) | 87.36% (2.21) | 87.81% (0.78) | 87.98% (0.76) ✓ | 98.52% (1.12) | 794 | 675 |
| | EM $(\alpha, \beta) = (0.4, 0.1)$ | 89.62% (0.73) | 86.43% (2.32) | 87.87% (0.62) | 88.71% (0.91) ✓ | 98.86% (0.98) | 1315 | 787 |
| | EM $(\alpha, \beta) = (0.2, 0.05)$ | 89.62% (0.93) | 83.80% (2.18) | 86.05% (0.88) | 87.24% (0.77) ✓ | 96.60% (1.43) | 1596 | 648 |
| | EM $(\alpha, \beta) = (0.7, 0)$ | 89.42% (0.95) | 83.80% (2.40) | 84.47% (0.80) | 86.02% (0.72) ✓ | 98.10% (1.31) | 855 | 685 |
| | PSB $(\alpha, \beta) = (0.3, 2)$ | 89.49% (0.68) | 80.24% (2.10) | 80.35% (0.73) | 84.19% (0.94) ✓ | 94.23% (1.23) | 1188 | 537 |
| | PSB $(\alpha, \beta) = (0.4, 1)$ | 89.51% (0.91) | 83.17% (2.19) | 84.77% (0.83) | 86.14% (1.08) ✓ | 97.52% (0.92) | 1010 | 606 |
| | PSB $(\alpha, \beta) = (0.15, 4)$ | 89.57% (0.75) | 74.60% (2.47) | 75.96% (1.02) | 80.35% (1.14) ✓ | 90.15% (1.04) | 1409 | 203 |
| | inverse $\alpha = 1.0$ | 94.28% (0.78) | 87.90% (1.80) | 88.67% (0.94) | 89.88% (0.90) ✓ | 95.19%(1.21) | 909 | 727 |
| | inverse $\alpha = 5.0$ | 94.96% (0.82) | 84.43% (2.10) | 86.88% (0.88) | 88.58% (0.85) ✓ | 82.89%(2.47) | 1349 | 456 |
| | inverse $\alpha = 8.0$ | 95.15% (0.70) | 81.22% (1.95) | 83.09% (0.83) ✓ | 83.02% (0.78) | 80.38%(2.51) | 1443 | 423 |
| | linear $\alpha = 0.3$ | 94.46% (0.62) | 93.43% (2.00) | 93.77% (0.65) | 94.12% (0.70) ✓ | 95.24%(1.06) | 289 | 273 |
| | linear $\alpha = 0.8$ | 92.98% (0.71) | 88.06% (2.25) | 88.62% (0.78) ✓ | 88.62% (0.80) ✓ | 92.14%(1.83) | 712 | 585 |
| | linear $\alpha = 1.0$ | 93.34% (0.67) | 86.79% (1.80) | 87.79% (0.72) | 88.68% (0.78) ✓ | 91.84%(2.02) | 901 | 760 |
| Overlap and Biased | exponential $(\alpha, \beta) = (3.0, 0.6)$ | 91.74% (0.85) | 85.55% (2.20) | 85.99% (0.75) ✓ | 85.69% (0.78) | 89.42%(1.55) | 678 | 605 |
| | EM $(\alpha, \beta) = (0.4, 0.1)$ | 94.33% (0.64) | 86.82% (1.95) | 88.48% (0.82) ✓ | 89.44% (0.87) | 89.53%(1.55) | 1146 | 745 |
| | EM $(\alpha, \beta) = (0.4, 0.1)$ | 94.91% (0.74) | 83.84% (2.25) | 86.49% (0.68) | 87.31% (0.72) ✓ | 81.50%(2.70) | 1355 | 400 |
| | EM $(\alpha, \beta) = (0.4, 0.1)$ | 93.63% (0.60) | 88.35% (1.85) | 88.90% (0.77) | 89.45% (0.82) ✓ | 94.66%(1.06) | 910 | 693 |
| | PSB $(\alpha, \beta) = (0.3, 2)$ | 94.59% (0.70) | 86.27% (2.30) | 88.13% (0.84) | 89.34% (0.78) ✓ | 87.08%(2.47) | 1238 | 511 |
| | PSB $(\alpha, \beta) = (0.4, 1)$ | 93.90% (0.66) | 87.32% (1.95) | 88.54% (0.80) | 89.58% (0.85) ✓ | 90.91%(1.78) | 1065 | 594 |
| | PSB $(\alpha, \beta) = (0.15, 4)$ | 94.90% (0.75) | 80.24% (2.20) | 82.51% (0.69) | 84.21% (0.71) ✓ | 75.97%(3.09) | 1412 | 203 |

We compared six model combinations, each with a relabeling model (random forest (RF), logistic regression (LR), or multi-layer perceptron (MLP)) and a classifier (LR or MLP). The relabeling method was applied to unlabeled data, with RF, MLP, and LR relabeling 49.8%, 54.7%, and 8.9% of unlabeled data, respectively. While LR achieved the highest accuracy in relabeling, it relabeled the fewest data points. Both relabeling quantity and accuracy affected classifier performance, with logistic regression showing the lowest performance among the three methods. Details of relabeling performance are in Appendix B, Table 5.

Table 3 summarizes the accuracy improvements of the six relabeling methods for both standard relabeled sets and density-refined sets. For the standard setting ($\Delta_1$), MLP+MLP and RF+MLP achieved the largest gains (2.35% and 1.39%). After applying the density ratio approach ($\Delta_2$), RF+MLP markedly outperformed all others with an average improvement of 4.69%, followed by MLP+MLP (3.33%) and LR+MLP (1.12%). The strong performance of RF+MLP highlights its effectiveness in handling clinical data heterogeneity. Tree-based relabeling appears to better capture the relationship between features and probabilistic gap variation, leveraging hierarchical partitioning to improve sensitivity in low-density and boundary regions where uncertainty is higher.

**Table 3.** Compares classifier accuracy across original, labeled-set-only, relabeled, and relabeled with density ratio datasets.

| Relabel Model | Inference Model | Data Type | Disease (Accuracy) | | | | |
|---|---|---|---|---|---|---|---|
| | | | Stroke | Diabetes | Hypertension | Coronary Heart | Chronic Kidney |
| RF | LR | Original Set | 0.8177 | 0.9526 | 0.7865 | 0.7946 | 0.8692 |
| | | Labeled-set-only | 0.7877 | 0.9503 | 0.7519 | 0.7823 | 0.8473 |
| | | Relabeled Set | 0.7980✓ | 0.9523✓ | 0.7715✓ | 0.7831 | 0.8531 |
| | | + Density | 0.7525 | 0.9514 | 0.7530 | 0.7928✓ | 0.8510 |
| | MLP | Original Set | 0.9200 | 0.9558 | 0.8984 | 0.8688 | 0.9246 |
| | | Labeled-set-only | 0.7473 | 0.9357 | 0.7373 | 0.7650 | 0.8246 |
| | | Relabeled Set | 0.7540 | 0.9469✓ | 0.7596 | 0.7826 | 0.8365 |
| | | + Density | 0.8483✓ | 0.9377 | 0.8142✓ | 0.7980✓ | 0.8461✓ |
| LR | LR | Original Set | 0.8177 | 0.9526 | 0.7865 | 0.7946 | 0.8692 |
| | | Labeled-set-only | 0.7877 | 0.9503 | 0.7519 | 0.7823 | 0.8473 |
| | | Relabeled Set | 0.7883 | 0.9519✓ | 0.7734 | 0.7861✓ | 0.8538 |
| | | + Density | 0.7973✓ | 0.9519✓ | 0.7739✓ | 0.7842 | 0.8546✓ |
| | MLP | Original Set | 0.9200 | 0.9558 | 0.8984 | 0.8688 | 0.9246 |
| | | Labeled-set-only | 0.7473 | 0.9357 | 0.7373 | 0.7650 | 0.8246 |
| | | Relabeled Set | 0.7692✓ | 0.9373 | 0.7569 | 0.7712 | 0.8228 |
| | | + Density | 0.7278 | 0.9410✓ | 0.7692✓ | 0.8174✓ | 0.8326✓ |
| MLP | LR | Original Set | 0.8177 | 0.9526 | 0.7865 | 0.7946 | 0.8692 |
| | | Labeled-set-only | 0.7877 | 0.9503 | 0.7519 | 0.7823 | 0.8473 |
| | | Relabeled Set | 0.7938 | 0.9542 | 0.7653 | 0.7696 | 0.8573 |
| | | + Density | 0.8038✓ | 0.9546✓ | 0.7692✓ | 0.7876✓ | 0.8556 |
| | MLP | Original Set | 0.9200 | 0.9558 | 0.8984 | 0.8688 | 0.9246 |
| | | Labeled-set-only | 0.7473 | 0.9357 | 0.7373 | 0.7650 | 0.8246 |
| | | Relabeled Set | 0.7881 | 0.9476 | 0.7653 | 0.7765 | 0.8500 |
| | | + Density | 0.7958✓ | 0.9477✓ | 0.7846✓ | 0.7813✓ | 0.8670✓ |

We compared our probability-gap–driven relabeling method (Ours) with representative semi-supervised and positive-unlabeled (PU) learning approaches, including self-training (ST) [8], label propagation (LP) [12], and a PU learning method [6]. These methods were selected due to their common use in addressing PNU problems under selection bias. As reported in Appendix B Table 4, traditional semi-supervised approaches such as ST and LP showed limited performance, whereas our method outperformed all baselines across five datasets.

## 5 Conclusion

In this work, we addressed incomplete labeling with selection bias in clinical datasets by formulating a PNU learning framework. We introduced a probability-gap–driven relabeling strategy and incorporated the relabeled data into a pseudo-likelihood framework with KMM-based reweighting to correct for selection bias and distributional shift under the SAR assumption. Experiments on synthetic and real-world medical datasets show that our method mitigates selection bias and improves classification performance. Future work will explore modeling the labeling probability as a function of the probabilistic gap and extending the framework to multi-class and survival analysis with deep generative models for high-dimensional, multi-modal clinical data.

**Acknowledgments.** This research was supported by National Key Research and Development Program of China (Grant No. 2024YFC2511003) and Natural Science Foundation of Zhejiang Province (Project No. LDT23E05015A01).

## A Proof of Theorems

Proof of Theorem 1:

*Proof.*

$$P(\tilde{Y}_+ = +1 \mid X) = \rho_-(X) + (1 - \rho_-(X))P(Y = +1 \mid X),$$
$$P(\tilde{Y}_+ = -1 \mid X) = (1 - \rho_-(X))P(Y = -1 \mid X).$$

As a result, we obtain the following equation:

$$\Delta\tilde{P}_+(X) = P(\tilde{Y}_+ = +1 \mid X) - P(\tilde{Y}_+ = -1 \mid X) = \rho_-(X) + (1 - \rho_-(X))\Delta P(X). \tag{9}$$

From Eq. (9),

$$\Delta\tilde{P}_+(X) = \rho_-(X) + (1 - \rho_-(X))\Delta P(X),$$

And from $\rho_-(X) = P(\tilde{Y} = +1 \mid Y = -1, X) \in (0, 1)$,$1 - \rho_-(X) < 1$.

$$\Delta\tilde{P}_+(X) = (1 - \rho_-(X))(\Delta P(X) - 1) + 1.$$

Therefore, $\Delta\tilde{P}_+(X) - 1 = (1 - \rho_-(X))(\Delta P(X) - 1)$, $\Delta P(X) - 1 \leq 0$,

$$\Delta\tilde{P}_+(X) \geq \Delta P(X).$$

Additionally, when $\Delta P(X)$ decrease, $\rho_-(X)$ decrease, which leads to $\Delta\tilde{P}_+(X)$ decrease. The second part of the theorem has been proofed.

Proof of Theorem 3:

*Proof.* When $\Delta P(X) = 0$, we have $\Delta\tilde{P}_+(X) = l_+$. Therefore:

- If $\Delta\tilde{P}_+(X) > l_+$, then $\Delta P(X) > 0$;
- Otherwise when $\Delta\tilde{P}_+(X) \leq l_+$, $\Delta P(X) < 0$.

# B Performance on Real World Datasets

**Table 4.** Performance Comparison among different algorithms on the Real World Datasets

| Dataset | Model | Metrics | | | | |
|---|---|---|---|---|---|---|
| | | Acc | Precision | Recall | F1 | AUC |
| Stroke | ST | 0.606 ± 0.038 | 0.547 ± 0.044 | 0.906 ± 0.018 | 0.682 ± 0.035 | 0.691 ± 0.034 |
| | LP | 0.617 ± 0.037 | 0.576 ± 0.042 | 0.941 ± 0.015 | 0.715 ± 0.032 | 0.693 ± 0.033 |
| | PU | 0.768 ± 0.027 | 0.701 ± 0.035 | 0.919 ± 0.017 | 0.795 ± 0.025 | 0.879 ± 0.021 |
| | Ours | 0.848 ± 0.028 | 0.938 ± 0.012 | 0.742 ± 0.031 | 0.829 ± 0.024 | 0.924 ± 0.011 |
| Diabetes | ST | 0.593 ± 0.039 | 0.545 ± 0.044 | 0.948 ± 0.014 | 0.692 ± 0.034 | 0.736 ± 0.031 |
| | LP | 0.628 ± 0.037 | 0.587 ± 0.042 | 0.935 ± 0.016 | 0.721 ± 0.032 | 0.705 ± 0.033 |
| | PU | 0.913 ± 0.016 | 0.861 ± 0.024 | 0.974 ± 0.009 | 0.914 ± 0.015 | 0.978 ± 0.008 |
| | Ours | 0.955 ± 0.008 | 0.985 ± 0.005 | 0.932 ± 0.021 | 0.958 ± 0.018 | 0.978 ± 0.003 |
| Hypertension | ST | 0.575 ± 0.041 | 0.582 ± 0.041 | 0.221 ± 0.041 | 0.320 ± 0.043 | 0.633 ± 0.037 |
| | LP | 0.610 ± 0.039 | 0.873 ± 0.023 | 0.852 ± 0.025 | 0.869 ± 0.021 | 0.967 ± 0.009 |
| | PU | 0.750 ± 0.031 | 0.667 ± 0.038 | 0.900 ± 0.018 | 0.765 ± 0.029 | 0.827 ± 0.027 |
| | Ours | 0.814 ± 0.026 | 0.710 ± 0.037 | 0.877 ± 0.021 | 0.785 ± 0.028 | 0.877 ± 0.022 |
| Coronary Heart Disease | ST | 0.553 ± 0.043 | 0.622 ± 0.039 | 0.182 ± 0.038 | 0.281 ± 0.041 | 0.664 ± 0.036 |
| | LP | 0.603 ± 0.039 | 0.555 ± 0.044 | 0.907 ± 0.018 | 0.689 ± 0.034 | 0.679 ± 0.035 |
| | PU | 0.720 ± 0.032 | 0.664 ± 0.038 | 0.856 ± 0.025 | 0.748 ± 0.030 | 0.819 ± 0.026 |
| | Ours | 0.817 ± 0.026 | 0.847 ± 0.012 | 0.743 ± 0.037 | 0.792 ± 0.022 | 0.874 ± 0.018 |
| Chronic Kidney Disease | ST | 0.575 ± 0.041 | 0.537 ± 0.044 | 0.871 ± 0.022 | 0.664 ± 0.036 | 0.626 ± 0.038 |
| | LP | 0.618 ± 0.038 | 0.581 ± 0.042 | 0.939 ± 0.016 | 0.718 ± 0.033 | 0.684 ± 0.035 |
| | PU | 0.784 ± 0.027 | 0.744 ± 0.032 | 0.837 ± 0.025 | 0.788 ± 0.026 | 0.864 ± 0.023 |
| | Ours | 0.867 ± 0.016 | 0.874 ± 0.016 | 0.842 ± 0.022 | 0.858 ± 0.022 | 0.916 ± 0.009 |

**Table 5.** Relabel Statistics for Each Disease

| Disease | Relabel Model | Relabel Accuracy | #Relabeled | #Unlabeled |
|---|---|---|---|---|
| Stroke | RF | 0.7756 | 11899 | 20339 |
| | LR | 0.9412 | 1326 | 20339 |
| | MLP | 0.8148 | 10780 | 20339 |
| Diabetes | RF | 0.9476 | 143426 | 367291 |
| | LR | 0.9852 | 8433 | 367291 |
| | MLP | 0.9513 | 208995 | 367291 |
| Hypertension | RF | 0.7677 | 72265 | 145424 |
| | LR | 0.8049 | 7808 | 145424 |
| | MLP | 0.7131 | 64427 | 145424 |
| Coronary Heart | RF | 0.8004 | 39356 | 73940 |
| | LR | 0.7857 | 15163 | 73940 |
| | MLP | 0.7939 | 40386 | 73940 |
| Chronic Kidney | RF | 0.8172 | 22299 | 46112 |
| | LR | 0.8889 | 4494 | 46112 |
| | MLP | 0.8212 | 29930 | 46112 |

## References


1. Amini, M.R., Feofanov, V., Pauletto, L., Hadjadj, L., Devijver, E., Maximov, Y.: Self-training: a survey. Neurocomputing **616**, 128904 (2025). https://doi.org/10.1016/j.neucom.2024.128904
2. Bekker, J., Davis, J.: Learning from positive and unlabeled data: a survey. Mach. Learn. **109**(4), 719–760 (2020). https://doi.org/10.1007/s10994-020-05877-5
3. Bekker, J., Robberechts, P., Davis, J.: Beyond the selected completely at random assumption for learning from positive and unlabeled data. In: Joint European Conference on Machine Learning and Knowledge Discovery in Databases, pp. 71–85. Springer (2019)
4. Berlinet, A., Thomas-Agnan, C.: Reproducing Kernel Hilbert Spaces in Probability and Statistics. Springer (2011)
5. Chen, H., et al.: Softmatch: addressing the quantity-quality trade-off in semi-supervised learning. arXiv preprint arXiv:2301.10921 (2023)
6. He, F., Liu, T., Webb, G.I., Tao, D.: Instance-dependent PU learning by Bayesian optimal relabeling. arXiv preprint arXiv:1808.02180 (2018)
7. Kato, M., Teshima, T., Honda, J.: Learning from positive and unlabeled data with a selection bias. In: International Conference on Learning Representations (2019)
8. Kumar, A., Ma, T., Liang, P.: Understanding self-training for gradual domain adaptation. In: International Conference on Machine Learning, pp. 5468–5479. PMLR (2020)
9. Luo, Y., Luo, G., Duan, G., Chen, A.: Label as equilibrium: a performance booster for graph neural networks on node classification. Neural Netw. **186**, 107284 (2025)
10. Teisseyre, P., Mielniczuk, J., Łazęcka, M.: Different strategies of fitting logistic regression for positive and unlabelled data. In: Computational Science – ICCS 2020, vol. 12140, pp. 3–17. Springer (2020). https://doi.org/10.1007/978-3-030-50423-6_1
11. Tian, S., Wei, H., Wang, Y., Feng, L.: Crosel: cross selection of confident pseudo labels for partial-label learning. In: Proceedings of the IEEE/CVF Conference on Computer Vision and Pattern Recognition, pp. 19479–19488 (2024)
12. Wang, H., Leskovec, J.: Unifying graph convolutional neural networks and label propagation. arXiv preprint arXiv:2002.06755 (2020)
13. Wang, Y., et al.: Freematch: self-adaptive thresholding for semi-supervised learning. arXiv preprint arXiv:2205.07246 (2022)
14. Ward, G., Hastie, T., Barry, S., Elith, J., Leathwick, J.R.: Presence-only data and the EM algorithm. Biometrics **65**(2), 554–563 (2009). https://doi.org/10.1111/j.1541-0420.2008.01116.x
15. Wei, Q., Sun, H., Lu, X., Yin, Y.: Self-filtering: a noise-aware sample selection for label noise with confidence penalization. In: European Conference on Computer Vision, pp. 516–532. Springer (2022)
16. Wilson, G., Cook, D.J.: A survey of unsupervised deep domain adaptation. ACM Trans. Intell. Syst. Technol. (TIST) **11**(5), 1–46 (2020). https://doi.org/10.1145/3400066
17. Łazecka, M.: Bayesian logistic model for positive and unlabeled data. In: International Conference on Modeling Decisions for Artificial Intelligence, pp. 157–168. Springer (2023). https://doi.org/10.1007/978-3-031-33498-6_11

# HERGC: Heterogeneous Experts Representation and Generative Completion for Multimodal Knowledge Graphs

Yongkang Xiao and Rui Zhang(✉)

University of Minnesota, Minneapolis, MN, USA
{xiao0290,ruizhang}@umn.edu

**Abstract.** Multimodal knowledge graphs (MMKGs) enrich traditional knowledge graphs (KGs) by incorporating diverse modalities such as images and text. Multimodal knowledge graph completion (MMKGC) seeks to exploit these heterogeneous signals to infer missing facts, thereby mitigating the intrinsic incompleteness of MMKGs. Existing MMKGC methods typically leverage only the information contained in the MMKGs under the closed-world assumption and adopt discriminative training objectives, which limits their reasoning capacity during completion. Recent large language models (LLMs), empowered by massive parameter scales and pretraining on vast corpora, have demonstrated strong reasoning abilities across various tasks. However, their potential in MMKGC remains largely unexplored. To bridge this gap, we propose **HERGC**, a flexible **H**eterogeneous **E**xperts **R**epresentation and **G**enerative **C**ompletion framework for MMKGs. HERGC first deploys a Heterogeneous Experts Representation Retriever that enriches and fuses multimodal information and retrieves a compact candidate set for each incomplete triple. It then uses a Generative LLM Predictor, implemented via either in-context learning or lightweight fine-tuning, to accurately identify the correct answer from these candidates. Extensive experiments on three standard MMKG benchmarks demonstrate HERGC's effectiveness, achieving superior performance over existing methods.

**Keywords:** Knowledge Graph Completion · Multimodal Knowledge Graph

## 1 Introduction

Knowledge graphs (KGs) represent real-world facts as triples of entities and their relations, offering a structured semantic representation [7,15]. Multimodal knowledge graphs (MMKGs) [4] extend traditional KGs by incorporating additional modalities such as images and text, thereby enriching the contextual information of entities and enhancing the expressiveness of the graph. Both KGs and MMKGs have been widely adopted in various AI systems [19].


R. C.-W. Wong et al. (Eds.): PAKDD 2026, LNAI 16597, pp. 327–339, 2026.
https://doi.org/10.1007/978-981-92-1300-9_26

Like traditional KGs, MMKGs suffer from incompleteness, often due to missing facts in the underlying data sources or facts that have yet to be discovered by humans. Unlike unimodal knowledge graph completion (KGC), which primarily leverages the graph's structure and local neighborhood information, multimodal knowledge graph completion (MMKGC) [4] introduces additional complexity through the incorporation of multimodal signals. Recent MMKGC methods have mainly focused on modality fusion [10] and modality information representation [32], typically constructing joint embeddings or employing ensemble strategies to score candidate facts. While these approaches offer valuable insights, their reasoning capabilities during the completion process remain limited due to their reliance solely on closed-world triples and available multimodal information.

Meanwhile, recent advances in unimodal KGC have introduced generative completion approaches powered by large language models (LLMs) [11,25]. These approaches achieve strong performance by exploiting the extensive factual knowledge and reasoning capabilities of LLMs. However, this paradigm remains largely underexplored in the MMKGC setting due to its inherent limitations. First, these methods typically rely on existing knowledge graph embedding (KGE) models (e.g., RotatE [20], as used in KICGPT [24]) for candidate retrieval, which ignore multimodal signals and yield low-recall candidate sets. Second, their LLM-based predictors are designed to process only textual inputs, excluding visual and structural modalities that are essential for multimodal reasoning. These limitations motivate the development of novel mechanisms that seamlessly integrate multimodal information into both retrieval and generation, enabling more accurate, context-aware reasoning within constrained candidate spaces.

To address these challenges and fill the gap in generative MMKGC, we propose HERGC, a novel and flexible generative framework that overcomes the limitations of closed-world reasoning based solely on MMKG contents and introduces a multimodal-aware generative completion paradigm. Inspired by the LLM-based reranking idea, HERGC comprises two core components: the Heterogeneous Experts Representation Retriever (HERR) and the Generative LLM Predictor (GLP). HERR employs a Mixture of Heterogeneous Experts Network (MoHE) to enrich each modality's embeddings from multiple and hierarchical perspectives and a Relation-aware Gated Multimodal Unit (RaGMU) to obtain high-quality fused embeddings, which are then used to score and retrieve candidate entities. GLP supports (1) directly using powerful closed-source LLMs (e.g., GPT-4) via APIs to make predictions with in-context learning, or (2) injecting the fused multimodal embeddings into open-source LLMs (e.g., LLaMA) and performing LoRA fine-tuning on minimal instruction data. This flexible design allows GLP to adapt to diverse resource conditions while enabling the LLMs to accurately select the correct entity from the retrieved candidates. We conduct comprehensive experiments on three public MMKG benchmarks to validate the effectiveness of HERGC. Our contributions are summarized as follows:

- We propose HERGC, the first, to the best of our knowledge, MMKGC framework based on the generative paradigm. It features a flexible GLP module

that supports both open-source and closed-source LLMs, enabling effective integration of external knowledge for multimodal structural reasoning.
- We design a novel representation and retrieval module, HERR, to extract multi-perspective signals from heterogeneous and distributionally diverse modalities, and to adaptively fuse them based on relation types, producing high-quality fused embeddings and candidate sets.
- We conduct extensive experiments on three standard MMKGC benchmarks, demonstrating that HERGC consistently outperforms strong baselines.

## 2 Related Work

### 2.1 Unimodal Knowledge Graph Completion

Unimodal KGC primarily focuses on embedding entities and relations into continuous vector spaces to predict triples. Most of them leverage KG's structure to learn the embeddings by maximize the positive and negative samples score difference, such as Translational-Distance approaches (TransE [2] and RotatE [20]) and Semantic-Matching approaches (DistMult [28], ComplEX [21], and Tucker [1]). To improve the embedding quality, graph neural network based (GNN-based) methods have been proposed, such as R-GCN [16]. Besides structural information, KGs naturally carry text information. Therefore, the text-based methods, which encodes text information in KGs through pretrained language models (PLMs), have also emerged, such as KG-Bert [29]. With the recent development of LLM, the novel generative methods have come into view. They mainly use the rich external knowledge and powerful reasoning capabilities of LLMs to complete KGC in a sequence-to-sequence form, including KICGPT [24] and DIFT [12].

### 2.2 Multimodal Knowledge Graph Completion

While recent text-based and generative KGC approaches have started to incorporate both structural and textual information, they often lack tight coordination between these modalities during inference. Moreover, the emergence of KGs enriched with additional modalities, such as images, audio, and video, further raises the bar for the design of dedicated MMKGC models. Initial MMKGC models, like IKRL [26], extract visual features from entities using pre-trained visual encoders and combine these with structural embeddings. Extensions such as TransAE [23] and TBKGC [14] incorporate both textual and visual features, enhancing entity representations Fusion-oriented methods, including OTKGE [3], employ sophisticated strategies like optimal transport and modality-specific representations to achieve effective multimodal integration. IMF [10] utilizes an interactive fusion framework, training separate models for each modality to collaboratively infer missing links. Furthermore, MMKRL [13] employs adversarial training but focuses specifically on robustness against modality-specific perturbations. Meanwhile, approaches like MyGO [32] leverage fine-grained contrastive learning to enhance the granularity of multimodal embeddings. Also, there are methods that use multi-perspective ideas to enhance modal representation, such as MoMoK [31] that uses a mixture of expert model and information decoupling and MCKGC [6] that integrates information in a mixed curvature space.

## 3 Preliminary

In this work, we focus on the most common form of MMKG with visual and textual modalities. An MMKG can be represented as a directed multigraph with modal attributes, denoted as $\mathcal{G} = (\mathcal{E}, \mathcal{R}, \mathcal{T}, \mathcal{V}, \mathcal{D})$, where $\mathcal{E}$ is the set of entities, $\mathcal{R}$ is the set of relations, and $\mathcal{T} = \{(h, r, t)|h, t \in \mathcal{E}, t \in \mathcal{R}\}$ is the set of triples. The $\mathcal{V}$ and $\mathcal{D}$ are the collection of visual images and descriptive text associated with entities. MMKGC aims to leverage the observed triples $\mathcal{T}$ together with the visual and textual attributes of entities ($\mathcal{V}$ and $\mathcal{D}$) to infer missing triples. The set of potential facts is defined as $\{(h', r', t') \mid h', t' \in \mathcal{E},\ r' \in \mathcal{R}\}$, where $(h', r', t') \notin \mathcal{T}$ represents missing triples in the MMKG. In this work, we formulate MMKGC as the task of completing incomplete query triples of the form $(?, r_q, t_q)$ and $(h_q, r_q, ?)$, corresponding to head prediction and tail prediction, respectively. Here, $h_q$ or $t_q$ denotes the query entity, and $r_q$ denotes the query relation.

## 4 Methodology

In this section, we present the HERGC framework and provide a detailed description of each module. An overview of HERGC is shown in Fig. 1.

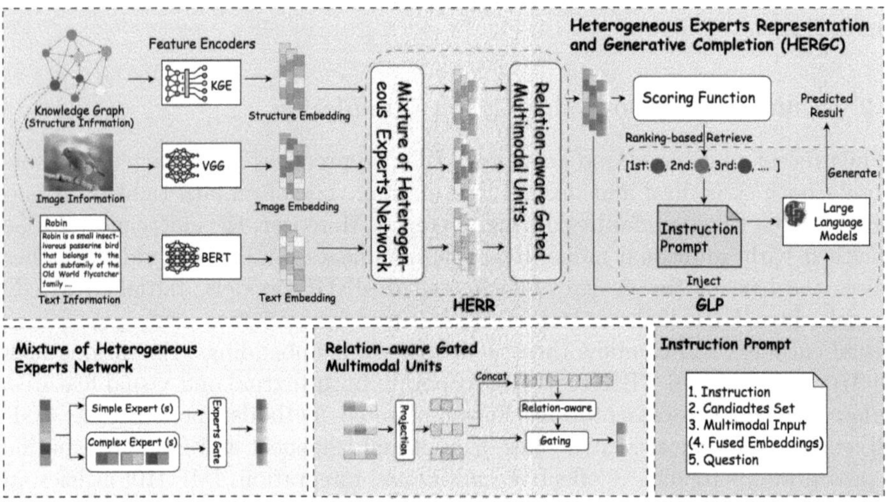


**Fig. 1.** Overview of the HERGC framework.

### 4.1 Multimodal Information Embedding

To enable effective fusion of multimodal information, we first encode each entity's visual and textual modalities into embedding representations, denoted as $\mathbf{e}_{\mathcal{V}}$ and $\mathbf{e}_{\mathcal{D}}$. Additionally, we embed each entity's structural information from the KG into a structural representation $\mathbf{e}_{\mathcal{S}}$ to capture graph contextual information.

**Image and Text Embedding.** We utilize pre-trained models to encode the visual and textual information associated with each entity. To ensure a fair comparison, we maintain consistency with recent baselines [6,31] by adopting BERT [5], an encoder-only transformer model trained on large-scale textual corpora, for text embeddings, and VGG [18], a convolutional neural network trained on large-scale image datasets, for visual feature extraction. Each entity's descriptive text and image are processed through BERT and VGG, respectively, yielding the initial modality-specific embeddings $\mathbf{e}_{\mathcal{D}}$ and $\mathbf{e}_{\mathcal{V}}$.

**Structure Embedding.** To encode structural information from the MMKG, we adopt TuckER [1], a lightwight KGE model that learns entity and relation embeddings via tensor factorization. TuckER is also employed as the scoring function in the retrieval module. The resulting embedding is taken as the structural representation $\mathbf{e}_{\mathcal{S}}$ for each entity.

### 4.2 Heterogeneous Experts Representation Retriever

We propose the Heterogeneous Experts Representation Retriever (HERR) to obtain high-quality joint representations and retrieval candidate sets in the complex multimodal setting. HERR first enhances modality-specific features through a Mixture of Heterogeneous Experts network, then fuses them using a Relation-Aware Gated Multimodal Unit to obtain joint embeddings. Finally, HERR employs a scoring function to compute triple scores and generate a coarse-ranked list of candidate entities.

**Mixture of Heterogeneous Experts Module.** To obtain representative and multi-perspective embeddings for each modality we design a heterogeneous experts layer composed of both simple and complex experts. Given the inherent heterogeneity and distinct distributions of modality-specific embeddings, our goal is to align them across modalities while preserving their unique characteristics. Thus, we introduce the Mixture of Heterogeneous Experts (MoHE) module.

MoHE extends the standard Mixture of Experts (MoE) [17] architecture by combining diverse expert types. Given the input feature vector of the modality, MoHE outputs the weighted sum of top-$\kappa$ experts outputs:

$$\mathbf{h}_{i,m} = \sum_{\kappa \in S} G_{\kappa}(\mathbf{x}_{i,m}) E_{\kappa}(\mathbf{x}_{i,m}), \tag{1}$$

where $\mathbf{x}_{i,m}$ and $\mathbf{h}_{i,m}$ denote the input and output embeddings of the $i$-th entity in modality $m$, the $\mathbf{x}_{i,m}$ comes from $\mathbf{e}_m$, $E(\cdot)$ is the expert network, and $S = \text{Top}\kappa(G_\kappa(\mathbf{x}_{i,m}))$ denotes the selected expert indices based on gating weights.

The gating weight $G_\kappa(\cdot)$ for the corresponding expert $E_\kappa$ is computed as:

$$G_\kappa(\mathbf{x}_{i,m}) = \text{softmax}(\frac{\mathbf{W}_{gate}\mathbf{x}_{i,m} + \mathbf{W}_\epsilon \mathbf{x}_{i,m}}{\tau}), \tag{2}$$

where $\mathbf{W}_{gate}$ is the gate weight parameter matrix, $\mathbf{W}_\epsilon$ injects noise for exploration, and $\tau$ is the gate temperature hyperparameter.

The simple expert in the MoHE layer performs a linear transformation and feature whitening. To better adapt to heterogeneous modalities and capture richer cross-dimensional interactions, MoHE also incorporates complex PHM experts inspired by Block-Hypercomplex Linear Transformations [30]. Specifically, the input $\mathbf{x} \in \mathbb{R}^d$ will be partitioned into $n$ sub-blocks of size $d/n$:

$$\mathbf{x} = [\mathbf{x}^{(1)}; \mathbf{x}^{(2)}; ...; \mathbf{x}^{(n)}], \mathbf{x}^{(j)} \in \mathbb{R}^{d/n}, \tag{3}$$

then each sub-block is transformed by a shared weight matrix $\mathbf{W}_{block} \in \mathbb{R}^{\frac{d}{n} \times \frac{d}{n}}$ and a per-block weight matrix $\mathbf{H}_j \in \mathbb{R}^{\frac{d}{n} \times \frac{d}{n}}$:

$$\mathbf{h}^{(j)} = \mathbf{H}_j \mathbf{W}_{block} \mathbf{x}^{(j)}. \tag{4}$$

Finally, the PHM expert output is obtained by concatenating all transformed sub-blocks.

$$E_{PHM}(\mathbf{x}) = [\mathbf{h}^{(1)}; \mathbf{h}^{(2)}; ...; \mathbf{h}^{(n)}] \in \mathbb{R}^d. \tag{5}$$

**Relation-Aware Gated Multimodal Units.** In MMKGC tasks, the importance of each modality can vary across relation types. To address this, we propose the Relation-aware Gated Multimodal Unit (RaGMU), which dynamically adjusts fusion weights based on the relations.

Specifically, each modality embedding $\mathbf{x}_m$ is projected into a shared latent space:

$$\mathbf{h}_m = \tanh(\mathbf{W}_{proj,m}\mathbf{x}_m + \mathbf{b}_{proj,m}) \tag{6}$$

where $\mathbf{W}_{proj,m}$ and $\mathbf{b}_{proj,m}$ are the projection matrix and bias of RaGMU projector for modality $m$.

Next, the gate vector can be calculated by:

$$\mathbf{z} = \text{softmax}(g_r(\mathbf{r}) \odot (\mathbf{W}_z \mathbf{h}_{concat} + \mathbf{b}_z)) \tag{7}$$

where $\mathbf{h}_{concat} = [\mathbf{h}_1; \mathbf{h}_2; ...; \mathbf{h}_M]$ is the concatenation of all projected modality embeddings, the $\mathbf{W}_z$ and $\mathbf{b}_z$ are the gating weight matrix and bias, $\odot$ denotes Hadamard product, and $g_r(\mathbf{r})$ is a relation-aware modulation function that generates a scaling vector from the relation embedding $\mathbf{r}$.

Finally, the fused multimodal embedding is computed by applying the gate vector to each modality's hidden projection $\mathbf{h}_m$:

$$\mathbf{h}_{fuse} = \sum_m \mathbf{z}_m \odot \mathbf{h}_m, \tag{8}$$

where $\mathbf{z}_m$ is the $m$-th element gate vector corresponding to modality $m$.

**Score Function.** After getting the fused multimodal embeddings, we compute triple plausibility scores using the TuckER [1] scoring function:

$$S(h, r, t) = \mathbf{W}_{tucker} \times_1 \mathbf{h}_h \times_2 \mathbf{r} \times_3 \mathbf{h}_t. \tag{9}$$

where $\mathbf{h}_h$ and $\mathbf{h}_t$ denote the fused embeddings of the head $h$ and tail $t$, $\mathbf{r}_r$ denotes the embedding of the relation $r$, and $\times_n$ denotes the $n$-mode tensor product.

To optimize HERR, we adopt a binary classification objective that encourages higher scores for positive triples and lower scores for negative ones. Negative samples are generated via uniform negative sampling. The loss function is defined as:

$$\mathcal{L} = -\sum[y \log \sigma(S) + (1-y)\log(1-\sigma(S))], \tag{10}$$

where $y \in \{0, 1\}$ is the label indicating whether the triple is positive or negative, and $\sigma(\cdot)$ is the sigmoid function.

### 4.3 Generative LLM Predictor

The Generative LLM Predictor (GLP) aims to predict the correct entity from a set of candidates given an incomplete query triple. Each query is reformulated as a natural language question derived from the query entity and relation. We adopt an instruction prompt that directly asks the LLM to complete an incomplete triple by choosing the most suitable entity.

**Prompt Template.** Taking the tail prediction scenario as an example, we first use HERR to retrieve the ranking of all candidates based on the query $(h_q, r_q, ?)$, ensuring that the resulting triples do not already exist in the MMKG. We then select the top-$k$ candidates, denoted as $C = [e_1, e_2, \ldots, e_k]$. A natural language question $Q$ is generated based on the query relation $r_q$ and entity $h_q$. Finally, we construct a prompt $P$ by combining the instruction $I$, the question $Q$, the candidate list $C$, and the entity descriptions $D$ (including text descriptions for $h_q$ and each $e \in C$, and the image for $h_q$ when using multimodal LLMs such as LLaMA-3-Vision):

$$P = [I, Q, C, D]. \tag{11}$$

Using this prompt, closed-source LLMs can perform prediction via in-context learning without additional training.

**LoRA Fine-tuning.** For open-source LLM, We perform Low-Rank Adaptation (LoRA) fine-tuning on a small number of query–answer pairs. In this setting, we inject the fused embedding into the LLM through a learnable adapter layer, which maps it to the LLM embedding space and replaces the embeddings of placeholder tokens in the prompt. The prompt template becomes:

$$P = [I, Q, C, D, E], \tag{12}$$

where $E$ denotes the fused embeddings of $h_q$ and each $e \in C$. This lightweight adaptation enables the model to follow our completion instruction while largely relying on its pretrained knowledge. The injected fused features provide additional grounding signals, guiding the model toward more accurate predictions.

## 5 Experiments

### 5.1 Experiment Setup

**Dataset.** We evaluate our proposed method on three benchmark MMKG datasets, MKG-Y [27], MKG-W [27] and DB15K [12]. Table 1 presents the statistical details of these three datasets.

**Table 1.** Statistics of the datasets.

| Datasets | Entities | Relations | Training | Validation | Testing |
|---|---|---|---|---|---|
| MKG-W | 15,000 | 169 | 34,196 | 4,276 | 4,274 |
| MKG-Y | 15,000 | 28 | 21,310 | 2,665 | 2,663 |
| DB15K | 12,842 | 279 | 79,222 | 9,902 | 9,904 |

**Baseline Methods.** For MMKGC, we consider both unimodal and multimodal methods as baselines, including: TransE [2], RotatE [20], DistMult [28], ComplEx [21] and TuckER [1], IKRL [26], TBKGC [14], TransAE [23], MMKRL [13], RSME [22], OTKGE [3], IMF [10], QEB [9], VISTA [9], MyGO [32], MoMoK [31], MCKGC [6]. The baseline comparisons in this paper are based on the reported performance values of these methods.

**Implementation Details.** For modality-specific feature extraction, we use bert-base-uncased to encode text and VGG-16 to encode images. For HERR training, we tune the embedding dimension from $\{200, 300, 400\}$ and set the batch size to $\{512, 1024\}$. We use the Adam optimizer [8], with the learning rate selected from $\{0.005, 0.001, 0.0005\}$. The MoHE module is configured with 2 simple experts and 2 complex PHM experts. The number of retrieved candidate entities is selected from $\{10, 20, 30, 40\}$. For the GLP, we employ GPT-4, LLaMA-3-8B and LlaMA-3.2-11B-Vision. We set the LoRA hyperparameters to $r = 64$, $\alpha = 16$, a dropout rate of 0.1, and a learning rate of 0.0002. Model performance is evaluated using standard ranking-based metrics: Mean Reciprocal Rank (MRR), and Hits@1, Hits@3, and Hits@10, under the "filtered" setting [2].

All experiments were conducted on an AMD EPYC 7763 64-Core CPU, an NVIDIA A100-SXM4-40GB GPU, and Rocky Linux 8.10.

**Table 2.** Main results of the comparison between HERGC with different LLMs and the baselines on MKG-W, MKG-Y and DB15K. For each metric, the best performance is highlighted in **bold**, and the second-best is underlined.

| Methods | MKG-W | | | | MKG-Y | | | | DB15K | | | |
|---|---|---|---|---|---|---|---|---|---|---|---|---|
| | MRR | Hits@1 | Hits@3 | Hits@10 | MRR | Hits@1 | Hits@3 | Hits@10 | MRR | Hits@1 | Hits@3 | Hits@10 |
| **Unimodal Methods** | | | | | | | | | | | | |
| TransE | 29.19 | 21.06 | 33.20 | 44.23 | 30.73 | 23.45 | 35.18 | 43.37 | 24.86 | 12.78 | 31.48 | 47.07 |
| RotatE | 33.67 | 26.80 | 36.68 | 46.73 | 34.95 | 29.10 | 38.35 | 45.30 | 29.28 | 17.87 | 36.12 | 49.66 |
| DistMult | 20.99 | 15.93 | 22.28 | 30.86 | 25.04 | 19.33 | 27.80 | 35.95 | 23.03 | 14.78 | 26.28 | 39.59 |
| ComplEx | 24.93 | 19.09 | 26.69 | 36.73 | 28.71 | 22.26 | 32.12 | 40.93 | 27.48 | 18.37 | 31.57 | 45.37 |
| TuckER | 30.39 | 24.44 | 32.91 | 41.25 | 37.05 | 34.59 | 38.43 | 41.45 | 33.86 | 25.33 | 37.91 | 50.38 |
| **Multimodal Methods** | | | | | | | | | | | | |
| IKRL | 32.36 | 26.11 | 34.75 | 44.07 | 33.22 | 30.37 | 34.28 | 38.26 | 26.82 | 14.09 | 34.93 | 49.09 |
| TBKGC | 31.48 | 25.31 | 33.98 | 43.24 | 33.99 | 30.47 | 35.27 | 40.07 | 28.40 | 15.61 | 37.03 | 49.86 |
| TransAE | 30.00 | 21.23 | 34.91 | 44.72 | 28.10 | 25.31 | 29.10 | 33.03 | 28.09 | 21.25 | 31.17 | 41.17 |
| MMKRL | 30.10 | 22.16 | 34.09 | 44.69 | 36.81 | 31.66 | 39.79 | **45.31** | 26.81 | 13.85 | 35.07 | 49.39 |
| RSME | 29.23 | 23.36 | 31.97 | 40.43 | 34.44 | 31.78 | 36.07 | 39.09 | 29.76 | 24.15 | 32.12 | 40.29 |
| OTKGE | 34.36 | 28.85 | 36.25 | 44.88 | 35.51 | 31.97 | 37.18 | 41.38 | 23.86 | 18.45 | 25.89 | 34.23 |
| IMF | 34.50 | 28.77 | 36.62 | 45.44 | 35.79 | 32.95 | 37.14 | 40.63 | 32.25 | 24.20 | 36.00 | 48.19 |
| QEB | 33.38 | 25.47 | 35.06 | 45.32 | 34.37 | 29.49 | 37.00 | 42.30 | 28.18 | 14.82 | 36.67 | 51.55 |
| VISTA | 32.91 | 26.12 | 35.38 | 45.61 | 30.45 | 24.87 | 32.39 | 41.53 | 30.42 | 22.49 | 33.56 | 45.94 |
| MyGO | 36.10 | 29.78 | 38.54 | 47.75 | 38.44 | 35.01 | 39.84 | 44.19 | 37.72 | 30.08 | 41.26 | 52.21 |
| MoMoK | 38.89 | 30.38 | 37.54 | 46.31 | 37.91 | 35.09 | 39.20 | 43.20 | 39.54 | 32.38 | 43.45 | 54.14 |
| MCKGC | 36.88 | 31.32 | 38.92 | 47.43 | 38.92 | 35.49 | 40.57 | 45.21 | 39.79 | 31.92 | 43.80 | 54.66 |
| **HERGC** | | | | | | | | | | | | |
| – GPT-4 | 38.28 | 32.03 | **41.80** | 47.82 | 39.23 | 35.69 | **42.22** | 45.09 | 39.70 | 31.22 | 45.09 | **55.38** |
| – LLaMA-3 | **39.12** | **33.65** | 41.67 | 48.12 | 39.82 | **36.73** | 41.42 | 44.84 | **40.95** | **33.47** | **45.66** | 55.12 |
| – LLaMA-3-Vision | 38.76 | 33.01 | 41.43 | **48.54** | **40.26** | 36.31 | 40.91 | 45.22 | 40.28 | 32.30 | 45.20 | 54.67 |

## 5.2 Main Results

Table 2 presents the main results of our proposed HERGC compared with advanced unimodal and multimodal KGC methods. HERGC consistently achieves the best overall performance on three datasets across most evaluation metrics. Notably, HERGC improves Hits@1 on MKG-Y, MKG-W, and DB15K by 7.44%, 3.94%, and 3.37%, respectively, over the strongest baseline on each dataset.

We also assess the impact of different LLM predictors within GLP. LLaMA-3, after fused embeddings injection and lightweight LoRA tuning, yields consistently strong results, whereas LLaMA-3-Vision offers only marginal gains, likely because the images do not directly carry the discriminant information of the current relations. Furthermore, comparisons with GPT-4 show that, by injecting fused embeddings and applying lightweight fine-tuning, the open-source model can outperform the powerful closed-source model, indicating that LLM's structural reasoning ability can be enhanced through multimodal integration.

## 5.3 Ablation Studies

To verify the rationality of the HERGC design, we conduct an ablation study consisting of: (1) ablation of modality-specific inputs to assess the contribution of each modality and the model's ability to leverage multimodal information; (2) ablation of key components within HERGC. The results are shown in Table 3.

**Table 3.** Ablation study results on three datasets.

| Setting | MKG-W | | MKG-Y | | DB15K | |
|---|---|---|---|---|---|---|
| | MRR | Hits@1 | MRR | Hits@1 | MRR | Hits@1 |
| **Modality Information (w/o)** | | | | | | |
| Image Modality | 36.83 | 31.19 | 38.57 | 34.96 | 39.41 | 30.74 |
| Text Modality | 36.17 | 30.59 | 38.42 | 34.72 | 39.59 | 31.18 |
| Structure Modality | 37.98 | 32.34 | 39.09 | 36.48 | 40.17 | 32.35 |
| **Model Components (w/o)** | | | | | | |
| Complex Experts | 37.95 | 32.26 | 39.04 | 35.51 | 40.26 | 32.30 |
| GMU | 37.02 | 31.41 | 38.96 | 35.19 | 39.97 | 31.28 |
| Relation-awareness | 37.56 | 32.02 | 39.21 | 36.14 | 40.34 | 32.41 |
| Embedding Injection | 37.94 | 32.26 | 39.00 | 35.84 | 39.05 | 31.16 |
| GLP | 36.22 | 30.56 | 38.42 | 35.11 | 38.76 | 30.67 |
| $HERGC_{LLaMA-3}$ | **39.12** | **33.65** | **39.82** | **36.73** | **40.95** | **33.47** |

For modality ablation, we individually remove the textual, visual, and structural information. In all cases, performance declines, indicating that each modality contributes meaningfully to the model's predictions and that HERGC effectively integrates multimodal information. For component ablation, we examine the impact of removing complex PHM experts, the RaGMU module, and relation-awareness mechanism, embedding-injection for GLP, and GLP itself. Removing any of these components results in performance degradation, highlighting their importance. Notably, omitting the embedding injection also leads to a performance drop, indicating that incorporating exogenous fused multimodal embeddings indeed enhances the LLM's structural reasoning capability.

### 5.4 Further Analyses

**Representation Visualization.** We use t-SNE to visualize the entity representations learned by HERR on DB15K and compare them with individual modality embeddings. As shown in Fig. 2, the fused embeddings form almost perfectly separated clusters for each entity type, with clear inter-type boundaries and uniform intra-type distributions. By contrast, image-only embeddings exhibit highly entangled regions; structure-only embeddings fail to distinguish the "Language" cluster and yield a diffuse "Writer" grouping; and text-only embeddings conflate "Writer" and "Singer" entities—likely due to their lexical similarity (e.g., names). These observations confirm that HERR effectively integrates multimodal signals.

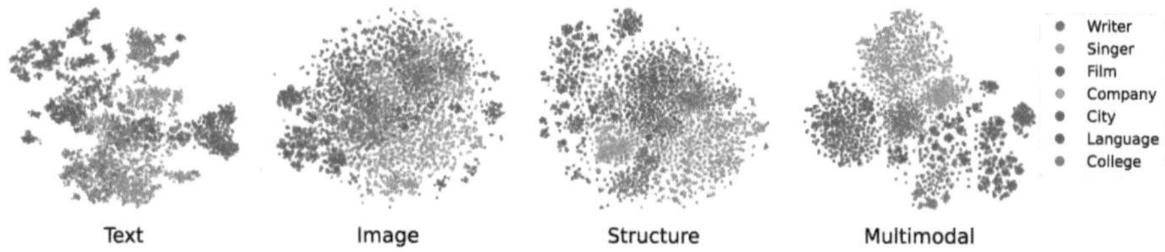


**Fig. 2.** t-SNE visualization of the entity representations on the DB15K dataset.

**LLM Predictor Exploration.** We further investigate the GLP when using open-source LLMs by examining the effect of varying the candidate set size $k$ on model performance and fine-tuning time. Figure 3(a) shows the trends in time consumption and ranking-based metrics as $k$ varies. As expected, inference time increases approximately linearly with larger $k$ values due to longer prompts and more complex candidate sets. However, the performance gains are marginal beyond $k = 20$; only increasing $k$ from 10 to 20 yields a noticeable improvement in MRR. Balancing effectiveness and efficiency, we set $k = 20$ in all experiments.

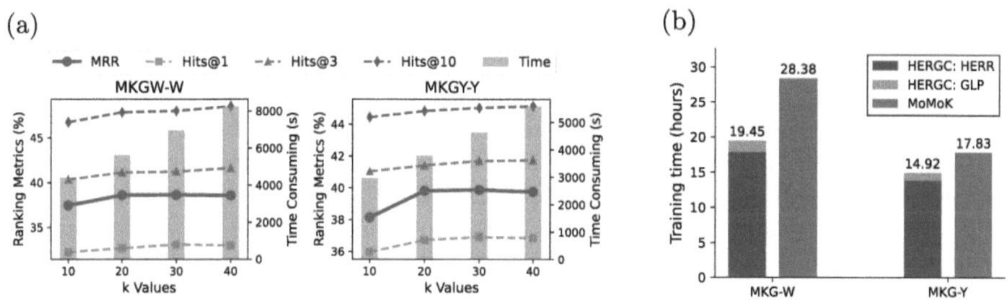


**Fig. 3.** Further analysis: (a) Performance and time consumption of the GLP with LlaMA-3 on MKG-W and MKG-Y under different candidate set sizes $k$. (b) Training time comparison between HERGC and MoMoK.

**Efficiency Analysis.** We further compare HERGC's efficiency with that of MoMoK. Since HERGC consists of a retrieval stage and a generation stage, and requires no additional training when using GPT-4 in GLP, we consider the scenario in which fine-tuning with the LLaMA-3 model incurs additional time. The results are presented in Fig. 3(b), confirming HERGC's overall efficiency.

## 6 Conclusion

In this paper, we present HERGC, a novel generative framework for MMKGC. HERGC consists of HERR, which fuses multimodal signals into high-quality embeddings and retrieves candidate sets, and GLP, which predicts the correct entity from candidates and supports both open- and closed-source LLMs. Extensive experiments on three standard MMKGC benchmarks demonstrate that HERGC achieves state-of-the-art performance. HERGC bridges the generative paradigm and MMKGC, providing a generalizable solution for future research.

**Acknowledgments.** This work was supported by the U.S. National Institutes of Health (NCCIH: R01AT009457; NIA: R01AG078154; NCI: R01CA287413) and the Food & Drug Administration (U01FD008720). The content is solely the responsibility of the authors and does not represent the official views of the NIH or FDA.

## References

1. Balažević, I., Allen, C., Hospedales, T.: Tucker: tensor factorization for knowledge graph completion. In: EMNLP-IJCNLP (2019)
2. Bordes, A., Usunier, N., Garcia-Duran, A., Weston, J., Yakhnenko, O.: Translating embeddings for modeling multi-relational data. In: NeurIPS, vol. 26 (2013)
3. Cao, Z., Xu, Q., Yang, Z., He, Y., Cao, X., Huang, Q.: Otkge: multi-modal knowledge graph embeddings via optimal transport. In: NeurIPS, vol. 35, pp. 39090–39102 (2022)
4. Chen, Z., et al.: Knowledge graphs meet multi-modal learning: a comprehensive survey. arXiv:2402.05391 (2024)
5. Devlin, J., Chang, M.W., Lee, K., Toutanova, K.: Bert: pre-training of deep bidirectional transformers for language understanding. In: NAACL-HLT (2019)
6. Gao, Y., Zhang, F., Zhang, Z., Min, X., Zhuang, F.: Mixed-curvature multi-modal knowledge graph completion. In: AAAI (2025)
7. Ji, S., Pan, S., Cambria, E., Marttinen, P., Yu, S.: A survey on knowledge graphs: representation, acquisition, and applications. IEEE TNNLS **33**(2), 494–514 (2021)
8. Kingma, D.P., Ba, J.: Adam: a method for stochastic optimization. arXiv:1412.6980 (2017)
9. Lee, J., Chung, C., Lee, H., Jo, S., Whang, J.: Vista: visual-textual knowledge graph representation learning. In: EMNLP Findings (2023)
10. Li, X., Zhao, X., Xu, J., Zhang, Y., Xing, C.: IMF: interactive multimodal fusion model for link prediction. In: WWW (2023)
11. Liu, Y., Tian, X., Sun, Z., Hu, W.: Finetuning generative large language models with discrimination instructions for knowledge graph completion. In: ISWC. (2024)
12. Liu, Y., Li, H., Garcia-Duran, A., Niepert, M., Onoro-Rubio, D., Rosenblum, D.S.: MMKG: multi-modal knowledge graphs. In: ESWC (2019)
13. Lu, X., Wang, L., Jiang, Z., He, S., Liu, S.: MMKRL: a robust embedding approach for multi-modal knowledge graph representation learning. Appl. Intell. (2022)
14. Mousselly-Sergieh, H., Botschen, T., Gurevych, I., Roth, S.: A multimodal translation-based approach for knowledge graph representation learning. In: SEM (2018)
15. Nickel, M., Murphy, K., Tresp, V., Gabrilovich, E.: A review of relational machine learning for knowledge graphs. Proc. IEEE **104**(1), 11–33 (2015)
16. Schlichtkrull, M., Kipf, T.N., Bloem, P., Van den Berg, R., Titov, I., Welling, M.: Modeling relational data with graph convolutional networks. In: ESWC (2018)
17. Shazeer, N., et al.: Outrageously large neural networks: the sparsely-gated mixture-of-experts layer. In: ICLR (2017)

18. Simonyan, K., Zisserman, A.: Very deep convolutional networks for large-scale image recognition. arXiv:1409.1556 (2014)
19. Sun, R., et al.: Multi-modal knowledge graphs for recommender systems. In: CIKM (2020)
20. Sun, Z., Deng, Z.H., Nie, J.Y., Tang, J.: Rotate: knowledge graph embedding by relational rotation in complex space. In: ICLR (2019)
21. Trouillon, T., Welbl, J., Riedel, S., Gaussier, É., Bouchard, G.: Complex embeddings for simple link prediction. In: ICML (2016)
22. Wang, M., Wang, S., Yang, H., Zhang, Z., Chen, X., Qi, G.: Is visual context really helpful for knowledge graph? A representation learning perspective. In: ACM MM (2021)
23. Wang, Z., Li, L., Li, Q., Zeng, D.: Multimodal data enhanced representation learning for knowledge graphs. In: IJCNN (2019)
24. Wei, Y., Huang, Q., Zhang, Y., Kwok, J.: Kicgpt: large language model with knowledge in context for knowledge graph completion. In: EMNLP Findings (2023)
25. Xiao, Y., et al.: DRKGC: dynamic subgraph retrieval-augmented LLMs for knowledge graph completion across general and biomedical domains. In: EMNLP Findings (2025)
26. Xie, R., Liu, Z., Luan, H., Sun, M.: Image-embodied knowledge representation learning. In: IJCAI (2017)
27. Xu, D., Xu, T., Wu, S., Zhou, J., Chen, E.: Relation-enhanced negative sampling for multimodal knowledge graph completion. In: ACM MM (2022)
28. Yang, B., Yih, S.W.T., He, X., Gao, J., Deng, L.: Embedding entities and relations for learning and inference in knowledge bases. In: ICLR (2015)
29. Yao, L., Mao, C., Luo, Y.: Kg-bert: bert for knowledge graph completion. arXiv:1909.03193 (2019)
30. Zhang, A., et al.: Beyond fully-connected layers with quaternions: parameterization of hypercomplex multiplications with $1/n$ parameters. In: ICLR (2017)
31. Zhang, Y., et al.: Multiple heads are better than one: mixture of modality knowledge experts for entity representation learning. In: ICLR (2025)
32. Zhang, Y., et al.: Tokenization, fusion, and augmentation: towards fine-grained multi-modal entity representation. In: AAAI (2025)

# EXCODER: EXplainable Classification Of DiscretE Time Series Representations

Yannik Hahn(✉), Antonin Königsfeld, Hasan Tercan, and Tobias Meisen

Institute for Technologies and Management of Digital Transformation, University of Wuppertal, Rainer-Gruenter-Straße 21, 42119 Wuppertal, Germany
{yhahn,koenigsfeld,tercan,meisen}@uni-wuppertal.de

**Abstract.** Deep learning has significantly improved time series classification, yet the lack of explainability in these models remains a major challenge. While Explainable AI (XAI) techniques aim to make model decisions more transparent, their effectiveness is often hindered by the high dimensionality and noise present in raw time series data. In this work, we investigate whether transforming time series into discrete latent representations (DLR) not only preserves but enhances explainability by reducing redundancy and focusing on the most informative patterns. We show that applying XAI methods to these compressed representations leads to concise and structured explanations that maintain faithfulness without sacrificing classification performance. Additionally, we propose Similar Subsequence Accuracy (SSA), a novel metric that quantitatively assesses the alignment between XAI-identified salient subsequences and the label distribution in the training data. SSA provides a systematic way to validate whether the features highlighted by XAI methods are truly representative of the learned classification patterns. Our findings demonstrate that discrete latent representations not only retain the essential characteristics needed for classification but also offer a pathway to more compact and interpretable explanations in time series analysis.

## 1 Introduction

Time series analysis is fundamental to numerous domains, including healthcare, finance, and industrial monitoring. As the volume and complexity of temporal data continue to grow, accurate and interpretable time series classification (TSC) becomes essential for deriving actionable insights. Recent advances in deep learning have led to state-of-the-art performance in TSC by capturing complex, non-linear dependencies directly from raw data, often outperforming traditional approaches. However, despite their predictive power, these models remain largely opaque, making it difficult to understand how decisions are made which is crucial for ensuring robustness and reliability.

XAI methods address this challenge by making model decisions more transparent, enabling practitioners to identify key features, diagnose errors, and

Y. Hahn and A. Königsfeld—Equal contribution.


R. C.-W. Wong et al. (Eds.): PAKDD 2026, LNAI 16597, pp. 340–352, 2026.
https://doi.org/10.1007/978-981-92-1300-9_27

improve model trustworthiness. However, applying XAI to time series data introduces unique challenges due to the high dimensionality, temporal dependencies, and noise inherent in raw time series. These factors have driven a growing need for meaningful and structured explanations specifically designed for time series analysis. Theissler et al. [26] proposed a taxonomy that categorizes XAI methods for time series into point-based approaches assessing the importance of individual time steps, and subsequence-based approaches, which identify meaningful temporal segments. However, both categories come with inherent limitations: point-based methods may fail to consider broader temporal context, while subsequence-based approaches might lack the fine-grained resolution needed to capture short but critical variations in the data. Beyond the adaptation of existing XAI methods, a complementary approach to improve interpretability in TSC lies in learning structured representations of time series data. Models like the vector quantised variational autoencoder (VQ-VAE) [21] and discrete variational autoencoder (DVAE) [28] map continuous time series into a lower-dimensional space using a finite set of discrete codes, leading to more structured and compact representations. Vahdat et al. [28] highlight several advantages of this approach, including noise reduction, improved robustness, and particularly enhanced feature interpretability. Hence, in this work, we explore the potential of discrete latent representations (DLRs) to enhance the explainability of TSC models. Specifically, we propose a patching mechanism, inspired by Nie et al. [20], that maps each discrete latent code back to its corresponding subsequence in the original time series. This ensures a direct and interpretable connection between abstract latent representations and the original input data.

However, evaluating explanations in a discrete latent space calls for a different approach than traditional XAI evaluation methods, which are typically designed for high-dimensional continuous input data. To address this, we introduce similar subsequence accuracy (SSA), a novel metric designed to quantify how well an XAI method identifies meaningful patterns corresponding to the training data within the latent space, thereby providing a quantitative framework for assessing the consistency and relevance of explanations in discretized representations. To evaluate this, we explore the application of XAI methods on DLRs of time series and compare their effectiveness to traditional approaches that operate directly on the original, continuous input space. Summarized, the main contributions of this paper are:

1. We adapt and implement a suite of XAI methods to operate effectively on DLRs of time series.
2. We show that XAI methods applied to discrete representations not only maintain performance comparable to those applied to the original input space but also produce significantly more compact explanations.
3. We introduce SSA, a novel metric for comparing XAI results across latent input spaces.

## 2 Related Work

Existing XAI methods for TSC can be broadly categorized into gradient-based, perturbation-based, and attention-based approaches, each providing distinct insights into model behavior and feature importance. Prominent gradient-based approaches include Gradient explanations [4], Saliency Map (SM) [24] and Integrated Gradients (IG) [25]. However, gradient-based methods can struggle to capture long-range temporal dependencies and complex non-linear relationships within time series data. Modern deep learning architectures, particularly transformer-based models, inherently utilize attention mechanisms, which can be leveraged to provide explainability. In this work, we refer to these techniques as Attention Maps (ATM). Post-hoc model-independent methods provide an alternative approach to explainability by generating local explanations without requiring modifications to the model itself. Techniques such as Local Interpretable Model-agnostic Explanations (LIME) [23] and Randomized Input Sampling for Explanation of Black-box Models (RISE) [22] achieve this by perturbing input features and analyzing the corresponding changes in the model's predictions. These methods are particularly useful for gaining instance-specific insights but can struggle with the nature of time series data. While widely used, applying these approaches to time series is challenging due to high dimensionality and temporal dependencies, potentially leading to fragmented or overly localized attributions that miss broader sequential patterns. This motivates exploring DLRs as a structured alternative for enhanced explainability. Our adapted methods operate directly in the latent space, where temporal dependencies are inherently preserved through the encoding process of discrete representations. This enables the application of standard XAI techniques without requiring task-specific perturbation strategies, thereby simplifying the explainability process while maintaining structured, sequence-aware explanations.

Various approaches exist for evaluating XAI methods and comparing their explanations. Adadi et al. [1] discuss evaluation as part of a broader review of XAI techniques, while Chromik et al. [7] focus on user studies as an evaluation framework. However, as Doshi-Velez et al. [9] emphasize, user studies often favor explanations that pass an initial 'face-validity' test—appearing intuitive to humans—without necessarily ensuring their correctness or generalizability. Similarly, anecdotal evaluations risk cherry-picking examples that seem plausible but lack rigorous validation. To ensure a more systematic and objective evaluation, we adopt quantitative evaluation methods as proposed by Zhou et al. [31] and especially Nauta et al. [19]. The latter introduced the "Co-12" properties, a framework for assessing explanation quality along multiple dimensions. These properties include:

1. Correctness, which measures how faithfully an explanation reflects the model's reasoning, assessed via randomization checks for model parameters [2] and explanations [17], as well as perturbation methods [5].
2. Output-completeness, evaluated using Preservation and Deletion Checks [8].
3. Coherence, assessed through agreement between different XAI methods [3].

In this paper, we apply Perturbation Analysis (PA) through incremental deletion and XAI method agreement. These evaluation methods are adaptable to DLRs and cover a broad spectrum of available Co-12 properties, such as correctness, completeness, consistency, and coherence of explanations. We detail the evaluation techniques further in Sect. 3.2.

# 3 Methodology

## 3.1 Discrete Representations for Time Series Classification

As previously discussed, we leverage DLRs to enhance interpretability in time series classification. To achieve this, we focus on VQ-VAE [21] as well as DVAE [28] and added SAX [16] as a baseline. SAX divides the time series into equal-sized segments (piecewise aggregate approximation), then maps these segments to discrete symbols using breakpoints that produce equiprobable regions under a Gaussian curve, allowing a lower-bounding distance measure to be defined. VQ-VAE and DVAE learn a structured, discrete representation of time series data but differ in their quantization strategy: VQ-VAE maps continuous latent vectors to a learned codebook of discrete codes, while DVAE models a categorical distribution over latent codes, enabling sampling via the Gumbel-Softmax trick [13]. In both approaches, a raw time series $x$ is transformed into a sequence of discrete tokens $z \in \{0, \dots, K-1\}^T$, where $K$ is the codebook size and $T$ the sequence length. A central aspect of both VQ-VAE and DVAE is their reliance on a reconstruction objective, where the models learn to reconstruct the original time series from their compressed, discrete latent codes, thereby minimizing the information loss from this quantization step. To retain temporal structure, we incorporate a patching mechanism, inspired by [20]. This technique associates each latent code $z$ with a 25-timestep segment of the original time series, balancing the need to capture fine-grained local features with the requirement for descriptive, explainable temporal features. In addition to reconstruction, the learned discrete representations are directly used for classification. Instead of processing raw time series data, the classifier operates on the sequence of discrete latent codes, providing a structured representation that may facilitate interpretability by reducing complexity and highlighting key temporal patterns.

## 3.2 Adaptation of XAI Methods and Evaluation Methods

As previously discussed, existing XAI methods are primarily designed for continuous input data and require adaptations to remain effective in the discrete latent space. While LIME and attention map (ATM) require no modifications, this section outlines how we adapted SM, IG, and RISE to ensure meaningful explanations within this transformed representation. A key challenge in this adaptation is the concept of masking, which plays a crucial role in IG and RISE but must be redefined for discrete token-based representations. Since each token in the latent space corresponds to specific features of the original time series, the standard approach of using an uninformative baseline—such as zero-padding in image data—is not directly applicable. Instead, we introduce an *unknown* token, analogous to the *MASK* token in BERT [15], to serve as an additional embedding and neutral placeholder, ensuring minimal disruption to the learned representations while still allowing us to probe model behavior.

**SM:** For SM, we adapted the method to operate on the discrete tokens by summing the gradients associated with each embedding vector. Let $e_i \in \mathbb{R}^d$ be the embedding vector corresponding to the $i$-th token with a dimension of $d$. The model's output for a given class c is denoted as $S_c(z)$. The saliency value $s_i$ for the $i$-th token is computed as follows:

$$s_i = \sum_{j=1}^{d} \left| \frac{\partial S_c(z)}{\partial e_{i,j}} \right| \tag{1}$$

where $e_{i,j}$ is the $j$-th element of the embedding vector $e_i$.

**IG:** IG requires a neutral baseline to accumulate gradients along a path from the baseline to the input token. We introduce the *unknown* token during training, replacing tokens with a probability of $p_{unk} \in \{0.1, 0.15\}$, similar to noise-injection strategies in NLP models [15]. Let $z'$ denote the baseline sequence where all tokens are replaced with the *unknown* token, and $e'$ its corresponding embedding representation. We compute the distances $\delta(e_i, e')$ between token embedding $e_i$ and $e'$. These distances guide the path for integrating gradients. For each token, we define a path with $m$ steps, $\alpha = 1, \ldots, m$, to the token embedding $e_i$ using linear interpolation. The integrated gradient $I_i$ for the $i$-th token is then:

$$IG_i = (e_i - e') \cdot \sum_{\alpha=1}^{m} \frac{\partial S_c(z^{(\alpha)})}{\partial z^{(\alpha)}} \times \frac{1}{m} \tag{2}$$

with $z^{(\alpha)} = e' + \frac{\alpha}{m}(e_i - e')$ where $\frac{\partial S_c(z^{(\alpha)})}{\partial z^{(\alpha)}}$ represents the gradient of the model's output with respect to the interpolated embedding at step $\alpha$ [25].

**RISE:** For RISE, we extend the masking mechanism using the *unknown* token to randomly perturb representations. Let $M \in \{0, 1\}^T$ be a binary mask where $M_i = 0$ indicates replacement of the $i$-th token with the *unknown* token. The RISE importance score $r_i$ for token $i$ is:

$$R = \frac{1}{|\mathcal{M}|} \sum_{M \in \mathcal{M}} P(c|z \odot M) \cdot M \tag{3}$$

where $\mathcal{M}$ is a set of random masks, $z \odot M$ represents the element-wise multiplication of the input sequence and the mask, and $P(c|z \odot M)$ is the probability of the class c given the masked input [22].

**Perturbation Analysis (PA):** To ensure a fair comparison between models operating on latent and non-latent representations, we adapt the evaluation procedures accordingly. Perturbation Analysis (PA) is used to validate whether the features identified by XAI methods are truly important for the model's decision-making process. By systematically altering the input and measuring the impact on predictions, we assess how strongly the model relies on the highlighted features. In non-latent models perturbation is applied to the raw time series by replacing a segment with a designated baseline value (e.g., zero values). In latent models we replace a sequence of discrete tokens with the *unknown* token, ensuring structural consistency within the representation.To establish a baseline for comparison, we introduce random perturbation (RND), where perturbations occur at randomly selected positions. For targeted perturbations, we rank features by importance according to each XAI method and incrementally perturb the highest-ranked features and compute the corresponding AUC score $\text{AUC}_{XAI}$ based on the F1 score across all classes, reflecting the model's robustness to arbitrary perturbations. To quantify the relative effectiveness of XAI-guided perturbations, we compute $\text{AUC}_{XAI-RND} = \text{AUC}_{RND} - \text{AUC}_{XAI}$. Thereby, a higher $\text{AUC}_{XAI-RND}$ indicates that the XAI method successfully identifies features with a strong influence on model decisions.

**XAI Method Agreement:** To evaluate explanation stability, we test for XAI method agreement [19], which measures explanation consistency across methods using cosine similarity. While disagreement is not necessarily negative, as different methods can offer valuable perspectives, explainability that relies more on the choice of XAI method than the model's actual process indicates instability in feature attribution.

**Similar Subsequence Accuracy (SSA):** While the previously introduced evaluation methods assess different aspects of explainability, they do not explicitly validate whether the identified features correspond to class-discriminative subsequences in the training data. The use of DLRs now enables a direct comparison between model explanations and the structured patterns learned from the data. To leverage this, we introduce SSA, a metric designed to evaluate the alignment between XAI explanations and recurring patterns in the training data. SSA quantifies how often the most influential subsequences identified by an XAI method reappear in training instances of the same class. This enables a structured assessment of whether model explanations capture genuine decision-relevant patterns or are influenced by spurious correlations. To compute SSA for a given XAI method, we follow the steps in Algorithm 1.

A high SSA indicates that the subsequence identified as important by the XAI method is frequently present in time series of the same class within the training data, and rarely present in time series of other classes. This suggests that the subsequence is a strong indicator of the class and that the XAI method has successfully identified a meaningful pattern in the latent space. Conversely, a low SSA suggests that the identified subsequence is not a reliable indicator of

**Algorithm 1.** SSA Algorithm

**Require:** Dataset $D^{train}_{latent}$ and $D^{test}_{latent}$, trained model $M$, XAI method $XAI$, subsequence length $l$, set of Classes $C$

**for** $c \in C$ **do** ▷ iterate all classes
  $Matches_c, Neighborhoods_c \leftarrow 0, 0$
  $Z_c \leftarrow$ all instances of $D^{test}_{latent}$ that belong to class c.
  **for** $(z, c) \in Z_c$ **do**
    $i \leftarrow \text{argmax}(XAI(M, z))$ ▷ argmax of importance
    $N(z) \leftarrow \emptyset$ ▷ initialize neighborhood
    **for** $(z', y') \in D^{train}_{latent}$ **do**
      **if** $z_{i:i+l} \in z'$ **then** ▷ check matching sequence
        $N(z) \leftarrow N(z) \cup \{z'\}$
        **if** $y' = c$ **then** $N(z)_{match} \leftarrow N(z)_{match} \cup \{z'\}$
        **end if**
      **end if**
    **end for**
    $Matches_c \mathrel{+}= |N(z)_{match}|$ ; $Neighborhoods_c \mathrel{+}= |N(z)|$
  **end for**
  $\text{SSA}_{\text{class}}(c) \leftarrow \frac{Matches_c}{Neighborhoods_c}$
**end for**
**return** $\text{SSA} = \frac{1}{|C|} \cdot \sum_{i=1}^{|C|} \text{SSA}_{\text{class}}(c_i)$

the class, either because it is not prevalent in the training data or because it is also common in other classes. Similarly to the PA, we additionally compute the score on a random baseline highlighting arbitrary subsequences. Due to the heavy preprocessing and temporal dependencies in our datasets, we only focus on neighbors that feature similar subsequences at the same positions.

In essence, SSA provides a quantitative measure of how well the DLRs capture class-specific patterns that are consistent with the historical data. With regards to the Co-12 properties defined by Nauta et al. [19], SSA measures explanation correctness by taking the alignment between true labels and model predictions into account, as well as consistency by ensuring that similar subsequences in the latent space lead to consistent predictions and explanations.

## 4 Experiments

We use three TSC datasets from industrial and medical domains to evaluate our approach. These datasets provide a diverse set of challenges, covering both univariate and multivariate time series and classification tasks with varying class imbalances. The Welding dataset [10] provides multivariate current and voltage time series from arc welding processes, labeled as either "substandard" (43%) or "satisfactory" (57%) welds. The CNC dataset [27] presents multivariate tri-axial accelerometer data from CNC machine operations with a moving average filter applied to reduce its size, followed by segmentation into smaller chunks. The ECG dataset is a combination of the MIT-BIH Arrhythmia [18] and PTB Diagnostic ECG [6] databases, preprocessed and segmented into individual heartbeats

**Table 1.** Combined results for all datasets. The table is structured with AUC metrics in the top rows and XAI methods as subcolumns within each model. For each dataset, the highest values are highlighted in bold.

| Dataset | Model Name | $\mathbf{AUC_{XAI-RND}}$ | | | | | **CS** |
|---|---|---|---|---|---|---|---|
| | | SM | IG | RISE | LIME | ATM | Agreement |
| CNC | DLinear | $0.0 \pm 0.0$ | $0.0 \pm 0.0$ | $-0.01 \pm 0.0$ | $-0.0 \pm 0.0$ | – | $0.84 \pm 0.0$ |
| | DVAE MLP | $0.03 \pm 0.03$ | $-0.03 \pm 0.05$ | $0.06 \pm 0.03$ | $0.03 \pm 0.02$ | – | $0.88 \pm 0.01$ |
| | DVAE Tr | $-0.01 \pm 0.01$ | $0.0 \pm 0.0$ | $0.0 \pm 0.0$ | $-0.0 \pm 0.0$ | $-0.01 \pm 0.02$ | $0.78 \pm 0.01$ |
| | MLP | $-0.0 \pm 0.01$ | $0.03 \pm 0.0$ | $-0.0 \pm 0.01$ | $0.01 \pm 0.01$ | – | $0.85 \pm 0.0$ |
| | TS Tr | $0.0 \pm 0.0$ | $0.0 \pm 0.0$ | $0.0 \pm 0.0$ | $0.0 \pm 0.0$ | $0.0 \pm 0.0$ | $0.83 \pm 0.01$ |
| | TimesNet | $0.0 \pm 0.01$ | $0.04 \pm 0.01$ | $0.01 \pm 0.0$ | $0.02 \pm 0.01$ | – | $0.75 \pm 0.02$ |
| | SAX MLP | $0.01 \pm 0.01$ | $0.01 \pm 0.01$ | $0.01 \pm 0.01$ | $0.04 \pm 0.01$ | – | $\mathbf{0.89 \pm 0.01}$ |
| | VQ-VAE MLP | $\mathbf{0.08 \pm 0.03}$ | $-0.03 \pm 0.05$ | $0.06 \pm 0.01$ | $-0.02 \pm 0.02$ | – | $0.8 \pm 0.01$ |
| | VQ-VAE Tr | $-0.05 \pm 0.01$ | $0.01 \pm 0.01$ | $0.0 \pm 0.01$ | $-0.02 \pm 0.0$ | $-0.01 \pm 0.05$ | $0.78 \pm 0.02$ |
| ECG | DLinear | $0.06 \pm 0.01$ | $0.03 \pm 0.01$ | $0.12 \pm 0.01$ | $0.06 \pm 0.0$ | – | $0.69 \pm 0.0$ |
| | DVAE MLP | $0.11 \pm 0.01$ | $0.03 \pm 0.03$ | $0.14 \pm 0.0$ | $0.17 \pm 0.01$ | – | $0.87 \pm 0.01$ |
| | DVAE Tr | $0.06 \pm 0.01$ | $0.07 \pm 0.01$ | $0.07 \pm 0.02$ | $0.08 \pm 0.02$ | $0.02 \pm 0.06$ | $0.73 \pm 0.05$ |
| | MLP | $0.19 \pm 0.0$ | $\mathbf{0.21 \pm 0.01}$ | $0.2 \pm 0.01$ | $0.12 \pm 0.0$ | – | $0.71 \pm 0.02$ |
| | TS Tr | $0.09 \pm 0.02$ | $0.12 \pm 0.02$ | $0.04 \pm 0.03$ | $0.01 \pm 0.03$ | $0.0 \pm 0.12$ | $0.7 \pm 0.06$ |
| | TimesNet | $0.07 \pm 0.01$ | $0.09 \pm 0.01$ | $0.11 \pm 0.0$ | $0.01 \pm 0.01$ | – | $0.69 \pm 0.0$ |
| | SAX MLP | $0.07 \pm 0.01$ | $0.02 \pm 0.01$ | $0.07 \pm 0.0$ | $0.07 \pm 0.01$ | – | $0.87 \pm 0.01$ |
| | VQ-VAE MLP | $0.15 \pm 0.05$ | $0.04 \pm 0.07$ | $0.12 \pm 0.02$ | $0.17 \pm 0.02$ | – | $\mathbf{0.9 \pm 0.02}$ |
| | VQ-VAE Tr | $0.11 \pm 0.05$ | $0.08 \pm 0.01$ | $0.14 \pm 0.02$ | $0.11 \pm 0.02$ | $0.06 \pm 0.04$ | $0.76 \pm 0.01$ |
| Welding | DLinear | $0.12 \pm 0.01$ | $0.12 \pm 0.01$ | $\mathbf{0.25 \pm 0.01}$ | $0.02 \pm 0.0$ | – | $0.69 \pm 0.0$ |
| | DVAE MLP | $0.07 \pm 0.03$ | $0.01 \pm 0.01$ | $0.1 \pm 0.02$ | $0.12 \pm 0.03$ | – | $0.87 \pm 0.01$ |
| | DVAE Tr | $0.04 \pm 0.04$ | $0.03 \pm 0.01$ | $0.11 \pm 0.01$ | $0.16 \pm 0.02$ | $0.07 \pm 0.04$ | $0.78 \pm 0.04$ |
| | MLP | $0.04 \pm 0.02$ | $0.03 \pm 0.02$ | $0.22 \pm 0.02$ | $0.03 \pm 0.01$ | – | $0.82 \pm 0.0$ |
| | TS Tr | $0.13 \pm 0.04$ | $0.14 \pm 0.03$ | $0.03 \pm 0.02$ | $-0.03 \pm 0.06$ | $0.07 \pm 0.05$ | $0.65 \pm 0.09$ |
| | TimesNet | $0.05 \pm 0.02$ | $0.03 \pm 0.01$ | $0.21 \pm 0.01$ | $-0.01 \pm 0.04$ | – | $0.66 \pm 0.03$ |
| | SAX MLP | $0.02 \pm 0.02$ | $0.0 \pm 0.02$ | $0.1 \pm 0.01$ | $0.1 \pm 0.03$ | – | $\mathbf{0.88 \pm 0.01}$ |
| | VQ-VAE MLP | $0.09 \pm 0.06$ | $0.05 \pm 0.03$ | $0.14 \pm 0.01$ | $0.09 \pm 0.02$ | | $0.87 \pm 0.01$ |
| | VQ-VAE Tr | $0.02 \pm 0.03$ | $0.03 \pm 0.01$ | $0.11 \pm 0.02$ | $0.09 \pm 0.01$ | $0.05 \pm 0.04$ | $0.76 \pm 0.01$ |

[14] classified into five categories. Each time series is standardized to a uniform length, determined by the longest sequence within each dataset.

We evaluate the impact of XAI methods on both raw time series data and learned DLRs. For the non-latent models, we evaluated DLinear [30] TimesNet [29], a multilayer perceptron (MLP) and a decoder-based time series Transformer. We utilized VQ-VAE [21], DVAE [28], and SAX [16] to generate DLRs. Classification on these latent spaces was performed using a Transformer architecture (following [11,12]) and an MLP for comparison. We report the mean and standard deviation of explainability metrics across five runs to ensure robustness and reproducibility. To achieve a broad evaluation, we employed SM, IG, LIME,

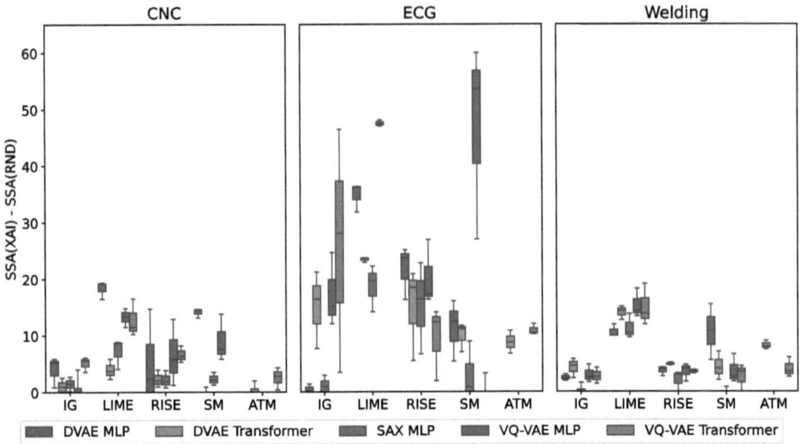


**Fig. 1.** Distribution of SSA improvement over random baselines across the CNC, ECG, and Welding datasets.

RISE, and ATM. The best hyperparameters and the full code are available on GitHub[1].

The experiments aim to evaluate the robustness of latent and non-latent models under perturbations, analyze the agreement between different XAI methods, and assess the stability of explanations across multiple training runs. Table 1 presents the results of the PA revealing that the effectiveness of a given XAI method varies depending on both the model architecture and the dataset. Notably, VQ-VAE combined with an MLP using SM achieves the highest score on the CNC dataset, while the MLP with IG outperforms other models on the ECG dataset, and DLinear with RISE achieves the best performance on the Welding dataset. RISE and LIME appear to perform particularly well with latent models, achieving the best scores for this model type on the ECG and Welding datasets. These findings indicate that the efficacy of XAI methods is not uniform but instead depends on the interplay between the model and the data. The last column of Table 1 illustrates the agreement between different XAI methods when applied to the same model and dataset. The latent models, particularly DVAE with an MLP on the CNC and Welding datasets and VQ-VAE with an MLP on the ECG dataset, exhibit the highest degree of agreement. Specifically, the VQ-VAE MLP model consistently shows high agreement among different XAI methods, indicating that the choice of XAI method is less critical when using this particular model and that the model's predictions is more robust.

**SSA Results:** The mean number of neighbors per instance (Sect. 3.2) drastically decreases as subsequence length increases, which directly affects SSA validity. Therefore, to evaluate the SSA results, we consider no subsequences longer

[1] https://github.com/tmdt-buw/EXCODER.

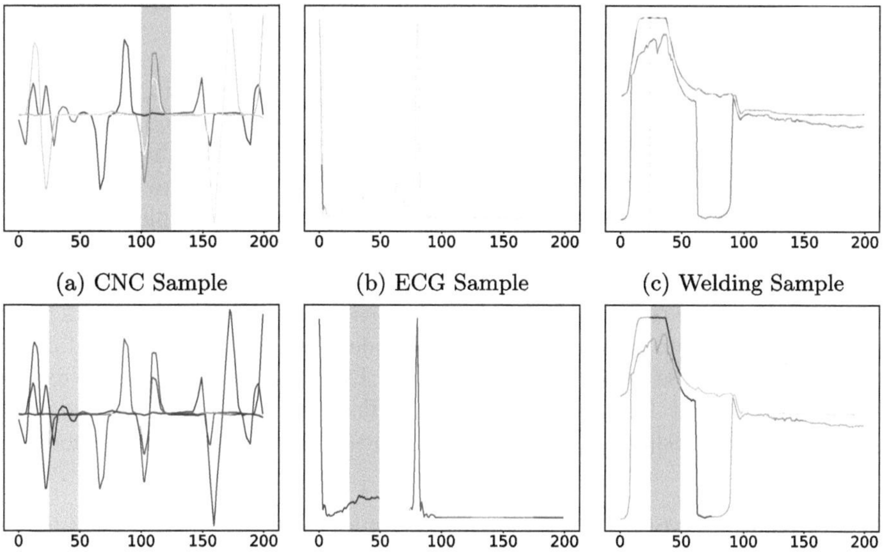


(a) CNC Sample (b) ECG Sample (c) Welding Sample

(d) SSA: 93.82% with 1941 neighbors (e) SSA: 98.9% with 1464 neighbors (f) SSA: 83.3% with 2322 neighbors

**Fig. 2.** Explanations for random samples from CNC (a, d), ECG (b, e), and Welding (c, f) datasets. Top row (a–c): Explanations from best-performing models (via perturbation), with darker colors indicating higher feature importance. Bottom row (d–f): Explanations with highest SSA, highlighting the subsequence used for SSA calculation.

than three tokens. In order to provide an overview for all tested model architectures and datasets, we compare a mean over three different subsequence lengths for each method to the random baseline. Figure 1 indicates that according to the mean difference $SSA(XAI) - SSA(RND)$, explanations provided by LIME consistently outperform other XAI methods for all datasets and model architectures. Using LIME, the VQ-VAE MLP has the best SSA score for the Welding and ECG dataset, while the DVAE MLP works best for the CNC dataset. Figure 2 demonstrates the advantages of subsequence-based explanations over point-wise approaches. While point-based methods often fail to clearly identify crucial features, the patch-based representations highlight relevant subsequences with greater precision. This allows for a more accurate identification of important features within the time series. Moreover, the highlighted regions, combined with the SSA percentage (representing the prevalence of the same subsequence and label combination in the training data), provide a two-fold benefit: they indicate which part of the time series is important and offer insight into the reliability of the prediction based on the training data distribution.

## 5 Conclusion

To overcome the limitations of applying XAI to raw time series data, we proposed an approach that utilizes DLRs to enhance the interpretability and effectiveness of XAI methods. We demonstrated that the adapted XAI methods achieve comparable performance to traditional XAI approaches operating on raw time series, while producing significantly more compact and structured explanations. We introduced SSA, a novel metric designed to assess whether the features identified by an XAI method genuinely represent class-discriminative information in the training data rather than misleading correlations. The experimental results highlight that there is no universally superior model-XAI combination, reinforcing the importance of dataset-specific evaluations. Furthermore, latent space experiments suggest that feature attributions in discrete representations are more stable with LIME displaying the best SSA score for every dataset and model.

## References

1. Adadi, A., Berrada, M.: Peeking inside the black-box: a survey on explainable artificial intelligence (xai). IEEE Access **6**, 52138–52160 (2018)
2. Adebayo, J., Muelly, M., Liccardi, I., Kim, B.: Debugging tests for model explanations. In: Proceedings of the 34th International Conference on Neural Information Processing Systems, NIPS '20. Curran Associates Inc., Red Hook (2020)
3. Ancona, M., Oztireli, C., Gross, M.: Explaining deep neural networks with a polynomial time algorithm for Shapley value approximation. In: International Conference on Machine Learning, pp. 272–281. PMLR (2019)
4. Baehrens, D., Schroeter, T., Harmeling, S., Kawanabe, M., Hansen, K., Müller, K.R.: How to explain individual classification decisions. J. Mach. Learn. Res. **11**, 1803–1831 (2010)
5. Bhatt, U., Weller, A., Moura, J.M.: Evaluating and aggregating feature-based model explanations. arXiv preprint arXiv:2005.00631 (2020)
6. Bousseljot, R., Kreiseler, D., Schnabel, A.: Nutzung der ekg-signaldatenbank cardiodat der ptb über das internet (1995)
7. Chromik, M., Schuessler, M.: A taxonomy for human subject evaluation of black-box explanations in xai. Exss-atec@ iui **1**, 1–7 (2020)
8. Dhurandhar, A., et al.: Explanations based on the missing: towards contrastive explanations with pertinent negatives. In: Advances in Neural Information Processing Systems, p. 31 (2018)
9. Doshi-Velez, F., Kim, B.: Considerations for evaluation and generalization in interpretable machine learning. In: Explainable and Interpretable Models in Computer Vision and Machine Learning, , pp. 3–17 (2018)
10. Hahn, Y., et al.: Metal arc welding (2025). 10/g98352
11. Hahn, Y., et al.: Quality prediction in arc welding: Leveraging transformer models and discrete representations from vector quantised-vae. In: Proceedings of the 33rd ACM International Conference on Information and Knowledge Management, CIKM '24, pp. 4522–4529. ACM (2024). 10.1145/3627673.3680031
12. Hahn, Y., Voets, J., Königsfeld, A., Tercan, H., Meisen, T.: Out of distribution detection for efficient continual learning in quality prediction for arc

welding. In: Proceedings of the 34th ACM International Conference on Information and Knowledge Management, CIKM '25, pp. 5699–5706. ACM (2025). 10.1145/3746252.3761535
13. Jang, E., Gu, S., Poole, B.: Categorical reparameterization with gumbel-softmax. arXiv preprint arXiv:1611.01144 (2016)
14. Kachuee, M., Fazeli, S., Sarrafzadeh, M.: ECG heartbeat classification: a deep transferable representation. In: 2018 IEEE International Conference on Healthcare Informatics (ICHI), pp. 443–444. IEEE (2018)
15. Kenton, J.D.M.W.C., Toutanova, L.K.: Bert: pre-training of deep bidirectional transformers for language understanding. In: Proceedings of NAACL-HLT, Minneapolis, Minnesota, vol. 1 (2019)
16. Lin, J., Keogh, E., Wei, L., Lonardi, S.: Experiencing sax: a novel symbolic representation of time series. Data Min. Knowl. Disc. **15**, 107–144 (2007)
17. Mohankumar, A.K., Nema, P., Narasimhan, S., Khapra, M.M., Srinivasan, B.V., Ravindran, B.: Towards transparent and explainable attention models. arXiv preprint arXiv:2004.14243 (2020)
18. Moody, G.B., Mark, R.G.: The impact of the MIT-BIH arrhythmia database. IEEE Eng. Med. Biol. Mag. **20**(3), 45–50 (2001)
19. Nauta, M., et al.: From anecdotal evidence to quantitative evaluation methods: a systematic review on evaluating explainable ai. ACM Comput. Surv. **55**(13s), 1–42 (2023)
20. Nie, Y., Nguyen, H., Sinthong, N., Kalagnanam, P., J.: A time series is worth 64 words: long-term forecasting with transformers. In: International Conference on Learning Representations (2023)
21. van den Oord, A., Vinyals, O., Kavukcuoglu, K.: Neural discrete representation learning. In: Proceedings of the 31st International Conference on Neural Information Processing Systems, NIPS'17, pp. 6309–6318. Curran Associates Inc., Red Hook (2017)
22. Petsiuk, V.: Rise: randomized input sampling for explanation of black-box models. arXiv preprint arXiv:1806.07421 (2018)
23. Ribeiro, M.T., Singh, S., Guestrin, C.: "why should i trust you?": explaining the predictions of any classifier. In: Proceedings of the 22nd ACM SIGKDD International Conference on Knowledge Discovery and Data Mining, KDD '16, pp. 1135–1144. Association for Computing Machinery, New York 2(016). https://doi.org/10.1145/2939672.2939778
24. Simonyan, K., Vedaldi, A., Zisserman, A.: Deep inside convolutional networks: visualising image classification models and saliency maps. In: Bengio, Y., LeCun, Y. (eds.) 2nd International Conference on Learning Representations, ICLR 2014, Banff, AB, Canada, 14–16 April 2014, Workshop Track Proceedings (2014). https://doi.org/10.48550/arXiv.1312.6034
25. Sundararajan, M., Taly, A., Yan, Q.: Axiomatic attribution for deep networks. In: International Conference on Machine Learning, pp. 3319–3328. PMLR (2017)
26. Theissler, A., Spinnato, F., Schlegel, U., Guidotti, R.: Explainable ai for time series classification: a review, taxonomy and research directions. IEEE Access **10**, 100700–100724 (2022)
27. Tnani, M.A., Feil, M., Diepold, K.: Smart data collection system for brownfield cnc milling machines: a new benchmark dataset for data-driven machine monitoring. Procedia CIRP **107**, 131–136 (2022). https://doi.org/10.1016/j.procir.2022.04.022
28. Vahdat, A., Andriyash, E., Macready, W.G.: Dvae#: discrete variational autoencoders with relaxed Boltzmann priors. In: Proceedings of the 32nd International

Conference on Neural Information Processing Systems, NIPS'18, pp. 1869–1878. Curran Associates Inc., Red Hook (2018)
29. Wu, H., Hu, T., Liu, Y., Zhou, H., Wang, J., Long, M.: Timesnet: temporal 2d-variation modeling for general time series analysis. arXiv preprint arXiv:2210.02186 (2022)
30. Zeng, A., Chen, M., Zhang, L., Xu, Q.: Are transformers effective for time series forecasting? In: Proceedings of the AAAI Conference on Artificial Intelligence, vol. 37, pp. 11121–11128 (2023)
31. Zhou, J., Gandomi, A.H., Chen, F., Holzinger, A.: Evaluating the quality of machine learning explanations: a survey on methods and metrics. Electronics **10**(5), 593 (2021)

# Neural-Symbolic Logic Query Answering in Non-euclidean Space

Lihui Liu(✉)

Wayne State University, Detroit, MI, USA
hw6926@wayne.edu

**Abstract.** Answering complex first-order logic (FOL) queries on knowledge graphs is essential for reasoning. Symbolic methods offer interpretability but struggle with incomplete graphs, while neural approaches generalize better but lack transparency. Neural-symbolic models aim to integrate both strengths but often fail to capture the hierarchical structure of logical queries, limiting their effectiveness. We propose HYQNET, a neural-symbolic model for logic query reasoning that fully leverages hyperbolic space. HYQNET decomposes FOL queries into relation projections and logical operations over fuzzy sets, enhancing interpretability. To address missing links, it employs a hyperbolic GNN-based approach for knowledge graph completion in hyperbolic space, effectively embedding the recursive query tree while preserving structural dependencies. By utilizing hyperbolic representations, HYQNET captures the hierarchical nature of logical projection reasoning more effectively than Euclidean-based approaches. Experiments on three benchmark datasets demonstrate that HYQNET achieves strong performance, highlighting the advantages of reasoning in hyperbolic space.

## 1 Introduction

A knowledge graph organizes entities and their relationships, making it a valuable tool for applications like question answering [19], recommendation systems [6], and computer vision [8]. Reasoning over knowledge graphs helps infer new knowledge and answer queries based on existing data, with answering complex First-Order Logic (FOL) queries being a key challenge. FOL queries involve logical operations such as existential quantifiers ($\exists$), conjunction ($\wedge$), disjunction ($\vee$), and negation ($\neg$). For example, the question "Which universities do Turing Award winners in deep learning work at?" can be formulated as an FOL query, as shown in Fig. 1.

Most existing methods rely on neural networks to model logical operations and learn query embeddings to find answers. Approaches like [5,7,20,21] represent queries using boxes, beta distributions, or points, but their reasoning process is often opaque since neural networks function as black boxes. On the other hand, symbolic methods provide clear interpretability by explicitly deriving answers from stored facts or using subgraph matching. However, they struggle with incomplete knowledge graphs, which limits their applicability in real-world scenarios.

Recent research has explored combining neural reasoning with symbolic techniques to achieve both strong generalization and interpretability. For example, GNN-QE [28]


R. C.-W. Wong et al. (Eds.): PAKDD 2026, LNAI 16597, pp. 353–365, 2026.
https://doi.org/10.1007/978-981-92-1300-9_28

employs a graph neural network (GNN) for projection operations and integrates symbolic graph matching to enhance robustness and interpretability. However, GNN-QE encodes local subtrees in Euclidean space, which is suboptimal for capturing hierarchical relationships. Hyperbolic embeddings provide a more natural representation for such structures.

In this paper, we propose a hyperbolic GNN for answering complex First-Order Logic (FOL) queries on knowledge graphs. Unlike Euclidean-based approaches, our model leverages hyperbolic embeddings with learned curvature at each layer, enabling it to more effectively capture hierarchical query structures. Building upon prior work [28], we decompose FOL queries into expressions over fuzzy sets, where relation projections are modeled using the hyperbolic GNN with learned curvature. This design improves reasoning accuracy. By operating in hyperbolic space, our model naturally preserves the tree-like structure of logical projections, facilitating more effective representation learning. Additionally, the learned curvature allows the GNN to adapt to the geometry of the knowledge graph, improving entity projections while maintaining interpretability through fuzzy set-based logic operations. We evaluate our method on three standard datasets for FOL queries. Our experiments demonstrate that our approach consistently outperforms existing baselines across various query types, achieving state-of-the-art performance on all datasets.

In summary, we make the following contributions:

- We propose a novel method that leverages hyperbolic embeddings with learned curvature to better capture hierarchical structures in logical projections. By learning curvature at each layer, our model gains increased flexibility, leading to improved reasoning performance.
- We demonstrate that our method achieves state-of-the-art performance on three standard FOL query datasets.

## 2 Problem Definition

In this section, we introduce the background knowledge of FOL queries on knowledge graphs and fuzzy sets.

### 2.1 First-Order Logic Queries on Knowledge Graphs

Given a set of entities $V$ and a set of relations $R$, a knowledge graph $G = (V, E, R)$ is a collection of triplets $E = \{(h_i, r_i, t_i)\} \subseteq V \times R \times V$, where each triplet is a fact from head entity $h_i$ to tail entity $t_i$ with the relation type $r_i$.

A FOL query on a knowledge graph is a formula composed of constants (denoted with English terms), variables (denoted with $a$, $b$, $c$), relation symbols (denoted with $R(a, b)$), and logic symbols ($\exists$, $\wedge$, $\vee$, $\neg$). In the context of knowledge graphs, each constant or variable is an entity in $V$. A variable is *bounded* if it is quantified in the expression, and *free* otherwise. Each relation symbol $R(a, b)$ is a binary function that indicates whether there is a relation $R$ between a pair of constants or variables. For logic symbols, we consider queries that contain conjunction ($\wedge$), disjunction ($\vee$), negation ($\neg$), and existential quantification ($\exists$)[1].

Figure 1 illustrates the FOL query for the natural language question "where did Canadian citizens with Turing Award graduate?". Given a FOL query, the goal is to find answers to the free variables, such that the formula is true.

$$q = V_?, \exists V: Win(TuringAward, V) \wedge Citizen(Canada, V) \wedge Graduate(V, V_?)$$

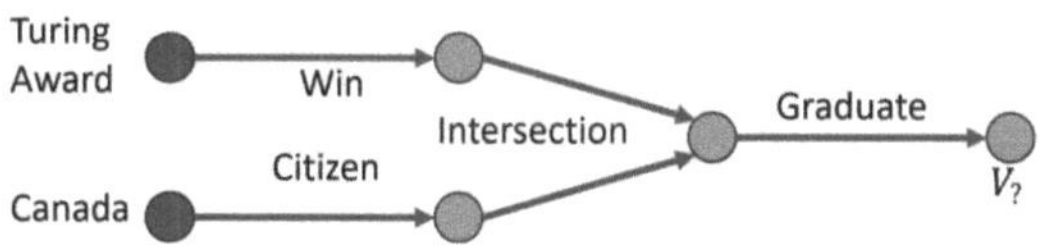


**Fig. 1.** An example of logical query.

## 2.2 Fuzzy Logic Operations

Fuzzy sets [10] are a continuous relaxation of sets whose elements have degrees of membership. A fuzzy set $A = (U, x)$ contains a universal set $U$ and a membership function $x : U \rightarrow [0, 1]$. For each $u \in U$, the value of $x(u)$ defines the degree of membership (i.e., probability) for $u$ in $A$. Similar to Boolean logic, fuzzy logic defines three logic operations: *AND*, *OR*, and *NOT*, over the real-valued degree of membership. There are several alternative definitions for these operations, such as product fuzzy logic, Gödel fuzzy logic, and Łukasiewicz fuzzy logic.

In this paper, fuzzy sets are used to represent the assignments of variables in FOL queries, where the universe $U$ is always the set of entities $V$ in the knowledge graph. Since the universe is a finite set, we represent the membership function $x$ as a vector $x$. We use $x_u$ to denote the degree of membership for element $u$. For simplicity, we abbreviate a fuzzy set $A = (U, x)$ as $x$ throughout the paper.

## 2.3 Hyperbolic Space and Its Embedding

Hyperbolic space is a geometric setting where the parallel postulate of Euclidean geometry does not hold. Instead of parallel lines remaining equidistant, they diverge exponentially. This space, often denoted as $H^n$, exhibits a constant negative curvature, leading to unique geometric properties such as rapid expansion of volume with distance and altered notions of angles and distances. These properties make hyperbolic space particularly useful for representing hierarchical structures and relational data, where entities grow exponentially as one moves outward.

[1] For simplicity, we focus on this subset of FOL operations in this work.

**Mathematical Representation via the Poincaré Model.** One of the most practical ways to work with hyperbolic space is through the Poincaré ball model, which maps it onto the interior of a unit sphere in $\mathbb{R}^n$. Given a point $x$ within this unit ball, distances are not measured using standard Euclidean norms but rather through a modified metric that accounts for hyperbolic curvature. Specifically, the hyperbolic distance $d(x, y)$ between two points $x, y$ is computed as:

$$d(x,y) = \operatorname{arccosh}\left(1 + 2\frac{\|x-y\|^2}{(1-\|x\|^2)(1-\|y\|^2)}\right).$$

This metric ensures that distances grow exponentially as points move outward, mirroring the behavior of many complex networks and hierarchical systems.

**Tangent Space and Exponential Map.** Although hyperbolic space is fundamentally different from Euclidean space, computations can be simplified by leveraging the tangent space at a given point. The tangent space at a point $x \in D^n$ is a local Euclidean approximation of hyperbolic space, allowing for operations such as optimization and vector transformations. The exponential map $\exp_x : T_x H^n \to H^n$ projects vectors from the tangent space back onto the manifold, while its inverse, the logarithmic map $\log_x : H^n \to T_x H^n$, maps points from the manifold to the tangent space:

$$\exp_x(v) = x + \tanh(\|v\|)\frac{v}{\|v\|}$$

$$\log_x(y) = \frac{\operatorname{arctanh}(\|y-x\|)}{\|y-x\|}(y-x).$$

## 3 Methodology

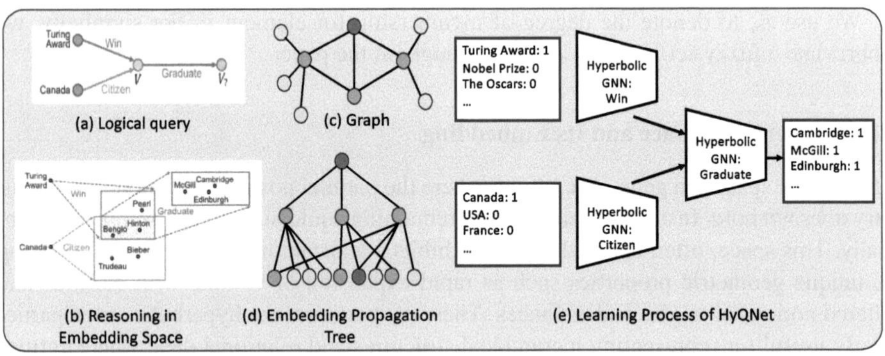


**Fig. 2.** Overview of HYQNET.

Here we present our model HYQNET. The high-level idea of HYQNET is to first decompose a FOL query into an expression of four basic operations (relation projection, conjunction, disjunction, and negation) over fuzzy sets, then parameterize the relation projection with a hyperbolic GNN with learned curvature adapted from KG completion, and instantiate the logic operations with product fuzzy logic operations.

Given a FOL query, the first step is to transform it into a structured computation process, where the query is broken down into a sequence of fundamental operations. We represent the query as a directed computation graph, starting from anchor entities and progressively applying transformations until the final answer set is obtained. The decomposition follows a structured process. First, relation projections identify the entities that satisfy specific relational constraints. Given an initial fuzzy set of head entities, the projection operation maps them to potential tail entities through the corresponding relation. Once relation projections establish the core structure, logical operations refine the intermediate results. Conjunction enforces conditions that must be met simultaneously, effectively taking the intersection of fuzzy sets. Disjunction broadens the set of possible answers by considering alternative paths, while negation eliminates entities that contradict the query constraints. By iteratively applying these operations, we propagate information through the computation graph, ultimately producing a fuzzy set representing the final answer. This structured execution ensures interpretability at each step while maintaining the flexibility needed for reasoning in knowledge graphs. We introduce the details of each logical operation below.

**4.2. Hyperbolic Neural Relation Projection.** To solve complex queries on incomplete knowledge graphs, we train a neural model to perform relation projection, defined as $y = P_q(x)$, where $x$ is the input fuzzy set, $q$ is the given relation, and $y$ is the output fuzzy set. The goal of relation projection is to generate the potential answer set $y$, consisting of entities that could be connected to $x$ via relation $q$. This task aligns with the knowledge graph completion problem, where missing links must be inferred. Specifically, the neural relation projection model aims to predict the fuzzy set of tail entities $y$ given the fuzzy set of head entities $x$ and the relation $q$, even in the presence of incomplete knowledge.

Many methods have been proposed to model projection operations, including [7, 20, 21]. Recently, [28] introduced a GNN-based framework for knowledge graph completion. When applying GNN models, it has been shown that their underlying mechanism is equivalent to learning node embeddings based on a recursive tree structure. An example is illustrated in Fig. 2 (d), where the goal is to predict a suitable answer for the query (Blue node, $r$, ?), where $r$ represents an arbitrary relation. However, the corresponding recursive tree structure grows exponentially with the number of propagation steps. Although naively applying GNNs as in [28] is feasible, it is suboptimal because standard GNNs struggle to effectively capture the exponential growth of node neighborhoods. In contrast, hyperbolic embeddings are better suited for encoding hierarchical structures, as they naturally capture the exponential growth of node neighborhoods. To address this, we propose leveraging hyperbolic embeddings with learned curvature for relation projection. Inspired by this idea, we develop a scalable hyperbolic GNN framework to enhance the effectiveness of relation projection in knowledge graph completion.

**Hyperbolic Relation Projection.** Our goal is to design a hyperbolic GNN model that predicts a fuzzy set of tail entities given a fuzzy set of head entities and a relation. Given a head entity $u$ and a projection relation $q$, we use the following iteration to compute a representation $h_v$ for each entity $v \in V$ with respect to the source entity $u$ according to the recursive tree structure. Given the source entity set $x_v$, we initalize its embedding by $h_v^{(0)} \leftarrow x_v q$, where the $x_v$ is the probability of entity $v$ in $x$. For each layer in the tree, the learned process can be denoted as

$$h_v^{(t)} \leftarrow \text{AGGREGATE}(\{\text{MESSAGE}(h_z^{(t-1)}(z, r, v)) \mid (z, r, v) \in E(v)\})$$

When utilizing hyperbolic embeddings, we iteratively update node embeddings in the knowledge graph. The update process is given by:

$$h_v^{(t+1)} = \sigma\left(\exp_c^0\left(\tilde{A}\,\mathbf{W}_t \log_c^0(h_z^{(t)})\right)\right), \qquad (z, r, v) \in E(v).$$

Here, the exponential map $\exp_c^x$ maps a tangent vector $v \in T_x D_c^n$ to the hyperbolic manifold:

$$\exp_c^x(v) = x \oplus_c \tanh\left(\sqrt{c}\frac{\|v\|}{\sqrt{c}}\right)\frac{v}{\|v\|}.$$

Conversely, the logarithmic map $\log_c^x$ transforms a point on the hyperbolic manifold back to the tangent space:

$$\log_c^x(y) = \frac{2\sqrt{c}\,\text{arctanh}(\sqrt{c}\|-x \oplus_c y\|)}{\|-x \oplus_c y\|}(-x \oplus_c y).$$

In practice, $\exp_c^0$ and $\log_c^0$ are used to efficiently convert between Euclidean vectors and their hyperbolic representations in the Poincaré ball model.

To apply the hyperbolic GNN framework for relation projection, we propagate the representations for $T$ layers. Then, we take the representations in the last layer and pass them into a multi-layer perceptron (MLP) $f$ followed by a sigmoid function $\sigma$ to predict the fuzzy set of tail entities: $P_q(x) = \sigma\left(f(h^{(T)})\right)$.

## 3.1 Fuzzy Logic Operations

In knowledge graph reasoning, operations like $C(x, y)$, $D(x, y)$, and $N(x)$ play an essential role in combining the results of multiple relation projections, thus forming the input fuzzy set for the next projection. These operations must ideally satisfy foundational logical principles, such as commutativity, associativity, and non-contradiction. While many prior approaches [7,20,21] have introduced geometric operations to model these logic operations within the embedding space, such neural operators are not always reliable in adhering to logical laws. As a result, chaining these operators can introduce errors.

Building on the work of [28], we employ product fuzzy logic operations to model conjunction, disjunction, and negation. Specifically, given two fuzzy sets $x, y \in [0, 1]^V$, we define the operations as:

$$C(x, y) = x \odot y \qquad \text{(Conjunction)}$$

$$D(x, y) = x + y - x \odot y \qquad \text{(Disjunction)}$$

$$N(x) = 1 - x \qquad \text{(Negation)}$$

In these equations, $\odot$ denotes element-wise multiplication, and 1 is a vector of ones (representing the universe).

### 3.2 Model Training

In terms of model training, we follow the approach used in previous studies [20,21], aiming to minimize the binary cross-entropy loss function, defined as:

$$L = -\frac{1}{|A_Q|} \sum_{a \in A_Q} \log p(a \mid Q) - \frac{1}{|V \setminus A_Q|} \sum_{a' \in V \setminus A_Q} \log\left(1 - p(a' \mid Q)\right)$$

Here, $A_Q$ refers to the set of correct answers for query $Q$, while $V \setminus A_Q$ represents the set of entities in the knowledge graph that are not part of $A_Q$. The terms $p(a \mid Q)$ and $p(a' \mid Q)$ indicate the predicted probabilities for $a$ and $a'$ being correct answers, respectively.

**Theorem 1.** *The proposed* HyQNet *consistently outperforms—or performs on par with—existing GNN-based logical query reasoning models that rely on Euclidean embeddings.*

*Proof.* Euclidean embeddings correspond to a special case of HyQNet where the learned curvature approaches zero. As HyQNet generalizes Euclidean embeddings by allowing variable curvature, it is capable of capturing a broader range of geometric structures. Consequently, HyQNet consistently matches or surpasses the performance of models using Euclidean embeddings.

## 4 Experiment

We evaluate HyQNet on three widely used knowledge graph datasets: FB15k, FB15k-237, and NELL995 [28]. To ensure consistency with prior studies, we adopt the benchmark FOL queries provided by BetaE [21], which include 9 EPFO query structures and 5 additional queries incorporating negation. Our model is trained on 10 query types (1p, 2p, 3p, 2i, 3i, 2in, 3in, inp, pni, pin), following the setup used in previous research [20,21,28]. For evaluation, we test HyQNet on both the 10 training query types and 4 unseen query types (ip, pi, 2u, up) to assess its generalization ability.

**Evaluation Protocol.** We follow the evaluation protocol of [20], where the answers to each query are divided into two categories: easy answers and hard answers. For test (or validation) queries, easy answers are entities that can be directly reached in the

**Table 1.** Test MRR results (%) on answering FOL queries.

| Model | $avg_p$ | $avg_n$ | 1p | 2p | 3p | 2i | 3i | pi | ip | 2u | up | 2in | 3in | inp | pin | pni |
|---|---|---|---|---|---|---|---|---|---|---|---|---|---|---|---|---|
| FB15k | | | | | | | | | | | | | | | | |
| GQE | 28.0 | – | 54.6 | 15.3 | 10.8 | 39.7 | 51.4 | 27.6 | 19.1 | 22.1 | 11.6 | – | – | – | – | – |
| Q2B | 38.0 | – | 68.0 | 21.0 | 14.2 | 55.1 | 66.5 | 39.4 | 26.1 | 35.1 | 16.7 | – | – | – | – | – |
| BetaE | 41.6 | 11.8 | 65.1 | 25.7 | 24.7 | 55.8 | 66.5 | 43.9 | 28.1 | 40.1 | 25.2 | 14.3 | 14.7 | 11.5 | 6.5 | 12.4 |
| CQD-CO | 46.9 | – | 89.2 | 25.3 | 13.4 | 74.4 | 78.3 | 44.1 | 33.2 | 41.8 | 21.9 | – | – | – | – | – |
| CQD-Beam | 58.2 | – | **89.2** | 54.3 | 28.6 | 74.4 | 78.3 | 58.2 | 67.7 | 42.4 | 30.9 | – | – | – | – | – |
| GNN-QE | 73.7 | 38.3 | 87.7 | 68.8 | 58.7 | 79.7 | 83.5 | 68.9 | **68.1** | 74.4 | 60.2 | **44.5** | 41.7 | 41.7 | 29.4 | 34.3 |
| HYQNET | **74.2** | **38.6** | 87.8 | **69.2** | **59.4** | **81.0** | **84.8** | **71.1** | 65.8 | **74.9** | **60.9** | 44.3 | **44.4** | **42.1** | **29.8** | **34.4** |
| FB15k-237 | | | | | | | | | | | | | | | | |
| GQE | 16.3 | – | 35.0 | 7.2 | 5.3 | 23.3 | 34.6 | 16.5 | 10.7 | 8.2 | 5.7 | – | – | – | – | – |
| Q2B | 20.1 | – | 40.6 | 9.4 | 6.8 | 29.5 | 42.3 | 21.2 | 12.6 | 11.3 | 7.6 | – | – | – | – | – |
| BetaE | 20.9 | 5.5 | 39.0 | 10.9 | 10.0 | 28.8 | 42.5 | 22.4 | 12.6 | 12.4 | 9.7 | 3.5 | 3.4 | 5.1 | 7.9 | 7.4 |
| CQD-CO | 21.8 | – | 46.7 | 9.5 | 6.3 | 31.2 | 40.6 | 23.6 | 16.0 | 14.5 | 8.2 | – | – | – | – | – |
| CQD-Beam | 22.3 | – | **46.7** | 11.6 | 8.0 | 31.2 | 40.6 | 21.2 | 18.7 | 14.6 | 8.4 | – | – | – | – | – |
| FuzzQE | 24.0 | 7.8 | 42.8 | **12.9** | 10.3 | 33.3 | 46.9 | 26.9 | 17.8 | **14.6** | 10.3 | **8.5** | 11.6 | 7.8 | 5.2 | 5.8 |
| GNN-QE | 26.1 | 8.4 | 40.1 | 12.1 | 10.2 | 35.5 | 51.6 | **29.2** | 16.9 | 13.4 | 11.2 | 7.7 | 14.4 | 8.3 | 5.7 | 5.9 |
| HYQNET | **26.5** | **8.9** | 40.5 | 11.9 | **10.3** | **36.3** | **52.1** | 28.8 | **17.9** | 14.3 | **11.2** | 7.6 | **15.8** | **8.9** | **6.3** | **5.9** |
| NELL995 | | | | | | | | | | | | | | | | |
| GQE | 18.6 | – | 32.8 | 11.9 | 9.6 | 27.5 | 35.2 | 18.4 | 14.4 | 8.5 | 8.8 | – | – | – | – | – |
| Q2B | 22.9 | – | 42.2 | 14.0 | 11.2 | 33.3 | 44.5 | 22.4 | 16.8 | 11.3 | 10.3 | – | – | – | – | – |
| BetaE | 24.6 | 5.9 | 53.0 | 13.0 | 11.4 | 37.6 | 47.5 | 24.1 | 14.3 | 12.2 | 8.5 | 5.1 | 7.8 | 10.0 | 3.1 | 3.5 |
| CQD-CO | 28.8 | – | 60.4 | 17.8 | 12.7 | 39.3 | 46.6 | **30.1** | 22.0 | 17.3 | **13.2** | – | – | – | – | – |
| CQD-Beam | 28.6 | – | **60.4** | **20.6** | 11.6 | 39.3 | 46.6 | 25.4 | **23.9** | **17.5** | 12.2 | – | – | – | – | – |
| FuzzQE | 27.0 | 7.8 | 47.4 | 17.2 | **14.6** | 39.5 | 49.2 | 26.2 | 20.6 | 15.3 | 12.6 | 7.8 | 9.8 | 11.1 | 4.9 | 5.5 |
| GNN-QE | 28.7 | 8.6 | 51.3 | 17.0 | 12.9 | 39.1 | 49.3 | 28.1 | 17.4 | 14.2 | 9.0 | **8.8** | 12.7 | 10.9 | 5.0 | 5.5 |
| HYQNET - | **28.9** | **8.9** | 50.3 | 18.0 | 14.1 | **39.7** | **49.9** | 27.4 | 17.5 | 14.5 | 10.8 | 8.5 | **13.5** | **11.4** | **5.8** | **5.5** |

validation (or train) graph using symbolic relation traversal. In contrast, hard answers require reasoning over predicted links, meaning the model must infer them rather than retrieve them directly. To evaluate performance, we rank each hard answer against all non-answer entities and use mean reciprocal rank (MRR) and HITS@K (H@K) as evaluation metrics. Specifically, we use MRR and HITS@1, as done in GNN-QE.

**Baselines.** We compare HYQNET against both embedding methods and neural-symbolic methods. The embedding methods include GQE [7], Q2B [20], BetaE [21] and FuzzQE [5]. The neural-symbolic methods include CQD-CO, CQD-Beam [1] and GNN-QE [28].

**Table 2.** Test H@1 results (%) on answering FOL queries.

| Model | $avg_p$ | $avg_n$ | 1p | 2p | 3p | 2i | 3i | pi | ip | 2u | up | 2in | 3in | inp | pin | pni |
|---|---|---|---|---|---|---|---|---|---|---|---|---|---|---|---|---|
| FB15k | | | | | | | | | | | | | | | | |
| GQE | 16.6 | – | 34.2 | 8.3 | 5.0 | 23.8 | 34.9 | 15.5 | 11.2 | 11.5 | 5.6 | – | – | – | – | – |
| Q2B | 26.8 | – | 52.0 | 12.7 | 7.8 | 40.5 | 53.4 | 26.7 | 16.7 | 22.0 | 9.4 | – | – | – | – | – |
| BetaE | 31.3 | 5.2 | 52.0 | 17.0 | 16.9 | 43.5 | 55.3 | 32.3 | 19.3 | 28.1 | 16.9 | 6.4 | 6.7 | 5.5 | 2.0 | 5.3 |
| CQD-CO | 39.7 | – | 85.8 | 17.8 | 9.0 | 67.6 | 71.7 | 34.5 | 24.5 | 30.9 | 15.5 | – | – | – | – | – |
| CQD-Beam | 51.9 | – | **85.8** | 48.6 | 22.5 | 67.6 | 71.7 | 51.7 | 62.3 | 31.7 | 25.0 | – | – | – | – | – |
| GNN-QE | 68.6 | 21.7 | 85.4 | 63.5 | 52.5 | 74.8 | 79.9 | 63.2 | **62.5** | 67.1 | 53.0 | 32.3 | 30.9 | 32.7 | 17.8 | 21.8 |
| HyQNet | **69.5** | **21.9** | 85.4 | **64.6** | **53.8** | **76.8** | **80.4** | **63.2** | 62.0 | **70.5** | **54.3** | **32.4** | **30.9** | **32.7** | **18.2** | **22.2** |
| FB15k-237 | | | | | | | | | | | | | | | | |
| GQE | 8.8 | – | 22.4 | 2.8 | 2.1 | 11.7 | 20.9 | 8.4 | 5.7 | 3.3 | 2.1 | – | – | – | – | – |
| Q2B | 12.3 | – | 28.3 | 4.1 | 3.0 | 17.5 | 29.5 | 12.3 | 7.1 | 5.2 | 3.3 | – | – | – | – | – |
| BetaE | 13.4 | 2.8 | 28.9 | 5.5 | 4.9 | 18.3 | 31.7 | 14.0 | 6.7 | 6.3 | 4.6 | 1.5 | 7.7 | 3.0 | 0.9 | 0.9 |
| CQD-CO | 14.7 | – | 36.6 | 4.7 | 3.0 | 20.7 | 29.6 | 15.5 | 9.9 | 8.6 | 4.0 | – | – | – | – | – |
| CQD-Beam | 15.1 | – | **36.6** | 6.3 | 4.3 | 20.7 | 29.6 | 13.5 | 12.1 | **8.7** | 4.3 | – | – | – | – | – |
| GNN-QE | 18.4 | 3.2 | 29.9 | 6.4 | 5.5 | 24.6 | 41.5 | **20.7** | 11.2 | 7.8 | 6.2 | 2.7 | 6.1 | 3.6 | 1.9 | 1.9 |
| HyQNet | **18.9** | **3.8** | 30.3 | **6.7** | **5.9** | **25.6** | **42.2** | 20.5 | **12.2** | 8.0 | **6.4** | **3.1** | **7.7** | **4.1** | **2.5** | **2.1** |
| NELL995 | | | | | | | | | | | | | | | | |
| GQE | 9.9 | – | 15.4 | 6.7 | 5.0 | 14.3 | 20.4 | 10.6 | 9.0 | 2.9 | 5.0 | – | – | – | – | – |
| Q2B | 14.1 | – | 23.8 | 8.7 | 6.9 | 20.3 | 31.5 | 14.3 | 10.7 | 5.0 | 6.0 | – | – | – | – | – |
| BetaE | 17.8 | 2.1 | 43.5 | 8.1 | 7.0 | 27.2 | 36.5 | 17.4 | 9.3 | 6.9 | 4.7 | 1.6 | 2.2 | 4.8 | 0.7 | 1.2 |
| CQD-CO | 21.3 | – | 51.2 | 11.8 | **9.0** | 28.4 | 36.3 | **22.4** | 15.5 | 9.9 | **7.6** | – | – | – | – | – |
| CQD-Beam | 21.0 | – | **51.2** | **14.3** | 6.3 | 28.4 | 36.3 | 18.1 | **17.4** | **10.2** | 7.2 | – | – | – | – | – |
| GNN-QE | 21.5 | 3.3 | 41.0 | 12.4 | 9.3 | **29.2** | **40.0** | 20.2 | 11.7 | 8.2 | 6.4 | 2.5 | 5.3 | 5.5 | 1.5 | 1.7 |
| HyQNet | **21.5** | **3.7** | 42.3 | 12.0 | 8.4 | 28.6 | 39.4 | 21.1 | 12.1 | 7.7 | 5.4 | **3.1** | 4.6 | **8.1** | 0.9 | **1.9** |

## 4.1 Complex Query Answering

**Table 3.** Spearman's rank correlation between the model prediction and the number of ground truth answers on FB15k. avg is the average correlation on all 12 query types in the table.

| Model | avg | 1p | 2p | 3p | 2i | 3i | pi | ip | 2in | 3in | inp | pin | pni |
|---|---|---|---|---|---|---|---|---|---|---|---|---|---|
| Q2B | – | 0.301 | 0.219 | 0.262 | 0.331 | 0.270 | 0.297 | 0.139 | – | – | – | – | – |
| BetaE | 0.494 | 0.373 | 0.478 | 0.472 | 0.572 | 0.397 | 0.519 | 0.421 | 0.622 | 0.548 | 0.459 | 0.465 | 0.608 |
| GNN-QE | 0.952 | 0.971 | 0.967 | 0.926 | 0.987 | 0.936 | 0.937 | 0.923 | 0.992 | 0.985 | 0.880 | 0.940 | 0.991 |
| HyQNet | **0.958** | **0.971** | **0.970** | **0.941** | **0.988** | **0.938** | **0.939** | **0.934** | 0.990 | 0.984 | **0.908** | **0.948** | **0.991** |

Table 1 shows the MRR results of different models for answering FOL queries. GQE, Q2B, CQD-CO, and CQD-Beam do not support queries with negation, so the corresponding entries are empty. We observe that HYQNET achieves the best result most of the time for both EPFO queries and queries with negation on all 3 datasets. Notably, HYQNET achives the best average performance and better overall performance most of the time compared with baselines, which shows the effectiveness of the proposed hyperbolic embedding model. Hyperbolic embedding can better capture the recursive learning tree structure, Thus, the learned embedding is more expressive compared with the euclidean embedding. Additionally, HYQNET consistently achieves the highest average MRR, particularly excelling in complex queries such as intersection (2i, 3i) and negation-based queries (2in, 3in, pin, pni). This suggests that these models effectively capture relational structures and logical patterns. CQD-Beam also performs well, especially on FB15k, benefiting from beam search strategies that enhance multi-hop reasoning. However, these gains are less pronounced in FB15k-237 and NELL995, likely due to the increased sparsity and relational diversity in these datasets. Overall, the results indicate that neural-symbolic models (e.g., GNN-QE, HYQNET) and query embedding-based approaches (e.g., CQD variants) are more effective at generalizing across different logical structures compared to earlier embedding-based methods like GQE and Q2B.

Table 2 shows the Hits@1 (H@1) results of different methods on FOL queries, demonstrating a consistent pattern of model performance across various datasets (FB15k, FB15k-237, and NELL-995). HYQNET consistently outperforms other models, achieving the highest average Hits@1 scores across all datasets. CQD-Beam also performs well, particularly on the FB15k dataset, benefiting from beam search strategies that enhance multi-hop reasoning. In contrast, models such as GQE and Q2B exhibit relatively lower performance, especially on more complex queries and larger datasets like FB15k-237 and NELL995. Compared to GNN-QE, HYQNET shows superior performance. Overall, the results suggest that the hyperbolic embedding model is better at capturing the recursive tree structure in embeddings.

### 4.2 Answer Set Cardinality Prediction

Similar to GNN-QE, HYQNET can predict the cardinality of the answer set (i.e., the number of answers) without explicit supervision. The cardinality of a fuzzy set is computed as the sum of entity probabilities exceeding a predefined threshold. We use 0.5 as the threshold, as it naturally aligns with our binary classification loss. Previous studies [20,21] have observed that the uncertainty in Q2B and BetaE is positively correlated with the number of answers. Following their approach, we report Spearman's rank correlation between our model's predictions and the ground truth. As shown in Table 3, HYQNET significantly outperforms existing methods most of the time.

## 5 Related Work

**Knowledge Graph Reasoning.** Knowledge graph reasoning has been studied for a long time [12–15]. Embedding-based approaches [3,22,24] map entities and relations into low-dimensional vectors to capture graph structure. Reinforcement learning methods [11,25,27] use agents to explore paths and predict missing links, while rule-based techniques [9,18,26] generate logical rules for link prediction.

**Graph Neural Networks for Relational Learning.** Graph neural networks (GNNs) [23,28] have gained popularity in learning entity representations for knowledge graph completion. In contrast, our approach adapts hyperbolic GNNs for relation projection and extends the task to complex logical query answering, which introduces more challenges than traditional graph completion.

**Hyperbolic Geometry in Knowledge Graphs.** Hyperbolic space is particularly effective for modeling hierarchical structures due to its exponential growth properties. Early works such as Poincaré embeddings [17] demonstrated the benefits of hyperbolic geometry, which were later expanded in Hyperbolic Graph Neural Networks (HGNNs) [16] for better relational reasoning. Models like MuRP [2] and AttH [4] have shown improved performance in hierarchical link prediction tasks, making hyperbolic embeddings valuable for relational reasoning in large-scale knowledge graphs. Our work builds on these models to optimize hyperbolic representations for logical reasoning and query answering.

**Complex Logical Query Handling Complex Logical Queries.** Traditional knowledge graph completion techniques focus on predicting missing links, but more complex tasks involve answering logical queries that combine operations like conjunction, disjunction, and negation. Methods such as GQE [7] introduced compositional training to extend embedding-based approaches for path queries. GQE applies a geometric intersection operator for conjunctive queries ($\wedge$), while Query2Box [20] and BetaE [21] generalize this approach to handle more complex logical operators like $\vee$ and $\neg$. FuzzQE [5] incorporates fuzzy logic to align embeddings with classical logic, providing a more nuanced representation. However, these approaches often prioritize computational efficiency, using nearest neighbor search or dot product for query decoding, which sacrifices interpretability and makes intermediate reasoning opaque.

**Combining Symbolic and Neural Approaches.** To address the limitations of pure embedding methods, some approaches combine neural networks with symbolic reasoning. CQD [1] extends pretrained embeddings to answer complex queries, with CQD-CO employing continuous optimization and CQD-Beam utilizing beam search. While our method shares similarities with CQD-Beam in combining symbolic reasoning with graph completion, it stands apart by enabling direct training on complex queries. Unlike CQD-Beam, GNN-QE bypasses the need for exhaustive search or pretrained embeddings, offering a more efficient and interpretable solution for answering complex logical queries.

## 6 Conclusion

We proposed a hyperbolic graph neural network (GNN) for answering complex First-Order Logic (FOL) queries on knowledge graphs. By leveraging hyperbolic embeddings with learned curvature, our model effectively captures hierarchical query structures, improving reasoning accuracy and interpretability. Experiments on three benchmark datasets show that our method outperforms existing baselines, demonstrating the benefits of hyperbolic space for logical reasoning. Future work includes extending our approach to more expressive logical operations and exploring adaptive curvature learning for greater flexibility.

## References

1. Arakelyan, E., Daza, D., Minervini, P., Cochez, M.: Complex query answering with neural link predictors. arXiv preprint arXiv:2011.03459 (2020)
2. Balažević, I., Allen, C., Hospedales, T.: Multi-relational Poincaré Graph Embeddings. Curran Associates Inc., Red Hook (2019)
3. Bordes, A., Usunier, N., Garcia-Duran, A., Weston, J., Yakhnenko, O.: Translating embeddings for modeling multi-relational data. Adv. Neural. Inf. Process. Syst. **26** (2013)
4. Chami, I., Wolf, A., Juan, D.C., Sala, F., Ravi, S., Ré, C.: Low-dimensional hyperbolic knowledge graph embeddings. In: Proceedings of the 58th Annual Meeting of the Association for Computational Linguistics. Association for Computational Linguistics, Online (2020)
5. Chen, X., Hu, Z., Sun, Y.: Fuzzy logic based logical query answering on knowledge graphs. In: Proceedings of the AAAI Conference on Artificial Intelligence, vol. 36, pp. 3939–3948 (2022)
6. Guo, Q., et al.: A survey on knowledge graph-based recommender systems (2020). https://arxiv.org/abs/2003.00911
7. Hamilton, W., Bajaj, P., Zitnik, M., Jurafsky, D., Leskovec, J.: Embedding logical queries on knowledge graphs. Adv. Neural Inf. Process. Syst. **31** (2018)
8. He, K., Zhang, X., Ren, S., Sun, J.: Deep residual learning for image recognition (2015). https://arxiv.org/abs/1512.03385
9. Ho, V.T., Stepanova, D., Gad-Elrab, M.H., Kharlamov, E., Weikum, G.: Rule learning from knowledge graphs guided by embedding models. In: Vrandečić, D., Bontcheva, K., Suárez-Figueroa, M.C., Presutti, V., Celino, I., Sabou, M., Kaffee, L.-A., Simperl, E. (eds.) ISWC 2018. LNCS, vol. 11136, pp. 72–90. Springer, Cham (2018). https://doi.org/10.1007/978-3-030-00671-6_5
10. Klir, G.J., Yuan, B.: Fuzzy Sets and Fuzzy Logic: Theory and Applications. Prentice-Hall Inc., Upper Saddle River (1994)
11. Lin, X., Socher, R.: Multi-hop knowledge graph reasoning with reward shaping. In: EMNLP 2018 (2018)
12. Liu, L.: Hyperkgr: knowledge graph reasoning in hyperbolic space with graph neural network encoding symbolic path. In: Proceedings of the 2025 Conference on Empirical Methods in Natural Language Processing, pp. 25188–25199 (2025)
13. Liu, L.: Monte Carlo tree search for graph reasoning in large language model agents. In: Proceedings of the 34th ACM International Conference on Information and Knowledge Management, pp. 4966–4970 (2025)

14. Liu, L., Ding, J., Mukherjee, S., Yang, C.J.: Mixrag: mixture-of-experts retrieval-augmented generation for textual graph understanding and question answering. arXiv preprint arXiv:2509.21391 (2025)
15. Liu, L., Du, B., Xu, J., Xia, Y., Tong, H.: Joint knowledge graph completion and question answering. In: Proceedings of the 28th ACM SIGKDD Conference on Knowledge Discovery and Data Mining, pp. 1098–1108 (2022)
16. Liu, Q., Nickel, M., Kiela, D.: Hyperbolic graph neural networks (2019). https://arxiv.org/abs/1910.12892
17. Nickel, M., Kiela, D.: Poincaré embeddings for learning hierarchical representations. In: Guyon, I., et al. (eds.) Advances in Neural Information Processing Systems, vol. 30. Curran Associates, Inc. (2017). https://proceedings.neurips.cc/paper_files/paper/2017/file/59dfa2df42d9e3d41f5b02bfc32229dd-Paper.pdf
18. Qu, M., Chen, J., Xhonneux, L.P., Bengio, Y., Tang, J.: Rnnlogic: learning logic rules for reasoning on knowledge graphs. arXiv preprint arXiv:2010.04029 (2020)
19. Radford, A., Wu, J., Child, R.: Language models are unsupervised multitask learners (2018)
20. Ren, H., Hu, W., Leskovec, J.: Query2box: reasoning over knowledge graphs in vector space using box embeddings. arXiv preprint arXiv:2002.05969 (2020)
21. Ren, H., Leskovec, J.: Beta embeddings for multi-hop logical reasoning in knowledge graphs. Adv. Neural. Inf. Process. Syst. **33**, 19716–19726 (2020)
22. Sun, Z., Deng, Z.H., Nie, J.Y., Tang, J.: Rotate: knowledge graph embedding by relational rotation in complex space. arXiv preprint arXiv:1902.10197 (2019)
23. Teru, K., Denis, E., Hamilton, W.: Inductive relation prediction by subgraph reasoning. In: International Conference on Machine Learning, pp. 9448–9457. PMLR (2020)
24. Trouillon, T., Welbl, J., Riedel, S., Gaussier, É., Bouchard, G.: Complex embeddings for simple link prediction. In: International conference on machine learning, pp. 2071–2080. PMLR (2016)
25. Xiong, W., Hoang, T., Wang, W.Y.: Deeppath: A reinforcement learning method for knowledge graph reasoning. arXiv preprint arXiv:1707.06690 (2017)
26. Yang, F., Yang, Z., Cohen, W.W.: Differentiable learning of logical rules for knowledge base completion. CoRR arxiv:1702.08367 (2017)
27. Zhang, Q., Weng, X., Zhou, G., Zhang, Y., Huang, J.X.: ARL: an adaptive reinforcement learning framework for complex question answering over knowledge base. Inf. Process. Manag. **59**(3), 102933 (2022). https://doi.org/10.1016/j.ipm.2022.102933. https://www.sciencedirect.com/science/article/pii/S0306457322000565
28. Zhu, Z., Galkin, M., Zhang, Z., Tang, J.: Neural-symbolic models for logical queries on knowledge graphs. In: International Conference on Machine Learning, pp. 27454–27478. PMLR (2022)

# LAPTG: Length-Aligned Attentive Prefix-Target Graph for Sequential Recommendation

Zhiqiang Yang[1], Pengfei Wang[2], Yanxin Zhang[3], Boying Li[3], Danlu Chen[4], Xinguang Xiang[1(✉)], and Xiaoyu Du[1]

[1] School of Computer Science and Engineering, Nanjing University of Science and Technology, Nanjing, China
{zhiqiangyang,xgxiang,duxy}@njust.edu.cn
[2] GienTech Technology Co., Ltd., Beijing, China
pengfei.wang14@gientech.com
[3] University of Wisconsin-Madison, Madison, USA
yzhang2879@wisc.edu
[4] China Tobacco Jiangsu Industrial Co., Ltd., Nanjing, China

**Abstract.** Sequential recommender systems aim to predict the next item a user will interact with based on historical behaviors by modeling the correlations between prefix interactions and target items. Recent studies construct Prefix-Target graphs to represent these correlations and employ graph models to capture inherent dependencies through their connections. However, the varying lengths of prefix interactions result in uneven information content in prefix representations, leading to biased correlation modeling and degraded representation quality. To address this issue, we propose a **L**ength-aligned **A**ttentive **P**refix-**T**arget **G**raph framework (LAPTG) that supplements short prefixes with relevant, unvisited sequences derived from observed interactions, enabling more balanced and informative prefix representations. To mitigate the noise introduced by augmented data, we devise an attentive graph-based sequential recommendation approach that selectively filters and aggregates useful information. Moreover, we identify that recent experimental setups may unintentionally leak user ID information, as sequences from the same user share identical prefixes. To address this issue and make fair evaluation, we redesign the data pre-processing pipeline and conduct extensive experiments on four real-world datasets. Experimental results demonstrate that LAPTG consistently outperforms state-of-the-art sequential recommendation models, achieving superior accuracy and robustness in capturing user intent.

**Keywords:** Sequential Recommendation · Prefix-Target Graph · Graph Attention Network · Length Alignment


R. C.-W. Wong et al. (Eds.): PAKDD 2026, LNAI 16597, pp. 366–378, 2026.
https://doi.org/10.1007/978-981-92-1300-9_29

# 1 Introduction

Sequential recommendation (SR) aims to predict the next item a user will interact with by modeling historical behavior sequences. Unlike traditional collaborative filtering methods [7,20], which rely on static interaction matrices, SR explicitly captures dynamic preference evolution, modeling both long-term interests and short-term intents within interaction sequences. In typical SR settings, the observed interactions are treated as a prefix, and the goal is to predict the subsequent target item conditioned on this prefix.

To model sequential dependencies, recurrent neural networks [8] and Transformer-based architectures [34] have been widely adopted. Representative models such as SASRec [13] and BERT4Rec [32] utilize self-attention mechanisms to adaptively assign importance to items in the prefix, enabling more accurate modeling of evolving user interests. In addition, multimodal methods [6,18,28] incorporate complementary contextual signals from different modalities to enhance item representations.

Recently, Prefix-Target (PT) graph-based methods have been proposed to explicitly model correlations between prefix interactions and target items [2]. By employing Graph Convolutional Networks (GCN), these approaches aggregate information among nodes that share similar prefixes or suffixes, providing a structured representation of sequential dependencies. Although effective, existing PT graph models still face two limitations. First, user sequences vary significantly in length, leading to imbalanced information across prefixes. Short prefixes often lack sufficient contextual signals, while long prefixes dominate representation learning. Second, during graph propagation, noisy or less relevant signals may be aggregated indiscriminately, which can negatively affect representation quality.

To alleviate these issues, we propose a **L**ength-aligned **A**ttentive **P**refix-**T**arget **G**raph (LAPTG) framework. We first compute pairwise sequence similarity and identify the most behaviorally relevant sequence for each short prefix. Unvisited yet informative items from the similar sequence are then selected and prepended to the short prefix, constructing a length-aligned PT graph that mitigates information imbalance caused by varying prefix lengths.

Furthermore, instead of using standard GCN for propagation, we adopt a Graph Attention Network (GAT) to adaptively weight neighboring nodes. This attentive aggregation mechanism helps emphasize informative dependencies while reducing the influence of noisy signals, resulting in more robust prefix representations and improved prediction performance.

We conduct extensive experiments on four real-world datasets. We observe that commonly used experimental data processing procedures may inadvertently cause user identity leakage, since sequences from the same user can share identical prefixes across splits. To ensure fair evaluation, we re-process the datasets to prevent user-level leakage and provide unbiased comparisons.

In summary, our contributions are threefold:

- We propose a Length-aligned Attentive Prefix-Target Graph framework (LAPTG) to alleviate prefix information imbalance by supplementing short prefixes with relevant unvisited sequences.

- We introduce an attentive graph aggregation mechanism to better distinguish informative signals from noise during propagation.
- We conduct extensive experiments under a leakage-free evaluation protocol and demonstrate the effectiveness of LAPTG.

# 2 Related Work

## 2.1 Graph-Based Recommendation Approaches

Recommendation data is naturally graph-structured, making Graph Neural Networks (GNNs) [4,16,24] suitable for representation learning. NGCF [35] integrates collaborative filtering signals with graph structural information. LightGCN [11] simplifies the GNN architecture by removing feature transformations and nonlinear activations, achieving competitive performance with improved efficiency. PinSage [41] and GraphSAGE [9] further extend GNN modeling to heterogeneous and large-scale graphs, while Bipartite influential community search [42] focuses on identifying cohesive subgraphs.

GNNs have also been applied to sequential recommendation. SR-GNN [38] models session-based sequences as graphs; GCE-GNN [37] integrates contrastive learning; SURGE [1] constructs intent-enhanced graphs via clustering; IHGCN [40] captures item similarities derived from shared prefix sequences; and ReducedGCN [14] improves efficiency by filtering less relevant interactions.

Despite these developments, many existing approaches focus primarily on item-level relations or simplified intent modeling. Methods leveraging shared prefixes, such as PTSR [2], partially capture high-order relationships among prefixes and targets. However, modeling complex and multidimensional user intents remains challenging, leaving room for further improvement.

## 2.2 Length-Aligned for Sequential Recommendation

Alignment strategies in SR differ according to sequence length. For long sequences, segmentation methods such as sliding windows [33] decompose complex behaviors into shorter segments. For medium-length sequences, generative expansion techniques—including reordering [39], masking [32], and reverse generation (e.g., ASReP [22], BARec [12], NSIR [26])—enrich contextual information and enhance robustness. For short sequences with insufficient signals, extension methods such as Llama4Rec [23] leverage large language models to expand behavioral context.

Although these strategies address sequence length from different perspectives, ensuring that aligned or augmented data faithfully reflects user intent without introducing additional noise remains a key challenge.

# 3 Method

This section presents the proposed Length-aligned Attentive Prefix-Target Graph (LAPTG) framework. Built upon the Prefix-Target Graph (PTG) architecture, LAPTG introduces two main enhancements: (1) a length-aligned data strategy to mitigate sequence length imbalance, and (2) the adoption of Graph Attention Network (GAT) for more adaptive graph propagation.

## 3.1 Problem Definition

Let $\mathcal{I} = \{i_1, i_2, \ldots, i_{|\mathcal{I}|}\}$ be the set of all items, where $|\mathcal{I}|$ is the number of unique items. A given interaction history is represented as a chronologically ordered sequence $s_u = [i_1^u, i_2^u, \ldots, i_L^u]$, where each element $i_t^u \in \mathcal{I}(1 \leq t \leq L)$ is an item interacted with at the $t$-th position, and $L$ denotes the sequence length. The goal of Sequential Recommendation is to predict the next item the user will interact with at the $L + 1$ step. This can be formulated as modeling the probability distribution over all items in $\mathcal{I}$ given the sequence $s_u$:

$$\arg\max_{i \in \mathcal{I}} P(i_{L+1}^u = i \mid s_u). \tag{1}$$

## 3.2 Length-Aligned Prefix-Target Graph

To address the data sparsity and modeling bias caused by limited information in short sequences, we propose a Length-Aligned Attentive Prefix-Target Graph (LAPTG) framework. As illustrated in Fig. 1, unlike traditional methods that simply pad sequences with null values, our framework enhances short prefixes by supplementing them with relevant, unvisited sequences derived from observed interactions. This approach not only alleviates the challenges of data sparsity but also helps construct more informative and enriched prefix representations.

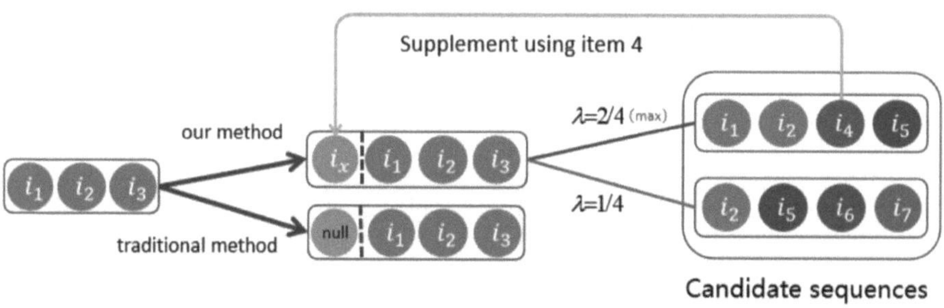


Fig. 1. Comparison of sequence length alignment process.

**Sequence Length Alignment.** For sequences with length less than $K$, we propose a length alignment mechanism to supplement information.

First, we compute the similarity $\lambda_{ij}$ between any two sequences $(p_i, t_i)$ and $(p_j, t_j)$ extracted from the original sequence as follows:

$$\lambda_{ij} = \frac{|p_i \cap p_j|}{|p_i \cup p_j|}. \tag{2}$$

For a short sequence $(p_i, t_i)$ requiring length alignment (where $|p_i|+|t_i| < K$), we find the most similar sequence:

$$(p_j, t_j) = Sim(p_i, t_i) = \mathop{\arg\max}_{(p_j,t_j)\in\mathcal{D},\forall i\neq j} \lambda_{ij}, \tag{3}$$

where $Sim(\cdot,\cdot)$ indicates the similarity function, $\mathcal{D}$ is the set of all sequences extracted from the original sequence. We then select items from $p_j$ not in $p_i$ for supplementation until the sequence reaches length $K$ or no items remain. This length alignment ensures uniform sequence lengths while preserving semantic relevance.

### 3.3 Attentive Graph-Based Aggregation

Through CDDRec [36], we observe significant noise in the sequence. Therefore, we build a global PTG and apply a Graph Attention Network for information propagation to filter out noisy information in the sequence.

**GAT Layer.** We employ Graph Attention Network to adaptively learn neighbor importance weights. For node $p$ and its neighbor $i \in \mathcal{N}_p$, the attention weight is computed as:

$$\alpha_{pi} = \frac{\exp\left(\text{LeakyReLU}\left(\mathbf{a}^\top[\mathbf{W}\mathbf{e}_p^{(l)}\|\mathbf{W}\mathbf{e}_i^{(l)}]\right)\right)}{\sum_{k\in\mathcal{N}_p}\exp\left(\text{LeakyReLU}\left(\mathbf{a}^\top[\mathbf{W}\mathbf{e}_p^{(l)}\|\mathbf{W}\mathbf{e}_k^{(l)}]\right)\right)}, \tag{4}$$

where $\mathbf{W}$ is the learnable weight matrix, $\mathbf{a}$ is the attention vector, and $\|$ represents vector concatenation operation. The embeddings are then updated via:

$$\mathbf{e}_p^{(l+1)} = \sigma\left(\sum_{i\in\mathcal{N}_p}\alpha_{pi}\mathbf{W}\mathbf{e}_i^{(l)}\right), \quad \mathbf{e}_i^{(l+1)} = \sigma\left(\sum_{p\in\mathcal{N}_i}\alpha_{ip}\mathbf{W}\mathbf{e}_p^{(l)}\right), \tag{5}$$

where $\sigma$ is an activation function and $\alpha_{ip}$ is computed analogously to $\alpha_{pi}$.

**Node Initialization and Final Representation.** Item nodes are initialized via a trainable embedding layer. Prefix nodes are initialized by averaging their constituent item embeddings. After multi-layer GAT propagation, we obtain enhanced embeddings that capture high-order transition relationships.

### 3.4 Training and Prediction

Enhanced item embeddings are fed into a sequential model (e.g., SASRec [13]). For a sequence $s$, the model takes prefix embeddings $[\mathbf{e}_{i_1}, \mathbf{e}_{i_2}, \ldots, \mathbf{e}_{i_{n-1}}]$ and outputs a prediction vector $\hat{\mathbf{e}}_{i_n}$. The probability for candidate item $i$ is:

$$P = \frac{\mathbf{e}_i \cdot \hat{\mathbf{e}}_{i_n}}{\sum_{j \in \mathcal{I}} \mathbf{e}_j \cdot \hat{\mathbf{e}}_{j_n}} \times 100\%, \tag{6}$$

where $\cdot$ denotes the vector inner product. The model is trained end-to-end using cross-entropy loss.

## 4 Experiments

In this section, we conduct comprehensive experiments to evaluate our model from various aspects and address the following issues:

- (RQ1): How does the performance of our model compare to existing models?
- (RQ2): What is the impact of each individual component in our model?
- (RQ3): What are the effects of the key hyperparameters, specifically the number of GAT layers and the PTG connection counts, on the performance of our model?

**Table 1.** Dataset Statistics

| Statistics | Sports | Beauty | Toys | ML-1M |
|---|---|---|---|---|
| # Users | 35,598 | 22,363 | 19,412 | 6,040 |
| # Items | 18,357 | 12,101 | 11,924 | 3,416 |
| # Actions | 296,337 | 198,502 | 167,597 | 999,611 |
| # Avg. Actions/User | 8.3 | 8.8 | 8.6 | 165.4 |
| # Avg. Actions/Item | 16.1 | 16.4 | 14.0 | 292.6 |
| Sparsity (%) | 99.95% | 99.93% | 99.93% | 95.16% |

### 4.1 Experimental Settings

**Datasets.** We evaluate the effectiveness of our proposed model on four datasets from two real-world applications:

- Amazon[1] [27]: This dataset comprises user reviews from Amazon.com. We select three distinct product categories—Sports, Beauty, and Toys—for experiments, each characterized by relatively short interaction sequences.

[1] https://jmcauley.ucsd.edu/data/amazon/.

- MovieLens[2] [10]: This dataset contains user behavior logs for movies, featuring notably long interaction sequences. Our experiments utilize the ML-1M version with one million ratings.

To ensure data quality, we remove users or items with fewer than five interactions [13,32]. Each user's interactions are aggregated and sorted chronologically. Following [3,30], we split the data into training, validation, and test sets based on timestamps: the last interaction for testing, the second-last for validation, and all prior interactions for training. The detailed statistics of the processed datasets are summarized in Table 1.

**Data Processing.** According to previous studies [2,29], we identify a *Potential ID Information Leakage* issue that using dynamic sliding windows may inadvertently lead to user identity leakage, as sequences from the same user typically share the same prefix. For instance, the raw sequence $s_0 = [i_1, i_2, i_3, i_4, i_5]$ derives sequences $s_1 = [i_1, i_2, i_3]$, $s_2 = [i_1, i_2, i_3, i_4]$ and $s_3 = [i_1, i_2, i_3, i_4, i_5]$ through a dynamic sliding window mechanism, which controls a fixed starting position but varies the window size. These sequences share the same prefix $[i_1, i_2, i_3]$, potentially exposing the fact that they originate from the same user sequence. To address this issue, we fix the size of sliding window as $K$, and perform the following operations for a sequence of length $L$:

- If $L \leq K$, use $s_u$ directly.
- If $L > K$, slide the window to obtain sequences $s_{uk} = [i_k^u, \ldots, i_{k+K-1}^u]$ for $k = 1, \ldots, L - K + 1$.

For each sequence $s = [i_1, \ldots, i_{|s|}]$, the prefix is $p = [i_1, \ldots, i_{|s|-1}]$ and the target is $t = i_{|s|}$. The global set of prefix-target pairs is $\mathcal{D} = \{(p, t)\}$.

**Baseline Model.** We will compare our model with the following representative SR models:

- **Non sequential model**: BPR [31] first uses Bayesian Personalized Sorting (BPR) loss to optimize the matrix factorization model.
- **Universal sequence model**: SASRec [13] first models the sequence using attention mechanism, greatly improving the performance of SR. BERT4Rec [32] first introduced BERT [5] to model sequences, which utilizes MIP (Mask Item Prediction) tasks to capture potential relationships between items and sequences. CoSeRec [21] is a contrastive learning framework based on the Transformer encoder that proposes two informative robust data enhancement operators based on CL4SRec [39].
- **Intention guided sequential model**: IOCRec [19] disentangles the latent intents from the user behavior sequence and uses the intent to guide the item representation learning. DuoRec [30] utilizes the randomness of dropout in the model to form different views for unsupervised contrastive learning and uses contrastive learning on sequences with the same next item. DSSRec [25]

[2] https://grouplens.org/datasets/movielens/.

proposes a seq2seq model and introduces a variable representing user intent. ICSRec [29] uses subsequences as training data and applies contrastive learning between similar subsequences to enhance representation. PTSR [2] constructs a PT graph and uses a GNN network to capture the multivariate relationships between prefixes of the same target and targets with the same prefix.

**Evaluation Metrics.** We follow [17,35] to rank the entire project set without negative sampling and use two widely used evaluation metrics to evaluate the model, including Hit Ratio @$k$ (HR@k) and Normalized Discount Cumulative Gain @$k$ (NDCG@k). Among them, $k \in \{5, 10, 20\}$.

**Implementation Details.** We use raw code with baseline original settings to reproduce the results on four benchmarks. We use Adam [15] optimizer with batch size set to 512. This model has 2 Transformer layers, 2 GAT layers, and a potential embedding size of 64 bits. Set the number of self attention blocks and attention heads to 2. The embedding dimension of the item is 64, the window length is 53, and PTG amplification is performed once per batch.

**Table 2.** Performance comparisons of different methods. The results of the best baseline are underlined in each row and bold values denote the best results. The last column is the relative improvements compared with the best baseline results.

| Dataset | Metric | BPR | SASRec | BERT4Rec | CoSeRec | IOCRec | DuoRec | ICSRec | PTSR | LAPTG | Impv. |
|---|---|---|---|---|---|---|---|---|---|---|---|
| Sports | HR@5 | 0.0127 | 0.0201 | 0.0197 | 0.0201 | 0.0206 | 0.0210 | 0.0207 | 0.0230 | **0.0236** | 2.61% |
| | HR@10 | 0.0195 | 0.0256 | 0.0263 | 0.0273 | 0.0280 | 0.0285 | 0.0270 | 0.0308 | **0.0325** | 5.52% |
| | HR@20 | 0.0251 | 0.0356 | 0.0345 | 0.0382 | 0.0399 | 0.0415 | 0.0411 | 0.0431 | **0.0460** | 6.73% |
| | NDCG@5 | 0.0087 | 0.0129 | 0.0119 | 0.0128 | 0.0134 | 0.0141 | 0.0134 | 0.0154 | **0.0164** | 6.49% |
| | NDCG@10 | 0.0093 | 0.0132 | 0.0124 | 0.0135 | 0.0157 | 0.0157 | 0.0151 | 0.0178 | **0.0192** | 7.87% |
| | NDCG@20 | 0.0124 | 0.0176 | 0.0181 | 0.0166 | 0.0183 | 0.0182 | 0.0179 | 0.0207 | **0.0226** | 9.18% |
| Beauty | HR@5 | 0.0212 | 0.0340 | 0.0339 | 0.0327 | 0.0341 | 0.0369 | 0.0352 | 0.0376 | **0.0402** | 6.91% |
| | HR@10 | 0.0255 | 0.0483 | 0.0485 | 0.0492 | 0.0507 | 0.0517 | 0.0501 | 0.0530 | **0.0556** | 4.91% |
| | HR@20 | 0.0351 | 0.0601 | 0.0612 | 0.0605 | 0.0623 | 0.0685 | 0.0658 | 0.0701 | **0.0745** | 6.28% |
| | NDCG@5 | 0.0118 | 0.0223 | 0.0205 | 0.0221 | 0.0237 | 0.0253 | 0.0240 | 0.0269 | **0.0289** | 7.43% |
| | NDCG@10 | 0.0157 | 0.0265 | 0.0231 | 0.0246 | 0.0276 | 0.0307 | 0.0303 | 0.0327 | **0.0338** | 3.36% |
| | NDCG@20 | 0.0154 | 0.0295 | 0.0286 | 0.0302 | 0.0325 | 0.0342 | 0.035 | 0.0365 | **0.0386** | 5.75% |
| Toys | HR@5 | 0.0169 | 0.0312 | 0.0301 | 0.0309 | 0.0323 | 0.0351 | 0.0334 | 0.0356 | **0.0384** | 7.87% |
| | HR@10 | 0.0214 | 0.0418 | 0.0412 | 0.0421 | 0.0430 | 0.0433 | 0.0428 | 0.0453 | **0.0486** | 7.28% |
| | HR@20 | 0.0261 | 0.0507 | 0.0508 | 0.0485 | 0.0522 | 0.0562 | 0.0531 | 0.0601 | **0.0614** | 2.16% |
| | NDCG@5 | 0.0117 | 0.0213 | 0.0212 | 0.0219 | 0.0227 | 0.0237 | 0.0226 | 0.0266 | **0.0288** | 8.27% |
| | NDCG@10 | 0.0132 | 0.0242 | 0.0232 | 0.0251 | 0.0275 | 0.0281 | 0.0282 | 0.0302 | **0.0321** | 6.29% |
| | NDCG@20 | 0.0154 | 0.0281 | 0.0285 | 0.0298 | 0.0315 | 0.0322 | 0.0327 | 0.0336 | **0.0353** | 5.06% |
| ml-1 m | HR@5 | 0.0798 | 0.1638 | 0.1486 | 0.1623 | 0.1702 | 0.1653 | 0.1726 | 0.1808 | **0.1892** | 4.65% |
| | HR@10 | 0.1326 | 0.2590 | 0.2445 | 0.2547 | 0.2356 | 0.2431 | 0.2632 | 0.2717 | **0.2748** | 1.14% |
| | HR@20 | 0.1855 | 0.3621 | 0.3017 | 0.3129 | 0.3485 | 0.3597 | 0.3695 | 0.3757 | **0.3839** | 2.18% |
| | NDCG@5 | 0.4630 | 0.1025 | 0.1096 | 0.1107 | 0.1153 | 0.1147 | 0.115 | 0.1216 | **0.1280** | 5.26% |
| | NDCG@10 | 0.0864 | 0.1307 | 0.1254 | 0.1365 | 0.1298 | 0.1289 | 0.1329 | 0.1510 | **0.1553** | 2.85% |
| | NDCG@20 | 0.0985 | 0.1695 | 0.1548 | 0.1685 | 0.1721 | 0.1723 | 0.1736 | 0.1746 | **0.1829** | 4.75% |

**Table 3.** The HR@20 and NDCG@20 performances achieved by LAPTG variants and PTSR on four datasets.

| Model | Dataset | | | | | | | |
|---|---|---|---|---|---|---|---|---|
| | Sports | | Beauty | | Toys | | ml-1 m | |
| | HR@20 | NDCG@20 | HR@20 | NDCG@20 | HR@20 | NDCG@20 | HR@20 | NDCG@20 |
| LAPTG | 0.0460 | 0.0226 | 0.0745 | 0.0386 | 0.0614 | 0.0353 | 0.3839 | 0.1829 |
| w/o GAT | 0.0449 | 0.0217 | 0.0736 | 0.0369 | 0.0605 | 0.0340 | 0.3821 | 0.1806 |
| w/o Enhanced | 0.0436 | 0.0210 | 0.0713 | 0.0381 | 0.0607 | 0.0344 | 0.3785 | 0.1788 |
| PTSR | 0.0431 | 0.0207 | 0.0701 | 0.0365 | 0.0601 | 0.0336 | 0.3757 | 0.1746 |

## 4.2 Performance Comparison (RQ1)

We evaluate LAPTG against multiple baselines, with overall results in Table 2. Key observations are:

- LAPTG significantly outperforms all baselines across datasets and metrics. By incorporating GAT and enhanced sequence prefixes in the PT graph, it achieves substantial gains. On Beauty, Sports, Toys, and ML-1M, LAPTG improves HR@20 over the best baseline by 6.73%, 6.28%, 2.16%, and 2.18%, respectively. Averaged over six metrics, it surpasses the previous SOTA, PTSR, by 5.77%, 6.40%, 6.15%, and 3.47% on these datasets.
- BPR performs worse than sequential models because it ignores temporal order and historical behavior. In contrast, SASRec [13], CoSeRec [21] and BERT4Rec [32] demonstrate that modeling sequences improves performance.
- Intent-guided models [2,19,25,29,30] further advance performance by capturing user intents behind item transitions. Similarly, LAPTG uses GAT to model prefix-target relations and enhances sequence prefixes to better reflect user intent, achieving state-of-the-art results.

## 4.3 Ablation Study (RQ2)

In order to evaluate the contribution of each component in LAPTG, we conduct ablation studies by excluding the following key elements related to the LAPTG mechanism:

- W/o Length-Aligned: The prefixes of sequences are not length-aligned after extraction, and are directly used for composition and training.
- W/o GAT: The GAT module is removed for item representation learning, and replaced with a simple GNN.
- W/o Both: Both components are removed, reducing the model to functional equivalence with PTSR [2].

As shown in Table 3, both the use of length-aligned sequence prefixes and the GAT-based embedding learning individually improve recommendation performance over the PTSR [2] baseline. Moreover, their combined use leads to further gains, demonstrating complementary benefits.

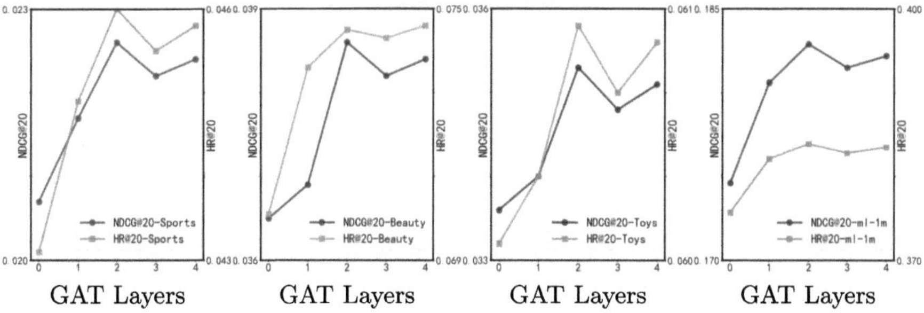


**Fig. 2.** The impact of different numbers of GAT Layers on four datasets.

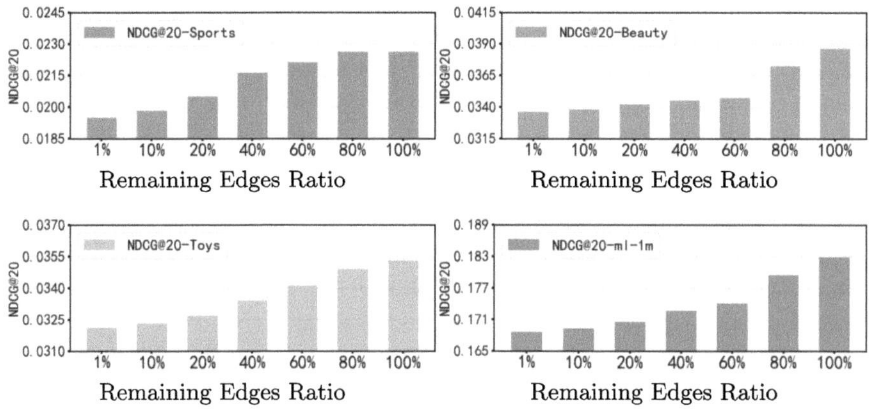


**Fig. 3.** The impact of the observed edges ratio in the Prefix-Target Graph on model performance.

## 4.4 Further Analysis (RQ3)

We conducted multiple other experiments on our model to explore some details of the model.

**Impact of GAT Layers.** As illustrated in Fig. 2, the model achieves optimal performance with two GAT layers. Beyond this point, additional layers lead to gradual performance degradation. This trend suggests that while deeper architectures can capture higher-order interactions, excessive depth may cause overfitting to the training data, consequently reducing generalization capability.

**Impact of the Observed Edges Ratio.** We investigate the impact of the observed edges ratio in PTG by randomly masking edges. As shown in Fig. 3, performance improves with higher edge ratios, indicating that more observed edges provide richer information for learning item-sequence relationships. LAPTG effectively utilizes this information to capture user intention transitions. Performance plateaus at sufficient edge ratios, demonstrating LAPTG's strong generalization ability with moderately sparse structures.

## 5 Conclusion

In this paper, we tried to address the challenge of sequence length imbalance in sequential recommendation by proposing LAPTG, a framework that integrates length-aligned data enhancement with attentive graph learning. Unlike conventional approaches, LAPTG constructs prefix-target pairs through fixed-size window sampling and enhances shorter sequences via similarity-based length alignment. The framework adopts Graph Attention Networks to model relationships between prefixes and targets, adaptively learning the importance of node connections. Experiments on four real-world datasets demonstrate that LAPTG achieves consistent improvements over strong baselines, indicating its effectiveness in modeling diverse sequence patterns and user intention transitions. In future work, we plan to explore more advanced alignment strategies and incorporate temporal dynamics to further enhance recommendation performance.

**Acknowledgments.** This work has been supported in part by the National Natural Science Foundation of China (Nos. 62272230).

## References

1. Chang, J., Gao, C.: Sequential recommendation with graph neural networks. In: SIGIR (2021)
2. Chen, J., Du, X., Pan, Y., Tang, J.: PTSR: prefix-target graph-based sequential recommendation. In: CIKM (2024)
3. Chen, Y., Liu, Z., Li, J., McAuley, J., Xiong, C.: Intent contrastive learning for sequential recommendation. In: WWW (2022)
4. Dai, L., Du, X., Tang, X., J.: TRGA: reconsidering the application of graph neural networks in two-view correspondence pruning. In: MM (2024)
5. Devlin, J., Chang, M.W., Lee, K., Toutanova, K.: Bert: pre-training of deep bidirectional transformers for language understanding. arXiv preprint arXiv:1810.04805 (2018)
6. Du, X., Wu, X., Feng, Z., He, F., Tang, X., J.: Invariant representation learning for multimedia recommendation. In: MM (2022)
7. Du, X., He, X., Yuan, F., et al.: Modeling embedding dimension correlations via convolutional neural collaborative filtering. TOIS (2019)
8. Grossberg, S.: Recurrent neural networks. Scholarpedia **8**(2), 1888 (2013)
9. Hamilton, W., Ying, R., Leskovec, J.: Inductive representation learning on large graphs. In: NeurIPS (2017)
10. Harper, F.M., Konstan, J.A.: The movielens datasets: history and context. TiiS **5**(4), 1–19 (2015)
11. He, X., Deng, K., Wang, X.: Lightgcn: simplifying and powering graph convolution network for recommendation. In: SIGIR (2020)

12. Jiang, J., Luo, Y., Kim, J.B., Zhang, K., Kim, S.: Sequential recommendation with bidirectional chronological augmentation of transformer. arXiv preprint arXiv:2112.06460 (2021)
13. Kang, W.C., McAuley, J.: Self-attentive sequential recommendation. In: ICDM (2018)
14. Kim, E., et al.: Reducedgcn: learning to adapt graph convolution for top-n recommendation. In: PAKDD (2025)
15. Kingma, D.P., Ba, J.: Adam: a method for stochastic optimization. arXiv preprint arXiv:1412.6980 (2014)
16. Kipf, T., Welling, M.: Semi-supervised classification with graph convolutional networks. arXiv preprint arXiv:1609.02907 (2016)
17. Krichene, W., Rendle, S.: On sampled metrics for item recommendation. In: SIGKDD (2020)
18. Lee, S., Yeo, K., Kim, E.: Mixture of conditional attention for multimodal fusion in sequential recommendation. In: PAKDD (2025)
19. Li, X., et al.: Multi-intention oriented contrastive learning for sequential recommendation. In: WSDM (2023)
20. Liu, L., Zhou, W., Wen, J., Zhang, Y., Wang, Y., Zhang, H.: LCAN: light cross-attention network for collaborative filtering recommendation. In: PAKDD (2022)
21. Liu, Z., Chen, Y., Li, J.: Contrastive self-supervised sequential recommendation with robust augmentation. arXiv preprint arXiv:2108.06479 (2021)
22. Liu, Z., Fan, Z., Wang, Y., Yu, P.S.: Augmenting sequential recommendation with pseudo-prior items via reversely pre-training transformer. In: SIGIR (2021)
23. Luo, S., Yao, Y., He, B.: Integrating large language models into recommendation via mutual augmentation and adaptive aggregation. arXiv preprint arXiv:2401.13870 (2024)
24. Ma, Y., et al.: Multicbr: multi-view contrastive learning for bundle recommendation. In: TOIS (2024)
25. Ma, J., Zhou, C., Yang, H., Cui, P., Wang, X., Zhu, W.: Disentangled self-supervision in sequential recommenders. In: SIGKDD (2020)
26. Ma, H., et al.: Negative sampling in recommendation. TOIS (2024)
27. McAuley, J., Targett, C., Shi, Q., Van Den Hengel, A.: Image-based recommendations on styles and substitutes. In: SIGIR (2015)
28. Niu, W., Li, Y., Liu, J., et al.: Parallel graph convolutional network for multi-modal recommendation. In: PAKDD (2025)
29. Qin, X., Yuan, H., Zhao, P., Liu, G., Zhuang, F., Sheng, V.S.: Intent contrastive learning with cross subsequences for sequential recommendation. In: WSDM (2024)
30. Qiu, R., Huang, Z., Yin, H., Wang, Z.: Contrastive learning for representation degeneration problem in sequential recommendation. In: WSDM (2022)
31. Rendle, S., Freudenthaler, C., Gantner, Z., Schmidt-Thieme, L.: BPR: Bayesian personalized ranking from implicit feedback. arXiv preprint arXiv:1205.2618 (2012)
32. Sun, F., et al.: Bert4rec: sequential recommendation with bidirectional encoder representations from transformer. In: CIKM (2019)
33. Tang, J., Wang, K.: Personalized top-n sequential recommendation via convolutional sequence embedding. In: WSDM (2018)
34. Vaswani, A., et al.: Attention is all you need. In: NeurIPS (2017)
35. Wang, X., He, X., Wang, M., Feng, F., Chua, T.S.: Neural graph collaborative filtering. In: SIGIR (2019)
36. Wang, Y., Liu, Z., Yang, L., Yu, P.S.: Conditional denoising diffusion for sequential recommendation. In: PAKDD (2024)

37. Wang, Z., Wei, W., Cong, G., Li, X.L., Mao, X.L., Qiu, M.: Global context enhanced graph neural networks for session-based recommendation. In: SIGIR (2020)
38. Wu, S., Tang, Y., Zhu, Y., Wang, L., Xie, X., Tan, T.: Session-based recommendation with graph neural networks. In: AAAI (2019)
39. Xu, X., Fei, S., Li, Z., Wang, S., Gao, J.: Contrastive learning for sequential recommendation. In: ICDE (2022)
40. Yan, Z., Xiang, X., Li, Z.: Item correlation modeling in interaction sequence for graph convolution session recommendation. Scientia Sinica (Informationis) **52**(6), 1069–1082 (2022)
41. Ying, R., He, R., Chen, K., Eksombatchai, P., Hamilton, W.L., Leskovec, J.: Graph convolutional neural networks for web-scale recommender systems. In: SIGKDD (2018)
42. Zhang, Y., Hua, Z., Yuan, L., Chen, Z.: Influential community search in bipartite graphs. arXiv preprint arXiv:2412.06216 (2024)

# Dual Nonlinear Sparse Feature Selection Method

Pan Xie[1], Cong Lei[2], Shanwen Zhang[1], and Shichao Zhang[1(✉)]

[1] Guangxi Normal University, Guilin, China
zhangsc@mailbox.gxnu.edu.cn
[2] Sydney AI Centre, School of Computer Science, The University of Sydney, Sydney, Australia
clei9249@uni.sydney.edu.au

**Abstract.** With the advancement of information technology, the growth in data dimensions poses challenges for data mining and machine learning. Feature selection serves as a critical countermeasure, reducing dimensionality while preserving core information to lay the groundwork for subsequent downstream applications. Existing feature selection methods often focus solely on single-level nonlinear modeling, struggling to capture nonlinear correlations between features and thus limiting their representational capabilities. To address this, this paper proposes DNSFS, a sparse feature selection method that simultaneously accounts for nonlinearity at both the sample and feature levels. This approach utilizes a kernel matrix to capture nonlinear dependencies among features and uncover their inherent association patterns. Concurrently, it constructs a hypergraph structure to model complex relationships between samples, thereby preserving nonlinear features at the sample level. Furthermore, a reasonable regularization strategy balances the weighting of nonlinear information across sample and feature levels, enabling sparse feature selection. Experiments demonstrate that DNSFS achieves superior classification accuracy compared to existing methods. Additionally, ablation studies further validate the effectiveness of the dual nonlinear design.

**Keywords:** Feature Selection · Hypergraph Representation · Low-Rank Constraint · Attribute Self-Representation · Kernel Function

## 1 Introduction

Information technology has brought tremendous convenience to people while driving explosive growth in the scale and dimensions of data. Not all features in raw data are relevant to the target task, and some features may be redundant [1]. Feature selection can filter out key features, reduce data dimensions, and effectively alleviate this issue. Through feature selection algorithms, irrelevant and redundant features can be removed, allowing the model to focus on valuable information [2].


R. C.-W. Wong et al. (Eds.): PAKDD 2026, LNAI 16597, pp. 379–390, 2026.
https://doi.org/10.1007/978-981-92-1300-9_30

In machine learning and data mining, feature selection involves filtering out the most information-rich subset of features from the raw data for model training or task objectives. Existing methods can be broadly categorized into linear and nonlinear feature selection based on how they model data association patterns. Linear feature selection methods, exemplified by Lasso regularization [3], achieve feature sparsity while uncovering linear relationships between features and target variables. Nonlinear feature selection methods leverage nonlinear functions or structures to capture complex nonlinear dependencies among features, such as LLE [4] utilizing manifold learning concepts. Since core relationships in real-world data often exhibit nonlinear characteristics rather than simple linear dependencies, nonlinear methods hold significant advantages in uncovering complex data structures [5]. Therefore, this paper focuses on the direction of nonlinear feature selection.

Despite numerous explorations into nonlinear feature selection methods, existing research still exhibits significant limitations: most approaches model nonlinear relationships in a single dimension, either emphasizing feature correlations or focusing on sample-level structures. Previous work has neglected the synergistic consideration of nonlinearity at both the sample and feature levels, failing to fully uncover the complex nonlinear structures coexisting in sample interactions and feature correlations. This results in constrained feature representation capabilities, ultimately impacting the performance of downstream tasks.

## Our Contribution

To screen for more valuable feature subsets, we propose the Dual Nonlinear Sparse Feature Selection (DNSFS) method. By synergistically considering nonlinear correlations between samples and features, it enhances the representational power of feature subsets, addressing the shortcomings of existing approaches. We achieve this through the following aspects:

- Using a nuclear matrix instead of the original feature matrix in the feature selection process aims to uncover nonlinear relationships among features. We employ hypergraph structures to model the structural relationships between similar samples. Leveraging the hypergraph's ability to represent multidimensional relationships, this approach enables nonlinear considerations at the sample level.
- Through sparse feature selection via attribute self-expression after low-rank decomposition, this self-expressive structure enables feature selection to focus more on the structural information inherent to the features themselves, thereby identifying more important feature subsets.

## 2 Related Work

The essence of feature selection lies in filtering out the most representative subset of features from the original feature set. From a procedural perspective, the feature selection process can be decomposed into two key stages: first, preprocessing the feature matrix of the original data; second, designing the search strategy for the optimal subset.

**Processing of the Original Feature Matrix.** Feature extraction is a common preprocessing method for raw feature matrices, generating entirely new low-dimensional features through transformation. Traditional approaches often utilize the unprocessed original matrix directly. Furthermore, while linear transformation methods such as PCA and its derivatives IPCA [6] and Sparse PCA [7] can accommodate large-scale data or enhance interpretability, they struggle to capture nonlinear relationships. Nonlinear transformation methods like LLE [4] and Isomap [8], while preserving feature information, tend to lose some inter-sample correlations.

**Optimal Subset Screening Strategy.** In recent years, meta-heuristic optimization algorithms such as WOA [9] and PSO [10] have emerged as mainstream feature selection strategies, enhancing subset optimization performance through various biomimetic and optimization mechanisms. However, these methods exhibit certain limitations: First, they consider feature information in a singular manner, struggling to coordinate deep-level correlations between data and features, resulting in feature subsets that fail to reflect the essence of the data. Second, biomimetic mechanisms are constrained by population interaction simulations, making it difficult to capture the diverse and complex correlations across different datasets, thereby limiting the effectiveness of the resulting subsets.

## 3 Preliminary Concepts

**Kernel Function.** For any two samples $X_1$ and $X_2$ in an arbitrary space, if there exists a mapping $\phi : \mathcal{X} \to \mathcal{H}$ from the low-dimensional space to the high-dimensional space (where $\mathcal{X}$ is the low-dimensional input space and $\mathcal{H}$ is the high-dimensional feature space), such that $k(x_1, x_2) = \langle \phi(x_1), \phi(x_2) \rangle$, then $k(x_1, x_2)$ is called a kernel function [11]. Here, $\langle \cdot, \cdot \rangle$ denotes the inner product operation in the high-dimensional space $\mathcal{H}$. The Gaussian kernel, the most widely used nonlinear kernel function, possesses strong nonlinear capabilities and can accommodate most complex data distributions:

$$K(x_1, x_2) = \exp\left(-\frac{\|x_1 - x_2\|^2}{2\sigma^2}\right) \tag{1}$$

**Attribute Self-expression.** For a given feature set $X = [x_1, x_2, \ldots, x_s] \in R^{s \times d}$ of a dataset, any sample $X_i (i \in N^+)$ can be represented as a linear combination of other samples $X_j (j \in N^+$ and $j \neq i)$ within the feature set $X$. This set of linear relationships can be represented by a column vector $\mu_i \in R^d \times 1$, where $\mu_i$ is termed the feature self-expression coefficient [12]. Furthermore, the self-expression structure of sample $X_i$ requires consideration of an error term. In summary, the feature set $X$ should be expressed as:

$$x_i = \mu_i X + b \tag{2}$$

**Hypergraph.** Hypergraphs are generalized graph structures that extend ordinary graph edges into hyperedges capable of connecting any number of vertices, and represent complex relationships through sets of vertices and hyperedges [13] (Fig. 1).

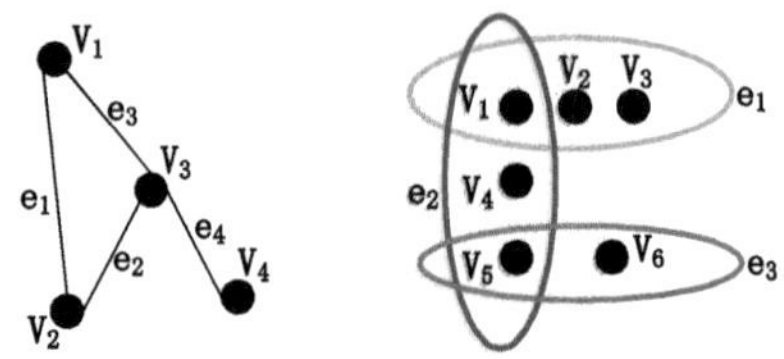


**Fig. 1.** Simple Graph (right) and HyperGraph (left)

$$L_{\text{norm}} = \mathbf{I} - \mathbf{D}_e^{-1/2}\mathbf{H}^\top \mathbf{D}_v^{-1}\mathbf{H}\mathbf{W}\mathbf{D}_e^{-1/2} \tag{3}$$

The normalized hypergraph feature similarity matrix $L_{\text{norm}}$ encapsulates the associative structural information between nodes and features within the hypergraph, serving as a standardized similarity metric [14]. $\mathbf{I}$ is a $d \times d$ identity matrix; $\mathbf{D}_e^{-1/2}$ (inverse square root of the hyperedge degree matrix) balances feature degree differences, and $\mathbf{D}_v^{-1}$ (inverse node degree matrix) normalizes node scale influence.

# 4 Our Approach

## 4.1 Method Objective Design

For an feature selection method with a regularization strategy, its objective function takes the following form:

$$\min_W \|Y - XW\|_F^2 + \alpha\|W\|_{2,p} + \beta\phi(W) \tag{4}$$

This function is commonly used in regularized feature selection problems. As a typical form of the embedded method, it combines a loss function and regularization terms to achieve feature selection [15].

The regularization term is $\alpha\|\mathbf{W}\|_{2,p}$, where $\alpha$ is a regularization parameter with $\alpha \geq 0$, used to balance the importance of the loss function and the regularization term. This term serves to constrain the weight matrix $\mathbf{W}$. By restricting $\mathbf{W}$ with different norms, the weights of certain features are driven toward 0. When the weight corresponding to an feature equals 0, it is equivalent to eliminating that feature from the model, thereby achieving feature selection.

The regularization term $\beta\phi(\mathbf{W})$ is a user-defined regularization function, which can be designed according to specific requirements and problems. $\beta$ is used to control the strength of this custom regularization term.

## 4.2 Method Description

Given a dataset $X \in R^{s\times d}$ (where $s$ is the number of samples and $d$ is the number of features), it is necessary to process the samples based on features. For this purpose, $X$ is split into $d$ column vectors $\mathbf{f}_i(x_i^1, x_i^2, x_i^3, \ldots, x_i^s)$ $(i = 1, 2, \ldots, d)$ according to the $d$ features. We select $X_i \in R^{s\times 1}$ $(i \in N^+)$ and perform kernel function transformation on each element between $f_i$.

The Gaussian kernel function used in this paper is $k(x_1, x_2) = e^{-\frac{\|x_1-x_2\|^2}{2\sigma^2}}$. Through the kernel transformation process $k(f_1, f_1), k(f_1, f_2), k(f_1, f_3), \ldots$, the original feature set $f_i$ is transformed and assembled into the following kernel matrix:

$$K = \begin{bmatrix} k(f_1, f_1) & k(f_1, f_2) & \ldots & k(f_1, f_d) \\ k(f_2, f_1) & k(f_2, f_2) & \ldots & k(f_2, f_d) \\ \vdots & \vdots & \ddots & \vdots \\ k(f_d, f_1) & k(f_d, f_2) & \ldots & k(f_d, f_d) \end{bmatrix} \tag{5}$$

As can be seen from Matrix (5), the kernel matrix satisfies the transpose relationship $K^T = K$. Subsequently, the kernel matrix $K$ becomes a new representation form of the feature set. The connections between features are obtained based on the attribute Self-Representation matrix:

$$K = KQ + eb \tag{6}$$

where $Q \in R^{d\times d}$ is the attribute Self-Representation matrix; $e \in R^{d\times 1}$ is a unit column vector (a $d$ dimensional column vector with all elements being 1); and $b \in R^{1\times d}$ is the error bias term.

In this paper, the mixed $\ell_{2,p}$-norm constraint is adopted, and the loss function term is formulated as $\min_{Q,b} \|K - KQ - eb\|_{2,p}$. Considering that the attribute Self-Representation matrix may lead to redundancy and outlier data issues, a reconstruction loss term $\Omega(\boldsymbol{Q})$ is added.

$$\Omega(\boldsymbol{Q}) = \|\boldsymbol{Q}\|_{2,p} = \left[\sum_{i=1}^{m}\left(\sum_{j=1}^{m} Q_{ij}^2\right)^{\frac{p}{2}}\right]^{\frac{1}{p}} \quad (0 < p < 2) \tag{7}$$

The kernel matrix $\mathbf{K}$ embeds the original feature set into a high-dimensional space through a global nonlinear mapping, revealing the data's global structure and latent relationships. Hypergraphs establish hyperedge connections based on samples, capturing nonlinear associations between adjacent samples. During feature selection, hypergraph data accounts for the proximity of samples, ensuring that hierarchical relationships among samples in the low-dimensional space are not diluted after feature selection. This benefits subsequent classification tasks. Consequently, we also introduce a hypergraph feature similarity matrix regularization term $\theta(\boldsymbol{Q})$:

$$\theta(\boldsymbol{Q}) = \operatorname{tr}\left(\boldsymbol{Q}^{\mathrm{T}}\boldsymbol{K}^{\mathrm{T}}\boldsymbol{L}_{\mathrm{norm}}\boldsymbol{K}\boldsymbol{Q}\right) \tag{8}$$

Furthermore, since $\boldsymbol{K}^{\mathrm{T}} = \boldsymbol{K}$, the following function is obtained:

$$\min_{Q,b} \|\boldsymbol{K} - \boldsymbol{K}\boldsymbol{Q} - \boldsymbol{e}\boldsymbol{b}\|_{2,p} + \alpha\|\boldsymbol{Q}\|_{2,p} + \beta\operatorname{tr}\left(\boldsymbol{Q}^{\mathrm{T}}\boldsymbol{K}\boldsymbol{L}_{\mathrm{norm}}\boldsymbol{K}\boldsymbol{Q}\right) \tag{9}$$

In Zhi Zhong's research [16], complex structural information is extracted by imposing low-rank constraints on the feature matrix, enabling data restoration and reconstruction. And it demonstrates that reasonable low-rank constraints preserve core relationships within the feature matrix while effectively eliminating noise and redundant interference.

This paper introduces a low-rank constraint to the feature self-expression matrix Q (rank $r \leq \min(s, d)$) to identify more critical and strongly correlated features within the low-rank space. The low-rank constraint is implemented as: $Q = AB$, where $A \in R^{dxr}$ and $B \in R^{rxd}$. This decomposition of the $Q$ matrix intuitively frees up storage space and enhances computational efficiency. In summary, the objective function is modeled as follows:

$$\min_{A,B,b} \|\boldsymbol{K} - \boldsymbol{K}\boldsymbol{A}\boldsymbol{B} - \boldsymbol{e}\boldsymbol{b}\|_{2,p} + \alpha\|\boldsymbol{A}\boldsymbol{B}\|_{2,p} + \beta\operatorname{tr}\left(\boldsymbol{B}^{\mathrm{T}}\boldsymbol{A}^{\mathrm{T}}\boldsymbol{K}\boldsymbol{L}_{\mathrm{norm}}\boldsymbol{K}\boldsymbol{A}\boldsymbol{B}\right)$$
$$\text{s.t. } \operatorname{rank}(\boldsymbol{A}\boldsymbol{B}) \leq \min(s, d) \tag{10}$$

The method flow is as follows:

**Algorithm 1:** DNSFS: Given the original dataset and hyperparameters $\alpha, \beta, r$, 10-fold cross-validation is adopted to output the classification accuracy of the SVM.

**Input:** Dataset $X \in R^{s \times d}$, Hyperparameters $\alpha, \beta, r$
**Output:** Classification accuracy

$K$ = MapKernel(Att($X$)) (5);
$L_{\mathrm{norm}}$ = CalcWeightMat($K$) and $Q$ = SelfExprAttr($K$) (3)(6);
$AB$ = LowRankFact($Q, \alpha, \beta, r$), rank($Q$) $\leq \min(s, d)$ (10);
**for** *each row of matrix AB* **do**
  row_abs_sum = sum of absolute values of elements in current row;
  threshold = 0.1 × mean(sum of absolute values of all elements);
  **if** *row_abs_sum ≤ threshold* **then**
    remove elements of the current row;
  **end**
**end**
Evaluate via SVM classification using the matrix AB;
**return** Classification accuracy;

### 4.3 Convergence Analysis of Objective Functions

TThis algorithm iteratively updates variables $A, B$ using an alternating optimization strategy [17]. The alternating optimization approach optimizes a multivariable objective function by fixing some variables and minimizing with respect to the others, alternating until convergence.

After eliminating the bias term $b$ using the centering matrix

$$H = I - \frac{1}{d} e e^T, \tag{11}$$

the objective function (10) becomes

$$\min_{A,B} \mathcal{J}' = \|HK - HKAB\|_{2,p} + \alpha \|AB\|_{2,p} + \beta \cdot \mathrm{tr}\left(B^T A^T K L_{\mathrm{norm}} K A B\right). \tag{12}$$

Let $A^{(t)}, B^{(t)}$ and $A^{(t+1)}, B^{(t+1)}$ denote the variables at iteration $t$ and $t+1$, respectively. We prove

$$\mathcal{J}'\left(A^{(t+1)}, B^{(t+1)}\right) \leq \mathcal{J}'\left(A^{(t)}, B^{(t)}\right). \tag{13}$$

The objective $\mathcal{J}'$ consists of three terms:

(1) Reconstruction error $\mathcal{J}_1 = \|HK - HKAB\|_{2,p}$. Let $D = HK - HKAB$, then

$$\mathcal{J}_1 = \sum_{i=1}^{d} \left(\|D_i\|_2^2\right)^{p/2}. \tag{14}$$

Since $A^{(t+1)}$ is optimal for fixed $B^{(t)}$ and $B^{(t+1)}$ optimal for fixed $A^{(t+1)}$, and the $\ell_{2,p}$ norm ($p \in [1,2]$) is convex, we have

$$\mathcal{J}_1^{(t+1)} \leq \mathcal{J}_1^{(t)}. \tag{15}$$

(2) Sparsity regularization $\mathcal{J}_2 = \alpha\|AB\|_{2,p}$. Let $Q = AB$, then

$$\mathcal{J}_2 = \alpha \sum_{i=1}^{d} \left(\|Q_i\|_2^2\right)^{p/2}. \tag{16}$$

By the same alternating optimality,

$$\mathcal{J}_2^{(t+1)} \leq \mathcal{J}_2^{(t)}. \tag{17}$$

(3) Hypergraph similarity $\mathcal{J}_3 = \beta \cdot \mathrm{tr}\left(Q^T K L_{\mathrm{norm}} K Q\right)$. Since $K$ and $L_{\mathrm{norm}}$ are positive semidefinite, $\mathcal{J}_3$ is a nonnegative quadratic form. The alternating minimization gives

$$\mathcal{J}_3^{(t+1)} \leq \mathcal{J}_3^{(t)}. \tag{18}$$

Combining the three terms yields

$$\mathcal{J}'^{(t+1)} = \mathcal{J}_1^{(t+1)} + \mathcal{J}_2^{(t+1)} + \mathcal{J}_3^{(t+1)} \leq \mathcal{J}_1^{(t)} + \mathcal{J}_2^{(t)} + \mathcal{J}_3^{(t)} = \mathcal{J}'^{(t)}. \tag{19}$$

Thus, the objective function is monotonically decreasing and bounded below by zero. By the monotone convergence theorem, $\mathcal{J}'$ converges to a finite limit, guaranteeing algorithm stability.

**Table 1.** Number of samples, features, and classes for each dataset

| Dataset | Samples | Features | Classes |
|---|---|---|---|
| Lung discrete | 73 | 325 | 7 |
| Ecoli | 336 | 343 | 8 |
| Sonar | 208 | 60 | 2 |
| Movements | 360 | 90 | 15 |
| Arrhythmia | 452 | 279 | 13 |
| Yale | 165 | 1024 | 15 |

# 5 Evaluation

**Datasets:** The feature selection method proposed in this paper will be tested on 6 datasets, all of which are derived from UCI [18]. The number of samples, number of features, and number of classes for each dataset are listed below (Table 1):

**Comparative Method:** The baseline method directly utilizes the raw feature matrix, while the remaining comparison methods encompass typical strategies in the feature selection domain: AdbetWOS [19] employs heuristic greedy search, and AAPSO [20] achieves global optimization of feature subsets through iterative particle swarm optimization algorithms. Both belong to optimal subset search methods that do not introduce nonlinear mechanisms; KPCA [21] employs kernel techniques to achieve global nonlinear consideration at the feature level, preserving overall data distribution information. KLPP [22] maintains sample neighborhood relationships to achieve local nonlinear embedding at the sample level. VHSICLasso [23] combines a single nonlinear method with variational inference to achieve sparse feature selection.

**Evaluation Criteria:** All methods employed 10-fold cross-validation for SVM classification experiments. To control variables, SVM classifiers were run using non-linear function. Classification accuracy served as the evaluation metric. After completing all experiments, results from 10 runs were averaged, with standard deviation added or subtracted to assess the performance of various methods.

## 5.1 Comparative Experiment

**Parameter Settings:** Low-rank constraint parameter $r \in \{5, 10, 15, 20, 25, 30\}$ (satisfying $r \leq \min(s, d)$, determined by sample size and feature dimension). Candidate values for parameters $\alpha$, $\beta$ are $\{10^{-3}, 10^{-2}, 10^{-1}, 10, 10^2, 10^3\}$.

DNSFS achieved an average classification accuracy improvement of 5.01% over the baseline and a maximum classification accuracy improvement of 8.25%. Compared to the baseline, the DNSFS method demonstrates significant advantages in classification tasks. Across all datasets, the 10-fold average classification accuracy of DNSFS outperformed all comparison methods. Compared to Adbet-WOS, DNSFS achieved an average improvement of 2.95%; compared to AAPSO, it achieved an average improvement of 3.29%; compared to KPCA by 2.62% on average, and compared to KLPP by 2.92% on average. The second-highest classification accuracy was achieved by VHSICLasso, which combines nonlinear methods with a sparse search strategy based on variational inference. The DNSFS method achieves nonlinear considerations at both the sample and feature levels, outperforming the single-nonlinear VHSICLasso by 1.72% (Table 2).

**Table 2.** Comparative experiment: statistical results of accuracy (mean ± mean square error)

| Method | Lung discrete | Ecoli | Sonar | Movements | Arrhythmia | Yale | Average Value |
|---|---|---|---|---|---|---|---|
| baseline | 80.29% ± 1.68% | 82.13% ± 0.47% | 82.51% ± 1.59% | 85.39% ± 0.68% | 66.52% ± 1.15% | 68.75% ± 1.11% | 77.60% ± 1.11% |
| AdbetWOS | 82.16% ± 1.33% | 83.30% ± 0.80% | 83.29% ± 1.78% | 87.72% ± 1.08% | 67.12% ± 0.89% | 74.34% ± 1.14% | 79.66% ± 1.17% |
| AAPSO | 81.64% ± 1.17% | 83.98% ± 0.63% | 83.95% ± 0.75% | 87.83% ± 0.92% | 67.36% ± 0.90% | 71.19% ± 0.92% | 79.32% ± 0.88% |
| KPCA | 83.44% ± 1.42% | 84.22% ± 0.38% | 83.72% ± 1.33% | 87.88% ± 1.43% | 68.29% ± 0.66% | 72.54% ± 0.95% | 80.01% ± 1.03% |
| KLPP | 84.21% ± 1.03% | 83.52% ± 0.21% | 86.08% ± 1.56% | 86.17% ± 0.71% | 67.39% ± 0.49% | 70.79% ± 0.37% | 79.69% ± 0.73% |
| VHSICLasso | 85.14% ± 1.02% | 84.56% ± 0.92% | 86.27% ± 0.23% | 86.64% ± 1.01% | 68.85% ± 0.77% | 73.88% ± 0.78% | 80.89% ± 0.80% |
| DNSFS | **88.54% ± 1.66%** | **85.43% ± 0.65%** | **87.60% ± 0.88%** | **88.42% ± 1.12%** | **70.08% ± 1.66%** | **75.58% ± 0.92%** | **82.61% ± 1.15%** |

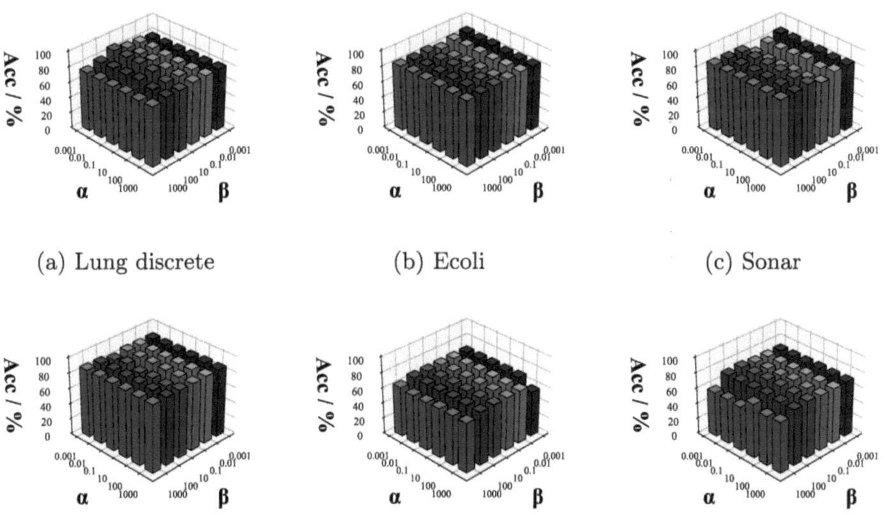


(a) Lung discrete (b) Ecoli (c) Sonar

(d) Movements (e) Arrhythmia (f) Yale

**Fig. 2.** 3D Bar Chart of DNSFS Performance with Different ($\alpha$, $\beta$) Parameter Combinations on Datasets.

Figure 2 shows the performance of the DNSFS with different regularization parameters ($\alpha$, $\beta$) across various datasets. Specifically, the optimal parameter combinations ($\alpha$, $\beta$) for DNSFS on the corresponding datasets are $(10^{-3}, 10)$, $(10^{-3}, 10^{-3})$, $(10^{3}, 10)$, $(10^{2}, 10^{-3})$, $(10^{-1}, 10^{-2})$ and $(10, 10)$ respectively.

### 5.2 Ablation Experiment

To validate the effectiveness of core components, this paper conducts ablation studies focusing on the kernel matrix $K$ and hypergraph structure. By removing components and comparing performance, the contribution of each component is clarified. Four model variants were tested: the full method DNSFS-full, DNSFS-w/o HyperGraph (without hypergraph), DNSFS-w/o Kernel (without kernel matrix), and the model DNSFS-base (omitting both). All models underwent 10-fold cross-validation, with classification accuracy mean serving as the evaluation metric. Optimal parameter combinations and dataset consistency were maintained across all tests to ensure reliable results.

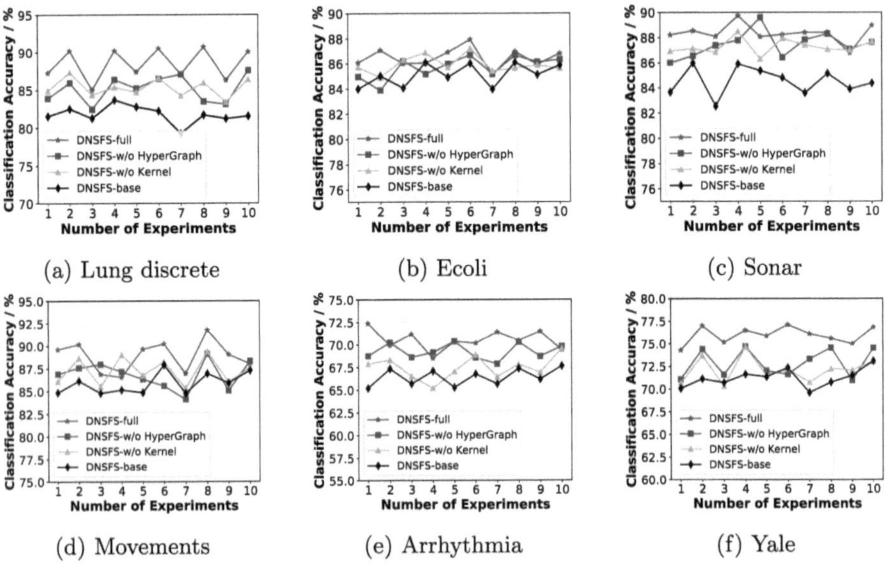


(a) Lung discrete (b) Ecoli (c) Sonar

(d) Movements (e) Arrhythmia (f) Yale

**Fig. 3.** Ablation Study Results of DNSFS Method.

Analysis of Fig. 3 reveals that DNSFS-full achieves an average classification accuracy improvement of 2% to 3% compared to single-component methods, representing a significant enhancement over DNSFS-base. Classification accuracy exhibits a marked decline from the full method to single-component ablation and ultimately to complete component removal. This fully demonstrates the necessity and effectiveness of the dual nonlinear design of the kernel matrix and hypergraph structure.

## 6 Conclusion

We propose a dual nonlinear feature selection method, DNSFS, that considers both sample and feature levels to select more valuable feature subsets across different datasets, thereby improving classification accuracy. Compared to linear and single nonlinear approaches, our method fully accounts for nonlinear correlations between features and samples. Through its dual nonlinear design, it captures deeper latent relationships among features. Comparative experiments on multiple public benchmark datasets demonstrate that DNSFS achieves significantly higher classification accuracy than existing methods. Ablation studies further validate that the dual nonlinearity design in DNSFS is crucial for enhancing the representational power of the feature subset. We believe our work provides a novel approach and strategy for feature selection, contributing to future feature selection fusion applications.

## References

1. Tosi, D., Kokaj, R., Roccetti, M.: 15 years of big data: a systematic literature review. J. Big Data **11**(1), 73 (2024)
2. Ramos-Pérez, I., Barbero-Aparicio, J.A., Canepa-Oneto, A., Arnaiz-González, Á., Maudes-Raedo, J.: An extensive performance comparison between feature reduction and feature selection preprocessing algorithms on imbalanced wide data. Information **15**(4), 223 (2024)
3. Tibshirani, R.: Regression shrinkage and selection via the lasso. J. R. Stat. Soc. Ser. B Stat Methodol. **58**(1), 267–288 (1996)
4. Roweis, S.T., Saul, L.K.: Nonlinear dimensionality reduction by locally linear embedding. Science **290**(5500), 2323–2326 (2000)
5. Basile, L., Acevedo, S., Bortolussi, L., Anselmi, F., Rodriguez, A.: Intrinsic dimension correlation: uncovering nonlinear connections in multimodal representations. arXiv preprint arXiv:2406.15812 (2024)
6. Aboalsamh, H.A., Mathkour, H.I., Assassa, G.M., Mursi, M.F.: Face recognition using incremental principal components analysis. In: 2009 International Conference on Computing, Engineering and Information, pp. 39–43. IEEE (2009)
7. Zou, H., Hastie, T., Tibshirani, R.: Sparse principal component analysis. J. Comput. Graph. Stat. **15**(2), 265–286 (2006)
8. Tenenbaum, J.B., Silva, V.D., Langford, J.C.: A global geometric framework for nonlinear dimensionality reduction. Science **290**(5500), 2319–2323 (2000)
9. Mafarja, M., Mirjalili, S.: Whale optimization approaches for wrapper feature selection. Appl. Soft Comput. **62**, 441–453 (2018)
10. Chen, K., Zhou, F.Y., Yuan, X.F.: Hybrid particle swarm optimization with spiral-shaped mechanism for feature selection. Expert Syst. Appl. **128**, 140–156 (2019)
11. Du, K.L., Jiang, B., Lu, J., Hua, J., Swamy, M.: Exploring kernel machines and support vector machines: principles, techniques, and future directions. Mathematics **12**(24), 3935 (2024)
12. Zhang, S., You, C., Vidal, R., Li, C.G.: Learning a self-expressive network for subspace clustering. In: Proceedings of the IEEE/CVF Conference on Computer Vision and Pattern Recognition, pp. 12393–12403 (2021)

13. Cheng, X., Zhu, Y., Song, J., Wen, G., He, W.: A novel low-rank hypergraph feature selection for multi-view classification. Neurocomputing **253**, 115–121 (2017)
14. Wei, W., Dai, H., Liang, W.: Regularized least squares locality preserving projections with applications to image recognition. Neural Netw. **128**, 322–330 (2020)
15. Bai, L., Li, H., Gao, W., Xie, J., Wang, H.: A joint multiobjective optimization of feature selection and classifier design for high-dimensional data classification. Inf. Sci. **626**, 457–473 (2023)
16. Zhong, Z.: Adaptive graph learning and low-rank constraint for supervised spectral feature selection. Neural Comput. Appl. **32**(11), 6503–6512 (2020)
17. Latafat, P., Themelis, A., Patrinos, P.: On the convergence of adaptive first order methods: proximal gradient and alternating minimization algorithms. In: 6th Annual Learning for Dynamics & Control Conference, pp. 197–208. PMLR (2024)
18. UCI repository of machine learning datasets (2025). http://archive.ics.uci.edu/ml. Accessed 09 Sept 2025
19. Bhattacharya, A., Saha, B., Chattopadhyay, S., Sarkar, R.: Deep feature selection using adaptive $\beta$-hill climbing aided whale optimization algorithm for lung and colon cancer detection. Biomed. Signal Process. Control **83**, 104692 (2023)
20. Pramanik, R., Sarkar, S., Sarkar, R.: An adaptive and altruistic PSO-based deep feature selection method for pneumonia detection from chest X-rays. Appl. Soft Comput. **128**, 109464 (2022)
21. Wang, Y., Yu, H., Li, X.: Efficient iterative dynamic kernel principal component analysis monitoring method for the batch process with super-large-scale data sets. ACS Omega **6**(15), 9989–9997 (2021)
22. Lin, C., Jiang, J., Zhao, X., Pang, M., Ma, Y.: Supervised kernel optimized locality preserving projection with its application to face recognition and palm biometrics. Math. Probl. Eng. **2015**(1), 421671 (2015)
23. Koyama, K., Kiritoshi, K., Okawachi, T., Izumitani, T.: Effective nonlinear feature selection method based on HSIC lasso and with variational inference. In: International Conference on Artificial Intelligence and Statistics, pp. 10407–10421. PMLR (2022)

# HEIR-Nets: Hierarchical Ensemble with Inheritance-Based Refinement Networks for Sperm Target Detection with Noisy Labels

Huiliang Lv[1], Yang Zhao[1(✉)], Xuan Yang[1], Jihong Pei[2], and Jiahui Wu[3]

[1] College of Computer Science and Software Engineering, Shenzhen University, Shenzhen 518060, China
2023150122@email.szu.edu.cn, zhaoyang1990@szu.edu.cn

[2] College of Electronics and Information Engineering, Shenzhen University, Shenzhen 518060, China
jhpei@szu.edu.cn

[3] College of Life Sciences and Oceanography, Shenzhen University, Shenzhen 518060, China
wujiahui@szu.edu.cn

**Abstract.** Sperm motility assessment in Assisted Reproductive Technology critically relies on accurate detection of biomedical targets within electron microscopic sperm images. However, annotating numerous small targets per image poses significant challenges, inevitably introducing noisy labels. These label errors severely degrade the learning efficacy of existing deep architectures, limiting high-precision sperm detection and classification. Therefore, we propose the Hierarchical Ensemble with Inheritance-based Refinement Networks (HEIR-Nets) featuring three synergistic strategies: 1) Self-Correcting Pseudo-label Inheritance (SPI) performs dual-action correction by analyzing pseudo-label confidence from the current branch model, while incorporating random target dropout intervention to generate self-corrected sample sets for inheritance by subsequent branches. 2) Multi-model Knowledge Inheritance (M2KI) initializes current branch models using predecessor branch model parameters, while iteratively training them on inherited purified datasets. 3) Multi-branch Fusion (MBF) integrates dual-weighted decisions by combining designed gamma-encoded model credibility weights with sample confidence weights. The performance of HEIR-Nets is verified on the self-built sperm image dataset with noisy labels of different proportions and the public sperm image dataset. The results validate HEIR-Nets' superiority and robustness. In particular, on the public dataset with 20% noisy labels, the F1 score and accuracy of HEIR-Nets exceeded those of the state-of-the-art methods by 4.56% and 7.54%, respectively.

**Keywords:** Noisy Labels · Sperm Target Detection · Self-Correcting Pseudo-label · Multi-branch Fusion


R. C.-W. Wong et al. (Eds.): PAKDD 2026, LNAI 16597, pp. 391–403, 2026.
https://doi.org/10.1007/978-981-92-1300-9_31

# 1 Introduction

In assisted reproductive technology (ART), sperm quality assessment is an extremely critical step [4]. Sperm medical target detection and classification are the foundation of downstream processes such as sperm morphology analysis, vitality assessment, and optimization strategy formulation [6]. Therefore, building a robust and high-precision sperm detection system is the core foundational task for promoting the intelligent process of reproductive diagnosis and treatment.

With the widespread adoption of deep learning in medical image analysis, automatic sperm image detection and classification have attracted increasing attention. Convolutional neural network (CNN)-based methods enable end-to-end feature extraction and prediction, among which YOLO series models are widely applied in medical object detection due to their real-time performance and flexible deployment [1,8]. YOLOv11 focuses on lightweight design, reconstructing the detection head through reparameterized convolution and channel attention module, optimizing the detection head to improve the balance between performance and parameter quantity [2]. YOLOv12 departs from traditional CNN-based designs by adopting a pure attention backbone with residual gating, significantly improving sensitivity to small targets and robustness in complex scenes [9]. In the sperm medical object detection task, the targets to be detected are predominantly small-scale objects. Thus, YOLOv12 is selected as the baseline model for our proposed architecture.

It's worth noting that sperm images have characteristics such as dense target distribution, small individual size, and complex backgrounds, resulting in uneven image quality, as shown in Fig. 1. Manual annotation is prone to problems such as confused categories, missing labels, and blurred boundaries. Although the mainstream YOLO models have excellent small target detection capability, it is difficult to resist the cumulative impact of noisy labels [7]. Several robust detection methods have been proposed to mitigate the impact of noisy labels, such as Meta Refine Net (MRNet) [11] and object-aware multiple instance learning (OA-MIL) [3]. While these approaches alleviate noise effects to some extent, they either require complex training procedures or exhibit limited stability under diverse noise distributions, making them less suitable for fine-grained medical object detection.

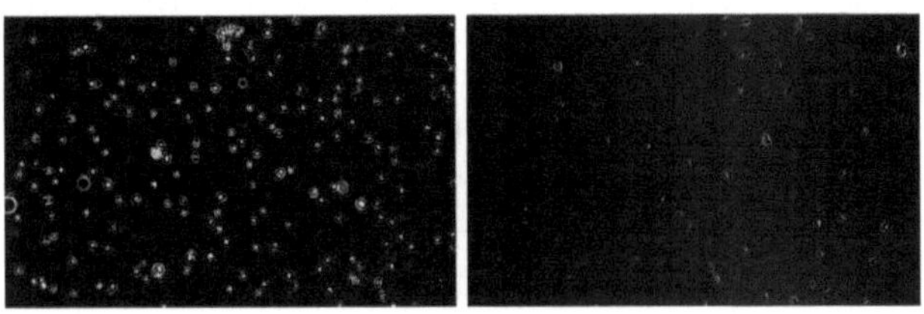

**Fig. 1.** Partial sperm electron microscopy images.

Overall, the existing methods have failed to effectively utilize data iteration mechanisms and ignore knowledge transfer between models when facing noisy labels in medical micro object detection, making it difficult to meet the practical needs of label diversity and complexity in sperm image detection.

To address the above issues, this paper proposes a multi object detection network framework for sperm images with noisy labels, namely **H**ierarchical **E**nsemble with **I**nheritance-based **R**efinement Networks (HEIR-Nets). The proposed HEIR-Nets adopts a multi-level inheritance approach for training and label self correction. Dynamically modify the original labels using the predicted results of the branch model itself. Inheriting the optimized labels and model parameters from the previous branch in each branch, gradually improving the quality of pseudo labels and the discriminative ability of the model. In the inference phase, a cross branch model fusion structure is constructed. The designed normalized gamma encoding function (NGEF) is used to obtain the credibility weight of the branch model. Finally, the dual fusion weight of model credibility weight and sample credibility weight is used to obtain the final multi branch fusion decision. The proposed network architecture fully integrates the predictive advantages of various stage models, while suppressing low-quality predictions, extracting more consistent and robust detection outputs from them, thereby improving overall detection accuracy. The main innovations and contributions are as follows:

1) **Dual-Path Hierarchical Inheritance Network Framework**: The self-correcting pseudo-label inheritance (SPI) strategy has been proposed at the data-level. The Multi-Model Knowledge Inheritance (M2KI) strategy at the parameter-level has been proposed. This network framework tightly couples data label self-correction with the model iteration process, enabling the model to quickly obtain stable features in the early stages of training and continuously improve label quality in the later stages, achieving a gradual evolution from "coarse labels" to "fine labels".
2) **Hierarchical Multi-Branch Fusion Decision Structure**: A dual weighted decision structure is designed that includes model credibility weights and sample credibility weights. By integrating the target detection results of various branch models and introducing double weighted decision-making to integrate the judgments of various models. This fusion mechanism not only effectively resists local misjudgment caused by noisy labels, but also enhances the robust adaptability of the model to complex situations such as target overlap and blurry morphology in sperm images.

## 2 Problem Statement

Electron microscopy images of sperm are characterized by dense target distributions, small object sizes, and complex backgrounds, as shown in Fig. 1. These factors make manual annotation highly error-prone, leading to noisy labels such as missing annotations, incorrect categories, and imprecise boundaries. When used for training, such noisy labels significantly degrade detection performance.

**Fig. 2.** The mislabeled targets correctly identified by YOLOv12 trained on noisy labels.

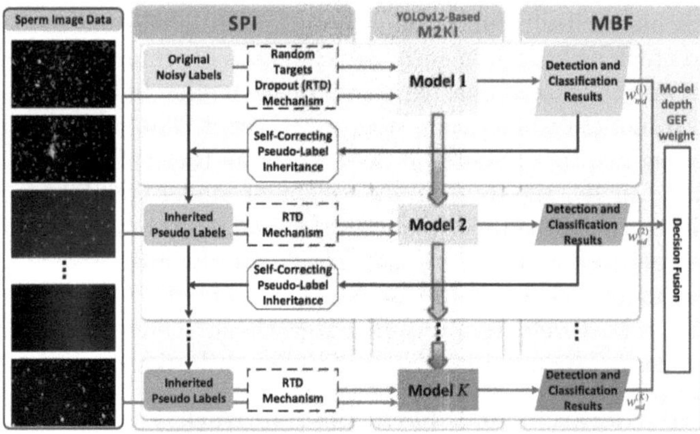


**Fig. 3.** Overall framework of the proposed HEIR-Nets. (Color figure online)

Although manual annotation inherently carries subjective errors, AI-based automated sperm detection methods have emerged as the predominant trend in recent years. Our preliminary experiments demonstrate that YOLOv12, when trained on sperm image datasets with noisy labels, can provide reasonable corrections for partially mislabeled samples and identify previously unannotated targets on the training dataset, as shown in Fig. 2. These observations indicate that the trained YOLOv12 model possesses a certain capability to rectify label noise. However, we also note that a standalone YOLOv12 model exhibits limited error tolerance when handling noisy labels - it remains vulnerable to annotation errors, resulting in unstable predictions characterized by increased false positive rates or missed small targets.

Intuitively, if the model training process can be iteratively coupled with label cleaning - progressively eliminating erroneous labels during model updates - this synergistic approach would significantly enhance both performance and robustness in noisy environments. Building upon YOLOv12's demonstrated capability to correctly identify mislabeled samples and inspired by the concept of coupling model training with label cleaning, this paper proposes the Hierarchical Ensemble with Inheritance-based Refinement Networks (HEIR-Nets) for target detection with noisy labels in sperm images.

# 3 Hierarchical Ensemble with Inheritance-Based Refinement Networks

## 3.1 Overall Framework of HEIR-Nets

The overall framework of HEIR-Nets is shown in Fig. 3. This framework integrates 3 core strategies: Self-Correcting Pseudo-Label Inheritance (SPI), Multi-model Knowledge Inheritance (M2KI), and Multi-Branch Fusion (MBF).

**SPI Strategy at Data-Level.** This strategy dynamically performs dual-action correction on the dataset by analyzing the pseudo label confidence of the current branch model, including removing low confidence targets and updating high confidence targets, as shown in Fig. 3 (red box). Meanwhile, a target random dropout intervention is designed to generate a self correcting sample set for subsequent branch models to inherit and train. The dataset after random target dropout operation has a certain probability of removing noise labels and reducing the possibility of model overfitting.

**M2KI Strategy at Parameter-Level.** The current branch model inherits the parameters of the previous branch model as initialization parameters, as illustrated in Fig. 3 (green box). Meanwhile, These models are then further trained on the self-correcting training dataset derived from previous branch model with SPI. This approach enables progressive knowledge accumulation while reducing error propagation. Through this process, the branch models gradually improve in robustness and generalization ability.

**MBF Structure at Decision-Level.** The detection results from all trained branch models are utilized for multi-branch fusion decision-making, as depicted in Fig. 3 (blue box). A normalized gamma encoding function (NGEF) is applied based on model depth to derive model credibility weights. The final multi-branch fusion decision is obtained through dual-weight fusion incorporating both model credibility weights and sample credibility weights. This hierarchical dual-weighting mechanism significantly enhances cross-branch synergy in object detection tasks, effectively mitigating misjudgments.

## 3.2 Dual-Path Hierarchical Inheritance Framework

The multi-level progressive inheritance proposed in this article includes two aspects: SPI strategy and M2KI strategy.

**SPI Strategy.** The SPI strategy leverages the prediction results from the current branch model to inversely evaluate label quality. Let the initial training set be $D^{(0)} = \{(x_i, y_i^{(0)}), i = 1, 2, ..., N\}$, where $x_i$ represents the $i$th sperm image,

$y_i^{(0)} = \{(b_{ij}^{(0)}, c_{ij}^{(0)}, s_{ij}^{(0)}), j = 1, 2, ..., n_i\}$, $b_{ij}^{(k)}$, $c_{ij}^{(k)}$, and $s_{ij}^{(k)}$ represent the coordinates, category, and confidence score of the $j$th target in the $i$th image, and these labels generally have noise. For the $k$th branch model $M_k$, its prediction result for the current training set $D^{(k)} = \{(x_i, y_i^{(k)})\}_{i=1}^N$ is represented as a pseudo-label set:

$$\hat{y}_i^{(k)} = \{(\hat{b}_{ij}^{(k)}, \hat{c}_{ij}^{(k)}, \hat{s}_{ij}^{(k)}), j = 1, 2, ..., n_i\}, \tag{1}$$

where $\hat{b}_{ij}^{(k)}$, $\hat{c}_{ij}^{(k)}$, and $\hat{s}_{ij}^{(k)}$ represent the coordinates, category, and confidence score of the $j$th predicted box in the $i$th image, respectively. Based on the prediction results, this paper designs a credibility driven dual-action label correction mechanism aimed at gradually purifying training labels. When the confidence level of the target obtained by the current branch model is lower, the label is considered unreliable and deleted, resulting in the target set to be removed:

$$D_{re}^{(k)} = \{(x_i, y_i^{(k)}) | (x_i, y_i^{(k)}) \in D^{(k)}, s_{ij}^{(k)} \leq \tau_{re}\}. \tag{2}$$

When the confidence level of the target obtained by the current branch model is higher, the predicted category of the current branch model is used to replace the original category, and the target set to be updated is obtained:

$$D_{cor}^{(k)} = \{(x_i, \hat{y}_i^{(k)}) | (x_i, y_i^{(k)}) \in D^{(k)}, s_{ij}^{(k)} \geq \tau_{cor}\}, \tag{3}$$

where $\hat{y}_i$ is obtained through Eq. (1). Then, the purified training sample set is:

$$D_{clean}^{(k)} = (D^{(k)} \triangleright D_{cor}^{(k)}) \ominus D_{re}^{(k)}, \tag{4}$$

where the operation $D^{(k)} \triangleright D_{cor}^{(k)}$ denotes that for duplicate targets in sets $D^{(k)}$ and $D_{cor}^{(k)}$, the label information of the duplicated targets in $D^{(k)}$ will be replaced by the corresponding information from $D_{cor}^{(k)}$. The operation $\ominus$ indicates that for duplicate targets in both sets, duplicate targets will be eliminated in the former. In addition, considering the potential interference left by the inability of each branch to remove all noisy labels, this paper designs a random target dropout (RTD) mechanism to randomly discard some retained labels, further suppressing the propagation of noise and the risk of overfitting. Then, the current training set can be updated to form a new pseudo labeled dataset:

$$D^{(k+1)} = RTDropout(D_{clean}^{(k)}, \rho), \tag{5}$$

where $RTDropout(\cdot, \rho)$ represents randomly removing targets from the set with probability $\rho$.

**M2KI Strategy.** To inherit the knowledge from the previous branch model, this section proposes the M2KI strategy. For the $(k+1)$th branch model, The parameters $\theta_*^{(k)}$ learned by the previous branch model $M_k$ are initialized as the starting parameters $\theta_0^{(k+1)}$ for the current branch model $M_{k+1}$, $\theta_0^{(k+1)} = \theta_*^{(k)}$ Subsequently, the current branch model $M_{k+1}$ is trained using the purified training set $D^{(k+1)}$ generated by the previous branch model. The optimization objective at this stage consists of the classification loss and the localization loss:

$$\mathcal{L}_{det} = \mathcal{L}_{cls} + \lambda \mathcal{L}_{loc}, \tag{6}$$

where $\mathcal{L}_{cls}$ represents the cross-entropy loss, $\mathcal{L}_{loc}$ denotes the localization error, and $\lambda$ serves as the balancing coefficient between these two components. The optimal model parameters can be mathematically expressed as:

$$\theta_*^{(k+1)} = \arg\min_{\theta} \mathbb{E}_{(x,y)\sim D^{(k+1)}} [\mathcal{L}_{det}(M_{k+1}(x;\theta), y)], \tag{7}$$

where $(x,y) \sim D^{(k+1)}$ represents the image and pseudo label pairs from $D^{(k+1)}$, $\mathcal{L}_{det}(M_{k+1}(x;\theta), y)$ represents the detection loss.

### 3.3 Hierarchical Multi-branch Fusion Decision Structure

The hierarchical multi-branch fusion decision consists of three main steps: multi-branch prediction merging, target clustering based on intersection over union (IOU), and dual weighted fusion. For the predicted results set extracted from all images by each trained branch model:

$$\mathcal{Y}^{(k)} = \{(b_{ij}^{(k)}, c_{ij}^{(k)}, s_{ij}^{(k)}), j = 1, 2, ..., n_i, i = 1, 2, ..., N\}, \tag{8}$$

where $b_{ij}^{(k)}$ is the predicted box coordinate, $c_{ij}^{(k)}$ is the predicted category, and $s_{ij}^{(k)}$ is the corresponding confidence level. To ensure fusion quality, the confidence threshold $\tau_{pre}$ is applied for filtering to obtain the overall set of candidate boxes:

$$\mathcal{Y}_{ca} = \bigcup_{k=1}^{K} \{(b,c,s) | (b,c,s) \in \mathcal{Y}^{(k)}, s \geq \tau_{pre}\}, \tag{9}$$

Then, the candidate box set $\mathcal{Y}_{ca}$ is spatially clustered to generate the target cluster set:

$$\mathcal{G} = \{\mathcal{G}_1, \mathcal{G}_2, ..., \mathcal{G}_J\}, \tag{10}$$

where $J$ represents the total number of medical targets in all sperm images. Each cluster $\mathcal{G}_j$ points to the spatially overlapping prediction box of the same real target, achieving spatial consistency integration and reducing redundant detection. Finally, in the decision fusion stage, a double-weighted fusion decision approach is designed for each target cluster $\mathcal{G}_j$ to obtain the final category attribution. Based on the normalized gamma encoding function (NGEF), a model credibility weight vector $w_{md}$ regarding the model depth is designed:

$$w_{md} = \frac{1}{\sum_{k=1}^{K} k^{\gamma}} [1^{\gamma}, 2^{\gamma}, ..., l^{\gamma}]^T, \tag{11}$$

where $\gamma \geq 0$, and $K$ is the number of branch models. When the value of $\gamma$ is larger, the credibility difference between models is greater, and more attention is

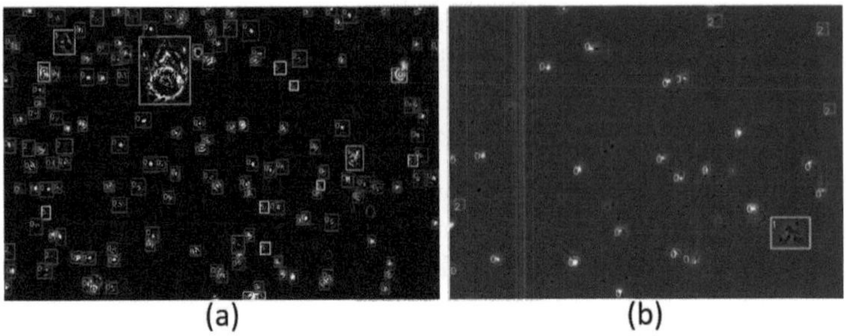


**Fig. 4.** Partially sperm images from self-built datasets and public datasets.

paid to deep branch models. Combining the credibility weight $w_{md}^{(k)}$ of the multi-branch models and the confidence degree $s_{ij}^{(k)}$ of the predicted target category, a dual weighted decision is constructed:

$$\hat{c}_{ij} = \arg\max_{\varsigma \in \mathcal{C}} \sum_{k=1}^{K} w_{md}^{(k)} s_{ij}^{(k)} \delta(c_{ij}^{(k)}, \varsigma), \tag{12}$$

where $\varsigma \in \mathcal{C}$ is the collection of categories defined by the task, $w_{md} \equiv [w_{md}^{(1)}, w_{md}^{(2)}, ..., w_{md}^{(K)}]$ is calculated through Eq. (11), and $\delta(\cdot, \cdot)$ is an indicator function:

$$\delta(c_{ij}^{(k)}, \varsigma) = \begin{cases} 1, & c_{ij}^{(k)} = \varsigma, \\ 0, & c_{ij}^{(k)} \neq \varsigma. \end{cases} \tag{13}$$

This hierarchical MBF decision structure effectively integrates the diverse information of multi-branch models.

# 4 Experimental Results and Analyses

## 4.1 Datasets and Experimental Setup

To evaluate the performance and robustness of the proposed algorithm under noisy label scenarios, experiments were conducted based on two data sources: a self-built clinical dataset (SCD) and the public dataset VISEMTracking (PDV).

The self-built dataset provided by a collaborating hospital comprises 101 images with a resolution of 1024 × 640 pixels, as shown in Fig. 4(a). Each image was annotated with 6 classes. There are a total of 7747 annotated bounding boxes. The public VISEM-Tracking dataset contains 20 human sperm microscopy videos, each lasting 30 s, totaling 29196 frames with an image resolution of 640 × 480 pixels, as shown in Fig. 4(b). Each frame was annotated with 3 classes.

To simulate label confusion caused by subjective labeling errors in real scenarios, this experiment designs four training data environments with noisy labels. 1) SDir-15: On the SCD, 15% of the normal sperm labels were randomly replaced

**Table 1.** Experimental results (%) of target detection and classification on the sperm image dataset PDir-20 (20% noisy labels).

| Model | Precision | Recall | F1 | Accuracy |
|---|---|---|---|---|
| Faster R-CNN | 63.76 ± 1.64 | 91.51 ± 0.77 | 75.14 ± 1.14 | 60.19 ± 1.47 |
| RT-DETR | 83.31 ± 2.62 | 93.39 ± 3.10 | 88.05 ± 2.65 | 78.74 ± 4.20 |
| YOLOv10 | 76.16 ± 2.87 | 86.80 ± 2.34 | 81.13 ± 2.56 | 68.32 ± 3.60 |
| YOLOv11 | 76.06 ± 3.34 | 90.17 ± 2.14 | 82.51 ± 2.81 | 70.31 ± 4.11 |
| YOLOv12 | 80.01 ± 2.95 | 92.38 ± 1.86 | 85.74 ± 2.47 | 75.11 ± 3.74 |
| HEIR-Nets | **91.16**±1.50 | **94.12**±2.01 | **92.61**±1.71 | **86.28**±2.91 |

with bubble impurity labels. 2) SDir-20: On the SCD, 20% of the normal sperm labels were randomly replaced with bubble impurity labels. 3) SRnd-20: On the SCD, 20% of the normal sperm labels randomly replaced with the labels of the other five categories. 4) PDir-20: On the PDV, 20% of the normal sperm target labels randomly replaced with small-headed sperm target labels. All experiments were repeated ten times and the average value was taken as the final result.

### 4.2 Experiments on Sperm Target Detection with Noise Labels

To validate the robustness and generalization capability of the proposed method under weak annotation conditions, systematic comparisons were conducted against mainstream detection models across different noise perturbation settings. Evaluated models included the two-stage detector Faster R-CNN [5], the YOLO series (YOLOv10 [10], YOLOv11 [2], YOLOv12 [9]) and the transformer-based RT-DETR [12]. Experiments were performed on three self-built sperm dataset (SDir-15, SDir-20, SRnd-20) and the public sperm dataset (PDir-20).

HEIR-Nets achieved state-of-the-art performance across all metrics on the public dataset, with particularly significant margins in F1-score and Accuracy over the second-best model (RT-DETR), as shown in Table 1. This demonstrates strong generalization beyond private medical data to broader domains. The results further validate the synergistic efficacy of the SPI, M2KI, and MBF strategies across diverse data distributions and noise structures. Notably, RT-DETR outperformed YOLO variants on this dataset, potentially due to its transformer- based global modeling being better suited for low-density, low-category detection tasks. With only three target classes and sparser annotations than the private dataset, RT-DETR's architectural advantages became more pronounced.

On the SDir-15 and SDir-20 datasets with different noise levels, HEIR-Nets outperformed all comparison models in Precision, F1, and Accuracy, as shown in Table 2. As the level of label noise increased, HEIR-Nets maintained a leading position in accuracy and precision, while maintaining a competitive F1 score. This indicates that HEIR-Nets reliably provides supervisory signals under severe label contamination, exhibiting exceptional fault tolerance. A slight decrease in

**Table 2.** Results (%) of target detection and classification on the sperm image datasets SDir-15 (15% noisy labels), SDir-20 (20% noisy labels) and SRnd-20 (20% noisy labels).

| Datasets | Model | Precision | Recall | F1 | Accuracy |
|---|---|---|---|---|---|
| SDir-15 | Faster R-CNN | 83.64 ± 4.79 | 71.25 ± 1.76 | 76.90 ± 2.39 | 62.52 ± 3.12 |
| | RT-DETR | 79.23 ± 1.81 | **91.65** ± 1.52 | 84.98 ± 1.56 | 73.91 ± 2.34 |
| | YOLOv10 | 84.13 ± 3.79 | 87.20 ± 2.29 | 85.58 ± 2.13 | 74.85 ± 3.25 |
| | YOLOv11 | 86.02 ± 2.56 | 89.73 ± 1.92 | 87.81 ± 1.57 | 78.30 ± 2.45 |
| | YOLOv12 | 85.62 ± 2.53 | 90.23 ± 2.03 | 87.85 ± 1.99 | 78.38 ± 3.14 |
| | HEIR-Nets | **89.87** ± 1.79 | 88.46 ± 2.04 | **89.15** ± 1.77 | **80.47** ± 2.83 |
| SDir-20 | Faster R-CNN | 80.87 ± 2.36 | 68.67 ± 2.40 | 74.25 ± 1.95 | 59.08 ± 2.45 |
| | RT-DETR | 79.25 ± 2.44 | **90.99** ± 2.59 | 84.71 ± 2.41 | 73.54 ± 3.63 |
| | YOLOv10 | 81.18 ± 4.00 | 87.69 ± 2.76 | 84.28 ± 3.10 | 72.94 ± 4.70 |
| | YOLOv11 | 85.42 ± 1.94 | 89.82 ± 1.78 | 87.55 ± 1.50 | 77.89 ± 2.37 |
| | YOLOv12 | 85.56 ± 1.72 | 90.02 ± 1.99 | 87.72 ± 1.59 | 78.16 ± 2.55 |
| | HEIR-Nets | **90.61** ± 1.58 | 87.22 ± 1.97 | **88.87** ± 1.50 | **80.00** ± 2.45 |
| SRnd-20 | Faster R-CNN | 84.78 ± 2.88 | 72.46 ± 2.47 | 77.85 ± 3.51 | 63.82 ± 4.77 |
| | RT-DETR | 79.42 ± 2.08 | **90.93** ± 2.02 | 84.78 ± 1.99 | 73.63 ± 2.99 |
| | YOLOv10 | 86.74 ± 3.34 | 86.17 ± 2.89 | 86.43 ± 2.63 | 76.18 ± 4.02 |
| | YOLOv11 | 85.82 ± 2.28 | 89.56 ± 1.78 | 87.64 ± 1.83 | 78.05 ± 2.88 |
| | YOLOv12 | 85.71 ± 2.14 | 90.13 ± 1.87 | 87.86 ± 1.88 | 78.39 ± 2.98 |
| | HEIR-Nets | **89.57** ± 1.95 | 87.94 ± 2.68 | **88.73** ± 1.91 | **79.79** ± 3.08 |

Recall compared to models like YOLOv12 was observed, likely attributable to the conservative confidence filtering mechanism in the MBF module, which prioritizes prediction quality over coverage. This trade-off remains justified since precision and stability are paramount for target applications. Table 2 also shows the experimental results on the SRnd-20 dataset with random noise labels. HEIR-Nets consistently led in precision, F1, and accuracy under random label errors. This confirms the method's effectiveness against both systematic bias and stochastic annotation inaccuracies, highlighting its generalized noise resilience.

### 4.3 Parameter Experiments

This section is an experimental analysis of the influence of four parameters in HEIR-Nets on the stability and final performance of the method.

The parameter $\rho$ controls random pseudo-label discarding during data inheritance to mitigate overfitting to erroneous labels, as shown in Fig. 5(a). Lower $\rho$ values suppress noise propagation and reduce overfitting, whereas excessive discarding critically diminishes supervisory signals causing underlearning.

The parameter $\tau_{cor}$ governs pseudo-label replacement, where predictions exceeding $\tau_{cor}$ supersede original labels for subsequent training, as illustrated

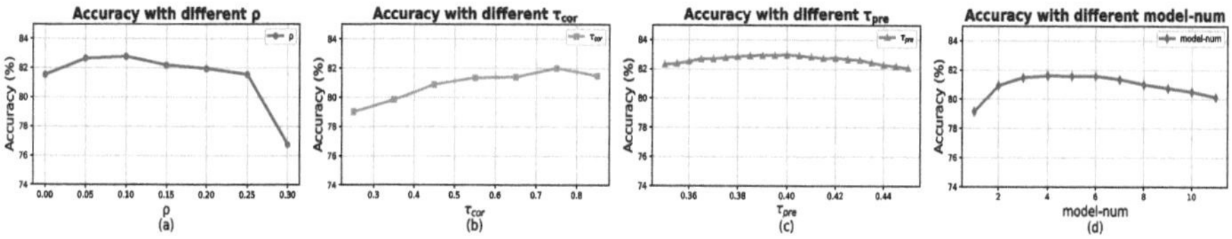


**Fig. 5.** Experimental results of different parameters in the HEIR-Nets.

**Table 3.** Ablation experiment results (%) on HEIR-Nets

| Model | RTD | M2KI | MBF | Precision | Recall | F1 Score | Accuracy |
|---|---|---|---|---|---|---|---|
| HEIR-Nets | × | × | × | 86.69 | 91.45 | 89.00 | 80.19 |
| | ✓ | × | × | 88.01 | 91.26 | 89.59 | 81.15 |
| | × | ✓ | × | 86.67 | 90.70 | 88.63 | 79.61 |
| | ✓ | ✓ | × | 88.37 | 91.08 | 89.69 | 81.31 |
| | × | ✓ | ✓ | 88.10 | **91.63** | 89.82 | 81.53 |
| | ✓ | ✓ | ✓ | **90.36** | 90.83 | **90.58** | **82.78** |

in Fig. 5(b). Higher $\tau_{cor}$ values increase pseudo-label reliability but narrow the replacement scope, while lower values risk pseudo-label contamination. Optimal performance is achieved when tuning $\tau_{cor}$ within [0.7, 0.8].

The parameter $\tau_{pre}$ sets the confidence lower bound for retaining detection boxes during inference, as shown in Fig. 5(c). Low $\tau_{pre}$ retain excessive low confidence predictions, elevating recall but introducing false positives. $\tau_{pre}$ beyond 0.4 excessively eliminate correct detections, causing performance degradation. $\tau_{pre} \in [0.38, 0.41]$ optimally balances recall and precision.

The branch count $K$ determines refinement depth and ensemble diversity, as depicted in Fig. 5(d). Validation accuracy rose continuously from $K = 1$ to $K = 4$ (peak: 81.62%), plateaued, then declined noticeably when $K > 6$. However, excessive branches ($K > 5$) risk error accumulation and overfitting to predecessor predictions, reducing diversity and generalization.

## 4.4 Ablation Experiments

To evaluate the independent and synergistic effects of key components in HEIR-Nets, ablation experiments were conducted on the self-built sperm dataset, as shown in Table 3. The study examines the impact of three strategies: RTD, M2KI strategy, and MBF.

When solely enabling RTD, the accuracy increased from 80.19% to 81.15%. This supports RTD's efficacy in reducing negative impacts from low-confidence pseudo-labels while enhancing stability and generalization under pseudo-label supervision. Combining M2KI with RTD significantly boosted accuracy, indicating that multi-model structures broaden feature learning scope and synergize with RTD to improve overall performance. The MBF strategy is pivotal for performance optimization. Activating MBF achieved peak accuracy, substantially outperforming other combinations. This confirms MBF successfully integrates advantages from multiple sub-models while suppressing individual model errors.

HEIR-Nets achieved the highest accuracy, significantly surpassing the baseline single model. This indicates that the synergistic use of various modules in HEIR-Nets can effectively handle the sperm target detection with noisy labels.

## 5 Conclusion

This paper proposes the HEIR-Nets, a novel framework designed to address the sperm target detection with noisy labels. HEIR-Nets effectively suppresses label noise while significantly enhancing sperm detection performance. The SPI module dynamically corrects and filters pseudo-labels through RTD intervention mechanism to construct purified and robust training sets. This substantially improves model stability and generalization in weakly supervised environments. The M2KI module facilitates knowledge inheritance across diverse structures. This enhances perception and representation of varied sperm morphologies while enriching discriminative feature extraction. The MBF module employs a dual-weighted fusion strategy, effectively integrating multi-branch predictions to boost detection accuracy and consistency while mitigating individual model errors. This study systematically evaluates HEIR-Nets on 4 datasets with varying levels of label noise, demonstrating its superior robustness against annotation errors. Experimental results show that HEIR-Nets outperforms state-of-the-art models in sperm target detection and classification tasks, exhibiting exceptional resistance to systematic labeling inaccuracies.

**Acknowledgments.** This work was supported in part by the Guangdong Basic and Applied Basic Research Foundation (2024A1515010977, 2025A1515060013), Scientific Foundation for Youth Scholars of Shenzhen University.

**Disclosure of Interests.** The authors have no competing interests to declare that are relevant to the content of this article.

## References

1. Huang, S., et al.: Yolo-med: multi-task interaction network for biomedical images. In: ICASSP 2024-2024 IEEE International Conference on Acoustics, Speech and Signal Processing (ICASSP), pp. 2175–2179. IEEE (2024)
2. Jocher, G., Qiu, J.: Ultralytics yolo11 (2024). https://github.com/ultralytics/ultralytics
3. Liu, C., Wang, K., Lu, H., Cao, Z., Zhang, Z.: Robust object detection with inaccurate bounding boxes. In: European Conference on Computer Vision, pp. 53–69. Springer, Cham (2022). https://doi.org/10.1007/978-3-031-20080-9_4
4. Organization, W.H.: Who Laboratory Manual for the Examination and Processing of Human Semen (2021)

5. Ren, S., He, K., Girshick, R., Sun, J.: Faster R-CNN: towards real-time object detection with region proposal networks. IEEE Trans. Pattern Anal. Mach. Intell. **39**(6), 1137–1149 (2016)
6. Sakkas, D., Alvarez, J.G.: Sperm DNA fragmentation: mechanisms of origin, impact on reproductive outcome, and analysis. Fertil. Steril. **93**(4), 1027–1036 (2010)
7. Song, H., Kim, M., Park, D., Shin, Y., Lee, J.G.: Learning from noisy labels with deep neural networks: a survey. IEEE Trans. Neural Networks Learn. Syst. **34**(11), 8135–8153 (2023)
8. Soni, A., Rai, A.: Yolo for medical object detection (2018–2024). In: 2024 IEEE 3rd International Conference on Electrical Power and Energy Systems (ICEPES), pp. 1–7. IEEE (2024)
9. Tian, Y., Ye, Q., Doermann, D.: Yolo12: attention-centric real-time object detectors. arXiv preprint arXiv:2502.12524 (2025)
10. Wang, A., et al.: Yolov10: real-time end-to-end object detection. In: Advances in Neural Information Processing Systems, vol. 37, pp. 107984–108011 (2024)
11. Xu, Y., Zhu, L., Yang, Y., Wu, F.: Training robust object detectors from noisy category labels and imprecise bounding boxes. IEEE Trans. Image Process. **30**, 5782–5792 (2021)
12. Zhao, Y., et al.: DETRs beat YOLOs on real-time object detection. In: Proceedings of the IEEE/CVF Conference on Computer Vision and Pattern Recognition, pp. 16965–16974 (2024)

# SEEDS: Curriculum-Driven Adaptive Regression Contrastive Learning for Robust Evidence-Based Depression Severity Classification

Tu-Phuong Mai[1], Duy-Cat Can[2], and Hoang-Quynh Le[1(✉)]

[1] VNU University of Engineering and Technology, Hanoi, Vietnam
21020552@vnu.edu.vn, lhquynh@vnu.edu.vn
[2] Lausanne University Hospital, Lausanne, Switzerland
duy-cat.can@chuv.ch

**Abstract.** Depression detection from social media faces critical challenges: sparse signals obscured by noisy content, severe class imbalance biasing models toward non-depressive text, and arbitrary feature extraction in end-to-end neural models causing overfitting on small datasets. We propose `SEEDS`–***S****equential* ***E****vidence-based* ***E****ncoder for* ***D****epression* ***S****everity classification*–which integrates evidence-based encoding with curriculum-driven adaptive contrastive learning. `SEEDS` extracts structured clinical features aligned with `BDI-II` symptoms and emotional states, replacing arbitrary extraction with interpretable evidence. A multi-task curriculum architecture progressively filters relevant content, providing balanced training data for severity estimation. We introduce adaptive regression contrastive learning to extend contrastive optimization to continuous prediction while dynamically balancing loss convergence. Evaluation on CLEF eRisk, Deptweet, and DepSeverity datasets demonstrates state-of-the-art macro F1 performance, validating `SEEDS`'s effectiveness across user-level and post-level depression detection. Source code is available at https://github.com/candleMind/SEEDS.

**Keywords:** depression · contrastive learning · curriculum learning

## 1 Introduction

While depression is projected to become the primary global disease burden by 2030 [29], a critical gap remains in clinical screening due to stigma and limited access to mental health screening services. Consequently, nearly 60% of suicide victims historically go undiagnosed despite exhibiting symptoms [6]. To bridge this gap, we exploit social media as a non-invasive resource where users freely express thoughts and emotions, thus possibly revealing cognitive fluctuations. This work aims for *depression risk detection from social media text*, by analyzing mental states at multiple granularities: from individual posts with consistent mental states to post sequences exhibiting complex temporal cognitive patterns.


R. C.-W. Wong et al. (Eds.): PAKDD 2026, LNAI 16597, pp. 404–416, 2026.
https://doi.org/10.1007/978-981-92-1300-9_32

Detecting depression from social media text presents four fundamental challenges with corresponding solutions. *First*, depression signals are inherently sparse, as users engage in diverse topics where potential indicators are obscured by noisy, unrelated discussions [22]. We address this through *evidence-based encoding*, which extracts structured clinical features aligned with `Beck Depression Inventory-II (BDI-II)` [2] depressive symptoms and emotional states to filter noise. *Second*, this sparsity creates severe class imbalance in training data, causing models to bias toward non-depressive content, while standard resampling techniques (random oversampling, undersampling, SMOTE) treat all samples uniformly, neglecting critical borderline instances near decision boundaries [26]. We employ *curriculum learning* using a filtering model to identify relevant content, providing balanced, borderline-enriched training data for severity estimation. *Third*, end-to-end deep learning models extract random features from limited data, overfitting on spurious correlations rather than generalizable depression indicators [27,28]. Our *evidence-based approach* replaces random extraction with targeted clinical feature learning on multiple datasets, producing interpretable evidence. *Fourth*, contrastive learning shows promise for this problem for its discriminative feature space construction and strong transfer learning performance [14], yet adapting it to regression tasks poses challenges due to traditional reliance on discrete positive-negative pair definitions incompatible with continuous severity scores, and combining multiple loss functions introduces convergence speed disparities where faster-converging losses dominate gradients [15]. We propose *adaptive regression contrastive learning* that extends contrastive optimization to continuous prediction without explicit pair definitions while dynamically balancing loss convergence.

In summary, we propose **SEEDS**–***S****equential* ***E****vidence-based* ***E****ncoder for* ***D****epression* ***S****everity classification*–integrating *evidence-based encoding*, *curriculum learning*, and *adaptive regression contrastive learning*. Comprehensive evaluation on CLEF eRisk 2022, Deptweet, and DepSeverity datasets demonstrates state-of-the-art `macro F1-score` performance while providing interpretable evidence of depression patterns.

## 2 Related Work

**Depression Detection on Social Media** has been formulated at both post-level and user-level granularities [23], and evolved through several methodological stages. Early approaches combined sentiment features with topic modeling like LDA [32], later advancing to hybrid neural models (e.g., FastText, CNN, LSTM) for richer text representation [24]. The field was significantly advanced by transformers, with BERT-based models capturing deep contextual signals [10] and LLMs being applied for both binary and severity classification [20]. A recent shift towards temporal modeling aims to capture the dynamic nature of psychological states [3]. *To overcome the limitations of purely semantic models*, recent studies have explored enriching inputs with emotional cues [25]. A key challenge in this domain is the *sparsity and noisiness of genuine depressive signals*, which are often obscured by unrelated user discussions [22]. While some methods attempt to filter this noise, they lack a trainable, symptom-based mechanism to discriminately identify relevant depressive text.

**Multi-task and Curriculum Learning** offer strategic frameworks for model training. Multi-task learning improves generalization by jointly optimizing a primary task with auxiliary objectives, such as sentiment analysis, though it requires careful loss balancing to prevent one task from dominating [13]. Curriculum learning, inspired by human education, increases the complexity of experience during the training process at the data-level or model-level. This strategy improves generalization and convergence speed across various machine learning domains [21]. Its application in sentiment analysis has shown promise, with task-specific curricula leading to better performance than general strategies [13,30].

**Contrastive Learning (CL)** has proven powerful for learning discriminative representations by pulling positive pairs together and pushing negative pairs apart in an embedding space. While traditionally applied to classification tasks where positive and negative pairs can be clearly defined [7,14], this requirement has limited its use in regression. Recent work has attempted to adapt these principles to regression through similarity thresholds or pseudo-labeling, yet still depending on discrete, pre-defined pairings [12]. This fundamental limitation motivates our proposed adaptive regression contrastive approach, which eliminates the need for explicit pair definitions through dynamic similarity learning.

# 3 Methodology

**Problem Formulation.** We formalize depression detection as sequential text classification with two granularities. Given a text sequence represented as $\{s_i\}_{i=1}^{|S|}$ where $s_i$ denotes the $i$-th sentence, we address: (1) *Post-level classification*, determining depression severity among C classes {non-depressed/minimal, mild, moderate, severe} for individual posts exhibiting consistent mental states; (2) *User-level classification*, predicting binary depression status {non-depressed, depressed} from longitudinal post sequences capturing temporal cognitive fluctuations. This dual problem validates `SEEDS`'s effectiveness and flexibility across diverse text granularities and task natures.

**Model Overview.** The `SEEDS` architecture (Fig. 1) comprises two core components that produce evidence-based sequential embeddings applicable to diverse downstream tasks. *First,* an Evidence-based Sentence Encoder (ESE) extracts clinically-informed features—`BDI-II` symptom relevance, severity levels, and emotional states—from each sentence. The ESE training employs *curriculum learning* through a multi-task architecture and the *new adaptive regression contrastive learning* to enhance feature discrimination while balancing loss convergence. *Second,* an LSTM aggregates sentence embeddings sequentially to produce the final evidence-based sequential representation $\mathbf{z}$. This representation serves as input to *task-specific classifiers* (detailed in Sect. 3.2), whose architectures vary by task requirements.

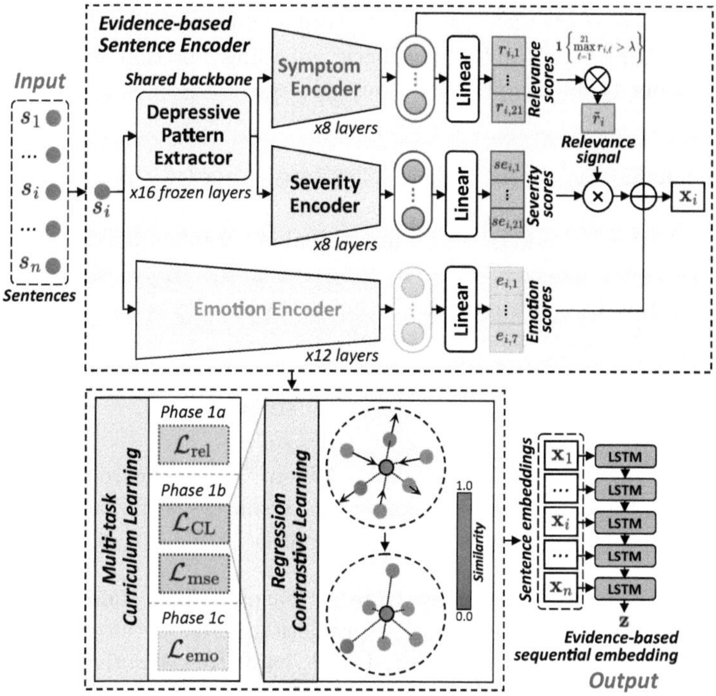


**Fig. 1.** Overview of the proposed SEEDS model architecture.

### 3.1 Sequential Evidence-Based Encoder

**Evidence-Based Sentence Encoder.** The ESE transforms each sentence $s_i$ into a clinically-informed embedding $\mathbf{x}_i$ by extracting three evidence types: BDI-II symptom and severity, and emotional states.

*a. Symptom and Severity Encoders.* Based on BDI-II's 21 depressive symptoms, each rated across four severity levels (0–3, where 0 indicates positive connotation and 3 indicates severe manifestation), we define two tasks: *symptom relevance detection* determines whether a sentence discusses the author and relates to any symptom, while *symptom severity estimation* quantifies the severity level for each relevant symptom.

We employ DepRoBERTa[1] as backbone, pre-trained on depression-specific social media data for effective transfer of depressive linguistic patterns. To optimize resources, a shared Depressive Pattern Extractor (first 16 Transformer layers) produces token representations $\mathbf{h}_i^{(16)} = \texttt{DepRoBERTa}^{1-16}(s_i)$.

[1] Available at https://huggingface.co/rafalposwiata/deproberta-large-depression.

Both encoders share identical architecture: eight fine-tuned Transformer layers processing $\mathbf{h}_i^{(16)}$. We define a generic encoding function where $\tau \in \{r, s\}$ denotes encoder type (symptom or severity):

$$\mathbf{v}_i^{(\tau)} = (\texttt{DepRoBERTa}_\tau^{17-24}(\mathbf{h}_i^{(16)}))_0 \in \mathbb{R}^{d_{\text{bert}}}, \quad \mathbf{o}_i^{(\tau)} = \sigma(\mathbf{v}_i^{(\tau)}\mathbf{W}_\tau + \mathbf{b}_\tau) \in [0,1]^{21}, \tag{1}$$

where $\mathbf{W}_\tau \in \mathbb{R}^{d_{\text{bert}} \times 21}$ and $\mathbf{b}_\tau \in \mathbb{R}^{21}$ are trainable parameters. For $\tau = r$, outputs are symptom relevance scores $\mathbf{r}_i = \mathbf{o}_i^{(r)}$; for $\tau = s$, severity scores $\mathbf{se}_i = \mathbf{o}_i^{(s)}$.

The symptom branch conditionally activates severity computation with $\tilde{r}_i \in \{0, 1\}$ denote the binary relevance signal:

$$\tilde{r}_i = \mathbf{1}\left\{\max_{\ell=1}^{21} r_{i,\ell} > \lambda\right\}, \quad \mathbf{se}'_i = \mathbf{se}_i \times \tilde{r}_i, \tag{2}$$

where $\lambda = 0.5$ is a relevance threshold. When $\tilde{r}_i = 0$ (no related symptom), severity computation is bypassed, conserving resources while balancing severity encoder training by filtering irrelevant sentences.

*b. Emotion Encoder.* We employ transfer learning from `RoBERTa`-based pretrained model for multi-label emotion prediction across seven categories {Neutral, Surprise, Fear, Sad, Joy, Anger, Love}, capturing affective states directly related to psychological conditions. For sentence $s_i$, the encoder outputs $\mathbf{e}_i \in [0, 1]^7$ representing emotion probabilities.

*c. Sentence Embedding Fusion.* The evidence-based sentence embedding $\mathbf{x}_i = [\mathbf{v}_i^{(\tau)}, (\mathbf{se}'_i \oplus \mathbf{e}_i)\mathbf{W}_{\text{f}}]\mathbf{W}_{\text{p}} \in \mathbb{R}^{d_s}$ integrates all three encoders, where $\oplus$ denotes concatenation, $\mathbf{W}_{\text{f}} \in \mathbb{R}^{28 \times d_{\text{f}}}$ and $\mathbf{W}_{\text{p}} \in \mathbb{R}^{(d_{\text{bert}}+d_{\text{f}}) \times d_s}$ are projection matrices. This hierarchical design preserves symptom-guided semantics from symptom encoding while incorporating severity and emotional dimensions through secondary fusion.

**Hierarchical Sequential Aggregation.** Direct sequence modeling over entire texts encounters computational inefficiency and information loss for overlengthy sequences. *Our hierarchical approach–vectorizing sentence-level units before sequential aggregation*–optimizes resources compared to full token-level attention while addressing information retention. Furthermore, existing post-level datasets provide only severity labels, lacking comprehensive symptom-based annotations available in sentence-level datasets. Aggregating clinically-enriched sentence embeddings from ESE produces informative sequential representations efficiently. Therefore, an LSTM processes sentence embeddings $\{\mathbf{x}_i\}_{i=1}^{|S|}$ in discourse order, producing the final evidence-based sequential embedding:

$$\mathbf{h}_i, \mathbf{c}_i = \text{LSTM}(\mathbf{x}_i, \mathbf{h}_{i-1}, \mathbf{c}_{i-1}), \quad \mathbf{z} = \mathbf{h}_{|S|}, \tag{3}$$

where $\mathbf{h}_i$ and $\mathbf{c}_i$ are hidden and cell states at position $i$, initialized as zero vectors, and $\mathbf{z} \in \mathbb{R}^{d_h}$ is the sequential representation.

### 3.2 Downstream Tasks

**Post-level Severity Classification.** For relatively small datasets, fine-tuning DepRoBERTa alone risks overfitting and forgetting pre-trained features. However, SEEDS alone is suboptimal as sentences within a post contain strong inter-sentence references, and moderate post lengths benefit from full attention. We therefore combine both representations:

$$\begin{aligned} \mathbf{h}_{\text{bert}} &= \texttt{DepRoBERTa}_{\text{finetuned}}(\text{post}), \quad \mathbf{z}_{\text{combined}} = [\mathbf{z}; \mathbf{h}_{\text{bert}}], \\ \mathbf{p}_{\text{sev}} &= \text{softmax}(\text{MLP}(\mathbf{z}_{\text{combined}})), \quad \hat{y}_{\text{post}} = \arg\max(\mathbf{p}_{\text{sev}}), \end{aligned} \tag{4}$$

where $\mathbf{z}$ is SEEDS output, $\mathbf{h}_{\text{bert}}$ is the CLS token from fine-tuned DepRoBERTa (trained for few epochs), and $\mathbf{p}_{\text{sev}} \in \mathbb{R}^C$ contains class probabilities.

**User-Level Depression Detection.** For extensive longitudinal sequences, indiscriminate processing of all texts imposes computational costs and dilutes depression signals with irrelevant content. We *filter sentences* by retaining only those identified as *symptom-relevant* (sentence $i$ with $\tilde{r}_i = 1$) or containing diagnostic keywords (e.g., "depression", "diagnosed"). Retained sentences are processed through SEEDS to obtain sequential embedding $\mathbf{z}$ and the predicted class $\hat{y}_{\text{user}} = \mathbf{1}_{\text{sigmoid}(\text{MLP}(\mathbf{z}))>\delta}$, with $\delta = 0.55$ is the selected classification threshold.

### 3.3 Model Training and Optimization

Training follows a consecutive two-stage process: Stage 1 establishes foundational clinical semantics at the sentence level; Stage 2 refines sequence-level understanding for downstream tasks.

#### Stage 1: Evidence-Based Sentence Encoder Training

***Multi-task Curriculum Learning.*** We employ a curriculum-based multi-task learning strategy that trains branches sequentially: first, train the symptom encoder with $\mathcal{L}_{\text{rel}}$ until convergence, freeze it, then train the severity encoder with $\mathcal{L}_{\text{sev}}$, and finally train the emotion encoder with $\mathcal{L}_{\text{emo}}$. This approach serves three purposes: (1) *Preventing suboptimal performance* caused by disparate convergence rates of different losses during simultaneous training; (2) *Knowledge preservation and efficiency* by sharing the frozen Depressive Pattern Extractor, which retains pre-trained depressive knowledge while preventing overfitting on small datasets and conserving computational resources; (3) *Progressive label balancing* where symptom encoder output activates severity encoder training, filtering irrelevant sentences to address class imbalance while retaining critical borderline instances, enabling the severity encoder to effectively distinguish levels among symptoms. The sequential training also ensures the severity encoder receives stable symptom predictions necessary for proper activation and noise-free learning.

*a. Symptom Relevance Encoder.* We train on CLEF eRisk 2023 and 2024 Task 1 datasets [18] (28,025 annotated sentences, including 17,414 control sentences, directly collected from Reddit user writings) for multi-label classification of 21 `BDI-II` symptoms using binary cross-entropy (BCE) with ground truth $y_\ell$ and predicted relevance $r_\ell$ for symptom $\ell$:

$$\mathcal{L}_{\text{rel}} = -\frac{1}{21}\sum_{\ell=1}^{21}\left[y_\ell \log(r_\ell) + (1-y_\ell)\log(1-r_\ell)\right], \tag{5}$$

*b. Symptom Severity Encoder.* We generate pseudo-labels for four severity levels (0–3) from CLEF eRisk data using `Gemini 2.5 Pro` and prompting with `BDI-II` questionnaire, then we manually check and relabel the trained model's errors. Labels are normalized to $[0, 1]$ for regression training with mean squared error:

$$\mathcal{L}_{\text{mse}} = \frac{1}{21}\sum_{\ell=1}^{21}(se'_\ell - \hat{se}_\ell)^2, \tag{6}$$

where $\hat{se}_\ell$ is ground truth normalized severity and $se'_\ell$ is prediction.

***Adaptive Regression Contrastive Learning.*** Contrastive learning is challenging to adapt to regression due to traditional reliance on discrete positive-negative pairs, which cannot capture continuous severity scores. We *propose a regression contrastive learning approach* inheriting the principle that similar-labeled instances should cluster while dissimilar ones separate. We define similarity through two factors: *overlap ratio of relevant symptom sets*, and *severity score proximity for overlapping symptoms.* For sentence $i$, we compute contrastive loss on projected low-dimensional severity embeddings $\mathbf{v}_i^{(s,\text{low})} = \mathbf{v}_i^s \mathbf{W}_{\text{low}} + \mathbf{b}_{\text{low}} \in \mathbb{R}^{128}$, as cosine similarity $(\text{sim}(*,*))$ become indifferent in high dimensions:

$$\mathcal{L}_{\text{CL}_i} = -\log \frac{\sum_{j\in\mathcal{A}} w_{ij}\cdot \exp(\text{sim}(\mathbf{v}_i^{(s,\text{low})}, \mathbf{v}_j^{(s,\text{low})})/\tau)}{\sum_{k\in\mathcal{A}} \exp(\text{sim}(\mathbf{v}_i^{(s,\text{low})}, \mathbf{v}_k^{(s,\text{low})})/\tau)}, \tag{7}$$

where $\mathcal{A}$ is a fixed set of randomly sampled anchors across all symptoms and severities from the train set (minimizing computation), $\tau > 0$ is temperature ($\tau \approx 0.26$), and weight $w_{ij} \in [0, 1]$ combines symptom overlap and severity proximity:

$$w_{ij} = \frac{|\mathcal{S}_i \cap \mathcal{S}_j|}{|\mathcal{S}_i \cup \mathcal{S}_j|} \cdot \frac{\beta}{\frac{1}{|\mathcal{S}_i\cap\mathcal{S}_j|}\sum_{\ell\in\mathcal{S}_i\cap\mathcal{S}_j} |se_{i,\ell} - se_{j,\ell}| + \beta}, \tag{8}$$

where $\mathcal{S}_i \subseteq \{1, \ldots, 21\}$ denotes relevant symptoms for sentence $i$, $se_{i,\ell} \in [0, 1]$ is severity for symptom $\ell$, and $\beta \in [0, 1]$ controls severity difference influence ($\beta = 0.2$ to ensure the balance among four severity levels).

*To address disparate convergence rates when combining losses*, we propose adaptive loss weighting that assigns a high weight to contrastive loss early (accelerating convergence), then gradually reduces that weight to prevent overfitting:

$$\mathcal{L}_{\text{sev}_t} = \mathcal{L}_{\text{mse}} + \mathcal{L}_{\text{CL}} \cdot \text{Ratio}_t, \quad \text{Ratio}_t = \frac{|\mathcal{L}_{\text{mse}}|}{|\mathcal{L}_{\text{CL}}|} \cdot (\text{rate}_{\text{init}})^{t-1}, \tag{9}$$

where $t$ is epoch, $\text{rate}_{\text{init}} \in (0,1)$ controls decay speed ($\text{rate}_{\text{init}} = 0.8$), and $\text{Ratio}_t$ balances magnitudes dynamically to prevent either loss dominating.

*c. Emotion Encoder.* We fine-tune `RoBERTa-base-emotion`[2] on multi-labeled SuperEmotion[3] dataset (439,000+ sentences with seven emotion labels) using BCE loss $\mathcal{L}_{\text{emo}} = -\frac{1}{7}\sum_{\ell=1}^{7}[y_\ell \log(e_\ell) + (1-y_\ell)\log(1-e_\ell)]$.

**Stage 2: Downstream Task Training**

*a. Post-level Classification.* We employ cross-entropy loss for multi-class severity prediction $\mathcal{L}_{\text{post}} = -\sum_{c=1}^{C} y_c \log(p_c)$, where $y_c \in \{0,1\}$ is ground truth and $p_c$ is predicted probability for class $c$.

*b. User-Level Classification.* To provide severity information beyond binary labels, we define reference severity scores based on depression-relevant sentence counts (sentence $i$ with $\tilde{r}_i = 1$), where more relevant sentences indicate higher severity while maintaining binary classification when score $> 0.5$:

$$\text{score}_{\text{ref}} = y_{\text{class}} \cdot \min\left(1.0, 0.51 + \frac{\log(n_{\text{rel}})}{2\log(n_{\text{max}})}\right), \; \mathcal{L}_{\text{user}} = (\text{score}_{\text{pred}} - \text{score}_{\text{ref}})^2, \tag{10}$$

where $y_{\text{class}} \in \{0,1\}$ is binary label, $n_{\text{rel}}$ is relevant sentence count, $n_{\text{max}}$ is a normalization constant, and $\text{score}_{\text{pred}} = \text{sigmoid}(\text{MLP}(\mathbf{z}))$ is predicted score.

# 4 Experimental Results

## 4.1 Experimental Setup

**Datasets and Metrics.** We evaluate `SEEDS` on three benchmarks spanning user-level and post-level depression detection (Table 1). **eRisk** [17] (CLEF eRisk 2022 Task 2) performs binary user-level classification from longitudinal post sequences, predicting whether users exhibit depression from their social media history. We train on eRisk 2017 and 2018 data, and test on eRisk 2022 Task 2, with performance measured by `Precision` (`P`), `Recall` (`R`), and `F1-score` (`F`). **Deptweet** [11] contains 40,191 posts for four-class severity classification while **DepSeverity** [16] provides 3,553 posts also across four severity classes. For post-level tasks, we split 80:20 of the dataset stratified by label for the train-test sets, and reserve 10% of the train data for validation. Post-level metrics include `macro F1` (`Mac-F`), `weighted F1` (`Wei-F`), `AUC-ROC` (`AUC`), and `Accuracy` (`Acc`), reported according to each dataset's standard evaluation protocol. All datasets exhibit *severe class imbalance* with predominant non-depressed samples, challenging model generalization on minority classes.

[2] Available at https://huggingface.co/bhadresh-savani/roberta-base-emotion.
[3] Available at https://huggingface.co/datasets/cirimus/super-emotion.

**Table 1.** Statistics of three benchmarks

| Metrics | eRisk | | Deptweet | | DepSeverity | |
|---|---|---|---|---|---|---|
| | Train set | Test set | Train set | Test set | Train set | Test set |
| Number of objects | 1707 users | 1400 users | 32152 posts | 8039 posts | 2841 posts | 712 posts |
| Label ratio | 88:12 *(non:dep)* | 93:7 *(non:dep)* | 81:13:4:2 *(non:mild:mod:sev)* | | 73:8:11:8 *(min:mild:mod:sev)* | |
| Avg. sentences/object | 1594.0 | 1618.8 | 2.2 | | 4.8 | |
| Avg. words/sentence | 12.0 | 11.5 | 16.5 | | 18.3 | |

*non: non-depressed, dep: depressed, min: minimum, mod: moderate, sev: severe.*

**Table 2.** Comparative results of our SEEDS and other top-performance methods (%)

| eRisk | | | | Deptweet | | | | DepSeverity | | | |
|---|---|---|---|---|---|---|---|---|---|---|---|
| Methods | P | R | F | Methods | Mac-F | Wei-F | AUC | Methods | Mac-F | Wei-F | Acc |
| DCWE[a] [8] | – | – | 35.00 | RoBERTa[b] [5] | 32.04 | 80.14 | – | BiLSTM[a] [31] | – | 42.80 | 58.00 |
| Random Forest[a] [4] | – | – | 57.00 | MentalBERT[b] [5] | 36.04 | 79.72 | – | GPT-4[c] [31] | – | 51.80 | 53.00 |
| BLUE[b](#2) [17] | 39.50 | **89.80** | 54.80 | 3M-Health[b] [5] | 46.43 | 83.09 | – | InterMind[c] [31] | 36.18 | 54.30 | 59.20 |
| NLPGroup[†](#1) [17] | 68.20 | 74.50 | 71.20 | DistilBERT[b] [11] | 48.60 | – | 79.75 | MentalBert[b] [31] | – | 56.40 | 62.70 |
| TAM-SenticNet [9] | 66.50 | 88.10 | 75.80 | Soft-voting[b] [1] | 63.32 | 85.03 | 93.26 | Few-Shot CoT[c] [19] | 41.20 | – | 64.20 |
| SEEDS (Our) | **69.67** | 86.73 | **77.27** | SEEDS (Our) | **65.79** | **85.42** | **93.40** | SEEDS (Our) | **54.62** | **75.64** | **76.40** |

P: Precision, R: Recall, F: F1-score, Mac: macro, Wei: weighted, AUC: AUC-ROC, Acc: Accuracy. #1,2: *eRisk shared task ranking.*

[a]*: traditional models,* [b]*: domain-specialized BERT models,* [c]*: LLM models. –: results not reported. Best scores in* ***bold***.

**Compared Methods.** We compare SEEDS against top-performing methods across three paradigms: traditional machine learning approaches, transformer-based models including domain-specialized variants and their ensemble methods, and large language models with few-shot prompting. Additional methods include symbolic reasoning with sentiment analysis (TAM-SenticNet) and a generated multimodal multi-teacher approach (3M-Health).

## 4.2 Results and Analysis

**Performance Comparison.** Table 2 demonstrates that SEEDS achieves state-of-the-art performance across all benchmarks. *On eRisk 2022 Task 2*, SEEDS attains 77.27% F1-score, outperforming TAM-SenticNet by 1.47 points with superior precision-recall balance. *For Deptweet*, SEEDS surpasses Soft-voting in all metrics, with higher 2.47 points macro F1, demonstrating robust discrimination despite extreme imbalance. *On DepSeverity*, SEEDS substantially improves macro F1 to 54.62% over Few-Shot CoT (41.20%), while achieving the highest weighted F1 (75.64%) and accuracy (76.40%). The consistent improvements across vastly different characteristics—from extensive user timelines to concise posts—validate SEEDS' flexibility and the effectiveness of evidence-based encoding with curriculum and adaptive contrastive learning.

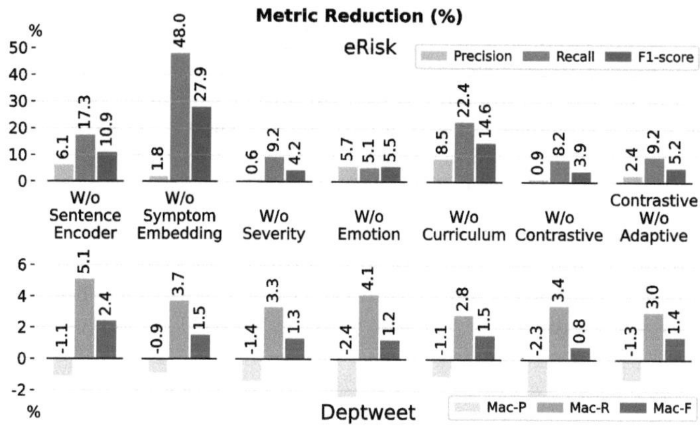


**Fig. 2.** Ablation results of SEEDS showing the performance drop (%) compared to the full model when a given component is removed (*W/o: Without*).

**Ablation Study.** Figure 2 demonstrates that each component contributes critically to SEEDS's performance on both downstream tasks: user-level (on eRisk dataset) and post-level (on Deptweet dataset) classification. Unlike the post-level task, user-level classification relies entirely on SEEDS without fine-tuned full-attention support, causing the performance gap. *Removing the sentence encoder* (using end-to-end post-level-based models) or individual evidence encoders *(symptom, severity, emotion)* substantially degrades metrics, confirming the importance of hierarchical evidence-based encoding. *Removing symptom embedding* causes the most severe drop with 27.9% F1 reduction on eRisk, and moderate 1.5% macro F1 drop on Deptweet. *Removing curriculum learning* substantially degrades performance across datasets (14.6% eRisk's F1 drop), indicating noisy symptom and severity information without curriculum learning. While removing *contrastive learning* reduces performance moderately (3.9% eRisk's F1 drop), *removing the adaptive mechanism* causes larger degradation (5.2% drop), demonstrating that effective contrastive learning requires adaptive weighting to prevent overfitting.

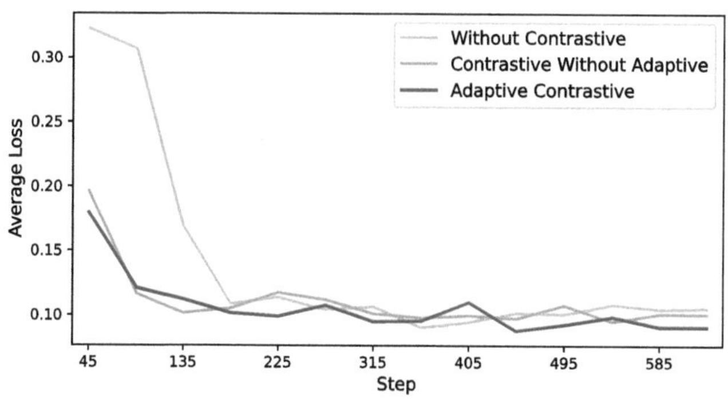


**Fig. 3.** Training dynamics of the sentence encoder's severity loss $\mathcal{L}_{mse}$ under different contrastive learning settings.

**Impact of Adaptive Regression Contrastive Learning.** Figure 3 analyzes the effect of adaptive regression contrastive learning on severity loss (averaged across labels for accurate evaluation under imbalanced data). *Without contrastive learning*, the model converges slowly with loss remaining high throughout training. *Contrastive learning (without adaptive weighting)* accelerates initial convergence, with loss dropping rapidly in early steps, but fails to reach optimal minima and exhibits instability—evidence of overfitting due to loss imbalance where contrastive gradients dominate. *Our adaptive mechanism* addresses both issues: it inherits the fast early convergence from contrastive learning while dynamically rebalancing losses, enabling the model to reach lower loss values and maintain stable convergence in later epochs.

## 5 Conclusion

We presented SEEDS, a novel framework addressing depression detection challenges through evidence-based encoding, curriculum learning, and adaptive regression contrastive learning. By extracting structured BDI-II-aligned features and employing curriculum filtering, SEEDS handles class imbalance while maintaining interpretable evidence. Our adaptive contrastive mechanism successfully extends contrastive optimization to severity estimation with dynamic loss balancing. Experimental results demonstrate the best performance across three benchmarks, validating flexibility across diverse text granularities.

Despite these achievements, SEEDS currently focuses on detection rather than early prediction, analyzing existing content without modeling temporal progression patterns signaling depression onset. Future work will develop early detection capabilities by incorporating temporal dynamics modeling in longitudinal sequences, enabling timely intervention.

**Acknowledgments.** This work was supported by the Vietnam National Foundation for Science and Technology Development (NAFOSTED) under the project 'Research and Development of a Personalized Machine Learning System for Early AlzheimerâĂŹs Disease Diagnosis from Adaptive Multimodal Biomarkers' (Grant No. 102.05-2025.54).

## References

1. Ahmed, T., Ivan, S., Munir, A., Ahmed, S.: Decoding depression: analyzing social network insights for depression severity assessment with transformers and explainable AI. Nat. Lang. Process. J. (2024)
2. Beck, A.T., Steer, R.A., Brown, G.: Beck depression inventory—II (BDI–II). Psychol. Assessment (1996)
3. Bucur, A.M., Cosma, A., Rosso, P., Dinu, L.P.: It's just a matter of time: detecting depression with time-enriched multimodal transformers. In: Proceedings of ECIR (2023)
4. Burgueño Paz, L.H., Zareei, M., de León Languré, A.: Assessing the effect of writing style features in detecting depression on social media. In: Proceedings of ICITS (2025)

5. Cabral, R.C., Luo, S., Poon, J., Han, S.C.: 3M-health: multimodal multi-teacher knowledge distillation for mental health detection. In: Proceedings of CIKM (2024)
6. Chaudhary, S., et al.: Youth suicide and preceding mental health diagnosis. JAMA Netw. Open (2024)
7. Chen, T., Kornblith, S., Norouzi, M., Hinton, G.: A simple framework for contrastive learning of visual representations. In: Proceedings of ICML (2020)
8. Couto, M., Perez, A., Parapar, J., Losada, D.E.: Temporal word embeddings for early detection of psychological disorders on social media. J. Healthc. Inf. Res. (2025)
9. Dou, R., Kang, X.: TAM-SenticNet: a neuro-symbolic AI approach for early depression detection via social media analysis. Comput. Electr. Eng. (2024)
10. Helmy, A., Nassar, R., Ramdan, N.: Depression detection for twitter users using sentiment analysis in English and Arabic tweets. AIDS Patient Care STDS (2024)
11. Kabir, M., et al.: DEPTWEET: a typology for social media texts to detect depression severities. Comput. Hum. Behav. (2023)
12. Karagoz, B., Ashraf, A., Khan, S.S.: Supervised sequential contrastive regression: improving performance on imbalanced rehabilitation exercises. Biomed. Signal Process. Control (2026)
13. Lange, N., Frasincar, F., Trușcă, M.M.: Curriculum learning for a hybrid approach for aspect-based sentiment analysis. Expert Syst. Appl. (2025)
14. Li, W., et al.: System log anomaly detection based on contrastive learning and retrieval augmented. Sci. Rep. (2025)
15. Lin, B., et al.: Dual-balancing for multi-task learning. Neural Networks (2025)
16. Naseem, U., Dunn, A.G., Kim, J., Khushi, M.: Early identification of depression severity levels on reddit using ordinal classification. In: Proceedings of WWW (2022)
17. Parapar, J., Martín-Rodilla, P., Losada, D.E., Crestani, F.: eRisk 2022: pathological gambling, depression, and eating disorder challenges. In: Proceedings of ECIR (2022). Springer, Cham. https://doi.org/10.1007/978-3-030-99739-7_54
18. Parapar, J., Martin-Rodilla, P., Losada, D.E., Crestani, F.: Overview of eRisk 2024: early risk prediction on the internet. In: Proceedings of CLEF (2024)
19. Patil, A., Kour Gedhu, A.: Cognitive-mental-LLM: evaluating reasoning in large language models for mental health prediction via online text. arXiv e-prints (2025)
20. Shah, S.M., Gillani, S.A., Baig, M.S.A., Saleem, M.A., Siddiqui, M.H.: Advancing depression detection on social media platforms through fine-tuned large language models. OSNEM (2025)
21. Soviany, P., Ionescu, R.T., Rota, P., Sebe, N.: Curriculum learning: a survey. Int. J. Comput. Vis. (2022)
22. Su, Y., et al.: Semantic distillation and enhanced diagnostic alignment: a novel approach for depression detection in social media. Expert Syst. Appl. (2025)
23. Ta, P., Tran, N., Nguyen, H., Nguyen, H.D.: Detecting signs of depression on social media: a machine learning analysis and evaluation. Sustain. Futures (2025)
24. Tejaswini, V., Sathya Babu, K., Sahoo, B.: Depression detection from social media text analysis using natural language processing techniques and hybrid deep learning model. TALLIP (2024)
25. Turcan, E., Muresan, S., McKeown, K.: Emotion-infused models for explainable psychological stress detection. In: Proceedings of the 2021 Conference of the North American Chapter of the Association for Computational Linguistics: Human Language Technologies (2021)
26. Wainer, J.: An empirical evaluation of imbalanced data strategies from a practitioner's point of view. Expert Syst. Appl. (2024)

27. Wen, H., Shi, J., Chen, W., Liu, Z.: Transferring and generalizing deep-learning-based neural encoding models across subjects. Neuroimage (2018)
28. Zhang, C., Bengio, S., Hardt, M., Recht, B., Vinyals, O.: Understanding deep learning requires rethinking generalization. In: Proceedings of ICLR (2017)
29. Zhang, Y., Jia, X., Yang, Y., Sun, N., Shi, S., Wang, W.: Change in the global burden of depression from 1990-2019 and its prediction for 2030. J. Psychiatr. Res. (2024)
30. Zhao, F., Li, C., Wu, Z., Ouyang, Y., Zhang, J., Dai, X.: M2DF: multi-grained multi-curriculum denoising framework for multimodal aspect-based sentiment analysis. In: Proceedings of EMNLP (2023)
31. Zhou, Z., Liu, J., Wang, S., Hao, S., Guo, Y., Hong, R.: InterMind: doctor-patient-family interactive depression assessment empowered by large language models. In: Proceedings of ACM MM (2025)
32. Zogan, H., Razzak, I., Jameel, S., Xu, G.: DepressionNet: learning multi-modalities with user post summarization for depression detection on social media. In: Proceedings of ACM SIGIR (2021)

# Dual Self-expression Subspace Clustering with Multi-scale Features and Structural Consistency Alignment

Yuhshan Peng, Ji Feng(✉), Mingxun Xi, and Yan Xu

College of Computer and Information Science, Chongqing Normal University, Chongqing, Chongqing, China
fengji@cqnu.edu.cn

**Abstract.** In recent years, deep clustering has shown strong potential for unsupervised representation learning on high-dimensional and complex data. Nevertheless, many existing approaches struggle to simultaneously capture fine-grained visual details across different scales and the intrinsic structural relationships embedded in data manifolds, which limits their clustering performance. To address these challenges, this paper proposes a Dual Self-Expression Subspace Clustering method with Multi-Scale Features and Structural Consistency Alignment (DMSSC). In DMSSC, EfficientNet is employed to extract hierarchical multi-scale features that balance local texture information and global semantic cues, while Chebyshev graph convolution effectively models structural dependencies and mitigates the over-smoothing issue commonly observed in standard GCNs. Furthermore, dual self-expression modules are designed to learn complementary content and structural affinities, which are aligned through a consistency constraint to produce more robust and discriminative subspace representations. Extensive experiments on four benchmark datasets validate the effectiveness and stability of DMSSC, achieving 87.58% ACC and 82.12% NMI on STL10. The implementation is publicly available at https://github.com/DMSSC-123/DeepSubspaceClustering.

**Keywords:** Deep clustering · Subspace clustering · EfficientNet · Unsupervised learning

## 1 Introduction

Subspace clustering focuses on revealing the underlying subspace structures of high-dimensional data via self-expression or low-rank constraints, thereby realizing automatic data grouping. Traditional methods like Sparse Subspace Clustering (SSC) [5] and Low-Rank Representation (LRR) [12] rely on linear feature spaces, limiting their ability to capture the nonlinearities in high-dimensional visual data.


R. C.-W. Wong et al. (Eds.): PAKDD 2026, LNAI 16597, pp. 417–428, 2026.
https://doi.org/10.1007/978-981-92-1300-9_33

The deep subspace clustering methods leverage deep neural networks to learn nonlinear features and model subspaces end-to-end. Representative works include the Deep Subspace Clustering Network (DSC-Net) [8], which uses an autoencoder for low-dimensional latent features and self-expression constraints in the latent space, and LARE [3], which combines autoencoders with linear subspace clustering. Multi-view methods, such as MvDSCN [22], fuse multi-scale or multi-modal information to improve stability, but still suffer from limitations. These methods rely on single-scale feature representations, neglecting important shallow-layer features, which weakens the completeness and discriminative power of the learned representations. Furthermore, they fail to effectively model structural relationships using graph structures, hindering their ability to maintain consistency between content and structure.

To address these issues, we propose DMSSC. DMSSC combines multi-scale feature extraction with structural learning and consistency-based self-expression fusion. EfficientNet [16] extracts multi-scale features, while Chebyshev Graph Convolutional Network (ChebNet) [4] learns structural relationships. A content and structure self-expression module is used to learn coefficient matrices, and a consistency constraint fuses these representations for enhanced clustering. The main contributions of this work are summarized as follows:

- We integrate the strengths of EfficientNet and the Chebyshev Graph Convolutional Network (ChebNet) to achieve multi-scale feature extraction and structure-aware representation learning, thereby improving feature learning efficiency and subspace discriminability.
- We propose a dual self-expression consistency fusion mechanism that aligns the content and structural domains through a consistency constraint, resulting in more unified and discriminative subspace embeddings.
- We conducted extensive experiments on multiple benchmark datasets, and the results demonstrate that the proposed method achieves highly competitive performance in terms of clustering accuracy(ACC), normalized mutual information(NMI), and adjusted Rand index(ARI).

## 2 Related Work

### 2.1 Deep Clustering

To address the limitations of linear models, the Deep Clustering (DC) framework integrates feature learning and clustering optimization using deep neural networks. Representative methods include Deep Embedded Clustering (DEC) [19], which uses an autoencoder optimized with a K-means objective, and its improved variant, IDEC [7], which adds a reconstruction constraint for structural consistency. VaDE [10] combines a variational autoencoder with a Gaussian mixture model to model cluster distributions. Methods like DCCM [18] and IIC [9] introduce consistency constraints or mutual information maximization to improve feature discriminability and robustness. However, most deep clustering methods rely on single-scale features, neglecting shallow-layer details like edges and texture, limiting their multi-level representation ability.

### 2.2 Graph-Based Clustering

Graph-based clustering approaches exhibit prominent capability in modeling nonlinear data relationships, which is achieved by building adjacency graphs or similarity matrices to depict pairwise sample correlations and using Graph Convolutional Networks (GCN) to propagate node features. Chebyshev Graph Convolutional Network (ChebNet) extends this by using polynomial approximations to improve graph signal flexibility and expand feature propagation. However, GCN suffers from over-smoothing, reducing feature discriminability. Methods like GAT [17] and C3 [14] address this with attention mechanisms and contrastive consistency constraints to optimize content and structure features. Despite advancements, these methods lack a unified framework to align content and structure representations, limiting clustering performance. This paper proposes a Dual Self-Expression Subspace Clustering with Multi-Scale Features and Structural Consistency Alignment (DMSSC) framework that enhances clustering by achieving cross-domain consistency while maintaining feature diversity.

## 3 Proposed Method

This section presents the proposed method for efficient clustering by integrating visual content features with graph structural information. A pretrained EfficientNet extracts multi-layer image features, and ChebNet learns structural representations on a K-nearest neighbor (KNN) graph. A self-expression mechanism captures pairwise relationships, while a consistency constraint fuses content and structural information. Finally, spectral clustering is applied to the fused matrix to obtain clustering results.

To validate the method, we visualize the affinity matrices on STL10 and CIFAR10 (Fig. 1). The clear block-diagonal patterns, with strong intra-cluster and weak inter-cluster connections, indicate that the model effectively captures the underlying subspace structure.

### 3.1 Multi-scale Feature and Content Learning

We use EfficientNet as the backbone to extract multi-scale visual features, balancing local details and global semantics by taking features from intermediate MBConv blocks (3rd, 6th, 9th for shallow, middle, deep features) and the final layer. Each feature map is flattened into a vector via Global Average Pooling, forming a multi-scale feature set:

$$\mathcal{Z} = \{\boldsymbol{z}^{(1)}, \boldsymbol{z}^{(2)}, \ldots, \boldsymbol{z}^{(L)}\}, \quad \boldsymbol{z}^{(l)} \in \mathbb{R}^{d_l} \tag{1}$$

These features are used for structural modeling to construct KNN graphs, capturing local structural relationships among samples. Meanwhile, the global semantic feature from EfficientNet's final layer is used as the visual content representation $\mathbf{X}$, which encodes global semantic information and provides a more discriminative visual representation for the subsequent content self-expression modeling (Fig. 2).

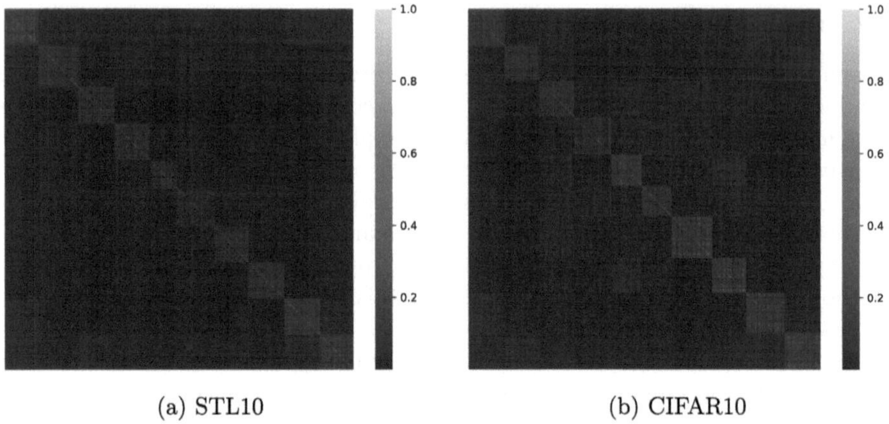


**Fig. 1.** Fusion self-expression affinity matrices on (a) STL10 and (b) CIFAR10 datasets.

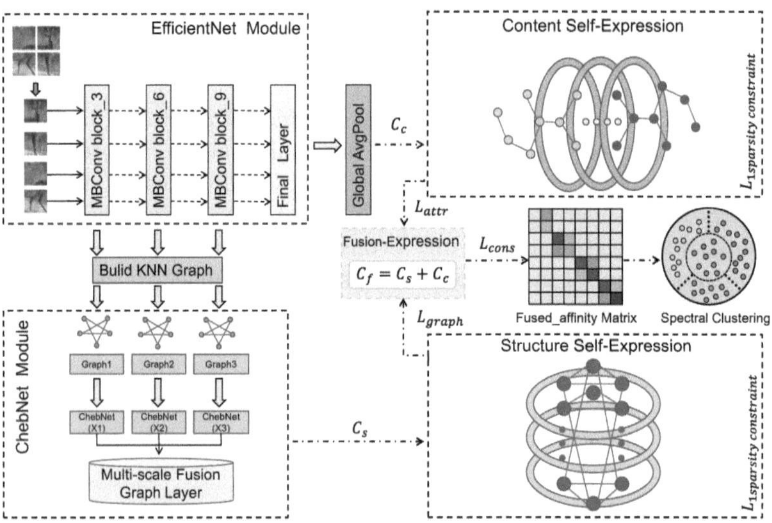


**Fig. 2.** Our model framework consists of three main components: multi-scale feature extraction (EfficientNet), structural feature learning (ChebNet), and dual self-expression fusion.

## 3.2 Graph Structure and Structural Learning

After obtaining the multi-scale features, we further model the structural relationships among samples to capture the underlying manifold topology. For each scale feature $\boldsymbol{z}^{(l)}$, a K-nearest neighbor (KNN) sparse graph is constructed based on the Euclidean distance. The Euclidean metric is adopted because the semantic features extracted by CNN exhibit strong local manifold preservation in Euclidean space, allowing for a more stable reflection of pairwise similarity among samples. To avoid the over-smoothing problem caused by repeated Laplacian multiplications in standard GCN, this paper employs Chebyshev polynomial approximation-based spectral graph convolution, whose general formulation can be expressed as follows:

$$X^{(l+1)} = \sum_{k=0}^{K} T_k(\tilde{L}) X^{(l)} W_k \tag{2}$$

where $\tilde{L} = \frac{2L}{\lambda_{\max}} - I_N$ is the normalized graph Laplacian, and $L = I_N - D^{-\frac{1}{2}} A D^{-\frac{1}{2}}$ the standard Laplacian ($A$: adjacency matrix, $D$: degree matrix). $W_k$ denotes the learnable weight matrix for the $k$-th polynomial, and $T_k(\cdot)$ the $k$-th Chebyshev polynomial (defined recursively) is defined as follows:

$$T_0(x) = 1, \quad T_1(x) = x, \quad T_k(x) = 2xT_{k-1}(x) - T_{k-2}(x) \tag{3}$$

This formulation approximates spectral filtering via polynomial expansion, enabling graph convolution to model K-hop node dependencies. When $K = 2$, the network captures interactions between nodes and their second-order neighbors, preserving local smoothness and enhancing structural distinctions. $K$-order Chebyshev approximation-based spectral graph convolution is applied to extract structural features $\widetilde{\mathbf{X}}^{(l)}$:

$$Z^{(l)} \xrightarrow{\text{ChebConv}} \widetilde{X}^{(l)} \in R^{N \times h} \tag{4}$$

where $h$ denotes the hidden dimension (set to 16 in our experiments). The three groups of structural features extracted from graphs at different scales, $\tilde{\boldsymbol{X}}^{(1)}, \tilde{\boldsymbol{X}}^{(2)}, \tilde{\boldsymbol{X}}^{(3)}$, are then projected onto a unified 384-dimensional space through three linear layers and fused across scales by element-wise summation:

$$S = \sum_{l=1}^{3} W^{(l)} \widetilde{X}^{(l)}, \quad W^{(l)} : R^h \to R^{384} \tag{5}$$

The final fused structural representation $S$ simultaneously encodes both local and global structural information across different scales, serving as the input representation for subsequent structural self-expression modeling.

### 3.3 Training Objective and Optimization

The core idea of subspace clustering lies in the assumption that samples within the same subspace can be approximately represented as a linear combination of other samples from the same subspace. Therefore, we separately learn self-expression coefficient matrices in the visual content space and the graph-structural space:

$$X \approx C_c X, \quad S \approx C_s S \tag{6}$$

In the implementation, $X$ and $S$ denote the visual feature matrix and the graph-structural feature matrix obtained from the previous two sections, while

$C_c$ and $C_s$ are content self-expression matrix and the structural self-expression matrix. To enhance interpretability and stability, an $L_1$ regularization is applied to the coefficient matrices to suppress noisy connections and promote sparse structures. The self-expression losses for the content and structural domains are defined as follows:

$$\mathcal{L}_{\text{attr}} = \|X - C_c X\|_F^2, \quad \mathcal{L}_{\text{graph}} = \|S - C_s S\|_F^2 \tag{7}$$

An additional $L_1$ sparsity regularization is imposed to promote strong connections within clusters while reducing connections between different clusters:

$$\|C_c\|_1, \|C_s\|_1 \tag{8}$$

To enhance the consistency between the content and structural relation graphs, a simple yet effective $L_2$ consistency term is introduced:

$$\mathcal{L}_{\text{cons}} = \|C_c - C_s\|_F^2 \tag{9}$$

This term encourages the self-expressive relationships in the content and structural domains to remain consistent, thereby aligning the two similarity structures during optimization. After obtaining $C_c$ and $C_s$, we combine them to obtain the final fused self-expression matrix $C_f$. In this work, an interpretable additive fusion strategy is adopted:

$$C_f = C_c + C_s \tag{10}$$

An additional $L_1$ regularization is imposed on the fused matrix $C_f$:

$$\|C_c + C_s\|_1 \tag{11}$$

To promote the sparsity and separability of the affinity matrix, this design intuitively reflects the complementarity between "content and structure", while avoiding the additional uncertainty and hyperparameter overhead introduced by complex attention or gating mechanisms. Therefore, our total loss function is defined as follows:

$$\begin{aligned} \mathcal{L}_{\text{total}} = & \lambda_1 \|X - C_c X\|_F^2 + \|C_c\|_1 \\ & + \lambda_2 \|S - C_s S\|_F^2 + \|C_s\|_1 \\ & + \lambda_3 \|C_c - C_s\|_F^2 + \|C_c + C_s\|_1 \end{aligned} \tag{12}$$

# 4 Experiments

## 4.1 Datasets and Evaluation Metrics

Datasets: Experiments were conducted on four publicly available datasets: CIFAR-10, CIFAR-100, Fashion-MNIST, and STL-10. CIFAR-10 contains 60,000 images from 10 categories with a resolution of 32×32, while CIFAR-100 is an extended version with 100 categories. Fashion-MNIST consists of 70,000 grayscale clothing images from 10 categories (28×28 resolution), and STL-10 includes 13,000 RGB natural images from the same categories as CIFAR-10 (96×96 resolution).

Evaluation Metrics: In this study, three commonly used clustering metrics are employed to evaluate performance: Clustering Accuracy (ACC), Normalized Mutual Information (NMI), and Adjusted Rand Index (ARI). These metrics assess the consistency between the clustering results and the ground-truth labels from different perspectives.

### 4.2 Experimental Setup

In the experiments, EfficientNet-B3 was adopted as the backbone network, initialized with pretrained parameters. Multi-scale features were extracted from the intermediate layers (MBConv-3, MBConv-6, and MBConv-9) as well as from the final convolutional layer. All input images were resized to 224×224 and normalized before processing. During the clustering stage, a K-nearest neighbor $KNN$ graph was constructed using the extracted features, with $Knn = 20$. Subsequently, a Chebyshev Graph Convolutional Network (ChebNet) was employed to learn structural features, which can efficiently model high-order dependencies on graphs through polynomial approximation. In our implementation, the order of the Chebyshev polynomial was set to $K = 2$, enabling the model to capture second-order neighborhood interactions while avoiding excessive oversmoothing. The model was optimized using the Adam optimizer with a batch size of 128, trained for 1000 epochs. The learning rate was set to $10^{-4}$, except for the Fashion-MNIST dataset, where it was adjusted to $10^{-5}$. Unlike most deep clustering methods, our approach does not rely on additional pretraining or multi-stage optimization strategies. Instead, it achieves end-to-end joint optimization of multiple loss functions, enabling unified learning of content and structural subspaces.

### 4.3 Experimental Results and Analysis

We compare the proposed DMSSC with the following methods: Deep Embedded Clustering (DEC) [19], Improved Deep Embedded Clustering (IDEC) [7], Deep K-means (DKM) [6], Variational Deep Embedding (VaDE) [10], Deep Adaptive Clustering (DAC) [2], Deep Comprehensive Correlation Mining (DCCM) [18], Invariant Information Clustering (IIC) [9], Contrastive Clustering (CC) [11], Multi-Scale Graph Contrastive Clustering (IcicleGCN) [20], Cross-Sample Contrastive Clustering (C3) [14], Hierarchical Contrastive Clustering (CoHiClust) [23], Graph Attention-based Contrastive Learning for Clustering (GACL) [13], Efficient Deep Embedded Subspace Clustering (EDESC) [1], Contrastive Deep Embedded Clustering (CDEC) [15] and Self-Expressive Network (SENet) [21] (Fig. 3).

**Table 1.** Clustering performance of ACC(%), NMI(%), and ARI(%). Best and second-best results are highlighted.(Since the implementation links of some methods are unavailable or invalid, we mark their results with a dash '–'. All the presented results are derived from publicly reported data in the original literature or reproduced based on the authors' official implementations.)

| Method | CIFAR-10 | | | STL-10 | | | Fashion-MNIST | | | CIFAR-100 | | |
|---|---|---|---|---|---|---|---|---|---|---|---|---|
| | ACC | NMI | ARI | ACC | NMI | ARI | ACC | NMI | ARI | ACC | NMI | ARI |
| DEC | 30.10 | 25.70 | 16.10 | 35.90 | 27.60 | 18.60 | 51.27 | 54.58 | 43.98 | 18.50 | 13.06 | 05.00 |
| IDEC | 36.99 | 32.53 | 23.41 | 32.53 | 18.85 | 11.09 | 52.90 | 55.70 | 43.98 | 19.16 | 14.58 | 02.51 |
| DKM | 35.26 | 26.12 | – | 32.61 | 29.12 | – | 51.31 | 55.57 | – | 18.14 | 12.30 | – |
| VaDE | 29.10 | 24.50 | – | 28.10 | 20.00 | – | 50.39 | 59.63 | – | 15.20 | 10.08 | – |
| DAC | 52.20 | 39.60 | 30.60 | 47.00 | 36.60 | 25.70 | 61.50 | 63.20 | 50.20 | 23.80 | 18.50 | 08.80 |
| DCCM | 62.30 | 49.60 | 40.80 | 48.20 | 37.60 | 26.20 | – | – | – | 32.70 | 28.50 | 17.30 |
| IIC | 61.70 | 51.30 | 41.10 | 49.90 | 43.10 | 29.50 | 65.70 | 63.70 | **52.30** | – | 25.70 | – |
| CC | 79.00 | 70.50 | 63.70 | 85.00 | 76.40 | **72.60** | – | – | – | 42.90 | 43.10 | 26.60 |
| EDESC | 67.20 | 46.40 | – | 74.50 | 68.70 | – | 63.10 | **67.00** | – | 38.50 | 37.00 | – |
| CDEC | 56.40 | 57.78 | – | 73.28 | 71.83 | – | 53.51 | 54.04 | – | – | – | – |
| IcicleGCN | 80.70 | 72.90 | 66.00 | – | – | – | – | – | – | 46.10 | 45.90 | 31.10 |
| C3 | 74.30 | **83.60** | 70.30 | – | – | – | – | – | – | 43.50 | 45.60 | 27.40 |
| CoHiClust | 77.90 | **83.90** | **73.10** | 58.40 | 61.30 | 47.40 | – | – | – | 46.70 | 43.70 | 29.90 |
| SENet | 76.50 | 65.50 | – | 82.32 | 75.41 | – | **69.70** | 66.30 | – | – | – | – |
| GACL | **87.50** | 79.30 | **75.30** | **86.30** | **78.30** | **74.40** | – | – | – | **48.80** | **49.60** | **32.10** |
| OURS | **82.50** | 73.49 | 66.45 | **87.58** | **82.12** | 70.46 | **68.30** | **68.52** | **55.37** | **49.50** | **66.72** | **30.42** |

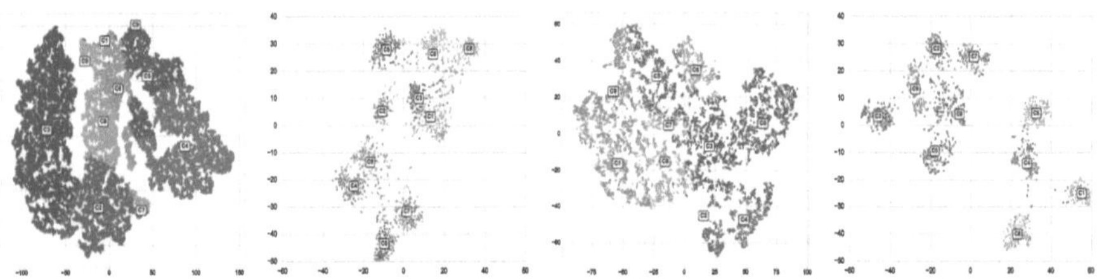

(a) DEC on CIFAR10 (b) OURS on CIFAR10 (c) DEC on STL10 (d) OURS on STL10

**Fig. 3.** t-SNE visualization of clustering results for CIFAR10 and STL10 datasets.

As observed from Table 1, our method exhibits stable and competitive performance across the four datasets. On CIFAR-10 and STL-10, it performs favorably compared with most existing approaches, while on the more challenging CIFAR-100 dataset, it achieves relatively strong results in terms of ACC and NMI. These observations suggest that incorporating multi-scale feature representations and graph structural information through the proposed consistency-aware subspace learning framework can effectively enhance clustering performance.

### 4.4 Ablation Study

To verify the effectiveness of the proposed method, this section conducts ablation experiments from two perspectives: network architecture and method components, including comparisons of backbone networks and analyses of key modules.

**Table 2.** Ablation study on different EfficientNet backbones across four datasets(B0–B4 denote EfficientNet-b0–b4). Best and second-best percolumn are highlighted.

| Network | STL-10 | | | Fashion-MNIST | | | CIFAR-10 | | | CIFAR-100 | | |
|---|---|---|---|---|---|---|---|---|---|---|---|---|
| | ACC | NMI | ARI | ACC | NMI | ARI | ACC | NMI | ARI | ACC | NMI | ARI |
| B0 | 85.31 | 79.00 | 66.93 | 62.90 | 61.30 | 49.11 | 55.16 | 50.12 | 39.49 | 39.23 | 60.13 | 20.30 |
| B1 | **88.59** | **83.08** | **74.19** | 65.50 | 63.59 | 50.37 | 74.92 | 67.26 | 52.03 | 43.43 | 62.45 | 23.91 |
| B2 | 85.62 | 80.94 | 66.93 | **68.60** | **66.28** | **52.37** | **78.12** | **68.66** | **58.67** | 44.87 | 64.16 | 19.60 |
| B3 | **87.58** | **82.12** | **70.46** | **68.30** | **68.52** | **55.37** | **82.50** | **73.49** | **66.45** | **49.50** | **66.72** | **30.42** |
| B4 | 51.41 | 45.87 | 30.23 | 63.10 | 61.43 | 48.42 | 57.58 | 48.83 | 32.79 | **45.57** | **66.92** | **27.83** |

As shown in Table 2, the clustering performance first improves and then declines as the model capacity increases. EfficientNet-B3 achieves the best or second-best results across all four datasets, indicating that a medium-sized backbone strikes a favorable balance between feature abstraction and detail preservation, whereas overly shallow (B0) or overly deep (B4) models tend to suffer from degraded performance due to insufficient or excessively abstract representations.

**Table 3.** Ablation study on different loss components of the proposed model.

| Dataset | $\mathcal{L}_{attr}$ | | | $\mathcal{L}_{attr} + \mathcal{L}_{graph}$ | | | $\mathcal{L}_{attr} + \mathcal{L}_{graph} + \mathcal{L}_{cons}$ | | |
|---|---|---|---|---|---|---|---|---|---|
| | ACC | NMI | ARI | ACC | NMI | ARI | ACC | NMI | ARI |
| Fashion-MNIST | 66.50 | 67.25 | 53.59 | 67.10 | 68.33 | 54.81 | **68.30** | **68.52** | **55.37** |
| STL10 | 83.59 | 80.05 | 62.73 | 84.38 | 80.09 | 63.22 | **87.58** | **82.12** | **70.46** |
| CIFAR10 | 80.80 | 71.35 | 62.76 | 80.63 | 71.51 | 63.29 | **82.50** | **73.49** | **66.45** |
| CIFAR100 | 46.77 | 65.45 | 21.18 | 46.93 | 65.72 | 28.90 | **49.50** | **66.72** | **30.42** |

Table 3 reports the ablation results under three settings: content-only($\mathcal{L}_{attr}$), content +structure($\mathcal{L}_{attr} + \mathcal{L}_{graph}$) and the full model with consistency($\mathcal{L}_{attr} + \mathcal{L}_{graph} + \mathcal{L}_{cons}$ ) Performance improves steadily as each component is added, showing that structural modeling and consistency alignment jointly enhance feature discriminability. On STL-10, for example, ACC rises from 0.8438 to 0.8758 and NMI from 0.8009 to 0.8212, demonstrating the effectiveness of the proposed consistency constraint mechanism.

### 4.5 Effect of Graph Convolution Type

To ensure a fair comparison, both ChebNet and GCN adopt the same two-layer architecture and ReLU activation. As presented in Table 4, ChebNet achieves better performance than GCN on all datasets. On CIFAR100, for instance, ACC increases from 0.4337 to 0.4950, confirming the benefits of high-order spectral modeling in reducing over-smoothing.

**Table 4.** Comparison of clustering performance between GCN and ChebNet.

| Method | Fashion-MNIST | | | STL10 | | | CIFAR10 | | | CIFAR100 | | |
|---|---|---|---|---|---|---|---|---|---|---|---|---|
| | ACC | NMI | ARI | ACC | NMI | ARI | ACC | NMI | ARI | ACC | NMI | ARI |
| GCN | 66.80 | 67.42 | 54.00 | 86.33 | 80.45 | 68.45 | 79.14 | 71.54 | 57.77 | 43.37 | 63.22 | 27.19 |
| ChebNet | **68.30** | **68.52** | **55.37** | **87.58** | **82.12** | **70.46** | **82.50** | **73.49** | **66.45** | **49.50** | **66.72** | **30.42** |

### 4.6 Parameter Sensitivity Analysis

We further examined the impact of ChebNet's neighbor count (K) and loss coefficients $\lambda_1$, $\lambda_2$, and $\lambda_3$ on experimental results. Default settings were K=20, $\lambda_1 = \lambda_2 = 1$, and $\lambda_3 = 0.5$; other parameters remained fixed when testing one. We evaluated K $\in 5, 10, 15, 20, 25, 30$ and $\lambda_1, \lambda_2, \lambda_3 \in 0.01, 0.1, 0.5, 1, 10, 100$. Results indicate our method is largely insensitive to K, with $\lambda_2$ and $\lambda_3$ exerting minor effects on clustering performance. $\lambda_1 = 1$ is dominant excessively large values degrade performance on CIFAR-100 and Fashion-MNIST while $\lambda_1$=1 yields stable, competitive results across all datasets (Fig. 4).

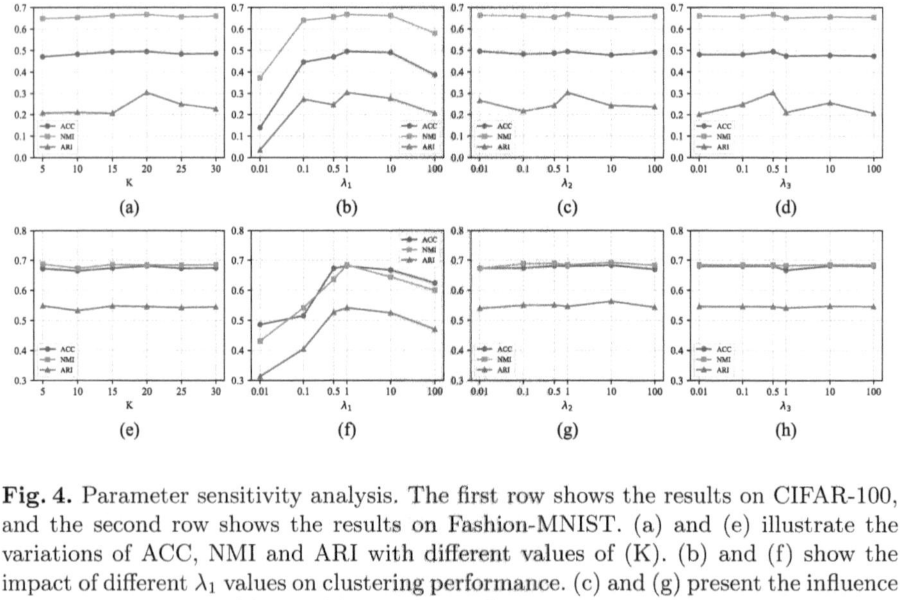


**Fig. 4.** Parameter sensitivity analysis. The first row shows the results on CIFAR-100, and the second row shows the results on Fashion-MNIST. (a) and (e) illustrate the variations of ACC, NMI and ARI with different values of (K). (b) and (f) show the impact of different $\lambda_1$ values on clustering performance. (c) and (g) present the influence of different $\lambda_2$ values. (d) and (h) depict the effect of varying $\lambda_3$ values.

### 4.7 Convergence Analysis

As shown in Fig. 5, the model converges stably on CIFAR-10 and Fashion-MNIST, with slight fluctuations on CIFAR-100. This is mainly due to its larger number of classes and fewer samples per category, which make similarity estimation noisier. Overall, the model demonstrates strong robustness and stable convergence even in complex multi-class settings.

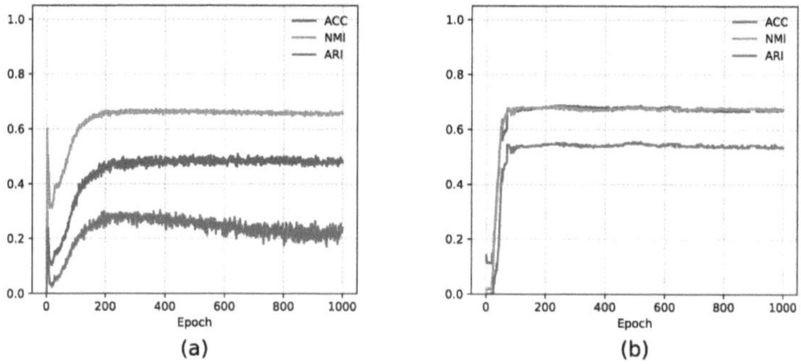


**Fig. 5.** Clustering convergence analysis. (a) shows the variations of ACC, NMI, and ARI with training epochs on the CIFAR-100 dataset, while (b) presents the corresponding changes on the Fashion-MNIST dataset.

## 5 Conclusion

This paper proposes a Consistency-Aligned Dual Self-Expressive Deep Subspace Clustering method that integrates content and structural feature learning through consistency alignment. Experimental results show that DMSSC achieves strong clustering performance on four benchmark datasets. Future work will focus on cross-modal feature fusion and adaptive graph structure learning to enhance generalization and robustness.

## References

1. Cai, J., Fan, J., Guo, W., Wang, S., Zhang, Y., Zhang, Z.: Efficient deep embedded subspace clustering. In: Proceedings of the IEEE/CVF Conference on Computer Vision and Pattern Recognition, pp. 1–10 (2022)
2. Chang, J., Wang, L., Meng, G., Xiang, S., Pan, C.: Deep adaptive image clustering. In: Proceedings of the IEEE International Conference on Computer Vision, pp. 5879–5887 (2017)
3. Chen, Y., Zhang, L., Yi, Z.: Subspace clustering using a low-rank constrained autoencoder. Inf. Sci. **424**, 27–38 (2018)
4. Defferrard, M., Bresson, X., Vandergheynst, P.: Convolutional neural networks on graphs with fast localized spectral filtering. In: Advances in Neural Information Processing Systems, vol. 29 (2016)

5. Elhamifar, E., Vidal, R.: Sparse subspace clustering: algorithm, theory, and applications. IEEE Trans. Pattern Anal. Mach. Intell. **35**(11), 2765–2781 (2013)
6. Fard, M.M., Thonet, T., Gaussier, E.: Deep k-means: jointly clustering with k-means and learning representations. Pattern Recogn. Lett. **138**, 185–192 (2020) clustering with k-means and learning representations. Pattern Recogn. Lett. **138**, 185–192 (2020)
7. Guo, X., Gao, L., Liu, X., Yin, J.: Improved deep embedded clustering with local structure preservation. IJCAI **17**, 1753–1759 (2017)
8. Ji, P., Zhang, T., Li, H., Salzmann, M., Reid, I.: Deep subspace clustering networks. In: Advances in Neural Information Processing Systems, vol. 30 (2017)
9. Ji, X., Henriques, J.F., Vedaldi, A.: Invariant information clustering for unsupervised image classification and segmentation. In: Proceedings of the IEEE/CVF International Conference on Computer Vision, pp. 9865–9874 (2019)
10. Jiang, Z., Zheng, Y., Tan, H., Tang, B., Zhou, H.: Variational deep embedding: an unsupervised and generative approach to clustering. arXiv preprint arXiv:1611.05148 (2016)
11. Li, Y., Hu, P., Liu, Z., Peng, D., Zhou, J.T., Peng, X.: Contrastive clustering. In: Proceedings of the AAAI Conference on Artificial Intelligence, vol. 35, pp. 8547–8555 (2021)
12. Liu, G., Lin, Z., Shuicheng Yan, J., Sun, Y.Y., Ma, Y.: Robust recovery of subspace structures by low-rank representation. IEEE Trans. Pattern Anal. Mach. Intell. **35**(1), 171–184 (2012)
13. Liu, M., et al.: Deep clustering by graph attention contrastive learning. Electronics **12**(11), 2489 (2023)
14. Sadeghi, M., Hojjati, H., Armanfard, N.: C3: cross-instance guided contrastive clustering. arXiv preprint arXiv:2211.07136 (2022)
15. Sheng, G., Wang, Q., Pei, C., Gao, Q.X.: Contrastive deep embedded clustering. Neurocomputing **514**, 13–20 (2022)
16. Tan, M., Le, Q.: EfficientNet: rethinking model scaling for convolutional neural networks. In: International Conference on Machine Learning, pp. 6105–6114. PMLR (2019)
17. Veličković, P., Cucurull, G., Casanova, A., Romero, A., Lio, P., Bengio, Y.: Graph attention networks. arXiv preprint arXiv:1710.10903 (2017)
18. Wu, J., et al.: Deep comprehensive correlation mining for image clustering. In: Proceedings of the IEEE/CVF International Conference on Computer Vision, pp. 8150–8159 (2019)
19. Xie, J., Girshick, R., Farhadi, A.: Unsupervised deep embedding for clustering analysis. In: International Conference on Machine Learning, pp. 478–487. PMLR (2016)
20. Yuankun, X., Huang, D., Wang, C.-D., Lai, J.-H.: Deep image clustering with contrastive learning and multi-scale graph convolutional networks. Pattern Recogn. **146**, 110065 (2024)
21. Zhang, S., You, C., Vidal, R., Li, C.-G.: Learning a self-expressive network for subspace clustering. In: Proceedings of the IEEE/CVF Conference on Computer Vision and Pattern Recognition, pp. 12393–12403 (2021)
22. Zhu, P., Yao, X., Wang, Yu., Hui, B., Du, D., Hu, Q.: Multi-view deep subspace clustering networks. arXiv preprint arXiv:1908.01978 (2019)
23. Znalezniak, M., Rola, P., Kaszuba, P., Tabor, J., Śmieja, M.: Contrastive hierarchical clustering. In: Joint European Conference on Machine Learning and Knowledge Discovery in Databases, pp. 627–643. Springer, Cham (2023). https://doi.org/10.1007/978-3-031-43412-9_37

# AnomalyFilter: Selective Denoising Diffusion Model for Time Series Anomaly Detection

Kohei Obata[1(✉)], Zheng Chen[1], Yasuko Matsubara[1], Lingwei Zhu[2], and Yasushi Sakurai[1]

[1] SANKEN, The University of Osaka, Osaka, Japan
{obata88,chenz,yasuko,yasushi}@sanken.osaka-u.ac.jp
[2] Great Bay University, Dongguan, China
zhulingwei@gbu.edu.cn

**Abstract.** Time series anomaly detection (TSAD) has been an important area of research for decades, with reconstruction-based methods, mostly based on generative models, gaining popularity and demonstrating success. Diffusion models have recently attracted attention due to their advanced generative capabilities. Existing diffusion-based methods for TSAD rely on a conditional strategy, which reconstructs input instances from white noise with the aid of the conditioner. However, this poses challenges in accurately reconstructing the normal parts, resulting in suboptimal detection performance. In response, we propose a novel diffusion-based method, named AnomalyFilter, which acts as a selective filter that only denoises anomaly parts in the instance while retaining normal parts. To build such a filter, we mask Gaussian noise during the training phase and conduct the denoising process without adding noise to the instances. The synergy of the two simple components greatly enhances the performance of naive diffusion models. Extensive experiments on five datasets demonstrate that AnomalyFilter achieves notably low reconstruction error on normal parts, providing empirical support for its effectiveness in anomaly detection.

**Keywords:** Time series · Anomaly detection · Diffusion model

## 1 Introduction

Time series anomaly detection (TSAD) is an important task for real-world applications, including robot-assisted systems [14] and cybersecurity [21]. The goal of TSAD is to detect unexpected/unusual patterns/values within sequences that are significantly dissimilar to the majority. Since anomalies are typically rare and labeling them from quantities of data is laborious, TSAD is generally approached as an unsupervised learning problem in which detection relies on a normality assumption [12], namely most observations are normal.

A prevalent approach for TSAD involves reconstruction-based methods. These methods learn to reconstruct normal samples during training and assume


R. C.-W. Wong et al. (Eds.): PAKDD 2026, LNAI 16597, pp. 429–441, 2026.
https://doi.org/10.1007/978-981-92-1300-9_34

that normal samples are better reconstructed than anomalous samples during inference. Then, anomalies can be detected by thresholding the difference between the input and its reconstruction. Hence, the ideal reconstruction exhibits low reconstruction error on normal parts and high error on anomalous parts. Many methods, such as those based on GAN [32] and Transformer [31], are categorized as encoder-decoder models because they reconstruct samples from embeddings in latent space. However, controlling the balance between concise representation (i.e., regularization) and representation power (i.e., reconstruction) remains challenging [5]. This issue is unavoidable when reconstructing from latent space.

Recently, diffusion models [6] have emerged as a promising approach due to their advanced generative capabilities. Unlike encoder-decoder models, diffusion models operate directly in data space, gradually reconstructing data from noisy inputs through an iterative denoising process. Current diffusion models for TSAD primarily focus on developing an effective conditional strategy to introduce inductive bias into the denoising process [3,30]. However, reconstructing from white noise proves challenging, even with the aid of the conditioner, due to the unavailability of original information. Consequently, while it denoises anomalous parts well and the reconstruction of anomalous parts deviates from the original input, it limits the accurate reconstruction of the normal parts.

Thus, a natural research question arises: *How can we accurately reconstruct the normal parts while exploiting the denoising capability of diffusion models?* In this work, we propose AnomalyFilter, a new diffusion-based method designed for TSAD, which overcomes the limitations of the existing diffusion-based methods regarding reconstruction quality. Specifically, we build a selective filter that denoises only the anomalous parts in the sample while retaining the normal parts. While alternative noise strategies for diffusion models have been explored in the image domain [29], the impact of noise design on time series remains largely underexplored. Thus, we investigate the utility of *masked Gaussian noise*, which is generated through a combination of Gaussian noise and a Bernoulli mask. During the training, the model learns to predict the noise added to a sample similar to general diffusion models, enabling the model to act as a filter that selectively denoises nonmasked parts (i.e., erases Gaussian noise) while passing through masked parts (i.e., retains input). Furthermore, we observe that adding noise to the sample during the denoising process merely degrades reconstruction quality, leading to poor detection accuracy. To address this, we propose *noiseless inference*, which takes a scaled noiseless sample as input and performs denoising without any noise, aligning seamlessly with the masked Gaussian noise. Neither masked Gaussian noise nor noiseless inference alone contributes significantly to improving the performance of the vanilla DDPM. However, the synergy of the two simple components allows the model to act as an effective filter, achieving substantial performance improvements over the vanilla DDPM.

In summary, this paper makes the following contributions:

- We investigate the limitations of the existing diffusion-based methods for TSAD from the perspective of noise design and empirically demonstrate their impact on reconstruction quality and anomaly detection performance.

- We introduce AnomalyFilter[1], a novel diffusion-based method designed for TSAD. It achieves ideal reconstruction, i.e., a low reconstruction error for normal parts and relatively high reconstruction error for anomalous parts.
- We show the superiority of AnomalyFilter by comparing it with state-of-the-art baselines on five datasets. Notably, by combining two simple components, masked Gaussian noise and noiseless inference, we achieve a 45.1% improvement in VUS-PR compared with vanilla DDPM.

## 2 Related Work

**Diffusion Model for Time Series Anomaly Detection.** TSAD differs from forecasting or imputation tasks in that all values are known. Overly powerful conditioners risk carrying anomaly information, which can lead to the accurate reconstruction of even anomalous parts. Therefore, the current mainstream focus is on a unique conditional strategy [15,27]. For example, IMDiffusion [3] is inspired by imputation techniques, where part of the sample is masked, and the unmasked parts serve as a condition to impute the masked parts. DiffAD [30] tackles the case where anomalous points concentrate in certain parts, which is called anomaly concentration, and incorporates a density ratio-based point selection strategy for masking anomalous parts. However, since they rely on partially observed input as a condition, which often loses information needed for reconstructing normal parts, their reconstruction errors are relatively high, resulting in a failure to detect anomalies that differ slightly from normality.

**Denoising Model for Anomaly Detection.** Denoising models, including diffusion models, perform anomaly detection by taking corrupted inputs, learning to denoise them, and generating clean reconstructions. Similar to diffusion models, a denoising autoencoder (DAE) [26] adds noise to the input but differs in that it directly predicts the clean input rather than using a stepwise denoising process. The selection of noise is critical in denoising models. Recent studies have explored deterministic noise [2,4], and non-Gaussian noise [11] for diffusion models. In image anomaly detection, [7] shows that DAE trained with coarse noise outperforms diffusion models at Brain MRI, and AnoDDPM [29] demonstrates the effectiveness of Simplex noise. Despite these advancements in the image domain, where noise has been extensively studied and often operates on 2D data, noise tailored for time series remains largely unexplored. This highlights a significant gap in understanding the impact of noise design in TSAD.

## 3 Preliminaries

### 3.1 Problem Definition

In this paper, we focus on unsupervised anomaly detection for time series. We adopt a local contextual window to model their temporal dependency, which

[1] The code is available at https://github.com/KoheiObata/AnomalyFilter.

is a common practice with most deep TSAD methods. Consider an instance $\boldsymbol{X} = \{\boldsymbol{x}_1, \boldsymbol{x}_2, \ldots, \boldsymbol{x}_L\} \in \mathbb{R}^{K \times L}$ to be a multivariate time series with $K$ features and $L$ timesteps. The objective of TSAD is to identify anomalous timesteps in a test set $\mathcal{D}^{test}$ by training the model in an unlabeled training set $\mathcal{D}^{train}$ assumed to contain mostly normal instances. Thus, the model aims to assign an anomaly score to each timestep, where higher scores are more anomalous. In a reconstruction-based approach, the model outputs a reconstruction $\hat{\boldsymbol{X}}$ of the given instance $\boldsymbol{X}$, and the anomaly score is calculated by comparing them by a score function $\mathcal{A}(\boldsymbol{x}_l, \hat{\boldsymbol{x}}_l) : \mathbb{R}^K \to \mathbb{R}$ (for simplicity, we consider the MSE score). To solve TSAD efficiently, we need a powerful method that achieves ideal reconstruction, i.e., has a low reconstruction error on the normal parts and a high reconstruction error on the anomalous parts of the instance.

### 3.2 Denoising Diffusion Probablistic Model

Our AnomalyFilter is based on the denoising diffusion probabilistic model (DDPM) [6], a well-known diffusion model. DDPM consists of a forward diffusion process and a reverse denoising process. The forward diffusion process $q(\boldsymbol{X}_t \mid \boldsymbol{X}_{t-1})$ gradually corrupts data from some target distribution $q(\boldsymbol{X}_0)$ into a normal distribution, where $\boldsymbol{X}_t$ denotes corrupted data indexed by a diffusion step $t$. Thus, in practice, $\boldsymbol{X}_0$ is $\boldsymbol{X}$. Then, a learned reverse process $p_\theta(\boldsymbol{X}_{t-1} \mid \boldsymbol{X}_t)$ generates data by turning noise into data from $q(\boldsymbol{X}_0)$.

**Forward Process.** This process can be formally described as Markovian:

$$q(\boldsymbol{X}_t \mid \boldsymbol{X}_{t-1}) = \mathcal{N}(\boldsymbol{X}_t; \sqrt{1-\beta_t}\boldsymbol{X}_{t-1}, \beta_t \boldsymbol{I}), \tag{1}$$

for $t = \{1, \ldots, T\}$, where $T$ is a forward diffusion step. The variance schedule $\beta_t \in (0, 1)$ is a hyperparameter defined by a monotonically increasing sequence $\beta_1 < \beta_2 < \cdots < \beta_T$ such that the data become sequentially corrupted.

To sample from an arbitrary $t$, namely without having to find intermediate steps $\boldsymbol{X}_{t-1}, \ldots, \boldsymbol{X}_1$, the variance schedule is simplified to $\alpha_t = 1 - \beta_t$ and $\bar{\alpha}_t = \prod_{i=1}^{T} \alpha_i$, as follows:

$$q(\boldsymbol{X}_t \mid \boldsymbol{X}_0) = \mathcal{N}(\boldsymbol{X}_t; \sqrt{\bar{\alpha}_t}\boldsymbol{X}_0, (1-\bar{\alpha}_t)\boldsymbol{I}), \tag{2}$$

$$\boldsymbol{X}_t = \bar{\alpha}_t \boldsymbol{X}_0 + \sqrt{1-\bar{\alpha}_t}\epsilon_t, \tag{3}$$

where $\epsilon_t \sim \mathcal{N}(\boldsymbol{0}, \boldsymbol{I})$ is the same size as $\boldsymbol{X}_0$.

**Reverse Process.** The generative model, parameterized by $\theta$, is a learned reverse process to denoise $\boldsymbol{X}_T$ and reconstruct $\boldsymbol{X}_0$. This is accomplished by iteratively computing the following Gaussian transitions:

$$p_\theta(\boldsymbol{X}_{t-1} \mid \boldsymbol{X}_t) = \mathcal{N}(\boldsymbol{X}_{t-1}; \mu_\theta(\boldsymbol{X}_t, t), \tilde{\beta}_t \boldsymbol{I}), \tag{4}$$

**Algorithm 1.** Naive Training

1: **Input:** Training set $\mathcal{D}^{train}$, Forward diffusion step $T$;
2: **repeat**
3: Sample data $\boldsymbol{X}_0 \sim \mathcal{D}^{train}$;
4: Sample diffusion step $t \sim Uniform(\{1, \ldots, T\})$;
5: Sample noise $\epsilon_t \sim \mathcal{N}(\mathbf{0}, \boldsymbol{I})$;
6: Corrupt $\boldsymbol{X}_t = \bar{\alpha}_t \boldsymbol{X}_0 + \sqrt{1-\bar{\alpha}_t}\epsilon_t$;
7: Calculate $\mathcal{L} = \|\epsilon_t - \epsilon_\theta(\boldsymbol{X}_t, t)\|$;
8: **until** Converged;

**Algorithm 2.** Naive Inference

**Input:** Data $\boldsymbol{X}_0 \in \mathcal{D}^{test}$, Reverse diffusion step $\lambda$, Trained denoising function $\epsilon_\theta$;
**Output:** Reconstruction $\hat{\boldsymbol{X}}_0$ (i.e., $\hat{\boldsymbol{X}}$);
Corrupt $\hat{\boldsymbol{X}}_\lambda = \bar{\alpha}_\lambda \boldsymbol{X}_0 + \sqrt{1-\bar{\alpha}_\lambda}\epsilon_\lambda$;
**for** $t = \lambda : 1$ **do**
Sample $\epsilon_t \sim \mathcal{N}(\mathbf{0}, \boldsymbol{I})$ if $t > 1$ else $\epsilon_t = 0$;
Denoise $\hat{\boldsymbol{X}}_{t-1} = \frac{1}{\sqrt{\alpha_t}}(\hat{\boldsymbol{X}}_t - \frac{\beta_t}{\sqrt{1-\bar{\alpha}_t}}\epsilon_\theta(\hat{\boldsymbol{X}}_t, t)) + \tilde{\beta}_t \epsilon_t$;
**end for**
**return** $\hat{\boldsymbol{X}}_0$;

for $t = \{T, \ldots, 1\}$, where $\tilde{\beta}_t = \frac{1-\bar{\alpha}_{t-1}}{1-\bar{\alpha}_t}$. $\mu_\theta$ can, for example, be implemented with U-net-like architectures [16], and it can be learned by setting

$$\mu_\theta(\boldsymbol{X}_t, t) = \frac{1}{\alpha_t}\big(\boldsymbol{X}_t - \frac{\beta_t}{\sqrt{1-\bar{\alpha}_t}}\epsilon_\theta(\boldsymbol{X}_t, t)\big). \tag{5}$$

**Objective Function.** [6] shows that optimizing the following simplified training objective leads to better generation quality:

$$\mathcal{L} = \mathbb{E}_{\boldsymbol{X}_0 \sim q(\boldsymbol{X}_0), \epsilon_t \sim \mathcal{N}(\mathbf{0}, \boldsymbol{I}), t}\big[\|\epsilon_t - \epsilon_\theta(\boldsymbol{X}_t, t)\|\big], \tag{6}$$

where $\epsilon_t$ is the noise used to obtain $\boldsymbol{X}_t$ from $\boldsymbol{X}_0$ in Eq. (3) at step $t$, and $\|\cdot\|$ denotes the Frobenius norm.

### 3.3 Anomaly Detection with Diffusion Model

Here, we focus on general anomaly detection tasks that are not restricted to time series. The diffusion model is a reconstruction-based approach. Thus, hereafter, we describe the way in which the diffusion model reconstructs $\hat{\boldsymbol{X}}$.

**Training Phase.** During the training phase, the model $\epsilon_\theta$ learns to predict the noise $\epsilon_t$ added to $\boldsymbol{X}_0 \in \mathcal{D}^{train}$ given $\boldsymbol{X}_t$ and $t$. The training procedure is shown as Algorithm 1. In each iteration of the training (Line 3–7), a diffusion step $t$ is randomly sampled, at which point the diffusion process Eq. (3) is applied. Unlike the other tasks, in anomaly detection, $\boldsymbol{X}_T$ does not need to be white noise. The loss Eq. (6) is calculated afterward.

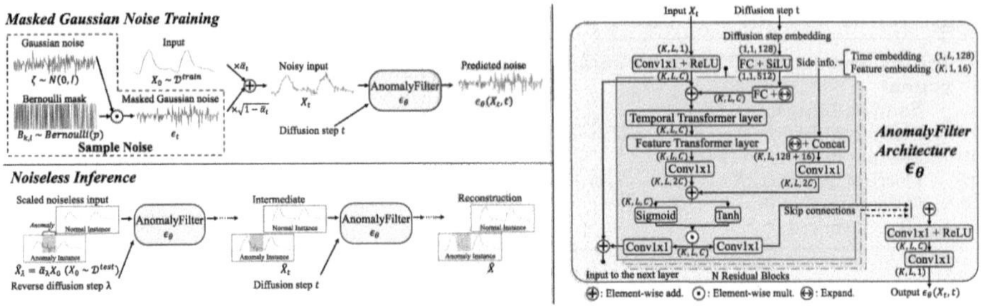


**Fig. 1.** An overview of AnomalyFilter framework and its architecture.

**Inference Phase.** In the inference phase, the model $\epsilon_\theta$ generates a reconstruction $\hat{\boldsymbol{X}}$ given the input instance $\boldsymbol{X}_0 \in \mathcal{D}^{test}$. The inference procedure is shown in Algorithm 2. It first corrupts $\boldsymbol{X}_0$ into $\hat{\boldsymbol{X}}_\lambda$ (Line 3):

$$\hat{\boldsymbol{X}}_\lambda = \bar{\alpha}_\lambda \boldsymbol{X}_0 + \sqrt{1-\bar{\alpha}_\lambda}\epsilon_\lambda, \tag{7}$$

where $\lambda$ is a reverse diffusion step that can be shorter than $T$. In general, relatively weak noise is added to form $\hat{\boldsymbol{X}}_\lambda$ so that it retains some of the input features [15,29]. This can be seen as a conditional DDPM on the input instance. Then, in each step (Line 6), $\hat{\boldsymbol{X}}_\lambda$ is gradually denoised to reconstruct $\hat{\boldsymbol{X}}$.

$$\hat{\boldsymbol{X}}_{t-1} = \frac{1}{\sqrt{\alpha_t}}\big(\hat{\boldsymbol{X}}_t - \frac{\beta_t}{\sqrt{1-\bar{\alpha}_t}}\epsilon_\theta(\hat{\boldsymbol{X}}_t, t)\big) + \tilde{\beta}_t\epsilon_t. \tag{8}$$

## 4 Proposed AnomalyFilter

In this section, we propose AnomalyFilter, a novel diffusion-based method designed for TSAD, which functions as a selective filter that removes only the anomalous parts while retaining the normal parts.

**Overview.** An overview of the AnomalyFilter framework is in Fig. 1 (left). It consists of *masked Gaussian noise training* and *noiseless inference*. Masked Gaussian noise makes the model learn to allow normal parts to pass through selectively by masking the Gaussian noise. To address the reconstruction quality issue, we propose noiseless inference, which eliminates the addition of noise during the inference. This aligns with the learned masked Gaussian noise and achieves the accurate reconstruction of normal parts.

**Architecture.** The architecture of AnomalyFilter is shown in Fig. 1 (right) and is based on CSDI [22], which is a modification of DiffWave [9] employing transformer layers. It is composed of multiple residual layers with a latent dimension $C$. Diffusion step $t$ is embedded as a 128-dimensional encoding vector with positional encoding [25] and input into the model together with data $\boldsymbol{X}_t$. To incorporate temporal and inter-variable dependencies beyond $L$ timesteps and $K$ features, we employ a temporal and feature transformer layer. The positional encoding of time and feature is added afterward. As is typical for diffusion models, the output dimensions match the input dimensions.

### 4.1 Masked Gaussian Noise

Our objective is to build a filter with two functionalities: (1) removing the anomalous parts and (2) allowing the normal parts to pass through. We design masked Gaussian noise for this purpose. Specifically, we represent anomalous parts with Gaussian noise, as in typical DDPM. To enable the model to allow normal parts to pass through, we randomly mask the noise. The stochastic selection of the noise to mask enables the filter to selectively denoise anomalous parts. Formally, our masked Gaussian noise $\epsilon_t$ is defined as follows:

$$\epsilon_t = \boldsymbol{B} \circ \zeta, \tag{9}$$

**Algorithm 3.** Masked Gaussian Noise Training

1: **Input:** Training set $\mathcal{D}^{train}$, Forward diffusion step $T$, Mask ratio $p$;
2: **repeat**
3: Sample data $\boldsymbol{X}_0 \sim \mathcal{D}^{train}$;
4: Sample diffusion step $t \sim Uniform(\{1, \ldots, T\})$;
5: Sample noise $\epsilon_t = \boldsymbol{B} \circ \zeta$ ,where $\zeta \sim \mathcal{N}(\mathbf{0}, \boldsymbol{I}), \boldsymbol{B}_{k,l} \sim Bernoulli(p)$;
6: Corrupt $\boldsymbol{X}_t = \bar{\alpha}_t \boldsymbol{X}_0 + \sqrt{1 - \bar{\alpha}_t} \epsilon_t$;
7: Calculate $\mathcal{L} = \|\epsilon_t - \epsilon_\theta(\boldsymbol{X}_t, t)\|$;
8: **until** Converged;

where $\circ$ is element-wise multiplication, $\zeta \sim \mathcal{N}(\mathbf{0}, \boldsymbol{I})$ is Gaussian noise, and $\boldsymbol{B}$ is a mask with elements drawn from a Bernoulli distribution, $\boldsymbol{B}_{k,l} \sim Bernoulli(p)$. The hyperparameter $p$ controls the mask ratio; the larger the value, the larger the ratio of Gaussian noise.

### 4.2 Masked Gaussian Noise Training

The simplified DDPM loss function Eq. (6) can be interpreted as training the model to predict the noise added to the data. Similarly, we train our model $\epsilon_\theta$ to predict the masked Gaussian noise $\epsilon_t$.

**Algorithm.** The training procedure for AnomalyFilter is summarized in Algorithm 3, where the differences from the conventional algorithm (Algorithm 1) are highlighted in blue. We follow the conventional training procedure for the diffusion model except for the noise $\epsilon_t$ (Line 5). Specifically, given data $\boldsymbol{X}_0$, we sample noisy data $\boldsymbol{X}_t$ and train $\epsilon_\theta$ by minimizing the loss function in Eq. (6).

**Objective Function.** Although our loss function is similar to that of DDPM, with the introduction of masked Gaussian noise, it can be divided into two parts.

$$\begin{aligned}\mathcal{L} &= \mathbb{E}[\|\epsilon_t - \epsilon_\theta(\boldsymbol{X}_t, t)\|] \\ &= \underbrace{\mathbb{E}[\boldsymbol{B} \circ \|\zeta - \epsilon_\theta(\boldsymbol{X}_t, t)\|]}_{\text{Noisy part}} + \underbrace{\mathbb{E}[(\boldsymbol{1} - \boldsymbol{B}) \circ \|\boldsymbol{0} - \epsilon_\theta(\boldsymbol{X}_t, t)\|]}_{\text{Noiseless part}} \\ &= \mathcal{L}_{NonMask} + \mathcal{L}_{Mask}.\end{aligned} \tag{10}$$

Thus, it enables learning a filter with two functionalities: (1) the removal of anomalous parts ($\mathcal{L}_{NonMask}$) and (2) the retention of normal parts ($\mathcal{L}_{Mask}$).

### 4.3 Noiseless Inference

To achieve low reconstruction errors for normal parts, we introduce noiseless inference, which inputs the data as they are and does not add noise at each denoising step. Algorithm 4 is our proposed noiseless inference. Instead of inputting corrupted data $\hat{\boldsymbol{X}}_\lambda = \bar{\alpha}_\lambda \boldsymbol{X}_0 + \sqrt{1 - \bar{\alpha}_\lambda}\epsilon_\lambda$ into the model, we input the following scaled noiseless data (Line 3):

$$\hat{\boldsymbol{X}}_\lambda = \bar{\alpha}_\lambda \boldsymbol{X}_0. \tag{11}$$

**Algorithm 4.** Noiseless Inference

1: **Input:** Data $\boldsymbol{X}_0 \in \mathcal{D}^{test}$, Reverse diffusion step $\lambda$, Trained denoising function $\epsilon_\theta$;
2: **Output:** Reconstruction $\hat{\boldsymbol{X}}_0$ (i.e., $\hat{\boldsymbol{X}}$);
3: Scale $\hat{\boldsymbol{X}}_\lambda = \bar{\alpha}_\lambda \boldsymbol{X}_0$ ;
4: **for** $t = \lambda : 1$ **do**
5: Denoise $\hat{\boldsymbol{X}}_{t-1} = \frac{1}{\sqrt{\alpha_t}}(\hat{\boldsymbol{X}}_t - \frac{\beta_t}{\sqrt{1-\bar{\alpha}_t}}\epsilon_\theta(\hat{\boldsymbol{X}}_t, t))$;
6: **end for**
7: **return** $\hat{\boldsymbol{X}}_0$;

Then, at each denoising step, $\boldsymbol{X}_\lambda$ is gradually denoised to reconstruct $\hat{\boldsymbol{X}}$. This step is conducted without any addition of noise (Line 5):

$$\hat{\boldsymbol{X}}_{t-1} = \frac{1}{\sqrt{\alpha_t}}\left(\hat{\boldsymbol{X}}_t - \frac{\beta_t}{\sqrt{1 - \bar{\alpha}_t}}\epsilon_\theta(\hat{\boldsymbol{X}}_t, t)\right). \tag{12}$$

Thanks to the masked Gaussian noise, the model functions as a filter. Thus, noiseless inference presumes that the model outputs data unchanged when given normal data and removes anomalous parts when given anomaly data.

## 5 Experiments

### 5.1 Experiment Setup

**Datasets.** We use five datasets in our paper: (1) UCR anomaly archive (UCR) [28], (2) AIOps, (3, 4) Yahoo real and Yahoo bench, and (5) Server Machine Dataset (SMD) [20]. We trained and tested separately for each of the subdatasets, and the average results of five runs are presented.

**Table 1.** Anomaly detection performance. Best results are in **bold**, and second-best results are underlined (higher is better).

| | UCR | | | | AIOps | | | Yahoo real | | | Yahoo bench | | | SMD | | |
|---|---|---|---|---|---|---|---|---|---|---|---|---|---|---|---|---|
| | V-R | V-P | RF | Acc. | V-R | V-P | RF | V-R | V-P | RF | V-R | V-P | RF | V-R | V-P | RF |
| IF | 0.641 | 0.106 | 0.041 | 0.236 | 0.798 | 0.169 | 0.061 | 0.699 | 0.358 | 0.000 | 0.611 | 0.295 | 0.000 | 0.735 | 0.245 | 0.093 |
| OCSVM | 0.645 | 0.167 | 0.092 | 0.300 | 0.848 | 0.336 | 0.152 | 0.715 | 0.443 | 0.000 | 0.608 | 0.289 | 0.000 | 0.698 | 0.239 | 0.156 |
| BeatGAN | **0.746** | 0.205 | 0.137 | 0.351 | 0.889 | 0.487 | **0.239** | 0.751 | 0.503 | 0.060 | 0.720 | 0.373 | 0.004 | 0.819 | 0.356 | 0.232 |
| LSTM-VAE | 0.739 | 0.141 | 0.113 | 0.217 | 0.864 | 0.359 | 0.185 | 0.899 | 0.612 | 0.076 | 0.709 | 0.329 | 0.003 | 0.711 | 0.223 | 0.190 |
| USAD | 0.674 | 0.156 | 0.129 | 0.272 | 0.889 | 0.445 | 0.227 | 0.830 | 0.571 | 0.075 | 0.612 | 0.305 | 0.000 | 0.508 | 0.145 | 0.088 |
| AT | 0.704 | 0.132 | 0.103 | 0.202 | 0.881 | 0.347 | 0.157 | 0.872 | 0.557 | 0.051 | 0.710 | 0.291 | 0.005 | 0.861 | 0.420 | 0.252 |
| TranAD | 0.579 | 0.072 | 0.059 | 0.157 | 0.762 | 0.230 | 0.160 | 0.774 | 0.496 | 0.061 | 0.589 | 0.270 | 0.000 | 0.636 | 0.194 | 0.144 |
| DADA | 0.585 | 0.028 | 0.015 | 0.036 | 0.839 | 0.255 | 0.161 | 0.860 | 0.548 | 0.095 | 0.745 | 0.309 | 0.007 | 0.875 | 0.419 | 0.262 |
| DiffAD | 0.605 | 0.063 | 0.039 | 0.095 | 0.859 | 0.395 | 0.191 | 0.844 | 0.520 | 0.077 | 0.768 | 0.346 | 0.004 | 0.606 | 0.169 | 0.157 |
| IMDiffusion | 0.630 | 0.047 | 0.028 | 0.060 | 0.866 | 0.358 | 0.126 | 0.860 | 0.552 | 0.062 | 0.725 | 0.300 | 0.002 | 0.850 | 0.341 | 0.196 |
| D3R | 0.580 | 0.071 | 0.060 | 0.157 | 0.839 | 0.396 | 0.213 | 0.839 | 0.560 | 0.099 | 0.668 | 0.315 | 0.000 | 0.828 | 0.330 | 0.235 |
| **AnomalyFilter** | 0.726 | **0.238** | **0.204** | **0.399** | **0.932** | **0.597** | 0.230 | **0.900** | **0.625** | **0.109** | **0.793** | **0.416** | **0.011** | **0.891** | **0.477** | **0.273** |

**Table 2.** Ablation study.

| | V-R | V-P | RF |
|---|---|---|---|
| DAE | 0.760 (↓ 2.8%) | 0.321 (↓ 10.6%) | 0.125 (↓ 8.2%) |
| DAE w/ **M**ask | 0.722 (↓ 7.6%) | 0.264 (↓ 26.6%) | 0.093 (↓ 31.8%) |
| DAE w/ **N**oiseless | 0.756 (↓ 3.3%) | 0.300 (↓ 16.4%) | 0.116 (↓ 14.9%) |
| DAE w/ M&N | 0.731 (↓ 6.5%) | 0.287 (↓ 20.1%) | 0.103 (↓ 24.1%) |
| DDPM | 0.706 (↓ 9.7%) | 0.197 (↓ 45.1%) | 0.050 (↓ 63.0%) |
| DDPM w/ M | 0.700 (↓ 10.4%) | 0.182 (↓ 49.4%) | 0.044 (↓ 67.9%) |
| DDPM w/ N | 0.740 (↓ 5.3%) | 0.265 (↓ 26.1%) | 0.090 (↓ 33.8%) |
| **AnomalyFilter** (DDPM w/M&N) | **0.782** | **0.359** | **0.136** |

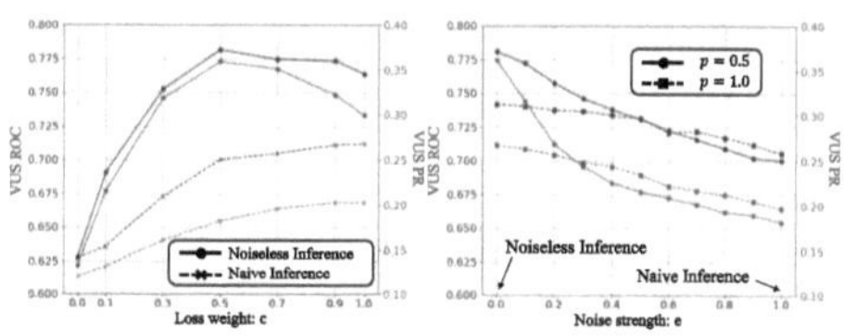


(a) Loss ratio $c$ (b) Noise strength $e$

**Fig. 2.** Effect of loss weights and noise strength.

**Baselines.** We compare our method with the following 13 baselines in four categories: (1) Classical machine-learning methods, Isolation Forest (IF) [10] and OCSVM [18], (2) GAN, VAE, and AE-based methods, BeatGAN [32], LSTM-VAE [14], and USAD [1], (3) Transformer-based methods, AT [31], TranAD [24], and DADA [19], and (4) Denoising-based methods, DiffAD [30], IMDiffusion [3], D3R [27], DAE [26] and DDPM [6]. We include DAE and DDPM in our ablation study, since they share the same architecture as AnomalyFilter.

**Evaluation Metrics.** It is claimed that the traditional metric for TSAD, namely the F1 score calculated with a protocol named point adjustment (PA), overestimates the model performance [8,17]. Thus, we adopt threshold-independent measures: Range F-score [23], volume under the surface (VUS)-ROC, and VUS-PR [13]. For UCR, we also employ accuracy as a metric, where we consider that the model detects the anomaly if the highest anomaly score is located inside the given anomaly range, since it only has a single anomaly in a sequence.

### 5.2 Anomaly Detection Performance

Table 1 presents the performance of AnomalyFilter and the baseline methods for five datasets. While the second-best method varies depending on the dataset and metric, the proposed AnomalyFilter achieves the highest detection accuracy across most of the datasets, demonstrating an average improvement of 4.1% in VOC-ROC, 14.4% in VOC-PR, and 29.0% in Range F-score compared with the second-best method. Overall, reconstruction-based methods exceed classic machine-learning methods. BeatGAN struggles with Yahoo, which has limited training data, and SMD, which is multivariate data. In contrast, AT performs well on SMD, but tends to overfit on simple patterns, and underperforms on UCR. These results highlight the difficulty of appropriately balancing the information bottleneck across diverse datasets. DADA is a zero-shot anomaly detection method that identifies relative anomalies within a window. However, when most values in a window are anomalous (as in UCR), it fails to detect them accurately, resulting in lower performance. Diffusion-based methods, DiffAD and IMDiffusion, face challenges in UCR, which contains many pattern-wise anomalies, reflecting limitations in the reconstruction quality. These results demonstrate the effectiveness of AnomalyFilter.

### 5.3 Ablation Studies

**Effect of Key Components.** Table 2 shows quantitative results across datasets w/o each component with DAE and DDPM as a base method. We can see that the noiseless inference enhances the performance of the vanilla DDPM. While the Bernoulli mask alone does not independently improve the accuracy of DDPM, its combination with noiseless, namely AnomalyFilter, leads to substantial performance gains. This synergy underscores the importance of jointly employing these two simple components in improving anomaly detection capabilities. The failure of the mask to work effectively on its own is because, in naive inference, noise is added at each denoising step, resulting in a mismatch with what the model has learned. Interestingly, when DAE is used as the base method, neither the Bernoulli mask nor noiseless inference contributes to performance improvement. This is probably because the direct prediction framework of DAE makes it challenging to learn a complex filter capable of both retaining and denoising, while DDPM has demonstrated remarkable capabilities in learning the filter.

**Effect of Loss Weights.** During our experiments, we give the same importance to each regularizer, i.e., the coefficients $c$ in $\mathcal{L} = c\mathcal{L}_{NonMask} + (1-c)\mathcal{L}_{Mask}$ are set at 0.5. Here, we vary $c$ from $\{0.1, \ldots 1.0\}$, using $p = 0.5$ for training, where $c = 1.0$ focused solely on denoising and $c = 0.0$ focused solely on retention. The results, shown in Fig. 2 (a), indicate that $c = 0.5$ achieved the best performance, aligning with $p = 0.5$ and suggesting that explicit weighting is unnecessary. However, under naive inference, $c = 1.0$ performed best, demonstrating that denoising is critical when inference includes noise, whereas both processes are significant when inference is noise-free.

**Effect of Noise Strength During Inference.** We modify the noise added during naive inference (Algorithm 2, Lines 3 and 6) from $\epsilon_t$ to $e\epsilon_t$ and gradually

**Table 3.** Reconstruction quality performance. $MSE_a$ and $MSE_n$ are the MSE values of anomalous and normal parts during the inference. Not necessary, but $MSE_a$ should be higher, and $MSE_n$ should be lower. The higher the ratio $\frac{MSE_a}{MSE_n}$ is, the easier it is to detect anomalies.

| | UCR | | | AIOps | | | Yahoo real | | | Yahoo bench | | | SMD | | |
|---|---|---|---|---|---|---|---|---|---|---|---|---|---|---|---|
| | $MSE_a$ | $MSE_n$ | $\frac{MSE_a}{MSE_n}$ | $MSE_a$ | $MSE_n$ | $\frac{MSE_a}{MSE_n}$ | $MSE_a$ | $MSE_n$ | $\frac{MSE_a}{MSE_n}$ | $MSE_a$ | $MSE_n$ | $\frac{MSE_a}{MSE_n}$ | $MSE_a$ | $MSE_n$ | $\frac{MSE_a}{MSE_n}$ |
| BeatGAN | 0.0153 | 0.0030 | 5.1 | 0.0251 | 0.0005 | 50.6 | 9.6410 | 0.3539 | 27.2 | 0.0660 | 0.0296 | 2.2 | 0.0257 | 0.0053 | 4.8 |
| LSTM-VAE | 0.0171 | 0.0030 | 5.7 | 0.0361 | 0.0014 | 25.5 | 10.6639 | 0.3497 | 30.5 | 0.1656 | 0.1131 | 1.5 | 0.0387 | 0.0130 | 3.0 |
| USAD | 0.0394 | 0.0152 | 2.6 | 0.0715 | 0.0068 | 10.5 | 12.6297 | 0.7915 | 16.0 | 0.1593 | 0.1265 | 1.3 | 0.0036 | 0.0015 | 2.3 |
| AT | 0.0039 | **0.0004** | 10.3 | 0.0091 | 0.0002 | 50.7 | 9.6449 | 0.2966 | 32.5 | 0.0266 | 0.0133 | 2.0 | 0.0239 | 0.0042 | 5.6 |
| TranAD | 0.0447 | 0.0328 | 1.4 | **0.0892** | 0.0187 | 4.8 | **12.6501** | 0.8379 | 15.1 | **0.2679** | 0.2464 | 1.1 | 0.0183 | 0.0032 | 5.8 |
| DADA | 0.0090 | 0.0050 | 1.8 | 0.0072 | 0.0003 | 23.0 | 2.2592 | 0.0204 | **110.8** | 0.0195 | **0.0059** | 3.3 | 0.0029 | **0.0006** | 4.8 |
| DiffAD | 0.0373 | 0.0244 | 1.5 | 0.0436 | 0.0051 | 8.5 | 10.4784 | 0.2780 | 37.70 | 0.0844 | 0.0362 | 2.3 | **0.0490** | 0.0314 | 1.6 |
| IMDiffusion | 0.0532 | 0.0273 | 1.9 | 0.0501 | 0.0110 | 4.6 | 10.0372 | 0.2937 | 34.2 | 0.1038 | 0.0528 | 2.0 | 0.0150 | 0.0031 | 4.8 |
| D3R | **0.0540** | 0.0440 | 1.2 | 0.0476 | 0.0094 | 5.0 | 10.0916 | 0.1463 | 69.0 | 0.0636 | 0.0252 | 2.5 | 0.0142 | 0.0028 | 5.0 |
| **AnomalyFilter** | 0.0108 | 0.0008 | **13.9** | 0.0344 | **0.0001** | **257.3** | 0.3551 | **0.0075** | 47.2 | 0.0276 | 0.0070 | **4.0** | 0.0093 | 0.0010 | **9.5** |
| DAE | 0.0103 | 0.0016 | 6.7 | 0.0374 | 0.0006 | 66.7 | 10.3125 | 0.3423 | 30.1 | 0.0642 | 0.0382 | 1.7 | 0.0037 | **0.0006** | 6.0 |
| DDPM | 0.0357 | 0.0170 | 2.1 | 0.0415 | 0.0053 | 7.8 | 0.5271 | 0.0716 | 7.4 | 0.0917 | 0.0472 | 1.9 | 0.0155 | 0.0053 | 3.0 |
| DDPM w/ N | 0.0261 | 0.0080 | 3.3 | 0.0392 | 0.0026 | 15.1 | 0.4989 | 0.0401 | 12.4 | 0.0757 | 0.0311 | 2.4 | 0.0139 | 0.0046 | 3.0 |
| DDPM w/ M | 0.0339 | 0.0201 | 1.7 | 0.0472 | 0.0111 | 4.3 | 0.4564 | 0.0690 | 6.6 | 0.0837 | 0.0511 | 1.6 | 0.0250 | 0.0169 | 1.5 |

reducing $e$ from $\{1.0, \dots, 0.0\}$, even achieving noiseless inference. This is equivalent to sampling Gaussian noise $\zeta$ from $\mathcal{N}(\mathbf{0}, e^2\boldsymbol{I})$. Figure 2 (b) shows the results for models trained with masked Gaussian noise ($p = 0.5$) and Gaussian noise ($p = 1.0$). In both cases, the performance improved as $e$ decreased, with the best results achieved at $e = 0.0$ (i.e., noiseless inference), indicating that noise is unnecessary for inference in TSAD. The model trained on Gaussian noise outperformed masked Gaussian noise at around $e \geq 0.4$, suggesting that models trained with Gaussian noise are more suitable when inference includes higher levels of noise, aligning with their training paradigm.

### 5.4 Reconstruction Quality

We show the MSE of anomalous and normal parts and their ratio in Table 3. Compared to baselines and ablation settings, AnomalyFilter achieves a significantly lower $MSE_n$, resulting in a larger ratio $\frac{MSE_a}{MSE_n}$. This demonstrates that AnomalyFilter is well-suited for effective anomaly detection. Although AT achieves a very small $MSE_n$ on UCR and AIOps, its $MSE_a$ is similarly small, resulting in a mediocre ratio. This causes false positives by even slight misalignments in normal regions. DADA, a zero-shot anomaly detector, shows favorable ratios on datasets like Yahoo, where the number of training samples is limited and point anomalies are dominant, but struggles in UCR, which contains longer anomalous parts than the window size. Diffusion-based methods such as DiffAD, IMDiffusion, and DDPM exhibit high values in both $MSE_a$ and $MSE_n$, indicating the limitation of diffusion models. DDPM w/ Mask yields results similar to the vanilla DDPM, suggesting that the Bernoulli mask by itself is not sufficient for improving detection ability. In contrast, DDPM w/ Noiseless setting shows a notable reduction in $MSE_n$, leading to improved ratio. These findings are consistent with the anomaly detection performance reported in Table 2 and further support the effectiveness of the proposed approach, particularly the combined contribution of the Bernoulli mask and noiseless inference.

## 6 Conclusion

We proposed a novel diffusion-based method called AnomalyFilter for unsupervised TSAD. To resolve the low reconstruction quality of existing reconstruction-based methods, we built a selective filter that accurately denoises anomalous parts while retaining normal parts. Specifically, we investigated the utility of noise in diffusion models and devised masked Gaussian noise and noiseless inference. Extensive experiments on five datasets demonstrated the effectiveness of the method. Ablation studies confirmed the contribution of our design to anomaly detection in a quantitative manner and demonstrated that the synergy of the proposed two simple components strongly enhances vanilla DDPM. Overall, we believe that our work paves the way for a new diffusion-based approach towards TSAD.

**Acknowledgments.** The authors would like to thank the anonymous referees for their valuable comments and helpful suggestions. This work was supported by JST CREST JPMJCR23M3, JST START JPMJST2553, JST CREST JPMJCR20C6, JST K Program JPMJKP25Y6, JST COI-NEXT JPMJPF2009, JST COI-NEXT JPMJPF2115, the Future Social Value Co-Creation Project - Osaka University.

## References

1. Audibert, J.: USAD: unsupervised anomaly detection on multivariate time series. In: KDD, pp. 3395–3404 (2020)
2. Bansal, A.: Cold diffusion: inverting arbitrary image transforms without noise. In: NeurIPS, vol. 36 (2024)
3. Chen, Y., et al.: Imdiffusion: imputed diffusion models for multivariate time series anomaly detection. VLDB **17**(3), 359–372 (2023)
4. Daras, G.: Soft diffusion: score matching for general corruptions. arXiv preprint arXiv:2209.05442 (2022)
5. Gong, D., et al.: Memorizing normality to detect anomaly: memory-augmented deep autoencoder for unsupervised anomaly detection. In: ICCV, pp. 1705–1714 (2019)
6. Ho, J., et al.: Denoising diffusion probabilistic models. In: NeurIPS, vol. 33, pp. 6840–6851 (2020)
7. Kascenas, A., et al.: The role of noise in denoising models for anomaly detection in medical images. Med. Image Anal. **90**, 102963 (2023)
8. Kim, S.: Towards a rigorous evaluation of time-series anomaly detection. In: AAAI, vol. 36, pp. 7194–7201 (2022)
9. Kong, Z.: Diffwave: a versatile diffusion model for audio synthesis. In: ICLR (2021)
10. Liu, F.T.: Isolation forest. In: ICDM, pp. pp. 413–422 (2008)
11. Nachmani, E.: Non gaussian denoising diffusion models. arXiv preprint arXiv:2106.07582 (2021)

12. Obata, K.: Robust and explainable detector of time series anomaly via augmenting multiclass pseudo-anomalies. In: KDD (2025)
13. Paparrizos, J., et al.: Volume under the surface: a new accuracy evaluation measure for time-series anomaly detection. VLDB **15**(11), 2774–2787 (2022)
14. Park, D., et al.: A multimodal anomaly detector for robot-assisted feeding using an LSTM-based variational autoencoder. IEEE Robot. Autom. Lett. **3**(3), 1544–1551 (2018)
15. Pintilie, I.: Time series anomaly detection using diffusion-based models. In: 2023 IEEE International Conference on Data Mining Workshops (ICDMW), pp. 570–578. IEEE (2023)
16. Ronneberger, O., et al.: U-Net: convolutional networks for biomedical image segmentation. In: Medical Image Computing and Computer-Assisted Intervention–MICCAI 2015: 18th International Conference, Munich, Germany, October 5–9, 2015, Proceedings, Part III 18, pp. 234–241. Springer, Cham (2015). https://doi.org/10.1007/978-3-319-24574-4_28
17. Sarfraz, M.S.: Position: quo vadis, unsupervised time series anomaly detection? In: ICML (2024)
18. Schölkopf, B.: Support vector method for novelty detection. In: NeurIPS, vol. 12 (1999)
19. Shentu, Q.: Towards a general time series anomaly detector with adaptive bottlenecks and dual adversarial decoders. In: ICLR (2025)
20. Su, Y., et al.: Robust anomaly detection for multivariate time series through stochastic recurrent neural network. In: KDD, pp. 2828–2837 (2019)
21. Tan, S.C., et al.: Fast anomaly detection for streaming data. In: IJCAI'11, AAAI, pp. 1511–1516. AAAI Press (2011)
22. Tashiro, Y., et al.: CSDI: conditional score-based diffusion models for probabilistic time series imputation. In: NeurIPS, vol. 34, pp. 24804–24816 (2021)
23. Tatbul, N., et al.: Precision and recall for time series. In: NIPS'18, NeurIPS, pp. 1924–1934. Curran Associates Inc., Red Hook, NY, USA (2018)
24. Tuli, S., et al.: TranAD: deep transformer networks for anomaly detection in multivariate time series data. Proc. VLDB **15**(6), 1201–1214 (2022)
25. Vaswani, A.: Attention is all you need. In: NeurIPS, vol. 30 (2017)
26. Vincent, P.: Stacked denoising autoencoders: learning useful representations in a deep network with a local denoising criterion. J. Mach. Learn. Res. **11**(12) (2010)
27. Wang, C.: Drift doesn't matter: dynamic decomposition with diffusion reconstruction for unstable multivariate time series anomaly detection. In: NeurIPS, vol. 36 (2024)
28. Wu, R., Keogh, E.J.: Current time series anomaly detection benchmarks are flawed and are creating the illusion of progress. IEEE Trans. Knowl. Data Eng. **35**(3), 2421–2429 (2021)
29. Wyatt, J.: AnoDDPM: anomaly detection with denoising diffusion probabilistic models using simplex noise. In: CVPR, pp. 650–656 (2022)
30. Xiao, C., et al.: Imputation-based time-series anomaly detection with conditional weight-incremental diffusion models. In: KDD, pp. 2742–2751 (2023)
31. Xu, J.: Anomaly transformer: time series anomaly detection with association discrepancy. In: ICLR (2022)
32. Zhou, B., et al.: BeatGAN: anomalous rhythm detection using adversarially generated time series. In: IJCAI, vol. 2019, pp. 4433–4439 (2019)

# Nonlinear Characteristic-Driven Partial Multi-label Learning

Fuyu Qi[1], Linqing Huang[2], Jian Feng[1], Zhiyong Li[1], Qingkai Bu[1], Wenpeng Lu[3,4], and Jinfu Fan[1(✉)]

[1] School of Computer Science and Technology, Qingdao University, Qingdao, China
{qifuyu,fengjian,lizhiyong}@qdu.edu.cn, fan_jinfu@163.com

[2] School of Computer Science, Shanghai Jiao Tong University, Shanghai, China

[3] Key Laboratory of Computing Power Network and Information Security, Ministry of Education, Shandong Computer Science Center (National Supercomputer Center in Jinan), Qilu University of Technology (Shandong Academy of Sciences), Jinan, China
wenpeng.lu@qlu.edu.cn

[4] Shandong Provincial Key Laboratory of Computing Power Internet and Service Computing, Shandong Fundamental Research Center for Computer Science, Jinan, China

**Abstract.** *Partial Multi-label Learning* (PML) addresses challenges where each instance is associated with multiple candidate labels, which include both relevant and irrelevant labels. Traditional label disambiguation strategies often overlook the significance of data lying on nonlinear subspace. Assuming that data points closely adhere to multiple linear subspaces is restrictive and may fail in many applications. Linear subspace clustering algorithms frequently struggle with data residing on multiple nonlinear manifolds, as they inherently focus solely on the global linear associations between data points. To address this gap, we propose a novel approach called *Nonlinear Characteristic-Driven Partial Multi-Label Learning* (NLPML). This method introduces a kernel low-rank representation technique to learn the instance relationship matrix within a nonlinear subspace, effectively identifying instances with complex nonlinear structures. Simultaneously, we introduce the Jaccard distance to quantify label relevance by constructing a label affinity matrix. By jointly optimizing instance-level and label-level affinity matrices, NLPML leverages complementary information for effective label denoising. Experimental results demonstrate that NLPML significantly outperforms existing methods.

**Keywords:** Partial multi-label learning · Weakly supervised learning

## 1 Introduction

In the domain of supervised machine learning, *Multi-label learning* (MLL) deals with scenarios where a single entity can be assigned to multiple categories [16,21,23,40]. Standard MLL algorithms require exhaustive and precise label


R. C.-W. Wong et al. (Eds.): PAKDD 2026, LNAI 16597, pp. 442–454, 2026.
https://doi.org/10.1007/978-981-92-1300-9_35

annotations for training data [40], but obtaining such annotations is time-consuming and labor-intensive [13,43]. This leads to incomplete labels with both correct and redundant information, a scenario defined as *Partial Multi-label Learning* (PML) [24,31]. Existing solutions for PML include multi-label classification techniques like BR [3], ML-kNN [39], and CPLST [5], but noisy labels in candidate sets may distort training [9,35]. Recent PML methods mostly exploit feature and label space properties such as low-rank or sparse decomposition [25,26,32] or binary decomposition [20] but share critical limitations. They overlook nonlinear subspace characteristics of real-world data (forcing linear modeling of nonlinear features) and underexplore the joint role of instances and candidate labels [42], compromising prediction accuracy. To address these issues, we propose a novel PML method named NLPML. While low-rank representation (LRR) [17] performs well for linear subspace clustering, its effectiveness degrades considerably when the data lie on nonlinear subspace [36]. Motivated by the advantages of nonlinear subspace learning in improving similarity estimation [1], we adopt *Robust Kernel Low-Rank Representation* (RKLRR) [30] to map data into a feature domain where nonlinear subspaces transform into linear ones. We then construct an instance affinity matrix to group same-label instances. Additionally, we use Jaccard distance to measure candidate label similarity for a label affinity matrix, integrating both matrices to harness complementary information for label disambiguation. This research offers two key contributions:

1. We propose a novel method for learning instance affinity matrices by exploring nonlinear subspace data and candidate label information. Kernel-based high-dimensional mapping effectively identifies same-label instances even when original data resides in nonlinear subspaces.
2. We utilize Jaccard distance to generate a label affinity matrix from candidate labels, and integrate it with the instance affinity matrix to disambiguate labels via complementary information (Fig. 1).

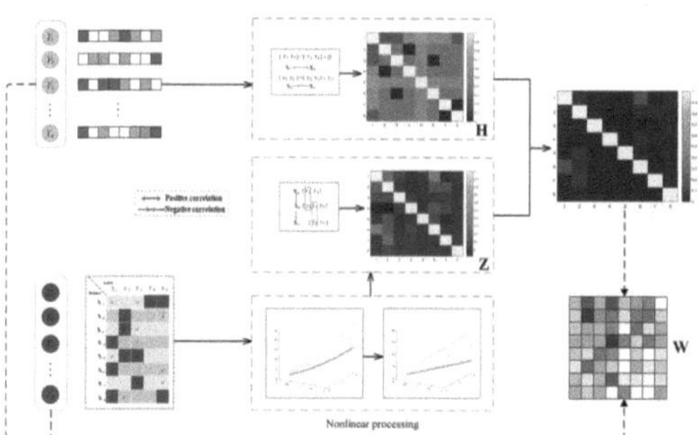


**Fig. 1.** The model of NLPML

## 2 Related Work

Partial Label Learning (PLL) [18] is a weak supervision paradigm distinct from supervised and unsupervised learning, with the core objective of identifying the unique ground-truth label for each sample from a candidate label set, where each set contains one true label and multiple noisy labels [8]. Existing PLL methods can be categorized into three strategies. Label disambiguation [22] includes mean-based methods such as convex optimization for averaging candidate label model outputs [6], vote aggregation on neighbor samples in candidate sets [11], and recognition-based methods that model ground-truth as latent variables to infer the highest-probability labels via parametric models [27]. Label transformation [38] aims to mitigate false-positive label interference in candidate sets. Theory-driven methods [33] focus on the theoretical foundations of PLL, such as statistical consistency.

PML increases the complexity because, unlike PLL, the true label count in each candidate set is unknown. A straightforward approach treats all candidate labels as correct and applies existing multi-label algorithms such as ML-KNN [39]. These methods are prone to being misled by incorrect candidate labels, which reduces performance. To address this, dedicated PML methods have been developed. Xie et al. [31] designed PML-lc and PML-fp to identify true labels via confidence metrics. Sun et al. [26] proposed PML-LRS based on low-rank and sparse structures. Fang et al. [37] introduced PARTICLE, which detects trustworthy labels through diffusion before robust classification. Gong et al. [12] developed MILI-PML to identify genuine labels via mutual information. Other works integrate feature information. Yu et al. [34] proposed fPML, which leverages label-feature synergy to separate valid labels from noise. Xie et al. [32] proposed PML-NI, attributing label noise to instance ambiguity and partitioning the prediction matrix to identify true labels. Wu et al. [29] proposed FIMAN, which enhances feature representation and label accuracy via multi-view data geometry. Hao et al. [15] proposed PML-FSSO, focusing on key feature identification by exploring label and feature spaces. Liu et al. [20] proposed PAMB, which applies binary decomposition to PML. Hang et al. [14] proposed PARD, adopting probabilistic graphical models. Wang et al. [28] proposed PML-ED, which uses KNN-based label attention and conditional layer normalization to reduce noise impact.

Despite advancements, existing methods have two gaps: (1) Most assume linear subspaces, while real-world data exhibit complex nonlinear patterns. (2) Joint utilization of label correlations and instance relationships is insufficient. To address these, we propose NLPML, integrating kernel-based nonlinear feature low-rank representation and Jaccard distance-driven label disambiguation.

## 3 The Proposed Approach

We denote the feature matrix as $\mathbf{X} = [\mathbf{x_1}, \mathbf{x_2}, \ldots, \mathbf{x_n}] \in \mathbb{R}^{d\times n}$ (where $d$ is the feature dimension and $n$ is the number of instances) and the label matrix as

$\mathbf{Y} = [\mathbf{y_1}, \mathbf{y_2}, \ldots, \mathbf{y_n}] \in \mathbb{R}^{m \times n}$ (where $m$ is the number of labels). Here, $y_{ji} = 1$ means instance $\mathbf{x}_i$ is annotated with the $j$-th label, and $y_{ji} = 0$ otherwise.

In multi-label learning, we train a projection matrix $\mathbf{W} = [\mathbf{w}_1, \mathbf{w}_2, \ldots, \mathbf{w}_d] \in \mathbb{R}^{m \times d}$ to map features to the label space. The optimization of $\mathbf{W}$ minimizes the quadratic loss with a regularization term to avoid overfitting:

$$\min_{\mathbf{W}} \frac{1}{2} \|\mathbf{Y} - \mathbf{WX}\|_F^2 + \beta \|\mathbf{W}\|_F^2 \tag{1}$$

where $\beta$ is the balancing parameter, and the squared Frobenius norm $\|\mathbf{W}\|_F^2$ serves as the regularizer.

### 3.1 Learn Nonlinear Relationships

The equality constraint of LRR is given by $\mathbf{E} = \mathbf{X} - \mathbf{XZ}$, and its optimization problem is equivalent to the following:

$$\min_{\mathbf{Z}} \|\mathbf{Z}\|_* + \lambda \|\mathbf{X} - \mathbf{XZ}\|_{2,1} \tag{2}$$

By introducing the auxiliary matrix $\mathbf{P} = \mathbf{I} - \mathbf{Z}$ (where $\mathbf{p}_i$ denotes the $i$-th column of $\mathbf{P}$), $\|\mathbf{X} - \mathbf{XZ}\|_{2,1}$ can be transformed into $\sum_{i=1}^{n} \sqrt{\mathbf{p}_i^T \mathbf{X}^T \mathbf{X} \mathbf{p}_i}$. Accordingly, Eq. 2 is reformulated as:

$$\begin{aligned} &\min_{\mathbf{Z},\mathbf{P}} \|\mathbf{Z}\|_* + \lambda \sum_{i=1}^{n} \sqrt{\mathbf{p}_i^T \mathbf{X}^T \mathbf{X} \mathbf{p}_i} \\ &\text{s.t.} \mathbf{P} = \mathbf{I} - \mathbf{Z} \end{aligned} \tag{3}$$

Given that $\mathbf{X}^T\mathbf{X}$ only encodes inner product relationships, we introduce the kernel mapping $\Psi : \mathbb{R}^d \to \mathcal{F}$ (with $\mathcal{F}$ as the feature space) and define $\Psi(\mathbf{X}) = [\Psi(\mathbf{x}_1), \ldots, \Psi(\mathbf{x}_n)]$, where the kernel matrix $\mathbf{K} = \Psi(\mathbf{X})^T \Psi(\mathbf{X})$. Let $g\{\mathbf{P}\} = \sum_{i=1}^{n} \sqrt{\mathbf{p}_i^T \mathbf{K} \mathbf{p}_i}$. The nonlinear extension of LRR is then expressed as:

$$\min_{\mathbf{Z}} \|\mathbf{Z}\|_* + \lambda \|\Psi(\mathbf{X}) - \Psi(\mathbf{X})\mathbf{Z}\|_{2,1} \tag{4}$$

which can be further rewritten as:

$$\begin{aligned} &\min_{\mathbf{Z},\mathbf{P}} \|\mathbf{Z}\|_* + \lambda g(\mathbf{P}) \\ &\text{s.t.} \mathbf{P} = \mathbf{I} - \mathbf{Z} \end{aligned} \tag{5}$$

For the linear kernel $\mathbf{K} = \mathbf{X}^T\mathbf{X}$, Eq. 5 reduces to the original LRR formulation. This study adopts the Gaussian kernel $K(\mathbf{x}, \mathbf{y}) = e^{-\gamma\|\mathbf{x}-\mathbf{y}\|^2}$, which maps data into a reproducing kernel Hilbert space to excavate nonlinear patterns and capture instance similarity. The higher $z_{ij}$ value in the low-rank representation indicates a stronger correlation between $\mathbf{x}_i$ and $\mathbf{y}_j$.

### 3.2 Learning Label Affinity Matrix

In PML, utilizing inter-instance relationships improves label clarification accuracy. Similarity assessment based solely on feature distance is limited, because two instances $\mathbf{x}_i$ and $\mathbf{x}_j$ may appear similar [10,41] but belong to distinct classes ($S_i \cap S_j = \emptyset$, where $S_i, S_j$ denote their potential label sets), while others may share labels but have dissimilar features.

To address this, we adopt the Jaccard similarity index to measure the similarity between potential label sets:

$$J_{ac}(S_i, S_j) = \frac{|S_i \cup S_j| - |S_i \cap S_j|}{|S_i \cup S_j|} \in [0, 1] \tag{6}$$

We define the label affinity matrix $\mathbf{H} = [h_{ij}]_{n \times n}$ with $h_{ij} = 1 - J_{ac}(S_i, S_j)$, where $h_{ij}$ represents the likelihood that $\mathbf{x}_i$ and $\mathbf{x}_j$ share identical labels.

### 3.3 Predictor

Specifically, the intrinsic value of potential label data improves subspace clustering precision. For example, instances $\mathbf{x}_i$ and $\mathbf{x}_j$ are distinct if they share no labels ($S_i \cap S_j = \emptyset$), so their corresponding $z_{ij}$ values should be negligible (near zero or zero).

Guided by this, we incorporate the term $\|\mathbf{Z} \circ \mathbf{H}\|_1^2$ [2] to integrate potential label data, where $\mathbf{H} \in \mathbb{R}^{n \times n}$ is the label affinity matrix.

Combining the above, the proposed NLPML is formulated as:

$$\begin{aligned} &\min_{\mathbf{W},\mathbf{Z},\mathbf{P}} \|\mathbf{Z}\|_* + \lambda g(\mathbf{P}) + \frac{1}{2}\|\mathbf{Y} - \mathbf{WXZ}\|_F^2 + \alpha \|\mathbf{Z} \circ \mathbf{H}\|_1^2 + \beta\|\mathbf{W}\|_F^2 \\ &\text{s.t.} \mathbf{P} = \mathbf{I} - \mathbf{Z} \end{aligned} \tag{7}$$

where $\alpha$ is the parameter controlling the weak supervision constraint.

### 3.4 Solving the Optimization Problem

We introduce auxiliary variables $\mathbf{J}$, $\mathbf{U}$, and $\mathbf{V}$, rewriting Eq. 7 as:

$$\min_{\mathbf{W},\mathbf{V},\mathbf{J},\mathbf{Z},\mathbf{U},\mathbf{P}} \|\mathbf{J}\|_* + \lambda g(\mathbf{P}) + \alpha\|\mathbf{U} \circ \mathbf{H}\|_1^2 + \frac{1}{2}\|\mathbf{Y} - \mathbf{WXV}\|_F^2 + \beta\|\mathbf{W}\|_F^2, \quad \text{s.t. } \mathbf{P} = \mathbf{I} - \mathbf{Z}. \tag{8}$$

Following the ALM framework [19], the augmented Lagrangian becomes:

$$\begin{aligned} \mathcal{L} = &\|\mathbf{J}\|_* + \lambda g(\mathbf{P}) + \alpha\|\mathbf{U} \circ \mathbf{H}\|_1^2 + \tfrac{1}{2}\|\mathbf{Y} - \mathbf{WXV}\|_F^2 + \beta\|\mathbf{W}\|_F^2 \\ &+ \tfrac{\mu}{2}\Big(\|\mathbf{I} - \mathbf{Z} - \mathbf{P} + \tfrac{\mathbf{Y}_1}{\mu}\|_F^2 + \|\mathbf{Z} - \mathbf{J} + \tfrac{\mathbf{Y}_2}{\mu}\|_F^2 + \|\mathbf{Z} - \mathbf{U} + \tfrac{\mathbf{Y}_3}{\mu}\|_F^2 + \|\mathbf{Z} - \mathbf{V} + \tfrac{\mathbf{Y}_4}{\mu}\|_F^2\Big) + \frac{1}{2\mu}\sum_{k=1}^{4}\|\mathbf{Y}_k\|_F^2, \end{aligned} \tag{9}$$

where $\mathbf{Y}_1$–$\mathbf{Y}_4$ are Lagrange multipliers and $\mu > 0$.

We update variables alternately as follows.

1) Updating $\mathbf{W}$

$$\min_{\mathbf{W}} \tfrac{1}{2}\|\mathbf{Y}-\mathbf{W}\mathbf{X}\mathbf{V}\|_F^2+\beta\|\mathbf{W}\|_F^2, \tag{10}$$

with closed-form solution:

$$\mathbf{W}=(\mathbf{Y}\mathbf{V}^T\mathbf{X}^T)\left(\mathbf{X}\mathbf{V}\mathbf{V}^T\mathbf{X}^T+2\beta\mathbf{I}_d\right)^{-1}. \tag{11}$$

2) Updating $\mathbf{V}$

$$\mathbf{V}=\left[(\mathbf{W}\mathbf{X})^T(\mathbf{W}\mathbf{X})+\mu\mathbf{I}\right]^{-1}\left[(\mathbf{W}\mathbf{X})^T\mathbf{Y}+\mu\mathbf{Z}+\mathbf{Y}_4\right]. \tag{12}$$

3) Updating $\mathbf{J}$

$$\min_{\mathbf{J}} \|\mathbf{J}\|_*+\tfrac{\mu}{2}\left\|\mathbf{Z}-\mathbf{J}+\tfrac{\mathbf{Y}_2}{\mu}\right\|_F^2, \tag{13}$$

equivalently:

$$\min_{\mathbf{J}} \tfrac{1}{2}\left\|\mathbf{J}-(\mathbf{Z}+\tfrac{\mathbf{Y}_2}{\mu})\right\|_F^2+\tfrac{1}{\mu}\|\mathbf{J}\|_*, \tag{14}$$

solved via SVD with singular value soft-thresholding.

4) Updating $\mathbf{Z}$

$$\mathbf{Z}=\tfrac{1}{4}(\mathbf{I}-\mathbf{P}+\mathbf{J}+\mathbf{U}+\mathbf{V})+\tfrac{1}{4\mu}(\mathbf{Y}_1-\mathbf{Y}_2-\mathbf{Y}_3-\mathbf{Y}_4). \tag{15}$$

5) Updating $\mathbf{U}$

$$\mathbf{U}_{k+1}=\arg\min_{\mathbf{U}} \frac{\alpha}{\mu_k}\|\mathbf{U}\circ\mathbf{H}\|_1+\tfrac{1}{2}\left\|\mathbf{U}-(\mathbf{Z}_{k+1}+\tfrac{\mathbf{Y}_{3,k}}{\mu_k})\right\|_F^2, \tag{16}$$

with closed-form elementwise soft-thresholding:

$$\mathbf{U}_{k+1}=\Psi_{\frac{\alpha\mathbf{H}}{\mu_k}}\left(\mathbf{Z}_{k+1}+\tfrac{\mathbf{Y}_{3,k}}{\mu_k}\right), \tag{17}$$

where $\Psi$ performs elementwise $\max(|\mathbf{X}|-\frac{\alpha}{\mu_k}\mathbf{H},0)\circ\operatorname{sgn}(\mathbf{X})$.

6) Updating $\mathbf{P}$. The problem is:

$$\min_{\mathbf{P}} \lambda g(\mathbf{P})+\frac{\mu_k}{2}\|\mathbf{P}-\mathbf{C}_{k+1}\|_F^2 \tag{18}$$

where $\mathbf{C}_{k+1}=\mathbf{I}-\mathbf{Z}_{k+1}+\mathbf{Y}_{1,k}/\mu_k$. Let $\tau=\mu_k/\lambda$, then Eq. 18 can be reformulated by normalizing with $\lambda$ as:

$$\min_{\{\mathbf{p}_i\}_{i=1}^n} \sum_{i=1}^n \sqrt{\mathbf{p}_i{}^T\mathbf{K}\mathbf{p}_i}+\frac{\tau}{2}\sum_{i=1}^n \|\mathbf{p}_i-\mathbf{c}_i\|^2 \tag{19}$$

where $\mathbf{c}_i$ ($\mathbf{p}_i$) is the $i$-th column of $\mathbf{C}$ ($\mathbf{P}$). This decomposes into $n$ independent subtasks:

$$\min_{\mathbf{p}_i} \sqrt{\mathbf{p}_i{}^T\mathbf{K}\mathbf{p}_i}+\frac{\tau}{2}\|\mathbf{p}_i-\mathbf{c}_i\|^2 \tag{20}$$

*Lemma 1*: Considering a positive scalar $\tau$, a constant vector $\mathbf{b} \in \mathbb{R}^q$ (where $q$ is a positive integer), and a diagonal matrix $\mathbf{S} = \text{diag}(\{s_i\}_{1\leq i\leq q}) \in \mathbb{R}^{q\times q}$ with $\text{diag}(\{s_i\}_{1\leq i\leq q})$ being positive scalars in descending order, the optimal solution $\mathbf{p}^*$ for the problem below can be derived as:

$$\min_{\mathbf{p}\in\mathbb{R}^q} \sqrt{\mathbf{p}^T\mathbf{S}^2\mathbf{p}} + \frac{\tau}{2}\|\mathbf{p}-\mathbf{b}\|^2 \tag{21}$$

is given by

$$\mathbf{p}^* = \begin{cases} \left(\frac{\mathbf{S}^2}{\tau\delta} + \mathbf{I}\right)^{-1}\mathbf{b}, & \text{if} \|\mathbf{S}^{-1}\mathbf{b}\| > \frac{1}{\tau} \\ \mathbf{0}_q, & \text{otherwise} \end{cases} \tag{22}$$

where the inverse of $\mathbf{S}$, denoted as $\mathbf{S}^{-1}$, is given by the diagonal matrix $\text{diag}(\{s_i^{-1}\}_{1\leq i\leq q})$, and the positive scalar $\delta$ meets the following condition:

$$\mathbf{b}^T\text{diag}\left(\left\{\frac{s_i^2}{(\tau\delta + s_i^2)^2}\right\}_{1\leq i\leq q}\right)\mathbf{b} = \frac{1}{\tau^2} \tag{23}$$

Specifically, if $\|\mathbf{S}^{-1}\mathbf{b}\|$ exceeds $1/\tau$, then Eq. 23 (pertaining to $\delta$) possesses a unique positive solution, retrievable via the bisection technique as detailed in [4]. Leveraging Lemma 1, we derive the subsequent theorem.

*Theorem 1*: The optimal solution $\mathbf{p}_\mathbf{i}^*$ for problem Eq. 23 (with $\tau > 0$) can be expressed as:

$$\mathbf{p}_\mathbf{i}^* = \begin{cases} \hat{\mathbf{p}}, & \text{if} \left\| \left[\frac{1}{\sigma_1}, \ldots, \frac{1}{\sigma_{r_K}}\right]^T \circ \tilde{\mathbf{b}}_\mathbf{u} \right\| > \frac{1}{\tau} \\ \mathbf{c}_\mathbf{i} - \mathbf{G}_K\tilde{\mathbf{b}}_\mathbf{u}, & \text{otherwise} \end{cases} \tag{24}$$

in which $\tilde{\mathbf{b}}_\mathbf{u} = \mathbf{G}_K^T\mathbf{c}_\mathbf{i} \in \mathbb{R}^{r_K}$, where $\mathbf{G}_K \in \mathbb{R}^{n\times r_K}$ is formed by the first $r_K$ columns of $\mathbf{G}$, and the vector $\hat{\mathbf{p}}$ is defined as

$$\hat{\mathbf{p}} = \mathbf{c}_\mathbf{i} - \mathbf{G}_K\left(\left[\frac{\sigma_1^2}{\tau\delta + \sigma_1^2}, \ldots, \frac{\sigma_{r_K}^2}{\tau\delta + \sigma_{r_K}^2}\right]^T \circ \tilde{\mathbf{b}}_\mathbf{u}\right) \tag{25}$$

where $\delta$ is a positive scalar that satisfies:

$$\tilde{\mathbf{b}}_u^\mathrm{T}\text{diag}\left(\left\{\frac{\sigma_i^2}{(\tau\delta + \sigma_i^2)^2}\right\}_{1\leq i\leq r_K}\right)\tilde{\mathbf{b}}_u = \frac{1}{\tau^2} \tag{26}$$

Specifically, if the norm $\|[1/\sigma_1, \ldots, 1/\sigma_{r_K}]^T \circ \tilde{\mathbf{b}}_\mathbf{u}\|$ surpasses $1/\tau$, then Eq. 26 (concerning $\delta$) possesses a singular positive solution, retrievable using the bisection approach as referenced in [4].

**Table 1.** Comparison of NLPML with State-of-the-art PML methods on six evaluation metrics, where the best performances are shown in bold face. ($r = 3$, pairwise t-test at 0.05 significance level)

| Data | NLPML | ML-KNN | PML-*lc* | PARTICLE | PAMB | PML-NI | NATAL | PLAIN |
|---|---|---|---|---|---|---|---|---|
| Hamming Loss (the lower the better) | | | | | | | | |
| Bibtex | .015 ± .000 | .998 ± .018 | .019 ± .012 | .021 ± .032 | .025 ± .063 | .016 ± .073 | .987 ± .014 | **.013 ± .072** |
| Birds | **.066 ± .006** | .996 ± .035 | .079 ± .077 | .212 ± .042 | .235 ± .023 | .067 ± .047 | .935 ± .028 | .069 ± .027 |
| Emotions | .335 ± .040 | .813 ± .035 | .696 ± .018 | .599 ± .032 | **.257 ± .037** | .557 ± .026 | .816 ± .029 | .533 ± .026 |
| Image | **.184 ± .008** | .751 ± .058 | .225 ± .023 | .793 ± .022 | .312 ± .047 | .274 ± .065 | .776 ± .067 | .984 ± .036 |
| Medical | **.012 ± .001** | .989 ± .032 | .017 ± .081 | .018 ± .016 | .038 ± .012 | .015 ± .024 | .981 ± .041 | .022 ± .074 |
| Scene | **.139 ± .006** | .874 ± .034 | .733 ± .047 | .274 ± .043 | .353 ± .043 | .278 ± .022 | .858 ± .026 | .525 ± .022 |
| Ranking Loss (the lower the better) | | | | | | | | |
| Bibtex | .137 ± .004 | .243 ± .071 | .198 ± .019 | .136 ± .014 | .405 ± .057 | .141 ± .064 | .176 ± .041 | **.127 ± .026** |
| Birds | .202 ± .024 | .473 ± .021 | .323 ± .016 | .237 ± .013 | .479 ± .022 | .219 ± .036 | .441 ± .054 | **.114 ± .055** |
| Emotions | **.248 ± .016** | .409 ± .029 | .398 ± .047 | .344 ± .014 | .291 ± .014 | .333 ± .077 | .374 ± .045 | .323 ± .047 |
| Image | **.182 ± .016** | .343 ± .028 | .271 ± .033 | .192 ± .032 | .394 ± .017 | .319 ± .018 | .386 ± .071 | .376 ± .022 |
| Medical | **.030 ± .004** | .083 ± .063 | .038 ± .066 | .036 ± .024 | .167 ± .034 | .040 ± .024 | .074 ± .022 | .076 ± .017 |
| Scene | **.124 ± .005** | .184 ± .107 | .288 ± .045 | .205 ± .034 | .832 ± .038 | .222 ± .032 | .253 ± .076 | .302 ± .057 |
| One Error (the lower the better) | | | | | | | | |
| Bibtex | **.388 ± .012** | .632 ± .021 | .455 ± .077 | .393 ± .065 | .859 ± .094 | .395 ± .045 | .509 ± .036 | .478 ± .061 |
| Birds | **.391 ± .043** | .736 ± .062 | .715 ± .094 | .533 ± .028 | .752 ± .031 | .493 ± .054 | .609 ± .013 | .754 ± .064 |
| Emotions | **.383 ± .020** | .546 ± .074 | .579 ± .072 | .582 ± .028 | .407 ± .092 | .426 ± .046 | .455 ± .031 | .550 ± .039 |
| Image | **.339 ± .024** | .562 ± .011 | .523 ± .014 | .369 ± .043 | .612 ± .033 | .509 ± .048 | .565 ± .023 | .415 ± .021 |
| Medical | **.160 ± .009** | .332 ± .054 | .194 ± .098 | .215 ± .017 | .653 ± .038 | .212 ± .045 | .288 ± .064 | .347 ± .027 |
| Scene | **.320 ± .016** | .384 ± .029 | .629 ± .024 | .463 ± .055 | .938 ± .046 | .502 ± .064 | .408 ± .034 | .585 ± .048 |
| Coverage (the lower the better) | | | | | | | | |
| Bibtex | .247 ± .005 | .912 ± .005 | **.242 ± .022** | .251 ± .022 | .529 ± .088 | .253 ± .046 | .612 ± .094 | .928 ± .037 |
| Birds | **.145 ± .020** | .514 ± .077 | .297 ± .038 | .194 ± .026 | .789 ± .044 | .163 ± .027 | .495 ± .031 | .458 ± .054 |
| Emotions | **.374 ± .017** | .474 ± .039 | .488 ± .031 | .444 ± .019 | .381 ± .045 | .459 ± .024 | .567 ± .047 | .576 ± .043 |
| Image | **.197 ± .016** | .990 ± .071 | .268 ± .041 | .254 ± .057 | .389 ± .045 | .315 ± .071 | .954 ± .065 | .345 ± .043 |
| Medical | **.047 ± .004** | .456 ± .035 | .054 ± .012 | .095 ± .043 | .216 ± .076 | .056 ± .066 | .328 ± .026 | .359 ± .046 |
| Scene | **.117 ± .002** | .986 ± .071 | .261 ± .034 | .185 ± .061 | .715 ± .064 | .198 ± .037 | .466 ± .094 | .589 ± .035 |
| Average Precision (the higher the better) | | | | | | | | |
| Bibtex | **.520 ± .007** | .306 ± .037 | .473 ± .042 | .518 ± .054 | .146 ± .029 | .517 ± .024 | .472 ± .034 | .461 ± .045 |
| Birds | **.602 ± .038** | .258 ± .034 | .389 ± .018 | .352 ± .061 | .339 ± .071 | .541 ± .018 | .415 ± .067 | .293 ± .035 |
| Emotions | **.716 ± .010** | .595 ± .034 | .590 ± .012 | .608 ± .046 | .709 ± .034 | .662 ± .029 | .622 ± .061 | .621 ± .049 |
| Image | **.782 ± .016** | .626 ± .081 | .674 ± .031 | .767 ± .051 | .617 ± .084 | .655 ± .014 | .613 ± .041 | .393 ± .048 |
| Medical | **.870 ± .006** | .720 ± .067 | .836 ± .097 | .819 ± .051 | .471 ± .037 | .829 ± .061 | .802 ± .024 | .723 ± .027 |
| Scene | **.802 ± .008** | .720 ± .042 | .592 ± .066 | .701 ± .053 | .228 ± .034 | .678 ± .073 | .801 ± .073 | .573 ± .021 |
| microF1 (the higher the better) | | | | | | | | |
| Bibtex | .389 ± .025 | .247 ± .022 | .365 ± .032 | .379 ± .025 | .269 ± .019 | **.400 ± .004** | .365 ± .019 | .357 ± .041 |
| Birds | **.329 ± .020** | .122 ± .024 | .217 ± .019 | .204 ± .033 | .186 ± .056 | .325 ± .012 | .163 ± .024 | .263 ± .046 |
| Emotions | **.545 ± .015** | .468 ± .032 | .492 ± .022 | .515 ± .021 | .492 ± .024 | .499 ± .012 | .496 ± .028 | .495 ± .051 |
| Image | **.613 ± .003** | .425 ± .051 | .473 ± .031 | .483 ± .049 | .465 ± .052 | .422 ± .018 | .458 ± .032 | .324 ± .021 |
| Medical | **.700 ± .010** | .475 ± .048 | .675 ± .021 | .678 ± .019 | .397 ± .029 | .681 ± .017 | .538 ± .025 | .617 ± .015 |
| Scene | **.636 ± .012** | .499 ± .034 | .529 ± .047 | .542 ± .017 | .253 ± .024 | .463 ± .029 | .467 ± .031 | .497 ± .047 |

**Table 2.** Friedman Test Statistics $F_F$ and Critical Values (0.05 significance level, $k = 8$ algorithms, $N = 21$ datasets)

| Evaluation metric | $F_F$ | Critical value |
|---|---|---|
| Hamming Loss | 27.8088 | |
| Ranking Loss | 18.9746 | |
| One Error | 16.0883 | |
| Coverage | 41.1131 | 2.0756 |
| Average Precision | 27.2242 | |
| microF1 | 32.0354 | |

# 4 Experiments

## 4.1 Experimental Setup

**Datasets.** We evaluate NLPML on three real world PML datasets *music_ emotion*, *music_ style*, and *mirflickr*. In addition, six multi label datasets *Bibtex*, *Birds*, *Emotions*, *Image*, *Medical*, and *Scene* are used to construct synthetic PML datasets by injecting $r \in \{1, 2, 3\}$ false positive labels for each instance, resulting in eighteen synthetic datasets. All experiments are conducted on these datasets.

**Evaluation Metrics.** We adopt five widely used evaluation metrics *Ranking loss*, *Hamming loss*, *One error*, *Coverage*, and *Average precision* [40]. Lower values indicate better performance for the first four metrics, while higher values indicate better performance for Average Precision. We also report the microF1 score [20], under which NLPML consistently achieves the best or near best performance.

## 4.2 Performance Comparison

Table 1 summarizes the results. NLPML achieves the strongest performance on most datasets across all metrics. Table 2 presents the Friedman statistics. The null hypothesis is rejected, demonstrating significant differences among methods. We further employ the Bonferroni Dunn test [7] to examine performance differences among the eight compared algorithms. NLPML is used as the benchmark, and its rank differences relative to the other methods are evaluated by the critical difference (CD). A method is considered significantly different from NLPML when the difference in average ranks exceeds the CD value, which is 2.0334 in this study with $k = 8$ algorithms and $N = 21$ datasets. The CD results for each metric are shown in Fig. 2. In this figure, lower ranks appear on the right, and a bold line links NLPML with a method when their rank difference is within the CD range.

From Fig. 2, it can be observed that: 1)NLPML achieves the best overall rank on all metrics and shows a clear advantage in the ranking loss metric. 2) NLPML

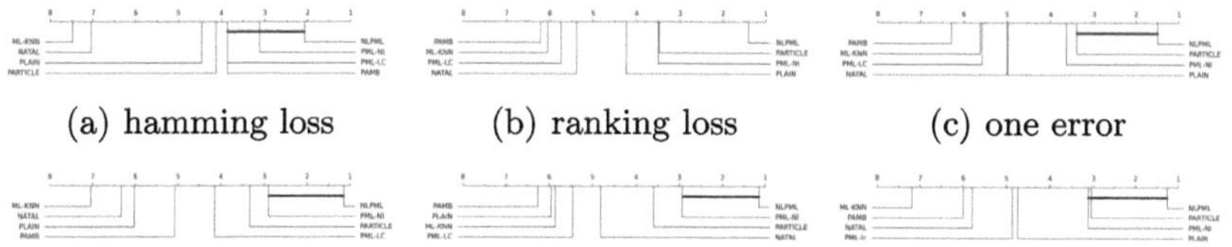

(a) hamming loss (b) ranking loss (c) one error

(d) coverage (e) average precision (f) microF1

**Fig. 2.** NLPML compared with seven algorithms using the Bonferroni-Dunn test (CD = 2.0756, significance level 0.05).

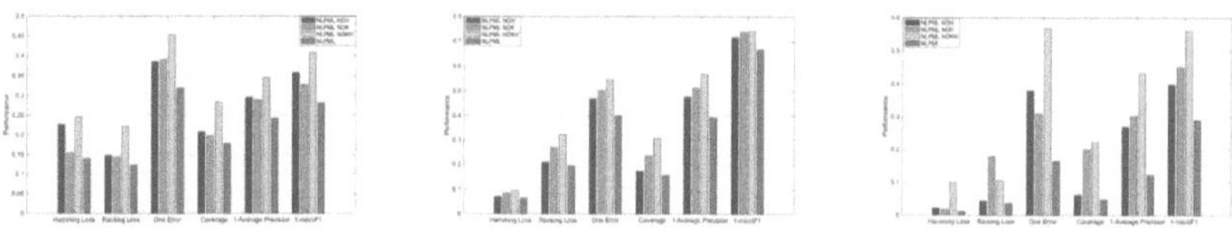

(a) music_style (b) Birds_3 (c) Medical_3

**Fig. 3.** Performance of NLPML-NOH, NLPML-NOK, NLPML-NOKH, and NLPML.

outperforms all methods except PML NI in coverage and Average Precision. 3) NLPML shows clear gains over most competitors in one error except PARTICLE, and surpasses PML NI, PAMB, and PML lc in hamming loss. 4) In microF1, NLPML noticeably outperforms all methods except PARTICLE and PML NI.

### 4.3 Ablation Study

We evaluate three simplified variants of NLPML: (1) NLPML-NOK: removes nonlinear term $g(\mathbf{P})$. (2) NLPML-NOH: removes the label affinity term $\alpha\|\mathbf{Z} \circ \mathbf{H}\|_1^2$. (3) NLPML-NOKH: removes both components.

Figure 3 reports results on *music_style*, *Birds_3*, and *Medical_3*. NLPML consistently outperforms all variants. NLPML-NOKH performs worst, confirming the necessity of combining nonlinear modeling with label affinity.

### 4.4 Parameter Analysis

We analyze four hyperparameters $(\alpha, \beta, \gamma, \lambda)$. Figure 4 shows that: (1) $\alpha$ and $\gamma$ have stable trends. (2) $\lambda$ is stable except at very small values. (3) $\beta$ exhibits a rise–fall behavior, with optimal values between 1 and 10.

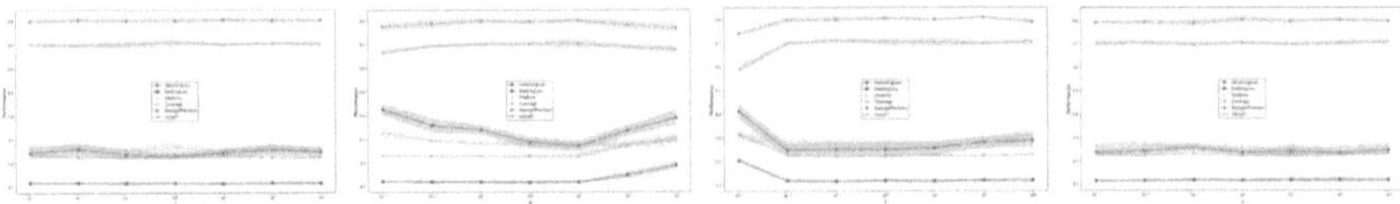

(a) $\alpha$ (b) $\beta$ (c) $\lambda$ (d) $\gamma$

**Fig. 4.** Performance of NLPML on the *mirflickr* dataset across different parameter settings.

## 5 Conclusion

The proposed NLPML framework addresses nonlinear partial multi-label learning by employing kernel low-rank representation to capture nonlinear data structures and identify complex instances often missed by linear methods. It further exploits Jaccard distance to model label correlations and integrates nonlinear affinity matrices to better suppress label noise. Experiments demonstrate that NLPML consistently outperforms existing methods, highlighting the importance of nonlinear feature modeling and label dependency exploitation. Future work will extend NLPML to broader application scenarios.

**Acknowledgments.** This research was supported by the National Natural Science Foundation of China (under Project No. 62401309), the Natural Science Foundation of Shandong Province (under Project No. ZR2024QF119), the Qingdao Natural Science Foundation Project (under Project No. 24-4-4-zrjj-89-jch), the China Postdoctoral Science Foundation (under Project No. 2025M771692), the Open Fund Project of Key Laboratory of Computing Power Internet and Information Security, Ministry of Education (under Project No. 2024PY030), and the National Natural Science Foundation of China (under Project No. 62376130).

## References

1. Abdolali, M., Gillis, N.: Beyond linear subspace clustering: a comparative study of nonlinear manifold clustering algorithms. Comput. Sci. Rev. **42**, 100435 (2021)
2. Auricchio, G., Brigati, G., Giudici, P., Toscani, G.: Multivariate gini-type discrepancies. Math. Models Methods Appl. Sci. **35**(05), 1267–1296 (2025)
3. Boutell, M.R., Luo, J., Shen, X., Brown, C.M.: Learning multi-label scene classification. Pattern Recogn. **37**(9), 1757–1771 (2004)
4. Burden, R., Faires, J.: Numerical differentiation & integration numerical differentiation I. Numer. Anal. 174–184 (2011)
5. Chen, Y.N., Lin, H.T.: Feature-aware label space dimension reduction for multi-label classification. In: Advances in Neural Information Processing Systems, vol. 25 (2012)
6. Cour, T., Sapp, B., Taskar, B.: Learning from partial labels. J. Mach. Learn. Res. **12**, 1501–1536 (2011)
7. Demšar, J.: Statistical comparisons of classifiers over multiple data sets. J. Mach. Learn. Res. **7**, 1–30 (2006)

8. Feng, L., An, B.: Partial label learning by semantic difference maximization. In: IJCAI, vol. 2019, p. 2294 (2019)
9. Fu, D., et al.: Graph relationship-driven label coded mapping and compensation for multi-label textile fiber recognition. Eng. Appl. Artif. Intell. **133**, 108484 (2024)
10. Gao, W., Hu, J., Li, Y., Zhang, P.: Feature redundancy based on interaction information for multi-label feature selection. IEEE Access **8**, 146050–146064 (2020)
11. Gong, X., Yang, J., Yuan, D., Bao, W.: Generalized large margin $k$NN for partial label learning. IEEE Trans. Multimedia **24**, 1055–1066 (2021)
12. Gong, X., Yuan, D., Bao, W.: Understanding partial multi-label learning via mutual information. In: Advances in Neural Information Processing Systems, vol. 34, pp. 4147–4156 (2021)
13. Han, Q., Hu, L., Gao, W.: Feature relevance and redundancy coefficients for multi-view multi-label feature selection. Inf. Sci. **652**, 119747 (2024)
14. Hang, J.Y., Zhang, M.L.: Partial multi-label learning with probabilistic graphical disambiguation. In: Advances in Neural Information Processing Systems, vol. 36 (2024)
15. Hao, P., Hu, L., Gao, W.: Partial multi-label feature selection via subspace optimization. Inf. Sci. **648**, 119556 (2023)
16. Huang, L., Fan, J., Wang, S., Liu, G., Liew, A.W.C.: Incomplete data classification via distribution alignment with evidence combination. Mach. Intell. Res. (2026)
17. Jiang, W., Liu, J., Qi, H., Dai, Q.: Robust subspace segmentation via nonconvex low rank representation. Inf. Sci. **340**, 144–158 (2016)
18. Li, W., Fan, L., Shao, S., Song, A.: Generalized contrastive partial label learning for cross-subject EEG-based emotion recognition. IEEE Trans. Instrum. Meas. (2024)
19. Lin, Z., Liu, R., Su, Z.: Linearized alternating direction method with adaptive penalty for low-rank representation. In: Advances in Neural Information Processing Systems, vol. 24 (2011)
20. Liu, B.Q., Jia, B.B., Zhang, M.L.: Towards enabling binary decomposition for partial multi-label learning. IEEE Trans. Pattern Anal. Mach. Intell. (2023)
21. Liu, D., et al.: Multi-directional attention: a lightweight attention module for slender structures. Intell. Robot. **5**(4), 827–43 (2025)
22. Liu, L., Dietterich, T.: A conditional multinomial mixture model for superset label learning. In: Advances in Neural Information Processing Systems, vol. 25 (2012)
23. Qian, W., Xiong, Y., Ding, W., Huang, J., Vong, C.M.: Label correlations-based multi-label feature selection with label enhancement. Eng. Appl. Artif. Intell. **127**, 107310 (2024)
24. Sun, F., Xie, M.K., Huang, S.J.: A deep model for partial multi-label image classification with curriculum-based disambiguation. Mach. Intell. Res. **21**(4), 801–814 (2024)
25. Sun, L., Feng, S., Liu, J., Lyu, G., Lang, C.: Global-local label correlation for partial multi-label learning. IEEE Trans. Multimedia **24**, 581–593 (2021)
26. Sun, L., Feng, S., Wang, T., Lang, C., Jin, Y.: Partial multi-label learning by low-rank and sparse decomposition. In: Proceedings of the AAAI Conference on Artificial Intelligence, vol. 33, pp. 5016–5023 (2019)
27. Wang, W., Zhang, M.L.: Partial label learning with discrimination augmentation. In: Proceedings of the 28th ACM SIGKDD Conference on Knowledge Discovery and Data Mining, pp. 1920–1928 (2022)
28. Wang, Z., Liu, F., Han, M., Tang, H., Wan, B.: PML-ED: a method of partial multi-label learning by using encoder-decoder framework and exploring label correlation. Inf. Sci. **661**, 120165 (2024)

29. Wu, J.H., Wu, X., Chen, Q.G., Hu, Y., Zhang, M.L.: Feature-induced manifold disambiguation for multi-view partial multi-label learning. In: Proceedings of the 26th ACM SIGKDD International Conference on Knowledge Discovery & Data Mining, pp. 557–565 (2020)
30. Xiao, S., Tan, M., Xu, D., Dong, Z.Y.: Robust kernel low-rank representation. IEEE Trans. Neural Netw. Learn. Syst. **27**(11), 2268–2281 (2015)
31. Xie, M.K., Huang, S.J.: Partial multi-label learning. In: Proceedings of the AAAI Conference on Artificial Intelligence, vol. 32 (2018)
32. Xie, M.K., Huang, S.J.: Partial multi-label learning with noisy label identification. IEEE Trans. Pattern Anal. Mach. Intell. **44**(7), 3676–3687 (2021)
33. Xu, N., Liu, B., Lv, J., Qiao, C., Geng, X.: Progressive purification for instance-dependent partial label learning. In: International Conference on Machine Learning, pp. 38551–38565. PMLR (2023)
34. Yu, G., et al.: Feature-induced partial multi-label learning. In: 2018 IEEE International Conference on Data Mining (ICDM), pp. 1398–1403. IEEE (2018)
35. Zhang, J., Jiang, X., Tian, N., Wu, M.: Label noise correction for crowdsourcing using dynamic resampling. Eng. Appl. Artif. Intell. **133**, 108439 (2024)
36. Zhang, J., et al.: Convolutional neural network aided movable antenna array design for channel estimation. Intell. Robot. **5**(4), 844–58 (2025)
37. Zhang, M.L., Fang, J.P.: Partial multi-label learning via credible label elicitation. IEEE Trans. Pattern Anal. Mach. Intell. **43**(10), 3587–3599 (2020)
38. Zhang, M.L., Yu, F., Tang, C.Z.: Disambiguation-free partial label learning. IEEE Trans. Knowl. Data Eng. **29**(10), 2155–2167 (2017)
39. Zhang, M.L., Zhou, Z.H.: ML-KNN: a lazy learning approach to multi-label learning. Pattern Recogn. **40**(7), 2038–2048 (2007)
40. Zhang, M.L., Zhou, Z.H.: A review on multi-label learning algorithms. IEEE Trans. Knowl. Data Eng. **26**(8), 1819–1837 (2013)
41. Zhang, P., Gao, W.: Feature relevance term variation for multi-label feature selection. Appl. Intell. **51**, 5095–5110 (2021)
42. Zhang, P., Gao, W., Hu, J., Li, Y.: Multi-label feature selection based on high-order label correlation assumption. Entropy **22**(7), 797 (2020)
43. Zhang, P., Gao, W., Hu, J., Li, Y.: A conditional-weight joint relevance metric for feature relevancy term. Eng. Appl. Artif. Intell. **106**, 104481 (2021)

# Partial Multi-label Learning via Label Anchor Graph

Jinfu Fan[1], Fuyu Qi[1], Linqing Huang[2], Jian Feng[1], Zhiyong Li[1], Qingkai Bu[1], and Wenpeng Lu[3,4](✉)

[1] School of Computer Science and Technology, Qingdao University, Qingdao, China
{qifuyu,fengjian,lizhiyong}@qdu.edu.cn
[2] School of Computer Science, Shanghai Jiao Tong University, Shanghai, China
[3] Key Laboratory of Computing Power Network and Information Security, Ministry of Education, Shandong Computer Science Center (National Supercomputer Center in Jinan), Qilu University of Technology (Shandong Academy of Sciences), Jinan, China
wenpeng.lu@qlu.edu.cn
[4] Shandong Provincial Key Laboratory of Computing Power Internet and Service Computing, Shandong Fundamental Research Center for Computer Science, Jinan, China

**Abstract.** Partial Multi-Label Learning (PML) addresses scenarios where each instance is associated with a candidate label set containing both relevant and irrelevant labels. The coexistence of noisy and missing labels poses significant challenges for reliable classifier learning and label disambiguation. Existing approaches often neglect the exploitation of co-occurring label structures and suffer from inaccurate feature-to-label mappings caused by linear projection assumptions. This paper presents Partial Multi-Label Learning via Label Anchor Graph (PML-LAG), a dynamic anchor-based framework that jointly refines semantic prototypes and classifier mappings under dual manifold constraints. Dynamic label anchors construct a semantic prototype graph capturing frequent label co-occurrences, while an anchor Laplacian preserves semantic smoothness among prototypes. In parallel, a neighborhood-preserving embedding enforces local feature consistency, and a collaborative optimization scheme aligns the feature and prototype spaces for robust label reconstruction. Extensive experiments on multiple real-world and synthetic datasets demonstrate that PML-LAG achieves superior noise robustness and label disambiguation performance compared with state-of-the-art partial multi-label learning methods.

**Keywords:** Partial multi-label learning · Weakly supervised learning

## 1 Introduction

Multi-label learning (MLL) addresses tasks where each instance is associated with multiple labels [9,13,19]. It captures complex semantic correlations among


R. C.-W. Wong et al. (Eds.): PAKDD 2026, LNAI 16597, pp. 455–467, 2026.
https://doi.org/10.1007/978-981-92-1300-9_36

labels and has been applied to text categorization [16], multimedia content analysis [2], and bioinformatics [4]. However, label annotation is expensive and error-prone [23], often yielding candidate label sets that contain both relevant and irrelevant labels. Noisy or missing labels mislead traditional MLL models [8,34,35], resulting in degraded generalization and unstable classifier training.

To alleviate this issue, Partial Multi-Label Learning (PML) [10,20,30] has been introduced as a weakly supervised paradigm, where each instance is associated with a candidate label set containing the ground-truth labels as well as false positives. Existing PML methods can be categorized into two groups: two-stage disambiguation methods (TDM) [25,30,36], which first identify trustworthy labels and then train classifiers, and unified framework methods (UFM) [12,21,22,27,31], which jointly perform label disambiguation and classifier optimization. Although these approaches have achieved progress, they still face two key limitations. First, they fail to exploit frequent co-occurrence label structures that can serve as semantic prototypes to guide disambiguation. Second, they rely on linear mappings from feature to label space, which ignore the intrinsic manifold geometry of the data and lead to suboptimal label alignment.

In practice, constructing direct instance–instance or label–label relationships in high-dimensional spaces is highly vulnerable to noise. Mislabelled samples can create incorrect neighborhood links, and these inaccurate connections may gradually accumulate, leading to significant error propagation across the relational graph. In addition, linear projection methods often distort the intrinsic manifold structure of the feature space, which hinders the preservation of local label dependencies. Consequently, most existing PML models exhibit substantial difficulty in jointly maintaining semantic smoothness in the label space and geometric consistency in the feature space.

To overcome these limitations, this paper proposes Partial Multi-Label Learning via Label Anchor Graph (PML-LAG), a dynamic anchor-based framework that jointly refines semantic prototypes and classifier mappings under dual manifold constraints. The method first applies k-means clustering in the initial label space to obtain anchors representing frequently co-occurring label patterns. These dynamic anchors construct a semantic prototype graph that encodes prototype–instance associations with a low-rank constraint, facilitating denoised label reconstruction. An anchor Laplacian preserves semantic smoothness among prototypes, encouraging semantically similar instances to cluster in the latent space. Meanwhile, a neighborhood-preserving embedding constrains the projection matrix to maintain the local manifold of the feature space, while semantic prototype alignment enforces consistency between projected features and prototype representations.

The main contributions of this study are summarized as follows:

1. A dynamic semantic anchor construction strategy that adaptively refines label prototypes and their associations during alternating optimization, capturing co-occurring label patterns and suppressing noise propagation.

2. An alternating multi-task learning framework integrating feature–prototype alignment and neighborhood-preserving embedding, improving projection accuracy and robustness.
3. A dual-manifold constraint that unifies feature geometry preservation and semantic smoothness, providing a cohesive geometric interpretation for partial multi-label learning.

## 2 Related Work

### 2.1 Partial Label Learning

Partial Label Learning (PLL) [5–7], a core branch of weakly supervised learning, aims to identify the correct label from a candidate set containing one ground-truth label and several false positives [6]. PLL methods are generally categorized into three families: disambiguation-based, transformation-based, and theory-driven approaches. Disambiguation strategies infer the true label by averaging or confidence estimation [17]. Transformation-based methods [18] reformulate the learning objective to reduce the influence of false positives. Theory-driven studies investigate the statistical consistency and generalization guarantees of PLL [14]. PLL has also been extended to various scenarios such as class imbalance [15,28], multi-view [37], online [26], multi-dimensional [24], and dimensionality reduction tasks [1,29]. While PLL focuses on selecting a single correct label per instance, PML generalizes this setting to multiple ground-truth labels within the candidate set, introducing higher ambiguity and modeling complexity.

### 2.2 Partial Multi-label Learning

Two-stage disambiguation methods (TDM) first identify reliable labels and then train multi-label classifiers based on disambiguated confidences. Xie et al. [30] proposed PML-*lc* and PML-*fp*, estimating label probabilities from correlations among similar instances. Zhang et al. [36] introduced PARTICLE, propagating credible labels iteratively with pairwise ranking. Wang et al. [25] developed DRAMA, learning a label confidence matrix from feature manifolds and fitting it with gradient boosting. Liu et al. [12] presented PAMB, decomposing PML into binary tasks via error-correcting codes and augmenting features with label correlations. TDM methods are prone to error propagation: mistakes in the first stage often transfer to the second, especially under heavy label noise, causing unstable confidence estimation.

Unified framework methods (UFM) jointly optimize label disambiguation and classifier training. Many exploit low-rank subspaces to capture label–feature relationships and suppress noise [21,22,33]. For instance, Sun et al. [22] proposed PML-LRS using low-rank and sparse decomposition, while [21] combined global label structures with local correlations. Yu et al. [33] represented labels and features in a unified low-rank space, and Xie et al. [31] applied trace norm and $l_1$ regularization for joint label recovery and noise identification. Wang et al. [27] adopted manifold-based models with gradient boosting.

Deep PML approaches, including Yan et al. [32] PML-GAN and Xie et al. [31] meta-disambiguation, integrate disambiguation and prediction via adversarial or online feedback mechanisms. The proposed framework leverages low-rank recovery for ground-truth estimation and introduces label semantic prototypes representing frequent co-occurring labels. By jointly optimizing feature manifolds, label manifolds, and prototype spaces, it overcomes linear mapping limitations and provides a lightweight, interpretable alternative that integrates semantic prototypes without heavy networks.

## 3 The Proposed Approach

### 3.1 Notations

In PML, each instance may be associated with only a subset of relevant labels, and some observed labels may be noisy. To facilitate model formulation, we introduce the following notations. Let $\mathbf{X} = [\mathbf{x}_1, \mathbf{x}_2, \ldots, \mathbf{x}_n] \in \mathbb{R}^{d\times n}$ denote the feature matrix, where $d$ is the feature dimensionality and $n$ is the number of instances. The partial multi-label annotations are represented by $\mathbf{Y} = [\mathbf{y}_1, \mathbf{y}_2, \ldots, \mathbf{y}_n] \in \mathbb{R}^{l\times n}$, where $l$ is the number of labels and $y_{ji} = 1$ indicates that instance $\mathbf{x}_i$ is associated with the $j$-th label. To capture label co-occurrence structures and semantic relationships, we further define a semantic prototype matrix $\mathbf{A} = [\mathbf{a}_1, \mathbf{a}_2, \ldots, \mathbf{a}_m] \in \mathbb{R}^{l\times m}$, where $m \ll n$, and each column $\mathbf{a}_j$ represents a label semantic prototype extracted from the label space.

### 3.2 The PML-LAG Framework

In conventional multi-label learning, a linear projection $\mathbf{W} = [\mathbf{w}_1, \mathbf{w}_2, \ldots, \mathbf{w}_l] \in \mathbb{R}^{d\times l}$ is learned to map features to the label space. This projection can be obtained by minimizing the squared reconstruction error with a regularization term:

$$\min_{\mathbf{W}} \frac{1}{2}\|\mathbf{Y} - \mathbf{W}^T\mathbf{X}\|_F^2 + \Psi(\mathbf{W}) \tag{1}$$

where $\Psi(\mathbf{W})$ is a model complexity regularizer that prevents overfitting.

In the PML setting, the observed label matrix $\mathbf{Y}$ may contain both correct and noisy candidate labels. We decompose it as

$$\mathbf{Y} = \mathbf{Y}^* + \mathbf{E}$$

where $\mathbf{Y}^*$ denotes the latent ground-truth labels and $\mathbf{E}$ represents the noise. To recover the true labels while handling noise, we impose a low-rank constraint on $\mathbf{Y}^*$ and a column-sparse constraint on $\mathbf{E}$, yielding the following optimization problem:

$$\begin{aligned} \min_{\mathbf{W},\mathbf{Y}^*,\mathbf{E}} \ & \frac{1}{2}\|\mathbf{Y}^* - \mathbf{W}^T\mathbf{X}\|_F^2 + \|\mathbf{Y}^*\|_* + \beta\|\mathbf{E}\|_{2,1} + \Psi(\mathbf{W}) \\ \text{s.t.} \ & \mathbf{Y} = \mathbf{Y}^* + \mathbf{E} \end{aligned} \tag{2}$$

where $\|\mathbf{Y}^*\|_*$ encourages low-rank structure in the true label matrix and $\beta > 0$ controls the sparsity of the noise matrix $\mathbf{E}$.

Because local neighborhoods in the feature space approximately lie on a linear manifold, each instance $\mathbf{x}_i$ can be reconstructed by its $k$ nearest neighbors. Let $\mathbf{G} \in \mathbb{R}^{n\times n}$ store the reconstruction weights, and define the reconstruction loss as

$$\varphi(\mathbf{G}) = \sum_{i=1}^{n} \left\| \mathbf{x}_i - \sum_{j=1}^{n} G_{ij}\mathbf{x}_j \right\|_2^2 \tag{3}$$

subject to $\sum_{j=1}^{n} G_{ij} = 1$ and $G_{ij} = 0$ if $\mathbf{x}_j$ is not a neighbor of $\mathbf{x}_i$. This reconstruction preserves local geometric structures in the feature space. Accordingly, the corresponding manifold regularizer in the label space is defined as

$$\mathrm{tr}\big(\mathbf{W}^T\mathbf{X}\mathbf{M}\mathbf{X}^T\mathbf{W}\big) \tag{4}$$

where $\mathbf{M} = (\mathbf{I}-\mathbf{G})(\mathbf{I}-\mathbf{G})^T$ encodes the local neighborhood information. Incorporating this term into the PML model leads to

$$\begin{aligned} \min_{\mathbf{W},\mathbf{Y}^*,\mathbf{E}} \ & \frac{1}{2}\|\mathbf{Y}^* - \mathbf{W}^T\mathbf{X}\|_F^2 + \mathrm{tr}(\mathbf{W}^T\mathbf{X}\mathbf{M}\mathbf{X}^T\mathbf{W}) + \|\mathbf{Y}^*\|_* + \beta\|\mathbf{E}\|_{2,1} \\ \text{s.t.} \ & \mathbf{Y} = \mathbf{Y}^* + \mathbf{E}. \end{aligned} \tag{5}$$

To capture label correlations, subspace clustering reconstructs each label vector using a linear combination of other labels:

$$\min_{\mathbf{Z}^*} \ \|\mathbf{Y} - \mathbf{Y}\mathbf{Z}^*\|_F^2 + \alpha f(\mathbf{Z}^*) \tag{6}$$

where $\mathbf{Z}^* \in \mathbb{R}^{n\times n}$ captures the relationships among labels, and $f(\mathbf{Z}^*)$ is a regularization function promoting desired structure (e.g., sparsity or low-rank) in $\mathbf{Z}^*$.

For the purpose of reducing error propagation caused by noisy labels and decreasing computational cost,anchor points are introduced to approximate label semantic prototypes:

$$\min_{\mathbf{A},\mathbf{Z}} \ \|\mathbf{Y}^* - \mathbf{A}\mathbf{Z}\|_F^2 + \alpha\|\mathbf{Z}\|_F^2 \tag{7}$$

where $\mathbf{A} \in \mathbb{R}^{l\times m}$ stores $m$ semantic prototypes obtained via $k$-means, and $\mathbf{Z} \in \mathbb{R}^{m\times n}$ denotes the associations between prototypes and instances. This prototype-based approximation effectively suppresses the influence of noisy candidate labels, since semantic structures are captured at a higher abstraction level and thus more robust to local label fluctuations.

For guide the feature projection toward meaningful semantic directions, we align the projected features with the prototype representations:

$$\gamma\sum_{i=1}^{n} \|\mathbf{W}^T\mathbf{x}_i - \mathbf{A}\mathbf{z}_i\|_2^2 = \gamma\|\mathbf{W}^T\mathbf{X} - \mathbf{A}\mathbf{Z}\|_F^2. \tag{8}$$

This alignment forces the learned projection to preserve semantic consistency, ensuring that instances mapped to similar prototype combinations remain close in the projected space. Such a constraint compensates for the distortion introduced by linear projections and improves the model's ability to maintain the underlying manifold structure.

Combining this alignment with the manifold regularization yields

$$\min_{\mathbf{W}} \operatorname{tr}\left(\mathbf{W}^T\mathbf{X}\mathbf{M}\mathbf{X}^T\mathbf{W} + \gamma(\mathbf{W}^T\mathbf{X} - \mathbf{A}\mathbf{Z})(\mathbf{W}^T\mathbf{X} - \mathbf{A}\mathbf{Z})^T\right). \tag{9}$$

Here, the manifold regularization preserves the intrinsic local geometry of the feature space, while the semantic alignment term introduces global semantic guidance through prototype-based supervision. Together, these constraints suppress the propagation of noisy neighborhood relations, enforce coherent feature–label interactions, and maintain semantic smoothness even under ambiguous or partially incorrect label annotations.

To further regularize prototype consistency, we compute pairwise cosine similarity between prototypes:

$$P_{ij} = \frac{\mathbf{A}_{:,i}^T \mathbf{A}_{:,j}}{\|\mathbf{A}_{:,i}\|_2 \, \|\mathbf{A}_{:,j}\|_2} \tag{10}$$

and enforce smoothness across similar prototypes using the Laplacian regularization:

$$\operatorname{tr}(\mathbf{Z}\mathbf{L}\mathbf{Z}^T) = \frac{1}{2}\sum_{i,j=1}^{m} P_{ij}\|\mathbf{Z}_{:,i} - \mathbf{Z}_{:,j}\|_2^2 \tag{11}$$

where $\mathbf{L} = \mathbf{D} - \mathbf{P}$ and $D_{ii} = \sum_j P_{ij}$.

Finally, integrating all the components above yields the PML-LAG objective:

$$\begin{aligned} \min_{\mathbf{Y}^*,\mathbf{A},\mathbf{Z},\mathbf{W},\mathbf{E}} \quad & \|\mathbf{Y}^* - \mathbf{A}\mathbf{Z}\|_F^2 + \alpha\|\mathbf{Z}\|_F^2 + \tfrac{1}{2}\|\mathbf{Y}^* - \mathbf{W}^T\mathbf{X}\|_F^2 \\ & + \operatorname{tr}\left(\mathbf{W}^T\mathbf{X}\mathbf{M}\mathbf{X}^T\mathbf{W} + \gamma(\mathbf{W}^T\mathbf{X} - \mathbf{A}\mathbf{Z})(\mathbf{W}^T\mathbf{X} - \mathbf{A}\mathbf{Z})^T\right) \\ & + \|\mathbf{Y}^*\|_* + \beta\|\mathbf{E}\|_{2,1} + \operatorname{tr}(\mathbf{Z}\mathbf{L}\mathbf{Z}^T), \\ \text{s.t.} \quad & \mathbf{Y} = \mathbf{Y}^* + \mathbf{E}. \end{aligned} \tag{12}$$

## 3.3 Optimization

To solve Eq. (12), auxiliary variables $\mathbf{U}$ and $\mathbf{Q}$ are introduced:

$$\begin{aligned} \min_{\mathbf{Y}^*,\mathbf{A},\mathbf{Z},\mathbf{W},\mathbf{E}} \quad & \|\mathbf{Y}^* - \mathbf{A}\mathbf{Z}\|_F^2 + \alpha\|\mathbf{Z}\|_F^2 + \tfrac{1}{2}\|\mathbf{Y}^* - \mathbf{W}^T\mathbf{X}\| \\ & + \operatorname{Tr}\left(\mathbf{W}^T\mathbf{X}\mathbf{M}\mathbf{X}^T\mathbf{W} + \gamma(\mathbf{W}^T\mathbf{X} - \mathbf{A}\mathbf{U})(\mathbf{W}^T\mathbf{X} - \mathbf{A}\mathbf{U})^T\right) \\ & + \|\mathbf{Y}^*\|_* + \beta\|\mathbf{E}\|_{2,1} + \operatorname{Tr}(\mathbf{Q}\mathbf{L}\mathbf{Q}^T) \\ & \text{s.t. } \mathbf{Y} = \mathbf{Y}^* + \mathbf{E},\ \mathbf{Z} = \mathbf{U},\ \mathbf{Z} = \mathbf{Q}. \end{aligned} \tag{13}$$

**Algorithm 1.** Optimization of PML-LAG

**Input**: $\mathbf{X}, \mathbf{Y}$ **Init**: $\mathbf{A}$ (k-means), $\mathbf{Z} = \mathbf{U} = \mathbf{Q} = 0$, $\mathbf{Y}^* = \mathbf{E} = 0$, $\mathbf{Y}_i = 0$; set $\alpha, \beta, \gamma, m, \mu_i, \rho, \mu_{\max}, \varepsilon_i$. **Output**: $\mathbf{W}$

**while** not converged **do**

$\quad \mathbf{W} = \left(2\mathbf{X}\mathbf{M}\mathbf{X}^T + (2\gamma+1)\mathbf{X}\mathbf{X}^T\right)^{-1}\left(2\gamma\mathbf{X}\mathbf{U}^T\mathbf{A}^T + \mathbf{X}\mathbf{Y}^*\right)$

$\quad \mathbf{Z} = (2\mathbf{A}^T\mathbf{A} + (2\alpha+\mu_2+\mu_3)\mathbf{I})^{-1}\left(2\mathbf{A}^T\mathbf{Y}^* + \mu_2(\mathbf{U} - \frac{\mathbf{Y}_2}{\mu_2}) + \mu_3(\mathbf{Q} - \frac{\mathbf{Y}_3}{\mu_3})\right)$

$\quad \mathbf{U} = (2\gamma\mathbf{A}^T\mathbf{A} + \mu_2\mathbf{I})^{-1}\left(2\gamma\mathbf{A}^T\mathbf{W}^T\mathbf{X} + \mu_2(\mathbf{Z} + \frac{\mathbf{Y}_2}{\mu_2})\right)$

$\quad \mathbf{Q} = \mu_3\left(\mathbf{Z} + \frac{\mathbf{Y}_3}{\mu_3}\right)(2\mathbf{L} + \mu_3\mathbf{I})^{-1}$

$\quad \mathbf{D} = \dfrac{2\mathbf{A}\mathbf{Z} + \mathbf{W}^T\mathbf{X} + \mu_1(\mathbf{Y} - \mathbf{E} + \mathbf{Y}_1/\mu_1)}{3 + \mu_1}$

$\quad \mathbf{Y}^* = \arg\min_{\mathbf{Y}^*} \frac{1}{2}\|\mathbf{Y}^* - \mathbf{D}\|_F^2 + \frac{1}{3+\mu_1}\|\mathbf{Y}^*\|_*$ (SVD+shrinkage)

$\quad \mathbf{E}_{k+1} = \Omega_{\beta/\mu_{1,k}}\left(\mathbf{Y} - \mathbf{Y}_k^* + \frac{\mathbf{Y}_{1,k}}{\mu_{1,k}}\right)$

$\quad \mu_i = \min(\rho\mu_i, \mu_{\max})$; $\mathbf{Y}_1 \leftarrow \mathbf{Y}_1 + \mu_1(\mathbf{Y} - \mathbf{Y}^* - \mathbf{E})$; $\mathbf{Y}_2 \leftarrow \mathbf{Y}_2 + \mu_2(\mathbf{Z} - \mathbf{U})$; $\mathbf{Y}_3 \leftarrow \mathbf{Y}_3 + \mu_3(\mathbf{Z} - \mathbf{Q})$

$\quad$ **if** iteration $\%T = 0$ **then**

$\quad\quad \mathbf{A} \leftarrow$ k-means($\mathbf{Y}^*$); update $\mathbf{L}$ and refresh $\mathbf{Z}$

$\quad$ **end if**

$\quad$ Check convergence

**end while**

The Alternating Linearized Minimization (ALM) framework [11] updates variables iteratively:

$$\begin{aligned} \min_{\mathbf{Y}^*,\mathbf{A},\mathbf{Z},\mathbf{W},\mathbf{E}} & \|\mathbf{Y}^* - \mathbf{A}\mathbf{Z}\|_F^2 + \alpha\|\mathbf{Z}\|_F^2 + \gamma\|\mathbf{W}^T\mathbf{X} - \mathbf{A}\mathbf{U}\|_F^2 \\ & + \mathrm{Tr}(\mathbf{W}^T\mathbf{X}\mathbf{M}\mathbf{X}^T\mathbf{W}) + \frac{1}{2}\|\mathbf{Y}^* - \mathbf{W}^T\mathbf{X}\|_F^2 \\ & + \|\mathbf{Y}^*\|_* + \beta\|\mathbf{E}\|_{2,1} + \mathrm{Tr}(\mathbf{Q}^T\mathbf{L}\mathbf{Q}) \\ & + \sum_{i=1}^{3}\frac{\mu_i}{2}\left\|\mathcal{C}_i + \frac{\mathbf{Y}_i}{\mu_i}\right\|_F^2. \end{aligned} \tag{14}$$

Here, the constraint residuals $\mathcal{C}_i$ correspond to $\mathcal{C}_1 = \mathbf{Y}^* + \mathbf{E} - \mathbf{Y}$, $\mathcal{C}_2 = \mathbf{Z} - \mathbf{U}$, and $\mathcal{C}_3 = \mathbf{Z} - \mathbf{Q}$. Semantic prototypes are refreshed periodically by k-means on $\mathbf{Y}^*$.

# 4 Experiments

## 4.1 Datasets

Experiments are conducted on three real-world datasets (*music_emotion*, *music_style*, *mirflickr*) and seven benchmark multi-label datasets (*Bibtex*, *Birds*, *Emotions*, *Image*, *Medical*, *Scene*, *Yeast*). To simulate partial multi-label settings, $r \in \{1, 2, 3\}$ false-positive labels are injected into each instance, yielding 21 synthetic PML datasets.

**Table 1.** PML-LAG is compared against leading PML methods using real-world PML datasets, highlighting the top-performing results in bold. (Employing a pairwise t-test with a significance level of 0.05).

| Hamming loss (the lower the better) | | | | | | | | |
|---|---|---|---|---|---|---|---|---|
| Data | **PML-LAG** | ML-KNN | PML-*lc* | PARTICLE | PAMB | PML-NI | NATAL | PLAIN |
| music_emotion | **.195 ± .001** | .889 ± .024 | .241 ± .006 | .243 ± .009 | .208 ± .021 | .257 ± .025 | .877 ± .027 | .281 ± .018 |
| music_style | **.113 ± .003** | .855 ± .028 | .153 ± .023 | .137 ± .035 | .123 ± .031 | .154 ± .028 | .493 ± .035 | .361 ± .011 |
| mirflickr | .219 ± .022 | .761 ± .115 | .209 ± .012 | .228 ± .022 | **.164 ± .035** | .228 ± .018 | .381 ± .014 | .236 ± .026 |
| Ranking loss (the lower the better) | | | | | | | | |
| Data | **PML-LAG** | ML-KNN | PML-*lc* | PARTICLE | PAMB | PML-NI | NATAL | PLAIN |
| music_emotion | **.222 ± .005** | .372 ± .012 | .389 ± .002 | .336 ± .014 | .234 ± .021 | .247 ± .028 | .294 ± .009 | .298 ± .034 |
| music_style | .137 ± .007 | .235 ± .017 | .699 ± .025 | .351 ± .024 | .152 ± .033 | **.133 ± .043** | .302 ± .051 | .260 ± .037 |
| mirflickr | **.092 ± .005** | .197 ± .022 | .199 ± .021 | .188 ± .018 | .136 ± .016 | .134 ± .025 | .251 ± .017 | .196 ± .014 |
| One error (the lower the better) | | | | | | | | |
| Data | **PML-LAG** | ML-KNN | PML-*lc* | PARTICLE | PAMB | PML-NI | NATAL | PLAIN |
| music_emotion | .414 ± .013 | .600 ± .005 | .623 ± .003 | .587 ± .005 | **.403 ± .008** | .453 ± .013 | .527 ± .012 | .453 ± .022 |
| music_style | **.337 ± .010** | .403 ± .021 | .495 ± .002 | .419 ± .031 | .367 ± .017 | .341 ± .019 | .578 ± .024 | .542 ± .028 |
| mirflickr | **.161 ± .004** | .459 ± .034 | .269 ± .005 | .165 ± .009 | .357 ± .014 | .275 ± .021 | .547 ± .025 | .157 ± .031 |
| Coverage (the lower the better) | | | | | | | | |
| Data | **PML-LAG** | ML-KNN | PML-*lc* | PARTICLE | PAMB | PML-NI | NATAL | PLAIN |
| music_emotion | **.386 ± .005** | .568 ± .015 | .521 ± .023 | .484 ± .009 | .412 ± .023 | .408 ± .025 | .492 ± .043 | .412 ± .019 |
| music_style | .195 ± .007 | .302 ± .007 | .467 ± .011 | .368 ± .005 | .213 ± .012 | **.191 ± .023** | .352 ± .013 | .405 ± .017 |
| mirflickr | **.208 ± .007** | .285 ± .009 | .224 ± .022 | .305 ± .032 | .237 ± .016 | .235 ± .032 | .317 ± .011 | .284 ± .014 |
| Average precision (the higher the better) | | | | | | | | |
| Data | **PML-LAG** | ML-KNN | PML-*lc* | PARTICLE | PAMB | PML-NI | NATAL | PLAIN |
| music_emotion | **.646 ± .008** | .525 ± .024 | .489 ± .003 | .542 ± .007 | .636 ± .013 | .612 ± .012 | .581 ± .021 | .612 ± .034 |
| music_style | .741 ± .008 | .654 ± .123 | .523 ± .052 | .621 ± .073 | .722 ± .023 | **.743 ± .059** | .548 ± .046 | .580 ± .037 |
| mirflickr | **.841 ± .006** | .685 ± .008 | .675 ± .034 | .690 ± .028 | .768 ± .028 | .781 ± .048 | .660 ± .062 | .768 ± .033 |

### 4.2 Experimental Setup

Performance is evaluated using five standard metrics: Ranking loss, Hamming loss, One error, Coverage, and Average precision. Lower values are preferred for the first four metrics, while higher values indicate better performance for Average precision.

### 4.3 Results Comparison

PML-LAG is compared against seven state-of-the-art methods (one multi-label and six partial multi-label) using the optimal settings reported in their original papers. Tables 1 and 2 show that PML-LAG achieves the best or near-best performance across most datasets. It performs especially well on *music_emotion* and *mirflickr*, while slightly inferior to PML-NI on *music_style* (see Fig. 1).

The Friedman test [3] rejects the null hypothesis of equal performance at the 0.05 level, confirming significant differences among all methods.

**Table 2.** THE performance of PML-LAG is juxtaposed with that of the state-of-the-art PML methods across six metrics, with superior results indicated in bold. (For $r = 3$, a pairwise t-test was conducted at a 0.05 significance level)

| Hamming loss (the lower the better) | | | | | | | | |
|---|---|---|---|---|---|---|---|---|
| Data | **PML-LAG** | ML-KNN | PML-*lc* | PARTICLE | PAMB | PML-NI | NATAL | PLAIN |
| Bibtex | .014 ± .001 | .998 ± .018 | .019 ± .012 | .021 ± .032 | .025 ± .063 | .016 ± .073 | .987 ± .014 | **.013 ± .072** |
| Birds | **.053 ± .007** | .996 ± .035 | .079 ± .077 | .212 ± .042 | .235 ± .023 | .067 ± .047 | .935 ± .028 | .069 ± .027 |
| Emotions | .290 ± .008 | .813 ± .035 | .696 ± .018 | .599 ± .032 | **.257 ± .037** | .557 ± .026 | .816 ± .029 | .533 ± .026 |
| Image | **.190 ± .012** | .751 ± .058 | .225 ± .023 | .793 ± .022 | .312 ± .047 | .274 ± .065 | .776 ± .067 | .984 ± .036 |
| Medical | **.011 ± .001** | .989 ± .032 | .017 ± .081 | .018 ± .016 | .038 ± .012 | .015 ± .024 | .981 ± .041 | .022 ± .074 |
| Scene | **.144 ± .004** | .874 ± .034 | .733 ± .047 | .274 ± .043 | .353 ± .043 | .278 ± .022 | .858 ± .026 | .525 ± .022 |
| Yeast | .236 ± .015 | .821 ± .145 | .278 ± .011 | .277 ± .034 | .474 ± .092 | **.216 ± .083** | .901 ± .045 | .304 ± .027 |
| Ranking loss (the lower the better) | | | | | | | | |
| Data | **PML-LAG** | ML-KNN | PML-*lc* | PARTICLE | PAMB | PML-NI | NATAL | PLAIN |
| Bibtex | **.118 ± .000** | .243 ± .071 | .198 ± .019 | .136 ± .014 | .405 ± .057 | .141 ± .064 | .176 ± .041 | .127 ± .026 |
| Birds | .195 ± .025 | .473 ± .021 | .323 ± .016 | .237 ± .013 | .479 ± .022 | .219 ± .036 | .441 ± .054 | **.114 ± .055** |
| Emotions | **.224 ± .035** | .409 ± .029 | .398 ± .047 | .344 ± .014 | .291 ± .014 | .333 ± .077 | .374 ± .045 | .323 ± .047 |
| Image | **.158 ± .018** | .343 ± .028 | .271 ± .033 | .192 ± .032 | .394 ± .017 | .319 ± .018 | .386 ± .071 | .376 ± .022 |
| Medical | **.034 ± .007** | .083 ± .063 | .038 ± .066 | .036 ± .024 | .167 ± .034 | .040 ± .024 | .074 ± .022 | .076 ± .017 |
| Scene | **.166 ± .013** | .184 ± .107 | .288 ± .045 | .205 ± .034 | .832 ± .038 | .222 ± .032 | .253 ± .076 | .302 ± .057 |
| Yeast | .189 ± .006 | .194 ± .025 | **.187 ± .044** | .239 ± .091 | .366 ± .041 | .209 ± .025 | .343 ± .031 | .291 ± .027 |
| One error (the lower the better) | | | | | | | | |
| Data | **PML-LAG** | ML-KNN | PML-*lc* | PARTICLE | PAMB | PML-NI | NATAL | PLAIN |
| Bibtex | **.389 ± .002** | .632 ± .021 | .455 ± .077 | .393 ± .065 | .859 ± .094 | .395 ± .045 | .509 ± .036 | .478 ± .061 |
| Birds | **.493 ± .030** | .736 ± .062 | .715 ± .094 | .533 ± .028 | .752 ± .031 | .493 ± .054 | .609 ± .013 | .754 ± .064 |
| Emotions | **.358 ± .087** | .546 ± .074 | .579 ± .072 | .582 ± .028 | .407 ± .092 | .426 ± .046 | .455 ± .031 | .550 ± .039 |
| Image | **.285 ± .023** | .562 ± .011 | .523 ± .014 | .369 ± .043 | .612 ± .033 | .509 ± .048 | .565 ± .023 | .415 ± .021 |
| Medical | **.166 ± .025** | .332 ± .054 | .194 ± .098 | .215 ± .017 | .653 ± .038 | .212 ± .045 | .288 ± .064 | .347 ± .027 |
| Scene | .400 ± .020 | **.384 ± .029** | .629 ± .024 | .463 ± .055 | .938 ± .046 | .502 ± .064 | .408 ± .034 | .585 ± .048 |
| Yeast | **.239 ± .010** | .255 ± .071 | .241 ± .081 | .528 ± .023 | .318 ± .027 | .288 ± .049 | .395 ± .017 | .363 ± .024 |
| Coverage (the lower the better) | | | | | | | | |
| Data | **PML-LAG** | ML-KNN | PML-*lc* | PARTICLE | PAMB | PML-NI | NATAL | PLAIN |
| Bibtex | **.215 ± .005** | .912 ± .005 | .242 ± .022 | .251 ± .022 | .529 ± .088 | .253 ± .046 | .612 ± .094 | .928 ± .037 |
| Birds | **.134 ± .030** | .514 ± .077 | .297 ± .038 | .194 ± .026 | .789 ± .044 | .163 ± .027 | .495 ± .031 | .458 ± .054 |
| Emotions | **.349 ± .006** | .474 ± .039 | .488 ± .031 | .444 ± .019 | .381 ± .045 | .459 ± .024 | .567 ± .047 | .576 ± .043 |
| Image | **.182 ± .016** | .990 ± .071 | .268 ± .041 | .254 ± .057 | .389 ± .045 | .315 ± .071 | .954 ± .065 | .345 ± .043 |
| Medical | **.050 ± .009** | .456 ± .035 | .054 ± .012 | .095 ± .043 | .216 ± .076 | .056 ± .066 | .328 ± .026 | .359 ± .046 |
| Scene | **.151 ± .012** | .986 ± .071 | .261 ± .034 | .185 ± .061 | .715 ± .064 | .198 ± .037 | .466 ± .094 | .589 ± .035 |
| Yeast | **.485 ± .010** | .879 ± .011 | .492 ± .076 | .511 ± .074 | .827 ± .049 | .501 ± .048 | .539 ± .021 | .571 ± .049 |
| Average precision (the higher the better) | | | | | | | | |
| Data | **PML-LAG** | ML-KNN | PML-*lc* | PARTICLE | PAMB | PML-NI | NATAL | PLAIN |
| Bibtex | **.544 ± .001** | .306 ± .037 | .473 ± .042 | .518 ± .054 | .146 ± .029 | .517 ± .024 | .472 ± .034 | .461 ± .045 |
| Birds | **.565 ± .010** | .258 ± .034 | .389 ± .018 | .352 ± .061 | .339 ± .071 | .541 ± .018 | .415 ± .067 | .293 ± .035 |
| Emotions | **.741 ± .044** | .595 ± .034 | .590 ± .012 | .608 ± .046 | .709 ± .034 | .662 ± .029 | .622 ± .061 | .621 ± .049 |
| Image | **.807 ± .016** | .626 ± .081 | .674 ± .031 | .767 ± .051 | .617 ± .084 | .655 ± .014 | .613 ± .041 | .393 ± .048 |
| Medical | **.856 ± .017** | .720 ± .067 | .836 ± .097 | .819 ± .051 | .471 ± .037 | .829 ± .061 | .802 ± .024 | .723 ± .027 |
| Scene | .747 ± .013 | .720 ± .042 | .592 ± .066 | .701 ± .053 | .228 ± .034 | .678 ± .073 | **.801 ± .073** | .573 ± .021 |
| Yeast | .747 ± .005 | .733 ± .041 | **.752 ± .082** | .657 ± .031 | .693 ± .023 | .719 ± .062 | .598 ± .011 | .648 ± .018 |

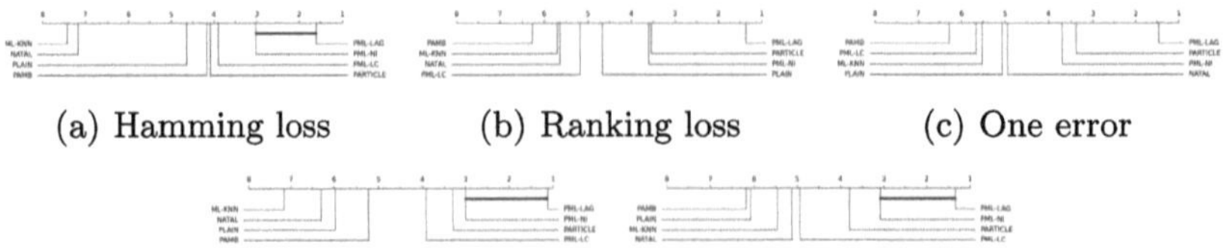

(a) Hamming loss (b) Ranking loss (c) One error

(d) Coverage (e) Average precision

**Fig. 1.** Bonferroni-Dunn comparison between PML-LAG (control) and eight baselines (CD = 1.9021, $\alpha = 0.05$).

Overall observations are summarized as follows:

- PML-LAG attains the lowest average ranking across all evaluation metrics, demonstrating consistently superior performance.
- The method shows notable improvements in *Ranking loss* and *One error*.
- In *Hamming loss*, *Coverage*, and *Average precision*, PML-LAG performs strongly, with the only close competitor being PML-NI, which relies on explicit noise identification. PML-LAG benefits from dynamic prototype refinement and anchor-based denoising.

### 4.4 Ablation Study

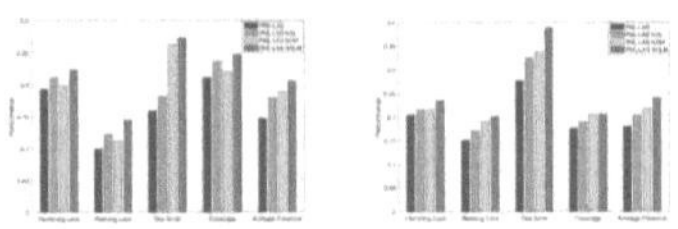

(a) *mirflickr* (b) *image_3*

**Fig. 2.** Ablation results on *mirflickr* and *image_3*.

To evaluate each component, we construct two degraded variants: **PML-LAG_NOL** removes the anchor Laplacian, and **PML-LAG_NOM** removes the feature-manifold constraint **M**.

As shown in Fig. 2, removing either constraint leads to clear degradation, confirming that each component contributes positively to the final performance.

### 4.5 Sensitive Analysis

We examine the influence of $\alpha$, $\beta$, $\gamma$, and the number of semantic prototypes $m$. Each parameter is varied while fixing the others. Figure 3 reports results on *mirflickr*. PML-LAG remains stable within a broad parameter range. The number of prototypes $m$ should not be overly small. A practical choice is at least 2× the number of labels.

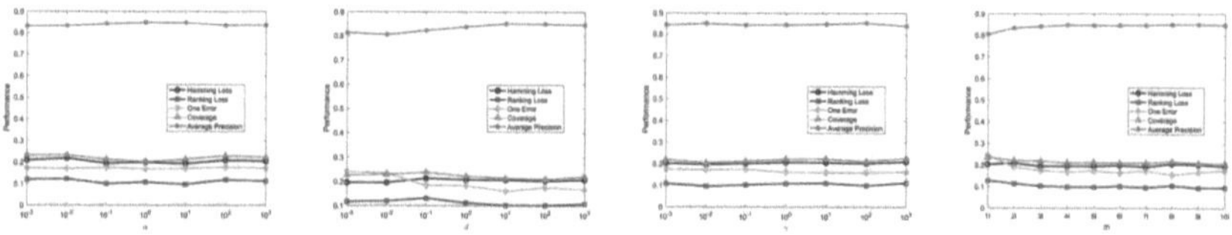

**Fig. 3.** Effect of trade-off parameters on *mirflickr*.

# 5 Conclusion

This paper presents PML-LAG, a prototype-driven partial multi-label framework that leverages dynamically updated anchor graphs to model instance–prototype associations while suppressing label noise. A dual-manifold constraint further integrates feature geometry and semantic smoothness, enabling stable and discriminative label reconstruction. Extensive experiments on real-world and synthetic datasets demonstrate the effectiveness of the proposed method. Future work will explore nonlinear prototype modeling and extensions to cross-modal partial multi-label scenarios.

**Acknowledgments.** This research was supported by the National Natural Science Foundation of China (under Project No.62401309), the Natural Science Foundation of Shandong Province (under Project No.ZR2024QF119), the Qingdao Natural Science Foundation Project (under Project No.24-4-4-zrjj-89-jch), the China Postdoctoral Science Foundation (under Project No.2025M771692), the Open Fund Project of Key Laboratory of Computing Power Internet and Information Security, Ministry of Education (under Project No.2024PY030), and the National Natural Science Foundation of China (under Project No.62376130).

# References

1. Bao, W.X., Hang, J.Y., Zhang, M.L.: Submodular feature selection for partial label learning. In: Proceedings of the 28th ACM SIGKDD Conference on Knowledge Discovery and Data Mining, pp. 26–34. (2022)
2. Briggs, F., Fern, X.Z., Raich, R.: Rank-loss support instance machines for MIML instance annotation. In: Proceedings of the 18th ACM SIGKDD International Conference on Knowledge Discovery and Data Mining, pp. 534–542 (2012)
3. Demšar, J.: Statistical comparisons of classifiers over multiple data sets. J. Mach. Learn. Res. **7**, 1–30 (2006)
4. Diplaris, S., Tsoumakas, G., Mitkas, P.A., Vlahavas, I.: Protein classification with multiple algorithms. In: Advances in Informatics: 10th Panhellenic Conference on Informatics, PCI 2005, Volas, Greece, 11–13 November 2005. Proceedings 10, pp. 448–456. Springer (2005)
5. Fan, J., Huang, L., Gong, C., You, Y., Gan, M., Wang, Z.: KMT-PLL: K-means cross-attention transformer for partial label learning. IEEE Trans. Neural Netw. Learn. Syst. (2024)

6. Fan, J., Yu, Y., Wang, Z.: Partial label learning with competitive learning graph neural network. Eng. Appl. Artif. Intell. **111**, 104779 (2022)
7. Feng, J., et al.: Partial label learning via mutual information representation learning. In: IEEE Transactions on Circuits and Systems for Video Technology (2025)
8. Fu, D., et al.: Graph relationship-driven label coded mapping and compensation for multi-label textile fiber recognition. Eng. Appl. Artif. Intell. **133**, 108484 (2024)
9. Huang, L., Fan, J., Wang, S., Liu, G., Liew, A.W.C.: Incomplete data classification via distribution alignment with evidence combination. Mach. Intell. Res. (2026). https://doi.org/10.1007/s11633-025-1565-1
10. Li, Z., Huang, L., Gu, T., Bu, Q., Qi, F., Fan, J.: Partial multi-label learning via K-means graph transformer. Knowl.-Based Syst. **325**, 114017 (2025)
11. Lin, Z., Liu, R., Su, Z.: Linearized alternating direction method with adaptive penalty for low-rank representation. In: Advances in Neural Information Processing Systems, vol. 24 (2011)
12. Liu, B.Q., Jia, B.B., Zhang, M.L.: Towards enabling binary decomposition for partial multi-label learning. IEEE Trans. Pattern Anal. Mach. Intell. (2023)
13. Liu, D., et al.: Multi-directional attention: a lightweight attention module for slender structures. Intell. Robot. **5**(4), 827–43 (2025)
14. Liu, L., Dietterich, T.: Learnability of the superset label learning problem. In: International Conference on Machine Learning, pp. 1629–1637. PMLR (2014)
15. Liu, W., Wang, L., Chen, J., Zhou, Y., Zheng, R., He, J.: A partial label metric learning algorithm for class imbalanced data. In: Asian Conference on Machine Learning, pp. 1413–1428. PMLR (2021)
16. Loza Mencía, E., Fürnkranz, J.: Efficient pairwise multilabel classification for large-scale problems in the legal domain. In: Joint European Conference on Machine Learning and Knowledge Discovery in Databases, pp. 50–65. Springer (2008)
17. Lyu, G., Feng, S., Huang, W., Dai, G., Zhang, H., Chen, B.: Partial label learning via low-rank representation and label propagation. Soft. Comput. **24**, 5165–5176 (2020)
18. Lyu, G., Feng, S., Wang, T., Lang, C., Li, Y.: GM-PLL: graph matching based partial label learning. IEEE Trans. Knowl. Data Eng. **33**(2), 521–535 (2019)
19. Qian, W., Xiong, Y., Ding, W., Huang, J., Vong, C.M.: Label correlations-based multi-label feature selection with label enhancement. Eng. Appl. Artif. Intell. **127**, 107310 (2024)
20. Sun, F., Xie, M.K., Huang, S.J.: A deep model for partial multi-label image classification with curriculum-based disambiguation. Mach. Intell. Res. **21**(4), 801–814 (2024)
21. Sun, L., Feng, S., Liu, J., Lyu, G., Lang, C.: Global-local label correlation for partial multi-label learning. IEEE Trans. Multimedia **24**, 581–593 (2021)
22. Sun, L., Feng, S., Wang, T., Lang, C., Jin, Y.: Partial multi-label learning by low-rank and sparse decomposition. In: Proceedings of the AAAI Conference on Artificial Intelligence, vol. 33, pp. 5016–5023 (2019)
23. Wang, C., et al.: Multitype label noise modeling and uncertainty-weighted label correction for concealed object detection. IEEE Trans. Instrum. Meas. **72**, 1–12 (2023)
24. Wang, H., Liu, W., Zhao, Y., Hu, T., Chen, K., Chen, G.: Learning from multi-dimensional partial labels. In: Proceedings of the Twenty-Ninth International Conference on International Joint Conferences on Artificial Intelligence, pp. 2943–2949 (2021)
25. Wang, H., Liu, W., Zhao, Y., Zhang, C., Hu, T., Chen, G.: Discriminative and correlative partial multi-label learning. In: IJCAI, pp. 3691–3697 (2019)

26. Wang, H., et al.: Online partial label learning. In: Machine Learning and Knowledge Discovery in Databases: European Conference, ECML PKDD 2020, Ghent, Belgium, 14–18 September 2020, Proceedings, Part II, pp. 455–470. Springer (2021)
27. Wang, H., et al.: Deep partial multi-label learning with graph disambiguation. arXiv preprint: arXiv:2305.05882 (2023)
28. Wang, J., Zhang, M.L.: Towards mitigating the class-imbalance problem for partial label learning. In: Proceedings of the 24th ACM SIGKDD International Conference on Knowledge Discovery & Data Mining, pp. 2427–2436 (2018)
29. Wu, J.H., Zhang, M.L.: Disambiguation enabled linear discriminant analysis for partial label dimensionality reduction. In: Proceedings of the 25th ACM SIGKDD International Conference on Knowledge Discovery & Data Mining, pp. 416–424 (2019)
30. Xie, M.K., Huang, S.J.: Partial multi-label learning. In: Proceedings of the AAAI Conference on Artificial Intelligence, vol. 32 (2018)
31. Xie, M.K., Huang, S.J.: Partial multi-label learning with noisy label identification. IEEE Trans. Pattern Anal. Mach. Intell. **44**(7), 3676–3687 (2021)
32. Yan, Y., Guo, Y.: Adversarial partial multi-label learning with label disambiguation. In: Proceedings of the AAAI Conference on Artificial Intelligence, vol. 35, pp. 10568–10576 (2021)
33. Yu, G., et al.: Feature-induced partial multi-label learning. In: 2018 IEEE International Conference on Data Mining (ICDM), pp. 1398–1403. IEEE (2018)
34. Zhang, J., Jiang, X., Tian, N., Wu, M.: Label noise correction for crowdsourcing using dynamic resampling. Eng. Appl. Artif. Intell. **133**, 108439 (2024)
35. Zhang, J., et al.: Convolutional neural network aided movable antenna array design for channel estimation. Intell. Robot. **5**(4), 844–58 (2025)
36. Zhang, M.L., Fang, J.P.: Partial multi-label learning via credible label elicitation. IEEE Trans. Pattern Anal. Mach. Intell. **43**(10), 3587–3599 (2020)
37. Zhao, L., Xiao, Y., Liu, B., Hao, Z.: Multi-view partial label machine. Inf. Sci. **586**, 310–325 (2022)

# Empowering Contactless Sleep Health Monitoring with Multi-task Learning

Zeyu Long[1,2], Beihong Jin[1,2(✉)], Siheng Li[1], Zhi Wang[1,2], Xiaoyong Ren[3], and Haiqin Liu[3]

[1] Institute of Software, Chinese Academy of Sciences, Beijing, China
Beihong@iscas.ac.cn
[2] University of Chinese Academy of Sciences, Beijing, China
[3] The Second Affiliated Hospital of Xi'an Jiaotong University, Xi'an, China

**Abstract.** Contactless sensing enables the simultaneous monitoring of multiple sleep indicators without disturbing the individual's sleep. However, complex relationships among multiple tasks pose challenges for joint monitoring. In the paper, we develop a contactless sleep health monitoring system named SHMIU, which analyzes IR-UWB signals reflected off the human body and predicts sleep stages, sleep apnea state, and blood oxygen saturation during sleep. For enhancing the performance of three tasks, we design a mixture of experts based multi-task learning model and dynamically reweight tasks via Nash bargaining to optimize multi-task learning. We conduct experiments on a 100-subject dataset, and the experimental results show that SHMIU achieves balanced performance across three tasks, with its performance comparable to or superior to that of existing two-task methods.

**Keywords:** Multi-task learning · Contactless sensing · Health monitoring · UWB signal

## 1 Introduction

Sleep is a fundamental physiological process of the human body, accounting for approximately one-third of a day. Tracking physiological indicators such as respiration rate, heart rate, and blood oxygen saturation during an individual's sleep can help diagnose whether the individual's sleep state is healthy, forming the basis for improving sleep quality. Until now, polysomnography (PSG) [10] has been regarded as the gold standard in sleep health monitoring. However, PSG often results in a poor user experience, as it requires the subject to wear specialized devices on different body parts and might disrupt natural sleep processes.

Recently, contactless sensing using wireless signals such as Wi-Fi, frequency modulated continuous wave (FMCW) and impulse-radio ultra-wideband (IR UWB) signals has emerged as a promising solution for health monitoring. Applying contactless sensing to sleep monitoring can avoid disturbing the monitored individual's sleep. It also has the potential to detect multiple sleep indicators


R. C.-W. Wong et al. (Eds.): PAKDD 2026, LNAI 16597, pp. 468–479, 2026.
https://doi.org/10.1007/978-981-92-1300-9_37

simultaneously. This is because contactless sensing can capture chest movements caused by cardiorespiratory activity, which are intrinsically correlated with these sleep indicators. For example, when a subject enters the non-rapid eye movement (NREM) sleep stages, chest movements are generally weaker than those during the rapid eye movement (REM) stage. When a subject experiences apnea, chest movements not only diminish but also lose the original periodic rhythm. Additionally, in a medical context, fluctuations in nocturnal blood oxygen saturation (e.g., drops) are closely linked to cardiac rhythm disorders (e.g., arrhythmia) and respiratory events (e.g., apnea), thus being indirectly associated with chest movements. However, relatively little research has been conducted in this regard, with only a few studies focusing on simultaneously detecting two indicators. We note that the reason behind this lies in the great complexity within and between tasks. As sleep stage, respiration, and blood oxygen saturation are concerned, these variations are not independent but mutually influence each other. For instance, sleep apnea, which typically occurs intermittently, will trigger transitory episodes of hypoxia, leading to decreased blood oxygen saturation and then causing arousal and heavy breathing. Consequently, the models constructed for these tasks are susceptible to negative transfer, where conflicts between tasks degrade performance compared to single-task baselines [16,20].

In this paper, we combine biomedical signal processing with a deep learning model to achieve contactless monitoring of sleep apnea, blood oxygen saturation ($SpO_2$), and sleep staging. Specifically, we develop a contactless sleep health monitoring system named SHMIU (Sleep Health Meter using IR-UWB signals). SHMIU adopts a UWB radar to transmit signals toward the human body and collect reflected signals, a choice driven by its high distance resolution and strong anti-interference capabilities [1]. In SHMIU, we take full advantage of the dual nature of wireless signals as both images and sequences, and design convolution operations on images and specific encoding for sequences to extract features from the amplitude and phase of the signals. Moreover, we discover that conflicts exist between task gradients and the average aggregation of task gradients, where the task gradient is the gradient of the loss function of the corresponding task. These gradient conflicts induce the direction of parameter updates to deviate from the optimal direction, thereby leading to performance imbalance. We thus adopt shared experts and task-specific experts in network structure, as well as a Nash bargaining strategy for gradient aggregation, to reduce the harmful impact of gradient conflicts on performance. Our contributions are summarized as follows:

- We are the first to employ IR-UWB signals to simultaneously achieve sleep staging, sleep apnea monitoring, and sleep $SpO_2$ estimation.
- We design a Mixture-of-Experts (MoE) based multi-task learning module to make task-specific predictions, while an adaptive loss reweighting strategy is proposed to mitigate the adverse effect of conflicting gradients.
- We conduct extensive experiments on a self-built dataset of 100 subjects. The experimental results indicate that SHMIU achieves balanced performance across all three tasks. Moreover, its performance is on par with, or even surpasses, that of existing two-task methods.

The rest of the paper is organized as follows. Section 2 introduces the related work. Section 3 describes the design of SHMIU in detail. Section 4 gives the experimental evaluation. Lastly, the paper is concluded in Sect. 5.

## 2 Related Work

**Contactless Sleep Health Monitoring.** Most existing studies focus on monitoring a single indicator [6,15,18], while studies on tracking two sleep indicators remain relatively limited, with noteworthy ones as follows. Hypnos [5] can identify different sleep stages as well as sleep apnea. It collects and cleans the UWB signals reflected off a subject, designs a CNN-Transformer based network as the encoder for the signal sequences, and combines it with contrastive learning. RF-Oximeter [3] is the first to achieve sleep blood oxygen monitoring via RF signals. Besides combining BERT and UNet, it incorporates a gating module, which first predicts sleep stages and then leverages the current sleep stage and the subject's gender information to dynamically adjust the estimation of the subject's blood oxygen saturation. Zhang et al. [19] employ an FMCW radar to collect multi-channel radar data, then feed them into a network that integrates an attention mechanism with UNet to estimate both blood oxygen saturation and duration of sleep apnea.

**Multi-task Learning.** The essence of multi-task learning (MTL) is to extract and leverage consensual and complementary information among tasks to enhance the data utilization efficiency and generalization of multi-task models.

As a promising network structure for MTL, MoE [4] combines the output of different experts to make predictions for specified tasks. Furthermore, MMoE (Multi-gate MoE) [8] explicitly introduces one gate for each task, resulting in enhanced scalability as the number of tasks increases, and CGC (Customized Gate Control) network [12] divides all experts into shared experts and task-specific experts in advance, with each task fixedly activating both shared experts and its dedicated experts. On the other hand, to mitigate the negative impact of inter-task gradient conflicts, some research delves into gradient correction methods, while others explore different gradient aggregation methods. For example, PCGrad (Projecting Conflicting Gradients) [17] projects a task gradient onto the normal plane of the gradient of any other task that has a conflicting gradient, thereby eliminating destructive interference from conflicting gradients. CAGrad (Conflict-Averse Gradient descent) [7] looks for the best update vector within a neighborhood of the average of task gradients that maximizes the worst local improvement of individual tasks. Aligned-MTL [11] is to align gradients using the principal components of the gradient matrix, and then the aligned gradients, which are orthogonal (non-conflicting) and of equal magnitude (non-dominant), are summed up with pre-defined task weights, resulting in a cumulative gradient. As to gradient aggregation, the simplest one, known as Linear Sum (LS), is to treat all tasks equally and sum their gradients. Nash-MTL [9] regards gradient aggregation as a cooperative bargaining game and obtains task gradient weights through the Nash bargaining solution.

Compared to existing work, we explore the potential of IR-UWB signals for sleep health monitoring. By combining multiple expert collaboration and dynamic gradient aggregation, we effectively monitor sleep apnea, sleep staging, and $SpO_2$ during sleep.

## 3 Our System

### 3.1 System Overview

Let $\mathbf{R} \in \mathbb{C}^{N_t \times N_\gamma}$ denote the matrix of reflected signals collected by the IR-UWB device. Here, $N_t = T \times f_t$ is the span of reflected signals on slow-time axis, $f_t$ denotes the frame rate (frames per second), and $T$ is the duration of the subject's sleep, $N_\gamma$ corresponds to the number of sampling points along the fast-time axis, which reflects the signal propagation delay and thus encodes the distance information between the radar and reflecting objects. The objective of our system is to map $\mathbf{R}$ to three indicators: sleep apnea events $\hat{y}_{\text{Apnea}}$, sleep stages $\hat{y}_{\text{Stage}}$, and blood oxygen saturation $\hat{y}_{\text{SpO}_2}$. Specifically, both apnea and $SpO_2$ are predicted once per second, while the sleep stage (i.e., NREM, REM or wake) is estimated once per sleep epoch (one sleep epoch corresponds to 30 s).

Our system comprises two parts: signal processing and sleep health prediction. In the first part, we perform signal extraction and refinement, and obtain the subject's overnight respiratory signal, denoted as $x = (x^a, x^p)$, where $x^a \in \mathbb{R}^{N_t}$ and $x^p \in \mathbb{R}^{N_t}$ represent the amplitude and phase of reflected signals, respectively. In the second part, we develop a deep learning model for sleep health prediction.

### 3.2 Signal Processing

We first divide the signal matrix $\mathbf{R}$ along the slow-time dimension into several submatrices $\{\mathbf{R}_1, \mathbf{R}_2, \ldots\}$ using a fixed time window of 30 s. Next, we apply a background subtraction algorithm [14] to each submatrix to suppress clutter caused by static objects in the environment, resulting in the background-subtracted submatrices $\{\mathbf{R}_1^b, \mathbf{R}_2^b, \ldots\}$. To locate the subject, we leverage the high intensity of human reflections. Specifically, we accumulate the signal amplitudes of each $\mathbf{R}_i^b$ along the slow-time axis to obtain intensity vectors $\{v_i\}$ ($v_i \in \mathbb{R}^{N_\gamma}$). The indices of the highest peaks in $\{v_i\}$ then serve as the initial candidate positions $D = \{d_i\}$.

The candidate set $D$ needs to be further refined by the periodicity of the reflected signals. For each position in $D$, we extract the corresponding one-dimensional reflected signal sequence $\{r_1, r_2, \ldots\}$ from the submatrices $\{\mathbf{R}_1, \mathbf{R}_2, \ldots\}$ along the slow-time axis and evaluate its periodicity using an auto-correlation function (ACF). If the signal exhibits strong periodicity (the ACF peak intensity exceeds a threshold $\epsilon$) and its dominant frequency falls within the range of normal respiration rates, we identify the presence of a valid subject and add the corresponding position to a validated set, denoted as $P$. The median value $p_{bed}$ of set $P$ is taken as the center position of the bed, and the bed area

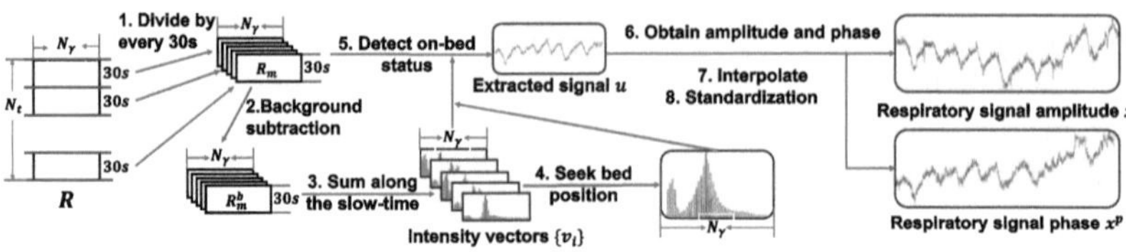


**Fig. 1.** Signal processing part

is defined as $[p_{bed} - w_{bed}/2, p_{bed} + w_{bed}/2]$, where $w_{bed}$ is the predefined width of the bed.

We constrain signal extraction to the defined bed area. For the signal intensity vector $v_i$ of the $i$-th time window, if a peak within the bed boundary exceeds a threshold $\mu$, the corresponding signal is extracted. This yields the overnight complex sequence $u \in \mathbb{C}^{N_t}$. Then, we derive the amplitude sequence $u^a$ and phase sequence $u^p$ from $u$. To address frame loss during data acquisition (assuming $N_{miss}$ missing points in an amplitude or phase sequence), we apply cubic spline ($N_{miss} < \eta_1$) or polynomial interpolation ($\eta_1 \leq N_{miss} < \eta_2$), masking segments where $N_{miss} \geq \eta_2$. Finally, z-score normalization is applied to obtain the processed signal $x = (x^a, x^p)$.

The whole signal processing procedure is shown in Fig. 1.

### 3.3 Sleep Health Prediction Model

Our model, as shown in Fig. 2, consists of three modules, i.e., a sequential encoding module, a contrastive learning module, and a multi-task learning module.

**Sequential Encoding Module.** This module incorporates an embedding encoder and a sequence encoder to capture local features and long-range dependencies in respiratory signals.

The embedding encoder $E$ is composed of seven stacked convolutional layers. Each odd layer shares the same structure, i.e., a 1D convolution followed by batch normalization and a LeakyReLU activation. Each even layer includes a 1D convolution followed by a LeakyReLU activation and max-pooling.

The preprocessed respiratory signal $x$ is first segmented into a series of consecutive windows, denoted as $x = \{x_1, x_2, \ldots, x_N\}$, where $x_i = (x_i^a, x_i^p)$ represents the amplitude-phase pair of the $i$-th segment, and $N$ is the total number of windows. Next, two sequences, composed of amplitudes and phases taken from $n$ consecutive windows, respectively, are denoted as $X^a$ and $X^p$ and fed into the embedding encoder $E$. Then, a positional embedding $e_p \in \mathbb{R}^{d\times(30n)}$ is injected into the output of the embedding encoder, forming $X_0 = E(X^a) + E(X^p) + e_p$, where $d$ is the embedding dimension. Finally, $X_0$ is fed into the sequence encoder.

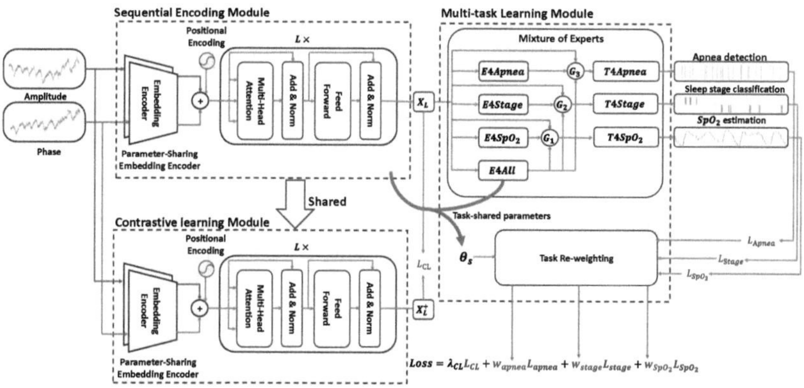


**Fig. 2.** Architecture of the proposed model

The sequence encoder adopts the multi-head self-attention mechanism [13] and contains $L$ self-attention blocks. The operations at the $l$-th self-attention block are as follows:

$$\mathbf{A}_l^i = \text{softmax}\left(\frac{\mathbf{X}_{l-1}\mathbf{W}_l^{Qi}\left(\mathbf{X}_{l-1}\mathbf{W}_l^{Ki}\right)^T}{\sqrt{d/h}}\right)\mathbf{X}_{l-1}\mathbf{W}_l^{Vi} \tag{1}$$

$$\mathbf{A}_l = \text{concat}\left(\mathbf{A}_l^1, \mathbf{A}_l^2, \ldots, \mathbf{A}_l^h\right) \tag{2}$$

$$\mathbf{B}_l = \text{LayerNorm}\left(\mathbf{X}_{l-1} + \text{Dropout}(\text{MLP}(\mathbf{A}_l))\right) \tag{3}$$

$$\mathbf{C}_l = \text{MLP}(\text{Dropout}(\text{ReLU}(\text{MLP}(\mathbf{B}_l)))) \tag{4}$$

$$\mathbf{X}_l = \text{LayerNorm}\left(\mathbf{B}_l + \text{Dropout}(\mathbf{C}_l)\right) \tag{5}$$

where $\mathbf{X}_{l-1}$ is the output of the layer $l-1$, $\mathbf{W}_l^{Qi}$, $\mathbf{W}_l^{Ki}$ and $\mathbf{W}_l^{Vi}$ are all learnable weight matrices, $h$ is the number of self-attention heads. The final output of the sequential encoding module is $\mathbf{X}_L \in \mathbb{R}^{d\times(30n)}$.

**Contrastive Learning Module.** To enhance discriminability at the sample level, we design a contrastive learning module to learn more generalizable feature representations. Specifically, we feed $X^a$ and $X^p$ into the sequential encoding module again, where the only difference between the two passes lies in the Dropout operations (due to different random seeds). This yields two outputs, $\mathbf{X}_L$ and $\mathbf{X}_L^+$, which share the same semantics and are regarded as a positive pair of samples. Meanwhile, the outputs of the sequential encoding module derived from other input data within the same batch are denoted as $\mathbf{X}_L^-$. $\mathbf{X}_L^-$ and $\mathbf{X}_L$ are called a negative pair of samples instead.

To learn semantically consistent representations, we employ a contrastive learning loss function that increases the similarity of positive pairs and decreases that of negative ones within each batch:

$$\mathcal{L}_{CL}\left(\mathbf{X}_L, \mathbf{X}_L^-, \mathbf{X}_L^+\right) = -\log \frac{\exp\left(\text{sim}\left(s, s^+\right)/\tau\right)}{\exp(\text{sim}\left(s, s^+\right)/\tau) + \sum_{s^- \neq s^+}^{2B} \exp(\text{sim}(s, s^-)/\tau)} \tag{6}$$

Here, $s, s^-, s^+$ are the results of the average pooling applied to $\mathbf{X}_L, \mathbf{X}_L^-, \mathbf{X}_L^+$, respectively, $\tau$ is the temperature parameter, $B$ denotes the batch size and $sim(\cdot)$ represents cosine similarity.

**Multi-task Learning Module.** This module adopts an MoE structure, consisting of expert clusters, gating networks, and task-specific tower networks, to produce task-specific predictions. In addition, this module dynamically reweights task losses to facilitate simultaneous convergence of loss functions.

We construct four expert clusters, denoted as $E4Apnea$, $E4Stage$, $E4SpO_2$, and $E4All$, where the first three clusters are dedicated to specific tasks and the last one is shared by all tasks. Each cluster contains $\phi$ experts. The $i$-th expert in a cluster is defined as:

$$Q_i(\mathbf{X}_L) = \mathbf{W}_{i_2} \cdot \text{Dropout}(\text{ReLU}(\mathbf{W}_{i_1}\mathbf{X}_L + b_{i_1})) + b_{i_2} \tag{7}$$

where $\mathbf{X}_L$ is the output of the sequential encoding module, $\mathbf{W}_{i_1} \in \mathbb{R}^{d_{eh} \times d}$ and $\mathbf{W}_{i_2} \in \mathbb{R}^{d_{eo} \times d_{eh}}$ are learnable weight matrices, $b_{i_1}$ and $b_{i_2}$ are bias terms.

Given the outputs of all experts in each expert cluster, task-specific gating networks are designed to selectively aggregate these expert outputs. Taking the sleep stage classification task as an example, the gating network first computes the normalized attention weights for relevant experts:

$$H_{\text{Stage}}(\mathbf{X}_L) = \text{Softmax}\left(\mathbf{W}_{H_{\text{Stage}}}\mathbf{X}_L + b_{H_{\text{Stage}}}\right) \tag{8}$$

where $\mathbf{W}_{H_{Stage}} \in \mathbb{R}^{(|E4Stage|+|E4All|) \times d}$ and $b_{H_{Stage}} \in \mathbb{R}^{|E4Stage|+|E4All|}$ are trainable parameters. Subsequently, the gating network produces the final task-specific representation by linearly combining the expert outputs according to these calculated weights:

$$\mathbf{Y}_{\text{Stage}} = \sum_{i \in E4Stage \cup E4All} [H_{\text{Stage}}(\mathbf{X}_L)]_i Q_i(\mathbf{X}_L) \tag{9}$$

Further, we design a tower network for each task, denoted as $T4Apnea$, $T4Stage$, and $T4SpO_2$. Each tower network adopts the same structure as the expert networks, consisting of two fully connected layers with intermediate ReLU activation and dropout regularization, but differs in the hidden and output dimensions. Tower networks output $\hat{y}_{\text{Apnea}}$, $\hat{y}_{\text{Stage}}$ and $\hat{y}_{\text{SpO}_2}$, where average pooling is specifically applied to generate the sleep epoch-level $\hat{y}_{\text{Stage}}$.

We employ cross-entropy loss for the classification tasks (a binary classification task for apnea detection and a ternary classification task for sleep staging), and the mean absolute error (MAE) loss for the regression task ($\text{SpO}_2$ estimation task).

$$\begin{aligned}
\mathcal{L}_{\text{Apnea}} &= \mathcal{L}_{\text{binary cross-entropy}}\left(y_{\text{Apnea}}, \hat{y}_{\text{Apnea}}\right), \\
\mathcal{L}_{\text{Stage}} &= \mathcal{L}_{\text{categorical cross-entropy}}\left(y_{\text{Stage}}, \hat{y}_{\text{Stage}}\right), \\
\mathcal{L}_{\text{SpO}_2} &= \mathcal{L}_{\text{MAE}}\left(y_{\text{SpO}_2}, \hat{y}_{\text{SpO}_2}\right),
\end{aligned} \tag{10}$$

To alleviate gradient conflicts in multi-task learning, we follow the idea of Nash-MTL [9] and obtain dynamic task weights as follows. First, let $\theta_{sh}$ denote the parameters of the shared networks (i.e., the sequential encoding module and shared experts), and compute the gradient of each task-specific loss w.r.t. $\theta_{sh}$, denoted as $g_k = \nabla_{\theta_{sh}} \mathcal{L}_k$ for $k \in \{\text{Apnea, Stage, SpO}_2\}$. These gradient vectors are then stacked column-wise to form the gradient matrix $G = [g_{\text{Apnea}},\ g_{\text{Stage}},\ g_{\text{SpO}_2}]$. Next, we treat each task as a player whose utility function is defined as $g_k^T \Delta\theta_{sh}$ and formulate the problem of finding the common descent direction as a bargaining game of maximizing the utilities of players. Then, from a Nash Bargaining Solution (NBS), we have that the parameter update direction is $\sum_k w_k g_k$, where $w = [w_k]$ is the solution to equation $G^\top G w = w'$, and $w'$ denotes the element-wise reciprocal of $w$. As we know, this equation has been addressed using the concave-convex procedure, which iteratively approximates it with convex subproblems that can be efficiently solved by standard convex optimization solvers such as ECOS [2]. Finally, $w$ is used as the optimized task weights to adaptively reweight the task losses:

$$\begin{aligned} \mathcal{L}_{MTL} = \{\mathcal{L}_{\text{Apnea}},\ \mathcal{L}_{\text{Stage}},\ \mathcal{L}_{\text{SpO}_2}\} \cdot w = \\ w_{\text{Apnea}}\mathcal{L}_{\text{Apnea}} + w_{\text{Stage}}\mathcal{L}_{\text{Stage}} + w_{\text{SpO}_2}\mathcal{L}_{\text{SpO}_2} \end{aligned} \tag{11}$$

Owing to the inherent invariance of NBS to affine transformations, the $w$ serves to regularize task gradients, regardless of the norms of these task gradients.

Finally, the final loss of our model can be represented as:

$$\mathcal{L} = \lambda_{CL}\mathcal{L}_{CL} + \mathcal{L}_{MTL} \tag{12}$$

where $\lambda_{CL}$ is the weight of the contrastive learning loss.

## 4 Experiments

### 4.1 Experimental Setup

**Implementation Details.** The system proposed in this paper is implemented in Python 3.8 with PyTorch 1.10. In the part of signal processing, the frame rate $f_t$ is set to 50 Hz, and the number of fast-time sampling points $N_\gamma$ is set to 99. The bed width $w_{bed}$ is set to 90cm. The periodicity threshold $\epsilon$ and the person presence threshold $\mu$ are set to 0.4 and 0.15, respectively. Two interpolation thresholds $\eta_1$ and $\eta_2$ are set to 12 and 80. In the sleep health prediction model, the number of windows contained in the sequences input into the model $n$ is set to 20. The dimension $d$ is 64. The number of self-attention blocks $L$ is 2 and the number of heads $h$ is 4. The temperature parameter $\tau$ is fixed to 1. Dropout rates are set to 0.1. $d_{eh}$ and $d_{eo}$, i.e., hidden and output dimensions of the expert network, are set to 64, while the hidden dimension of the tower network is set to 32. $\phi$ is 4 in the task-specific expert and 2 in shared expert cluster. $\lambda_{CL}$ is set to 0.1. Batch size $B$ is set to 32. Adam is adopted as the optimizer.

**Dataset.** The dataset used in this study was collected from hospital wards, where a UWB device was placed on the bedside table of each ward and oriented

**Table 1.** Performance comparison. The values in bold indicate the best ones of the corresponding columns.

| Task | Apnea | | Stage | | $SpO_2$ | |
|---|---|---|---|---|---|---|
| Metric | ACC | F1 | ACC | F1 | MAE | RMSE |
| Hypnos | 0.869 | 0.860 | **0.770** | **0.766** | N/A | N/A |
| RF-Oximeter | N/A | N/A | 0.684 | 0.618 | 1.45 | 1.61 |
| Zhang et al. | 0.795 | 0.796 | N/A | N/A | 1.46 | 1.61 |
| SHMIU | **0.880** | **0.868** | 0.763 | 0.753 | **1.42** | **1.58** |

towards the subject. It comprises 94,075 sleep epochs (~784 h) collected from 100 individuals. Notably, 28% of these subjects exhibit severe sleep apnea symptoms. Ground-truth annotations for sleep stages, apnea state, and $SpO_2$ values are obtained from synchronized PSG recordings. To ensure patient privacy and data security, all recordings are collected exclusively by certified clinicians without additional interventions. The data collection protocol has been reviewed and approved by the ethics committee of the hospital, and written informed consent has been obtained from all subjects before data acquisition.

The dataset is randomly partitioned into 10 subsets, each containing 10 subjects. Each time one subset is held out for testing, one of the remaining nine subsets is randomly selected for validation, and the other eight subsets are used for training. As a result, each subset serves once as the testing set. The final results are reported as the average performance across 10 folds.

**Metrics.** For the apnea detection and sleep stage classification tasks, accuracy (ACC) and F1-score (F1) are used as evaluation criteria. For the $SpO_2$ estimation task, we adopt MAE and root mean square error (RMSE) as regression metrics. We also employ an additional metric, mean rank (MR), to assess the overall performance of models using MTL. MR is defined as $\text{MR} = \frac{1}{M_t}\sum_{i=1}^{M_t} r_i$, where $M_t$ denotes the number of metrics for each model, $r_i$ represents the rank of the model on the $i$-th metric. A lower MR value indicates better overall performance across multiple tasks.

### 4.2 Performance Comparison

**Overall Performance.** We choose three contactless sensing methods that can monitor two sleep health indicators as competitors, i.e., Hypnos [5], RF-Oximeter [3], and the method of Zhang et al. [19]. We implement the models within these methods and train them using signals from our dataset but processed by the pipeline detailed in Sect. 3.2.

From the experimental results shown in Table 1, SHMIU outperforms all compared methods in both apnea detection and $SpO_2$ estimation tasks, and is slightly inferior to Hypnos in sleep stage classification. This is reasonable since Hypnos explicitly treats sleep staging as the primary task and leverages apnea

**Table 2.** Performance of variants with different components.

| Task | Apnea | | Stage | | $SpO_2$ | | |
|---|---|---|---|---|---|---|---|
| Metric | ACC | F1 | ACC | F1 | MAE | RMSE | MR |
| SHMIU | **0.880** | 0.868 | **0.763** | **0.753** | 1.42 | 1.58 | **1.50** |
| Variant A1 | 0.825 | 0.748 | 0.669 | 0.537 | 1.64 | 1.77 | 4.67 |
| Variant A2 | 0.873 | 0.858 | 0.721 | 0.699 | 1.64 | 1.82 | 3.67 |
| Variant A3 | 0.879 | **0.869** | 0.751 | 0.746 | 1.45 | 1.60 | 2.17 |
| Variant A4 | 0.874 | 0.857 | 0.718 | 0.687 | **1.39** | **1.55** | 2.83 |

**Table 3.** Performance of variants with different gradient update strategies.

| Task | Apnea | | Stage | | $SpO_2$ | | |
|---|---|---|---|---|---|---|---|
| Metric | ACC | F1 | ACC | F1 | MAE | RMSE | MR |
| SHMIU | **0.880** | **0.868** | **0.763** | 0.753 | **1.42** | **1.58** | **1.16** |
| Variant B1 (PCGrad) | 0.877 | 0.867 | 0.725 | 0.715 | 1.47 | 1.61 | 3.00 |
| Variant B2 (CAGrad) | 0.879 | 0.865 | 0.742 | 0.723 | 1.47 | 1.61 | 2.67 |
| Variant B3 (Aligned-MTL) | 0.877 | **0.868** | 0.760 | **0.756** | 1.50 | 1.64 | 2.50 |

detection as an auxiliary task to enhance its accuracy, whereas our system is designed to achieve balanced performance across all three tasks rather than optimizing one at the expense of others. In summary, across all metrics of the three tasks, SHMIU either outperforms the state-of-the-art methods or approaches their performance.

**Ablation Study.** To observe the impact of key components in SHMIU on performance, we construct four variants. Variant A1 replaces the embedding encoder in the sequential encoding module with an adaptive pool layer, variant A2 replaces the multi-head self-attention in the sequential encoding module with an identity mapping, variant A3 removes the contrastive learning module, and variant A4 replaces task reweighting in the MTL module with the LS method.

The results are shown in Table 2. As indicated by the MR column, all variants rank behind SHMIU, and differences in MR values between SHMIU and variants can reflect the relative importance of the corresponding components. Moreover, variants A1 and A2 perform worse than the original SHMIU across all metrics, and variant A4 exhibits a seesaw effect in performance, that is, it has minimal error in $SpO_2$ predictions but low accuracy in sleep stage classification. Its performance demonstrates the effectiveness of our dynamic reweighting strategy for multiple tasks.

### 4.3 MTL Strategy Analysis

**Comparison of Different Gradient Update Strategies.** We construct three variants (B1, B2, B3), where variant B1 replaces the task reweighting method in

**Table 4.** Performance of SHMIU with different configurations of experts. The first column shows the configuration of experts, where the first three numbers represent the number of task-specific experts for the three tasks, respectively, and the last number indicates the number of shared experts.

| Task | Apnea | | Stage | | $SpO_2$ | | |
|---|---|---|---|---|---|---|---|
| Metric | ACC | F1 | ACC | F1 | MAE | RMSE | MR |
| SHMIU (4-4-4-2) | **0.880** | 0.868 | **0.763** | **0.753** | **1.42** | 1.58 | **1.33** |
| Variant C1 (1-1-1-1) | 0.876 | 0.865 | 0.736 | 0.709 | **1.42** | **1.56** | 2.83 |
| Variant C2 (4-4-4-4) | 0.878 | **0.870** | 0.740 | 0.724 | 1.47 | 1.62 | 2.83 |
| Variant C3 (4-4-4-1) | **0.880** | 0.864 | 0.750 | 0.721 | 1.45 | 1.59 | 2.50 |

SHMIU with PCGrad [17], B2 with CAGrad [7], and B3 with Aligned-MTL [11], and perform an experiment to observe their impact on performance.

The experimental results are listed in Table 3, which in essence reflect the varying capabilities of different methods to mitigate the adverse impacts of gradient conflicts. Specifically, across our three tasks, PCGrad, CAGrad, and Aligned-MTL show a progressive upward trend in capability, while our system exhibits the strongest capability, as indicated by the MR column.

**Sensitivity Analysis of Different Expert Configurations.** We construct three variants (C1, C2, C3) with different total numbers of experts, as well as different numbers of shared experts and task-specific experts, and conduct an experiment to observe the impact of expert configurations on performance.

The results are listed in Table 4. It can be seen that SHMIU achieves the best overall performance (MR = 1.33), maintaining high accuracy in classification tasks and the lowest errors on the regression task. This confirms that an intermediate level of parameter sharing, i.e., combining sufficient task-specific capacity with limited but effective shared experts, strikes the best balance between specialization and generalization in the multi-task setting.

## 5 Conclusion

In this paper, we propose a contactless sleep health monitoring system, which can be deployed at home for monitoring sleep stages, sleep apnea state, and $SpO_2$ during sleep simultaneously. Through the adoption of expert collaboration and adaptive task reweighting strategies, our system mitigates inter-task conflicts and consequently enhances overall prediction performance. The experimental results show that our system can achieve balanced and excellent performance across these three tasks.

**Acknowledgements.** This work is supported by the National Natural Science Foundation of China (No. 62172394, No. 62072450) and the Basic Research Program of ISCAS (No. ISCAS-JCMS-202305).

## References


1. Cheraghinia, M., Shahid, A., et al.: A comprehensive overview on UWB radar: applications, standards, signal processing techniques, datasets, radio chips, trends and future research directions. IEEE Commun. Surv. Tutorials (2024)
2. Domahidi, A., Chu, E., Boyd, S.: ECOS: an SOCP solver for embedded systems. In: European Control Conference (2013)
3. He, H., Yuan, Y., Chen, Y.C., Cao, P., Katabi, D.: Contactless oxygen monitoring with radio waves and gated transformer. In: Machine Learning for Healthcare Conference, pp. 248–265. PMLR (2023)
4. Jacobs, R.A., Jordan, M.I., Nowlan, S.J., Hinton, G.E.: Adaptive mixtures of local experts. Neural Comput. **3**(1), 79–87 (1991)
5. Li, S., Jin, B., et al.: Hypnos: A contactless sleep stage monitoring system using UWB signals. Proc. ACM Inter., Mob., Wearable Ubiquit. Technol. **8**(3), 1–27 (2024)
6. Li, S., Jin, B., et al.: Leveraging attention-reinforced UWB signals to monitor respiration during sleep. ACM Trans. Sens. Netw. **20**(5), 1–28 (2024)
7. Liu, B., Liu, X., et al.: Conflict-averse gradient descent for multi-task learning. In: Advances in Neural Information Processing Systems, vol. 34, pp. 18878–18890 (2021)
8. Ma, J., Zhao, Z., et al.: Modeling task relationships in multi-task learning with multi-gate mixture-of-experts. In: Proceedings of the 24th ACM SIGKDD International Conference on Knowledge Discovery & Data Mining, pp. 1930–1939 (2018)
9. Navon, A., Shamsian, A.: Multi-task learning as a bargaining game. In: International Conference on Machine Learning, pp. 16428–16446. PMLR (2022)
10. Rundo, J.V., Downey III, R.: Polysomnography. Handb. Clin. Neurol. **160**, 381–392 (2019)
11. Senushkin, D., Patakin, N.: Independent component alignment for multi-task learning. In: Proceedings of the IEEE/CVF Conference on Computer Vision and Pattern Recognition, pp. 20083–20093 (2023)
12. Tang, H., Liu, J., Zhao, M., Gong, X.: Progressive layered extraction (PLE): a novel multi-task learning (MTL) model for personalized recommendations. In: Proceedings of the 14th ACM Conference on Recommender Systems, pp. 269–278 (2020)
13. Vaswani, A., Shazeer, N., Parmar, N., Uszkoreit, J., Jones, L.: Attention is all you need. In: Advances in Neural Information Processing Systems (2017)
14. Yim, D., Cho, S.H.: An equidistance multi-human detection algorithm based on noise level using mono-static IR-UWB radar system, pp. 131–134 (2015)
15. Yu, B., Wang, Y., et al.: WiFi-sleep: sleep stage monitoring using commodity Wi-Fi devices. IEEE Internet Things J. **8**(18), 13900–13913 (2021)
16. Yu, J., Dai, Y.: Unleashing the power of multi-task learning: a comprehensive survey spanning traditional, deep, and pretrained foundation model eras. arXiv preprint: arXiv:2404.18961 (2024)
17. Yu, T., Kumar, S., et al.: Gradient surgery for multi-task learning. In: Advances in Neural Information Processing Systems, vol. 33, pp. 5824–5836 (2020)
18. Zhang, F., Wu, C., et al.: SMARS: sleep monitoring via ambient radio signals. IEEE Trans. Mob. Comput. **20**(1), 217–231 (2019)
19. Zhang, H., Bo, S., et al.: Event-level identification of sleep apnea using FMCW radar. Bioengineering **12**(4), 399 (2025)
20. Zhang, Y., Yang, Q.: A survey on multi-task learning IEEE transactions on knowledge and data engineering. IEEE Trans. Knowl. Data Eng. (2021)

# Triangle Counting Under Edge Relationship Local Differential Privacy: The Case of Restricted Extended Local Views

Wenzheng Xia[1], Shuangqing Xu[1,2], Yifeng Zheng[2,3(✉)], Lei Xu[4], and Zhongyun Hua[1]

[1] Harbin Institute of Technology, Shenzhen, China
huazhongyun@hit.edu.cn
[2] The Hong Kong Polytechnic University, Hong Kong SAR, China
yifeng.zheng@polyu.edu.hk
[3] The Hong Kong Polytechnic University Shenzhen Research Institute, Shenzhen, China
[4] Nanjing University of Science and Technology, Nanjing, China

**Abstract.** Triangle counting is a fundamental primitive in graph analysis. However, in decentralized graph settings, directly aggregating users' local views would leak sensitive social connections. Ensuring edge privacy is challenging because users' local views are often correlated. Existing methods typically focus on the Extended Local View (ELV) model, which assumes that users fully disclose their neighbor lists to all their neighbors. However, in realistic social network applications where users may selectively disclose their connections, this full-visibility assumption breaks, rendering ELV-based approaches inadequate. In this paper, we explicitly capture such selective disclosure behavior and formalize it as the Restricted ELV (RELV) model. With this as a foundation, we propose SEPALS, a new framework for accurate triangle counting under RELV. In SEPALS, we develop a targeted neighbor-list collection strategy to recover unobservable structural information and propose a redundancy-aware weighting mechanism to unbiasedly aggregate contributions from incomplete local views. We formally prove that SEPALS satisfies the advanced notion of edge relationship local differential privacy. SEPALS significantly outperforms baselines that directly adapt existing ELV-based approaches for the RELV model in accuracy.

**Keywords:** Decentralized graph analysis · Edge relationship local differential privacy · Triangle counting

## 1 Introduction

Triangle counting is a fundamental task in graph analysis, as it enables analysts to compute important metrics like clustering coefficient [12] and serves as


R. C.-W. Wong et al. (Eds.): PAKDD 2026, LNAI 16597, pp. 480–492, 2026.
https://doi.org/10.1007/978-981-92-1300-9_38

a foundational primitive for many applications such as social network analysis [3,17,18] and community detection [14,20]. However, graphs usually contain rich information which could be sensitive, e.g., the private connections between different people/organizations. This raises critical privacy concerns, and makes it imperative to ensure graph data privacy when performing triangle counting. As a rigorous framework for privacy-preserving data analysis, differential privacy (DP) [2] is widely used to protect sensitive information in graph analysis. Initially, DP was studied in a centralized setting, where a trusted curator collects all user data, executes computations over it, and releases the results [5,16,24]. However, the assumption of a fully trusted curator is too strong in practice [9].

Local differential privacy (LDP) performs local perturbation and does not require a fully trusted curator [6]. It is much more suitable to support privacy-preserving data analysis, and a line of work has built on LDP to enable privacy-preserving analytics on decentralized graphs [8,9,11,17,22,23]. A central challenge in decentralized graph analysis is that users only locally maintain a limited *local view* of the global graph. A widely used abstraction for modeling users' partial knowledge is the Extended Local View (ELV) model [8,9,17], which assumes that each user can access the complete multi-hop subgraph revealed by her neighbors (with the two-hop ELV being the standard instantiation).

However, ELV assumes that every user fully discloses their neighbor list to all their neighbors, which fails to capture the *selective* privacy preferences observed in real-world social networks. For example, platforms such as Facebook allow users to hide or disclose their friend lists. Such selective behavior directly violates the full-disclosure assumption underlying ELV and fundamentally affects what triangles are observable across users. To rigorously model this realistic constraint, we introduce the Restricted Extended Local View (RELV) model, which provides the first formalization that captures users' selective disclosure of their connections in decentralized graph analysis. Unlike ELV, RELV explicitly models users' selective disclosure through customized per-user flags. Under RELV, whether a triangle is observable depends jointly on the disclosure choices of its three nodes. Counting triangles under RELV, however, presents challenges that do not arise in the idealized ELV setting. First, triangle visibility becomes inherently incomplete: some triangles appear in multiple users' RELVs, whereas others are entirely unobservable. Second, even for observable triangles, incomplete visibility results in inconsistent report redundancy, leading to unequal contributions to the global triangle count. These issues render existing ELV-based methods incapable of supporting accurate triangle counting under RELV.

In light of this, we propose SEPALS, a new framework designed for accurate and privacy-preserving triangle counting under RELV. We build SEPALS through a careful synergy of correlation estimation, targeted neighbor-list collection, and redundancy-aware weighting. To handle unobservable triangles, we introduce a targeted neighbor collection strategy that selectively queries noisy edges from non-disclosing users to reconstruct missing structural information. To correct the bias from incomplete visibility, we develop a redundancy-aware weighting mechanism that dynamically adjusts the contribution of each reported triangle based on its observation redundancy. Since edges may appear in multiple users' RELVs, the resulting local views are inherently correlated, violating the independence assumption required by standard LDP. Therefore, we adopt the recent advanced notion of Edge Relationship Local Differential Privacy (Edge-RLDP) [8,9] and ensure that SEPALS satisfies $(\epsilon, \delta)$-Edge-RLDP.

Our main contributions are summarized as follows: (1) For the first time, we formally introduce the *Restricted Extended Local View (RELV)* to model decentralized graphs where users selectively disclose their connections. (2) We design a redundancy-aware and targeted collection framework enabling accurate global triangle counting under $(\epsilon, \delta)$-Edge-RLDP. (3) Extensive experiments demonstrate that our method significantly improves accuracy over baselines that directly adapt state-of-the-art schemes in the ELV model [4,8].

## 2 Related Work

Differentially private graph analysis has been extensively studied in the centralized setting [1,13,24]. Prior work investigates private triangle counting under the central model of DP using techniques such as smooth sensitivity [13], empirical sensitivity [1], and ladder functions [24]. However, all these approaches rely on a trusted curator that centrally collects users' private graphs, which is challenging to satisfy in real-world applications. LDP has become an attractive alternative for privacy-preserving graph analysis [8,9,11,17,22,23]. For example, Ye et al. [22,23] propose LF-GDPR, a one-round protocol that estimates global statistics from noisy neighbor lists generated via Randomized Response (RR) [19]. More recently, Mendonça et al. [11] propose PEG, which releases edge-labeled graphs under LDP by leveraging partitioning and clustering to better allocate noise and preserve graph structure.

Among LDP-based approaches for graph analytics, several works [8,9,17] focus on private triangle counting. These works are built on the common assumption that each user can observe their two-hop ELV. Specifically, to mitigate the privacy risks arising from correlations in such extended views, Sun et al. [17] propose decentralized DP (DDP), which requires each user to protect not only her own information but also that of her neighbors. Liu et al. [8,9] later introduce Edge-RLDP, which provides formal guarantees even when correlated reports are jointly observed and thus addresses the limitations of DDP in handling multi-user dependency. However, these approaches all assume that each user can fully observe their two-hop ELV, implicitly requiring complete visibility of their neighbors' connections. In contrast, our work introduces a more realistic visibility model—the Restricted Extended Local View (RELV)—which captures selective connections disclosure. Building on Edge-RLDP, we develop algorithms that remain privacy-preserving while achieving accurate triangle counting under this more realistic constraint.

## 3 Preliminaries of Differentially Private Graph Analysis

Differential privacy (DP) [2] is a rigorous framework for protecting sensitive information in data analysis. Formally, DP is defined as follows.

**Definition 1 (Differential Privacy [2]).** *A randomized mechanism $\mathcal{M} : \mathcal{D} \to \mathcal{R}$ satisfies $(\epsilon, \delta)$-DP if, for any two neighboring datasets $\mathcal{D}$ and $\mathcal{D}'$ differing in a single record, and for any subset $\mathcal{S} \subseteq \mathcal{R}$:* $\Pr[\mathcal{M}(\mathcal{D}) \in \mathcal{S}] \leq e^{\epsilon} \cdot \Pr[\mathcal{M}(\mathcal{D}') \in \mathcal{S}] + \delta$.

Here, $\epsilon$ is the *privacy budget* that controls the level of privacy protection, with smaller values indicating stronger protection, while $\delta$ denotes failure probability. In centralized DP, a trusted curator directly collects users' raw data. In comparison, local differential privacy (LDP) [6] removes this assumption by requiring each user to randomize their data locally before reporting. A basic LDP mechanism is the Laplace mechanism [2], which achieves $(\epsilon, 0)$-DP by adding noise drawn from $\mathrm{Lap}(\Delta f/\epsilon)$, where $\Delta f$ is the $\ell_1$ sensitivity of $f$.

When applied to private graph analysis, DP can be defined under node-level LDP or edge-level LDP [23]. Following prior works [9,15], we focus on edge-level LDP (Edge-LDP) for graphs, which is defined as follows.

**Definition 2 (Edge Local Differential Privacy [15]).** *A mechanism $\mathcal{M} : \mathcal{G} \to \mathcal{R}$ satisfies $(\epsilon, \delta)$-Edge-LDP if, for any neighboring graphs $G_i, G_i'$ differing by one edge and any $S \subseteq \mathcal{R}$:* $\Pr[\mathcal{M}(G_i) \in S] \leq e^{\epsilon} \cdot \Pr[\mathcal{M}(G_i') \in S] + \delta$.

In this paper, we also use the recently advanced Relationship Local Differential Privacy (RLDP) [9], which extends LDP to handle the privacy leakage caused by highly correlated data. The Edge-RLDP is defined as:

**Definition 3 (Edge Relationship Local Differential Privacy [9]).** *Given a class of randomized mechanisms $\{\mathcal{M}_j \mid j \in [n]\}$, a mechanism $\mathcal{M}_i$ satisfies $(\epsilon, \delta)$-Edge-RLDP if, for any neighboring graphs $G_i$, $G_i' \in \mathcal{G}_i$ that differ by one edge, it holds:* $\Pr[\mathcal{M}_1(G_1) \in S_1, \ldots, \mathcal{M}_n(G_n) \in S_n] \leq e^{\epsilon} \cdot \Pr[\mathcal{M}_1(G_1') \in S_1, \ldots, \mathcal{M}_n(G_n') \in S_n] + \delta$.

## 4 Problem Statement

We consider private triangle counting in the emerging *decentralized* graph setting [9,10,17,22]. In this setting, each user is modeled as a node in an undirected graph, and edges represent connections between users (e.g., friendships). A data analyst, who has no access to the global graph $G$, aims to estimate the global triangle count of $G$ by collecting reports from users based on their local views (i.e., subgraphs of $G$). Formally, let $G = (V, E)$ be an undirected graph with $n = |V|$ nodes. Each node $v_i \in V$ corresponds to a user $u_i$, and each undirected edge $(v_i, v_j) \in E$ denotes a connection between $u_i$ and $u_j$.

Following prior works [8,9,17], we assume that each user can observe at most her two-hop neighborhood. But unlike the standard ELV model as assumed in prior work, we consider a more realistic model in which each user holds a

**Algorithm 1:** Our Privacy-Preserving Triangle Counting Framework

**Input**: Privacy budget $\epsilon$, failure probability $\delta$, budget split ratio $\beta$, sample rate $p$, RELVs $\mathcal{G} = \{G_1, \dots, G_n\}$

**Output**: Noisy global triangle count $c'$ satisfies $(\epsilon, \delta)$-Edge-RLDP

$\epsilon_1 \leftarrow \beta \cdot \epsilon$, $\epsilon_2 \leftarrow (1 - \beta) \cdot \epsilon$;

`/* Phase 1: Correlation Estimation */`

$k_s^* \leftarrow \textsc{GlobalCorrEst}(\epsilon_1, \delta, \mathcal{G})$;

`/* Phase 2: Flag-Aware Triangle Counting */`

**for** $v_i \in V \land \mathbb{I}_{v_i} = 0$ **do**

  $\mathcal{N}_i' \leftarrow \textsc{PerturbNeighbor}(\mathcal{N}_i, \epsilon_2/2, \mu)$;

  Upload $\mathcal{N}_i'$ to the data analyst;

$E' \leftarrow \{(v_j, v_k) \mid \mathcal{N}_j'(k) = 1, j < k\}$;

**for** $v_i \in V$ **do**

  $c'_{i,w,s} \leftarrow \textsc{LocalCount}(G_i, \epsilon_2, \mathbb{I}_{v_i}, E', p)$;

$c' \leftarrow \frac{1}{w \cdot p} \sum_{i=1}^{n} c'_{i,w,s}$;

**return** $c'$;

*restricted* ELV (RELV), denoted as $G_i = (V_i, E_i)$, where users may selectively disclose their connections. To capture this behavior, we associate each node $v_i$ with a binary disclosure flag $\mathbb{I}_{v_i} \in \{0, 1\}$. With $\mathbb{I}_{v_i} = 1$, user $u_i$ *discloses* her one-hop neighbor list to those neighbors who have direct connection to $u_i$ (i.e., to nodes in $\mathcal{N}_i$); when $\mathbb{I}_{v_i} = 0$, $u_i$ withholds her neighbor list. We formally define the two-hop RELV model as follows.

**Definition 4 (Two-Hop RELV).** *Given node $v \in V$, its two-hop RELV $G_v$ comprises: (1) the node $v$ and its one-hop neighbors: $\{v\} \cup \mathcal{N}_v$; (2) edges incident to $v$: $\{(v, u) \in E\}$; (3) $v$'s visible two-hop neighbors: $\{w \mid \exists u \in V, (u, v) \in E \land (u, w) \in E \land \mathbb{I}_u = 1\}$; (4) visible edges adjacent to $v$'s one-hop neighbors: $\{(u, w) \in E \mid u \in \mathcal{N}_v,\ \mathbb{I}_u = 1\}$.*

In this paper, we consider a semi-honest data analyst and assume that the analyst knows the set of users $V$ and their disclosure flags $\{\mathbb{I}_v \mid v \in V\}$, but not the edges among them.

## 5 Our Approach

### 5.1 Overview of SEPALS

SEPALS is aimed at enabling accurate estimation of the global triangle count under Edge-RLDP under the constraints of selective edge disclosure. The key challenge arises from the incomplete visibility caused by users' disclosure flags, which makes different nodes contribute unequally and introduces strong correlations across their reports. SEPALS addresses these issues through a combination of correlation estimation, targeted neighbor-list collection, and redundancy-aware weighting. At a high level, SEPALS involves two phases: Correlation Estimation and Flag-Aware Triangle Counting. In Phase 1, we first allocate a small portion of the privacy budget to estimate a global correlation metric $k_s^*$, following the edge-using-frequency formulation [8]. This metric quantifies how frequently an edge may appear across users' RELVs and plays a central role in calibrating noise in the second phase. Then, in Phase 2, we perform the estimation computation. To handle incomplete visibility, only nodes with $\mathbb{I}_v = 0$ contribute noisy neighbor lists, from which the analyst constructs an auxiliary noisy edge set $E'$. This auxiliary information compensates for the missing information caused by selective disclosure and reduces the variance of local estimates. Each node then computes a weighted and subsampled local triangle count using both its RELV and a subset of $E'$. The weighting corrects for inconsistent redundancy in RELVs, and the subsampling further improves the utility. Finally, each local count is privatized under Edge-RLDP and aggregated to produce an unbiased estimate of the global triangle count. The overall workflow is summarized in Algorithm 1.

### 5.2 Phase 1: Correlation Estimation

Recall from Definition 3 that the privacy loss of an edge $e_i$ is determined by all local reports whose outputs depend on whether $e_i$ is present. In the worst-case scenario, $e_i$ may affect the outputs of all $n$ users, which means that we should allocate at most $\epsilon/n$ privacy budget to each user to guarantee the overall $\epsilon$-Edge-RLDP. However, real-world graphs are typically sparse, and such worst cases rarely occur. Instead of splitting the privacy budget according to $n$, we follow [8] and allocate the budget according to the *maximum edge-using frequency* $k$, defined as the maximum number of users whose local views may simultaneously contain a common edge when counting triangles. Formally, for each node $v_i$, let $\Delta f_{c_i}$ denote the local sensitivity of its triangle count in $G_i$, and we define: $k = \max_{i=1}^{n}(\Delta f_{c_i} + 2)$. Intuitively, $k$ upper-bounds the correlation degree of any single edge across users. Therefore, if we divide the privacy budget according to $k$ instead of $n$, the per-user budget can be enlarged from $\epsilon/n$ to roughly $\epsilon/k$ ($k \ll n$), significantly reducing the amount of injected noise.

To privately estimate $k$ under $(\epsilon_1, \delta_1)$-Edge-RLDP, SEPALS uses a small fraction $\epsilon_1$ of the overall privacy budget $\epsilon$ to (i) privately identify users with high degrees and (ii) query their noisy local sensitivities. By combining these noisy statistics, it computes a differentially private estimate $k^*$ of the true maximum edge-using frequency $k$ (Details in Algorithm 2 in Appendix A of the full version [21]). Note that SEPALS performs triangle counting with subsampling in Phase 2 which reduces data correlations since the probability that a common edge appears in all $k^*$ RELVs becomes small. We therefore adjust $k^*$ to obtain a noisy upper bound after subsampling $k_s^* = \max\big((k^* - 2)p + 2 + r\sigma,\ 1\big)$, where $p$ is the sampling rate, $r$ is a positive constant, and $\sigma$ is the standard deviation of the normal distribution $N\big(p(k^* - 2),\ p(1 - p)(k^* - 2)\big)$. The correctness of this bound is guaranteed by Theorem 5 in [8]. The resulting $k_s^*$ is then used in Phase 2 to allocate the remaining privacy budget $\epsilon_2 = \epsilon - \epsilon_1$ to each user.

### 5.3 Phase 2: Flag-Aware Triangle Counting

Two-hop RELV provides each user with an incomplete local neighborhood: any node with disclosure flag $\mathbb{I}_v = 0$ hides its connection information from its one-hop neighbors. Consequently, some edges become invisible, and certain triangles may be unobservable in users' RELVs. This incomplete visibility makes triangle counting fundamentally more challenging under RELV.

A key insight is that the visibility of a triangle is determined solely by the disclosure flags of its three constituent nodes. Some triangles naturally appear in multiple users' RELVs, whereas others remain completely unobserved. We formalize this observability condition below.

**Proposition 1 (Observable Condition under RELV).** *Let $\langle u, v, w\rangle$ be a triangle in an undirected graph. In RELV setting, a triangle is observable in users' RELVs if and only if at least one of its three nodes has disclosure flag set to 1, i.e., $\mathbb{I}_u = 1$ or $\mathbb{I}_v = 1$ or $\mathbb{I}_w = 1$. Conversely, the triangle is unobservable in all users' RELVs if and only if all three nodes have disclosure flags equal to 0, i.e., $\mathbb{I}_u = \mathbb{I}_v = \mathbb{I}_w = 0$.*

Building on the above proposition, SEPALS adopts a flag-aware triangle counting strategy that consists of two components: (1) targeted neighbor-list collection, which reconstructs noisy edges that involve nodes with $\mathbb{I}_v = 0$, and (2) redundancy-aware weighting, which corrects the inconsistent reporting redundancy induced by incomplete visibility. Finally, we inject carefully calibrated Laplace noise into users' local reports to satisfy $(\epsilon_2, \delta_2)$-Edge-RLDP.

In Phase 2, the analyst first collects noisy one-hop neighbor lists from nodes with $\mathbb{I}_v = 0$, using the Asymmetric Randomized Response (ARR) mechanism [4]. ARR is parameterized by the privacy budget $\epsilon$ and a reporting probability $\mu \in [0, e^\epsilon/(e^\epsilon+1)]$, where $\mu$ regulates the trade-off between communication overhead and estimation accuracy. We denote this mechanism by $ARR_{\epsilon,\mu}$. Given a binary input $x \in \{0,1\}$ that indicates the presence or absence of an edge, the output $b \in \{0,1\}$ of $ARR_{\epsilon,\mu}(x)$ is drawn from the Bernoulli distribution $ARR_{\epsilon,\mu}(x) \sim \text{Bern}(\mu_x)$, where $\mu_x = \mu\,(\rho^{1-x})$ and $\rho = e^{-\epsilon}$. We omit the detailed algorithm of targeted neighbor-list collection which can be seen in Appendix A of the full version [21].

After collecting the targeted noisy neighbor lists, the analyst proceeds to construct auxiliary information that compensates for the missing edges needed to estimate triangles that are unobservable in any user's RELV. Guided by Proposition 1, we generate auxiliary edges for each node as follows. First, the analyst aggregates all noisy neighbor lists and discards edges incident to nodes with $\mathbb{I}_v = 1$. The remaining noisy edges between nodes with $\mathbb{I}_v = 0$ are retained to form an auxiliary edge set: $E' = \{(v_j, v_k) \mid \mathcal{N}'_k(j) = 1,\ j < k\}$. Next, for each node $v_i$, we construct a personalized subset $M_i \subseteq E'$ based on the 4-cycle trick [4]. Depending on the disclosure flag $\mathbb{I}_{v_i}$, we define

$$M_i = \begin{cases} \varnothing, & \text{if } \mathbb{I}_{v_i} = 1, \\ \{(v_j, v_k) \in E' \mid (v_i, v_k) \in E',\ j < k < i\}, & \text{if } \mathbb{I}_{v_i} = 0. \end{cases} \tag{1}$$

Because $M_i$ is derived solely from the perturbed one-hop neighbor lists and the public disclosure flags, its construction consumes no extra privacy budget.

Each node $v_i \in V$ then augments its RELV using $M_i$ to obtain $\widetilde{G}_i = G_i \cup M_i$, computes its local noisy triangle count $c'_{i,w,s}$, and contributes it to the aggregation of the global estimate $c'$. First, each node samples the triangles to be counted from all triangles involving itself in $\widetilde{G}_i$. Let $\mathcal{T}_i$ denote the set of all triangles involving $v_i$ in $\widetilde{G}_i$. Each triangle $\tau \in \mathcal{T}_i$ is independently included in the reporting set $\mathcal{T}_i'$ with probability $p$. Subsequent estimation is based solely on $\mathcal{T}_i'$, rather than on the entire set $\mathcal{T}_i$. Let $c_i = |\mathcal{T}_i|$ and $c_{i,s} = |\mathcal{T}_i'|$. The expected subsampled local count satisfies $\mathbb{E}[c_{i,s}] = \mathrm{p}c_i$, which implies that an unbiased estimate of $c_i$ can be obtained by scaling $c_{i,s}$ by $1/p$. Since subsampling reduces the number of triangles participating in the local counting process, it also decreases the local sensitivity $\Delta f_{c_i}$. Therefore, we re-estimate the local sensitivity after subsampling, denoted by $\Delta f_{c_{i,s}}$. Following the approach in [8,9], a noisy upper bound $\Delta f^*_{c_{i,s}}$ of $\Delta f_{c_{i,s}}$ is given by

$$\Delta f^*_{c_{i,s}} = \max\left(d_i^* + r\sqrt{k_s^* p(1-p)},\ 1\right), \tag{2}$$

where $p$ is the sampling rate, $r$ is a positive integer, and $d_i^*$ is the noisy upper bound of the node degree estimated in Phase 1. The correctness of this bound is formally established in [9].

Due to incomplete visibility, a triangle may not appear in the augmented RELVs of all three of its constituent nodes; its contribution to the global estimate can be counted more than once and is non-uniform. We refer to this multiplicity of contributions as the triangle's reporting redundancy, which is determined solely by the disclosure flags of its constituent nodes. Specifically, the reporting redundancy falls into the following three cases: (1) redundancy = 3, which occurs when *at least two* of the three nodes have $\mathbb{I}_v = 1$, in which case all three nodes observe and report the triangle; (2) redundancy = 2, which arises when exactly one node has $\mathbb{I}_v = 1$, so the triangle is reported only by the two nodes with $\mathbb{I}_v = 0$; and (3) redundancy = 1, which happens when all three nodes have $\mathbb{I}_v = 0$, in which case the triangle is assigned to a single node (the one with the largest index) via auxiliary information and is reported exactly once.

SEPALS applies a redundancy-aware weighting mechanism when estimating the local count, ensuring that each triangle has a uniform expected redundancy in the global aggregation. Let $c^j_{i,s}$ denote the count of triangles in user $u_i$'s reporting set $\mathcal{T}_i'$ with redundancy $j$. For $j = 2, 3$, $c^j_{i,s}$ is the number of triangles with redundancy $j$ in $\mathcal{T}_i'$. For $j = 1$, we have:

$$c^1_{i,s} = \begin{cases} 0, & \mathbb{I}_{v_i} = 1, \\ \dfrac{t_{i,s} - \mu^2 \rho s_{i,s}}{\mu^2(1-\rho)}, & \mathbb{I}_{v_i} = 0, \end{cases} \tag{3}$$

where the second case is an unbiased estimator [4] to correct the bias caused by the introduced noise. Here, $t_{i,s}$ denotes the number of sampled triangle candidates with redundancy 1 in $\mathcal{T}_i'$, and $s_{i,s}$ denotes the number of sampled wedges centered at $v_i$ (with both adjacent nodes having $\mathbb{I}_v = 0$), each retained independently with probability $p$. The corresponding weighting factor $\omega_j$ for each $j \in \{1, 2, 3\}$ is defined as $\omega_j = \omega/j$ where $\omega = \text{lcm}(1, 2, 3) = 6$ represents the uniform target redundancy that each triangle should contribute to the final estimate. Then the weighted local triangle count for user $u_i$ after subsampling is computed as $c_{i,w,s} = \omega_1 c^1_{i,s} + \omega_2 c^2_{i,s} + \omega_3 c^3_{i,s}$.

Since nodes with $\mathbb{I}_v = 0$ consume a portion of $\epsilon_2$ (i.e., $\epsilon_2/2$ in this work) to perturb their one-hop neighbor lists, whereas nodes with $\mathbb{I}_v = 1$ do not execute this step. Accordingly, nodes with $\mathbb{I}_v = 0$ are allocated a privacy budget $\epsilon'_{2,0}$, which takes into account privacy amplification by subsampling and applies the advanced composition theorem to enforce an $\epsilon_2/2$ privacy guarantee. Similarly, nodes with $\mathbb{I}_v = 1$ are allocated a privacy budget $\epsilon'_{2,1}$ to enforce an $\epsilon_2$ privacy guarantee. Because the weighting mechanism only rescales the contributions of reported triangles and does not affect which triangles are reported, the local sensitivity is affected solely by the subsampling. Following Eq. 2, the noisy upper bound of local sensitivity of $c_{i,w,s}$ is given by $\Delta f^*_{c_{i,w,s}} = \max\left(d^*_i + r\sqrt{k^*_s p(1-p)},\, 1\right)$. To preserve the privacy budget, we reuse the $d^*_i$ values estimated in Phase 1 rather than re-estimating them here.

Given the allocated privacy budgets $\epsilon'_{2,0}$ and $\epsilon'_{2,1}$, each node $v_i$ adds Laplace noise to $c_{i,w,s}$ as follows:

$$c'_{i,w,s} = \begin{cases} c_{i,w,s} + \text{Lap}(\frac{\Delta f^*_{c_{i,w,s}}}{\epsilon'_{2,0}}), & \mathbb{I}_{v_i} = 0, \\ c_{i,w,s} + \text{Lap}(\frac{\Delta f^*_{c_{i,w,s}}}{\epsilon'_{2,1}}), & \mathbb{I}_{v_i} = 1. \end{cases} \tag{4}$$

Finally, the analyst estimates the global triangle count $c$ by $c' = \frac{1}{\omega \cdot p} \sum_{i=1}^{n} c'_{i,w,s}$.

### 5.4 Privacy Analysis of SEPALS

**Theorem 1.** *SEPALS's approach for privacy-preserving triangle counting (as shown in Algorithm 1) satisfies $(\epsilon, \delta)$-Edge-RLDP, where $\epsilon = \epsilon_1 + \epsilon_2$ and $\delta = \delta_1 + 2\delta_2 + \delta_3$.*

We provide proof for Theorem 1 in Appendix B of the full version [21].

## 6 Experiments

### 6.1 Experimental Setup

We develop a prototype implementation of SEPALS using Python 3.9. All experiments are conducted on a workstation with Intel(R) Xeon(R) Gold 6240R CPU @ 2.40 GHz and 128 GB memory, and all reported results are averaged over 300 independent runs to ensure statistical robustness. Given the overall privacy budget $\epsilon$, we allocate $\epsilon_1 = 0.2\epsilon$ to Phase 1 and $\epsilon_2 = 0.8\epsilon$ to Phase 2. In addition, we simulate user disclosure willingness by independently assigning each node $v$

with a binary flag $\mathbb{I}_v \in \{0, 1\}$ sampled from $\text{Bern}(p_s)$, where $p_s = 0.6$. Additional parameter settings follow prior work and are provided in Appendix C of the full version [21].

**Datasets.** We use three real-world graph datasets from the *Stanford Large Network Dataset Collection (SNAP)* [7]: wiki-Vote, Facebook, and cit-HepTh. All graphs are converted into undirected versions. Detailed graph statistics are provided in the Appendix C of the full version [21].

**Baseline Methods.** To evaluate the effectiveness of our approach, we compare it against two state-of-the-art two-phase methods: ARROneNS [4] and PRIVET[1] [8]. We also consider two variants of SEPALS, denoted as SEPALS-A and SEPALS-B. SEPALS-A adopts a higher value of $\mu$ to match the communication overhead of ARROneNS while SEPALS-B disables the redundancy-aware weighting mechanism. All methods follow their original configurations and recommended settings.

**Metrics.** We adopt the Mean Relative Error (MRE) to evaluate the accuracy of global triangle count estimation. Let $\hat{f}_\triangle(G) \in \mathbb{R}$ denote the estimated number of global triangles, and let $f_\triangle(G) \in \mathbb{R}$ denote the ground truth. For any $f_\triangle(G) \neq 0$, the MRE is defined as $\text{MRE} = \frac{|\hat{f}_\triangle(G) - f_\triangle(G)|}{f_\triangle(G)}$. A smaller MRE indicates a higher estimation accuracy.

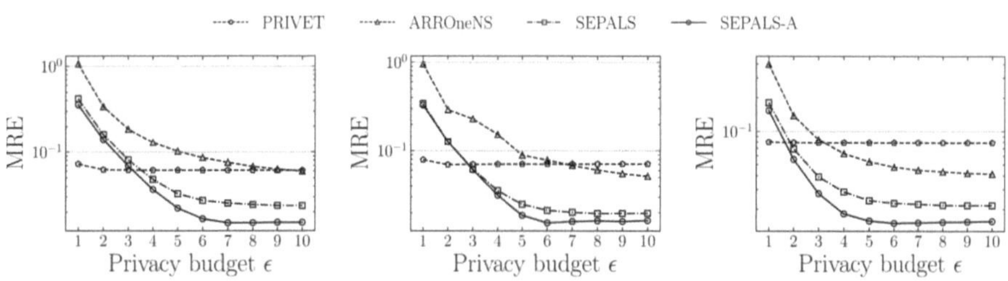


(a) wiki-Vote (b) cit-HepTh (c) Facebook

**Fig. 1.** MRE of global triangle counts with $\epsilon$ varying from 1 to 10 ($p_s = 0.6$).

## 6.2 Experimental Results

Our evaluation consists of two parts. First, we compare SEPALS and SEPALS-A against two baselines for a comprehensive performance assessment. Second, we perform an ablation study by evaluating SEPALS and SEPALS-A against SEPALS-B.

**Triangle Counts.** Figure 1 compares SEPALS, SEPALS-A, and the baselines in terms of the MRE of global triangle counting. Both SEPALS and SEPALS-A consistently outperform ARROneNS and achieve lower errors than PRIVET

[1] We apply the redundancy-aware weighting mechanism to ensure a fair comparison.

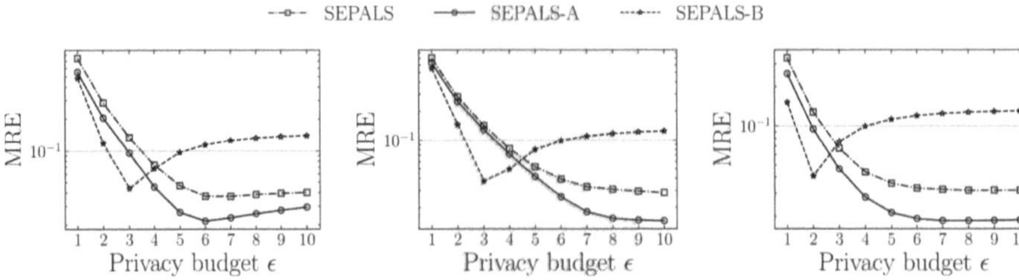


(a) wiki-Vote (b) cit-HepTh (c) Facebook

**Fig. 2.** MRE of global triangle counts with $\epsilon$ varying from 1 to 10 ($p_s = 0.5$).

when the privacy budget is moderate to large. Under small privacy budgets ($\epsilon \leq 3$), the noise required to infer unobservable triangles overwhelms the accuracy gains, making PRIVET, which ignores such triangles in the RELV setting, relatively competitive. As $\epsilon$ increases, our methods benefit from improved reconstruction accuracy, resulting in a clear reduction in MRE, whereas PRIVET remains largely insensitive to $\epsilon$.

**Effectiveness of the Weighting Mechanism.** Figure 2 evaluates the weighting mechanism in SEPALS. SEPALS and SEPALS-A outperform SEPALS-B once $\epsilon$ is sufficiently large ($\epsilon \geq 3$ on Facebook and $\epsilon \geq 5$ on wiki-Vote and cit-HepTh). Under small $\epsilon$, heavy perturbation introduces bias in estimating unobservable triangles, partially offsetting the missing-weight effect in SEPALS-B. However, as $\epsilon$ increases, this bias cancellation disappears, leading to unstable and significantly worse performance for SEPALS-B, while SEPALS and SEPALS-A continue to improve steadily.

## 7 Conclusion

This paper presents SEPALS, a framework for accurate triangle counting under the newly formulated RELV model, where users might selectively disclose their connections. SEPALS is built through a delicate synergy of correlation estimation, targeted neighbor-list collection, and redundancy-aware weighting. We formally prove that SEPALS satisfies $(\epsilon, \delta)$-Edge-RLDP. Extensive experiments show that SEPALS achieves substantially higher accuracy than baselines that directly adapt existing ELV-based schemes on real-world datasets while ensuring strong privacy protection.

**Acknowledgments.** This work was supported in part by National Cryptologic Science Fund of China (Grant 2025NCSF02033), by National Natural Science Foundation of China (Grants 62572150 and 62572243), by Shenzhen Science and Technology Program (Grant JCYJ20230807094411024), by Guangdong Basic and Applied Basic Research Foundation (Grant 2024A1515012299), and by Youth Science and Technology Talents Lifting Project of Jiangsu Association of Science and Technology JSTJ-2024-163.

## References


1. Chen, S., Zhou, S.: Recursive mechanism: towards node differential privacy and unrestricted joins. In: Proc. of SIGMOD (2013)
2. Dwork, C., Roth, A.: The algorithmic foundations of differential privacy. Found. Trends Theor. Comput. Sci. **9**(3–4), 211–407 (2014)
3. Gehrke, J., Lui, E., Pass, R.: Towards privacy for social networks: a zero-knowledge based definition of privacy. In: Proc. of TCC (2011)
4. Imola, J., Murakami, T., Chaudhuri, K.: Communication-efficient triangle counting under local differential privacy. In: Proc. of USENIX Security (2022)
5. Karwa, V., Raskhodnikova, S., Smith, A.D., Yaroslavtsev, G.: Private analysis of graph structure. Proc. VLDB Endow. **4**(11), 1146–1157 (2011)
6. Kasiviswanathan, S.P., Lee, H.K., Nissim, K., Raskhodnikova, S., Smith, A.D.: What can we learn privately? SIAM J. Comput. **40**(3), 793–826 (2011)
7. Leskovec, J., Krevl, A.: SNAP datasets: Stanford large network dataset collection. http://snap.stanford.edu/data
8. Liu, Y., Wang, T., Liu, Y., Chen, H., Li, C.: Edge-protected triangle count estimation under relationship local differential privacy. IEEE Trans. Knowl. Data Eng. **36**(10), 5138–5152 (2024)
9. Liu, Y., Zhao, S., Liu, Y., Zhao, D., Chen, H., Li, C.: Collecting triangle counts with edge relationship local differential privacy. In: Proc. of ICDE (2022)
10. Mar, J., Liu, J., Tang, Y., Chen, W., Sun, T.: Evaluating and analyzing reliability over decentralized and complex networks. In: Proc. of PAKDD (2018)
11. Mendonça, A.L.C., Brito, F.T., Machado, J.C.: PEG: local differential privacy for edge-labeled graphs. In: Proc. of EDBT (2025)
12. Newman, M.E.: Random graphs with clustering. Phys. Rev. Lett. **103**(5), 058701 (2009)
13. Nissim, K., Raskhodnikova, S., Smith, A.D.: Smooth sensitivity and sampling in private data analysis. In: Proc. of STOC (2007)
14. Pan, R., Wang, Y., Sun, J., Liu, H., Zhao, Y., Xia, J., Chen, W.: Simplifying social networks via triangle-based cohesive subgraphs. Vis. Informatics **7**(4), 84–94 (2023)
15. Qin, Z., Yu, T., Yang, Y., Khalil, I., Xiao, X., Ren, K.: Generating synthetic decentralized social graphs with local differential privacy. In: Proc. of CCS (2017)
16. Roth, E., Newatia, K., Ma, Y., Zhong, K., Angel, S., Haeberlen, A.: Mycelium: large-scale distributed graph queries with differential privacy. In: Proc. of SOSP (2021)
17. Sun, H., et al.: Analyzing subgraph statistics from extended local views with decentralized differential privacy. In: Proc. of CCS (2019)
18. Task, C., Clifton, C.: A guide to differential privacy theory in social network analysis. In: Proc. of ASONAM (2012)
19. Warner, S.L.: Randomized response: a survey technique for eliminating evasive answer bias. J. Am. Stat. Assoc. **60**(309), 63–69 (1965)
20. Wu, S., Wu, D., Cheuk, S., Chan, T.N., Lu, K.: GREAT: generalized reservoir sampling based triangle counting estimation over streaming graphs. Proc. VLDB Endow. **18**(7), 2031–2043 (2025)
21. Xia, W., Xu, S., Zheng, Y., Xu, L., Hua, Z.: Triangle counting under edge relationship local differential privacy: the case of restricted extended local views. CoRR (2026)
22. Ye, Q., Hu, H., Au, M.H., Meng, X., Xiao, X.: Towards locally differentially private generic graph metric estimation. In: Proc. of ICDE (2020)

23. Ye, Q., Hu, H., Au, M.H., Meng, X., Xiao, X.: LF-GDPR: a framework for estimating graph metrics with local differential privacy. IEEE Trans. Knowl. Data Eng. **34**(10), 4905–4920 (2022)
24. Zhang, J., Cormode, G., Procopiuc, C.M., Srivastava, D., Xiao, X.: Private release of graph statistics using ladder functions. In: Proc. of SIGMOD (2015)

# Leveraging Variational Information Bottleneck for Fine-Grained Urban Traffic Flow Inference

Qiang Ai[1,2] and Lixin Dong[1,2](✉)

[1] College of Computer, Qinghai Normal University, Xining 810008, China
202433331044@stu.qhnu.edu.cn, 2013040@qhnu.edu.cn

[2] The State Key Laboratory of Tibetan Intelligence, Xining 810008, China

**Abstract.** Accurately inferring fine-grained traffic flow from coarse-grained observations can significantly reduce traffic monitoring costs while enhancing urban traffic analysis and decision-making. Existing methods typically rely on direct deep-network fitting, which struggles to effectively handle noise and redundancy in multimodal inputs. To address this issue, we propose a novel multimodal fine-grained traffic flow inference framework based on the Variational Information Bottleneck (VIBFI). By imposing information-theoretic constraints, VIBFI explicitly compresses redundant content while preserving prediction-relevant factors, enabling the learning of more robust spatial distributions of fine-grained traffic flow. Specifically, we first construct multimodal representations encompassing road structure, external factors, and traffic flow; then model the spatial influence of temporal and meteorological conditions through discrete embeddings and grid-aligned mappings; and finally capture local flow patterns via residual convolutions. Importantly, we introduce the Variational Information Bottleneck (VIB) into each modality as an information regularizer to suppress modality-specific but prediction-irrelevant redundancy, thereby strengthening the most critical representation factors for fine-grained inference. After VIB compression, the fused multimodal features are processed through short-range convolutions to encode local dynamics and a Transformer to capture global long-range dependencies, producing structurally consistent and detail-preserving fine-grained flow maps. Extensive experiments on multiple real-world datasets demonstrate that VIBFI consistently outperforms state-of-the-art methods across diverse urban scenarios.

**Keywords:** Variational information bottleneck · Fine-grained urban flow inference · Multi-modal learning · Road network · Information theoretic

## 1 Introduction

Rapid urbanization and the surge in vehicle ownership have increased the demand for fine-grained traffic flow monitoring in intelligent transportation systems (ITS) [1,2]. Although dense sensor networks can provide detailed inflow,


R. C.-W. Wong et al. (Eds.): PAKDD 2026, LNAI 16597, pp. 493–505, 2026.
https://doi.org/10.1007/978-981-92-1300-9_39

outflow, and link-level information essential for congestion warning and traffic assessment, their large-scale deployment and long-term maintenance incur substantial financial and operational costs [3,4]. As ITS move toward fine-grained governance, fine-grained traffic flow inference has emerged as a practical alternative, aiming to recover high-resolution traffic flows from coarse-grained observations [5,6]. This approach reduces sensor density while supporting tasks such as congestion management, spatiotemporal forecasting, and traffic simulation. Nevertheless, the problem remains challenging due to strong spatial heterogeneity, high local variation, and complex cross-region dependencies characteristic of urban traffic systems.

Modern urban traffic systems typically involve multi-modal inputs, including road topology, external factors such as weather and temporal patterns, and traffic observations at different spatial resolutions [7,8]. These modalities are strongly coupled while containing substantial noise, bias, and modality-specific redundant features. Identifying the key factors truly relevant to fine-grained urban traffic flow inference while suppressing intra- and inter-modality redundancy remains a core challenge.

Recent advances in fine-grained urban flow inference have explored multiple directions to improve reconstruction quality under limited sensing resources. A prominent line of work incorporates structural priors from road networks. For example, Liu et al. [9] propose RATFM, which explicitly integrates road-topology knowledge. By extracting semantic road features via multidirectional 1-D convolutions and using these features as Transformer queries, RATFM effectively models both local road-relative patterns and long-range spatial dependencies, enabling structure-consistent fine-grained flow recovery. Another thread focuses on spatial heterogeneity. Zhang et al. [10] show that complex urban regions tend to accumulate larger reconstruction errors and design FocusNet to emphasize these "hard" areas. Through an adaptive masking strategy and an encoder–dual-decoder refinement, their framework improves inference accuracy in spatially diverse environments. External influencing factors (e.g., weather, temporal periodicity) also play a critical role. Pan et al. [11] propose MS-SCL, which performs multi-scale global modeling and adopts a multi-view mechanism so that each external factor is mapped to its dedicated feature representation. This design better captures nonlinear impacts of heterogeneous external conditions. Cross-city generalization has become another key focus. Yang et al. [12] introduce STDA to alleviate spatio-temporal distribution deviations and structural inconsistencies across cities, while Zheng et al. [13] develop AdaTM, which extracts spatial-temporal features using a Transformer-based encoder and adaptively fuses city-invariant and city-specific knowledge, yielding significant gains for data-scarce cities. In parallel, Abideen et al. [14] enhance fine-grained flow reconstruction for travel-time estimation via MSHM, which extracts multi-scale road semantics and leverages EDSR-based super-resolution to recover detailed spatial patterns, thereby supporting more accurate downstream predictions.

However, a common characteristic across existing methods is their reliance on fully end-to-end deep networks, which expand model capacity or architectural complexity to let the network automatically explore useful features in high-dimensional multi-modal inputs. While effective with abundant training data, these approaches do not explicitly optimize or constrain the intermediate latent variables. Consequently: (1) noise, bias, and redundancy across modalities are easily absorbed by the latent space; (2) modality-specific but task-irrelevant factors may be amplified; (3) cross-modality interactions lack explicit mechanisms for information selection and compression. As a result, current deep models often exhibit limited generalization and suboptimal stability under strong noise, high redundancy, or severe cross-region discrepancies.

Motivated by these challenges, we propose VIBFI, a **V**ariational **I**nformation **B**ottleneck-driven framework for **F**ine-grained urban traffic flow **I**nference. Unlike prior work that primarily improves expressive capacity through architectural design, VIBFI introduces an information-theoretic compression–selection mechanism on latent variables. Specifically, we apply a variational information bottleneck to each modality—road structure, external factors, and traffic flow—enabling the latent space to retain information relevant to the prediction target while implicitly filtering unstable or redundant components within each modality. The compressed latent features are subsequently fused and modeled through residual convolutions and Transformers [15] to capture both local detail patterns and global long-range dependencies, generating structurally consistent and detail-rich fine-grained traffic flow maps. Extensive experiments demonstrate that VIBFI substantially outperforms existing methods across multiple cities and under various noise conditions, particularly in complex, highly heterogeneous, or cross-region inference scenarios, achieving superior robustness and generalization.

## 2 Methodology

The proposed framework VIBFI is illustrated in Fig. 1. Unlike existing approaches that rely primarily on direct deep network fitting, our method is grounded in information theory. The goal is to obtain a latent representation $Z$ that is maximally predictive of the fine-grained target $Y$ while being minimally dependent on the coarse-grained input $X$. This compression–preservation principle allows the model to effectively suppress noise, reduce redundancy across modalities, and enhance robustness under complex urban traffic dynamics.

Specifically, the VIBFI mainly consists of five core components. (1) Road network representation: this module encodes the binary road topology map to extract geometric structure and connectivity patterns of the urban road network. (2) External factor representation: temporal and meteorological conditions are modeled through discrete embeddings and further mapped into both grid-aligned feature maps and low-dimensional vectors to capture their spatial and contextual influences. (3) Flow pattern representation: a short-range convolutional pathway is employed to extract localized spatial patterns from the coarse-grained traffic flow map. (4) Multi-Modal information regularizer with VIB: as the central mechanism of the framework, VIBFI incorporates a VIB into each modality to regulate the statistical structure of its latent representation. Concretely, the VIB mechanism simultaneously enhances the mutual information between

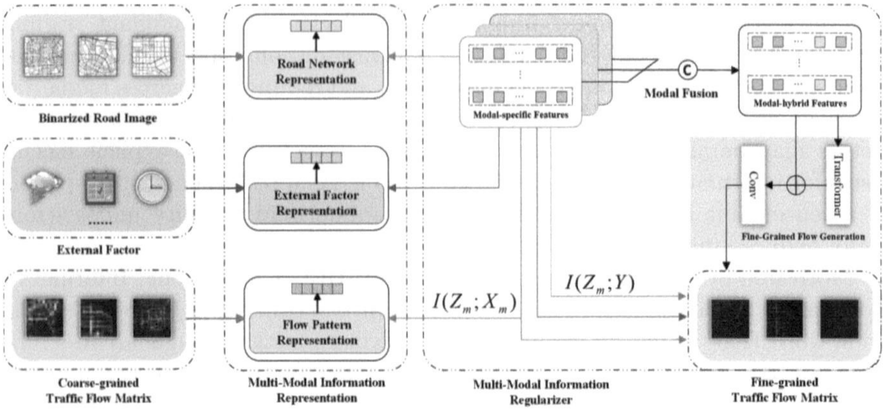


**Fig. 1.** Structure of VIBFI.

the latent variables and the prediction target, while suppressing their mutual information with the raw inputs. This enables the model to filter out modality-specific redundancies irrelevant to fine-grained traffic inference and selectively retain the most informative factors for prediction. The resulting latent variables become compact yet highly discriminative, providing a robust and generalizable representational basis for subsequent fine-grained flow generation. (5) Fine-grained flow generation: this final module integrates short-range local features with long-range dependencies. The global structural information captured by the Transformer is spatially aligned and fused with the locally extracted convolutional patterns; the fused representation is then transformed through several convolutional mappings to produce the fine-grained traffic flow map. This design simultaneously preserves local detail and global consistency, yielding complete and high-accuracy fine-grained traffic flow predictions.

Before elaborating on the architecture, we first define the task of fine-grained traffic flow inference. Following prior studies [9–14], we partition the city into a regular grid map based on geographic coordinates, instead of adopting graph-structured partitioning, to ensure spatial alignment with the Euclidean road network. A coarse-grained traffic map at time $t$ is expressed as

$$X_t = [x_t^{(1,1)}, \dots, x_t^{(i,j)}, \dots, x_t^{(I_c,J_c)}] \in \mathbb{R}^{I_c \times J_c \times S}, \tag{1}$$

where each $x_t^{(i,j)} \in \mathbb{R}^S$ contains $S$ traffic states, the symbol $S$ representing the number of traffic flow type. To achieve fine-grained inference, each coarse grid is subdivided into $N \times N$ cells, yielding

$$Y_t \in \mathbb{R}^{I_f \times J_f \times S}, \quad I_f = NI_c,\ J_f = NJ_c. \tag{2}$$

Within this problem setting, we incorporate VIB into the fusion of traffic flow features, road network structures, and external factors. Each modality is encoded by a variational encoder to produce information-constrained latent variables.

The VIB objective guides the encoders to discard irrelevant information and preserve only the components that contribute to predicting fine-grained traffic flow. The short-range convolutional module captures localized spatial patterns, while the Transformer module models global long-range interactions. Finally, the reconstructed latent representations from all modalities are fused to generate high-resolution traffic flow maps.

In summary, the central innovation of this work is introducing the VIB into fine-grained urban traffic flow inference. This enables principled information compression, improved generalization, robustness to noise, and more effective multi-modal fusion under complex urban environments.

### 2.1 Road Network Representation

This module encodes the binary road-topology map $M \in \{0,1\}^{H \times W}$ into a compact structural prior that preserves both geometric shapes and connectivity patterns of urban roads. Considering the extreme sparsity and elongated linear structures in road networks, we jointly leverage the local spatial modeling of a 2D convolutional encoder $\mathcal{E}_{\mathrm{2D}}(\cdot)$ to obtain low-level features $F'_m = \mathcal{E}_{\mathrm{2D}}(M)$, and a 1D structural enhancement module $\mathcal{E}_{\mathrm{1D}}(\cdot)$ that refines these features as $F_m = \mathcal{E}_{\mathrm{1D}}(F'_m)$, thereby aggregating directional cues along road-aligned patterns and improving sensitivity to long, narrow, and potentially fragmented segments.

To alleviate attention dilution caused by large homogeneous backgrounds, we introduce a sparsity-aware spatial attention mechanism. It takes both $F_m$ and a downsampled mask $\tilde{M}$ as inputs and generates an attention map $A = \sigma(\phi([F_m; \tilde{M}])) \in [0,1]^{1 \times H \times W}$, where $[\cdot;\cdot]$ denotes channel-wise concatenation and $\phi(\cdot)$ is a lightweight convolutional projector. Explicitly injecting $\tilde{M}$ constrains attention to road-relevant regions, reduces noise from background pixels, and helps preserve global topological continuity.

The structural features are further refined via a residual attention formulation $\mathbf{X}_R = F_m \odot (1 + A) + F_m$, which enhances road-aware responses without destabilizing the base representation. Finally, $\mathbf{X}_R$ is projected to a unified channel dimension $C_r$ and spatially aligned with the backbone, yielding $\mathbf{X}_R \in \mathbb{R}^{C_r \times H \times W}$ as a robust topology-preserving prior. This representation is subsequently fed into the VIB module, which selectively compresses redundant spatial details while retaining critical connectivity information essential for fine-grained traffic flow inference.

### 2.2 External Factor Representation

To model the influence of temporal and meteorological conditions on traffic dynamics, VIBFI incorporates an external factor representation module that encodes both discrete and numerical contextual variables. Discrete components such as day-of-week, time-of-day, and weather category are embedded into learnable vectors $e_{day}$, $e_{time}$, $e_{weather}$, while numerical indicators (e.g., holidays or special events) are concatenated with these embeddings to form a unified external vector $v_{ext} = [e_{day}; e_{time}; e_{weather}; \ldots] \in \mathbb{R}^{d_{ext}}$, which compactly summarizes temporal and environmental context at each time step.

To make these global contextual cues compatible with spatial processing, the external vector is projected through a lightweight transformation $\mathcal{P}(\cdot)$ to generate a grid-aligned map $\mathbf{X}_E = \mathcal{P}(v_{ext})$ whose spatial resolution matches that of the coarse-grained traffic tensor $X$. This spatialized representation is then fused with traffic features via channel-wise concatenation, i.e., $[X; \mathbf{X}_E]$, enabling convolutional layers to directly capture how external conditions shape localized traffic patterns.

This dual representation strategy allows the model to simultaneously exploit global context (via the compact vector) and localized interactions (via the spatial projection). The vector form can be efficiently encoded in the VIB latent space to regulate contextual dependencies, while the grid-aligned features expose external influences to spatial convolution, making the overall architecture more sensitive to temporal rhythms, weather variations, and event-driven anomalies in fine-grained traffic flow inference.

### 2.3 Flow Pattern Representation

The flow pattern representation module focuses on extracting localized traffic dynamics from the coarse-grained flow grid and its associated external features. Given the integrated input tensor $X$, a shallow convolutional layer is first applied to obtain the base feature map $F_{\text{conv}} = \text{Conv}(X) \in \mathbb{R}^{C \times H \times W}$, which captures short-range spatial correlations and fundamental flow variations across the grid.

To enrich these local representations, $F_{\text{conv}}$ is further processed by a sequence of residual blocks, formulated as $\mathbf{X}_F = \text{ResBlocks}(F_{\text{conv}})$. The residual structure preserves essential low-level information while enabling the network to model more complex interactions, such as localized congestion transitions and short-term flow fluctuations.

Through this combination of convolutional encoding and residual refinement, the module produces a stable and expressive feature map $\mathbf{X}_F$ that serves as the core modality describing fine-scale traffic dynamics. This representation is subsequently fed into the VIB module and the multimodal fusion stage, ensuring that local traffic patterns are accurately preserved and effectively integrated into the overall fine-grained inference process.

### 2.4 Multi-modal Information Regularizer

To refine the modality-specific representations for fine-grained traffic inference, VIBFI incorporates a VIB regularizer. For each modality $m$, we denote the input feature as $\mathbf{X}_m$ and the learned latent variable after VIB as $\mathbf{Z}_m$. Our goal is to encourage $\mathbf{Z}_m$ to retain predictive information with respect to the traffic target $Y$, while suppressing redundant information from the raw input $\mathbf{X}_m$. Concretely, we define the VIB regularizer for modality $m$ as:

$$\mathcal{L}_m = \beta_m I(\mathbf{Z}_m; \mathbf{X}_m) - I(\mathbf{Z}_m; Y), \tag{3}$$

where $B$ denotes the batch size, and $\beta_m$ is a scaling factor controlling the trade-off between minimality and predictiveness. The first term penalizes information in $\mathbf{Z}_m$ that is irrelevant to prediction (minimality), while the second term encourages $\mathbf{Z}_m$ to retain information relevant to $Y$ (predictive alignment). In this way, redundant modality-specific factors are suppressed, whereas the most informative components for fine-grained traffic inference are emphasized.

The overall modality-specific VIB regularization is given by:

$$\mathcal{L}_{\text{spec}} = \sum_{m \in \text{M}} \mathcal{L}_m, \quad \text{M} = \{\text{flow, ext, road}\}. \tag{4}$$

After VIB processing, each modality's latent variable $\mathbf{Z}_m$ is projected back to a spatial feature map $\mathbf{Z}_m^{VIB}$ and then fused with other modalities. Following the concatenation of all modality-specific VIB-enhanced features:

$$\mathbf{Z}_{\text{fused}} = \text{Concat}(\mathbf{Z}_{\text{flow}}^{\text{VIB}}, \mathbf{Z}_{\text{road}}^{\text{VIB}}, \mathbf{Z}_{\text{ext}}^{\text{VIB}}), \tag{5}$$

the fused representation serves as the input to subsequent residual and Transformer-based modules for final fine-grained flow generation. This formulation ensures that the downstream modules operate on compact yet highly discriminative representations, while irrelevant modality-specific redundancies are minimized.

### 2.5 Fine-Grained Flow Generation

The VIB-enhanced fused features $\mathbf{Z}_{\text{fused}}$ are combined with the global structural representation $\mathbf{Z}_{\text{global}}$ extracted by a Transformer. After spatially aligning $\mathbf{Z}_{\text{global}}$ to match $\mathbf{Z}_{\text{fused}}$, local and global information are integrated via element-wise addition:

$$\mathbf{Z}_{\text{merged}} = \mathbf{Z}_{\text{fused}} + \mathbf{Z}_{\text{global}}. \tag{6}$$

The merged features are then processed through several convolutional layers $\mathcal{C}(\cdot)$ to produce the final high-resolution traffic flow prediction $\hat{Y} = \mathcal{C}(\mathbf{Z}_{\text{merged}})$, preserving local fine-grained structures while maintaining global consistency. This design effectively balances short- and long-range traffic dynamics in urban flow inference.

### 2.6 Loss Function

The training objective of VIBFI combines the standard regression loss for traffic flow prediction with the VIB regularization across all modalities. Let $\mathcal{L}_{\text{pred}}$ denote the mean squared error (MSE) between the predicted flow $\hat{Y}$ and ground truth $Y$, and $\mathcal{L}_{\text{spec}}$ the sum of KL divergences for flow, road, and external feature latent variables. The overall loss is defined as

$$\mathcal{L} = \mathcal{L}_{\text{pred}} + \beta \mathcal{L}_{\text{spec}}, \tag{7}$$

where $\beta$ is a weighting coefficient controlling the strength of the VIB regularization. This design encourages the latent representations to retain predictive information while suppressing modality-specific redundancies irrelevant to fine-grained traffic flow inference.

# 3 Experiments

## 3.1 Datasets

We conduct experiments on three widely-used fine-grained urban traffic flow inference datasets: Xi'an [9], Chengdu [9] and TaxiBJ [16]. The detailed statistics of the three datasets are summarized in Table 1, Table 2 and Fig. 2.

**Table 1.** Overview of the Xi'an and Chengdu datasets.

| Property | Xi'an | Chengdu |
|---|---|---|
| City | Xi'an, China | Chengdu, China |
| Year Collected | 2016 | 2016 |
| Flow Type | Inflow/Outflow | Inflow/Outflow |
| Time Interval | 15 min | 15 min |
| Road Map Resolution | $928 \times 928$ | $952 \times 952$ |
| Coarse-grained Grid | $16 \times 16 \times 2$ | $16 \times 16 \times 2$ |
| Fine-grained Grid | $64 \times 64 \times 2$ | $64 \times 64 \times 2$ |
| Time Span | 10/01–11/30 | 10/01–10/31 |
| Training Samples | 10/01–11/02 | 10/01–10/21 |
| Validation Samples | 11/03–11/16 | 10/22–10/24 |
| Testing Samples | 11/17–11/30 | 10/25–10/31 |
| Day-of-Week Range | [0, 6] | [0, 6] |
| Hour-of-Day Range | [0, 23] | [0, 23] |
| Weather Categories | 14 types | 14 types |
| Temperature (°C) | [−5, 32.2] | [2.2, 32.2] |
| Wind Speed (mph) | [0, 18] | [0, 22] |

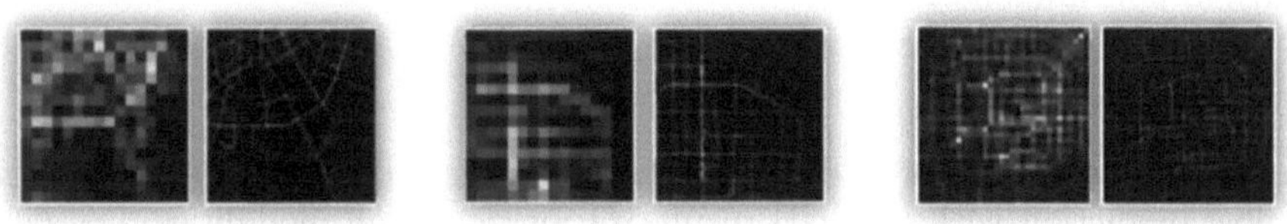

**Fig. 2.** Traffic patterns and the road network of Xi'an, Chengdu and TaxiBJ dataset.

## 3.2 Evaluation Metrics

We evaluate all methods using three widely adopted metrics: root mean square error (RMSE), mean absolute error (MAE), and mean absolute percentage error (MAPE). Specifically, RMSE is computed as $\sqrt{\frac{1}{n}\sum_{i=1}^{n}(\hat{Y}_i - Y_i)^2}$, emphasizing larger deviations due to its quadratic form. MAE is given by $\frac{1}{n}\sum_{i=1}^{n}|\hat{Y}_i - Y_i|$, measuring the average absolute prediction error. MAPE, defined as $\frac{1}{n}\sum_{i=1}^{n}\frac{|\hat{Y}_i - Y_i|}{|Y_i|}$, captures relative error and is suitable for comparing flows across different magnitude levels. Here, $n$ denotes the number of samples, $\hat{Y}_i$ is the predicted flow map, and $Y_i$ is the corresponding ground truth.

### 3.3 Comparison with Other Methods

Table 3 reports quantitative comparisons between VIBFI and representative baselines on TaxiBJ, Xi'an, and Chengdu. We focus on RMSE and MAE for clarity and first summarize overall trends before analyzing dataset-specific behavior.

**Table 2.** Overview of the TaxiBJ datasets.

| Property | TaxiBJ-P1 | TaxiBJ-P2 | TaxiBJ-P3 | TaxiBJ-P4 |
|---|---|---|---|---|
| City | Beijing, China | Beijing, China | Beijing, China | Beijing, China |
| Flow Type | Volume | Volume | Volume | Volume |
| Time Interval | 30 min | 30 min | 30 min | 30 min |
| Road Map Resolution | $1125 \times 1125$ | $1125 \times 1125$ | $1125 \times 1125$ | $1125 \times 1125$ |
| Coarse-grained Grid | $32 \times 32 \times 1$ | $32 \times 32 \times 1$ | $32 \times 32 \times 1$ | $32 \times 32 \times 1$ |
| Fine-grained Grid | $128 \times 128 \times 1$ | $128 \times 128 \times 1$ | $128 \times 128 \times 1$ | $128 \times 128 \times 1$ |
| Time Span | 07/01–10/31 | 03/01–03/30 | 03/01–06/30 | 11/01–04/10 |
| Training Samples | 07/01–08/31 | 03/01–03/20 | 03/01–04/30 | 11/01–12/31 |
| Validation Samples | 09/01–09/30 | 03/21–03/25 | 05/01–05/31 | 01/01–02/10 |
| Testing Samples | 10/01–10/31 | 03/26–03/30 | 06/01–06/30 | 02/11–04/10 |
| Day-of-Week Range | [0, 6] | [0, 6] | [0, 6] | [0, 6] |
| Hour-of-Day Range | [7, 21] | [7, 21] | [7, 21] | [7, 21] |
| Weather Categories | 16 types | 16 types | 16 types | 16 types |
| Temperature (°C) | [−24.6, 41.0] | [−24.6, 41.0] | [−24.6, 41.0] | [−24.6, 41.0] |
| Wind Speed (mph) | [0, 48.6] | [0, 48.6] | [0, 48.6] | [0, 48.6] |

Across all benchmarks, VIBFI achieves the lowest average RMSE and MAE, consistently outperforming prior methods. Compared with the strongest baseline DyBPN, VIBFI reduces RMSE by approximately 1%–2% and MAE by 3%–6% across datasets. Although the absolute margins may appear moderate, the improvements are systematic across cities with distinct traffic distributions. This consistency suggests that incorporating a variational information bottleneck effectively regularizes intermediate representations, suppresses redundant or noise-corrupted features, and enhances reconstruction fidelity under heterogeneous urban dynamics.

On TaxiBJ, which exhibits strong temporal periodicity and multi-scale mobility patterns, VIBFI achieves RMSE/MAE = 3.79 / 1.82, outperforming DyBPN

(3.83/1.94). This corresponds to approximately 1.0% improvement in RMSE and 6.2% improvement in MAE. The larger gain in MAE indicates that VIBFI substantially reduces systematic per-grid prediction bias. Since MAE is more sensitive to median-level deviations rather than extreme outliers, this result suggests that the proposed bottleneck mechanism improves overall stability and reduces consistent over-/under-estimation errors.

On Xi'an, characterized by dense urban structure and high spatial heterogeneity, VIBFI achieves RMSE/MAE = 15.41/5.15, surpassing DyBPN (15.73/5.38). The relative improvements reach 2.1% in RMSE and 4.3% in MAE, representing the largest margin among the evaluated datasets. This result indicates that information-constrained latent representations are particularly effective in complex spatial settings. By limiting mutual information between inputs and latent variables, VIBFI discourages overfitting to localized noise patterns and better captures essential spatial correlations within intricate road networks.

For Chengdu, which exhibits distinct congestion patterns and varying sensor coverage, VIBFI achieves RMSE/MAE = 19.84/7.23 compared to DyBPN's 20.19/7.46, yielding improvements of approximately 1.7% and 3.1%, respectively. The consistent gains across datasets with different spatial distributions demonstrate that VIB-based regularization improves generalization beyond a single-city regime. In particular, constraining representation capacity appears to enhance robustness against distributional variations in traffic intensity and spatial layout.

**Table 3.** Performance comparison of all methods.

| Model | TaxiBJ-P1 | | TaxiBJ-P2 | | TaxiBJ-P3 | | TaxiBJ-P4 | | Xi'an | | Chengdu | |
|---|---|---|---|---|---|---|---|---|---|---|---|---|
| | RMSE | MAE | RMSE | MAE | RMSE | MAE | RMSE | MAE | RMSE | MAE | RMSE | MAE |
| UrbanFM [16] | 4.08 | 2.10 | 4.33 | 2.22 | 4.50 | 2.32 | 3.50 | 1.82 | 17.42 | 6.28 | 21.83 | 8.46 |
| DeepLGR [17] | 4.26 | 2.10 | 4.85 | 2.37 | 4.98 | 2.48 | 3.87 | 1.92 | 17.10 | 6.11 | 21.45 | 8.29 |
| FODE [18] | 4.00 | 2.12 | 4.39 | 2.36 | 4.51 | 2.41 | 3.51 | 1.88 | 16.88 | 5.97 | 21.32 | 8.15 |
| UrbanPy [19] | 3.96 | 1.99 | 4.36 | 2.23 | 4.52 | 2.32 | 3.51 | 1.82 | 16.76 | 5.92 | 21.24 | 8.08 |
| CUFAR [20] | 3.88 | 1.96 | 4.27 | 2.19 | 4.39 | 2.24 | 3.42 | 1.81 | 16.59 | 5.84 | 21.10 | 8.02 |
| STCF [21] | 3.85 | 1.95 | 4.25 | 2.32 | 4.54 | 2.34 | 3.54 | 1.81 | 17.01 | 5.96 | 22.12 | 8.21 |
| UrbanODE [22] | 3.96 | 2.09 | 4.36 | 2.32 | 4.69 | 2.53 | 3.51 | 1.88 | 16.18 | 5.42 | 20.35 | 7.37 |
| RATFM [9] | 3.85 | 1.95 | 4.28 | 2.20 | 4.41 | 2.28 | 3.46 | 1.81 | 16.21 | 5.63 | 20.73 | 7.75 |
| MS-SCL [11] | 3.80 | 1.95 | 4.20 | 2.16 | 4.39 | 2.27 | 3.41 | 1.79 | 15.94 | 5.49 | 20.42 | 7.59 |
| DyBPN [23] | 3.83 | 1.94 | 4.20 | 2.15 | 4.36 | 2.23 | 3.38 | 1.75 | 15.73 | 5.38 | 20.19 | 7.46 |
| **VIBFI (ours)** | **3.79** | **1.82** | **4.16** | **2.13** | **4.29** | **2.18** | **3.34** | **1.63** | **15.41** | **5.15** | **19.84** | **7.23** |

### 3.4 Component Analysis

To verify the effectiveness of each component in our proposed model, we conduct ablation experiments on all datasets. Specifically, we remove three key variational bottlenecks: road network image ($\mathcal{B}_{\text{road}}$), external factors ($\mathcal{B}_{\text{ext}}$), and traffic flow ($\mathcal{B}_{\text{flow}}$)—to investigate their respective contributions. Tables 4 summarize the results on the TaxiBJ, Xi'an and Chengdu datasets, respectively.

**Table 4.** Ablation study of VIBFI.

| Model Variant | TaxiBJ-P1 | | TaxiBJ-P2 | | TaxiBJ-P3 | | TaxiBJ-P4 | | Xi'an | | Chengdu | |
|---|---|---|---|---|---|---|---|---|---|---|---|---|
| | RMSE | MAE | RMSE | MAE | RMSE | MAE | RMSE | MAE | RMSE | MAE | RMSE | MAE |
| **VIBFI (Full Model)** | **3.79** | **1.82** | **4.16** | **2.13** | **4.29** | **2.18** | **3.34** | **1.63** | **15.41** | **5.15** | **19.84** | **7.23** |
| w/o $\mathcal{B}_{\text{road}}$ | 3.95 | 1.90 | 4.34 | 2.22 | 4.46 | 2.26 | 3.49 | 1.70 | 15.92 | 5.43 | 20.28 | 7.45 |
| w/o $\mathcal{B}_{\text{ext}}$ | 3.88 | 1.88 | 4.27 | 2.19 | 4.39 | 2.24 | 3.43 | 1.68 | 15.68 | 5.32 | 20.09 | 7.36 |
| w/o $\mathcal{B}_{\text{flow}}$ | 3.84 | 1.86 | 4.23 | 2.17 | 4.35 | 2.22 | 3.39 | 1.66 | 15.56 | 5.28 | 19.97 | 7.31 |
| w/o All Bottlenecks | 4.15 | 2.02 | 4.55 | 2.35 | 4.69 | 2.43 | 3.65 | 1.82 | 16.14 | 5.57 | 20.66 | 7.68 |

From Table 4, it is evident that removing any of the three bottlenecks causes consistent degradation across all datasets.

(1) The road network bottleneck $\mathcal{B}_{\text{road}}$ contributes most significantly, as it encodes the spatial topology of the city road structure, which is critical for reconstructing fine-grained flow distributions. (2) The external factor bottleneck $\mathcal{B}_{\text{ext}}$ improves the model's adaptability to dynamic conditions such as weather, holidays, and temporal fluctuations. (3) The traffic flow bottleneck $\mathcal{B}_{\text{flow}}$ enhances temporal consistency, capturing the latent evolution of inflow and outflow patterns.

When all three are removed, the model degenerates into a basic encoder–decoder structure, leading to an average RMSE increase of about 0.3 on TaxiBJ and nearly 0.7–1.0 on other datasets, confirming that the proposed multi-bottleneck variational design is essential for robust and accurate urban flow inference.

## 4 Conclusion

This paper presents VIBFI, a fine-grained traffic flow inference framework that addresses spatial heterogeneity and feature redundancy through three complementary variational bottlenecks for road structure, external context, and traffic dynamics. By enforcing controllable information compression across modalities, VIBFI yields more compact and interpretable representations for urban traffic prediction. Experiments on three real-world city-level datasets show that VIBFI consistently surpasses state-of-the-art baselines, with ablation studies confirming the distinct contributions of each bottleneck to spatial accuracy, environmental adaptability, and temporal consistency. VIBFI offers a lightweight yet effective perspective on interpretable traffic modeling. Future extensions include cross-city transfer, integration of richer heterogeneous data, and real-time adaptive inference.

**Acknowledgments.** This research was supported by the Research Foundation from Qinghai Normal University (Grant No. 2023QZR002).

**Disclosure of Interests.** The authors have no competing interests to declare that are relevant to the content of this article.

## References

1. Wang, J., Liu, C., Wu, Z., Liao, R., Li, G., Lu, H.: The roadmap and strategy for prioritizing the development of public transport in China (2025)
2. Lu, Y., Shao, J., Yao, Y.: Data modeling of impact of green-oriented transportation planning and management measures on the economic development of small-and medium-sized cities. J. Adv. Transp. **2022**(1), 8676805 (2022)
3. Park, J.: Adaptive traffic signal control for urban traffic networks: simulation and experiments. Ph.D. thesis, New York University Tandon School of Engineering (2025)
4. Jiang, W., Luo, J.: Big data for traffic estimation and prediction: a survey of data and tools. Appl. Syst. Innov. **5**(1), 23 (2022)
5. Yuan, S., et al.: Fine-grained urban flow inference with multi-scale representation learning. arXiv preprint arXiv:2406.09710 (2024)
6. Li, J., Wang, S., Zhang, J., Miao, H., Zhang, J., Yu, P.S.: Fine-grained urban flow inference with incomplete data. IEEE Trans. Knowl. Data Eng. **35**(6), 5851–5864 (2022)
7. Han, X., Li, Z., Cao, H., Hou, B.: Multimodal spatio-temporal data visualization technologies for contemporary urban landscape architecture: a review and prospect in the context of smart cities. Land **14**(5), 1069 (2025)
8. Bhandari, P.: Spatio-temporal big data analysis for congestion mitigation in megacity transportation hubs. J. Digit. Transf. Cyber Resilience Infrastruct. Secur. **10**(1), 11–19 (2025)
9. Liu, L., Liu, M., Li, G., Wu, Z., Lin, J., Lin, L.: Road network-guided fine-grained urban traffic flow inference. IEEE Trans. Neural Netw. Learn. Syst. (2023)
10. Zhang, Y., Yu, W., Zhao, X., He, Z., Zhang, W., Lin, J.: Focus on hard areas of reconstruction: a fine-grained urban flow inference framework. Trans. GIS **29**(3), e70039 (2025)
11. Pan, L., Ren, Q., Li, Z., Zhao, C.: Multi-scale synchronous contextual network for fine-grained urban flow inference. Inf. Sci. **689**, 121459 (2025)
12. Yang, M., et al.: STDA: spatio-temporal deviation alignment learning for cross-city fine-grained urban flow inference. IEEE Trans. Knowl. Data Eng. (2025)
13. Zheng, Y., Wu, J., Cai, Z., Wang, S., Wang, J.: AdaTM: fine-grained urban flow inference with adaptive knowledge transfer across multiple cities. In: Proceedings of the 33rd ACM International Conference on Information and Knowledge Management, pp. 3424–3432 (2024)
14. Abideen, Z.U., Sun, X., Sun, C.: Multi-scale learning for fine-grained traffic flow-based travel time estimation prediction. Comput. Intell. **40**(5), e12693 (2024)
15. Vaswani, A., et al.: Attention is all you need. Adv. Neural Inf. Process. Syst. **30** (2017)
16. Liang, Y., et al.: UrbanFM: inferring fine-grained urban flows. In: Proceedings of the 25th ACM SIGKDD International Conference on Knowledge Discovery & Data Mining, pp. 3132–3142 (2019)

17. Liang, Y., et al.: Revisiting convolutional neural networks for citywide crowd flow analytics. In: Hutter, F., Kersting, K., Lijffijt, J., Valera, I. (eds.) ECML PKDD 2020. LNCS (LNAI), vol. 12457, pp. 578–594. Springer, Cham (2021). https://doi.org/10.1007/978-3-030-67658-2_33
18. Zhou, F., Li, L., Zhong, T., Trajcevski, G., Zhang, K., Wang, J.: Enhancing urban flow maps via neural odes. In: Proceedings of the Twenty-Ninth International Joint Conference on Artificial Intelligence, {IJCAI} 2020 (2020)
19. Ouyang, K., Liang, Y., Liu, Y., Tong, Z., Ruan, S., Zheng, Y., Rosenblum, D.S.: Fine-grained urban flow inference. IEEE Trans. Knowl. Data Eng. **34**(6), 2755–2770 (2020)
20. Yu, H., Xu, X., Zhong, T., Zhou, F.: Overcoming forgetting in fine-grained urban flow inference via adaptive knowledge replay. In: Proceedings of the AAAI Conference on Artificial Intelligence, vol. 37, pp. 5393–5401 (2023)
21. Xu, X., et al.: Spatial-temporal contrasting for fine-grained urban flow inference. IEEE Trans. Big Data **9**(6), 1711–1725 (2023)
22. Zhou, F., Jing, X., Li, L., Zhong, T.: Inferring high-resolutional urban flow with internet of mobile things. In: ICASSP 2021-2021 IEEE International Conference on Acoustics, Speech and Signal Processing (ICASSP), pp. 7948–7952. IEEE (2021)
23. Xu, M., Qi, G., Qi, G.: Fine-grained traffic flow inference model based on dynamic back projection network. J. Syst. Simul. **37**(3), 657–666 (2025)

# Self-weighted Contrastive Fusion for Deep Multi-view scRNA-Seq Clustering

Feiyu Chen[2], Zihao Feng[1], Zhongyang Zhou[1], Bin Tang[2], and Shangshang Zhao[1](✉)

[1] College of Computer and Information Science, Chongqing Normal University, Chongqing 401331, China
zs1999ism@163.com

[2] National Center for Applied Mathematics in Chongqing, Chongqing Normal University, Chongqing 401331, China

**Abstract.** Single-cell RNA sequencing data clustering is inherently challenged by ultra-high dimensionality, extreme sparsity, and technical noise, collectively presenting formidable computational burdens. Compounding this difficulty, the popular GNN-based clustering methods designed to address these issues are fundamentally limited by single-scale topology and the over-smoothing effect, which significantly hinders robust cell subtype identification in large datasets. To overcome these issues, we propose Self-Weighted Contrastive Learning (SCSW), a novel framework utilizing decoupled multi-view contrastive learning. SCSW constructs complementary dual-graph representations: the KNN graph captures local neighborhood structure, while the Diffusion Map graph models global manifold topology and long-range relationships. Our innovative decoupled encoder architecture separates representation learning from contrastive optimization, which bypasses GNN over-smoothing and integrates a cross-view contrastive mechanism to ensure feature discriminative consistency. The final robust embeddings are generated by a self-adaptive weighted fusion strategy that dynamically down-weights noisy or unreliable views. Extensive experiments on nine diverse scRNA-seq datasets confirm that SCSW consistently and significantly outperforms all tested on critical clustering metrics.

**Keywords:** scRNA-seq · Multi-view Clustering · Contrastive Learning

## 1 Introduction

Cellular heterogeneity is the cornerstone of biological organization and regulatory fidelity, with profound distinctions among cell subtypes driving diverse functions and specialized gene expression [1]. Understanding these differences is critical for dissecting developmental and homeostatic programs and elucidating pathogenic mechanisms in diseases. Historically, conventional bulk RNA sequencing obscured these critical, subtype-specific transcriptional signatures and continuous cellular state transitions by reporting only an ensemble-averaged


R. C.-W. Wong et al. (Eds.): PAKDD 2026, LNAI 16597, pp. 506–518, 2026.
https://doi.org/10.1007/978-981-92-1300-9_40

signal [2]. This fundamental limitation severely impeded the accurate identification of rare, often critical, cell populations and transient cellular states. The advent of single-cell RNA sequencing (scRNA-seq) technology represents a paradigm shift, effectively overcoming this bottleneck. By providing unprecedented single-cell resolution of genome-wide transcriptional landscapes, scRNA-seq has become a core technology accelerating critical advancements [3], including high-fidelity transcriptome-based cell type annotation, robust developmental trajectory reconstruction, and detailed mechanistic investigations into disease pathogenesis [4].

Within the analytical pipeline of scRNA-seq data, cell clustering remains a fundamental and often prerequisite task. Its primary objective is the identification of distinct cellular subpopulations and the revelation of novel biological states by grouping cells that share highly similar transcriptional patterns. However, the nature of scRNA-seq data itself imposes formidable computational hurdles: the intrinsic high dimensionality, severe data sparsity, and confounding complex technical noise collaboratively introduce significant challenges to robust clustering analysis [5]. Although the field has seen the proposal of several technically robust methods, including those leveraging regularization constraints, such as Rss NMF [6], and techniques based on subspace learning integrated with low-rank representation, such as NLRRC [7], a critical limitation persists in their capacity to effectively model the high-dimensional, non-linear manifold structure of single-cell data. Specifically, these conventional approaches often fail to capture the multi-scale cellular heterogeneity required for robust subtype discovery. These traditional or shallow machine learning models possess an inherent, constrained feature representation capacity [8]. As a direct consequence, they are markedly inefficient in fully exploring and capturing the highly non-linear and intricate underlying manifold structures that govern cellular relationships within increasingly large-scale and complex scRNA-seq datasets.

To robustly mitigate the representational constraints of conventional methods, deep clustering algorithms have emerged as the cornerstone of scRNA-seq analysis. Earlier milestones, such as scGNN [9] and scDSC [10], integrated ZINB-based autoencoders with GNNs to model gene distributions. However, these frameworks often rely on local structural priors, failing to capture the global manifold geometry that our SCSW addresses through Diffusion Maps. As the field pivoted towards self-supervised learning, ScCCL [11] and scDCCA [12] introduced contrastive paradigms for semantic consistency, while scMGCA [13] refined structural modeling via specialized decoders. Despite their success, these methods are frequently hampered by high computational overhead and GNN over-smoothing. In contrast, SCSW utilizes a decoupled Autoencoder architecture that bypasses cumbersome message-passing while maintaining superior representation power. More recently, research has expanded into multi-view and large-scale optimization. scMMN [14] and scDFN [15] explored feature fusion, while nsDCC [16] and scMCKC [17] introduced sampling and multi-constraint optimization for massive datasets. Nevertheless, these advanced models often utilize rigid fusion strategies that ignore feature quality heterogeneity. Our SCSW distinguishes itself through an Adaptive Robust Weighted Fusion mechanism with **Common Information Maximization**, ensuring that noise is suppressed while shared biological signals are adaptively prioritized.

In summary, SCSW bridges these research gaps through three core innovations:

- **Innovative Multi-View Data Construction Strategy.** We propose a computationally efficient Multi-View Data Construction Strategy, utilizing Raw Expression, the KNN graph, and Diffusion Map (DM). This approach assimilates multi-scale structural knowledge (gene expression, local topology, global manifold geometry) to yield embedding fidelity significantly superior to single-view models.
- **Highly Robust Decoupled Network Architecture.** To circumvent GNN efficiency and over-smoothing bottlenecks, SCSW employs a robust, decoupled Autoencoder (AE) network. Utilizing independent AE pathways, this architecture effectively bypasses cumbersome message passing, ensuring information decoupling, robust multi-view learning, and significantly reduced computational overhead.
- **Adaptive Robust Weighted Fusion with Common Information Maximization.** To overcome feature quality heterogeneity, we introduce an Adaptive Robust Weighted Fusion framework that dynamically assigns weights using a self-learning mechanism to suppress noise. Concurrently, a common information maximization module ensures the embedding retains shared latent representations, significantly enhancing clustering robustness.

# 2 Methodology

## 2.1 The Overall Architecture of the SCSW Framework

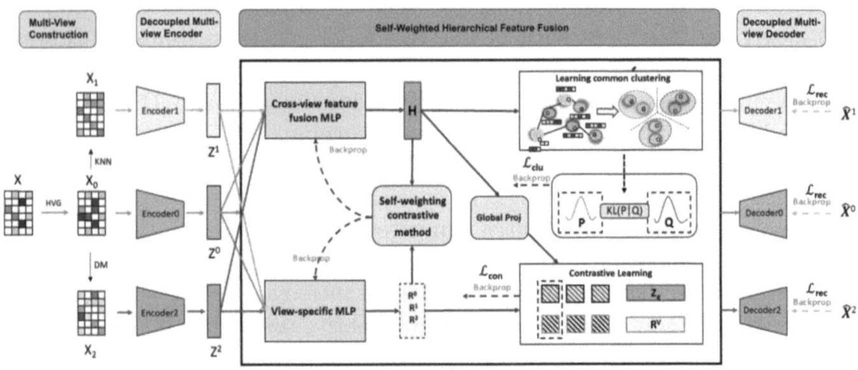


**Fig. 1.** Founded upon a Multi-View strategy and an integrated Decoupled Feature Embedding Network co-driven framework, HVG stands for Highly Variable Gene selection.

SCSW first preprocesses raw matrix $\mathbf{X}$ via Scanpy ($10^4$ normalization, $\log(1+x)$ transform) and selects 2,000–4,000 HVGs to yield filtered input $\mathbf{X}_0$. Building on $\mathbf{X}_0$, three complementary views are generated and encoded by independent AEs into latent representations $\mathbf{Z}^v$, which are then mapped to a shared contrastive space $\mathbf{R}^v$ to decouple feature learning from optimization. Finally, features are integrated via adaptive weighted fusion into a global embedding $\mathbf{H}$, jointly optimized by reconstruction ($\mathcal{L}_{\text{rec}}$), contrastive ($\mathcal{L}_{\text{con}}$), and clustering ($\mathcal{L}_{\text{clu}}$) losses to enhance discriminative power (Fig. 1).

### 2.2 Multi-view Data Construction

To effectively surmount the single-scale limitations in topological modeling, SCSW employs a unified, computationally efficient Multi-View Data Construction Strategy, distilling three complementary feature views from the expression matrix $\mathbf{X} \in \mathbb{R}^{N \times D}$. These views are: the Raw Feature View ($\mathbf{X}_0$) (transcriptional profile); the KNN Structure-Smoothed View ($\mathbf{X}_1$), generated by $\mathbf{X}_1 = \mathbf{A}_{\text{KNN}}\mathbf{X}_0$ to efficiently encode local topology and mitigate sparsity while bypassing GNN bottlenecks; and the Diffusion Map (DM) Projected View ($\mathbf{X}_2$), which extracts global non-linear manifold knowledge. This comprehensive strategy ensures robust capture of both local and long-range cellular structures:

$$\mathbf{X}_1 = \mathbf{A}_{\text{KNN}}\mathbf{X}_0, \tag{1}$$

$$\mathbf{X}_2 = \Phi(\mathbf{X}_0). \tag{2}$$

where $\Phi$ is the mapping function from the original data space $\mathbf{X}_0$ to the Diffusion Map (DM) projection space. The diffusion graph is constructed with a time step $t = 1$ and an adaptive kernel bandwidth $\sigma$ determined via the median heuristic, effectively balancing local manifold preservation with global connectivity.

This unified strategy guarantees that the final latent representation comprehensively and robustly captures both local fine-grained structures and global macroscopic cellular heterogeneity.

### 2.3 Decoupled Deep Feature Embedding Network

To effectively address the computational bottlenecks and over-smoothing issues of GNNs, SCSW employs a robust decoupled deep AE network with parameter-independent branches. This design achieves objective isolation: the view-specific latent representation $\mathbf{Z}^v$ is optimized solely by the reconstruction loss ($\mathcal{L}_{\text{rec}}$), ensuring the preservation of high-fidelity, view-specific information:

$$\mathbf{Z}^v = \mathbf{E}_v(\mathbf{X}^v), \tag{3}$$

Concurrently, a Cross-View Contrastive Projection mechanism maps $\mathbf{Z}^v$ to a new space $\mathbf{R}^v$.

$$\mathbf{R}^v = \mathbf{Projection}(\mathbf{Z}^v). \tag{4}$$

This critical projection decouples $\mathbf{Z}^v$ from $\mathbf{R}^v$, enabling the contrastive objective to maximize inter-view consistency and consensus discriminability. This dual optimization strategy mitigates representation collapse and significantly enhances clustering robustness without compromising reconstruction quality.

### 2.4 Self-weighted Hierarchical Feature Fusion

Though the contrastive projection enhances discriminability, the inherent heterogeneity and noise in scRNA-seq views persist, making simple fusion unreliable. To address this, we propose the innovative Adaptive Robust Weighted Fusion strategy, centered around a unique hierarchical fusion architecture, for generating the global embedding $\mathbf{H}$:

- **Adaptive Weight Calculation.** Specifically, we first employ the Softmax function to normalize the negative discrepancy between the contrastive projection embedding $\mathbf{R}^v$ and the global embedding $\mathbf{H}$, thereby yielding the adaptive weight $\alpha_v$:

$$\alpha^v = \frac{e^{-\mathcal{D}(\mathbf{R}^v,\mathbf{H})}}{\sum_{v'=1}^{V} e^{-\mathcal{D}(\mathbf{R}^{v'},\mathbf{H})}}, \tag{5}$$

  where $\mathcal{D}(\mathbf{R}^v, \mathbf{H})$ is the discrepancy measure. The closer $\mathbf{R}^v$ is to $\mathbf{H}$, the higher the weight $\alpha^v$ assigned to that view.
- **Maximum Mean Discrepancy.** Specifically, the discrepancy $\mathcal{D}(\mathbf{R}_v, \mathbf{H})$ is defined as the Squared ($\text{MMD}^2$). $\text{MMD}^2$ measures the distance between the mean embeddings of $\mathbf{R}^v$ and $\mathbf{H}$ in a Reproducing Kernel Hilbert Space (RKHS) $\mathcal{H}$. The empirical $\text{MMD}^2$ is formulated as the difference between the expected self-similarity within each view ($\mathcal{D}_{\text{self}}$) and the expected cross-similarity ($\mathcal{D}_{\text{cross}}$) [18].
  This term, $\mathcal{D}_{\text{self}}$, represents the sum of the expected internal similarity: the self-similarity of the view projection $\mathbf{R}^v$ and the global embedding $\mathbf{H}$:

$$\mathcal{D}_{\text{self}} = \frac{1}{N^2}\sum_{i=1}^{N}\sum_{i'=1}^{N} k(\mathbf{r}_i, \mathbf{r}_{i'}) + \frac{1}{N^2}\sum_{j=1}^{N}\sum_{j'=1}^{N} k(\mathbf{h}_j, \mathbf{h}_{j'}), \tag{6}$$

  The full empirical $\text{MMD}^2$ is then given by:

$$\mathcal{D}(\mathbf{R}^v, \mathbf{H}) = \mathcal{D}_{\text{self}} - \frac{2}{N^2}\sum_{i=1}^{N}\sum_{j=1}^{N} k(\mathbf{r}_i, \mathbf{h}_j), \tag{7}$$

  where $\mathbf{r}_i$ and $\mathbf{h}_j$ are the $i$-th and $j$-th cell samples from $\mathbf{R}^v$ and $\mathbf{H}$, $N$ is the total number of samples, and $k(\cdot,\cdot)$ is a characteristic kernel.
- **Global Feature Fusion Formula.** The global embedding $\mathbf{H}$ is then obtained as the result of a non-linear transformation, where the weighted sum of $\mathbf{Z}^v$ is fed into the cross-view feature fusion MLP ($\text{MLP}_{\text{fuse}}$):

$$\mathbf{H} = \text{MLP}_{\text{fuse}}\left(\sum_{v=1}^{V} \alpha^v \mathbf{Z}^v\right). \tag{8}$$

where $\mathbf{H}$ is the global embedding, $\sum \alpha^v \mathbf{Z}^v$ implements the robust weighting, and $\text{MLP}_{\text{fuse}}(\cdot)$ provides the non-linear feature integration.

### 2.5 Optimization Objective and Loss Function

The SCSW framework is driven by a multi-task joint optimization objective: a weighted summation of three core losses, ensuring an optimal balance among view-specific reconstruction, cross-view discriminative consistency, and final global clustering quality:

$$\mathcal{L}_{\text{total}} = \lambda \mathcal{L}_{\text{rec}} + \lambda_1 \mathcal{L}_{\text{con}} + \lambda_2 \mathcal{L}_{\text{clu}}, \tag{9}$$

where $\mathcal{L}_{\text{rec}}$, $\mathcal{L}_{\text{con}}$, and $\mathcal{L}_{\text{clu}}$ denote the reconstruction loss, the cross-view contrastive loss, and the global clustering loss, respectively. $\lambda$, $\lambda_1$ and $\lambda_2$ are positive hyperparameters.

- **Multi-View Feature Reconstruction Loss.** To ensure high-fidelity local information preservation, we strictly constrain the latent feature $\mathbf{Z}^v$ by minimizing the Mean Squared Error Loss (MSE) between the input $\mathbf{X}^v$ and its reconstruction $\hat{\mathbf{X}}^v$ for each view-specific encoder-decoder pair:

$$\mathcal{L}_{\text{rec}} = \sum_{v=1}^{V} \|\mathbf{X}^v - \hat{\mathbf{X}}^v\|_F^2, \tag{10}$$

  where $V$ denotes the number of views, $\mathbf{X}^v$ denotes the original input matrix of the $v$-th view, $\hat{\mathbf{X}}^v$ denotes the reconstructed output matrix of the $v$-th view after Encoder$^v$ and Decoder$^v$, $\|\cdot\|_F^2$ denotes the squared Frobenius norm, measuring the element-wise difference between two matrices.
- **Cross-View Contrastive Loss.** Because we maximize the discriminative consistency between the view projection $\mathbf{R}^v$ and the global embedding $\mathbf{H}$, the contrastive loss function uses normalized temperature-scale cross-entropy (NT-Xent). The total contrastive loss is the sum of all pairwise losses:

$$\mathcal{L}_{\text{con}} = \sum_{v=1}^{V} \sum_{v'=1, v' \neq v}^{V} \mathcal{L}_{\text{NT-Xent}}(\mathbf{R}^v, \mathbf{R}^{v'}) + \sum_{v=1}^{V} \mathcal{L}_{\text{NT-Xent}}(\mathbf{R}^v, \mathbf{Z}_g), \tag{11}$$

  The global representation $\mathbf{Z}_g \in \mathbb{R}^d$ is defined as $\mathbf{Z}_g = f_{\phi_g}(\mathbf{H})$, where $f_{\phi_g}(\cdot)$ is a global projection head consisting of a two-layer MLP with ReLU activation, mapping the integrated embedding $\mathbf{H}$ into a dedicated contrastive latent space.where NT-Xent loss is defined as:

$$\mathcal{L}_{\text{NT-Xent}}(\mathbf{z}_i, \mathbf{z}_j) = -\log \frac{\exp(\text{sim}(\mathbf{z}_i, \mathbf{z}_j)/\tau)}{\sum_{k \in \mathcal{A}} \exp(\text{sim}(\mathbf{z}_i, \mathbf{z}_k)/\tau)}, \tag{12}$$

  where $\text{sim}(\cdot,\cdot)$ represents the cosine similarity, $\tau$ is the temperature hyperparameter, and $\mathcal{A}$ is the set of all samples in the batch excluding $\mathbf{z}_i$.

- **Clustering Loss.** The soft assignment $q_{ij}$ and target distribution $p_{ij}$ are defined using Student's $t$-distribution kernel and cluster refinement respectively:

$$q_{ij} = \frac{(1 + \|\mathbf{h}_i - \boldsymbol{\mu}_j\|^2)^{-1}}{\sum_{j'}(1 + \|\mathbf{h}_i - \boldsymbol{\mu}_{j'}\|^2)^{-1}}, \quad p_{ij} = \frac{q_{ij}^2 / \sum_i q_{ij}}{\sum_{j'}(q_{ij'}^2 / \sum_i q_{ij'})} \tag{13}$$

where $\boldsymbol{\mu}_j$ is the $j$-th cluster centroid, and $\alpha = 1$ is assumed for the kernel.

$$\mathcal{L}_{\text{clu}} = \mathcal{D}_{\text{KL}}(P||Q) = \sum_{i=1}^{N}\sum_{c=1}^{C} P_{ic} \log \frac{P_{ic}}{Q_{ic}}. \tag{14}$$

where $N$ is the total number of samples, $C$ is the number of clusters, $Q_{ic}$ is the soft assignment probability of sample $i$ belonging to cluster $c$, and $P_{ic}$ is the auxiliary target distribution.

# 3 Experiments

## 3.1 Experimental Setup and Baselines

All models were implemented using the PyTorch deep learning framework and trained on a high-performance computing platform (Intel i9-14900KF, NVIDIA RTX4090, 256 GB RAM). Training was performed using the ADAM optimizer for a maximum of 400 epochs, with dataset-specific learning rates (e.g., 5e−4 for Klein, 1e−5 for 10X_PBMC) and applied weight decay to prevent overfitting. Upon convergence, the global embedding $\mathbf{H}$ was extracted, preprocessed, and subsequently subjected to the Spectral Clustering algorithm for final cell cluster assignments. Clustering performance was rigorously validated using four widely accepted external metrics: Accuracy (ACC), Normalized Mutual Information (NMI), Adjusted Rand Index (ARI), and Adjusted Mutual Information (AMI). For comprehensive evaluation, we compared SCSW against nine deep single-cell clustering methods, including scMMN [14], nsDCC [16], scDFN [15], scMCKC [17], ScCCL [11], scDCCA [12], scMGCA [13], scDSC [10], and scGNN [9]. All baseline implementations strictly adhered to the parameter settings specified in their original publications. Hyperparameter Tuning. For each dataset, we determined the optimal loss weights $\lambda$, $\lambda_1$, and $\lambda_2$ via a systematic grid search within the range $\{0.05, 0.1, 1\}$. Other parameters, such as the learning rate ($5 \times 10^{-4}$) and batch size (512), remained consistent across most experiments to ensure a fair comparison. Detailed optimal settings for all nine datasets are summarized in Table 1.

**Table 1.** Optimal loss weights across nine scRNA-seq datasets.

| Dataset | $\lambda$ | $\lambda_1$ | $\lambda_2$ |
|---|---|---|---|
| Wang_Lung/QS_Heart/Young/Chen/Plasschaert | 1.0 | 0.1 | 1.0 |
| QS_Trachea/QX_Spleen | 1.0 | 1.0 | 1.0 |
| 10X_PBMC/Klein | 1.0 | 0.05 | 1.0 |

**Table 2.** The characteristics of nine real scRNA-seq datasets.

| Datasets | Cells | Genes | Cell types |
|---|---|---|---|
| Klein | 2717 | 24047 | 4 |
| 10X_PBMC | 4721 | 16653 | 8 |
| Young | 5685 | 33658 | 11 |
| Chen | 12089 | 23284 | 46 |
| Plasschaert | 6977 | 28205 | 8 |
| QS_Trachea | 1350 | 23341 | 4 |
| QS_Heart | 4365 | 23341 | 8 |
| Wang_Lung | 9519 | 14561 | 2 |
| QX_Spleen | 9552 | 23341 | 5 |

### 3.2 Datasets

The datasets utilized in this study were derived from various sequencing platforms and inherently exhibit heterogeneous characteristics. For brevity and ease of reference, the Quake_Smart-seq2_Trachea and Quake_Smartseq2_Heart datasets are henceforth abbreviated as QS_Trachea and QS_Heart, respectively. Detailed statistics for all datasets, including cell count, gene count, and number of ground-truth clusters, are provided in Table 2.

### 3.3 Clustering Results

**Table 3.** Comparison based on **ACC**

| Dataset | scGNN | scDSC | scMGCA | scDCCA | scCCL | scMCKC | scDFN | nsDCC | scMMN | SCSW |
|---|---|---|---|---|---|---|---|---|---|---|
| Klein | 0.800 | 0.607 | 0.709 | 0.890 | 0.840 | 0.737 | 0.641 | 0.839 | 0.912 | **0.981** |
| 10X_PBMC | 0.663 | 0.766 | 0.749 | 0.772 | 0.825 | 0.689 | 0.736 | 0.785 | 0.746 | **0.827** |
| Young | 0.424 | 0.622 | 0.639 | 0.716 | 0.791 | 0.715 | 0.715 | 0.684 | 0.769 | **0.837** |
| Chen | 0.835 | 0.883 | 0.870 | 0.823 | 0.646 | 0.735 | 0.605 | 0.745 | 0.706 | **0.891** |
| Plasschaert | 0.714 | 0.749 | 0.632 | 0.739 | 0.797 | 0.812 | 0.695 | 0.683 | 0.704 | **0.973** |
| QS_Trachea | 0.557 | 0.706 | 0.678 | 0.690 | 0.890 | 0.827 | 0.759 | 0.706 | 0.842 | **0.990** |
| QS_Heart | 0.843 | 0.682 | 0.961 | 0.758 | 0.943 | 0.930 | 0.758 | 0.831 | 0.761 | **0.971** |
| WangLung | 0.773 | 0.705 | 0.805 | 0.796 | 0.992 | 0.554 | 0.992 | 0.890 | 0.986 | **0.993** |
| QX_Spleen | 0.756 | 0.782 | 0.866 | 0.818 | 0.938 | 0.766 | 0.719 | 0.819 | 0.929 | **0.978** |
| **Avg** | 0.707 | 0.722 | 0.768 | 0.778 | 0.851 | 0.752 | 0.736 | 0.776 | 0.817 | **0.938** |

**Table 4.** Comparison based on **NMI**

| Dataset | scGNN | scDSC | scMGCA | scDCCA | scCCL | scMCKC | scDFN | nsDCC | scMMN | SCSW |
|---|---|---|---|---|---|---|---|---|---|---|
| Klein | 0.701 | 0.586 | 0.696 | 0.874 | 0.776 | 0.716 | 0.772 | 0.781 | 0.912 | **0.943** |
| 10X_PBMC | 0.525 | 0.756 | 0.679 | 0.768 | 0.792 | 0.726 | 0.737 | 0.751 | 0.696 | **0.797** |
| Young | 0.359 | 0.545 | 0.590 | 0.643 | **0.801** | 0.731 | 0.737 | 0.691 | 0.713 | 0.787 |
| Chen | 0.812 | 0.828 | 0.834 | 0.788 | 0.756 | 0.752 | 0.757 | 0.770 | 0.771 | **0.843** |
| Plasschaert | 0.740 | 0.722 | 0.740 | 0.745 | 0.763 | 0.691 | 0.715 | 0.642 | 0.721 | **0.892** |
| QS_Trachea | 0.279 | 0.749 | 0.712 | 0.695 | 0.725 | 0.783 | 0.608 | 0.738 | 0.720 | **0.952** |
| QS_Heart | 0.822 | 0.655 | 0.901 | 0.702 | 0.900 | 0.861 | 0.820 | 0.798 | 0.682 | **0.968** |
| WangLung | 0.762 | 0.671 | 0.785 | 0.761 | 0.914 | 0.436 | 0.918 | 0.807 | 0.871 | **0.919** |
| QX_Spleen | 0.749 | 0.743 | 0.836 | 0.797 | 0.806 | 0.656 | 0.674 | 0.791 | 0.647 | **0.878** |
| **Avg** | 0.639 | 0.695 | 0.753 | 0.753 | 0.804 | 0.706 | 0.749 | 0.752 | 0.748 | **0.887** |

**Table 5.** Comparison based on **ARI**

| Dataset | scGNN | scDSC | scMGCA | scDCCA | scCCL | scMCKC | scDFN | nsDCC | scMMN | SCSW |
|---|---|---|---|---|---|---|---|---|---|---|
| Klein | 0.777 | 0.447 | 0.660 | 0.882 | 0.785 | 0.696 | 0.659 | 0.792 | 0.899 | **0.961** |
| 10X_PBMC | 0.677 | 0.770 | 0.699 | 0.769 | **0.794** | 0.635 | 0.693 | 0.769 | 0.625 | 0.772 |
| Young | 0.424 | 0.693 | 0.694 | **0.729** | 0.703 | 0.570 | 0.626 | 0.701 | 0.633 | 0.722 |
| Chen | 0.779 | 0.837 | 0.836 | 0.746 | 0.739 | 0.765 | 0.584 | 0.652 | 0.652 | **0.842** |
| Plasschaert | 0.656 | 0.683 | 0.612 | 0.696 | 0.731 | 0.685 | 0.633 | 0.664 | 0.658 | **0.934** |
| QS_Trachea | 0.468 | 0.767 | 0.727 | 0.779 | 0.829 | 0.866 | 0.485 | 0.883 | 0.724 | **0.980** |
| QS_Heart | 0.786 | 0.601 | 0.932 | 0.656 | 0.950 | 0.914 | 0.676 | 0.794 | 0.664 | **0.968** |
| WangLung | 0.701 | 0.623 | 0.746 | 0.720 | 0.965 | 0.262 | 0.967 | 0.829 | 0.941 | **0.967** |
| QX_Spleen | 0.683 | 0.712 | 0.815 | 0.754 | 0.908 | 0.674 | 0.505 | 0.780 | 0.549 | **0.944** |
| Avg | 0.661 | 0.681 | 0.747 | 0.748 | 0.823 | 0.674 | 0.648 | 0.763 | 0.705 | **0.899** |

**Table 6.** Comparison based on **AMI**

| Dataset | scGNN | scDSC | scMGCA | scDCCA | scCCL | scMCKC | scDFN | nsDCC | scMMN | SCSW |
|---|---|---|---|---|---|---|---|---|---|---|
| Klein | 0.863 | 0.518 | 0.678 | 0.878 | 0.776 | 0.715 | 0.766 | 0.781 | 0.912 | **0.926** |
| 10X_PBMC | 0.670 | 0.762 | 0.688 | 0.784 | 0.692 | 0.725 | 0.739 | 0.779 | 0.698 | **0.797** |
| Young | 0.410 | 0.643 | 0.681 | 0.706 | 0.763 | 0.730 | 0.736 | 0.682 | 0.622 | **0.786** |
| Chen | 0.827 | 0.831 | 0.842 | 0.766 | 0.749 | 0.750 | 0.751 | 0.749 | 0.717 | **0.843** |
| Plasschaert | 0.704 | 0.706 | 0.713 | 0.730 | 0.762 | 0.690 | 0.714 | 0.723 | 0.740 | **0.892** |
| QS_Trachea | 0.531 | 0.711 | 0.656 | 0.687 | 0.724 | 0.782 | 0.607 | 0.784 | 0.716 | **0.955** |
| QS_Heart | 0.821 | 0.635 | 0.901 | 0.688 | 0.899 | 0.864 | 0.819 | 0.797 | 0.678 | **0.931** |
| WangLung | 0.761 | 0.651 | 0.788 | 0.742 | 0.914 | 0.435 | 0.918 | 0.806 | 0.838 | **0.919** |
| QX_Spleen | 0.748 | 0.731 | 0.830 | 0.779 | 0.806 | 0.656 | 0.674 | 0.790 | 0.644 | **0.894** |
| Avg | 0.704 | 0.688 | 0.753 | 0.751 | 0.787 | 0.705 | 0.747 | 0.766 | 0.729 | **0.883** |

As shown in Table 3, 4, 5 and 6, SCSW achieves SOTA clustering performance across all nine scRNA-seq datasets, robustly surpassing baselines. This effectiveness is rooted in two core innovations: The self-weighted fusion mechanism adaptively suppresses noisy views, ensuring the global embedding $\mathbf{H}$ carries the highest consensus information. The synergistic application of the contrastive learning and KL objectives rigorously optimizes the discriminability of the global embedding $\mathbf{H}$, leading to a highly compact and cohesive intra-cluster configuration. Our method demonstrates superior robustness when applied to challenging, high-heterogeneity datasets. This notable stability is achieved through the integration of its multi-scale topological fusion module and direct clustering constraints, which together effectively stabilize the latent feature space and enable the precise separation of subtle cellular subpopulations.

### 3.4 Visualization

We employ the UMAP method for 2D visualization in the latent space to intuitively assess the quality of the global features learned by model and its effectiveness in preserving cell topological structure and achieving category separability. Figure 2 compares the visualization results of the original PCA features with the optimal global features learned by SCSW on four representative datasets. The results show that, compared to the original embeddings, Our method generates more compact and better separated clusters, with significantly clearer boundaries in datasets like Klein and Plasschaert. These visualizations, complementing quantitative metrics, confirm our method's stronger biological consistency and separability.

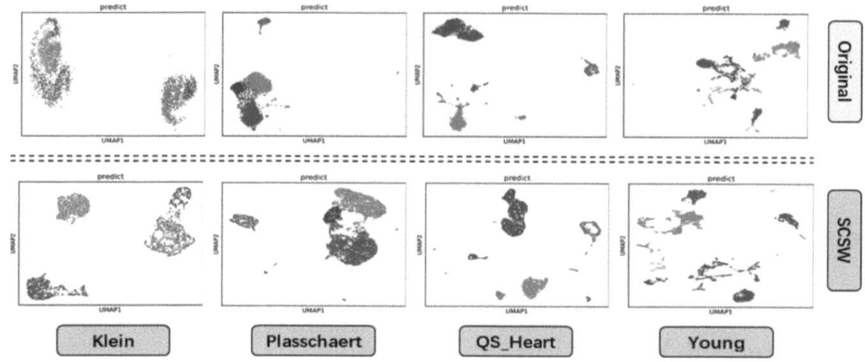


**Fig. 2.** UMAP visualization of cell embeddings from four scRNA-seq datasets, comparing the original data with the SCSW embeddings.

### 3.5 Ablation Experiments

To rigorously quantify the contribution of each core module, we performed comprehensive ablations on four datasets (Klein, 10X_PBMC, Young, Chen), assessing the average performance loss across variants. The variants involved removing or replacing the Multi-view construction (with single-view), the View projection head, and the Self-weighted fusion (with average fusion). Figure 3 visualizes the results via a radar chart, where the consistent and noticeable performance degradation upon the removal of any module validates their independent contributions. This analysis fundamentally emphasizes the synergistic and indivisible nature of the SCSW framework, confirming that every component is essential for its superior overall performance.

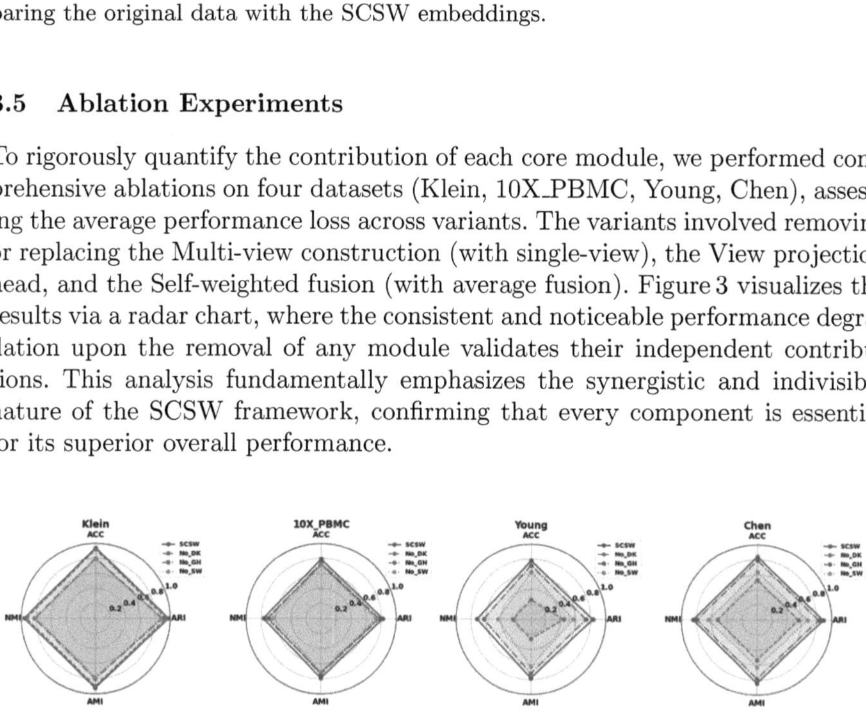


**Fig. 3.** Radar plots of ablation results for selected datasets showing Klein, 10X_PBMC, Young and Chen, NO_DK denotes removal of dual-graph topology, retaining only the raw view, NO_GH denotes exclusion of the global contrastive projection head, and NO_SW denotes replacement of the adaptive self-weighting mechanism with unweighted fusion.

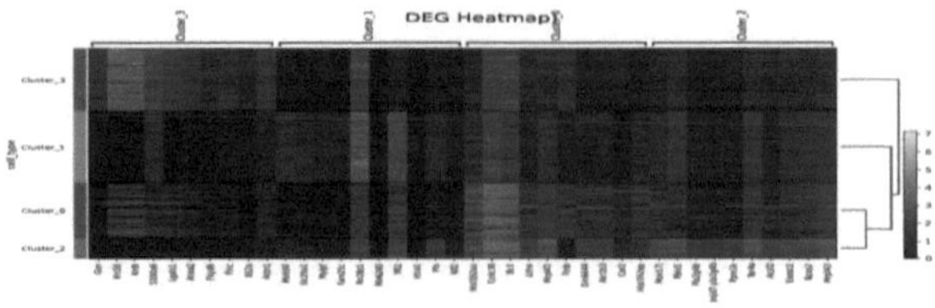


**Fig. 4.** DEG heatmap of the Top-10 marker genes for each SCSW cluster, highlighting the distinct and cluster-specific transcriptional profiles.

# 4 Biological Validation

## 4.1 Differential Expression Gene (DEG) Heatmap Analysis

We validated the biological plausibility of our clustering via Wilcoxon rank-sum DEG analysis (with BH correction) on the Klein dataset. Figure 4 confirms that every identified cluster exhibits a highly specific and non-overlapping transcriptional profile, where Top-10 DEG heatmap are expressed exclusively within their respective groups. This DEG heatmap demonstrates our method's superior ability to delineate subtle subpopulations, effectively preventing the erroneous merging of heterogeneous groups observed in baseline methods.

## 4.2 GO Analysis

To validate the functional relevance and biological homogeneity of Our method clusters, we performed GO enrichment analysis (Hypergeometric Test with FDR-corrected $P$-values) on the top DEG. Figure 5 displays the GO Biological Process (BP) terms for clusters C1-C4, which clearly reveal distinct, biologically meaningful functions for each subpopulation (e.g., C1: Developmental process; C3: Immune system process). These GO results confirm that SCSW identifies biologically homogeneous cell subpopulations, a success attributed to its self-weighted contrastive fusion mechanism which robustly processes noisy scRNA-seq data and effectively reinforces cluster boundaries.

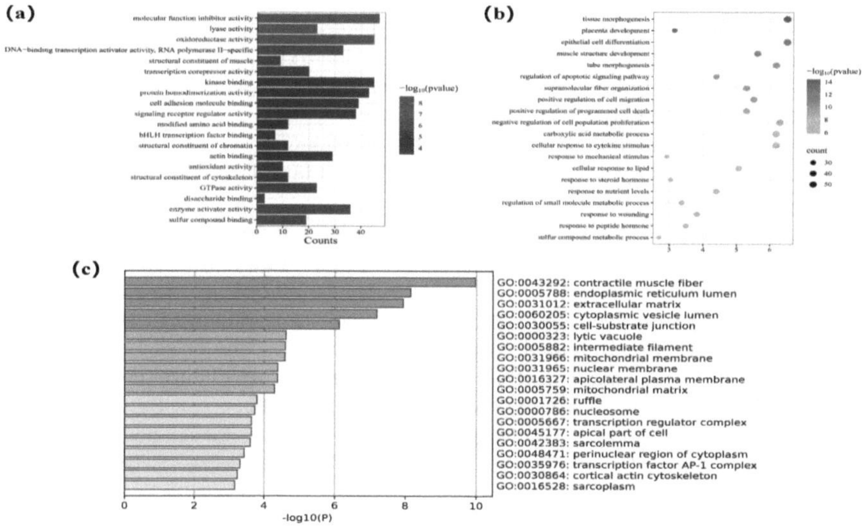


**Fig. 5.** GO enrichment analysis validates SCSW's functional relevance. The distinct GO Biological Process BP terms for clusters C1-C4 confirm the identification of biologically homogeneous cell subpopulations.

# 5 Conclusions

This paper introduces Self-Weighted Contrastive Fusion (SCSW), a novel deep multi-view framework mitigating inter-view conflict and representation degradation in scRNA-seq. SCSW achieves this by integrating three core innovations: multi-view construction, a decoupled feature network, and a self-weighted fusion mechanism. Extensive experiments on nine scRNA-seq datasets confirm SCSW significantly outperforms nine SOTA baselines and validates the discovery of cell clusters with high biological purity and functional specificity. To address the framework's current limitation with multi-modal data heterogeneity, future work will incorporate Generative Models (VAE/GAN) for robust weighting and imputation strategies.

# References

1. Li, H., McCarthy, D.J., Shim, H., Wei, S.: Trade-off between conservation of biological variation and batch effect removal in deep generative modeling for single-cell transcriptomics. BMC Bioinform. **23**(1), 460 (2022)
2. Dong, Y., Chen, Z., Yang, F., Wei, J., Huang, J., Long, X.: Prediction of immunotherapy responsiveness in melanoma through single-cell sequencing-based characterization of the tumor immune microenvironment. Transl. Oncol. **43**, 101910 (2024)
3. Hui-Qi, Q., Kao, C., Hakonarson, H.: Single-cell RNA sequencing technology landscape in 2023. Stem Cells **42**(1), 1–12 (2024)

4. Wang, X., Zhai, Y., Zheng, H.: Deciphering the cellular heterogeneity of the insect brain with single-cell RNA sequencing. Insect Sci. **31**(2), 314–327 (2024)
5. Amezquita, R.A., Lun, A.T.L., et al.: Orchestrating single-cell analysis with bioconductor. Nat. Methods **17**(2), 137–145 (2020)
6. Trapnell, C., Cacchiarelli, D., Grimsby, J., et al.: Pseudo-temporal ordering of individual cells reveals dynamics and regulators of cell fate decisions. Nat. Biotechnol. **32**(4), 381 (2014)
7. Wang, J., Wang, L.-P., Yuan, S.-S., et al.: NLRRC: a novel clustering method of jointing non-negative LRR and random walk graph regularized NMF for single-cell type identification. IEEE J. Biomed. Health Inform. **27**(10), 5199–5209 (2023)
8. Papalexi, E., Satija, R.: Single-cell RNA sequencing to explore immune cell heterogeneity. Nat. Rev. Immunol. **18**(1), 35–45 (2018)
9. Wang, J., Ma, A., Chang, Y., et al.: scGNN is a novel graph neural network framework for single-cell RNA-seq analyses. Nat. Commun. **12**(1), 1882 (2021)
10. Gan, Y., Huang, X., Zou, G.: Deep structural clustering for single-cell RNA-seq data jointly through autoencoder and graph neural network. Brief. Bioinform. **23**(2) (2022)
11. Linlin, D., Han, R., et al.: SCCCL: single-cell data clustering based on self-supervised contrastive learning. IEEE/ACM Trans. Comput. Biol. Bioinf. **20**(3), 2233–2241 (2023)
12. Wang, J., Xia, J., et al.: scDCCA: deep contrastive clustering for single-cell RNA-seq data based on auto-encoder network. Brief. Bioinform. **24**(1) (2023)
13. Yu, Z., Su, Y., et al.: Topological identification and interpretation for single-cell gene regulation elucidation across multiple platforms using scMGCA. Nat. Commun. **14**(1), 400 (2023)
14. Shu, Z., Xia, M., et al.: Multi-level multi-view network based on structural contrastive learning for scRNA-seq data clustering. Brief. Bioinform. **25**(6), bbae562 (2024)
15. Liu, T., Jia, C., et al.: scDFN: enhancing single-cell RNA-seq clustering with deep fusion networks. Brief. Bioinform. **25**(6), bbae486 (2024)
16. Wang, L., Li, W., et al.: nsDCC: dual-level contrastive clustering with nonuniform sampling for scRNA-seq data analysis. Brief. Bioinform. **25**(6), bbae477 (2024)
17. He, Y., Chen, X., et al.: Deep multi-constraint soft clustering analysis for single-cell RNA-seq data via zero-inflated autoencoder embedding. IEEE/ACM Trans. Comput. Biol. Bioinf. **20**(3), 2254–2265 (2023)
18. Lei, W., Ma, Z., et al.: Domain adaption based on source dictionary regularized RKHS subspace learning. Pattern Anal. Appl. **24**(4), 1513–1532 (2021)

# When +1% is Not Enough: A Paired Bootstrap Protocol for Evaluating Small Improvements

Wenzhang Du(✉)

Department of Computer Engineering, Mahanakorn University of Technology, International College (MUTIC), Bangkok, Thailand
dqswordman@gmail.com

**Abstract.** Recent machine learning papers often report one–two percentage point improvements from a single run. These gains are measured on standard benchmarks and are highly sensitive to random seeds, data ordering, and implementation details, yet are rarely accompanied by uncertainty estimates or significance tests. It is therefore unclear when a reported "+1–2%" reflects a real algorithmic advance versus noise.

We revisit this problem under realistic compute budgets, where only a few runs are affordable. We propose a simple, PC-friendly protocol based on paired multi-seed runs, bias-corrected and accelerated (BCa) bootstrap confidence intervals, and a sign-flip permutation test on per-seed deltas. The protocol is intentionally conservative and is meant as a guardrail against over-claiming.

We instantiate it on CIFAR-10, CIFAR-10N, and AG News using synthetic "no-improvement", small-gain, and medium-gain scenarios. Single runs and unpaired $t$-tests often suggest significant gains for 0.6–2.0 point improvements, especially on text. With only three seeds, our paired protocol never declares significance in these settings. We argue that such conservative evaluation is a safer default for small gains under tight budgets.

**Keywords:** evaluation protocol · bootstrap · permutation test · reproducibility

## 1 Introduction

Single numbers still dominate empirical reporting in machine learning. A typical paper trains a baseline and a new variant once, reports a difference of one or two percentage points on a benchmark, and concludes that the new method "significantly outperforms" the baseline. Yet a growing body of work has shown that performance can vary substantially across random seeds, data orderings, and even hardware and library versions [1–3, 8, 9]. Small improvements are particularly fragile: a different seed or slightly different environment may erase or reverse the observed gain.


R. C.-W. Wong et al. (Eds.): PAKDD 2026, LNAI 16597, pp. 519–531, 2026.
https://doi.org/10.1007/978-981-92-1300-9_41

Despite this, multi-seed reporting and principled uncertainty quantification remain the exception rather than the rule. When authors do repeat experiments, they often apply unpaired $t$-tests that assume independence and approximate normality, conditions that may not hold for small samples and skewed metric distributions [5–7]. Moreover, many practitioners can afford at most a handful of seeds (e.g., three), making classical asymptotic arguments even less convincing.

In this paper we do *not* propose yet another training trick. Instead, we turn the knife toward **evaluation itself** and ask:

> Given a small number of runs and a small apparent improvement, how can we avoid over-claiming success?

Our answer is a simple protocol that any lab can run on a single consumer GPU. For each of $k$ random seeds, we train both the baseline and the new variant under identical conditions (same data split, same seed, same number of epochs), record the per-seed metric difference, and then analyse these deltas with a BCa bootstrap interval and a sign-flip permutation test. A method is allowed to claim a "significant improvement" only if the BCa interval lies entirely above zero and the permutation $p$-value is below a chosen threshold (we use 0.05). Otherwise, the default recommendation is *not* to claim significance.

We study how this protocol behaves in realistic small-gain settings by constructing controlled scenarios on three benchmarks—CIFAR-10, CIFAR-10N, and AG News—with no, small, and medium modifications. We compare three evaluation schemes: (i) single-run deltas; (ii) unpaired Welch $t$-tests on independent runs; and (iii) our paired BCa+permutation procedure. Empirically we find that single-run comparisons and unpaired tests frequently report apparently significant gains, especially on noisy labels and text [14,15], while our protocol systematically refuses to endorse 0.5–2 percentage-point gains when only three seeds are available.

Our contributions are:

1. A simple, paired evaluation protocol for small improvements that combines per-seed deltas, BCa bootstrap confidence intervals, and sign-flip permutation tests into a conservative decision rule.
2. A controlled case study across three benchmarks and synthetic S0/S1/S2 scenarios, showing how standard practice overstates evidence for small gains, while our protocol behaves as a guardrail against over-claiming.
3. Practical guidance for authors and reviewers on interpreting small improvements under tight compute budgets.

## 2 Related Work

### 2.1 Reproducibility and Variance in ML Experiments

Multiple studies have documented substantial run-to-run variance in deep learning experiments, arising from random initialisation, mini-batch sampling, data ordering, and implementation quirks [1–3,8,14]. Such variance can reorder model rankings and invalidate conclusions drawn from single runs, especially on modest-size benchmarks.

Reproducibility initiatives and checklists for major conferences encourage authors to report seeds, hyperparameters, and environment details, and to release code when possible [4,9]. These efforts improve transparency but do not by themselves answer the question of how many runs are enough, or how to aggregate a small number of runs into a statistically grounded conclusion. Our work is complementary: we focus not on open-sourcing or environment management, but on the statistical layer on top of a handful of runs.

### 2.2 Statistical Comparison of Learning Algorithms

Classical work on comparing classifiers advocates a variety of tests, including cross-validated $t$-tests, the 5×2CV test, Wilcoxon signed-rank tests, and permutation methods [5,6,12]. These approaches typically assume reasonably large sample sizes (either many datasets or many folds) and i.i.d. conditions. In modern deep learning, we often have the opposite: a single dataset, few seeds, and complicated dependence structures.

Bootstrap methods have been proposed as more robust alternatives for non-Gaussian, skewed metrics. The BCa bootstrap adjusts for bias and skewness in the resampled distribution and enjoys good asymptotic properties [10,11]. Permutation tests provide finite-sample control of type-I error under symmetry assumptions [12]. Our protocol builds on these tools but adapts them to the **per-seed delta** setting, where we explicitly exploit pairing and accept that the number of deltas is tiny ($k \approx 3$).

### 2.3 Multi-seed Reporting and Paired Evaluation

In NLP and vision, several recent papers have argued for reporting score distributions over random seeds and for stronger statistical testing [7–9]. They show that rankings between models can flip when new seeds are added, and that standard practice often underestimates variance. These works typically use unpaired tests or cross-dataset comparisons. We extend this line by (i) formalising a paired delta + bootstrap + permutation pipeline, and (ii) empirically characterising its behaviour as a conservative filter in realistic small-gain scenarios.

## 3 Paired Bootstrap Evaluation Protocol

### 3.1 Problem Setting

We consider two models: a baseline $M_0$ and a variant $M_1$. Given a dataset $D$ and a performance metric $\theta(\cdot)$ (e.g., test accuracy), each training run produces a random outcome due to stochastic optimisation and data sampling. Let $X$ and $Y$ denote the metric for $M_1$ and $M_0$ respectively in one run.

The quantity of interest is the expected improvement

$$\Delta_{\text{true}} = \mathbb{E}[X - Y].$$

We aim to answer: Is $\Delta_{\text{true}}$ clearly positive, zero, or negative? Given only a small number of runs $k$, how cautious should we be in claiming $\Delta_{\text{true}} > 0$?

### 3.2 Paired Multi-seed Design

We adopt a paired design. For each seed $s_i$, $i = 1, \ldots, k$:

1. Fix all randomness (initialisation, data shuffling, augmentation) via seed $s_i$.
2. Train $M_0$ and evaluate $\theta_{0,i}$ on a fixed test set.
3. Under the same seed $s_i$, train $M_1$ and evaluate $\theta_{1,i}$.
4. Record the per-seed delta $\Delta_i = \theta_{1,i} - \theta_{0,i}$.

This mimics a counterfactual experiment: "holding the world defined by seed $s_i$ fixed, how much better is $M_1$ than $M_0$?"

If $X$ and $Y$ denote the random metrics in a paired run, basic variance algebra gives

$$\text{Var}(X - Y) = \text{Var}(X) + \text{Var}(Y) - 2\,\text{Cov}(X, Y).$$

In typical settings, $X$ and $Y$ are positively correlated: both models perform slightly better on easy splits and slightly worse on hard ones, so $\text{Cov}(X, Y) > 0$. Thus pairing reduces the variance of $X - Y$ relative to independent runs. We denote the sample mean of deltas by $\bar{\Delta} = \frac{1}{k}\sum_{i=1}^{k} \Delta_i$.

### 3.3 BCa Bootstrap Confidence Intervals

Given $\{\Delta_i\}$, we estimate uncertainty via the bias-corrected and accelerated (BCa) bootstrap [10,11]:

1. Generate $B$ bootstrap samples by resampling the $k$ deltas with replacement, each time computing the bootstrap mean $\bar{\Delta}^*$.
2. Compute the bias-correction and acceleration terms from the bootstrap and jackknife distributions.
3. Derive adjusted quantiles for the lower and upper bounds of the $(1 - \alpha)$ interval, $[L, U]$.

The BCa interval adjusts for bias and skewness in the empirical distribution of $\bar{\Delta}$ and is particularly useful when $k$ is small and the distribution is asymmetric. Under standard regularity conditions and large $B$, BCa intervals are asymptotically valid.

### 3.4 Sign-Flip Permutation Test

To test $H_0 : \Delta_{\text{true}} = 0$, we use a sign-flip permutation test [12]:

1. Compute the observed test statistic $T_{\text{obs}} = \bar{\Delta}$.
2. For each of $P$ permutations, sample random signs $\sigma_i \in \{-1, +1\}$ and compute $T^{(p)} = \frac{1}{k} \sum_{i=1}^{k} \sigma_i \Delta_i$.
3. Estimate the two-sided p-value as $p = \frac{1+\#\{p:|T^{(p)}| \geq |T_{\text{obs}}|\}}{1+P}$.

Under symmetry of the $\Delta_i$ distribution, this test controls type-I error exactly at finite $k$. In practice, it behaves conservatively when $k$ is very small, which we accept and even desire for our guardrail purpose.

### 3.5 Decision Rule

Our final decision rule is intentionally strict:

- Call "significant improvement" only if the BCa interval $[L, U]$ satisfies $L > 0$ and the permutation p-value $p < \alpha$ (we use $\alpha = 0.05$).
- In all other cases—including ambiguous intervals or borderline p-values—our default recommendation is: *Do not claim a statistically significant improvement.*

This rule is designed as a brake, not an accelerator: when evidence is weak or sample size is tiny, it pushes authors toward neutrality.

## 4 Experimental Setup

### 4.1 Datasets and Models

We evaluate the protocol on three benchmarks, chosen to cover both vision and text, and both clean and noisy labels.

*CIFAR-10.* 50k training and 10k test 32×32 colour images over 10 classes [13]. We train a standard ResNet-18 with SGD (momentum 0.9, batch size 128) and moderate data augmentation (random crop and horizontal flip).

*CIFAR-10N.* CIFAR-10 with human-annotated noisy labels [14]. We train the same ResNet-18 backbone but use noisy labels for training and the clean test set for evaluation.

*AG News.* A text classification dataset with 120k training and 7.6k test news headlines in four categories [15]. We join title and description, tokenise with a simple word tokenizer, build a 20k-word vocabulary, and train a lightweight TextCNN model, a standard convolutional architecture used in prior work on text classification and vision [15,16].

All experiments are run on a single consumer GPU (NVIDIA GeForce RTX 5090, CUDA 12.8) with PyTorch 2.8.0 under Windows 10, using Python 3.9.21 and `num_workers=0`. For each configuration we use exactly the same training schedule across protocols and seeds.

**Table 1.** Datasets and synthetic scenarios. Intended $\Delta$ is the rough expected gain in percentage points.

| Dataset | Modality | Scenario | New vs. old configuration | Intended $\Delta$ |
|---|---|---|---|---|
| CIFAR-10 | vision | S0 (no-improve) | Identical ResNet-18 baseline | $\approx 0$ |
| CIFAR-10 | vision | S1 (small gain) | + label smoothing ($\varepsilon = 0.05$) | 1–2 |
| CIFAR-10 | vision | S2 (medium gain) | + strong augmentation + label smoothing | $\approx 1$ |
| CIFAR-10N | vision/noisy | S1 (small gain) | Noisy labels, + label smoothing | $\approx 1$ |
| CIFAR-10N | vision/noisy | S2 (medium gain) | Noisy labels, + strong augmentation + LS | $\approx 1$ |
| AG News | text | S0 (no-improve) | Identical TextCNN baseline | $\approx 0$ |
| AG News | text | S1 (small gain) | + label smoothing ($\varepsilon = 0.05$) | 0.5–1 |
| AG News | text | S2 (medium gain) | + LS + word dropout (0.1) | $\approx 1$ |

### 4.2 Synthetic Scenarios S0/S1/S2

For each dataset we define scenarios that approximate no, small, and medium gains:

- **S0 (no-improve).** Baseline and variant are identical. This approximates $\Delta_{\text{true}} = 0$ and is used to measure false positive behaviour.
- **S1 (small gain).**
  - CIFAR-10 and CIFAR-10N: add label smoothing with $\varepsilon = 0.05$ to the cross-entropy loss.
  - AG News: TextCNN with label smoothing $\varepsilon = 0.05$.
- **S2 (medium gain).**
  - CIFAR-10 and CIFAR-10N: combine label smoothing with stronger augmentation.
  - AG News: label smoothing plus word dropout (10% tokens dropped to PAD during training).

Table 1 summarises the datasets and scenarios. All runs use $k = 3$ paired seeds.

### 4.3 Evaluation Protocols Compared

We compare three evaluation schemes:

1. **Single.** A single baseline run vs. a single variant run. We report only the difference in test accuracy, with no uncertainty measure.
2. **Unpaired *t*-test ($k = 3$).** We train 3 independent runs of the baseline and 3 independent runs of the variant, ignoring pairing, and apply Welch's *t*-test on the two sets of accuracies [5–7].
3. **Paired BCa+perm ($k = 3$).** Our proposed protocol: three paired runs, BCa confidence interval on deltas, and sign-flip permutation test, with the strict decision rule in Sect. 3.

**Table 2.** Summary of small-gain experiments (percentage-point test accuracy). $\Delta_{\text{single}}$ is the single-run difference. $\Delta_{\text{paired}}$ and the BCa interval come from our paired protocol (3 seeds). $p_{\text{perm}}$ is the paired sign-flip permutation p-value and $p_{\text{unpaired}}$ is Welch's unpaired $t$-test p-value.

| Dataset | Scenario | Type | $\Delta_{\text{single}}$ | $\Delta_{\text{paired}}$ (BCa 95% CI) | $p_{\text{perm}}$ | $p_{\text{unpaired}}$ |
|---|---|---|---|---|---|---|
| CIFAR-10 | S0 | no-improve | 0.00 | 0.00 [0.00, 0.00] | 1.00 | 1.00 |
| CIFAR-10 | S1 | small gain | 2.29 | 1.82 [0.37, 2.62] | 0.25 | 0.17 |
| CIFAR-10 | S2 | medium gain | 0.32 | 0.93 [0.32, 1.47] | 0.25 | 0.12 |
| CIFAR-10N | S1 | small gain | 1.88 | 1.02 [-1.26, 2.25] | 0.50 | 0.39 |
| CIFAR-10N | S2 | medium gain | 0.66 | 0.93 [0.35, 1.41] | 0.24 | 0.09 |
| AG News | S0 | no-improve | 0.00 | 0.00 [0.00, 0.00] | 1.00 | 1.00 |
| AG News | S1 | small gain | 0.46 | 0.64 [0.46, 0.79] | 0.24 | 0.03 |
| AG News | S2 | medium gain | 0.76 | 1.03 [0.76, 1.20] | 0.24 | 0.003 |

For each protocol we record per-scenario means, intervals, p-values, and decisions. Tables and figures in the paper are generated directly from the logged CSV files, in the spirit of transparent reporting advocated in prior work [8,9].

# 5 Results

## 5.1 Main Quantitative Results

Table 2 summarises the small-gain experiments across datasets and scenarios. Each row corresponds to one dataset–scenario pair and compares: (i) a single-run delta; (ii) the mean delta and BCa 95% interval under our paired protocol; and (iii) permutation and unpaired $t$-test p-values.

Several patterns emerge.

First, the S0 rows confirm a sanity check: in all no-improvement scenarios, the average delta is exactly zero, the BCa interval degenerates to $[0, 0]$, and both permutation and unpaired tests give $p = 1.0$. Our protocol does not introduce spurious differences.

Second, for S1 and S2 the single-run deltas frequently look attractive: between +0.46 and +2.29 percentage points. On AG News S2, for example, a single run shows a +0.76pp gain; on CIFAR-10 S1 some runs exceed +2pp.

Third, unpaired $t$-tests on three seeds tend to be optimistic. On AG News, unpaired p-values are 0.03 (S1) and 0.003 (S2), which many papers would describe as "significant" or even "highly significant". On CIFAR-10N S2 the unpaired p-value is 0.09, suggestive but not conclusive.

In contrast, our paired BCa+perm protocol is consistently conservative at $k = 3$. It never declares significance: all permutation p-values are either 0.24, 0.25, or larger, even when BCa intervals are strictly positive (e.g., AG News S2 with $[0.76, 1.20]$). This is a direct consequence of the small discrete sample of deltas and our strict decision rule, and we intentionally embrace this behaviour as a guardrail.

### 5.2 Effect Sizes and P-Values Across Tasks

Figure 1 visualises the mean improvements and BCa 95% intervals under the paired protocol. S0 bars cluster tightly around zero, confirming no systematic bias. S1/S2 bars lie mostly between +0.5 and +2.0pp, with CIFAR-10N showing the widest intervals due to label noise. Overall effect sizes are small and uncertainty is non-trivial, which supports our view that aggressive significance claims are risky at $k = 3$.

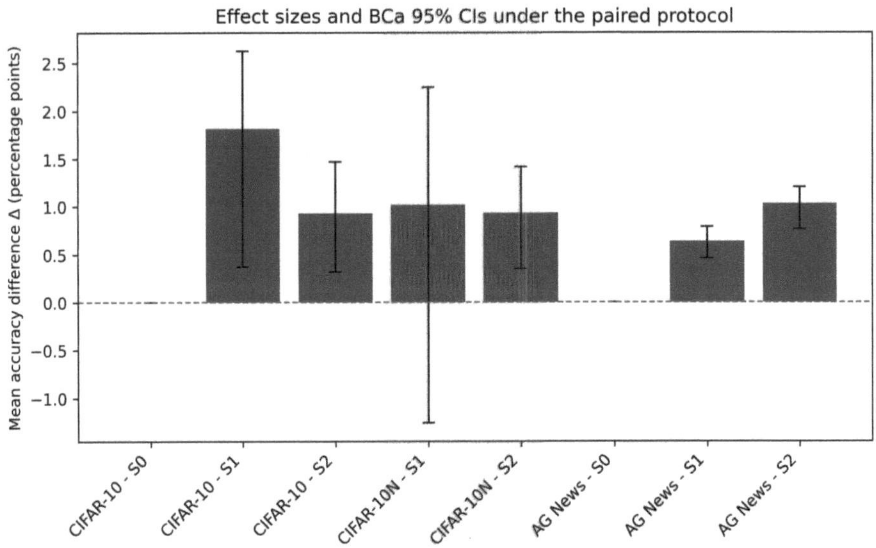


**Fig. 1.** Effect sizes and BCa 95% confidence intervals under the paired protocol. Bars show $\Delta_{\text{paired}}$; whiskers show BCa intervals; the dashed line marks zero.

Figure 2 compares unpaired $t$-test p-values (x-axis) with paired permutation p-values (y-axis) across S1/S2 scenarios. Points below the horizontal dashed line and to the left of the vertical line would correspond to cases where both tests call significance. In practice, all S1/S2 points lie in the region where unpaired p-values are small (sometimes $< 0.01$) but paired p-values remain around 0.24. Thus, at $k = 3$ our protocol systematically refuses to turn suggestive trends into "wins".

### 5.3 Case Study: CIFAR-10 Learning Curves

To illustrate how small gains emerge over training, Fig. 3 plots test accuracy vs. epoch for CIFAR-10 S1 (label smoothing), for three seeds and both baseline and variant. The variant consistently tracks slightly above the baseline after the early epochs, with end-of-training gaps between 1–3pp depending on the seed. However, the curves also show non-monotonicity and seed-to-seed variability: some seeds close the gap late, others diverge. Our protocol interprets this as suggestive but not conclusive evidence at $k = 3$.

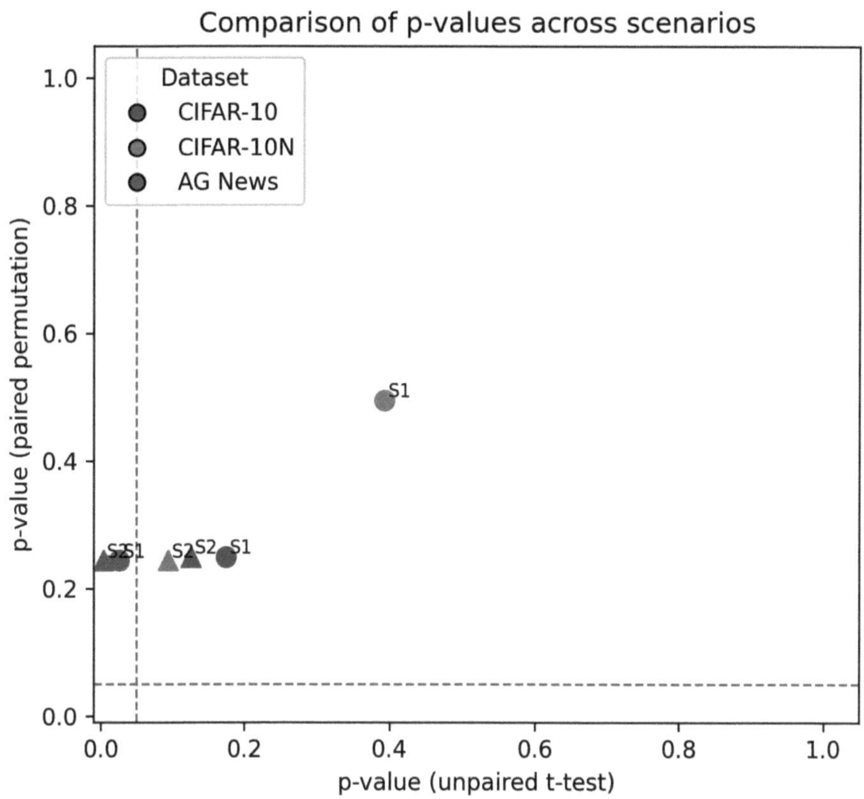


**Fig. 2.** Comparison of p-values across scenarios: unpaired $t$-test (x-axis) vs. paired permutation (y-axis). Colours denote datasets; markers denote S1/S2. Dashed lines mark the 0.05 threshold.

## 6 Discussion and Limitations

Our study argues for a simple but strong practical position: if your improvement is on the order of one to two percentage points and you only have three seeds, you should expect a careful paired bootstrap protocol to say: "This evidence is too weak to claim significance".

*When is the Protocol Most Useful?* The protocol is most useful when compute budgets limit you to very few seeds (3–5), effect sizes are small ($\leq$ 2pp), and you want a principled way to avoid fooling yourself or your readers. In such settings, our guardrail helps ensure that papers do not over-interpret fragile differences.

*Conservatism vs. Power.* At $k = 3$ the permutation test is deliberately conservative: the discrete null distribution makes it hard to obtain very small p-values. One could relax our decision rule by requiring only CI$>0$ *or* $p < 0.05$, which

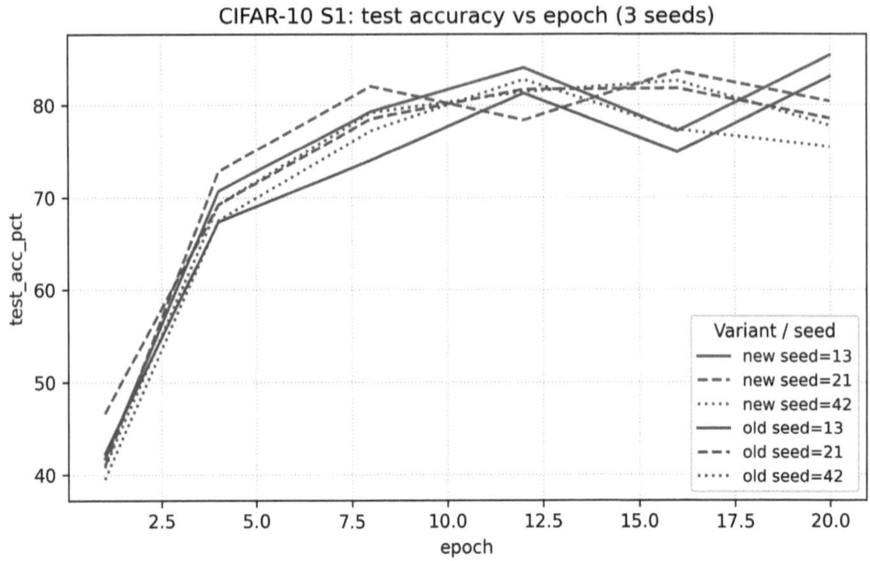


**Fig. 3.** CIFAR-10 S1 learning curves: test accuracy vs. epoch for three seeds, baseline (old) vs. variant (new).

would mark several S1/S2 scenarios as "significant". We intentionally adopt the stricter "CI$>0$ & $p < 0.05$" rule to prevent over-claiming at minimal $k$ and to encourage richer reporting. When larger budgets allow $k \geq 5$, the same protocol becomes less conservative and can detect moderate effect sizes with reasonable power.

*Limitations.* There are clear limitations. First, we only study small models on CIFAR-10/10N and AG News; larger-scale tasks may exhibit different variance structure. Second, we focus on accuracy; extending the protocol to fairness, robustness or calibration metrics is straightforward in principle but may raise issues with multiple testing and dependency. Third, the permutation test assumes approximate symmetry of the delta distribution; severe skewness may require alternative non-parametric tests or Bayesian treatment. Finally, our protocol does not replace domain judgment: a statistically significant +1% may or may not be practically important.

*Practical Recommendations.* Based on our findings, we recommend: (1) always report deltas across multiple seeds; (2) exploit pairing by training baseline and variant under the same seeds and data splits; (3) use bootstrap intervals and permutation p-values, not only parametric $t$-tests, especially when $k$ is small or metrics are skewed; (4) be explicit about decision rules; and (5) when in doubt, under-claim.

## 7 Ethical and Societal Considerations

Our work concerns evaluation methodology and uses only standard public datasets (CIFAR-10, CIFAR-10N, AG News) that contain no personally identifiable information. It does not involve human subjects or sensitive attributes. The primary societal impact is indirect and positive: if widely adopted, conservative small-gain evaluation could reduce publication of spurious improvements and encourage more rigorous empirical practice. At the same time, stricter evaluation might raise the bar for publishing modest but practically valuable tweaks. Statistical significance should not be the only criterion of value: effect sizes, computational cost, and domain needs should all be considered. Our protocol is meant as a tool to calibrate claims, not as a gatekeeping mechanism.

## 8 Reproducibility

All experiments are designed to be reproducible on a single consumer GPU. We fix random seeds at both the Python and framework level; log system fingerprints (OS, Python, PyTorch, CUDA, GPU model); store per-seed metrics and configuration files; and provide scripts to re-run experiments and regenerate tables and figures. Three key artifacts are: (i) data loading and training scripts for CIFAR-10/CIFAR-10N (ResNet-18) and AG News (TextCNN); (ii) an evaluation script that implements the paired BCa+perm protocol and outputs a summary CSV; and (iii) analysis code that reads the summary CSV and produces Figs. 1, 2 and 3 and the main tables. These artefacts allow other researchers to reproduce our findings, extend the protocol to new tasks, or stress-test it under different sample sizes and decision rules.

## 9 Conclusion

We revisited a mundane but pervasive situation: a new method shows a +1–2%-point improvement on a benchmark under a few random seeds. Should we believe this improvement? Should we call it statistically significant?

Our answer is deliberately cautious. Through a simple paired bootstrap protocol evaluated on CIFAR-10, CIFAR-10N, and AG News, we show that standard practices—single runs and unpaired $t$-tests—readily interpret such gains as significant, especially in noisy or textual settings. In contrast, our protocol acts as a guardrail: it never produces false positives in carefully constructed no-improvement scenarios and systematically declines to endorse small gains when only three seeds are available, even when p-values from unpaired tests look impressive. We view this conservatism not as a weakness, but as a feature: when evidence is weak, the default should be "no strong conclusion yet". We hope this work helps shift the culture around empirical claims in machine learning, especially for small improvements under tight budgets, and encourages the community to invest as much care in evaluation design as in model design.

**Disclosure of Interests.** The author has no competing interests to declare that are relevant to the content of this article.

## References


1. Bouthillier, X., Laurent, C., Vincent, P.: Unreproducible research is reproducible. In: Chaudhuri, K., Salakhutdinov, R. (eds.) Proceedings of the 36th International Conference on Machine Learning (ICML 2019). Proceedings of Machine Learning Research, vol. 97, pp. 725–734 (2019)
2. Bouthillier, X., et al.: Accounting for variance in machine learning benchmarks. Proc. Mach. Learn. Syst. **3**, 747–769 (2021)
3. Henderson, P., Islam, R., Bachman, P., Pineau, J., Precup, D., Meger, D.: Deep reinforcement learning that matters. In: Proceedings of the 32nd AAAI Conference on Artificial Intelligence (AAAI 2018), pp. 3207–3214. AAAI Press (2018)
4. Pineau, J., et al.: Improving reproducibility in machine learning research: a report from the NeurIPS 2019 reproducibility program. J. Mach. Learn. Res. **22**(164), 1–20 (2021)
5. Dietterich, T.G.: Approximate statistical tests for comparing supervised classification learning algorithms. Neural Comput. **10**(7), 1895–1923 (1998)
6. Demšar, J.: Statistical comparisons of classifiers over multiple data sets. J. Mach. Learn. Res. **7**, 1–30 (2006)
7. Dror, R., Baumer, G., Bogomolov, M., Reichart, R.: The hitchhiker's guide to testing statistical significance in natural language processing. In: Proceedings of the 56th Annual Meeting of the Association for Computational Linguistics (ACL 2018), pp. 1383–1392. Association for Computational Linguistics (2018)
8. Reimers, N., Gurevych, I.: Reporting score distributions makes a difference: performance study of LSTM-networks for sequence tagging. In: Proceedings of the 2017 Conference on Empirical Methods in Natural Language Processing (EMNLP 2017), pp. 338–348. Association for Computational Linguistics (2017)
9. Dodge, J., Gururangan, S., Card, D., Schwartz, R., Smith, N.A.: Show your work: improved reporting of experimental results in NLP. In: Proceedings of the 2019 Conference on Empirical Methods in Natural Language Processing and the 9th International Joint Conference on Natural Language Processing (EMNLP-IJCNLP 2019), pp. 2185–2194. Association for Computational Linguistics (2019)
10. Efron, B.: Better bootstrap confidence intervals. J. Am. Stat. Assoc. **82**(397), 171–185 (1987)
11. Efron, B., Tibshirani, R.J.: An Introduction to the Bootstrap. Chapman & Hall/CRC, New York (1994)
12. Good, P.I.: Permutation, Parametric, and Bootstrap Tests of Hypotheses. Springer Series in Statistics, 3rd edn. Springer, New York (2005)
13. Krizhevsky, A.: Learning multiple layers of features from tiny images. Technical report, Department of Computer Science, University of Toronto (2009)
14. Wei, J., Zhu, Z., Cheng, H., Liu, T., Niu, G., Liu, Y.: Learning with noisy labels revisited: a study using real-world human annotations. In: Proceedings of the 10th International Conference on Learning Representations (ICLR 2022) (2022)

15. Zhang, X., Zhao, J., LeCun, Y.: Character-level convolutional networks for text classification. In: Cortes, C., Lawrence, N.D., Lee, D.D., Sugiyama, M., Garnett, R. (eds.) Advances in Neural Information Processing Systems (NeurIPS 2015), vol. 28, pp. 649–657. Curran Associates, Inc. (2015)
16. Szegedy, C., Vanhoucke, V., Ioffe, S., Shlens, J., Wojna, Z.: Rethinking the Inception architecture for computer vision. In: Proceedings of the IEEE Conference on Computer Vision and Pattern Recognition (CVPR 2016), pp. 2818–2826. IEEE (2016)

# DDCG: Dual-granularity Dual-domain Collaborative Graph Neural Networks for Time Series Forecasting

Chenggang Xie, Ting Deng(✉), Yiqi Xiao, Ping Lu, and Renzhao Liang

Beihang University, Beijing, China
{xiechenggang,dengting,xiaoyiqi,luping,liangrenzhao}@buaa.edu.cn

**Abstract.** Multivariate time series forecasting has been widely applied in domains such as finance, traffic, energy, and healthcare. However, real-world time series often exhibit complex inter-channel dependencies. We propose DDCG, a Dual-granularity Dual-domain Collaborative Graph Neural Network, which models temporal structures at both global and patch levels. At the global scale, DDCG uses a multi-scale frequency-domain Mixture-of-Experts to enhance spectral representations and suppress noise, capturing the overall temporal characteristics and feature patterns of the series. At the patch scale, it employs a graph-based patch-level channeltemporal interaction mechanism to share channel graphs via a smooth-sharing strategy, effectively reducing computational complexity while preserving modeling precision. A fusion module integrates representations across both granularities, achieving cross-scale timefrequency collaboration and unified representation learning. Experiments on multiple real-world datasets show that DDCG achieves competitive performance, demonstrating its effectiveness and generality for complex multivariate time series modeling.

## 1 Introduction

Time series forecasting is a fundamental task in data mining, underpinning a wide range of applications such as finance [17], meteorology, transportation, and climate science [13]. Traditional statistical approaches (*e.g.,* ARIMA and exponential smoothing) benefit from interpretability and simplicity, but often fail to handle nonlinear, multivariate, and complex dynamic patterns. Deep learning architectures (*e.g.,* Recurrent Neural Networks (RNNs) [16], Convolutional Neural Networks (CNNs) [19], and Transformers [11]) have achieved remarkable progress by capturing long-term dependencies and complex temporal dynamics. However, these models still suffer from limited modeling granularity and high computational complexity, which hinder their ability to learn deep temporal representations.

Recently, patch-based and frequency-domain approaches [22] mitigate distributional drift via finer-grained modeling and spectral information. Nevertheless,


R. C.-W. Wong et al. (Eds.): PAKDD 2026, LNAI 16597, pp. 532–544, 2026.
https://doi.org/10.1007/978-981-92-1300-9_42

existing models lack dual-granularity and dual-domain collaboration: (1) At the global scale, frequency-domain modeling [4] lacks collaboration among frequency components and effective signal enhancement. (2) At the patch scale, channel-temporal interactions are modeled coarsely, failing to capture fine-grained dependencies or cross-scale relationships.

To overcome these limitations, we propose a **D**ual-granularity **D**ual-domain **C**ollaborative **G**raph neural network (DDCG) for time series forecasting, enabling multi-scale feature fusion and coordinated time-frequency modeling. At the global scale, DDCG introduces a global frequency-domain multi-scale expert module that decomposes signals into multiple frequency components [12] to capture global periodic patterns that complement local details, and enhances their representations through expert routing. At the patch scale, DDCG designs a patch-level channelinteraction graph module to model local dependencies and inter-channel relationships. Finally, a fusion module integrates representations from both granularities, enabling unified timefrequency coordination and multi-scale feature aggregation. Extensive experiments on multiple real-world datasets demonstrate that DDCG delivers strong and consistent performance, highlighting its effectiveness and generality for complex multivariate time series forecasting.

The main contributions of the paper are listed as follows:

- We propose DDCG, a dual-granularity dual-domain forecasting model that jointly models time series from global and local temporal perspectives.
- We design a two-level architecture, comprising a frequency-domain MoE module for multi-frequency enhancement and a graph-based patch module with shared channel graphs for fine-grained interactions.
- We propose a cross-domain fusion mechanism for coordinated time-frequency modeling and multi-scale integration, enhancing accuracy and robustness.

## 2 Related Work

**Deep Learning-Based Time Series Forecasting.** Time series forecasting has evolved from traditional approaches based on statistical learning [1] to neural network-based methods. Early deep models primarily rely on Recurrent Neural Networks (RNNs) [16] and Convolutional Neural Networks (CNNs) [2] to learn the dependencies between time series.

Recently, Transformer-based architectures and MLP methods have achieved good progress and become state-of-the-art approaches. Transformer-based methods like Informer [28] and PatchTST [14] improve efficiency via sparse attention and patching mechanisms. PathFormer [3] employs a dynamic routing mechanism to extract multi-scale temporal features from variable-length patches. ITransformer [11] further transposes the Transformer structure to capture the correlations between multivariate time series.

MLP structures have emerged as a lightweight alternative to Transformers. Lightweight MLP-based models such as DLinear [27] and SOFTS [6] focus on efficient temporal and channel-wise interactions.

**Frequency Learning and Temporal Learning.** Integrating signal processing and deep learning has advanced time series analysis by incorporating frequency-domain information and temporal-frequency interactions.

Frequency-domain techniques offer a distinct perspective, capturing pattern characteristics often overlooked by temporal methods. FreTS [26] employs a global frequency-domain view to learn temporal features and dependencies, FITS [22] enhances representation via frequency-domain low-pass filtering and complex spectral interpolation. Autoformer [20] introduces frequency-domain information to improve sequence decomposition. FEDformer [29] achieves efficient feature fusion via spectral random sampling and low-rank approximation, while Fan [24] proposes a novel frequency-domain normalization method to improve stationarity. Complementary to these, TimeFilter [8] and DUET [15] explore frequency-guided channel-temporal interactions using graph filtration and dual clustering.

From the temporal perspective, prior works focus on long-term temporal dependencies and cross-channel interactions. Some methods [3] prioritize modeling complex long-range dependencies along the time axis, while others [7] emphasize inter-channel correlations to capture sophisticated temporal-channel interactions.

**Novelty.** To the best of our knowledge, prior works typically focus on either dual-domain [26,29] or dual-granularity modeling [3,14]. In contrast, DDCG jointly unifies both within a collaborative framework, where a multi-scale frequency-domain MoE module augments time-domain features with frequency information, a patch-level channeltime graph is proposed to enrich global modeling with local features, and a cross-granularity fusion module integrates global and local features for timefrequency collaboration.

# 3 Methodology

We present a **D**ual-granularity **D**ual-domain **C**ollaboration **G**raph neural network (DDCG), a time series forecasting model that integrates global-local granularity modeling with synergistic time- and frequency-domain information to improve predictive performance. Given a sequence of historical observations $X \in \mathbb{R}^{C \times T}$, where $C$ is the number of channels and $T$ is the length of the look-back window, DDCG aims to predict future values $Y \in \mathbb{R}^{C \times F}$, where $F$ is the forecast horizon.

## 3.1 Overview of DDCG Framework

As illustrated in Fig. 1, we use Invertible Instance Normalization [11] to mitigate distributional drift and non-stationarity by normalizing inputs and denormalizing predictions. The backbone integrates the Global Frequency-domain Multi-scale Expert Module (FMR) and the Patch-level ChannelInteraction Graph Module (CIG) to extract dual-domain features, which are fused by the Cross-domain Dual-granularity Feature Fusion Module (CDF) to form high-resolution representations. Finally, a linear MLP outputs the forecasted values.

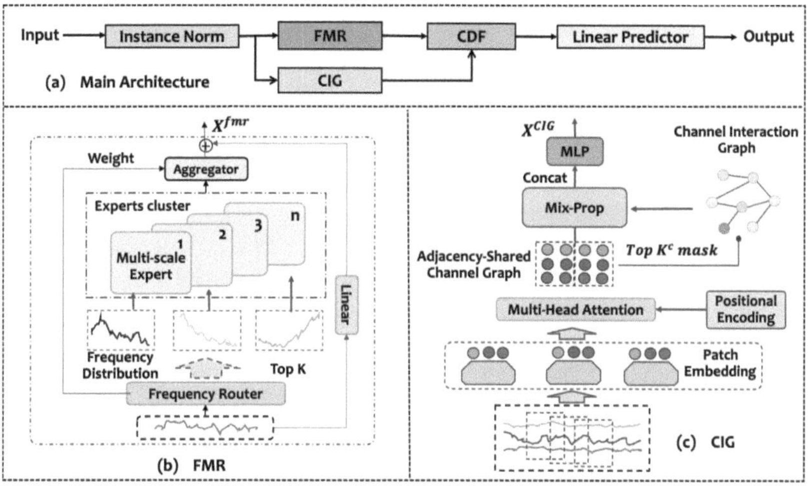


**Fig. 1.** The overall architecture of DDCG

### 3.2 Global Frequency-Domain Multi-scale Expert Module (FMR)

To fully exploit the heterogeneous frequency-domain characteristics across channels and enhance global frequency-domain representations, we develop a noise-balanced frequency-domain routing mechanism coupled with a multi-scale frequency-domain expert network.

**(1) Frequency-Domain Noise Routing.** To assign sequences with similar frequency distributions to each expert and balance expert load, we adopt the stochastic Mixture-of-Experts (sMoE) [18] technique. The procedure first computes the L2 norm of the FFT amplitudes for each normalized channel sequence, denoted as $Z = \|\mathbf{FFT}(X^{norm})\|_2$. We focus on amplitudes to capture energy distributions while avoiding the noise sensitivity often associated with phase information. The router generates expert scores augmented with stochastic noise via a reparameterization trick:

$$\mathbf{Router}(Z_i) = (Z_i \cdot W_g) + \epsilon \cdot \mathbf{Softplus}\left((Z_i \cdot W_{\mathbf{noise}})\right), \epsilon \sim \mathcal{N}(0, I). \quad (1)$$

Here the Softplus function ensures non-negative variance $\sigma_i = \mathbf{Softplus}(\cdot)$. This stabilizes score propagation and introduces stochastic perturbations. Then the top-$K$ sparsification is applied to the noisy routing scores by setting all values outside the top-$K$ to $-\infty$, followed by a softmax to produce the final routing weights:

$$\mathbf{TopK}(v_i, k) = \begin{cases} v_i, & \text{if } v_i \text{ is in the top } k \text{ elements of } v, \\ -\infty, & \text{otherwise,} \end{cases} \quad (2)$$
$$\mathbf{G}(Z) = \mathbf{Softmax}(\mathbf{TopK}(\mathbf{Router}(Z), K)).$$

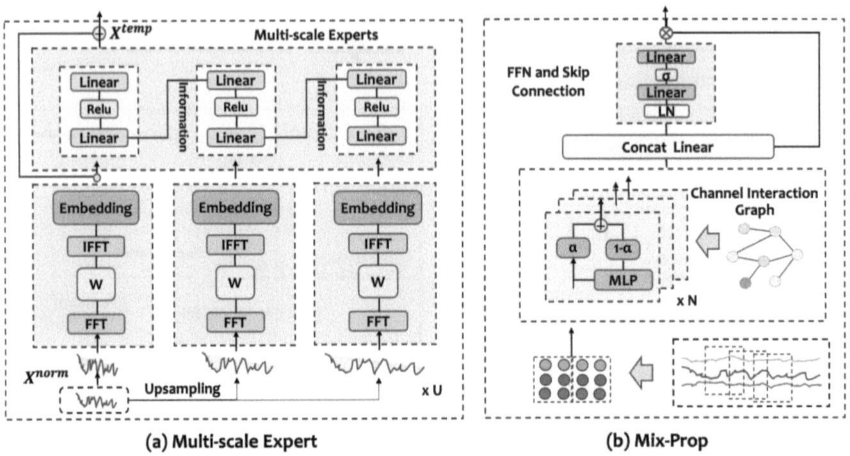


**Fig. 2.** (a) Frequency-domain multi-scale experts (Multi-scale Expert) incorporate upsampling and spectral enhancement, with temporal features extracted via feature fusion and residual connections. (b) Mix-Propagation Module (Mix-Prop) fuses multi-layer features via residual preservation to mitigate over-smoothing.

This design ensures that each expert focuses on sequences with similar frequency characteristics, improving feature extraction consistency.

**(2) Multi-scale Frequency-Domain Experts.** To enhance spectral signals and suppress noise, we design a multi-scale frequency-domain expert network (see Fig. 2(a)), where each expert focuses on specific frequency bands, and a multi-scale upsampling mechanism enriches frequency-domain features by applying FFT to upsampled sequences. The temporal signal is first hierarchically upsampled as

$$\chi = \mathbf{Upper}(X^{norm}, U), \quad \chi = \{\chi^1, \ldots, \chi^U\}, \tag{3}$$

where **Upper** denotes interpolation-based upsampling and $U$ is the number of upsampling levels. For each scale branch, the features are extracted via an independent learnable frequency-domain enhancer:

$$X_u^{emb} = \mathbf{Embedding}(\mathbf{FFT}^{-1}(\mathbf{FFT}(\chi^u) \odot W)), \tag{4}$$

where **Embedding** denotes the embedding layer.

We next adopt a top-down feature fusion architecture to integrate cross-scale features, via a top-down information propagation path. After dimensionality reduction, features from higher scales are fused hierarchically with lower-scale features, yielding a unified multi-scale representation:

$$\begin{aligned} &\mathbf{Trans}(X^{emb}) = \mathbf{MLP}(X^{emb}), \quad \mathbf{Trans} = \{\mathbf{Trans}_1, \dots \mathbf{Trans}_U\}, \\ &information = \begin{cases} \mathbf{Trans}_u(X_u^{emb}), & \text{if} \quad u = U \\ \mathbf{Trans}_u(X_u^{emb} + information), & \text{else} \end{cases}, \\ &X^{temp} = X_1^{emb} + information. \end{aligned} \tag{5}$$

Here, Eq. (5) denotes the fused features at level $u$ with multi-scale information extracted by lightweight MLPs and propagated hierarchically.

**(3) Aggregator.** Each multi-scale expert aggregates its temporal features using the routing-generated weights, followed by a linear projection layer to match the residual connection dimensions:

$$X_c^{temp} = \sum_{j=1}^{k} G(Z_c)_j \, X_c^{temp_j}, \; X^{fmr} = X^{temp} \oplus \mathbf{Linear}(X^{norm}), \tag{6}$$

where $G(Z_c)_j$ is the routing-generated weight for expert $j$ and channel $c$, $X_c^{temp_j}$ is the output from expert $j$, and $\oplus$ denotes residual fusion.

### 3.3 Patch-Level Channelinteraction Graph Module (CIG)

The CIG module extracts fine-grained temporal features by partitioning channel sequences into patches and building a patch-level temporal interaction graph. A shared-channel strategy reduces overfitting and enhances information sharing for global-scale modeling.

**(1) Patch-Level Temporal Feature Modeling.** Following PatchTST [14], we first divide the time series into multiple consecutive and overlapping patches. Formally, the normalized input sequence $X^{norm} \in \mathbb{R}^{C \times T}$ is partitioned into overlapping patches of length $P$ with overlap size $S$, forming $X_p \in \mathbb{R}^{C \times N \times D}$, where the number of patches is $N = \lfloor \frac{T-P}{S} \rfloor + 2$, and $D$ is the embedding dimension. Each patch is then projected into a new vector space with positional encoding:

$$X_{\text{p}} = X^{norm} \cdot W + P, \tag{7}$$

where $W$ denotes the linear mapping and $P$ is the cosine positional encoding. To capture patch-level temporal dependencies, a temporal attention layer is applied along the patch dimension:

$$X_{\text{tp}} = \text{LayerNorm}\left( X_{\text{p}} + \text{softmax}\left( \frac{X_Q X_K^{\top}}{\sqrt{d_k}} \right) X_V \right). \tag{8}$$

Here $X_Q$, $X_K^{\top}$, and $X_V$ are linear projections of $X_p$. This captures fine-grained temporal dynamics within patches, yielding high-quality features for cross-channel interactions.

**(2) Channel Graph Interaction Mechanism.** To model dynamic inter-channel dependencies, we design a Channel-Interaction Graph (CIG) module. For each patch, it constructs a channel-interaction graph, where nodes correspond to channels and edge weights are determined by the inter-channel feature similarity. This design enables information propagation and fusion across channels, through graph neural operations, to capture latent collaborative dynamics. To reduce complexity and suppress noise, only the top-$K^c$ most correlated channels are retained, forming a sparse adjacency matrix. Moreover, to mitigate overfitting and avoid per-patch recomputation, a shared-channel strategy is adopted such that every $r$ consecutive patches share the same channel-interaction graph, where $r$ is a hyperparameter to balance efficiency and locality.

Specifically, given the patch feature $X_{\text{tp}} \in \mathbb{R}^{C\times N\times D}$ computed using Formula 8, the number of required channel-interaction graphs is $M = \lceil \frac{N}{r} \rceil$, where $N$ is the number of patches. For each $m \in [1, M]$, inter-channel $m$-th channel-interaction graph is represented as the inter-channel similarity matrix $A_m$:

$$X_{tm}^m = \frac{1}{r} \sum_{q=m\times r}^{(m+1)\times r-1} X_{tp}^q, \quad \text{Mask}_{\text{Top-}K^c}(a_{ij}) = \begin{cases} a_{ij}, & \text{if } a_{ij} \in \text{Top-}K^c(a_i), \\ -\infty, & \text{otherwise,} \end{cases}$$
$$A_m(i,j) = \text{softmax}\left(\frac{1}{\tau} \text{Mask}_{\text{Top-}K^c}\left(\frac{X_{tm}^i \cdot X_{tm}^j}{\|X_{tm}^i\| \, \|X_{tm}^j\|}\right)\right), \tag{9}$$

Here, $X_{tm}^m$ is the mean of patch-level features that share the same adjacency matrix, averaged along the patch dimension to serve as the overall representation for the corresponding patch. Top-$K^c(a_i)$ is the top-$K^c$ element of the $i$-th row of $A_m$, and $\tau$ is a temperature parameter that controls the smoothness of the similarity distribution. Then information propagation is performed on each channel-interaction graph via a Mix-Prop graph convolution adapted from MTGNN [21](see Fig. 2(b)):

$$H_t^{(0)} = X_t, \quad H_t^{(l+1)} = \alpha X_t + (1-\alpha) A_t H_t^{(l)} W_g,$$
$$H_t = \text{Concat}(H_t^{(0)}, \ldots, H_t^{(g_{\text{dep}})}) W_{\text{mlp}}, \tag{10}$$

where $\alpha \in [0,1]$ controls residual preservation and $W_g$ is the graph convolution weight for propagation. The MixProp layer performs multi-step gated propagation, where each step updates node features through a residual combination of the original input and the graph-transformed representation. The outputs from all propagation steps are concatenated and projected via an MLP, followed by PreNorm and FFN refinement to enhance feature expressiveness.

After interaction, features from all patches are concatenated and projected back to the original space through an MLP:

$$X^{cig} = \text{MLP}(\text{Concat}(H_1, H_2, \ldots, H_N)). \tag{11}$$

### 3.4 Cross-Domain Dual-Granularity Feature Fusion Module (CDF)

This module aims to fuse multi-granularity representations from both frequency and temporal domains, enabling adaptive cross-domain feature fusion.

**(1) Adaptive Weighted Feature Fusion.** After obtaining multi-scale representations from the FMR and CIG modules, we design an adaptive weighting module to balance the contributions from different domains and granularities. Features are concatenated along the last dimension and aligned via lightweight MLP:

$$X^{concat} = \mathbf{MLP}([X^{fmr} \| X^{cig}]) = W_2\,\sigma(W_1[X^{fmr} \| X^{cig}] + b_1) + b_2, \quad (12)$$

where $W_1, W_2, b_1, b_2$ are learnable weight parameters. An adaptive weighting factor $\beta$ is then applied as a selective modulation signal to dynamically control information flow during cross-domain fusion, and then a residual connection $\oplus$ is used to prevent excessive information loss:

$$X^{adp} = (1 - \beta)\,X^{fmr} + \beta\,X^{cig} \oplus X^{fmr}. \quad (13)$$

**(2) Stochastic Pooling-Based Fusion Enhancement.** To improve robustness and generalization, a stochastic pooling mechanism [6] is applied to the initial fused representation, computed as:

$$Prob = \mathbf{MLP_{prob}}(X^{adp}), \quad X^{fused} = \mathbf{Pool}(Prob, X^{adp}), \quad (14)$$

where $\mathbf{MLP_{prob}}(\cdot)$ outputs the channel-wise fusion probabilities, and $\mathbf{Pool}(\cdot)$ performs stochastic selection and aggregation of features.

# 4 Experiments

We evaluate the performance of DDCG on eight benchmarks from multiple different domains, and conduct ablation studies to analyze the model's components.

## 4.1 Experimental Setup

**Datasets.** We conduct experiments on eight datasets covering weather, traffic and energy domains, including ETT [28], Electricity [20], Weather [20], Solar [9] and PEMS [10]. (see details of all datasets in Table 1). Following SOFTS [6], we apply the same data preprocessing and splitting: 6:2:2 for ETT and PEMS08, and 7:1:2 for the remaining datasets.

**Baselines.** We compare DDCG with eight state-of-the-art forecasting models, including (1) MLP-based models UniMixer [23], SOFTS [6], TimesNet [19], TSMixer [5] and DLinear [27]; (2) frequency-domain-based methods FilterNet [25] and (3) transformer-based models iTransformer [11] and PatchTST [14]. All baseline implementations are publicly available. For FilterNet, which provides two model variants, we report results using the better-performing architecture.

**Implementation Details.** To align with previous work, the length of the look-back window is fixed at 96. Prediction lengths are set to {12, 24, 48, 96} for

**Table 1.** Detailed dataset descriptions.

| Dataset | Domain | Channels | Frequency | Split |
|---|---|---|---|---|
| ETTh1, ETTh2 | Electricity | 7 | Hourly | 6:2:2 |
| ETTm1, ETTm2 | Electricity | 7 | 15min | 6:2:2 |
| Weather | Weather | 21 | 10min | 7:1:2 |
| ECL | Electricity | 321 | Hourly | 7:1:2 |
| Solar-Energy | Energy | 137 | 10min | 7:1:2 |
| PEMS08 | Transportation | 170 | 5min | 6:2:2 |

**Table 2.** Forecasting results (Best are in **bold** and the second best in underline).

| Models | | **DDCG** | | UniMixer | | SOFTS | | FilterNet | | iTransformer | | PatchTST | | TimesNet | | TSMixer | | DLinear | |
|---|---|---|---|---|---|---|---|---|---|---|---|---|---|---|---|---|---|---|---|
| Dataset | | MSE | MAE | MSE | MAE | MSE | MAE | MSE | MAE | MSE | MAE | MSE | MAE | MSE | MAE | MSE | MAE | MSE | MAE |
| ETTm1 | 96 | **0.312** | **0.353** | 0.320 | 0.358 | 0.328 | 0.364 | 0.318 | 0.359 | 0.334 | 0.368 | 0.329 | 0.365 | 0.338 | 0.375 | 0.323 | 0.363 | 0.345 | 0.372 |
| | 192 | **0.363** | **0.382** | 0.366 | 0.384 | 0.376 | 0.391 | 0.364 | 0.383 | 0.377 | 0.391 | 0.380 | 0.394 | 0.374 | 0.387 | 0.376 | 0.392 | 0.380 | 0.389 |
| | 336 | 0.403 | 0.408 | 0.400 | 0.408 | 0.418 | 0.417 | **0.395** | **0.406** | 0.426 | 0.420 | 0.400 | 0.410 | 0.410 | 0.411 | 0.407 | 0.413 | 0.413 | 0.413 |
| | 720 | 0.459 | 0.445 | 0.467 | 0.445 | 0.467 | 0.449 | **0.458** | **0.444** | 0.491 | 0.459 | 0.475 | 0.453 | 0.478 | 0.450 | 0.485 | 0.459 | 0.474 | 0.453 |
| | avg | **0.384** | **0.397** | 0.388 | 0.399 | 0.397 | 0.405 | **0.384** | 0.398 | 0.407 | 0.410 | 0.396 | 0.406 | 0.400 | 0.406 | 0.398 | 0.407 | 0.403 | 0.407 |
| ETTm2 | 96 | 0.174 | **0.256** | **0.172** | 0.258 | 0.182 | 0.262 | 0.175 | 0.257 | 0.180 | 0.264 | 0.184 | 0.264 | 0.187 | 0.267 | 0.182 | 0.266 | 0.193 | 0.292 |
| | 192 | **0.240** | **0.300** | 0.241 | 0.302 | 0.246 | 0.306 | **0.240** | 0.301 | 0.250 | 0.309 | 0.246 | 0.306 | 0.249 | 0.309 | 0.249 | 0.309 | 0.284 | 0.362 |
| | 336 | 0.302 | 0.339 | 0.306 | 0.341 | 0.325 | 0.356 | **0.295** | **0.338** | 0.311 | 0.348 | 0.308 | 0.346 | 0.321 | 0.351 | 0.309 | 0.347 | 0.369 | 0.427 |
| | 720 | 0.400 | 0.398 | 0.407 | 0.401 | 0.412 | 0.405 | **0.390** | **0.392** | 0.412 | 0.407 | 0.409 | 0.402 | 0.408 | 0.403 | 0.416 | 0.408 | 0.554 | 0.522 |
| | avg | 0.279 | 0.323 | 0.282 | 0.326 | 0.291 | 0.332 | **0.275** | **0.322** | 0.288 | 0.332 | 0.287 | 0.330 | 0.291 | 0.333 | 0.289 | 0.333 | 0.350 | 0.401 |
| ETTh1 | 96 | 0.380 | 0.399 | 0.394 | 0.413 | 0.382 | 0.401 | **0.378** | **0.398** | 0.385 | 0.404 | 0.394 | 0.406 | 0.384 | 0.402 | 0.401 | 0.412 | 0.386 | 0.400 |
| | 192 | **0.431** | **0.426** | 0.446 | 0.439 | 0.436 | 0.432 | 0.445 | 0.430 | 0.440 | 0.435 | 0.440 | 0.435 | 0.436 | 0.429 | 0.452 | 0.442 | 0.437 | 0.432 |
| | 336 | **0.470** | **0.447** | 0.480 | 0.454 | 0.479 | 0.452 | 0.487 | 0.451 | 0.486 | 0.457 | 0.491 | 0.462 | 0.491 | 0.469 | 0.492 | 0.463 | 0.481 | 0.459 |
| | 720 | **0.472** | **0.466** | 0.500 | 0.478 | 0.500 | 0.488 | 0.491 | 0.469 | 0.502 | 0.490 | 0.487 | 0.479 | 0.521 | 0.500 | 0.507 | 0.490 | 0.519 | 0.516 |
| | avg | **0.438** | **0.435** | 0.455 | 0.446 | 0.450 | 0.443 | 0.450 | 0.437 | 0.453 | 0.446 | 0.453 | 0.446 | 0.458 | 0.450 | 0.463 | 0.452 | 0.456 | 0.452 |
| ETTh2 | 96 | 0.290 | 0.346 | 0.297 | 0.346 | 0.297 | 0.348 | 0.292 | 0.343 | 0.297 | 0.349 | **0.288** | **0.340** | 0.340 | 0.374 | 0.319 | 0.361 | 0.333 | 0.387 |
| | 192 | **0.367** | **0.392** | 0.371 | 0.394 | 0.374 | 0.394 | 0.371 | 0.396 | 0.380 | 0.400 | 0.376 | 0.395 | 0.402 | 0.414 | 0.402 | 0.410 | 0.477 | 0.476 |
| | 336 | **0.412** | **0.426** | 0.415 | 0.428 | 0.419 | 0.431 | 0.421 | 0.432 | 0.428 | 0.432 | 0.440 | 0.451 | 0.452 | 0.452 | 0.444 | 0.446 | 0.594 | 0.541 |
| | 720 | 0.426 | **0.443** | **0.425** | 0.444 | 0.427 | 0.445 | 0.431 | 0.448 | 0.427 | 0.445 | 0.436 | 0.453 | 0.462 | 0.468 | 0.441 | 0.450 | 0.831 | 0.657 |
| | avg | **0.374** | **0.402** | 0.377 | 0.403 | 0.379 | 0.405 | 0.379 | 0.405 | 0.383 | 0.407 | 0.385 | 0.410 | 0.414 | 0.427 | 0.401 | 0.417 | 0.559 | 0.515 |
| ECL | 96 | **0.137** | **0.232** | 0.151 | 0.253 | 0.145 | 0.235 | 0.147 | 0.241 | 0.148 | 0.242 | 0.164 | 0.251 | 0.168 | 0.272 | 0.157 | 0.260 | 0.197 | 0.283 |
| | 192 | **0.156** | **0.250** | 0.173 | 0.273 | 0.162 | 0.253 | 0.163 | 0.254 | 0.162 | 0.253 | 0.173 | 0.262 | 0.184 | 0.289 | 0.173 | 0.274 | 0.196 | 0.286 |
| | 336 | **0.170** | **0.267** | 0.201 | 0.299 | 0.178 | 0.269 | 0.178 | 0.271 | 0.178 | 0.269 | 0.190 | 0.279 | 0.198 | 0.300 | 0.192 | 0.295 | 0.209 | 0.302 |
| | 720 | **0.200** | **0.293** | 0.225 | 0.315 | 0.219 | 0.310 | 0.228 | 0.318 | 0.225 | 0.318 | 0.230 | 0.313 | 0.220 | 0.320 | 0.223 | 0.318 | 0.245 | 0.334 |
| | avg | **0.166** | **0.261** | 0.188 | 0.285 | 0.176 | 0.267 | 0.179 | 0.271 | 0.178 | 0.271 | 0.189 | 0.276 | 0.192 | 0.295 | 0.186 | 0.287 | 0.212 | 0.301 |
| Weather | 96 | 0.162 | 0.205 | **0.156** | **0.203** | 0.170 | 0.211 | 0.165 | 0.212 | 0.175 | 0.214 | 0.176 | 0.217 | 0.172 | 0.220 | 0.166 | 0.210 | 0.195 | 0.255 |
| | 192 | 0.208 | **0.248** | **0.206** | 0.249 | 0.222 | 0.258 | 0.214 | 0.252 | 0.222 | 0.254 | 0.221 | 0.256 | 0.219 | 0.261 | 0.215 | 0.256 | 0.236 | 0.296 |
| | 336 | 0.265 | **0.293** | **0.263** | **0.293** | 0.280 | 0.299 | 0.272 | 0.298 | 0.279 | 0.296 | 0.275 | 0.296 | 0.280 | 0.306 | 0.287 | 0.300 | 0.282 | 0.335 |
| | 720 | **0.345** | **0.346** | 0.347 | 0.347 | 0.358 | 0.350 | 0.358 | 0.352 | 0.359 | 0.347 | 0.352 | **0.346** | 0.365 | 0.359 | 0.355 | 0.348 | **0.345** | 0.381 |
| | avg | 0.245 | **0.273** | **0.243** | **0.273** | 0.258 | 0.280 | 0.252 | 0.280 | 0.259 | 0.278 | 0.256 | 0.279 | 0.259 | 0.287 | 0.256 | 0.279 | 0.264 | 0.317 |
| Solar | 96 | **0.189** | **0.229** | 0.197 | 0.230 | 0.211 | 0.244 | 0.264 | 0.287 | 0.203 | 0.237 | 0.205 | 0.246 | 0.250 | 0.292 | 0.221 | 0.275 | 0.290 | 0.378 |
| | 192 | **0.229** | 0.263 | **0.229** | 0.272 | 0.233 | 0.265 | 0.326 | 0.316 | 0.233 | **0.261** | 0.237 | 0.267 | 0.296 | 0.318 | 0.268 | 0.306 | 0.320 | 0.398 |
| | 336 | **0.238** | **0.271** | 0.251 | 0.276 | 0.245 | 0.272 | 0.369 | 0.335 | 0.248 | 0.273 | 0.250 | 0.276 | 0.319 | 0.330 | 0.272 | 0.294 | 0.353 | 0.415 |
| | 720 | **0.243** | **0.273** | 0.254 | 0.281 | 0.252 | 0.279 | 0.362 | 0.331 | 0.249 | 0.275 | 0.252 | 0.275 | 0.338 | 0.337 | 0.281 | 0.313 | 0.356 | 0.413 |
| | avg | **0.225** | **0.259** | 0.233 | 0.265 | 0.235 | 0.265 | 0.330 | 0.317 | 0.233 | 0.262 | 0.236 | 0.266 | 0.301 | 0.319 | 0.260 | 0.297 | 0.330 | 0.401 |
| PEMS08 | 12 | **0.073** | **0.171** | 0.252 | 0.307 | 0.075 | **0.171** | 0.092 | 0.241 | 0.078 | 0.181 | 0.098 | 0.205 | 0.112 | 0.212 | 0.083 | 0.189 | 0.155 | 0.277 |
| | 24 | **0.097** | **0.193** | 0.356 | 0.325 | 0.105 | 0.201 | 0.142 | 0.249 | 0.114 | 0.218 | 0.162 | 0.266 | 0.141 | 0.238 | 0.117 | 0.226 | 0.248 | 0.353 |
| | 48 | **0.144** | **0.234** | 0.401 | 0.342 | 0.167 | 0.251 | 0.223 | 0.322 | 0.186 | 0.235 | 0.238 | 0.311 | 0.198 | 0.283 | 0.196 | 0.299 | 0.441 | 0.471 |
| | 96 | **0.174** | 0.257 | 0.508 | 0.361 | 0.210 | **0.252** | 0.380 | 0.423 | 0.220 | 0.266 | 0.303 | 0.318 | 0.320 | 0.351 | 0.266 | 0.331 | 0.675 | 0.566 |
| | avg | **0.122** | **0.214** | 0.379 | 0.334 | 0.139 | 0.219 | 0.209 | 0.309 | 0.149 | 0.225 | 0.200 | 0.275 | 0.193 | 0.271 | 0.165 | 0.261 | 0.378 | 0.415 |

PEMS08 and {96, 192, 336, 720} for the other datasets. All our experiments are conducted using PyTorch 2.0.1 on a single NVIDIA Tesla V100 GPU. We adopt the Adam optimizer with a cosine learning rate scheduler to control the learning rate decay. Hyperparameters were tuned via grid search over the following candidate sets: upsampling scale $U \in \{2,3,4\}$, patch sharing interval $r \in \{2,3\}$, overlap size $S \in \{8,12\}$, and patch length $P \in \{12,16,24\}$. Other hyperparameters, including the weight $\alpha$ and dropout rate, were fixed at 0.3 and 0.1, respectively. We report the results corresponding to the configuration with the best validation performance. During training, we employ an early stopping strategy based on validation performance and report results from the best checkpoint.

**Table 3.** Ablation study.

| Dataset | DDCG | | w/o FMR | | w/o CIG | | w/o CDF | |
|---|---|---|---|---|---|---|---|---|
| | MSE | MAE | MSE | MAE | MSE | MAE | MSE | MAE |
| ETTm1 | **0.384** | **0.397** | 0.393 | 0.402 | 0.408 | 0.406 | 0.388 | 0.403 |
| ETTh2 | **0.374** | **0.402** | 0.388 | 0.409 | 0.388 | 0.410 | 0.378 | 0.404 |
| ECL | **0.166** | **0.261** | 0.170 | 0.263 | 0.176 | 0.265 | 0.182 | 0.267 |
| Weather | **0.245** | **0.273** | 0.250 | 0.276 | 0.253 | 0.280 | 0.251 | 0.278 |

### 4.2 Main Results

Table 2 presents the results of DDCG on eight datasets, where lower MSE/MAE indicate higher prediction accuracy. DDCG achieves best or second-best results on nearly all datasets (57/80 cases optimal, 21/80 suboptimal).

(1) **Experiments on ETT.** The ETT dataset has fewer channels, so models that model channels independently (*e.g.*, PatchTST and FilterNet) perform well. Our model additionally performs global-scale modeling and incorporates frequency-domain information for enhancement, achieving optimal or near-optimal prediction performance even on datasets with fewer channels. Overall, DDCG yields competitive results on the four ETT subsets.
(2) **Experiments on ECL&Solar&Weather.** These datasets have more channels and exhibit more complex interdependencies than ETT. Baselines that model channel dependencies (*e.g.*, UniMixer and SOFTS) outperform channel-independent methods such as PatchTST. DDCG further enhances the modeling of dependencies by introducing more granular channel-temporal interaction graphs, achieving optimal or near-optimal results in most settings.
(3) **Experiments on PEMS08.** PEMS08 is a widely used traffic forecasting dataset with more fine-grained and complex channel-temporal interactions. DDCG builds a more detailed interaction graph and better captures the complex dependencies in the data, leading to a 12% improvement in average

MSE over the second-best model. This demonstrates the importance of feature fusion at two different granularities in capturing complex spatiotemporal dependencies.

### 4.3 Ablation Study

We perform an ablation study to assess the contributions of FMR, CIG and CDF. We compare DDCG with three variants: (1) w/o FMR, without the global frequency-domain multi-scale expert module; (2) w/o CIG, without the patch-level channelinteraction graph module; and (3) w/o CDF, without the cross-domain dual-granularity feature fusion module.

As shown in Table 3, removing any module degrades performance. Removing the FMR module leads to a loss of fine-grained temporal features and frequency domain information, causing a decline in prediction accuracy. This underscores the FMR module's role in temporal modeling and frequency enhancement. Disabling the CIG module results in a significant drop in metrics on ECL, indicating the importance of fine-grained spatiotemporal interaction modeling for handling data from multiple channels. Removing the CDF module affects fine-grained dual-domain information fusion, significantly reducing performance, particularly on ECL, highlighting its necessity for effective information fusion in modeling.

## 5 Conclusion

We propose DDCG, a dual-granularity dual-domain collaborative graph Neural Network, that integrates global and patch-level structures via a multi-scale MoE for spectral enhancement, and adopts a graph-based mechanism to reduce complexity while maintaining accuracy. The experiments on multiple datasets demonstrate that DDCG achieves competitive forecasting accuracy. The current framework still faces limitations in automatically selecting optimal hyperparameters. We leave this for future work.

**Acknowledgments.** This work is supported by the National Natural Science Foundation of China (Grants No U24B20143, No 62372030).

## References

1. Asteriou, D., Hall, S.G.: ARIMA models and the Box–Jenkins methodology. Appl. Econom. **2**(2), 265–286 (2011)
2. Bai, S., Kolter, J.Z., Koltun, V.: An empirical evaluation of generic convolutional and recurrent networks for sequence modeling. arXiv preprint arXiv:1803.01271 (2018)
3. Chen, P., et al.: Pathformer: multi-scale transformers with adaptive pathways for time series forecasting. arXiv preprint arXiv:2402.05956 (2024)

4. Chen, Y., Liu, S., Yang, J., Jing, H., Zhao, W., Yang, G.: A joint time-frequency domain transformer for multivariate time series forecasting. Neural Netw. **176**, 106334 (2024)
5. Ekambaram, V., Jati, A., Nguyen, N., Sinthong, P., Kalagnanam, J.: TSMixer: lightweight MLP-mixer model for multivariate time series forecasting. In: KDD, pp. 459–469 (2023)
6. Han, L., Chen, X.Y., Ye, H.J., Zhan, D.C.: SOFTS: efficient multivariate time series forecasting with series-core fusion. In: NeurIPS, pp. 64145–64175 (2024)
7. Hu, Y., Liu, P., Zhu, P., Cheng, D., Dai, T.: Adaptive multi-scale decomposition framework for time series forecasting. In: AAAI, vol. 39, pp. 17359–17367 (2025)
8. Hu, Y., et al.: TimeFilter: patch-specific spatial-temporal graph filtration for time series forecasting. In: ICML (2025)
9. Lai, G., Chang, W., Yang, Y., Liu, H.: Modeling long- and short-term temporal patterns with deep neural networks. In: SIGIR, pp. 95–104 (2018)
10. Liu, P., et al.: WFTNet: exploiting global and local periodicity in long-term time series forecasting. In: ICASSP, pp. 5960–5964 (2024)
11. Liu, Y., et al.: iTransformer: inverted transformers are effective for time series Forecasting. In: ICLR (2024)
12. Liu, Z.: FreqMoE: enhancing time series forecasting through frequency decomposition mixture of experts. In: AISTATS, pp. 3430–3438 (2025)
13. Mouatadid, S., et al.: SubseasonalClimateUSA: a dataset for subseasonal forecasting and benchmarking. In: NeurIPS (2023)
14. Nie, Y., Nguyen, N.H., Sinthong, P., Kalagnanam, J.: A time series is worth 64 words: long-term forecasting with transformers. In: ICLR (2023)
15. Qiu, X., Wu, X., Lin, Y., Guo, C., Hu, J., Yang, B.: DUET: dual clustering enhanced multivariate time series forecasting. In: KDD, pp. 1185–1196 (2025)
16. Salinas, D., Flunkert, V., Gasthaus, J., Januschowski, T.: DeepAR: probabilistic forecasting with autoregressive recurrent networks. Int. J. Forecast. **36**(3), 1181–1191 (2020)
17. Sezer, O.B., Gudelek, M.U., Ozbayoglu, A.M.: Financial time series forecasting with deep learning: a systematic literature review: 2005–2019. Appl. Soft Comput. **90**, 106181 (2020)
18. Shazeer, N., et al.: Outrageously large neural networks: the sparsely-gated mixture-of-experts layer. In: ICLR (2017)
19. Wu, H., Hu, T., Liu, Y., Zhou, H., Wang, J., Long, M.: TimesNet: temporal 2D-variation modeling for general time series analysis. In: ICLR (2023)
20. Wu, H., Xu, J., Wang, J., Long, M.: AutoFormer: decomposition transformers with auto-correlation for long-term series forecasting. In: NeurIPS, pp. 22419–22430 (2021)
21. Wu, Z., Pan, S., Long, G., Jiang, J., Chang, X., Zhang, C.: Connecting the dots: multivariate time series forecasting with graph neural networks. In: KDD, pp. 753–763 (2020)
22. Xu, Z., Zeng, A., Xu, Q.: FITS: modeling time series with 10k parameters. In: ICLR (2024)
23. Ye, J., Liu, C., Zhou, X., Shen, R., Wen, Y.: UniMixer: unified patch-wise and global inter-series dependency modeling for multivariate time series forecasting. In: DASFAA (2025)
24. Ye, W., Deng, S., Zou, Q., Gui, N.: Frequency adaptive normalization for non-stationary time series forecasting. In: NeurIPS (2024)
25. Yi, K., et al.: FilterNet: harnessing frequency filters for time series forecasting. In: NeurIPS (2024)

26. Yi, K., et al.: Frequency-domain MLPs are more effective learners in time series forecasting. In: NeurIPS (2023)
27. Zeng, A., Chen, M., Zhang, L., Xu, Q.: Are transformers effective for time series forecasting? In: AAAI, pp. 11121–11128 (2023)
28. Zhou, H., et al.: Informer: beyond efficient transformer for long sequence time-series Forecasting. In: AAAI, pp. 11106–11115 (2021)
29. Zhou, T., Ma, Z., Wen, Q., Wang, X., Sun, L., Jin, R.: FEDformer: frequency enhanced decomposed transformer for long-term series forecasting. In: ICML, pp. 27268–27286 (2022)

# A Transductive Model-Agnostic Contrastive Learning Framework for Few-Shot Learning

Tianyu Kang, Chengjie Zheng, Ping Chen(✉), and Wei Ding

University of Massachusetts Boston, Boston, MA 02125, USA
{tianyu.kang001,chengjie.zheng001,ping.chen,wei.ding}@umb.edu
https://www.umb.edu/

**Abstract.** Few-shot learning is critical in data mining systems where new or long-tail categories appear with only a few labeled examples. We propose **MACL**, a transductive and model-agnostic contrastive learning framework for robust few-shot classification. MACL contains two branches. The contrastive branch learns margin-aware embeddings through hard-negative mining, boundary-sensitive reweighting, and semi-positive construction. The transductive branch refines class weights by maximizing mutual information over the entire query set with support supervision. A lightweight gradient-based scheme adaptively balances the loss terms, and parameter-efficient tuning with adapters, LoRA, and visual prompts enables scalable adaptation across convolutional and transformer backbones. We construct controlled few-shot evaluations based on miniImageNet and tieredImageNet to assess robustness under spurious correlations, label noise, and open-set queries. Experiments show that MACL and its ViT-based variant consistently outperform strong inductive and transductive baselines in accuracy, calibration, and robustness. Overall, the results suggest that MACL benefits from a tightly coupled contrastive and transductive design in which margin-aware supervised contrastive learning shapes the posteriors optimized by the InfoMax objective under adaptive balancing. This cooperative mechanism goes beyond a simple combination of SupCon, TIM, and gradient reweighting and provides a unified and scalable solution for few-shot learning in practical data mining and knowledge discovery tasks.

**Keywords:** Few-shot learning · Contrastive learning · Transductive inference · Foundation models · Hard negatives

## 1 Introduction

Few-shot learning (FSL) aims to recognize new classes from only a handful of labeled examples, which is crucial when annotation is scarce. While

T. Kang and Ch. Zheng—Contributed equally as co-first authors.


R. C.-W. Wong et al. (Eds.): PAKDD 2026, LNAI 16597, pp. 545–556, 2026.
https://doi.org/10.1007/978-981-92-1300-9_43

episodic training and large-scale pretrained models (e.g., CLIP and SigLIP) have improved performance, current FSL systems still face limitations.

First, many methods underuse structure within training samples: supervised contrastive learning helps, but typical pipelines rely on random negatives and overlook hard negatives or boundary samples. Second, combining multiple objectives is unstable; fixed loss weights are dataset-dependent, while adaptive schemes remain cumbersome. Third, structure within the query set is often ignored—inductive inference treats queries independently, whereas transductive approaches exploit this structure but are rarely combined with contrastive learning or modern pretrained encoders.

We introduce MACL, a transductive and model-agnostic contrastive framework for heterogeneous and noisy few-shot scenarios. MACL couples margin-aware contrastive learning with query-level information maximization: the contrastive branch applies hard-negative mining, boundary-aware reweighting, and semi-positives to shape the embedding geometry, while the transductive branch refines class weights by maximizing mutual information over the query set. Both branches can be trained jointly or alternately and operate across CNNs, transformers, and visionlanguage models. A lightweight gradient-based balancer stabilizes multi-objective optimization, and parameter-efficient tuning (adapters, LoRA, visual prompts) supports scalable adaptation. We evaluate MACL on controlled benchmarks derived from miniImageNet and tieredImageNet, focusing on robustness, calibration, and the impact of spurious correlations and label noise. In this work, we deliberately focus on these controlled small-image benchmarks to enable fine-grained analysis under systematically varied nuisance factors. Conceptually, MACL departs from a naïve "SupCon + TIM" pipeline by jointly optimizing margin-aware geometry and InfoMax under a shared proto-softmax and unified gradient-norm balancing.

- We propose MACL, a unified few-shot learning framework that **jointly** optimizes a margin-aware supervised contrastive objective and a transductive InfoMax objective through a shared proto-softmax classifier and adaptive gradient-based loss balancing.
- We design a structured margin-aware contrastive branch, including hard-negative emphasis and boundary reweighting, that directly shapes the decision geometry used by transductive inference.
- We show that standard components such as SupCon-style contrastive learning, TIM-style InfoMax, and gradient-norm–based loss balancing are special cases of MACL, and that the coupled formulation improves robustness and calibration.
- Extensive evaluations on controlled benchmarks demonstrate superior accuracy, calibration, and robustness in scenarios with clutter, spurious correlations, label noise, and open-set queries.

## 2 Related Work

Few-shot learning has increasingly shifted from bespoke architectures to methods that build on large pretrained encoders. Vision–language models such as CLIP and SigLIP provide strong zero- and few-shot priors [4], motivating parameter-efficient adaptation through prompts, lightweight adapters, or LoRA-style refinements [3,10]. Similar strategies are adopted in vision-only backbones such as ViT and MAE [5]. These works focus on efficient adaptation of the encoder. In contrast, **MACL** keeps the backbone fixed and exploits episodic structure through contrastive and transductive objectives.

Contrastive learning plays a central role in representation learning. Supervised contrastive learning (SupCon) improves class separation by grouping positives [7], while later studies show that selecting informative negatives is crucial. Many few-shot pipelines, however, treat negatives uniformly. **MACL** builds on these insights by using structured sampling with hard negatives, boundary-aware reweighting, and semi-positives, and treats contrastive learning as a main driver of margin shaping.

Transductive inference provides complementary benefits in episodic classification. Classical methods such as ProtoNet and MAML use inductive inference [13], whereas transductive approaches jointly predict over the query set. TIM maximizes mutual information under support supervision [2]; structure-aware few-shot learners such as QSFormer and MetaDiff further exploit geometric information in the embedding space [14,15]. These works highlight the value of leveraging query-set structure, but typically do not integrate it tightly with margin-aware contrastive learning. **MACL** unifies both: the contrastive branch shapes the geometry on which InfoMax operates.

Finally, multi-objective training is an active topic in FSL. Pipelines often combine classification, contrastive, and transductive losses with fixed weights, which can be unstable across datasets and backbones. Approaches such as Auto-$\lambda$ and FAMO use gradient information to adapt loss weights [8,9]. Meanwhile, recent work such as CLAP emphasizes the need for broader evaluation and cleaner splits [12]. **MACL** incorporates these insights by using lightweight gradient normalization instead of meta-optimization and by evaluating not only accuracy but also calibration and robustness to nuisance factors.

Overall, prior work advances backbone adaptation, contrastive learning, transductive refinement, and dynamic loss balancing mostly in isolation. MACL offers a compact framework in which margin-aware contrastive learning and transductive InfoMax share the same proto-softmax parameterization and are jointly balanced, yielding more calibrated and robust few-shot classifiers.

## 3 Transductive Model-Agnostic Contrastive Framework

### 3.1 Problem Setup and MACL Overview

We follow the standard $K$-way $N$-shot episodic protocol for few-shot classification [13]. Each episode has a support set $\mathcal{S} = \{(x_s, y_s)\}_{s=1}^{K \cdot N}$ and a query set

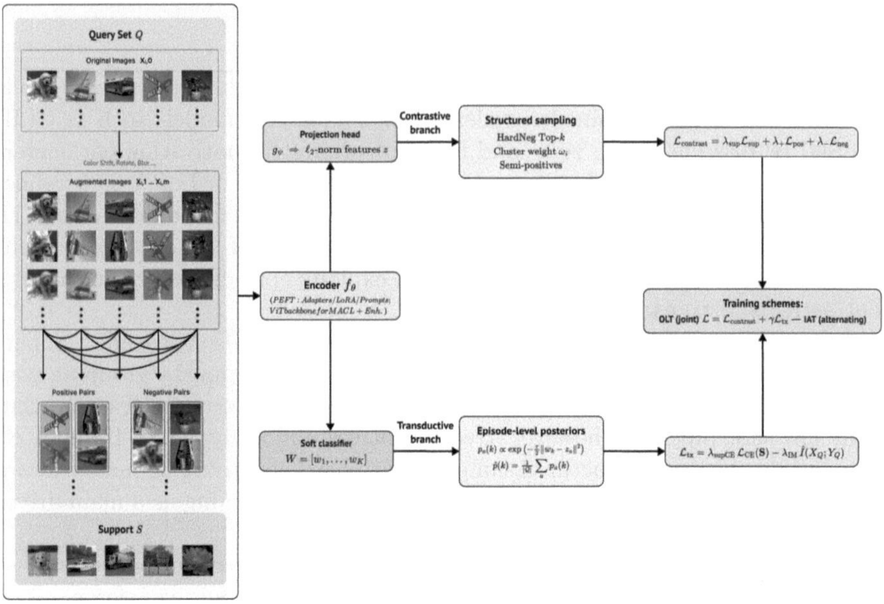


**Fig. 1. Overview of MACL.** A shared encoder $f_\theta$ and projection head $g_\psi$ produce embeddings for two branches. The contrastive branch forms structured positive and negative pairs and learns margin-aware representations. The transductive branch refines class weights $W$ using proto-softmax posteriors and an InfoMax objective. The two branches are trained with either one-loss training (OLT) or iterative alternating training (IAT), with adaptive weighting to balance objectives. Optional MAE and consistency losses can be added for improved robustness.

$\mathcal{Q} = \{x_q\}_{q=1}^{|\mathcal{Q}|}$, sampled from a novel label set $\mathcal{Y}_{\text{novel}}$ that does not overlap with the base labels used during pre-training, i.e., $\mathcal{Y}_{\text{base}} \cap \mathcal{Y}_{\text{novel}} = \emptyset$.

The backbone encoder is $f_\theta : \mathbb{R}^{H \times W \times C} \rightarrow \mathbb{R}^d$ and the projection head for contrastive learning is $g_\psi : \mathbb{R}^d \rightarrow \mathbb{R}^{d_z}$. We use the $L_2$-normalized embedding

$$z(x) = \frac{g_\psi(f_\theta(x))}{\|g_\psi(f_\theta(x))\|_2}, \tag{1}$$

so cosine similarity reduces to an inner product in $\mathbb{R}^{d_z}$.

For transductive inference we maintain class weights $W = [w_1, \dots, w_K] \in \mathbb{R}^{d_z \times K}$. At initialization, $w_k$ is the mean embedding of class $k$ in the support set, so $W$ starts as a prototype matrix and is later refined using query information. Given a query $x_a \in \mathcal{Q}$ with embedding $z_a = z(x_a)$, we use a distance-based softmax

$$p_a(k) \propto \exp\Big( -\tfrac{\tau}{2} \|w_k - z_a\|_2^2 \Big), \tag{2}$$

$$\hat{p}(k) = \frac{1}{|\mathcal{Q}|} \sum_{a \in \mathcal{Q}} p_a(k), \tag{3}$$

where $\tau > 0$ is a temperature. Here $p_a(k)$ is the posterior for $x_a$. We refer to Eq. (3) as **query pooling**, which aggregates query posteriors to form the empirical marginal label distribution $\hat{p}$ within an episode.

***Unified Contrastive–Transductive Framework.*** Figure 1 illustrates how the two branches interact. Augmented views of support and query images are encoded by $(f_\theta, g_\psi)$. The contrastive branch builds positive and negative pairs and learns a margin-aware embedding space through a unified contrastive loss. The transductive branch uses the same embeddings and the proto-softmax posteriors to refine $W$ by information maximization, while the cross-entropy loss on $\mathcal{S}$ keeps predictions anchored to the support labels. We use two training schemes: OLT optimizes a single combined loss, whereas IAT alternates contrastive and InfoMax updates inside each episode.

**MACL** is implemented as a lightweight head on top of a pretrained encoder and can work with CNNs, ResNets, transformers, and vision–language backbones. For transformer backbones we use an enhanced variant, MACL+Enh, which employs a ViT-B/16 encoder, inserts small adapters between transformer blocks, and feeds intermediate layers into the contrastive loss so that multi-scale information contributes to training. On selected layers we apply LoRA-style low-rank adaptation [6],

$$W'_\ell = W_\ell + A_\ell B_\ell^\top, \tag{4}$$

where $A_\ell, B_\ell \in \mathbb{R}^{d_\ell \times r}$ with $r \ll d_\ell$, and only $\{A_\ell, B_\ell\}$ and $g_\psi$ are updated. In this way MACL and MACL+Enh adapt across backbones with a modest number of trainable parameters.

### 3.2 Structured Contrastive Learning

***Structured Sampling and Margin-Aware Contrastive Objectives.*** The contrastive branch makes explicit use of episodic structure. For each labeled example $(x, y)$ we draw $m$ stochastic views $\{\tilde{x}^{(1)}, \ldots, \tilde{x}^{(m)}\}$ and compute normalized embeddings

$$z_i = \frac{g_\psi(f_\theta(\tilde{x}_i))}{\|g_\psi(f_\theta(\tilde{x}_i))\|_2},$$

with $\mathcal{I}$ the index set of all views in the batch. For an anchor $i \in \mathcal{I}$ we define

$$\mathcal{P}(i) = \{j \in \mathcal{I} \setminus \{i\} : y_j = y_i\}, \tag{5}$$

$$\mathcal{N}(i) = \{j \in \mathcal{I} : y_j \neq y_i\}. \tag{6}$$

Randomly sampling from $\mathcal{N}(i)$ produces many easy negatives. We instead keep the $k$ most similar negatives under cosine similarity,

$$\mathcal{N}_{\text{hard}}(i) = \text{Top-k}_{j \in \mathcal{N}(i)}(z_i^\top z_j), \tag{7}$$

so learning focuses on examples close to the decision boundary. Within each class we weight samples according to how close they are to other class centroids. Let

$\mu_c = \frac{1}{|\mathcal{I}_c|} \sum_{i \in \mathcal{I}_c} z_i$ be the prototype of class $c$. For a sample $i$ of class $y_i$ we set

$$\omega_i = \sigma\Big( - \alpha \cdot \min_{c \neq y_i} \left( \|z_i - \mu_c\|_2 - \|z_i - \mu_{y_i}\|_2 \right) \Big), \tag{8}$$

where $\sigma$ is a logistic function and $\alpha > 0$ is a scale factor. Samples with small margins obtain larger $\omega_i$ and contribute more to the loss.

Strong augmentations can create views that are distant in pixel space but still semantically similar. To preserve this signal, we define a soft-positive set

$$\widehat{\mathcal{P}}(i) = \{ j \in \mathcal{N}(i) : z_i^\top z_j \geq \delta \} \cup \{ j : \text{teacher predicts the same label as for } i \},$$

where $\delta$ is a similarity threshold and the teacher network comes from a teacher–student model. We then use the combined positive set $\widetilde{\mathcal{P}}(i) = \mathcal{P}(i) \cup \widehat{\mathcal{P}}(i)$ in the contrastive objectives.

Let $s_{ij} = z_i^\top z_j / \tau_z$ denote similarity with temperature $\tau_z > 0$. The supervised contrastive loss [7] is

$$\mathcal{L}_{\text{sup}} = \sum_{i \in \mathcal{I}} \frac{-1}{|\widetilde{\mathcal{P}}(i)|} \sum_{p \in \widetilde{\mathcal{P}}(i)} \log \frac{\exp(s_{ip})}{\sum_{a \in \mathcal{I} \setminus \{i\}} \exp(s_{ia})}. \tag{9}$$

To act directly on margins we add two logistic penalties,

$$\mathcal{L}_{\text{pos}} = \sum_i \frac{1}{|\widetilde{\mathcal{P}}(i)|} \sum_{p \in \widetilde{\mathcal{P}}(i)} \log\big(1 + \exp(-s_{ip})\big), \tag{10}$$

$$\mathcal{L}_{\text{neg}} = \sum_i \frac{1}{|\mathcal{N}_{\text{hard}}(i)|} \sum_{n \in \mathcal{N}_{\text{hard}}(i)} \log\big(1 + \exp(s_{in})\big), \tag{11}$$

which pull (soft-)positives closer and push hard negatives away.

The final contrastive loss is

$$\mathcal{L}_{\text{contrast}} = \sum_i \omega_i \Big( \lambda_{\text{sup}} \mathcal{L}_{\text{sup}}^{(i)} + \lambda_+ \mathcal{L}_{\text{pos}}^{(i)} + \lambda_- \mathcal{L}_{\text{neg}}^{(i)} \Big), \tag{12}$$

where $\omega_i$ is the boundary-aware weight and $\{\lambda_{\text{sup}}, \lambda_+, \lambda_-\}$ control the three components. By concentrating on hard positives and hard negatives and giving more emphasis to low-margin points, this objective produces a cleaner separation between classes and more stable decision boundaries.

### 3.3 Transductive InfoMax and Training

***Adaptive Multi-objective Training and Parameter-Efficient Tuning.*** The transductive branch uses an information-maximization objective on the query set. Using $p_a(k)$ and $\hat{p}(k)$ from Eqs. (2)–(3), we refine $W$ (and optionally $f_\theta$) with

$$\mathcal{L}_{\text{tx}} = \lambda_{\text{supCE}} \, \mathcal{L}_{\text{CE}}(\mathcal{S}) - \lambda_{\text{IM}} \, \hat{I}(X_{\mathcal{Q}}; Y_{\mathcal{Q}}), \tag{13}$$

where $\mathcal{L}_{\mathrm{CE}}$ is cross-entropy on the support set and $\hat{I}(\cdot;\cdot)$ estimates mutual information between query embeddings and predicted labels [2]. Concretely, we use a TIM-style estimator:

$$\hat{I}(X_{\mathcal{Q}};Y_{\mathcal{Q}}) = H(\hat{p}) - \frac{1}{|\mathcal{Q}|}\sum_{a\in\mathcal{Q}} H(p_a), \quad H(p) = -\sum_{k=1}^{K} p(k)\log p(k),$$

where the marginal $\hat{p}$ is computed by query pooling in Eq. (3). The weights $\lambda_{\mathrm{supCE}}$ and $\lambda_{\mathrm{IM}}$ control the balance between supervision and InfoMax.

In OLT we combine both branches in

$$\mathcal{L}_{\mathrm{OLT}} = \mathcal{L}_{\mathrm{contrast}} + \gamma\,\mathcal{L}_{\mathrm{tx}}, \tag{14}$$

with $\gamma > 0$. In IAT we alternate within each episode: contrastive updates on the augmented episodic batch (all views indexed by $\mathcal{I}$) using $\mathcal{L}_{\mathrm{contrast}}$, followed by InfoMax updates on $(\mathcal{S},\mathcal{Q})$ using $\mathcal{L}_{\mathrm{tx}}$ with a small number of inner iterations. In our main experiments we adopt OLT as the default training scheme because it is simpler to implement and tune; IAT is included as an alternative formulation and we leave a detailed empirical comparison of the two schemes to future work.

Fixed loss weights can behave very differently across datasets and backbones. To reduce this sensitivity, we adopt a gradient-based adaptive scheme inspired by Auto-$\lambda$ [9] and FAMO [8]. Let $\mathcal{G} = \{\mathcal{L}_{\mathrm{sup}}, \mathcal{L}_{\mathrm{pos}}, \mathcal{L}_{\mathrm{neg}}, \mathcal{L}_{\mathrm{CE}}, -\hat{I}\}$ be the loss set (we treat $-\hat{I}$ as the InfoMax loss), and let $\Lambda = \{\lambda_g\}_{g\in\mathcal{G}}$ denote their weights. In our notation, these correspond to $\{\lambda_{\mathrm{sup}}, \lambda_+, \lambda_-, \lambda_{\mathrm{supCE}}, \lambda_{\mathrm{IM}}\}$, respectively. We compute gradient norms with respect to the trainable parameters $\phi = \{\theta, \psi, W\}$ (or the corresponding trainable adapter/LoRA/prompt parameters when the backbone is frozen). For each $g \in \mathcal{G}$ we compute

$$u_g = \|\nabla_\phi \mathcal{L}_g\|_2, \quad \tilde{u}_g = \frac{u_g}{\frac{1}{|\mathcal{G}|}\sum_{g'} u_{g'}}, \tag{15}$$

$$\lambda_g \leftarrow \mathrm{Proj}_{[\lambda_{\min},\lambda_{\max}]}\Big(\lambda_g \cdot \exp(-\eta(\tilde{u}_g - 1))\Big), \tag{16}$$

with learning rate $\eta > 0$. This update keeps gradient scales of different objectives comparable and reduces manual tuning. The same mechanism applies when only adapters, LoRA blocks, or prompts in MACL+Enh are trainable on top of a frozen encoder.

Optimization uses Adam (learning rate $10^{-3}$, weight decay $10^{-5}$) with batches of 16 episodes and early stopping based on 200 validation episodes. Temperatures $(\tau, \tau_z)$ are tuned once per dataset seed and then fixed. The contrastive branch requires $O(B^2 d)$ pairwise similarities, where $B = |\mathcal{I}|$ is the number of views in the contrastive batch, and the transductive branch costs $O(|\mathcal{Q}|Kd)$ for distance and entropy computations. Unless stated otherwise, we use $\tau_z = 0.07$, $\tau = 15$, and $\gamma = 0.7$, with $\lambda_{\mathrm{sup}} = \lambda_+ = \lambda_- = 1$ at initialization. Hard negative mining uses $k = 16$ candidates, and boundary-aware weights $\omega_i$ are activated after a 100-episode warmup to mitigate instability when class prototypes are still noisy early in training. MAE-style auxiliary losses [5] are only used in the most challenging low-shot settings and do not affect the core contrastive–transductive design of MACL.

### 3.4 Intuition and Analysis

MACL jointly shapes both the *margin distribution* of the embedding space and the *posterior distribution* over classes. The supervised contrastive loss forms tight within-class clusters and pushes hard negatives away, while boundary-aware weights emphasize low-margin points near decision boundaries. Since the transductive posteriors $p_a(k)$ are exponential functions of squared distances to class weights, enlarging these margins sharpens the logits that enter the InfoMax objective.

The InfoMax term itself can be viewed as maximizing the entropy of the marginal label distribution while minimizing the conditional entropy of predictions. Query pooling provides the episode-level marginal $\hat{p}$ for the entropy term, which prevents collapsed assignments—where many queries map to a few classes—while still encouraging confident per-query decisions. Combined with cross-entropy on the support set, it produces calibrated decision regions whose effective volume reflects class frequency in the episode.

## 4 Experiments

We evaluate MACL and MACL+Enh. on miniImageNet and tieredImageNet using controlled 5-way 1/5-shot episodes (clutter, spurious correlations, label noise, open-set), reporting Top-1/Top-5 accuracy, 5-shot macro-F1, and calibration/robustness (mean ± 95% CI over 600 test episodes).

### 4.1 Experimental Setup and Baselines

We follow a 5-way episodic protocol with $N \in \{1, 5\}$ shots and 15 queries per class. We use miniImageNet with the standard 64/16/20 base–val–novel split and add clutter/foreground–background co-occurrence, and tieredImageNet with the original hierarchical split where base/val/novel classes come from disjoint superclasses. For controlled shifts, miniImageNet images are segmented with a saliency detector and composited with class-specific nuisance backgrounds with probability $\rho$=0.9 during training, and evaluated at $\rho/2$, 0, and $-\rho$; we additionally corrupt 10% of support labels per episode and evaluate open-set episodes where 20% of queries come from unseen classes and are classified among $K$+1 labels using a reject threshold tuned on validation data. We compare to ProtoNet [13], TIM [2], SNAIL [11], QSFormer [14], and TFSL-PLP [16], using publicly available implementations under comparable episodic/backbone settings (foundation-model-centric pipelines with substantially different pretraining/adaptation protocols are not included). All models share a PyTorch pipeline: CNN results use a 4-block convolutional backbone, while MACL+Enh. uses ViT-B/16 with a 2-layer MLP projection head and updates only adapters/LoRA/projection head with the backbone frozen; optimization uses Adam (lr $10^{-3}$, weight decay $10^{-5}$), 16 episodes per batch, early stopping on 200 validation episodes, and hyperparameters tuned on validation splits.

### 4.2 Results and Analysis

**In-Domain Performance.** Table 1 reports in-domain results on miniImageNet and tieredImageNet. Transductive methods outperform inductive ones, and MACL variants achieve the strongest performance among the compared methods under our protocol. In the 1-shot case, MACL improves Top-1 accuracy over TIM by more than four points on both datasets, and MACL+Enh. adds another two to three points on top of that. Compared with TIM, this corresponds to up to about 26% relative error reduction for MACL+Enh. The 5-shot setting shows the same pattern: on tieredImageNet, MACL raises Top-1 accuracy from 89.8 (TIM) to 92.8, and MACL+Enh. reaches 93.7, with macro-F1 improving in parallel. Compared with the strongest non-ours baseline TFSL-PLP, MACL+Enh. still yields consistent gains in Top-1 and other metrics. The performance gaps are larger than the corresponding 95% confidence intervals, indicating statistically meaningful improvements. Note that TIM can be viewed as an InfoMax-only special case that refines prototypes without an explicit supervised contrastive branch, while TFSL-PLP represents a strong transductive contrastive baseline without our margin-aware design. The consistent gains of MACL over TIM and TFSL-PLP therefore support the benefit of integrating margin-aware contrastive learning with transductive refinement.

**Robustness to Spurious Correlations.** To test robustness under distribution shift, we follow the IRM protocol [1] and couple content and nuisance subspaces with $\rho$=0.9 during training. At test time we weaken, remove, or invert this correlation ($\rho/2$, 0, and $-\rho$). Table 2 reports accuracy drops relative to a clean test set. MACL variants consistently suffer smaller degradations than all baselines. Under the strongest inversion shift ($-\rho$), ProtoNet loses almost twenty points, TIM about sixteen, while MACL and MACL+Enh. lose roughly eleven and fewer than ten points, respectively. When correlation is removed ($\rho$=0), MACL keeps 6–9 points more accuracy than ProtoNet and 3–5 points more than TIM, which matches the goal of reducing reliance on nuisance cues.

**Ablation and Calibration.** We ablate three components on miniImageNet (1-shot): (i) boundary-aware margin weighting (by setting $\alpha$=0), (ii) query pooling in the InfoMax branch, and (iii) parameter-efficient adapters in the enhanced variant. Table 3 reports Top-1 accuracy and expected calibration error (ECE; lower is better). Here "w/o Pooling" removes the marginal-entropy term computed from the episode-level marginal $\hat{p}$ (i.e., it disables query pooling in the InfoMax objective). "MACL+Enh. w/o Adapter" removes adapters from the enhanced variant. Removing margin weighting gives the largest drop and clearly hurts calibration. Without query pooling, accuracy decreases by almost three points and ECE increases. Adapters further improve both accuracy and calibration in the enhanced setting.

**Table 1. In-domain results** (% mean ± 95% CI) on miniImageNet and tieredImageNet. MACL improves over both inductive (ProtoNet, SNAIL) and transductive (TIM, TFSL-PLP) baselines, and MACL+Enh. further improves performance in the transformer setting (ViT-B/16 with parameter-efficient tuning). Best results are in **bold**, and the second-best are underlined.

| Method | Top-1 (1-shot) | Top-5 (1-shot) | Top-1 (5-shot) | Macro-F1 (5-shot) |
|---|---|---|---|---|
| miniImageNet | | | | |
| ProtoNet [13] | $62.4 \pm 0.3$ | $88.1 \pm 0.2$ | $79.5 \pm 0.3$ | $78.9 \pm 0.3$ |
| TIM [2] | $72.8 \pm 0.3$ | $93.5 \pm 0.2$ | $85.7 \pm 0.2$ | $85.1 \pm 0.3$ |
| SNAIL [11] | $74.5 \pm 0.2$ | $94.2 \pm 0.1$ | $86.3 \pm 0.2$ | $85.9 \pm 0.2$ |
| QSFormer [14] | $75.1 \pm 0.2$ | $94.7 \pm 0.1$ | $87.2 \pm 0.2$ | $86.8 \pm 0.2$ |
| TFSL-PLP [16] | $76.0 \pm 0.2$ | $94.9 \pm 0.1$ | $88.1 \pm 0.2$ | $87.5 \pm 0.2$ |
| **MACL (ours)** | $\underline{77.2 \pm 0.2}$ | $\underline{95.3 \pm 0.1}$ | $\underline{89.0 \pm 0.2}$ | $\underline{88.6 \pm 0.2}$ |
| **MACL+Enh. (ours)** | $\mathbf{79.9 \pm 0.2}$ | $\mathbf{96.3 \pm 0.1}$ | $\mathbf{91.2 \pm 0.2}$ | $\mathbf{90.9 \pm 0.2}$ |
| tieredImageNet | | | | |
| ProtoNet [13] | $66.1 \pm 0.3$ | $90.2 \pm 0.2$ | $81.0 \pm 0.3$ | $80.6 \pm 0.3$ |
| TIM [2] | $78.9 \pm 0.3$ | $95.1 \pm 0.1$ | $89.8 \pm 0.2$ | $89.4 \pm 0.2$ |
| SNAIL [11] | $79.7 \pm 0.2$ | $95.5 \pm 0.1$ | $90.1 \pm 0.2$ | $89.8 \pm 0.2$ |
| QSFormer [14] | $81.3 \pm 0.2$ | $96.2 \pm 0.1$ | $91.2 \pm 0.2$ | $90.7 \pm 0.2$ |
| TFSL-PLP [16] | $81.9 \pm 0.2$ | $96.3 \pm 0.1$ | $92.1 \pm 0.2$ | $91.8 \pm 0.2$ |
| **MACL (ours)** | $\underline{82.6 \pm 0.2}$ | $\underline{96.7 \pm 0.1}$ | $\underline{92.8 \pm 0.2}$ | $\underline{92.3 \pm 0.2}$ |
| **MACL+Enh. (ours)** | $\mathbf{85.0 \pm 0.2}$ | $\mathbf{97.4 \pm 0.1}$ | $\mathbf{93.7 \pm 0.2}$ | $\mathbf{93.4 \pm 0.2}$ |

**Table 2. Robustness to spurious correlation shift.** Accuracy deltas (%) relative to a clean test set; smaller magnitude is better.

| Method | $\rho/2$ | 0 | $-\rho$ |
|---|---|---|---|
| ProtoNet | −4.8 | −12.3 | −19.4 |
| TIM | −3.3 | −8.9 | −16.5 |
| TFSL-PLP | −2.8 | −6.5 | −13.2 |
| **MACL** | <u>−1.9</u> | <u>−5.4</u> | <u>−11.1</u> |
| **MACL+Enh.** | **−1.2** | **−3.6** | **−9.5** |

**Table 3. Ablation and calibration (ECE↓)** on miniImageNet (1-shot).

| Variant | Top-1 (%) | ECE |
|---|---|---|
| w/o Margin ($\alpha$=0) | 73.2 | 9.3 |
| w/o Pooling | 74.5 | 8.1 |
| MACL+Enh. w/o Adapter | 75.8 | 7.4 |
| **MACL** | <u>77.2</u> | <u>6.3</u> |
| **MACL+Enh.** | **79.9** | **4.9** |

**Table 4. Open-set and noisy FSL** on miniImageNet (1-shot). Higher is better for open-set F1; smaller magnitude drop is better under noise.

| Method | Open-set F1 (%) ↑ | Noisy $\Delta$Acc (↓) |
|---|---|---|
| ProtoNet | 62.8 | −6.3 |
| TIM | 69.1 | −4.7 |
| TFSL-PLP | 71.2 | −3.9 |
| **MACL** | 73.0 | −3.0 |
| **MACL+Enh.** | **75.5** | **−2.1** |

**Open-Set and Noisy Few-Shot Learning.** We also study two challenging regimes on miniImageNet (1-shot). In the open-set setting, 20% of queries belong to classes unseen in the support set and we measure macro-F1 over $K$+1 labels (the $K$ known classes plus a reject label). In the noisy-label setting, 10% of support labels are randomly corrupted and we report the resulting drop in Top-1 accuracy. Table 4 summarizes the results. For open-set recognition, MACL+Enh. improves macro-F1 by more than twelve points over ProtoNet and six points over TIM, with MACL also giving a clear gain. When labels are corrupted, ProtoNet loses more than six accuracy points, whereas MACL limits the drop to three points and MACL+Enh. to just over two. Combined with the correlation-shift and ablation results, this shows that MACL improves accuracy while also enhancing robustness to distribution shifts, label noise, and open-set uncertainty.

## 5 Conclusion

MACL unifies margin-aware contrastive geometry and query-set InfoMax under a shared proto-softmax classifier, yielding few-shot predictors that stay accurate, well-calibrated, and robust on controlled miniImageNet/tieredImageNet episodes with clutter, domain shift, spurious correlations, label noise, and open-set queries, while remaining lightweight and easy to adapt across CNN, transformer, and visionlanguage backbones via adapters, LoRA, and prompts. We intentionally focus on image-based benchmarks with controlled nuisance factors to make these effects measurable, and the same coupling principle naturally lends itself to broader domains and larger-scale settings where leveraging transductive structure and maintaining reliable margins are equally critical.

**Acknowledgment.** This material is based upon work partially supported by the National Science Foundation under NSF grants 2334665 and 2430699. Any opinions, findings, and conclusions or recommendations expressed in this material are those of the author(s) and do not necessarily reflect the views of the funding agencies.

## References


1. Arjovsky, M., Bottou, L., Gulrajani, I., Lopez-Paz, D.: Invariant risk minimization. arXiv preprint arXiv:1907.02893 (2019)
2. Boudiaf, M., Masud, Z.I., Rony, J., Dolz, J., Piantanida, P., Ben Ayed, I.: Transductive information maximization for few-shot learning. In: Advances in Neural Information Processing Systems (NeurIPS) (2020)
3. Ding, T., Li, W., Miao, Z., Pfister, H.: Tree of attributes prompt learning for vision-language models. In: International Conference on Learning Representations (ICLR) (2025)
4. Farina, M., Mancini, M., Iacca, G., Ricci, E.: Rethinking few-shot adaptation of vision-language models in two stages. In: Proceedings of the IEEE/CVF Conference on Computer Vision and Pattern Recognition (CVPR) (2025)
5. He, K., Chen, X., Xie, S., Li, Y., Dollár, P., Girshick, R.: Masked autoencoders are scalable vision learners. In: Proceedings of the IEEE/CVF Conference on Computer Vision and Pattern Recognition (CVPR) (2022)
6. Hu, E.J., et al.: LoRA: low-rank adaptation of large language models. In: International Conference on Learning Representations (ICLR) (2022)
7. Khosla, P., et al.: Supervised contrastive learning. In: Advances in Neural Information Processing Systems (NeurIPS) (2020)
8. Liu, B., Feng, Y., Stone, P., Liu, Q.: FAMO: fast adaptive multitask optimization. In: Advances in Neural Information Processing Systems (NeurIPS) (2023)
9. Liu, S., James, S., Davison, A.J., Johns, E.: Auto-$\lambda$: disentangling dynamic task relationships. Trans. Mach. Learn. Res. (2022)
10. Lu, H., et al.: UniAdapter: unified parameter-efficient transfer learning for cross-modal modeling. In: International Conference on Learning Representations (ICLR) (2024)
11. Mishra, N., Rohaninejad, M., Chen, X., Abbeel, P.: A simple neural attentive meta-learner. In: International Conference on Learning Representations (ICLR) (2018)
12. Silva-Rodríguez, J., Hajimiri, S., Ben Ayed, I., Dolz, J.: A closer look at the few-shot adaptation of large vision-language models. In: Proceedings of the IEEE/CVF Conference on Computer Vision and Pattern Recognition (CVPR), pp. 23681–23690 (2024)
13. Snell, J., Swersky, K., Zemel, R.: Prototypical networks for few-shot learning. In: Advances in Neural Information Processing Systems (NeurIPS) (2017)
14. Wang, X., Wang, X., Jiang, B., Luo, B.: Few-shot learning meets transformer: unified query-support transformers for few-shot classification. IEEE Trans. Circuits Syst. Video Technol. **33**(12), 7789–7802 (2023). https://doi.org/10.1109/TCSVT.2023.3282777
15. Zhang, B., et al.: MetaDiff: meta-learning with conditional diffusion for few-shot learning. In: Proceedings of the AAAI Conference on Artificial Intelligence (2024)
16. Zhu, H., Koniusz, P.: Transductive few-shot learning with prototype-based label propagation by iterative graph refinement. In: Proceedings of the IEEE/CVF Conference on Computer Vision and Pattern Recognition (CVPR) (2023)

# IAMRec: Intent-Adaptive Multimodal Recommendation with Collaborative–Modality Disentanglement

Jiayi Chen[1], Xin Zheng[2], Yixin Liu[3], Yi Li[1], Yongqiang Guo[1](✉), and Shirui Pan[3]

[1] Dalian University of Technology, Dalian, Liaoning, China
chenjy@mail.dlut.edu.cn, {liyi,guoyq}@dlut.edu.cn
[2] RMIT University, Melbourne, Australia
xin.zheng2@rmit.edu.au
[3] Griffith University, Brisbane, Australia
{yixin.liu,s.pan}@griffith.edu.au

**Abstract.** Multimodal recommender systems have become increasingly important to leverage rich semantic signals from textual and visual content for effective recommendation. Despite significant progress, most existing approaches still encounter two major challenges: (**C1**) Feature homogenization from collaborative-modality coupling, when collaborative filtering signals and modality-specific features are entangled within shared user embedding learner; and (**C2**) Inadequate modeling of historical–candidate semantic relevance, when existing uniform aggregation fails to capture dynamic user preference intention variations. To address these, we propose an **I**ntent-**A**daptive **M**ultimodal **REC**ommendation with collaborative-modality disentanglement, named **IAMREC**. Specifically, the proposed IAMREC contains a (1) *collaborative-modality representation disentangler* to separately capture structural collaborative signals and semantic modality features; and an (2) *intent-adaptive preference scorer* to build item-sensitive user preferences via relevant historical relevance modeling, where a collaborative-modality alignment strategy is also introduced to ensure cross-modal consistency. Extensive experiments on three public multimodal datasets could verify the consistently superior performance of our proposed method over existing baseline models. The code is available at https://github.com/jiayichen1121/IAMRec.

**Keywords:** Multimodal Recommendation · Collaborative Filtering · Representation Learning · Attention Mechanisms

## 1 Introduction

With the rapid growth of multimedia content (*e.g.*, images and text) on e-commerce and social platforms, multimodal recommender systems have emerged


R. C.-W. Wong et al. (Eds.): PAKDD 2026, LNAI 16597, pp. 557–569, 2026.
https://doi.org/10.1007/978-981-92-1300-9_44

to improve user engagement by incorporating rich item-side information. These systems aim to combine traditional collaborative filtering signals with semantic knowledge from multiple modalities such as visual images and textual descriptions to better capture user preferences. Earlier multimodal recommendation approaches [4,13] often rely on linear fusion or attention-based integration of item modality features with user-item interactions. While these methods yield certain gains, they typically fail to capture the complex structure of user behavior and multimodal semantics.

Recent advances in graph-based models [16,17,25] have established them as a mainstream approach in multimodal recommendation by enabling higher-order reasoning over user-item interactions and modality-aware structures. Despite the promising results, existing methods still face two critical challenges, as illustrated in Fig. 1: (**C1**) **Feature homogenization from collaborative-modality coupling**, where existing studies [15,17] typically use a shared feature propagator for both collaborative signals and modality-specific information. (**C2**) **Inadequate modeling of historical–candidate semantic relevance**, where most graph-based methods uniformly aggregate all interacted items when computing user representations, neglecting the semantic relevance between historical items and target candidate items [3,19,24].

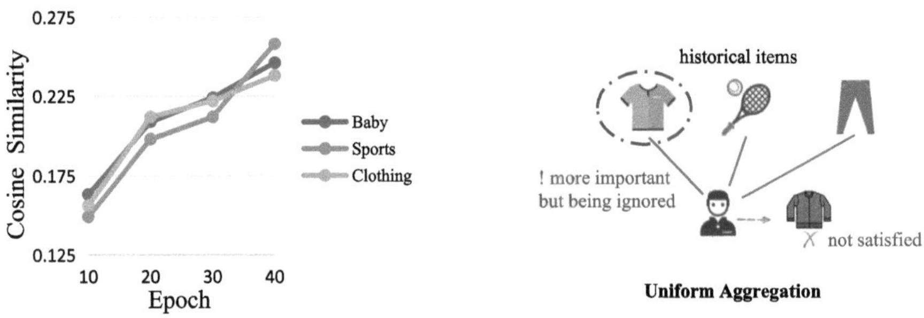


(a) Indistinguishable collaborative modality features.

(b) Inadequate item relevance modeling.

**Fig. 1.** Illustrations of two core challenges in multimodal recommendation.

To tackle the above challenges, we propose a novel Intent-Adaptive Multimodal **REC**ommendation method with collaborative–modality disentanglement, named **IAMREC**, to alleviate the feature homogenization issue and enhance the historical-candidate semantic relevance modeling for effective and accurate multimodal recommendation. Specifically, the proposed IAMREC contains two core modules: (1) *collaborative-modality representation disentangler*, which contains a collaborative signal propagator, a modality feature propagator, and a hybrid graph enhancer. Two parallel propagators capture collaborative relation patterns and modality-specific semantics without representation conflicts, and the enhancer integrates these patterns and semantics for stable parameter updates and discriminative item representations; And (2) *intent-adaptive preference scorer*, which dynamically learns ID-based user representations through a candidate-aware attention mechanism, so that the varying user intent can be well captured for more accurate preference matching. Meanwhile, we design a collaborative-modality alignment module to maintain cross-modal semantic consistency and stabilize representation learning.

In summary, the contributions of this work are listed as follows:

- We propose a novel multimodal recommendation method, named **IAMRec**, to achieve effective and accurate recommendation performance by disentangling collaborative and modality signals while adaptively modeling user intention.
- The proposed IAMRec contains two core modules: (1) *collaborative-modality representation disentangler* for distinguishable user embedding learning, and (2) *intent-adaptive preference scorer* for dynamic user intention capture with guaranteed cross-modal consistency.
- Extensive experiments on three real-world multimodal recommendation datasets demonstrate that IAMRec consistently outperforms competitive baseline models across multiple evaluation metrics.

## 2 Related Work

Graph Neural Networks (GNNs) are effective for modeling relational structures [6,20,21], a capability that underpins graph-based models in advancing multimodal recommendation by capturing complex user-item-content interactions [8,12]. Early models such as MMGCN [15] employed graph convolutional networks (GCNs) to model user-item relationships by fusing modality-specific features with ID embeddings. Building on this line of research, LATTICE [18] constructs an item–item graph to capture semantically correlated signals among items. Subsequent works [9,24] extended these designs for improved graph quality. Recent studies further explored graph fusion and denoising techniques. Models like EgoGCN [1] integrate inter-modal message aggregation and diffusion-aware representation learning. Techniques like DRAGON [22] tackle invariant learning and dual-graph construction for enhanced robustness. Furthermore, works like POWERec [2] explore prompt learning and disentanglement to improve multimodal recommendation. However, many existing models primarily capture local user preferences through user-item interactions, lacking a fine-grained modeling of intent-specific interests. Moreover, insufficient alignment across modalities often limits the effectiveness of multi-modal information integration.

## 3 Methodology

We consider a multimodal recommendation scenario with a user set $\mathcal{U}$ and an item set $\mathcal{I}$. Their historical interactions are represented as a bipartite graph

$\mathcal{G} = (\mathcal{V}, \mathcal{E})$ and encoded in an interaction matrix $\mathbf{R} \in \mathbb{R}^{|\mathcal{U}| \times |\mathcal{I}|}$. Each item $i \in \mathcal{I}$ is associated with multimodal features $\mathbf{x}_i^{(m)}$, typically extracted using pre-trained encoders such as ResNet or BERT. The goal of our work is to predict the top-$K$ non-interacted items for each user $u \in \mathcal{U}$. The overall architecture of the proposed model is depicted in Fig. 2.

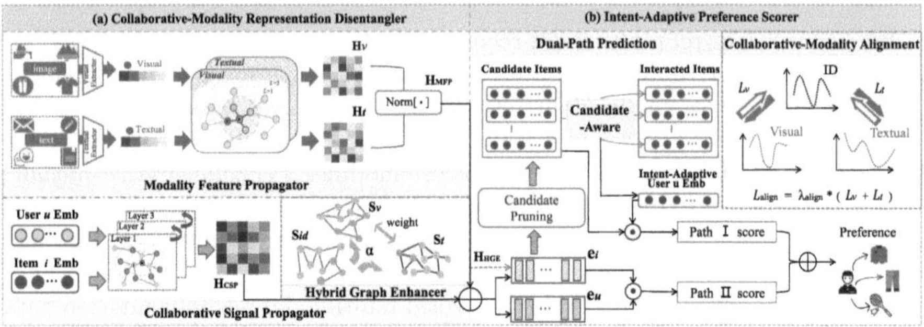


**Fig. 2.** Overview of our proposed IAMREC architecture.

### 3.1 Collaborative-Modality Representation Disentangler

**Collaborative Signal Propagator (CSP).** To capture the collaborative signals between users and items, we perform high-order message passing over the user-item interaction graph based on ID embeddings. At each layer $l$, the propagation of node embeddings follows:

$$\mathbf{H}^{(l+1)} = \mathbf{P}\mathbf{H}^{(l)}, \tag{1}$$

where $\mathbf{P} = \mathbf{D}^{-\frac{1}{2}}\mathbf{A}\mathbf{D}^{-\frac{1}{2}}$ is the normalized adjacency matrix. Here, $\mathbf{A}$ represents the binary interaction graph, and $D$ is a diagonal degree matrix. After $L$ propagation layers, we aggregate the representations from all layers to form the final collaborative embedding:

$$\mathbf{H}_{\text{CSP}} = \text{AveragePooling}\left(\left\{\mathbf{H}^{(0)}, \mathbf{H}^{(1)}, \ldots, \mathbf{H}^{(L)}\right\}\right), \tag{2}$$

where AveragePooling(·) denotes element-wise mean operation across different layers. The resulting $\mathbf{H}_{\text{CSP}}$ captures collaborative information over both direct and higher-order user-item relationships, which is independent of the modality-specific features.

**Modality Feature Propagator (MFP).** To capture semantically rich representations from different content modalities, we design a unified propagation module that operates on both user and item nodes. For each item $i$, we begin with its modality-specific features $\mathbf{x}_i^{(m)}$ and project them into a shared latent space via:

$$\tilde{\mathbf{x}}_i^{(m)} = \mathbf{W}_m^\top \mathbf{x}_i^{(m)}, \tag{3}$$

where $\mathbf{W}_m \in \mathbb{R}^{d_m \times d}$ is a modality-specific transformation matrix. Unlike items, users do not possess inherent modality features. To construct meaningful user embeddings for each modality, we aggregate the projected features of items the user has interacted with:

$$\tilde{\mathbf{x}}_u^{(m)} = \frac{1}{|\mathcal{I}(u)|} \sum_{i \in \mathcal{I}(u)} \tilde{\mathbf{x}}_i^{(m)}, \tag{4}$$

where $\mathcal{I}(u)$ denotes the set of items associated with user $u$. This design avoids introducing ID embeddings into modality modeling, preventing semantic interference. We then propagate the initial embeddings $\mathbf{H}^{(0)} = \{\tilde{\mathbf{x}}_u^{(m)}, \tilde{\mathbf{x}}_i^{(m)}\}$ over the user-item graph:

$$\mathbf{H}^{(k+1)} = \mathbf{P}\mathbf{H}^{(k)}, \tag{5}$$

where $\mathbf{P}$ is the normalized adjacency matrix. After $K$ propagation rounds, we obtain the final modality-aware representations $\mathbf{H}_{\text{MFP}} = \mathbf{H}^{(K)}$, which capture both high-order structural and semantic relations. These embeddings serve as a dedicated modality branch and are later integrated into the overall representation architecture.

**Hybrid Graph Enhancer (HGE).** To model fine-grained item-level semantic relations in multimodal content, we design a Hybrid Graph Enhancer that supplements collaborative signals with modality-aware item-item structures. Specifically, for each modality $m$, we compute item-item similarity using cosine similarity:

$$s_{ij}^{(m)} = \frac{\mathbf{x}_i^{(m)\top} \mathbf{x}_j^{(m)}}{\|\mathbf{x}_i^{(m)}\|_2 \|\mathbf{x}_j^{(m)}\|_2}. \tag{6}$$

We retain only the top-$k$ similar neighbors for each item to obtain a sparsified binary adjacency matrix $\tilde{\mathbf{S}}^{(m)}$, which is normalized as:

$$\hat{\mathbf{S}}^{(m)} = \mathbf{D}^{(m)-\frac{1}{2}} \tilde{\mathbf{S}}^{(m)} \mathbf{D}^{(m)-\frac{1}{2}}, \tag{7}$$

where $\mathbf{D}^{(m)}$ is the diagonal degree matrix with $\mathbf{D}_{ii}^{(m)} = \sum_j \tilde{\mathbf{S}}_{ij}^{(m)}$. To combine multiple modalities, we linearly fuse the normalized adjacency matrices:

$$\hat{\mathbf{S}} = \gamma \hat{\mathbf{S}}^{(v)} + (1-\gamma)\hat{\mathbf{S}}^{(t)}, \tag{8}$$

where $\gamma$ balances modality importance. The fused item-item graph $\hat{\mathbf{S}}$ is precomputed and fixed during training for efficiency. Unlike existing approaches that treat item-item graphs as isolated structures, we propose a hybrid propagation scheme that integrates collaborative and semantic signals at the item level. Given initial item embeddings $\mathbf{H}_{\mathrm{i}}^{(0)}$, the updated embeddings at layer $l+1$ are obtained by:

$$\mathbf{H}_{\mathrm{i}}^{(l+1)} = \alpha \mathbf{P}\mathbf{H}_{\mathrm{i}}^{(l)} + (1-\alpha)\hat{\mathbf{S}}\mathbf{H}_{\mathrm{i}}^{(l)}, \tag{9}$$

where $\mathbf{P}$ propagates collaborative signals while $\hat{\mathbf{S}}$ injects semantic relations. The resulting $\mathbf{H}_{\mathrm{HGE}}$ captures both interaction behavior and content-level priors, serving as an auxiliary representation that enhances the overall modeling alongside collaborative and modality-specific embeddings.

**Fusion Module.** To generate final representations for recommendation, we integrate three complementary components: collaborative embeddings $\mathbf{H}_{\mathrm{CSP}}$ encoding interaction structures, modality-aware features $\mathbf{H}_{\mathrm{MFP}}$ capturing content semantics, and hybrid graph embeddings $\mathbf{H}_{\mathrm{HGE}}$ encoding item-level structural relations. The fusion is formulated as:

$$\begin{aligned} \mathbf{e}_i &= \mathbf{H}_i^{\mathrm{CSP}} + \mathrm{Norm}[\mathbf{H}_i(v)^{\mathrm{MFP}} + \mathbf{H}_i(t)^{\mathrm{MFP}}] + \mathbf{H}_i^{\mathrm{HGE}}, \\ \mathbf{e}_u &= \mathbf{H}_u^{\mathrm{CSP}} + \mathrm{Norm}[\mathbf{H}_u^{\mathrm{MFP}}], \end{aligned} \tag{10}$$

where $\mathrm{Norm}(\cdot)$ denotes a scale normalization function to balance modules. Both user and item representations $(\mathbf{e}_u, \mathbf{e}_i)$ are subsequently used for scoring.

### 3.2 Intent-Adaptive Preference Scorer

In this section, we design an intent-adaptive preference scorer that overcomes the limitation of existing methods in modeling semantic relevance between historical and candidate items. Our approach constructs dynamic and candidate-aware user representations through target-conditioned attention to enhance recommendation accuracy.

**Candidate Pruning.** To balance personalization and efficiency, we perform relevance-based pruning to limit attention computation. For each user $u$, we compute similarity scores with all items, then select the Top-$K$ items as candidates:

$$\mathcal{C}_u(K) = \mathrm{Top}K_{i \in \mathcal{V}}(\mathbf{e}_u^{\top}\mathbf{e}_i). \tag{11}$$

This reduces memory overhead by $O(K/B)$ compared to full-batch modeling, and suppresses noise from irrelevant items.

**Adaptive Representation Construction.** For each candidate $v_t \in \mathcal{C}_u^{(K)}$, we construct a candidate-aware user representation by attending over historical items $\mathcal{N}(u)$. The attention score is computed as:

$$r_u^{t,i} = \frac{\exp(\psi(\mathbf{e}_{v_t}, \mathbf{e}_{v_i}))}{\sum_{v_j \in \mathcal{N}(u)} \exp(\psi(\mathbf{e}_{v_t}, \mathbf{e}_{v_j}))}, \tag{12}$$

where $\psi(\cdot)$ is a parametric compatibility function designed to capture semantic correlations between items. Specifically, $\psi$ is defined as:

$$\psi(\mathbf{x}, \mathbf{y}) = \frac{\mathbf{w}_{\psi}^{\top}\text{ELU}(\mathbf{W}_{\phi}\mathbf{x} + \mathbf{U}_{\phi}\mathbf{y} + \mathbf{b}_{\phi})}{\sqrt{d}}, \tag{13}$$

where $\mathbf{W}_{\phi}$ and $\mathbf{U}_{\phi}$ are learnable linear projections, $\mathbf{b}_{\phi}$ is a bias vector, and ELU$(\cdot)$ is an Exponential Linear Unit activation function introducing non-linearity to improve representation capacity. Finally, the candidate-aware user representation is computed as:

$$\tilde{\mathbf{e}}_u = \sum_{v_i \in \mathcal{N}(u)} r_u^{t,i} \cdot \mathbf{e}_{v_i}. \tag{14}$$

This adaptively emphasizes historically relevant items, improving personalization and relevance alignment.

**Dual-Path Prediction.** To capture both general preference and candidate-specific intent, we design a dual-path prediction mechanism combining collaborative matching and intent-adaptive scoring:

- **Path I (Collaborative Matching)**: We compute the base relevance score between a user $u$ and item $i$ via inner product:

$$\hat{y}_{u,i}^{(1)} = \mathbf{e}_u^{\top}\mathbf{e}_i. \tag{15}$$

- **Path II (Intent-Adaptive Scoring)**: Leveraging the candidate-aware user representation $\tilde{\mathbf{e}}_u$, we compute the intent-aware score as:

$$\hat{y}_{u,t}^{(2)} = \tilde{\mathbf{e}}_u^{\top}\mathbf{e}_{v_t}. \tag{16}$$

- **Final Prediction**: The two scores are combined via a convex combination controlled by a trade-off parameter $\beta$:

$$\hat{y}_{u,c} = (1-\beta) \cdot \hat{y}_{u,i}^{(1)} + \beta \cdot \hat{y}_{u,t}^{(2)}. \tag{17}$$

This prediction design preserves global preference patterns while adaptively emphasizing candidate-relevant intents for improved ranking precision.

### 3.3 Model Optimization

We first adopt the Bayesian Personalized Ranking (BPR) loss to train the model:

$$L_{\text{BPR}} = -\sum_{(u,i^+,i^-) \in D} \log \sigma(\hat{y}_{u,i^+} - \hat{y}_{u,i^-}), \tag{18}$$

where $D$ is the sampled interaction triplet set, and $\sigma(\cdot)$ is the sigmoid function.

However, introducing multimodal information may cause semantic inconsistency with ID signals, particularly under our decoupled propagation scheme that models them separately. To address this, we design a *collaborative-modality alignment loss* that leverages stable ID embeddings to align modality-specific representations. We first model the ID embedding $\mathbf{x}^{ID}$ and modality embeddings $\mathbf{x}^{v}$, $\mathbf{x}^{t}$ using Gaussian distributions:

$$\mathbf{x}^{ID} \sim \mathcal{N}(\boldsymbol{\mu}^{ID}, (\boldsymbol{\sigma}^{ID})^2), \quad \mathbf{x}^{v} \sim \mathcal{N}(\boldsymbol{\mu}^{v}, (\boldsymbol{\sigma}^{v})^2), \quad \mathbf{x}^{t} \sim \mathcal{N}(\boldsymbol{\mu}^{t}, (\boldsymbol{\sigma}^{t})^2), \tag{19}$$

where $\boldsymbol{\mu}$ and $\boldsymbol{\sigma}$ represent the mean and standard deviation parameters. We encourage the semantic consistency between modalities by minimizing the distributional distances between ID embeddings and each modality:

$$\begin{aligned} L_v &= \|\boldsymbol{\mu}^{\mathrm{ID}} - \boldsymbol{\mu}^{v}\|_1 + \|\boldsymbol{\sigma}^{\mathrm{ID}} - \boldsymbol{\sigma}^{v}\|_1, \\ L_t &= \|\boldsymbol{\mu}^{\mathrm{ID}} - \boldsymbol{\mu}^{t}\|_1 + \|\boldsymbol{\sigma}^{\mathrm{ID}} - \boldsymbol{\sigma}^{t}\|_1. \end{aligned} \tag{20}$$

The overall modality alignment loss is formulated as:

$$L_{\text{align}} = L_v + L_t. \tag{21}$$

Apart from the above two losses, we further regularize the model using $\ell_2$ weight decay. Finally, the overall learning objective is:

$$L = L_{\text{BPR}} + \lambda_{\text{reg}}\|\Theta\|_2^2 + \lambda_{\text{align}} L_{\text{align}}, \tag{22}$$

where $\lambda_{\text{reg}}$ and $\lambda_{\text{align}}$ are hyperparameters to control the contributions of two loss terms. The final objective encourages both discriminative ranking and semantic consistency across modalities.

## 4 Experiment

To evaluate the effectiveness of our proposed IAMREC, we conduct extensive experiments to answer the following research questions on three real-world datasets:

- RQ1: How does IAMREC perform compared to existing state-of-the-art multimodal recommendation baselines?
- RQ2: How does each key component in IAMREC individually contribute to the model's overall effectiveness?
- RQ3: How do critical hyperparameters impact the model's performance and stability?

### 4.1 Experimental Settings

*Datasets.* To assess the effectiveness of our proposed model, we conduct experiments on three benchmark datasets from Amazon: **Baby**, **Sports**, and **Clothing**. Following prior practice, we use 4096-dimensional image features for visual content and 384-dimensional text features for textual content. These feature sets are made publicly available in earlier works [23]. Dataset statistics are summarized in Table 1.

**Table 1.** Statistics of the experimental datasets.

| Dataset | # Users | # Items | # Interactions | Sparsity |
|---|---|---|---|---|
| Baby | 19,445 | 7,050 | 160,792 | 99.88% |
| Sports | 35,598 | 18,357 | 296,337 | 99.95% |
| Clothing | 39,387 | 23,033 | 278,677 | 99.97% |

*Evaluation and Parameter Settings.* We evaluate our approach using an 8:1:1 train/validation/test split, reporting Recall@N and NDCG@N averaged over all test users with $N$ set to 10 and 20, where these metrics respectively measure retrieval quality and ranking accuracy. Model parameters are initialized via Xavier uniform initialization, with hyperparameters tuned on the validation set. The number of collaborative and modality graph propagation layers $K$ is selected from $\{1, 2, 3, 4\}$, while the item-item GCN layers in the Hybrid Graph Enhancer are fixed at $L_{ii} = 1$. In our intent-adaptive module, the hyperparameter Top-$K$ is selected from the set 50, 100, 150. Additionally, key parameters including the fusion weight $\alpha$, alignment loss coefficient $\lambda_{\text{align}}$, and score integration factor $\beta$ are tuned within [0, 1] in increments of 0.2.

*Baselines.* We compare our proposed **IAMRec** with a diverse set of strong baselines across four categories:

- **Traditional CF models**: BPR [10];
- **Graph-based recommenders**: LightGCN [5], SGL [16], NCL [7];
- **Multimodal recommenders**: VBPR [4], MMGCN [15], GRCN [14], LATTICE [18], MMGCL [17], SLMRec [11], BM3 [25], MICRO [19], LGMRec [3].

All baselines are either re-implemented or evaluated using the authors' released code under the same experimental settings for fairness.

### 4.2 Experimental Results

**Overall Performance (RQ1).** We summarize the performance of all models across the benchmark datasets in Table 2, and draw the following observations: (1) **Overall effectiveness of our method.** Our model consistently outperforms all baseline methods under various metrics and datasets, demonstrating its strong generalization capability. (2) **Contribution of the collaborative-modality representation disentangler.** This module separately collaborative signals and modality features to address feature homogenization in conventional approaches. Combined with our hybrid graph enhancer that balances structural and semantic information, it achieves stable optimization and more discriminative representations. (3) **Impact of the intent-adaptive preference scorer.**

**Table 2.** The Comparison of different baselines and the IAMRec performance on three datasets. The best result is in **boldface** and the second best is underlined. The t-tests validate the significance of performance improvements with $p$-value $\leq 0.05$.

| Datesets | Baby | | | | Sports | | | | Clothing | | | |
|---|---|---|---|---|---|---|---|---|---|---|---|---|
| Metrics | R@10 | R@20 | N@10 | N@20 | R@10 | R@20 | N@10 | N@20 | R@10 | R@20 | N@10 | N@20 |
| BPR | 0.0379 | 0.0607 | 0.0202 | 0.0261 | 0.0452 | 0.0690 | 0.0252 | 0.0314 | 0.0211 | 0.0315 | 0.0118 | 0.0144 |
| LightGCN | 0.0464 | 0.0732 | 0.0251 | 0.0320 | 0.0553 | 0.0829 | 0.0307 | 0.0379 | 0.0331 | 0.0514 | 0.0181 | 0.0227 |
| SGL | 0.0532 | 0.0820 | 0.0289 | 0.0363 | 0.0620 | 0.0945 | 0.0339 | 0.0423 | 0.0392 | 0.0586 | 0.0216 | 0.0266 |
| NCL | 0.0538 | 0.0836 | 0.0292 | 0.0369 | 0.0616 | 0.0940 | 0.0339 | 0.0421 | 0.0410 | 0.0607 | 0.0228 | 0.0275 |
| VBPR | 0.0424 | 0.0663 | 0.0232 | 0.0284 | 0.0556 | 0.0854 | 0.0301 | 0.0378 | 0.0281 | 0.0412 | 0.0158 | 0.0191 |
| MMGCN | 0.0398 | 0.0649 | 0.0211 | 0.0275 | 0.0382 | 0.0625 | 0.0202 | 0.0263 | 0.0229 | 0.0363 | 0.0118 | 0.0152 |
| GRCN | 0.0531 | 0.0835 | 0.0291 | 0.0370 | 0.0610 | 0.0932 | 0.0335 | 0.0422 | 0.0431 | 0.0664 | 0.0262 | 0.0289 |
| LATTICE | 0.0536 | 0.0858 | 0.0287 | 0.0370 | 0.0616 | 0.0950 | 0.0337 | 0.0423 | 0.0459 | 0.0702 | 0.0253 | 0.0306 |
| MMGCL | 0.0522 | 0.0778 | 0.0289 | 0.0355 | 0.0635 | 0.0969 | 0.0360 | 0.0463 | 0.0443 | 0.0669 | 0.0223 | 0.0297 |
| SLMRec | 0.0540 | 0.0810 | 0.0296 | 0.0361 | 0.0676 | 0.1007 | 0.0324 | 0.0402 | 0.0452 | 0.0675 | 0.0247 | 0.0303 |
| BM3 | 0.0538 | 0.0857 | 0.0301 | 0.0378 | 0.0659 | 0.0979 | 0.0354 | 0.0437 | 0.0450 | 0.0669 | 0.0243 | 0.0295 |
| MICRO | 0.0570 | 0.0905 | 0.0310 | 0.0406 | 0.0675 | 0.1026 | 0.0365 | 0.0446 | 0.0496 | 0.0743 | 0.0264 | 0.0332 |
| LGMRec | 0.0637 | 0.0985 | 0.0338 | 0.0433 | 0.0711 | 0.1038 | 0.0381 | 0.0468 | 0.0543 | 0.0821 | 0.0292 | 0.0362 |
| **IAMRec (Ours)** | **0.0684*** | **0.1035*** | **0.0376*** | **0.0461*** | **0.0748*** | **0.1113*** | **0.0407*** | **0.0508*** | **0.0576*** | **0.0851*** | **0.0313*** | **0.0383*** |
| Improv. | 10.52% | 5.08% | 11.24% | 6.47% | 5.20% | 7.23% | 6.82% | 8.55% | 6.10% | 3.65% | 7.19% | 5.80% |

Our scoring module dynamically models historical–candidate semantic relevance to capture varying user intent, while collaborative–modality alignment preserves cross–modal semantic consistency for stable learning, thereby yielding more accurate prediction results. (4) **Comparison with related models.** Compared to existing approaches, VBPR and MMGCN either underutilize collaborative signals or suffer from over-smoothing during propagation, while MICRO and LGMRec fail to fully leverage historical interactions for user-candidate alignment. In contrast, our architecture maintains clear modality and interaction signals via a lightweight adaptive design.

**Ablation Study (RQ2).** We perform ablation studies by sequentially removing five core modules: Collaborative Signal Propagator (CSP), Modality Feature Propagator (MFP), Hybrid Graph Enhancer (HGE), Intent-Adaptive Preference Scorer (IAPS), and Collaborative-Modality Alignment (CMA). Table 3 summarizes the performance under each setting.

**Table 3.** Ablation results of removing each core component.

| Model Variant | Baby | | Sports | | Clothing | |
|---|---|---|---|---|---|---|
| Metrics | R@20 | N@20 | R@20 | N@20 | R@20 | N@20 |
| w/o MFP | 0.0978 | 0.0431 | 0.1073 | 0.0478 | 0.0811 | 0.0363 |
| w/o CSP | 0.0997 | 0.0433 | 0.1059 | 0.0468 | 0.0801 | 0.0360 |
| w/o HGE | 0.1009 | 0.0450 | 0.1077 | 0.0479 | 0.0819 | 0.0363 |
| w/o IAPS | 0.1007 | 0.0448 | 0.1083 | 0.0481 | 0.0824 | 0.0369 |
| w/o CMA | 0.1010 | 0.0449 | 0.1091 | 0.0489 | 0.0836 | 0.0373 |
| **IAMRec** | **0.1035** | **0.0461** | **0.1113** | **0.0508** | **0.0851** | **0.0383** |

The ablation study results demonstrate that each component contributes critically to the model performance. CSP and MFP serve as fundamental modules which capture collaborative signals and modality semantics. Furthermore, performance decreases observed in variants without HGE, IAPS, and CMA confirm that item-level semantic relations, adaptive user intent modeling, and cross-modal alignment effectively enhance recommendation accuracy and robustness.

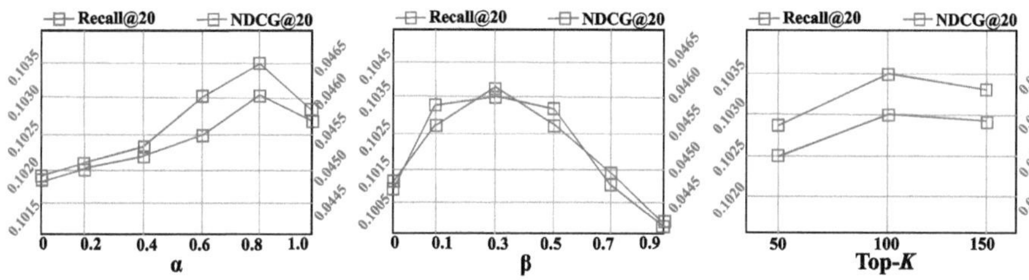


**Fig. 3.** Impact of parameter $\alpha$, $\beta$ and Top-$K$ on Baby.

**Hyper-parameter Sensitivity Analysis (RQ3).** As shown in Fig. 3, we evaluate three key hyper-parameters on the Baby dataset: the HGE fusion weight $\alpha$ achieves optimal performance when $\alpha > 0.5$, confirming the dominance of collaborative signals; the IAPS Top-$K$ candidate selection reaches peak effectiveness at $K = 100$, balancing diversity and noise reduction; and the score fusion weight $\beta$ performs best at $\beta = 0.3$, effectively integrating adaptive intents with collaborative filtering. These results demonstrate our model's stable performance under parameter variations.

## 5 Conclusion

In this work, we propose IAMRec, an intent-adaptive multimodal recommendation framework featuring collaborative-modality disentanglement. Our model employs a representation disentangler to separately capture collaborative signals and modality features, mitigating feature homogenization, while the hybrid graph enhancer fuses these signals to generate more discriminative item representations. The intent-adaptive preference scorer models historical-candidate relevance for candidate-aware prediction, supported by a collaborative-modality alignment strategy that ensures cross-modal consistency. Extensive experiments validate the effectiveness of our approach across multiple datasets and metrics.

**Acknowledgement.** This work is supported by the Fundamental Research Funds for the Central Universities (No. DUT25YG238).

## References


1. Chen, F., Wang, J., Wei, Y., Zheng, H.T., Shao, J.: Breaking isolation: multimodal graph fusion for multimedia recommendation by edge-wise modulation. In: Proceedings of the 30th ACM International Conference on Multimedia, pp. 385–394 (2022)
2. Dong, X., Song, X., Tian, M., Hu, L.: Prompt-based and weak-modality enhanced multimodal recommendation. Inf. Fusion **101**, 101989 (2024)
3. Guo, Z., Li, J., Li, G., Wang, C., Shi, S., Ruan, B.: LGMRec: local and global graph learning for multimodal recommendation. In: Proceedings of the AAAI Conference on Artificial Intelligence, vol. 38, pp. 8454–8462 (2024)
4. He, R., McAuley, J.: VBPR: visual Bayesian personalized ranking from implicit feedback. In: Proceedings of the AAAI Conference on Artificial Intelligence, vol. 30 (2016)
5. He, X., Deng, K., Wang, X., Li, Y., Zhang, Y., Wang, M.: Lightgcn: simplifying and powering graph convolution network for recommendation. In: Proceedings of the 43rd International ACM SIGIR Conference on Research and Development in Information Retrieval, pp. 639–648 (2020)
6. Ke, F., et al.: Explain before you answer: a survey on compositional visual reasoning. arXiv preprint arXiv:2508.17298 (2025)
7. Lin, Z., Tian, C., Hou, Y., Zhao, W.X.: Improving graph collaborative filtering with neighborhood-enriched contrastive learning. In: Proceedings of the ACM Web Conference 2022, pp. 2320–2329 (2022)
8. Liu, Y., Zheng, X., Li, Y., Guo, Y.: Test-time adaptation on recommender system with data-centric graph transformation. In: International Joint Conference on Artificial Intelligence (IJCAI) (2025)
9. Mu, Z., Zhuang, Y., Tan, J., Xiao, J., Tang, S.: Learning hybrid behavior patterns for multimedia recommendation. In: Proceedings of the 30th ACM International Conference on Multimedia, pp. 376–384 (2022)
10. Rendle, S., Freudenthaler, C., Gantner, Z., Schmidt-Thieme, L.: BPR: Bayesian personalized ranking from implicit feedback. arXiv preprint arXiv:1205.2618 (2012)
11. Tao, Z., et al.: Self-supervised learning for multimedia recommendation. IEEE Trans. Multimedia **25**, 5107–5116 (2022)
12. Wang, Z., et al.: Data-free model extraction for black-box recommender systems via graph convolutions. In: The Thirty-Ninth Annual Conference on Neural Information Processing Systems (2025)
13. Wei, Y., et al.: Contrastive learning for cold-start recommendation. In: Proceedings of the 29th ACM International Conference on Multimedia, pp. 5382–5390 (2021)
14. Wei, Y., Wang, X., Nie, L., He, X., Chua, T.S.: Graph-refined convolutional network for multimedia recommendation with implicit feedback. In: Proceedings of the 28th ACM International Conference on Multimedia, pp. 3541–3549 (2020)
15. Wei, Y., Wang, X., Nie, L., He, X., Hong, R., Chua, T.S.: MMGCN: multi-modal graph convolution network for personalized recommendation of micro-video. In: Proceedings of the 27th ACM International Conference on Multimedia, pp. 1437–1445 (2019)
16. Wu, J., et al.: Self-supervised graph learning for recommendation. In: Proceedings of the 44th International ACM SIGIR Conference on Research and Development in Information Retrieval, pp. 726–735 (2021)
17. Yi, Z., Wang, X., Ounis, I., Macdonald, C.: Multi-modal graph contrastive learning for micro-video recommendation. In: Proceedings of the 45th International ACM

SIGIR Conference on Research and Development in Information Retrieval, pp. 1807–1811 (2022)
18. Zhang, J., Zhu, Y., Liu, Q., Wu, S., Wang, S., Wang, L.: Mining latent structures for multimedia recommendation. In: Proceedings of the 29th ACM International Conference on Multimedia, pp. 3872–3880 (2021)
19. Zhang, J., Zhu, Y., Liu, Q., Zhang, M., Wu, S., Wang, L.: Latent structure mining with contrastive modality fusion for multimedia recommendation. IEEE Trans. Knowl. Data Eng. **35**(9), 9154–9167 (2022)
20. Zheng, X., Huang, H., Guo, Y., Wang, B., He, R.: BLAN: Bi-directional ladder attentive network for facial attribute prediction. Pattern Recogn. **100**, 107155 (2020)
21. Zheng, X., Huang, W., Zhou, C., Li, M., Pan, S.: Test-time graph neural dataset search with generative projection. In: Forty-Second International Conference on Machine Learning (2025)
22. Zhou, H., Zhou, X., Zhang, L., Shen, Z.: Enhancing dyadic relations with homogeneous graphs for multimodal recommendation. In: ECAI 2023, pp. 3123–3130. IOS Press (2023)
23. Zhou, X.: MMRec: simplifying multimodal recommendation. In: Proceedings of the 5th ACM International Conference on Multimedia in Asia Workshops, pp. 1–2 (2023)
24. Zhou, X., Shen, Z.: A tale of two graphs: freezing and denoising graph structures for multimodal recommendation. In: Proceedings of the 31st ACM International Conference on Multimedia, pp. 935–943 (2023)
25. Zhou, X., et al.: Bootstrap latent representations for multi-modal recommendation. In: Proceedings of the ACM Web Conference 2023, pp. 845–854 (2023)

# MPRL: Multi-Perspective Reinforcement Learning for Enhancing Format Adherence Capability of Large Language Models

Bo Qian[1], Yuting Wu[2](✉), Shuang Zeng[3], Ziming Wang[1], and Qiaochen Wang[3]

[1] School of Software Engineering, Beijing Jiaotong University, Beijing, China
[2] School of Cyberspace Science and Technology, Beijing Jiaotong University, Beijing, China
ytwu1@bjtu.edu.cn
[3] Tencent Inc., Beijing, China

**Abstract.** The ability to generate structured text like JSON is well-established in large language models (LLMs). However, a pronounced performance gap exists for smaller language models (SLMs) with less than 10B parameters, which often fall short in understanding and adhering to complex schema constraints provided in the input instruction. To systematically study this issue, we construct JSONBench, a benchmark dataset curated through a four-stage cleaning pipeline and covering diverse, complex schema constraints. By analyzing the errors occurring during JSON generation, we find that the majority of failures stem from semantic and structural constraints, such as symbol misuse, nesting errors, and type inconsistencies. Motivated by these observations, we propose a multi-perspective reinforcement learning framework. By modeling JSON schemas to extract intrinsic constraints, we devise a reward mechanism decomposed into syntax, structure, and content. This effectively aligns SLM outputs with strict schema requirements. Experimental results show that our approach significantly improves JSON generation accuracy across multiple model families, including Qwen and Llama, at various scales (3B, 7–8B, 14B). Moreover, unlike the performance degradation in supervised fine-tuning counterparts, our method demonstrates strong generalization on the FOFO benchmark and maintains the capabilities of SLMs on general tasks (GSM8K and MMLU).

**Keywords:** Large Language Models · Structured Text Generation · Format Adherence · Multi-Perspective Reinforcement Learning · JSON Schema

## 1 Introduction

Large language models (LLMs) have achieved remarkable progress in natural language processing. Models such as GPT-4 [1], DeepSeek-R1 [11], and Llama3


R. C.-W. Wong et al. (Eds.): PAKDD 2026, LNAI 16597, pp. 570–581, 2026.
https://doi.org/10.1007/978-981-92-1300-9_45

[10] are increasingly applied to tasks involving structured text generation, including code generation [4,5] and Text2SQL [19,20]. Since JSON is the most widely used format in modern web services and API interactions, the ability of LLMs to generate well-formed JSON has attracted significant attention.

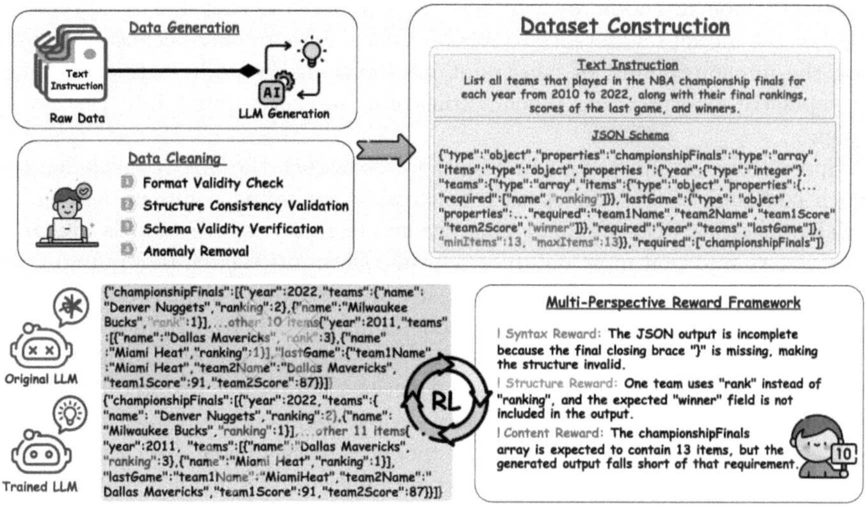


**Fig. 1.** Overview of the proposed framework for enhancing structured data generation in SLMs: The pipeline has two phases: (1) Data Construction: Creating the JSON-Bench dataset through a rigorous four-stage cleaning pipeline to ensure high-quality instruction-output pairs. (2) Reinforcement Learning: Introduce a *Multi-Perspective Reward Framework* using JSON schemas. Syntax, structure, and content rewards guide the model to fix errors and produce outputs that strictly follow schema rules.

Current state-of-the-art LLMs already demonstrate strong capabilities in generating JSON. However, research from FOFO [26] indicates that their ability to generate JSON-based to complex instructions still requires improvement, which is particularly pronounced in smaller models (under 10B parameters). Evaluations from SchemaBench [16] and DeepJSONEval [31] further highlight that models perform poorly when tasked with generating JSON strings under complex schema constraints. But these benchmarks mainly focus on whether the model's output fully complies with the schema as a whole, while lacking more fine-grained evaluations of how well the model adheres to the different constraints within the schema.

There are several approaches to enhancing the JSON generation capability of LLMs. Prompt engineering [3,23] is a simple yet effective technique that performs well on basic JSON generation tasks but struggles to handle complex schema constraints. In contrast, constrained decoding methods such as Outlines and XGrammar [8,9,25] improve model performance by modifying the decoding process to ensure outputs conform to JSON format and schema specifications. However, these methods introduce additional computational overhead during inference and may interfere with the generated JSON content, ultimately affecting the overall quality of JSON generation. In addition, several reinforcement learning approaches have been proposed, which construct reward functions based on the consistency between the generated JSON and the target JSON schema. Although these methods are effective to some extent, they fail to specifically improve the model's adherence to schema constraints, mainly because they overlook the interpretability of constraint information. We address this limitation by explicitly incorporating schema constraint interpretability into the reward design.

Specifically, as illustrated in Fig. 1, we construct the JSONBench dataset, which contains 7,511 samples with constraints including complex nesting and variable-length arrays. By analyzing schema elements where models are prone to errors, we find that most failures occur in constraints related to required fields, array length, and enumerated values. Motivated by these observations, we propose a **M**ulti-**P**erspective **R**einforcement **L**earning (MPRL) framework, which categorizes schema constraints into three perspectives—syntax, structure, and content—to construct targeted reward functions. Moreover, inspired by the idea of test-time scaling [11,17], we introduce a reasoning-aware reward mechanism that encourages models to perform deeper reasoning during JSON generation. Experimental results demonstrate that our method significantly enhances the model's ability to adhere to JSON format constraints and maintains strong generalization performance on the FOFO dataset. Unlike traditional supervised fine-tuning (SFT), it also preserves the model's capabilities on general tasks, such as MMLU [12] and GSM8K [7].

Our main contributions are summarized as follows:

- We construct the JSONBench[1] dataset, comprising 7,511 samples, each incorporating rich and complex schema constraints to evaluate models' JSON generation capabilities.
- We propose the MPRL[1] method, which enhances the model's adherence to JSON format constraints. Compared to SFT and SRL (schema reinforcement learning) [16], MPRL achieves substantially greater improvements.
- We conduct comprehensive evaluations on JSONBench, FOFO, MMLU, and GSM8K, validating the performance of our method in JSON generation (JSONBench, FOFO), general tasks (MMLU, GSM8K). The experimental results demonstrate that, through our method, the model not only significantly improves its adherence to the JSON format but also exhibits strong generalization capabilities while maintaining overall performance.

## 2 Related Work

Recent studies, such as SchemaBench [16] and DeepJSONEval [31], have actively explored the capability of large language models to generate outputs under

[1] Code, data, and other details are available at https://github.com/Tuziking/MPRL.

complex schema constraints. To improve schema adherence, several constrained decoding methods have been proposed, including Outlines, SGLang [30], and XGrammar. These approaches introduce constraint mechanisms into the decoding process to enhance decoding efficiency and strengthen compliance with format specifications. However, despite considering constraints during generation and improving efficiency, Lu et al.'s experiments in vLLM show that complex schemas can still trigger a NotSupportedError, indicating that existing methods still have limitations in handling complex schema constraints [16].

Prompt engineering has also been widely explored as a low-cost, model-agnostic approach. Few-shot prompting [3] and more advanced strategies such as thought chaining [24] or step-by-step guidance [23] improve adherence to complex instructions, while iterative and feedback-based methods like CLAIRIFY [22] and ReAct [28] leverage validator feedback to refine outputs. Despite their effectiveness, these approaches can increase interaction complexity and struggle when relevant JSON fields are distributed across long inputs [14].

Fine-tuning approaches for improving structured generation can be broadly categorized into SFT and reinforcement learning (RL)-based methods. Instruct-GPT [18] establishes a three-stage training paradigm that has since been widely adopted for LLM alignment. However, FOFO demonstrates that SFT alone provides limited improvement in format-following capabilities. Moreover, SFT typically requires large-scale, high-quality datasets and may negatively affect model performance on other downstream tasks. In contrast, RL-based methods, such as SRL [16] and ReFF [27], introduce reinforcement signals to enhance structural adherence. Although ReFF achieves promising results in format-following tasks, it focuses only on simple, non-nested structures. Similarly, SRL relies solely on binary schema validation rewards, overlooking the rich constraint information embedded within schemas, thereby limiting its ability to improve fine-grained schema adherence.

## 3 Dataset Construction

### 3.1 LLM-Assisted Generation

To enable precise evaluation and reinforcement learning optimization under complex schema constraints, we construct a dataset that is tailored for structured JSON generation. Its text instructions are sourced from the JSON-training dataset [2], while a powerful LLM is used to generate both the schemas and the corresponding ground truth outputs, ensuring high quality and consistency.

This entire process is mainly divided into two steps. First, we employ DeepSeek-v3 [13] to generate a JSON schema for each entry. This step establishes a unified standard for structural modeling and alignment of the format across the dataset. Second, we extract a subset of the original data and feed it into DeepSeek-r1 [21] to generate detailed and standardized outputs that include the model's reasoning process. Inspired by recent studies demonstrating that responses incorporating a chain of thought can enhance the performance of LLM on various tasks [6,15], we hypothesize that these reasoning-augmented outputs will improve the model's ability to adhere to formats. Through the two-step data construction process, we ultimately develop a high-quality dataset. Each sample in the dataset is paired with a corresponding JSON Schema, and a subset of samples further includes standardized outputs containing complete reasoning processes. All samples are subsequently subjected to a systematic data cleaning procedure to further enhance the overall quality and consistency of the dataset.

### 3.2 Dataset Cleaning

After distilling the data using DeepSeek-r1 and DeepSeek-v3, we curate the final dataset through a four-stage cleaning pipeline: (1) **Format Validity Check**: We use JSON and JSON Schema parsers to verify the syntax of raw samples, removing those that fail to parse as valid JSON and ensuring compliance with basic format standards. (2) **Structure Consistency Validation**: For samples that pass the format check, we analyze their hierarchical and field structures. Automated rule-matching and manual spot-checking are combined to eliminate samples with illogical or ambiguous hierarchies. (3) **Schema Validity Verification**: To ensure the reliability of the JSON Schemas generated by DeepSeek, we perform syntactic validity checks to confirm that all schemas can be correctly parsed. (4) **Anomaly Removal**: We detect and remove samples that contain semantic inconsistencies or abnormal values, even if they are syntactically and structurally valid, ensuring overall data quality and consistency. Typical cases include schemas with null fields or mismatches where the instruction specifies generating an array while the schema requires an object.

Following the data construction procedures described above, we curate a large-scale, high-quality JSON dataset, which serves as the basis for our evaluation benchmark. The dataset includes five types of JSON structures and is deliberately skewed towards complexity: 92.3% of schemas are complex-nested (nesting depth >3), 89.5% feature alternating array/object structures, 33.0% include array length constraints, and 6.5% contain enum constraints. Only 1.6% of schemas are simple, flat structures. Notably, these categories are not mutually exclusive, and a single schema may exhibit multiple complex attributes. The dataset is further partitioned for different experimental purposes: 4,810 samples are used for SFT, 2,990 samples for the MPRL method, and 7,511 samples for constructing the JSONBench.

## 4 Format Adherence Enhancement

### 4.1 Schema Formalization and Constraint Extraction

JSON Schema defines field types, required attributes, enumerated values, and array constraints for data objects, providing a standard foundation for data validation and automated processing. To precisely control LLM outputs, we compile JSON Schema specifications into a set of flattened, imperative constraints. Our algorithm starts from the root node (object or array) and recursively traverses all nested properties and array items in a depth-first manner. Field paths use a unified notation: nested object fields are represented with dots (e.g., `user.address.city`), and array elements with square brackets (e.g., `orders[].item_id`). During traversal, the algorithm extracts various constraints and organizes them into a structured dictionary for downstream tasks, supporting model guidance or output validation.

## 4.2 Multi-perspective Reward Framework

Our initial approach involves constructing a reward model using tree-structured architectures, such as Tree-LSTM, augmented with the attention mechanism. Although tree-based models naturally align with the hierarchical structure of JSON, this reward model yield suboptimal performance. Moreover, the high training costs and limited interpretability of the model motivate us to adopt rule-based RL. We find that structured information from JSON Schema can be seamlessly integrated into the GRPO framework [21]. Evaluating SLMs on JSONBench shows that most errors occurred in key schema elements, such as required fields and type consistency. To improve reward clarity, we decompose it into three perspectives—syntax, structure, and content—forming MPRL.

*Syntax Perspective.* This perspective ensures that the generated output is syntactically valid JSON and properly encapsulated within `<think>` and `<answer>` tags. The syntactic reward ($R_{syn}$) is computed in Eq. 1:

$$R_{syn} = \begin{cases} 1, & \text{if syntax is correct,} \\ 0, & \text{otherwise.} \end{cases} \tag{1}$$

*Structural Perspective.* This perspective evaluates whether the output strictly conforms to the structural framework mandated by the predefined JSON schema, such as field hierarchy and required properties. The structural reward ($R_{str}$) is computed in Eq. 2:

$$R_{str} = \begin{cases} 1, & \text{if structure is valid,} \\ 0, & \text{otherwise.} \end{cases} \tag{2}$$

*Content Perspective.* This perspective verifies the compliance of individual content within the content, ensuring they satisfy the specific constraints defined in the schema, such as correct data types, enumeration membership, and array length limits. The content perspective reward $S_{\text{con}}$ is computed as the sum of three sub-rewards: type correctness ($R_{type}$), enum validity ($R_{enum}$), and array length compliance ($R_{len}$). The total $R_{con}$ is computed in Eq. 3:

$$R_{con} = R_{type} + R_{enum} + R_{len}. \tag{3}$$

We use $\alpha$, $\beta$, and $\gamma$ as weights for the syntax, structure, and content to compute the total reward $R_t$ in Eq. 4:

$$R_t = \begin{cases} 0, & R_{syn} = 0, \\ \alpha R_{syn}, & R_{str} = 0, \\ \alpha R_{\text{syn}} + \beta R_{str} + \gamma R_{con}, & \text{otherwise.} \end{cases} \tag{4}$$

In our experiments, the optimal parameter settings are $\alpha = 3.5$, $\beta = 1$, and $\gamma = 1$. Unless specified otherwise, a base reward of 1 is granted for each perspective upon successful validation of its corresponding condition.

The final objective function for GRPO is defined as the per-sample loss:

$$\mathcal{J}_{\text{GRPO}}(\theta) = \mathbb{E}_{q,\{o_i\}\sim\pi_{\theta_{\text{old}}}}\left[\frac{1}{G}\sum_{i=1}^{G}\mathcal{L}_i(\theta)\right], \tag{5}$$

with the per-sample loss $\mathcal{L}_i(\theta)$ being a combination of the policy objective and a KL penalty:

$$\mathcal{L}_i(\theta) = \min\left(\rho_i(\theta)A_i^R,\ \rho_i'(\theta)A_i^R\right) \ - \ \lambda_{KL}\, D_{KL}, \tag{6}$$

where $\rho_i'(\theta)$ represents the clipped probability ratio:

$$\rho_i'(\theta) = \text{clip}\left(\rho_i(\theta), 1-\epsilon, 1+\epsilon\right), \tag{7}$$

and the advantage $A_i^R$ and the original probability ratio $\rho_i(\theta)$ are given by:

$$A_i^R = \frac{R_{t,i} - \text{mean}(\{R_{t,k}\}_{k=1}^G)}{\text{std}(\{R_{t,k}\}_{k=1}^G)}, \quad \rho_i(\theta) = \frac{\pi_\theta(o_i|q)}{\pi_{\theta_{\text{old}}}(o_i|q)}. \tag{8}$$

The total reward $r_i$ for each output $o_i$ is calculated using our multi-perspective reward system as Eq. 4.

# 5 Experiment

## 5.1 Experimental Setup

Our experiments[1] are conducted on a system composed of eight NVIDIA H20 GPUs, using the dataset constructed as described in Sect. 3. To mitigate the issues of unstable training and the generation of meaningless characters observed during model development, we introduce a curriculum learning, the details of which will be discussed later. To ensure the stability of the results, all metrics reported in this paper are the average of three independent experiments.

*Baseline.* Since SFT has been shown to improve task-specific performance and underlies most structural-consistency methods [29], we use it as our main baseline. We also evaluate the released Llama-based SRL checkpoint [16], as its training pipeline is unavailable.

*Assessment.* To comprehensively assess our method's effectiveness, we design a multi-faceted evaluation protocol: (1) **JSON Generation Assessment**: we utilize the JSONBench dataset and the JSON subset of the FOFO benchmark to evaluate JSON generation capability. We define the correct output as one that meets all three standards: valid syntax, correct structure, and accurate content. (2) **General Task Assessment**: We use the GSM8K and MMLU benchmarks to evaluate the retention of the model's general knowledge before and after training, in order to determine whether the model's fundamental capabilities are affected.

**Table 1.** Model performance on different benchmarks. SRL† refers to the model based on the open-source developed by Lu et al. [16](results on other LLMs are omitted as the official training code is unavailable). We group JSONBench and FOFO as JSON adherence tasks, while GSM8K and MMLU assess general knowledge.

| Model | Training | JSON (%) | | General (%) | | Avg (%) |
|---|---|---|---|---|---|---|
| | | JSONBench | FOFO | GSM8K | MMLU | |
| Llama-3.2-3B | Instruct | 28.62 | 24.67 | <u>77.58</u> | <u>60.61</u> | 47.87 |
| | +SRL† | 41.23 | **58.00** | **79.35** | 60.37 | 59.74 |
| | +SFT | <u>76.69</u> | 36.67 | 70.03 | 58.91 | <u>60.08</u> |
| | +MPRL | **88.93** | <u>52.67</u> | 73.74 | **60.58** | **69.00** |
| Qwen2.5-3B | Instruct | 39.06 | <u>52.00</u> | **84.61** | <u>66.39</u> | 60.52 |
| | +SFT | <u>86.88</u> | 44.67 | 74.60 | 64.96 | <u>67.78</u> |
| | +MPRL | **94.87** | **60.00** | <u>83.70</u> | **66.41** | **76.25** |
| Qwen2.5-7B | Instruct | 83.00 | <u>52.66</u> | <u>90.83</u> | <u>74.22</u> | 75.18 |
| | +SFT | <u>90.88</u> | 43.33 | 83.27 | 71.52 | <u>72.25</u> |
| | +MPRL | **97.19** | **64.00** | **90.95** | **74.38** | **81.63** |
| Qwen2.5-14B | Instruct | 90.95 | <u>54.00</u> | **93.63** | **79.96** | <u>79.64</u> |
| | +SFT | <u>92.00</u> | 49.33 | 90.62 | 77.44 | 77.35 |
| | +MPRL | **96.62** | **74.67** | <u>93.61</u> | <u>78.70</u> | **85.90** |
| Llama-3.1-8B | Instruct | 59.83 | 25.33 | **85.39** | **68.37** | <u>59.73</u> |
| | +SFT | <u>84.42</u> | <u>37.33</u> | 55.98 | 59.71 | 59.36 |
| | +MPRL | **92.70** | **57.33** | <u>77.81</u> | <u>66.96</u> | **73.70** |

### 5.2 Results and Analysis

As shown in Table 1, our method achieves the best overall performance across all datasets and delivers state-of-the-art results on most of them. Notably, the SFT baseline also achieves reasonably good scores on JSONBench, which we attribute to the high-quality distribution of our distilled training dataset. However, because SFT tends to overfit this specific distribution, it causes substantial degradation across all evaluated models in the general task assessment (GSM8K and MMLU). In contrast, our method effectively preserves the models' original capabilities, demonstrating superior performance and generalization in JSON generation while maintaining core competencies on downstream tasks.

However, we also observe that the Qwen2.5-14B model performs worse than its 7B counterpart. This outcome may stem from the fact that more capable models tend to have more deeply embedded pretraining knowledge. This knowledge can result in a less flexible policy during reinforcement learning, thereby restricting how effectively they can explore based on task-specific rewards.

Moreover, we observe that the SRL method outperforms ours on FOFO and GSM8K, which can largely be attributed to differences in the training data. The SRL training set contains a substantial amount of custom data featuring special formats such as phone numbers and passwords—formats that are also prevalent in FOFO. In contrast, our dataset includes relatively few schemas of this type, which likely contributes to our weaker performance on FOFO. Additionally, the escape subset in SRL specifically strengthens the model's ability to handle special character escaping, resulting in more robust answer extraction and thus providing an advantage on GSM8K. This explains why our method does not surpass SRL on GSM8K.

### 5.3 Curriculum Learning

**Table 2.** Impact of Curriculum Learning (CL) on the performance of Llama-3.1-8B-Instruct trained with MPRL.

| Method | Accuracy(%) | Performance(%) |
|---|---|---|
| Base Model | 59.83 | — |
| MPRL | 91.08 | ↑ 31.25 |
| **CL + MPRL** | **92.70** | **↑ 32.87** |

During the training of Llama-3.1-8B-Instruct, we encounter text degeneration and instability. To address this problem, we apply the CL approach, presenting samples in order of increasing difficulty so that the model first learns fundamental patterns before handling more complex structures. We implement this during data construction: while generating gold answers with DeepSeek-V3, samples are ranked by JSON structural complexity, and training proceeds from simpler to more complex structures. This approach significantly improves training stability, eliminates text degeneration, and boosts performance. As shown in Table 2, CL + MPRL achieves a final accuracy of 92.70%. All results reported in Table 1 are obtained using this CL + MPRL training strategy.

### 5.4 Ablation Study

**Table 3.** Results of the progressive ablation study.

| Method | Syntax(%) | Structure(%) | Content(%) | Error(%) | Acc(%) |
|---|---|---|---|---|---|
| Base Model | 19.85 | 17.83 | 2.49 | 40.17 | 59.83 |
| +syn. | 7.29 | 51.53 | 2.68 | 61.50 | 38.50 |
| +syn.+stru. | 7.01 | 2.01 | 9.56 | 18.58 | 81.42 |
| +syn.+stru. + cont. | 2.66 | 2.04 | 2.60 | 7.30 | 92.70 |

As shown in Table 3, Llama-3.2-3B-Instruct shows an overall error rate of 40.17%, mainly from syntax (19.85%) and structure (17.83%) errors. After adding the syntax perspective, syntax errors drop to 7.29%, but this exposes latent structural issues, increasing structure errors to 51.53%. Introducing the structure perspective drastically reduces structure errors to 2.01%, while slightly reducing syntax errors(7.01%). Finally, incorporating the content perspective minimizes all error types (syntax 2.66%, structure 2.04%, content 2.60%), lowering the total error rate to 7.30% with a 92.70% accuracy. These results demonstrate strong complementarity and synergy among the three components.

## 6 Conclusion

Our study introduces JSONBench, a benchmark designed to evaluate the ability of LLMs to adhere to complex JSON formatting requirements. We further propose the MPRL framework, which enhances LLMs' adherence with complex JSON by extracting multi-perspective rewards—covering syntax, structure, and content perspectives—from JSON schemas. Experimental results demonstrate that the MPRL method not only substantially improves JSON format adherence but also exhibits strong generalization capabilities. Results on GSM8K and MMLU further indicate that the method preserves the fundamental abilities of the models without degradation. Future work will build upon these findings to extend the framework to YAML, XML, and other structured formats, as they share the same hierarchical logic and schema-driven constraints as JSON.

**Acknowledgments.** This work is supported by Beijing Natural Science Foundation (No. L251028) and the National Natural Science Foundation of China under Grant 62406022.

## References

1. Achiam, J., et al.: GPT-4 technical report. arXiv preprint arXiv:2303.08774 (2023)
2. Azinn, C.: JSON-training dataset (2024). https://huggingface.co/datasets/ChristianAzinn/json-training. Accessed 30 July 2025
3. Brown, T., et al.: Language models are few-shot learners. Adv. Neural. Inf. Process. Syst. **33**, 1877–1901 (2020)
4. Chen, M., et al.: Evaluating large language models trained on code. arXiv preprint arXiv:2107.03374 (2021)
5. Chowdhery, A., et al.: Palm: scaling language modeling with pathways. J. Mach. Learn. Res. **24**(240), 1–113 (2023)
6. Chu, Z., et al.: Navigate through enigmatic labyrinth: a survey of chain of thought reasoning—advances, frontiers and future. In: Ku, L.W., Martins, A., Srikumar, V. (eds.) Proceedings of the 62nd Annual Meeting of the Association for Computational Linguistics (Volume 1: Long Papers), pp. 1173–1203. Association for Computational Linguistics (2024). https://doi.org/10.18653/v1/2024.acl-long.65
7. Cobbe, K., et al.: Training verifiers to solve math word problems. arXiv preprint arXiv:2110.14168 (2021)

8. Deutsch, D., Upadhyay, S., Roth, D.: A general-purpose algorithm for constrained sequential inference. In: Bansal, M., Villavicencio, A. (eds.) Proceedings of the 23rd Conference on Computational Natural Language Learning (CoNLL), pp. 482–492. Association for Computational Linguistics, Hong Kong, China (2019). https://doi.org/10.18653/v1/K19-1045. https://aclanthology.org/K19-1045/
9. Dong, Y., et al.: Xgrammar: flexible and efficient structured generation engine for large language models. Proc. Mach. Learn. Syst. **7** (2024)
10. Grattafiori, A., et al.: The llama 3 herd of models. arXiv preprint arXiv:2407.21783 (2024)
11. Guo, D., et al.: Deepseek-r1: incentivizing reasoning capability in LLMs via reinforcement learning. arXiv preprint arXiv:2501.12948 (2025)
12. Hendrycks, D., et al.: Measuring massive multitask language understanding (2021). https://arxiv.org/abs/2009.03300
13. Liu, A., et al.: Deepseek-v3 technical report. arXiv preprint arXiv:2412.19437 (2024)
14. Liu, N.F., et al.: Lost in the middle: how language models use long contexts. Trans. Assoc. Comput. Linguist. **12**, 157–173 (2024). https://doi.org/10.1162/tacl_a_00638. https://aclanthology.org/2024.tacl-1.9/
15. Lu, P., Qiu, L., Yu, W., Welleck, S., Chang, K.W.: A survey of deep learning for mathematical reasoning. In: Rogers, A., Boyd-Graber, J., Okazaki, N. (eds.) Proceedings of the 61st Annual Meeting of the Association for Computational Linguistics (Volume 1: Long Papers), pp. 14605–14631. Association for Computational Linguistics, Toronto, Canada (2023). https://doi.org/10.18653/v1/2023.acl-long.817. https://aclanthology.org/2023.acl-long.817/
16. Lu, Y., et al.: Learning to generate structured output with schema reinforcement learning. In: Che, W., Nabende, J., Shutova, E., Pilehvar, M.T. (eds.) Proceedings of the 63rd Annual Meeting of the Association for Computational Linguistics (Volume 1: Long Papers), pp. 4905–4918. Association for Computational Linguistics, Vienna, Austria (2025). https://doi.org/10.18653/v1/2025.acl-long.243. https://aclanthology.org/2025.acl-long.243/
17. OpenAI: learning to reason with LLMs (2024). https://openai.com/index/learning-to-reason-with-llms/
18. Ouyang, L., et al.: Training language models to follow instructions with human feedback. Adv. Neural. Inf. Process. Syst. **35**, 27730–27744 (2022)
19. Qi, J., et al.: RASAT: integrating relational structures into pretrained Seq2Seq model for text-to-SQL. In: Goldberg, Y., Kozareva, Z., Zhang, Y. (eds.) Proceedings of the 2022 Conference on Empirical Methods in Natural Language Processing, pp. 3215–3229. Association for Computational Linguistics, Abu Dhabi, United Arab Emirates (2022). https://doi.org/10.18653/v1/2022.emnlp-main.211. https://aclanthology.org/2022.emnlp-main.211/
20. Scholak, T., Schucher, N., Bahdanau, D.: PICARD: parsing incrementally for constrained auto-regressive decoding from language models. In: Moens, M.F., Huang, X., Specia, L., Yih, S.W.T. (eds.) Proceedings of the 2021 Conference on Empirical Methods in Natural Language Processing, pp. 9895–9901. Association for Computational Linguistics, Online and Punta Cana, Dominican Republic (2021). https://doi.org/10.18653/v1/2021.emnlp-main.779. https://aclanthology.org/2021.emnlp-main.779/
21. Shao, Z., et al.: Deepseekmath: pushing the limits of mathematical reasoning in open language models. arXiv preprint arXiv:2402.03300 (2024)
22. Skreta, M., et al.: Errors are useful prompts: instruction guided task programming with verifier-assisted iterative prompting. arXiv preprint arXiv:2303.14100 (2023)

23. Sun, T., et al.: Moss: an open conversational large language model. Mach. Intell. Res. **21**(5), 888–905 (2024)
24. Wei, J., et al.: Chain-of-thought prompting elicits reasoning in large language models. Adv. Neural. Inf. Process. Syst. **35**, 24824–24837 (2022)
25. Willard, B.T., Louf, R.: Efficient guided generation for large language models. arXiv preprint arXiv:2307.09702 (2023)
26. Xia, C., et al.: FOFO: a benchmark to evaluate LLMs' format-following capability. In: Ku, L.W., Martins, A., Srikumar, V. (eds.) Proceedings of the 62nd Annual Meeting of the Association for Computational Linguistics (Volume 1: Long Papers), pp. 680–699. Association for Computational Linguistics, Bangkok, Thailand (2024). https://doi.org/10.18653/v1/2024.acl-long.40. https://aclanthology.org/2024.acl-long.40/
27. Yao, J., et al.: Reff: reinforcing format faithfulness in language models across varied tasks. In: Proceedings of the AAAI Conference on Artificial Intelligence, vol. 39, pp. 25660–25668 (2025)
28. Yao, S., et al.: React: synergizing reasoning and acting in language models. In: International Conference on Learning Representations (ICLR) (2023)
29. Zhao, W.X., et al.: A survey of large language models. arXiv preprint arXiv:2303.18223, vol. 1, no. 2 (2023)
30. Zheng, L., et al.: Sglang: efficient execution of structured language model programs. In: Proceedings of the 38th International Conference on Neural Information Processing Systems. NIPS 2024. Curran Associates Inc, Red Hook, NY, USA (2024)
31. Zhou, Z., Li, J., Qiu, S., Huang, J., Qiu, L., Sun, Z.: Deepjsoneval: benchmarking complex nested json data mining for large language models (2025). https://arxiv.org/abs/2509.25922

# SegPPMTS: Unsupervised Segmentation for Pseudo-periodic Medical Time Series

Jinxi Wang(✉), Ling Luo, and Uwe Aickelin

School of Computing and Information Systems, The University of Melbourne, Melbourne, VIC 3053, Australia
{jinxi.wang.1,ling.luo,uwe.aickelin}@unimelb.edu.au

**Abstract.** Accurately segmenting medical time series into physiologically meaningful cycles is a critical preprocessing step to downstream tasks, such as classification. Unlike strictly periodic signals, medical time series, such as electrocardiograms and respiratory signals, exhibit pseudo-periodicity, characterised by variations in cycle length, amplitude, and morphology. The inherent variability of such medical signals poses challenges for existing segmentation approaches. To this end, we propose SegPPMTS, an unsupervised segmentation algorithm specifically designed for pseudo-periodic medical time series. SegPPMTS follows a three-step process: frequency-adaptive clustering to mitigate spectral leakage, periodicity hints validation to refine periodicity estimates, and dynamic time series segmentation to accommodate variations. It dynamically adjusts to pseudo-periodicity, enabling robust and adaptive segmentation across diverse signal types. Extensive experiments on four public medical datasets demonstrate that SegPPMTS outperforms existing segmentation methods across a comprehensive segmentation evaluation framework. Additionally, classification experiments validate our hypothesis that improved segmentation enhances classification performance, highlighting the importance of accurate pseudo-cycle segmentation of pseudo-periodic medical time series for downstream tasks. For clinical applications, our approach provides a promising solution for segmenting medical time series in scenarios where annotated segmentation ground truth is limited.

**Keywords:** Medical Time Series Segmentation · Pseudo-Periodic Signals · Medical Data Analytics

## 1 Introduction

Medical time series are fundamental to physiological monitoring and disease diagnosis. A core task is to classify recorded signals into physiological and pathological patterns [14]. Many medical time series, such as electrocardiogram (ECG) and respiratory signals, exhibit pseudo-periodic behaviour driven by underlying physiological cycles, such as heartbeats and breathing cycles. Accurately


R. C.-W. Wong et al. (Eds.): PAKDD 2026, LNAI 16597, pp. 582–596, 2026.
https://doi.org/10.1007/978-981-92-1300-9_46

segmenting such medical time series into physiologically meaningful pseudo-cycles is a critical preprocessing step, as improper segmentation can blur temporal dynamics across cycles and degrade the performance of classification models [4].

Compared with strictly periodic signals that repeat exactly with a fixed period, pseudo-periodic medical time series exhibit approximate cyclicity due to physiological variability and noise. Pseudo-periodicity manifests at intra-class and inter-class levels, as illustrated in Fig. 1. At the intra-class level, pseudo-cycles belonging to the same class share similar shapes but are not exactly identical, showing cycle-to-cycle variations in length, amplitude, and local morphology. At the inter-class level, pseudo-cycles from different classes exhibit differences in these properties. Such irregularities disrupt consistent temporal alignment across cycles, making it difficult for existing segmentation algorithms to identify the boundaries of pseudo-cycles.

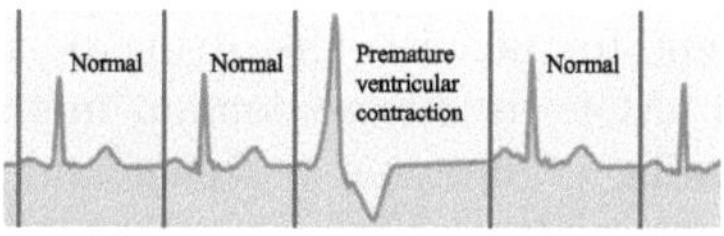


**Fig. 1.** An example of a pseudo-periodic ECG signal with a premature ventricular contraction [19]. Cycle-level segmentation is essential for the reliable identification and classification of abnormal events. Red vertical lines are the boundaries of cycles.

To this end, we propose SegPPMTS, an unsupervised segmentation algorithm designed to segment pseudo-periodic medical time series into pseudo-cycles. Leveraging the quasi-periodic nature, SegPPMTS follows a three-step process: frequency-adaptive clustering, periodicity hints validation, and dynamic time series segmentation. The contributions of this study are summarised below:

(1) We introduce SegPPMTS, an unsupervised segmentation algorithm tailored to identify the boundaries of pseudo-cycles within medical time series.
(2) We introduce a comprehensive evaluation framework that combines classification-related metrics with segmentation zone measures, enabling more precise assessment of segmentation quality.
(3) We validate SegPPMTS on four medical datasets, covering ECG, motion, and respiratory signals, highlighting its applicability and generalisability to a wide range of physiological signals.

## 2 Related Work

**Window-based** methods segment time series into fixed-length segments using either overlapping or non-overlapping schemes. However, these methods fail to adapt to natural cycle variations in medical signals. **Clustering-based**

approaches [8,15] attempt to group time series data based on feature similarity but fail to explicitly model periodicity. **Distance-based** methods, particularly those based on dynamic time warping (DTW) [3,10], align cycles of varying lengths by minimising temporal distortions. Despite effectively capturing local misalignments, DTW suffers from high computational complexity and sensitivity to noise. **Deep learning (DL)-based** methods have demonstrated state-of-the-art (SOTA) performance by automatically extracting features and identifying segmentation boundaries from raw data, such as graph neural networks [15] and U-Net-like models [11,16–18,21]. Despite the promising results, DL-based approaches require large amounts of labelled data for training and often exhibit poor generalisation across various physiological signals.

**Periodicity detection-based** segmentation methods aim to identify cyclic patterns by leveraging either frequency-domain or time-domain analysis. Fast Fourier transform (FFT)-based methods [4,22] estimate periodicity by analysing frequency components in medical signals. However, FFT-based methods assume signal stationarity and suffer from spectral leakage, where energy from one frequency spills into adjacent frequencies, distorting periodicity estimation. The autocorrelation function (ACF) is a time-domain method that measures periodicity based on the similarity between a signal and its lagged version [24]. However, when multiple periodic components exist, ACF generates ambiguous peaks, making it difficult to select the most suitable segmentation period.

# 3 Methodology

## 3.1 Problem Formulation

We consider a one-dimensional pseudo-periodic medical time series $\mathbf{x} = \{x_t\}_{t=1}^{N}$, which can be partitioned into a sequence of cycles $\{\phi_i\}_{i=1}^{T}$ by cycle boundaries $\{T_i\}_{i=1}^{N}$, as illustrated in Fig. 2. The goal of this work is to identify the boundary points of each cycle and thus segment the pseudo-periodic medical time series into physiologically meaningful pseudo-cycles.

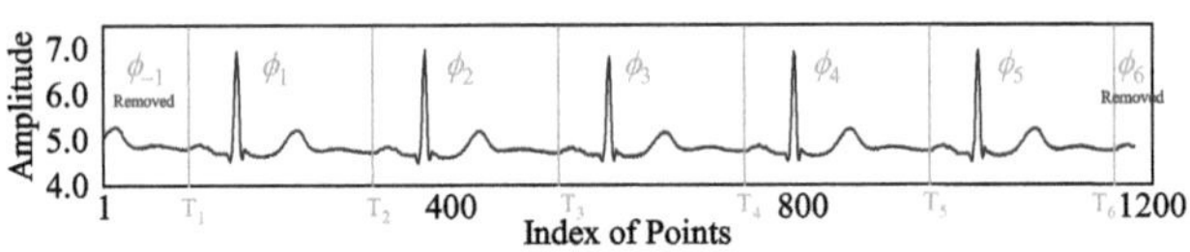


**Fig. 2.** A segment of ECG signal from the QT database [13]. The cyan vertical lines indicate the onsets of each cycle.

## 3.2 Method Overview

Our proposed segmentation method, SegPPMTS, is designed to segment pseudo-periodic medical time series into physiologically meaningful pseudo-cycles. It follows a three-step process: frequency-adaptive clustering, periodicity hints validation, and dynamic time series segmentation. The overall workflow is illustrated in Fig. 3, and the function of each step is detailed below:

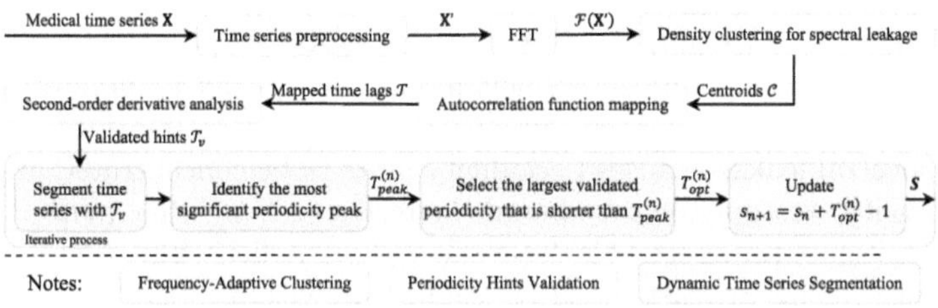


**Fig. 3.** Overall workflow of SegPPMTS.

(1) **Frequency-Adaptive Clustering:** Extracts dominant periodic components by detecting frequency peaks in the FFT spectrum while mitigating spectral leakage using a density-based clustering algorithm.
(2) **Periodicity Hints Validation:** Refines periodicity estimates and eliminates spurious periodicities through second-order derivative analysis of ACF.
(3) **Dynamic Time Series Segmentation:** Identifies the most appropriate period from validated hints and dynamically segments the entire time series from the beginning to the end, while accounting for both cycle variations and the coexistence of multiple periodic components.

### 3.3 Frequency-Adaptive Clustering

Before applying segmentation, we preprocess the pseudo-periodic medical time series $\mathbf{X}$ to ensure that the extracted periodic components are not affected by incomplete cycles at the beginning or end of the signal. Specifically, we remove any incomplete cycles that may exist at the start or end of the time series (e.g., $\phi_{-1}$ and $\phi_6$ in Fig. 2). Next, we normalise the signal to the range [0, 1] and remove the direct current (DC) component to enhance the frequency-domain analysis. The preprocessed signal $\mathbf{X}'$ is obtained.

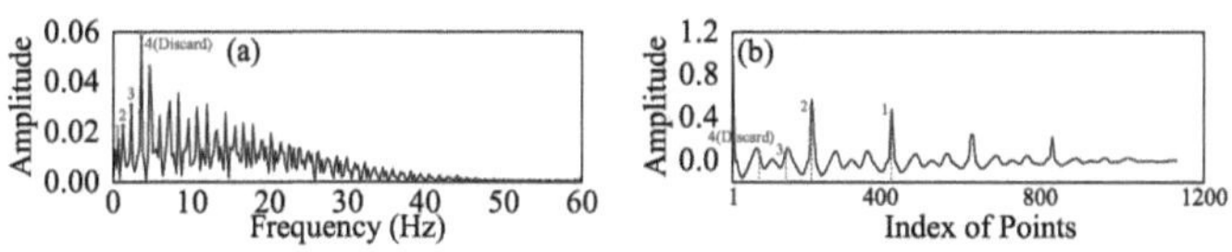


**Fig. 4.** FFT spectrum and autocorrelation function of the ECG shown in Fig. 2: (a) FFT spectrum and (b) autocorrelation spectrum.

We then apply FFT to convert the preprocessed time series $\mathbf{X}'$ into the frequency domain, enabling the extraction of periodicity hints from dominant frequency components. As shown in Fig. 4(a), the FFT spectrum $\mathcal{F}(\mathbf{X}')$ of the ECG signal in Fig. 2 reveals multiple frequency peaks. Initially, we identify local maxima in the FFT spectrum as periodicity hints, represented as the set $\mathcal{P} = \{f_i \mid \mathcal{F}(\mathbf{X}')(f_i) > \mathcal{F}(\mathbf{X}')(f_{i-1}), \mathcal{F}(\mathbf{X}')(f_i) > \mathcal{F}(\mathbf{X}')(f_{i+1})\}$. However, spectral leakage spreads signal energy across neighbouring frequencies, thereby distorting periodicity hints and causing inaccuracies in periodicity estimation.

Since spectral leakage causes periodicity hints to be concentrated around the true periodicity, we apply density-based clustering to $\mathcal{P}$ to refine periodicity estimates and mitigate the spectral leakage effect. A periodicity hint $f_i$ is classified as a core point if it has at least $MinPts$ neighbours within its $\varepsilon$-neighbourhood:

$$|\mathcal{N}_\varepsilon(f_i)| \geq MinPts. \tag{1}$$

The $\varepsilon$-neighbourhood of a potential core point $f_i$ is defined as:

$$\mathcal{N}_\varepsilon(f_i) = \{f_j \mid d(f_i, f_j) \leq \varepsilon_{N'/k}\}, \tag{2}$$

where $N'$ is the length of the preprocessed time series $\mathbf{X}'$, $f_i$ is a periodicity hint under evaluation, and $f_j$ represents the neighbouring points within distance $\varepsilon_{N'/k}$. The distance function $d(f_i, f_j)$ measures the Euclidean distance between frequency bins. If $f_i$ satisfies Eq. (1), it is classified as a core point.

To ensure that frequency components corresponding to shorter periodicities are not overwhelmed by longer ones, we sort candidate frequency peaks in ascending order before the clustering algorithm. The adaptive search radius $\varepsilon_{N'/k}$ dynamically adjusts based on frequency resolution and is defined as:

$$\varepsilon_{N'/k} = \frac{N'}{k-1} + 1, \tag{3}$$

where $k$ is the frequency bin index. The term $\frac{N'}{k-1}$ ensures that hints leaked into adjacent bins remain within the clustering range, adapting to frequency resolution. The additional constant $+1$ prevents numerical instability when adjacent bin differences are small, thereby improving clustering robustness.

Clusters are constructed iteratively by expanding from each core point, adding all density-reachable neighbours, and repeating this process until no additional points can be included.

$$C_i = \bigcup_{f_j \in C_i} \mathcal{N}_\varepsilon(f_j), \quad \text{for all core points } f_j. \tag{4}$$

Once clusters are established, the centroid of each cluster $c_i$ is computed to provide a refined periodicity estimate:

$$c_i = \frac{1}{|C_i|} \sum_{f_j \in C_i} f_j. \tag{5}$$

This clustering-based refinement mitigates spectral leakage and provides a more accurate estimate of periodicity hints compared to the raw periodicity hints corresponding to the FFT spectrum peaks. The refined periodicity estimates are represented as the set of centroids $\mathcal{C} = \{c_i\}_{i=1}^{M}$.

Figure 4(a) illustrates the result of density-based clustering on periodicity hints. The identified peaks (labelled as 1, 2, 3, and 4) serve as representative centroids of clustered periodicity hints, demonstrating the refinement of periodicity estimation while mitigating spectral leakage. Please note that these peaks are shown for illustrative purposes, and additional clustered periodicity hints may exist but are not displayed in Fig. 4(a).

### 3.4 Periodicity Hints Validation

To ensure that periodicity hints obtained from the frequency domain are effectively validated in the time domain, we first map the centroids of frequency clusters $\mathcal{C}$ to the corresponding time lags $\mathcal{T}$ in the ACF spectrum of the preprocessed time series $\mathbf{X}'$. The set of mapped time lags is given by:

$$\mathcal{T} = \{T_i \mid T_i = \frac{f_s}{c_i}, c_i \in \mathcal{C}\}, \tag{6}$$

where $f_s$ is the sampling frequency of the medical time series $\mathbf{X}$ or $\mathbf{X}'$. This transformation ensures that the periodic components identified in the FFT spectrum align with the time-lagged correlations observed in the ACF spectrum, enabling a consistent periodicity validation process.

To assess the validity of periodicity hints, we analyse the second-order derivative of ACF, denoted as $ACF''(\cdot)$. The underlying principle is as follows: (i) Local maxima in the ACF correspond to strong periodic correlations, indicating valid periodicity hints; (ii) Local minima in the ACF suggest weak correlations and should be disregarded. Based on this criterion, we identify periodicity hints as:

(1) Valid periodicity hints: If $ACF''(T_i) < 0$, the periodicity hint $T_i$ corresponds to a local maximum in the ACF, confirming it as a valid periodicity.
(2) Spurious periodicity hints: If $ACF''(T_i) > 0$, the periodicity hint $T_i$ corresponds to a local minimum in the ACF, indicating weak correlation and should therefore be discarded.

By filtering out spurious hints, we refine the segmentation process, ensuring that only periodicity hints corresponding to strong correlations are retained. As shown in Fig. 4(b), the periodicity hint labelled 4 is identified as spurious and removed. The set of validated periodicity hints is then given by:

$$\mathcal{T}_v = \{T_{vi} \mid T_{vi} \in \mathcal{T}, ACF''(T_{vi}) < 0\}. \tag{7}$$

### 3.5 Dynamic Time Series Segmentation

Now that we have obtained the validated periodicity hints $\mathcal{T}_v$ (in ascending order), we segment the time series iteratively based on validated periodicity hints. The optimal segment length should correspond to a single complete cycle while avoiding excessive overlap with adjacent cycles.

To determine the optimal segment length at iteration $n$, we evaluate the segments obtained from validated periodicity hints using ACF. Instead of selecting the periodicity hints with the highest autocorrelation coefficient directly (except zero-position), we define the optimal period $T_{\text{opt}}^{(n)}$ as follows:

(1) Identify the most significant periodicity peak $T_{\text{peak}}^{(n)}$ in the autocorrelation functions of the segments at iteration $n$, where $T_{\text{peak}}^{(n)}$ represents the length of the segment corresponding to the strongest periodicity peak:

$$T_{\text{peak}}^{(n)} = \arg\max_{T_{vi} \in \mathcal{T}_v} ACF(T_{vi}). \tag{8}$$

(2) Select the largest validated periodicity that is shorter than $T_{\text{peak}}^{(n)}$:

$$T_{\text{opt}}^{(n)} = \max\{T_{vi} \mid T_{vi} < T_{\text{peak}}^{(n)}, T_{vi} \in \mathcal{T}_v\}. \tag{9}$$

Using the selected $T_{\text{opt}}^{(n)}$, we iteratively update the segmentation points. Given the initial segmentation point $s_1 = T_1$, subsequent segmentation points are computed as:

$$s_{n+1} = s_n + T_{\text{opt}}^{(n)} - 1, \quad \text{until complete coverage of the signal } \mathbf{X}'. \tag{10}$$

Once all segmentation points are determined, the pseudo-periodic medical time series is partitioned accordingly, ensuring complete coverage of signal $\mathbf{X}'$.

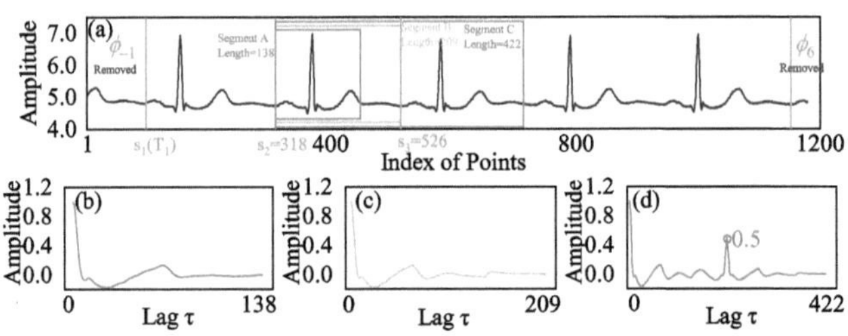


**Fig. 5.** Illustration of Step 3: (a) The ECG signal (shown in Fig. 2) with segmentation points indicated, and (b), (c), and (d) show the ACF spectra of Segments A, B, and C, respectively.

To illustrate the segmentation process, we consider the signal given in Fig. 2. The validated periodicity hints, identified as 138 points, 209 points, and 422 points, serve as candidate segment lengths for the iterative segmentation process.

At $n = 1$, we determined the segmentation point $s_2 = 318$. At $n = 2$, we tested three candidate segments: Segment A, Segment B, and Segment C, as shown in Fig. 5(a). We then evaluated the autocorrelation function for all three segments, as shown in Figs. 5(b), (c), and (d). The autocorrelation of segment C exhibits a significant peak (0.5), indicating the presence of multiple cycles within this segment. Since the periodic structure in Segment C extends beyond a single cycle, we select Segment B (209 points) as the optimal segment, as it is the shortest segment that preserves a complete cycle. Thus, the next segmentation point is: $s_3 = s_2 + T_{opt}^{(2)} = 318 + 209 - 1 = 526$. This process repeats iteratively to segment the entire time series.

**Algorithm 1:** SegPPMTS

**Input**: Pseudo-periodic medical time series **X**, $\varepsilon$ (adaptive), and $MinPts = 6$.
**Output**: Set of segmentation points **S**

```
/* Frequency-Adaptive Clustering, O(N log N) + O(M log M) (indexing
structure applied during clustering) or O(N log N) + O(M^2) (no
indexing structure used during clustering) */
Remove incomplete cycles at the start and end of X
Normalise X to range [0,1]
Remove DC component to obtain X'
Compute Fourier transform F(X')
Identify local maxima in F(X') as periodicity hints P
Apply density-based clustering to refine periodicity hints and compute centroids
C:
for each frequency hint f_i ∈ P do
    Compute ε_{N'/k} using Eq. (2)
    Determine ε-neighbourhood N_ε(f_i)
    Expand clusters iteratively for core points
    Compute cluster centroids using Eq. (5)
/* Periodicity Hints Validation, O(N log N)(based on the
WienerKhinchin theorem) */
Map centroids C to time lags T in ACF domain
Filter out spurious hints based on second-order derivative ACF''(T_i):
for each T_i ∈ T do
    if ACF''(T_i) < 0 then
        Add T_i to valid periodicity set T_v
/* Dynamic Time Series Segmentation, O(N) */
Initialise segmentation point s_1 = T_1
while s_n < N' do
    Compute the most significant periodicity peak using Eq. (8)
    Select the largest validated periodicity shorter than T_peak^(n) using Eq. (9)
    Update segmentation point using Eq. (10)
    if s_{n+1} ≥ N' then
        Break (Ensure complete coverage of the time series)
    n = n + 1
return Set of segmentation points S
```

### 3.6 Algorithm Implementation Description of SegPPMTS

We provide the step-by-step implementation of SegPPMTS in Algorithm 1. The algorithm takes a pseudo-periodic medical time series as input and outputs a set of segmentation points that divide the medical time series into pseudo-cycles.

The computational complexity of each step is also given in Algorithm 1. Given that $M \ll$ N empirically, the total computational complexity is dominated by the FFT term, i.e., $O(N \log N)$, which is compatible with real-time implementation.

## 4 Experiment

To evaluate the effectiveness of SegPPMTS, we conducted extensive experiments on four public medical datasets. The experiments were performed on a high-performance computing system equipped with an Intel Xeon Silver 4214 CPU (2.20 GHz) and an NVIDIA Tesla V100 SXM2 GPU, ensuring efficient computation and accelerated processing for deep learning tasks.

### 4.1 Datasets

We evaluate the proposed SegPPMTS method on four public healthcare datasets:

(1) QT Database (QTD) [13]: We use the 'MLII' channel for segmentation and 'pu0' for P wave onset annotation.
(2) Gait in Neurodegenerative Disease Dataset (GNDD) [7]: The 'left-foot' signal is used for segmentation, with gait cycle onset annotations extracted from Derived Time Series (.ts) files.
(3) Wrist Photoplethysmogram (PPG) during Exercise Dataset (WPED) [9]: Channel 'wrist_gyro_z'. Pseudo-cycle onsets are manually annotated.
(4) Ventilator Time Series Dataset (VTSD) [20]: Pressure signals are used for segmentation with breath-start annotations for segmentation evaluation.

### 4.2 Implementation Details for Time Series Segmentation

**Compared Segmentation Approaches.** The proposed approach is compared with the following five algorithms: fixed window, critical point clustering (CPC) [12], sub-sequence DTW (Sub-seq DTW) [2], trend filtering sliding discrete Fourier transform-based pseudo-periodic time series segmentation algorithm (EINSTEIN) [4], and U-Time [18]. The fixed window method partitions time series into non-overlapping segments, with segment length set to the average pseudo-cycle length.

**Hyperparameter Fine-Tuning.** Each dataset was partitioned into training and test sets with a ratio of 8:2. For hyperparameter fine-tuning, the training set was further split into 80% for training and 20% for validation. $F_1$-score was used as the criterion during this process.

**Comprehensive Segmentation Evaluation Framework.** Three classification-related metrics (precision, recall, and $F_1$-score) and three segmentation zone (SZ) measures (average segmentation count, ASC; absolute segmentation distance, ASD; and average direct tendency, ADT) are utilised to assess segmentation performance. More details can be found in Ref. [6].

### 4.3 Time Series Segmentation Result Analysis

**Time Series Segmentation Results.** Table 1 presents the results of the various segmentation algorithms across all datasets. Overall, SegPPMTS outperforms the other segmentation methods in four out of six evaluation metrics. Specifically, in terms of classification-related measures, SegPPMTS achieves the highest mean $F_1$-score of 0.9846, followed by U-Time (0.9780). It should be noted that U-Time is a supervised DL-based method that relies on labelled data for training, whereas SegPPMTS is unsupervised. This indicates the adaptability and label efficiency of SegPPMTS. When considering segmentation zone measures, it is evident that, on average, all algorithms predicted fewer segmentation points per SZ than required. Among them, CPC places segmentation points closest to the centre of the SZ. Furthermore, SegPPMTS, U-Time, and EINSTEIN tend to position segmentation points slightly before the SZ centre, while methods such as fixed window, CPC, and Sub-seq DTW place the segmentation points after the SZ centre.

**Table 1.** Comparison of segmentation results across datasets and methods.

| Dataset | Segmentation method | Evaluation measure | | | | | |
|---|---|---|---|---|---|---|---|
| | | Precision | Recall | $F_1$-score | ASC | ASD | ADT |
| QTD | Fixed window | 0.8214 | 0.8134 | 0.8174 | 0.8485 | 9.1039 | 0.7265 |
| | CPC | 0.9025 | 0.9601 | 0.9304 | 0.9837 | 6.8558 | 0.1583 |
| | Sub-seq DTW | 0.9913 | 0.9955 | 0.9934 | 0.9932 | 8.3521 | 0.8903 |
| | EINSTEIN | 0.9802 | 0.9828 | 0.9815 | 0.9877 | 7.0267 | 0.2206 |
| | U-Time (DL-based) | 0.9788 | 0.9647 | 0.9717 | 0.9631 | 6.0822 | 0.4025 |
| | SegPPMTS | 0.9803 | 0.9973 | 0.9887 | 0.9668 | 5.3543 | 0.3510 |
| GNDD | Fixed window | 0.7464 | 0.7028 | 0.7239 | 0.7455 | 8.8616 | 0.9204 |
| | CPC | 0.6575 | 0.5515 | 0.5999 | 0.4913 | 6.5501 | 0.2351 |
| | Sub-seq DTW | 0.9322 | 0.8637 | 0.8966 | 0.8696 | 3.2515 | 0.3171 |
| | EINSTEIN | 0.9614 | 0.9508 | 0.9561 | 0.9742 | 8.6593 | 0.2910 |
| | U-Time (DL-based) | 0.9705 | 0.9588 | 0.9646 | 0.9604 | 5.4298 | 0.3367 |
| | SegPPMTS | 0.9735 | 0.9679 | 0.9707 | 0.9750 | 7.8647 | 0.4594 |
| WPED | Fixed window | 0.8206 | 0.7860 | 0.8029 | 0.8248 | 15.6242 | 0.8492 |
| | CPC | 0.9820 | 0.8266 | 0.8976 | 0.6514 | 5.1177 | 0.8453 |
| | Sub-seq DTW | 0.9742 | 0.9906 | 0.9823 | 0.9900 | 12.2541 | 0.9525 |
| | EINSTEIN | 1.0000 | 1.0000 | 1.0000 | 0.9929 | 11.3582 | 0.2540 |
| | U-Time (DL-based) | 0.9923 | 0.9984 | 0.9953 | 0.9628 | 5.5683 | 0.2855 |
| | SegPPMTS | 1.0000 | 0.9835 | 0.9917 | 1.0000 | 9.6869 | 0.2183 |
| VTSD | Fixed window | 0.7088 | 0.6648 | 0.6861 | 0.6894 | 8.0538 | 0.7982 |
| | CPC | 0.6805 | 0.9708 | 0.8001 | 0.7549 | 4.5235 | 0.7909 |
| | Sub-seq DTW | 0.9933 | 0.9213 | 0.9559 | 0.8847 | 0.1108 | 0.2535 |
| | EINSTEIN | 0.9856 | 0.8887 | 0.9346 | 0.8111 | 3.3381 | 0.2665 |
| | U-Time (DL-based) | 0.9844 | 0.9766 | 0.9805 | 0.9708 | 5.8265 | 0.1876 |
| | SegPPMTS | 0.9924 | 0.9825 | 0.9874 | 0.9209 | 3.3520 | 0.5345 |
| Mean value | Fixed window | 0.7743 | 0.7418 | 0.7576 | 0.7771 | 10.4109 | 0.8236 |
| | CPC | 0.8056 | 0.8273 | 0.8070 | 0.7203 | 5.7618 | **0.5074** |
| | Sub-seq DTW | 0.9728 | 0.9428 | 0.9571 | 0.9344 | 5.9921 | 0.6034 |
| | EINSTEIN | 0.9818 | 0.9556 | 0.9681 | 0.9415 | 7.5956 | 0.2580 |
| | U-Time (DL-based) | 0.9815 | 0.9746 | 0.9780 | 0.9643 | **5.7267** | 0.3031 |
| | SegPPMTS | **0.9866** | **0.9828** | **0.9846** | **0.9657** | 6.5645 | 0.3908 |

**Parameter Sensitivity Testing and Ablation Studies.** We conducted one sensitivity analysis on the parameter $MinPts$ and two high-level ablation studies on the QTD dataset. For the ablation studies, we remove frequency-adaptive clustering or periodicity hints validation within the SegPPMTS framework. Since the parameter $\epsilon$ is adaptive, no sensitivity testing is applied to it. The results are given in Table 2. When $MinPts$ is too small, the algorithm tends to form spurious small clusters and include noise points as part of clusters, which slightly degrades the overall performance. However, when $MinPts$ becomes too large, the density threshold becomes strict, resulting in more points being labelled as noise and a gradual decrease in performance. We would recommend tuning this parameter, if necessary, to obtain better performance on specific datasets. Besides, our ablation studies demonstrate that both frequency-adaptive clustering and periodicity hints validation contribute to the performance of SegPPMTS.

### 4.4 Time Series Classification Result Analysis

We assume that an improvement in segmentation quality leads to an enhancement in classification performance. Therefore, classification experiments were conducted to validate our assumption.

**Classification Models.** Four classifiers, Light Gradient Boosting Machine (LightGBM) [23], Convolutional Neural Network (CNN) [1], Shapelet Transformer (ShapeFormer) [14], and Minimally Random Convolutional Kernel Transform (MiniRocket) [5], are employed to classify the pseudo-periodic cycles obtained by different segmentation strategies. Each experiment was repeated 10 times, and the reported results reflect (mean $\pm$ standard deviation) over these runs.

**Table 2.** Results of parameter sensitivity testing and ablation studies on QTD dataset.

| Segmentation method | Evaluation metrics | | | | | |
|---|---|---|---|---|---|---|
| | Precision | Recall | $F_1$-score | ASC | ASD | ADT |
| SegPPMTS ($MinPts = 6$, our choice) | 0.9803 | 0.9973 | 0.9887 | 0.9668 | 5.3543 | 0.3510 |
| SegPPMTS ($MinPts = 4$) | 0.9764 | 0.9611 | 0.9687 | 0.9575 | 4.9735 | 0.3329 |
| SegPPMTS ($MinPts = 8$) | 0.9733 | 0.9664 | 0.9698 | 0.9601 | 6.0392 | 0.5873 |
| SegPPMTS ($MinPts = 10$) | 0.9603 | 0.9529 | 0.9566 | 0.9438 | 6.9042 | 0.6743 |
| Our SegPPMTS w/o frequency-adaptive clustering | 0.8085 | 0.8325 | 0.8203 | 0.8431 | 6.5818 | 0.4552 |
| Our SegPPMTS w/o periodicity hints validation | 0.8964 | 0.9185 | 0.9073 | 0.8707 | 7.6525 | 0.5620 |

**Time Series Classification Results.** For the classification experiments, accuracy is employed as the evaluation metric, with the results presented in Table 3. When the segments generated by SegPPMTS are used as input across all classifiers and datasets, a consistently high overall classification accuracy of 92.48% is achieved. On the other hand, the U-Time segmentation method ranks second, yielding an average classification accuracy exceeding 90% across datasets, while all other segmentation methods result in lower average accuracies (below 90%). Together with the segmentation results in Table 1, these findings confirm that

**Table 3.** Comparison of accuracies (%) across datasets and methods.

| Dataset | Classifier | Segmentation method | | | | | |
|---|---|---|---|---|---|---|---|
| | | Fixed window | CPC | Sub-seq DTW | EINSTEIN | U-Time | SegPPMTS |
| QTD | LightGBM | $78.84 \pm 3.06$ | $80.38 \pm 2.72$ | $83.98 \pm 2.28$ | $84.56 \pm 1.69$ | $88.44 \pm 1.07$ | $90.34 \pm 0.85$ |
| | CNN | $80.22 \pm 2.46$ | $81.62 \pm 2.65$ | $86.29 \pm 1.73$ | $86.88 \pm 1.42$ | $89.62 \pm 1.24$ | $91.02 \pm 0.64$ |
| | ShapeFormer | $85.68 \pm 2.74$ | $88.02 \pm 1.69$ | $88.96 \pm 1.13$ | $92.84 \pm 1.25$ | $92.96 \pm 1.84$ | $94.25 \pm 0.23$ |
| | MiniRocket | $84.95 \pm 3.25$ | $86.41 \pm 0.62$ | $87.51 \pm 0.82$ | $92.10 \pm 1.15$ | $93.54 \pm 1.43$ | $95.68 \pm 0.73$ |
| GNDD | LightGBM | $77.04 \pm 2.89$ | $79.34 \pm 2.88$ | $82.38 \pm 1.53$ | $85.29 \pm 1.32$ | $88.06 \pm 0.76$ | $90.02 \pm 1.12$ |
| | CNN | $79.38 \pm 3.04$ | $82.67 \pm 1.98$ | $82.81 \pm 2.09$ | $85.86 \pm 1.49$ | $89.23 \pm 0.88$ | $91.96 \pm 0.94$ |
| | ShapeFormer | $81.56 \pm 1.97$ | $84.38 \pm 1.76$ | $85.62 \pm 2.19$ | $92.29 \pm 1.22$ | $92.69 \pm 0.51$ | $94.81 \pm 1.02$ |
| | MiniRocket | $81.55 \pm 3.70$ | $83.93 \pm 2.35$ | $87.52 \pm 1.41$ | $92.56 \pm 1.98$ | $93.15 \pm 0.74$ | $94.23 \pm 1.35$ |
| WPED | LightGBM | $80.55 \pm 2.14$ | $82.82 \pm 2.67$ | $84.08 \pm 1.67$ | $87.85 \pm 3.08$ | $89.71 \pm 2.15$ | $91.15 \pm 1.67$ |
| | CNN | $81.29 \pm 3.01$ | $83.01 \pm 1.98$ | $84.86 \pm 1.85$ | $88.38 \pm 0.55$ | $89.96 \pm 1.78$ | $92.58 \pm 0.64$ |
| | ShapeFormer | $83.54 \pm 2.65$ | $84.37 \pm 2.00$ | $86.54 \pm 1.36$ | $90.37 \pm 1.58$ | $90.86 \pm 1.07$ | $93.12 \pm 0.76$ |
| | MiniRocket | $83.59 \pm 1.52$ | $84.02 \pm 1.27$ | $85.34 \pm 1.02$ | $90.71 \pm 1.74$ | $91.82 \pm 0.54$ | $94.02 \pm 0.61$ |
| VTSD | LightGBM | $78.34 \pm 1.77$ | $80.55 \pm 3.47$ | $82.16 \pm 4.35$ | $84.29 \pm 2.87$ | $86.75 \pm 2.73$ | $90.65 \pm 1.66$ |
| | CNN | $79.43 \pm 2.67$ | $81.72 \pm 2.08$ | $83.68 \pm 3.08$ | $85.81 \pm 1.05$ | $86.94 \pm 1.89$ | $91.18 \pm 1.92$ |
| | ShapeFormer | $79.61 \pm 3.14$ | $83.98 \pm 3.78$ | $85.27 \pm 2.61$ | $88.29 \pm 2.15$ | $90.96 \pm 1.74$ | $92.64 \pm 1.16$ |
| | MiniRocket | $81.55 \pm 5.46$ | $83.55 \pm 4.05$ | $84.62 \pm 3.01$ | $87.51 \pm 2.67$ | $91.44 \pm 0.53$ | $91.97 \pm 2.11$ |
| Mean value | LightGBM | 78.69 | 80.77 | 83.15 | 85.50 | 88.24 | **90.54** |
| | CNN | 80.08 | 82.26 | 84.41 | 86.73 | 88.94 | **91.69** |
| | ShapeFormer | 82.60 | 85.19 | 86.60 | 90.95 | 91.87 | **93.71** |
| | MiniRocket | 82.91 | 84.48 | 86.25 | 90.72 | 92.49 | **93.98** |

improved segmentation quality consistently leads to higher classification accuracy. Moreover, the choice of classifier also impacts classification performance. Overall, MiniRocket and ShapeFormer outperform LightGBM and CNN in most scenarios, which aligns with the general expectation that more advanced classifiers tend to attain superior accuracy.

## 5 Conclusion and Future Work

In this work, we propose SegPPMTS, an unsupervised segmentation method that leverages the pseudo-periodic nature of medical time series to partition them into pseudo-cycles. Extensive experiments show that SegPPMTS consistently outperforms five baseline methods and achieves the best average rank on four out of six evaluation metrics, demonstrating its overall superiority. We further demonstrate that better segmentation quality improves classification performance. These results highlight the importance of accurate pseudo-cycle segmentation in medical time series analysis and position SegPPMTS as a possible solution in scenarios with limited boundary annotations.

Furthermore, the SegPPMTS framework can be readily extended to other domains, as it assumes the existence of quasi-periodic structures in local time windows rather than domain-specific characteristics. Under this assumption, SegPPMTS can, in principle, be applied to other domains that exhibit quasi-periodic behavior, such as environmental monitoring (e.g., daily temperature cycles and circadian patterns) and industrial sensor data (e.g., vibration signals from rotating machinery). Additional preprocessing pipelines may apply.

One limitation of our work is the potential uncertainty in segmentation boundaries resulting from manual annotations, as seen in the WPED dataset. It may affect the assessment reliability of segmentation when annotations are inconsistent. To minimise this uncertainty, we may refine the annotation process by involving multiple experts to annotate the same data or by introducing automatic validation mechanisms.

## References


1. Acharya, U.R., et al.: A deep convolutional neural network model to classify heartbeats. Comput. Biol. Med. **89**, 389–396 (2017)
2. Alshehri, M., Coenen, F., Dures, K.: Sub-sequence-based dynamic time warping. In: KDIR, pp. 274–281 (2019)
3. Chakraborty, G., Kamiyama, T., Takahashi, H., Kinoshita, T.: An efficient anomaly detection in quasi-periodic time series data—a case study with ECG. In: Time Series Analysis and Forecasting: Selected Contributions from ITISE 2017, pp. 147–157. Springer (2018)
4. Chaudhary, A., Luo, L., Aickelin, U.: Einstein: a fast online segmentation (pseudo-period extraction) algorithm for complex pseudo-periodic data streams. Int. J. Data Sci. Analytics (2025)
5. Dempster, A., Schmidt, D.F., Webb, G.I.: MINIROCKET: a very fast (almost) deterministic transform for time series classification. In: Proceedings of the 27th ACM SIGKDD Conference on Knowledge Discovery & Data Mining, pp. 248–257 (2021)
6. Gensler, A., Sick, B.: Novel criteria to measure performance of time series segmentation techniques. In: LWA, pp. 193–204 (2014)
7. Hausdorff, J.M., Lertratanakul, A., Cudkowicz, M.E., Peterson, A.L., Kaliton, D., Goldberger, A.L.: Dynamic markers of altered gait rhythm in amyotrophic lateral sclerosis. J. Appl. Physiol. (2000)
8. Huang, G., Zhang, Y., Cao, J., Steyn, M., Taraporewalla, K.: Online mining abnormal period patterns from multiple medical sensor data streams. World Wide Web **17**, 569–587 (2014)
9. Jarchi, D., Casson, A.J.: Description of a database containing wrist PPG signals recorded during physical exercise with both accelerometer and gyroscope measures of motion. Data **2**(1), 1 (2016)
10. Jiang, H., Zhang, K., Wang, J., Wang, X., Huang, P.: Anomaly detection and identification in satellite telemetry data based on pseudo-period. Appl. Sci. **10**(1), 103 (2019)
11. Joung, C., et al.: Deep learning based ECG segmentation for delineation of diverse arrhythmias. PLoS ONE **19**(6), e0303178 (2024)
12. Keogh, E., Chu, S., Hart, D., Pazzani, M.: An online algorithm for segmenting time series. In: Proceedings 2001 IEEE International Conference on Data Mining, pp. 289–296. IEEE (2001)
13. Laguna, P., Mark, R.G., Goldberg, A., Moody, G.B.: A database for evaluation of algorithms for measurement of qt and other waveform intervals in the ECG. In: Computers in Cardiology 1997, pp. 673–676. IEEE (1997)

14. Le, X.M., Luo, L., Aickelin, U., Tran, M.T.: ShapeFormer: Shapelet transformer for multivariate time series classification. In: Proceedings of the 30th ACM SIGKDD Conference on Knowledge Discovery and Data Mining, pp. 1484–1494 (2024)
15. Liu, F., et al.: Anomaly detection in quasi-periodic time series based on automatic data segmentation and attentional LSTM-CNN. IEEE Trans. Knowl. Data Eng. **34**(6), 2626–2640 (2020)
16. Matias, P., Folgado, D., Gamboa, H., Carreiro, A.: Time series segmentation using neural networks with cross-domain transfer learning. Electronics **10**(15), 1805 (2021)
17. Perslev, M., Darkner, S., Kempfner, L., Nikolic, M., Jennum, P.J., Igel, C.: U-Sleep: resilient high-frequency sleep staging. NPJ Digit. Med. **4**(1), 72 (2021)
18. Perslev, M., Jensen, M., Darkner, S., Jennum, P.J., Igel, C.: U-Time: a fully convolutional network for time series segmentation applied to sleep staging. In: Advances in Neural Information Processing Systems, vol. 32 (2019)
19. QALY: usage statistics of content languages for websites. https://www.qaly.co/post/premature-ventricular-contractions-pvcs. Accessed 21 Nov 2025
20. Rehm, G.B., Kuhn, B.T., Nguyen, J., Anderson, N.R., Chuah, C.N., Adams, J.Y.: Improving mechanical ventilator clinical decision support systems with a machine learning classifier for determining ventilator mode. In: MEDINFO 2019: Health and Wellbeing e-Networks for All, pp. 318–322. IOS Press (2019)
21. Sereda, I., Alekseev, S., Koneva, A., Kataev, R., Osipov, G.: ECG segmentation by neural networks: Errors and correction. In: 2019 International Joint Conference on Neural Networks (IJCNN), pp. 1–7. IEEE (2019)
22. Tarjano, C., Pereira, V.: An efficient algorithm for segmenting quasi-periodic digital signals into pseudo cycles: application in lossy audio compression. IEEE/ACM Trans. Audio, Speech, Lang. Process. **30**, 1694–1703 (2022)
23. Wang, C., Aickelin, U., Luo, L., Ristanoski, G., Howard, M.E., Berlowitz, D.: Patient-ventilator asynchrony detection via similarity search methods. J. Adv. Inf. Technol. **13**(1) (2022)
24. Wang, J., Zhang, F., Zhang, L., Jiang, M.: Maximum average impulse energy ratio deconvolution and its application for periodic fault impulses enhancement of rolling bearing. Adv. Eng. Inform. **53**, 101721 (2022)

# Prototype Fusion: A Training-Free Multi-layer Approach to OOD Detection

Shreen Gul[1], Mohamed Elmahallawy[2], Ardhendu Tripathy[1,2], and Sanjay Madria[1(✉)]

[1] Missouri University of Science and Technology, Rolla, MO 65401, USA
{sgchr,madrias}@mst.edu
[2] Washington State University, Richland, WA 99354, USA
mohamed.elmahallawy@wsu.edu

**Abstract.** Deep learning models are increasingly deployed in safety-critical applications, where reliable out-of-distribution (OOD) detection is essential to ensure robustness. Existing methods predominantly rely on the penultimate-layer activations of neural networks, assuming they encapsulate the most informative in-distribution (ID) representations. In this work, we revisit this assumption to show that intermediate layers encode equally rich and discriminative information for OOD detection. Based on this observation, we propose a simple yet effective model-agnostic approach that leverages internal representations across multiple layers. Our scheme aggregates features from successive convolutional blocks, computes class-wise mean embeddings, and applies $L_2$ normalization to form compact ID prototypes capturing class semantics. During inference, cosine similarity between test features and these prototypes serves as an OOD score—ID samples exhibit strong affinity to at least one prototype, whereas OOD samples remain uniformly distant. Extensive experiments on state-of-the-art OOD benchmarks across diverse architectures demonstrate that our approach delivers robust, architecture-agnostic performance and strong generalization for image classification. Notably, it improves AUROC by up to 4.41% and reduces FPR by 13.58%, highlighting multi-layer feature aggregation as a powerful yet underexplored signal for OOD detection, challenging the dominance of penultimate-layer-based methods. Our code is available at: https://github.com/sgchr273/cosine-layers.git.

**Keywords:** Deep Neural Networks · OOD Detection · Representation Learning

## 1 Introduction

Neural networks (NNs) have remarkable ability to perform complex tasks such as classification, detection, and segmentation with high accuracy [7]. However, despite their success, they exhibit a critical flaw—a tendency to be overconfident when presented with samples that lie outside the training distribution. This


R. C.-W. Wong et al. (Eds.): PAKDD 2026, LNAI 16597, pp. 597–611, 2026.
https://doi.org/10.1007/978-981-92-1300-9_47

overconfidence can lead to catastrophic failures in safety-critical applications such as autonomous driving and medical diagnostics [1]. Hence, it is imperative to design effective methods for out-of-distribution (OOD) detection to ensure the robust deployment of NNs in such domains [6].

A large body of work has focused on *leveraging signals from the penultimate layer of NNs* for OOD detection. For instance, Guan et al. [4] compute mean feature vectors in the penultimate feature space and exploit the observation that in-distribution (ID) samples exhibit higher similarity to these means than OOD samples. Similarly, KNN [20] measures the distance between test and training features, classifying a sample as OOD if the distance exceeds a threshold. NNGuide [16] refines OOD scores by averaging cosine similarities to the $k$ nearest ID neighbors in feature space, while NCI [14] defines a proximity score based on the norm of the projection of a centered feature onto its class weight vector, with larger norms indicating ID-like behavior. Although these methods achieve competitive results, they primarily depend on the penultimate representation and often fail to capture the richer class-level structure distributed across intermediate layers. As a result, their performance degrades substantially when faced with far-OOD samples [3].

Recent work shows that informative OOD signals are *not restricted to the penultimate layer* but also appear in *intermediate representations*. ESOOD [22], for instance, trains multiple *one-class* SVM detectors at different depths and uses a layer-selection policy for each test sample while using early stopping to terminate inference, but this can be computationally expensive due to the need to handle many layer-specific detectors (e.g., several SVMs per backbone in a ResNet). Ag-EBO [5] instead regularizes intermediate layers with an energy-based contrastive loss and aggregates several layers into a single OOD score, yet it does not identify which specific layers should be regularized to maximize performance. NMD [3] exploits batch-normalization statistics for OOD detection, a direction further supported by [4,12,13]; however, these methods do not explicitly encode class-specific information, which limits their discriminative power.

To address these limitations, we propose a method that combines *fine-grained class information, captured via class prototypes,* with comprehensive multi-layer feature utilization in NNs. Specifically, we construct $L_2$-normalized class prototypes from intermediate layers, which serve as compact, semantically rich representations of ID data. During inference, we compute the cosine similarity between a test sample and each class prototype. For each layer, we retain the maximum similarity across all classes, and a weighted average across layers produces the final confidence score. We observe that ID samples consistently exhibit higher cosine similarity than OOD samples, highlighting cosine similarity as a robust metric for OOD detection. Conceptually, our approach identifies anomalies by analyzing the angles in feature space: ID samples form smaller angles (higher similarity) with class prototypes, whereas OOD samples form larger angles.
Our main contributions are summarized as follows:

- We propose a novel, simple, and training-free OOD detection framework that jointly leverages the class-aware structure of ID data and the network's internal layers.
- Our method exploits geometric relationships in the feature space by computing cosine similarity between class prototypes and intermediate-layer activations. For each layer, the maximum class-wise similarity is retained, and a weighted aggregation across layers yields the final confidence score—enabling robust separation of ID and OOD samples.
- Extensive experiments across multiple ID datasets and diverse network architectures demonstrate that our approach consistently enhances OOD detection performance and generalizes effectively across domains. In particular, it improves AUROC by up to 4.7% and reduces FPR by 13.58%.

## 2 Related Work

**Neural-Collapse-Based Methods:** A recent thread of work leverages the geometric phenomena of neural collapse (NC) to improve OOD detection. Training-time strategies enforce *feature-space separation*: for example, Wu et al. [23] push OOD features into the subspace orthogonal to the ID weight span via a simple orthogonality loss under outlier exposure. Post-hoc alternatives avoid retraining by extracting NC-consistent signals from a trained model: Liu and Qin [14] score samples using the proximity of centered features to the predicted class weight together with feature norm, while NECO uses NC/PCA structure to distinguish ID from OOD without fine-tuning [1]. Complementary work by Harun et al. [8] studies how *controlling* the degree of neural collapse across layers trades off OOD detection and OOD generalization, proposing entropy regularization on the encoder plus an equiangular tight frame (ETF) projector on the head to balance the two objectives. Our method fits naturally within this line: we aggregate intermediate-layer activations into $L_2$-normalized class prototypes and score samples with cosine similarity, thereby capturing discriminative geometry that appears prior to full collapse while remaining post-hoc and model-agnostic.

**Output and Representation-Based OOD Detection methods:** Another large class of methods transforms a frozen network's outputs or representations into confidence scores. Output-based scores include the *maximum softmax probability* (MSP), which gauges peakiness, and *MaxLogit*, which uses the largest logit directly to mitigate softmax saturation. *Energy*-based scores replace the max with a log-sum-exp (negative free energy) aggregation to leverage the whole logit spectrum and often sharpen ID/OOD separation. Representation-based techniques instead exploit feature geometry: *ViM* decomposes penultimate features into an ID-aligned principal subspace plus a residual to form a virtual logit margin, while residual- or reconstruction-style methods score samples by reconstruction error or distance to a class subspace. Our method complements these approaches by extending geometric scoring beyond the penultimate layer: we build multi-stage, $L_2$-normalized class prototypes from intermediate features and apply cosine similarity, which captures early discriminative cues and yields a simple, effective post-hoc detector that is broadly applicable across architectures.

**Entropy-Based OOD Detection Methods:** The method [17] extends Maximum Concept Matching using intermediate-layer features from CLIP-like vision–language models. It applies an entropy-based rule to select and fuse informative layers for training-free OOD detection. This works best when the backbone has rich, diverse intermediate representations. Thus it becomes less effective and stable on models with flatter or redundant layers, such as MAE or Perception Encoder. In contrast, [24] models information flow by estimating conditional entropy between consecutive blocks. It defines a Conditional Entropy Ratio (CER) from ID and jigsaw-based pseudo-OOD images to locate the most discriminative block. However, EOOD needs jigsaw pseudo-OOD generation and k-NN–based entropy estimation, adding notable overhead. LaREx [2] proposes uncertainty-based scores LaRED and LaREM using dropout/DropBlock (zMCD) on latent features. It fits the entropy density of ID samples with kernel density estimation or a Gaussian model. Yet LaREx still requires Monte Carlo sampling and architecture-specific tuning of the noise layer and its size. So no universal, computation-light configuration exists, and nontrivial overhead remains.

## 3 Problem Statement and Proposed Methodology

Recent studies indicate that the hidden layers of NNs encode rich, discriminative information about the training data. Motivated by this, we leverage the activations of hidden convolutional layers to design our OOD detection method. To focus on class-specific information and reduce the influence of noise, we compute a mean feature (prototype) for each ID class at every hidden layer. At test time, each sample is compared to these per-layer prototypes, and the resulting similarities are aggregated across layers. This yields a simple, training-free OOD detector that fuses cosine similarities between samples and class prototypes computed from multiple intermediate layers of a pretrained network. A small ID calibration set is used to construct the prototypes, and OOD scores are obtained by taking the maximum cosine similarity to the prototypes, aggregated across layers (Fig. 1). The problem statement and details of our method are presented below.

### 3.1 Problem Statement

We consider a pretrained classifier $f \circ h : \mathcal{X} \to \mathcal{Y}$, where $\mathcal{X}$ denotes the input space (e.g., images), $\mathcal{Y} = \{1, \dots, K\}$ the set of class labels, $h : \mathcal{X} \to \mathbb{R}^d$ a feature extractor (backbone), and $f : \mathbb{R}^d \to \mathcal{Y}$ the classifier head. The model is trained on in-distribution (ID) data drawn from a distribution $P_{\text{in}}$.

In addition, we are given a small labeled calibration set $\mathcal{D}_{\mathrm{ID}}^{\mathrm{calib}} \subset \mathcal{X} \times \mathcal{Y}$, sampled from $P_{\mathrm{in}}$, which we use to construct class prototypes at a collection of hidden layers $\mathcal{L}$. At test time, a sample $x \in \mathcal{X}$ may originate either from the in-distribution $P_{\mathrm{in}}$ or from an unknown out-of-distribution $P_{\mathrm{out}}$.

Our goal is to design, using only the frozen network $f \circ h$ and the prototype bank built from $\mathcal{D}_{\mathrm{ID}}^{\mathrm{calib}}$, a scalar score function $S(x) \in \mathbb{R}$ that separates ID and OOD samples. The resulting detector is expressed as a decision function $G : \mathcal{X} \to \{0, 1\}$,

$$G(x) = \begin{cases} 1, & \text{if } S(x) \geq \tau \quad (x \text{ is ID}), \\ 0, & \text{if } S(x) < \tau \quad (x \text{ is OOD}), \end{cases} \tag{1}$$

where $\tau \in \mathbb{R}$ is a threshold chosen on a validation set.

### 3.2 Multi-layer Feature Representations

Let $f_\theta$ be a pretrained classifier with parameters $\theta$, trained on an ID dataset with $K$ classes. We select a set of internal layers as:

$$\mathcal{L} = \{\ell_1, \ell_2, \ldots, \ell_L\}, \tag{2}$$

which may include intermediate convolutional blocks. For an input image $x$, each layer $\ell \in \mathcal{L}$ produces an activation map as:

$$h_\ell(x) \in \mathbb{R}^{C_\ell \times H_\ell \times W_\ell}, \tag{3}$$

where $C_\ell$ is the number of output channels while $H_\ell$ and $W_\ell$ are the spatial height and width of the layer's feature map. To obtain a fixed-length representation for layer $\ell$, we apply global average pooling over the spatial dimensions $(H_\ell, W_\ell)$. For channel $c \in \{1, \ldots, C_\ell\}$, we define

$$z_\ell(x)_c = \frac{1}{H_\ell W_\ell} \sum_{i=1}^{H_\ell} \sum_{j=1}^{W_\ell} h_\ell(x)_{c,i,j}. \tag{4}$$

This yields a channel-wise pooled descriptor $z_\ell(x) \in \mathbb{R}^{C_\ell}$. We then apply $L_2$-normalization to obtain a unit-norm feature vector, which can be given as:

$$\tilde{z}_\ell(x) = \frac{z_\ell(x)}{\|z_\ell(x)\|_2} \in \mathbb{R}^{C_\ell}. \tag{5}$$

After this step, each input $x$ is represented by a collection of unit-normalized vectors, one per selected convolutional layer $\{\tilde{z}_\ell(x)\}_{\ell \in \mathcal{L}}$.

### 3.3 Class Prototypes in Feature Space

We construct a prototype representation for each ID class in each selected layer $\ell \in \mathcal{L}$. Let $\mathcal{D}_{\mathrm{ID}}^{\mathrm{calib}} = \{(x_n, y_n)\}_{n=1}^{N}, \quad y_n \in \{0, \ldots, K-1\}$ denotes a labeled

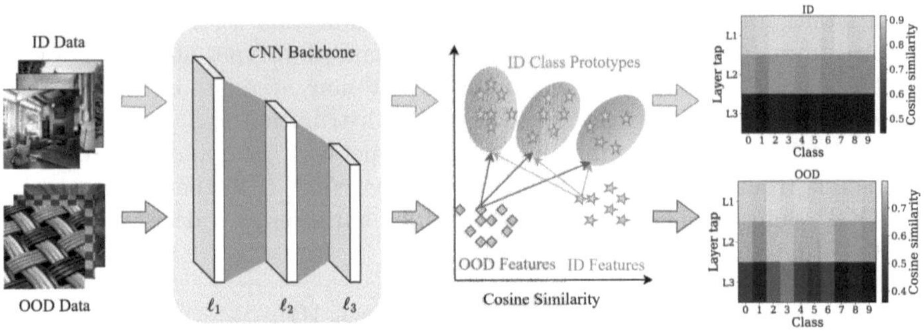


**Fig. 1.** Overview of the proposed training-free OOD detector. A pretrained CNN is tapped at layers $\ell_1$–$\ell_3$. ID calibration images are passed through the network, and per-layer features are global-average-pooled and $L_2$-normalized to form $K$ class prototypes (green regions). For a test image (ID or OOD), we compute cosine similarity between its features and each layer's prototypes (green arrows for ID, red arrows for OOD). At each layer, we retain the maximum class-wise similarity, and these per-layer maxima are averaged to produce an ID affinity score, with OODScore = 1 − affinity. The heatmaps on the right show that ID samples yield strong, consistent similarities across layers and classes, whereas OOD samples produce weaker and less coherent similarity patterns.(Color figure online)

calibration set of ID samples. For each layer $\ell \in \mathcal{L}$ and each class $c \in \{0, \ldots, K-1\}$, we define the empirical mean feature as:

$$\bar{p}_{\ell,c} = \frac{1}{N_{\ell,c}} \sum_{n:y_n=c} \tilde{z}_\ell(x_n), \quad N_{\ell,c} = \big|\{n : y_n = c\}\big|. \tag{6}$$

We then normalize this class mean to unit length $p_{\ell,c} = \frac{\bar{p}_{\ell,c}}{\|\bar{p}_{\ell,c}\|_2} \in \mathbb{R}^{C_\ell}$. The resulting matrix of class prototypes for each layer $\ell$ can be represented as:

$$P_\ell = \begin{bmatrix} p_{\ell,0}^\top \\ p_{\ell,1}^\top \\ \vdots \\ p_{\ell,K-1}^\top \end{bmatrix} \in \mathbb{R}^{K \times C_\ell}. \tag{7}$$

Thus, for every chosen layer $\ell$, we obtain a bank of $K$ unit-norm prototype vectors, one per ID class.

### 3.4 Cosine Similarity to Known-Class Manifolds

Given a test sample $x$, we extract its normalized feature vector $\tilde{z}_\ell(x)$ for each layer $\ell \in \mathcal{L}$, as described in Sect. 3.2. We then compute the cosine similarity between $\tilde{z}_\ell(x)$ and each class prototype $p_{\ell,c}$ in that layer:

$$\mathrm{sim}_\ell(x, c) = \tilde{z}_\ell(x)^\top p_{\ell,c}, \quad c = 0, \ldots, K-1. \tag{8}$$

**Algorithm 1:** Prototype-based multi-layer OOD scoring

**Data:** test sample $x$
**Inputs:** $f_\theta$ (pretrained net), $\mathcal{L}$ (selected layers), $\{p_{\ell,c}\}$ (class prototypes), $\{w_\ell\}$ (layer weights)
**Output:** OODScore($x$)

**foreach** $\ell \in \mathcal{L}$ **do**
  $h_\ell(x) \leftarrow$ activations of layer $\ell$ on $x$
  $z_\ell(x) \leftarrow \text{GlobalAvgPool}(h_\ell(x))$  // $\in \mathbb{R}^{C_\ell}$
  $\tilde{z}_\ell(x) \leftarrow z_\ell(x)/\|z_\ell(x)\|_2$
  $m_\ell(x) \leftarrow \max_c \tilde{z}_\ell(x)^\top p_{\ell,c}$  // max cosine sim

$s(x) \leftarrow \frac{\sum_{\ell\in\mathcal{L}} w_\ell\, m_\ell(x)}{\sum_{\ell\in\mathcal{L}} w_\ell}$
$\text{OODScore}(x) \leftarrow 1 - s(x)$
**return** OODScore($x$)

Next, for each layer, we retain the maximum similarity over classes as:

$$m_\ell(x) = \max_{c\in\{0,\ldots,K-1\}} \text{sim}_\ell(x, c). \tag{9}$$

The scalar $m_\ell(x)$ quantifies how well sample $x$ matches any known ID class at layer $\ell$.

### 3.5 Cross-Layer Aggregation and OOD Scoring

As different layers capture different forms of structure, we aggregate evidence across all selected layers. We assign a non-negative weight $w_\ell > 0$ to each layer $\ell \in \mathcal{L}$. For a test input $x$, we compute a normalized weighted average of the per-layer match scores as:

$$s(x) = \frac{\sum_{\ell\in\mathcal{L}} w_\ell\, m_\ell(x)}{\sum_{\ell\in\mathcal{L}} w_\ell}, \tag{10}$$

We interpret $s(x)$ as an ID affinity score, then define the final OOD detection score as:

$$\text{OODScore}(x) = 1 - s(x). \tag{11}$$

A higher OODScore($x$) indicates that the sample is less consistent with the set of ID class prototypes across all monitored layers, and is therefore more likely to be OOD. Algorithm 1 outlines the scoring procedure.

# 4 Performance Evaluation

## 4.1 Experimental Setup

**ID and OOD Datasets.** We evaluate our method on three standard benchmarks for OOD detection: CIFAR-10, CIFAR-100, and ImageNet-1k as the ID datasets, following the experimental protocols of prior work [4]. For CIFAR-10, we report results under two widely used evaluation protocols: (i) the small OOD benchmark suite consisting of iSUN, LSUN, iNaturalist, Textures, and Places, and (ii) the NECO protocol [1], where CIFAR-10 (resp. CIFAR-100) serves as the ID dataset and CIFAR-100 (resp. CIFAR-10) together with SVHN are used as OOD datasets. For ImageNet-1k as the ID dataset, we evaluate OOD detection performance against SUN, Textures, Places, and iNaturalist.

**Model Configuration.** Our method involves two key hyperparameters: (i) the weights assigned to the internal layers of the NN, and (ii) the number of samples selected from the ID training data to construct the calibration set ($\alpha$). We conduct an extensive analysis to study the impact of these hyperparameters on the overall OOD detection performance (see results in Figs. 2 and 4).

**Baselines.** We compare our method against *nine* widely used OOD detection baselines. All these methods are *post-hoc*, meaning they operate on pretrained networks without requiring access to model weights, gradients, or retraining losses. This makes them suitable baselines for our method, which is also post-hoc by design. The compared methods include MSP [10], MaxLogit [9], Energy [15], Mahalanobis [18], GradNorm [11], NNGuide [16], NECO [1], ReAct [19], and ViM [21].

**Evaluation Metrics.** We follow the same preprocessing, calibration, and evaluation procedures as in [4]: a small held-out ID calibration set is used to build prototypes, and all OOD scores are computed post-hoc on pretrained models. Performance is evaluated using the most widely adopted OOD detection metrics: (i) the Area Under the Receiver Operating Characteristic curve (AUROC) and (ii) the False Positive Rate at 95% True Positive Rate (FPR@95% TPR). Results are averaged across all OOD datasets unless stated otherwise. Higher AUROC and lower FPR@95% TPR indicate better OOD detection performance.

## 4.2 Performance Results

Tables 1, 2, and 3 report the results of our method compared to all baselines across multiple architectures. In all experiments, we use a uniform weighting of layers. In Table 1, for ResNet-18, our method improves AUROC by 2.34% over ESOOD and 3.66% over NNGuide, while reducing FPR by 10.88% and 7.25%, respectively. For DenseNet-100, our approach increases AUROC by 1.1% over ViM and 2.5% over Energy, and reduces FPR by 3.49% and 4.26%, respectively.

**Table 1.** OOD detection results on the **CIFAR-10** dataset. ↑/↓ denote that higher/lower values are better. The **first**, second, and third best results are highlighted in green, orange, and blue, respectively (values in %).

| Model | Method | iSUN | | LSUN | | Places | | iNat | | Textures | | Average | |
|---|---|---|---|---|---|---|---|---|---|---|---|---|---|
| | | AUROC↑ | FPR↓ | AUROC↑ | FPR↓ | AUROC↑ | FPR↓ | AUROC↑ | FPR↓ | AUROC↑ | FPR↓ | AUROC↑ | FPR↓ |
| ResNet-18 | MSP | 89.79 | 47.44 | 94.68 | 15.26 | 89.66 | 48.32 | 89.59 | 44.33 | 87.18 | 63.96 | 90.75 | 39.75 |
| | MaxLogit | 90.08 | 52.84 | 94.63 | 25.82 | 89.65 | 55.06 | 89.37 | 51.19 | 86.35 | 73.86 | 90.74 | 46.60 |
| | Energy | 88.42 | 63.00 | 92.34 | 51.48 | 87.97 | 62.87 | 87.46 | 59.14 | 84.24 | 79.10 | 89.02 | 57.46 |
| | Mahalanobis | 86.86 | 51.32 | 80.99 | 53.84 | 84.80 | 49.93 | 80.25 | 58.48 | 92.64 | 34.99 | 86.61 | 46.26 |
| | GradNorm | 83.44 | 70.92 | 91.86 | 37.20 | 79.30 | 74.98 | 90.81 | 44.29 | 81.20 | 82.75 | 86.56 | 57.97 |
| | NNGuide | 89.45 | 50.12 | 95.18 | 13.84 | 89.82 | 45.34 | 90.00 | 41.89 | 91.23 | 30.33 | 91.71 | 32.93 |
| | NECO | 84.69 | 54.92 | 85.32 | 53.14 | 81.87 | 61.78 | 83.80 | 56.51 | 91.74 | 42.23 | 87.08 | 50.14 |
| | ReAct | 85.78 | 75.53 | 91.70 | 50.88 | 84.21 | 79.63 | 84.66 | 73.83 | 80.95 | 92.19 | 86.49 | 69.73 |
| | ViM | 88.52 | 55.16 | 87.28 | 38.36 | 88.13 | 48.88 | 82.90 | 59.05 | 93.88 | 30.78 | 89.40 | 42.46 |
| | ESOOD | 96.70 | **8.36** | 96.26 | **12.00** | 84.51 | 65.92 | 84.40 | 59.85 | 96.30 | 18.37 | 92.66 | 29.30 |
| | LaREx | 89.00 | 39.62 | 86.57 | 48.32 | **94.70** | **22.68** | 89.94 | 45.76 | **98.63** | **5.83** | 91.77 | 32.44 |
| | **Ours** | **93.51** | 28.50 | **96.34** | 17.97 | 91.61 | 34.60 | **92.82** | **29.23** | 96.99 | 16.06 | **95.00** | **22.05** |
| DenseNet-100 | MSP | 96.13 | 14.18 | 92.64 | 20.07 | 89.94 | 33.02 | 88.10 | 40.26 | 90.86 | 31.39 | 91.86 | 26.11 |
| | MaxLogit | 98.68 | 6.92 | 95.95 | 16.26 | 92.53 | 33.27 | 87.74 | 50.95 | 91.29 | 42.66 | 93.73 | 27.37 |
| | Energy | 98.77 | 6.65 | 96.08 | 16.10 | **92.63** | 33.26 | 87.73 | 50.95 | 91.33 | 42.78 | 93.80 | 27.31 |
| | Mahalanobis | 97.12 | 15.90 | 96.52 | 15.13 | 70.75 | 77.46 | 75.42 | 71.10 | 95.97 | 25.11 | 88.83 | 36.69 |
| | GradNorm | 95.91 | 26.78 | 87.71 | 65.63 | 80.30 | 77.32 | 80.70 | 75.72 | 79.09 | 91.21 | 86.35 | 61.91 |
| | NNGuide | 95.49 | 18.68 | 93.28 | 19.24 | 88.94 | 42.79 | 89.81 | 34.08 | 92.92 | 24.66 | 92.37 | 26.34 |
| | NECO | 98.71 | 6.88 | 96.00 | 16.01 | 92.52 | 33.33 | 87.55 | 50.91 | 91.54 | 40.81 | 93.74 | 27.02 |
| | ReAct | 98.45 | 9.16 | 95.59 | 18.26 | 92.37 | **32.28** | 90.31 | 37.95 | 92.17 | 34.57 | 94.05 | 24.94 |
| | ViM | **99.50** | **2.25** | **98.62** | **6.59** | 90.91 | 38.35 | 85.72 | 54.51 | 96.76 | 19.01 | 95.13 | 20.68 |
| | ESOOD | 97.34 | 10.56 | 83.45 | 53.56 | 87.50 | 68.90 | 85.78 | 65.66 | 34.32 | 96.56 | 81.68 | 58.65 |
| | LaREx | 96.54 | 20.34 | 91.34 | 35.45 | 90.98 | 36.78 | 88.33 | 39.02 | 96.01 | 15.44 | 92.64 | 29.41 |
| | **Ours** | 98.24 | 9.81 | 97.67 | 10.73 | 90.07 | 40.08 | **94.14** | **23.59** | **97.92** | **12.41** | **96.13** | **17.19** |

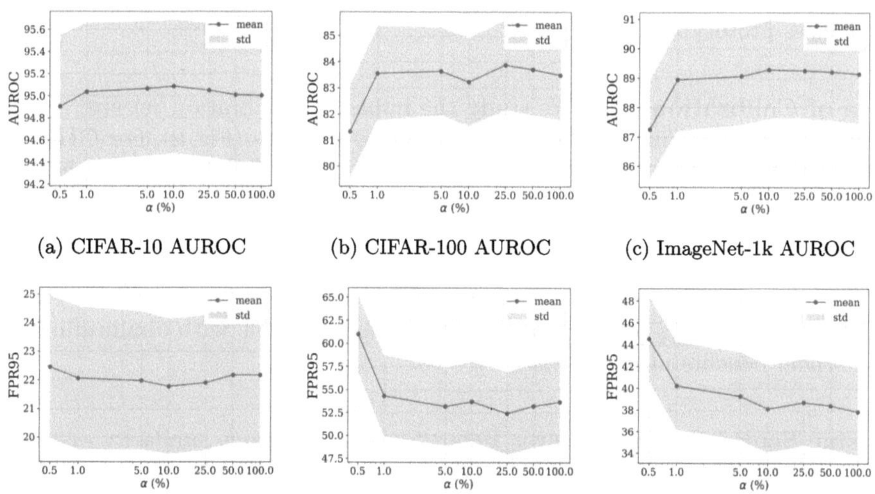


(a) CIFAR-10 AUROC (b) CIFAR-100 AUROC (c) ImageNet-1k AUROC

(d) CIFAR-10 FPR (e) CIFAR-100 FPR (f) ImageNet-1k FPR

**Fig. 2.** Effect of calibration set size on AUROC (top row) and false positive rate (bottom row) for CIFAR-10, CIFAR-100, and ImageNet-1k.

In Table 2 (ResNet-18, ID = CIFAR-10), our method improves AUROC over ESOOD and NNGuide by 0.85% and 0.69%, while reducing FPR by approximately 0.75% and 4.63%, respectively. For ID = CIFAR-100 on ResNet-18, our method improves AUROC by about 1.99% over ESOOD and 8.83% over NNGuide, and reduces FPR by roughly 6.97% and 19.03%. On DenseNet-100 with ID = CIFAR-10, our method improves AUROC by 0.5% over ViM and 1.31% over MSP, and reduces FPR by 3.5% and 5.68% relative to ViM and MSP, respectively. For CIFAR-100, ESOOD edges out our method in AUROC by 0.04%, but our method achieves a 5.09% lower FPR. ViM ranks third in AUROC, trailing ours by 0.05%, and second in FPR, with 2.63% higher than ours.

In Table 3 (ResNet-50), our method achieves the best Average score, improving AUROC by 3.38% over Mahalanobis and 4.47% over NNGuide, while reducing FPR by 7.9% and 14.53%, respectively. For RegNet, our method is the second-best performer after ReAct, with an AUROC lower by 3.52% and an FPR higher by 6.35%, whereas ReAct performs poorly on ResNet-50, highlighting that it is extrapolating the signal received by Energy. Compared to the third-place NNGuide, our approach improves AUROC by 4.41% and reduces FPR by 13.58%.

### 4.3 Ablation Studies

Our method builds an ID representation that captures variations across both classes and hierarchical depths of the backbone network. Accordingly, we analyze two key factors: (i) the stage-level features of the backbone used for prototype construction, and (ii) the size of the calibration set drawn from the ID data to form class prototypes.

**Size of Calibration Set.** We study the impact of calibration set size ($\alpha$) on performance, averaging results across four OOD datasets (Fig. 2). For **CIFAR-10**, AUROC improves modestly up to $\alpha \approx 10\%$ before plateauing, with FPR reaching its minimum near that point. For **CIFAR-100**, most gains occur by $\alpha \approx 5\%$, with incremental improvement up to $\alpha \approx 25\%$, where AUROC peaks and FPR is near its minimum. For **ImageNet-1K**, AUROC rises steadily up to $\alpha \approx 25\%$ and then levels off, while FPR decreases monotonically. Overall, small calibration sets ($\alpha \approx 5\%$–$10\%$) already perform well, with diminishing yet non-trivial benefits up to $\alpha \approx 25\%$.

**Cosine Scores Across Layers.** Figure 5 visualizes cosine similarity distributions for CIFAR-10 (ID) and four OOD datasets across the last three stages of ResNet-18. The ID scores remain sharply peaked near 1.0, while OOD scores cluster lower ($\sim 0.75$–$0.90$) and shift slightly leftward with depth, increasing the separation between ID and OOD samples. This behavior supports cosine similarity as an effective and stable criterion for OOD detection across intermediate layers.

**Table 2.** OOD detection results where CIFAR-10 is ID and OOD = CIFAR-100, SVHN) and ID = CIFAR-100 (OOD = CIFAR-10, SVHN). ↑/↓ denote that higher/lower is better. The **first**, second, and third best results are highlighted in green, orange, and blue respectively (values in %).

| Model | Method | ID = CIFAR-10 | | | | | | ID = CIFAR-100 | | | | | |
|---|---|---|---|---|---|---|---|---|---|---|---|---|---|
| | | CIFAR-100 | | SVHN | | Average | | CIFAR-10 | | SVHN | | Average | |
| | | AUROC↑ | FPR↓ | AUROC↑ | FPR↓ | AUROC↑ | FPR↓ | AUROC↑ | FPR↓ | AUROC↑ | FPR↓ | AUROC↑ | FPR↓ |
| ResNet-18 | MSP | 88.76 | 46.29 | 93.60 | 19.17 | 91.18 | 32.73 | 75.62 | **77.26** | 72.92 | 83.01 | 74.27 | 80.14 |
| | MaxLogit | 88.09 | 53.99 | 94.36 | 20.86 | 91.22 | 37.42 | 72.75 | 84.39 | 67.27 | 90.25 | 70.01 | 87.32 |
| | Energy | 85.04 | 64.90 | 93.69 | 29.20 | 89.37 | 47.05 | 66.38 | 86.81 | 58.26 | 91.68 | 62.32 | 89.25 |
| | Mahalanobis | 82.11 | 59.59 | 94.12 | 29.00 | 88.12 | 44.30 | 70.44 | 78.46 | 77.92 | 61.76 | 74.18 | 70.11 |
| | GradNorm | 76.00 | 77.34 | 92.76 | 37.65 | 84.38 | 57.50 | 27.77 | 97.37 | 22.41 | 98.05 | 25.09 | 97.71 |
| | NNGuide | **89.55** | **41.62** | 94.59 | 16.09 | 92.07 | 28.85 | 75.82 | 79.33 | 76.51 | 68.25 | 76.17 | 73.79 |
| | NECO | 76.86 | 67.72 | 95.04 | 32.26 | 85.95 | 49.99 | 39.79 | 95.27 | 45.93 | 91.06 | 42.86 | 93.16 |
| | ReAct | 80.06 | 82.33 | 91.64 | 46.31 | 85.85 | 64.32 | 28.57 | 95.95 | 36.60 | 92.76 | 32.59 | 94.36 |
| | ViM | 85.55 | 53.83 | 95.69 | 22.52 | 90.62 | 38.17 | 64.48 | 86.93 | 76.72 | 68.15 | 70.60 | 77.54 |
| | ESOOD | 84.56 | 62.34 | 98.19 | 8.67 | 91.38 | 35.51 | **79.78** | 80.65 | 90.19 | 35.45 | 84.99 | 58.05 |
| | LaREx | 88.20 | 53.45 | 95.32 | 18.50 | 91.76 | 35.98 | 73.11 | 82.45 | 90.45 | 36.09 | 81.78 | 59.27 |
| | **Ours** | 87.94 | 50.27 | **98.94** | **5.92** | **93.44** | **28.10** | 74.47 | 79.84 | **95.52** | **22.31** | **85.00** | **51.08** |
| DenseNet-100 | MSP | 86.78 | 47.06 | 93.46 | 17.71 | 90.12 | 32.39 | 75.72 | **62.15** | 79.82 | 52.74 | 77.77 | 57.45 |
| | MaxLogit | 86.41 | 59.11 | 96.17 | 14.18 | 91.29 | 36.65 | 77.10 | 63.57 | 83.45 | 48.09 | 80.28 | 55.83 |
| | Energy | 86.44 | 59.12 | 96.28 | 14.12 | 91.36 | 36.62 | 76.99 | 63.54 | 83.44 | 48.10 | 80.22 | 55.82 |
| | Mahalanobis | 78.99 | 67.97 | 97.18 | 15.42 | 88.09 | 41.70 | 40.51 | 95.53 | 89.23 | 43.25 | 64.87 | 69.39 |
| | GradNorm | 60.24 | 94.30 | 94.37 | 34.82 | 77.31 | 64.56 | 60.23 | 90.05 | 82.86 | 64.10 | 71.55 | 77.07 |
| | NNGuide | 87.20 | 46.60 | 93.78 | 18.59 | 90.49 | 32.60 | 72.79 | 83.51 | 83.04 | 41.86 | 77.92 | 62.69 |
| | NECO | 86.56 | 58.07 | 96.14 | 14.16 | 91.35 | 36.12 | **77.12** | 63.50 | 83.85 | 47.19 | 80.49 | 55.35 |
| | ReAct | 87.33 | 51.97 | 95.43 | 17.44 | 91.38 | 34.71 | 75.46 | 68.24 | 85.63 | 41.70 | 80.55 | 54.97 |
| | ViM | 86.65 | 57.02 | **99.29** | **3.39** | 92.97 | 30.21 | 71.68 | 69.70 | 93.66 | 28.37 | 88.85 | 37.74 |
| | ESOOD | 86.56 | 58.67 | 97.89 | 11.32 | 92.23 | 35.00 | 73.45 | 65.67 | 94.56 | 30.32 | **88.94** | 40.20 |
| | LaREx | 89.34 | 50.15 | 97.99 | 10.15 | **93.67** | 30.15 | 75.12 | 70.05 | 93.49 | 29.98 | 84.31 | 50.02 |
| | **Ours** | **88.34** | **45.81** | 98.74 | 6.53 | 93.54 | **26.17** | 67.57 | 79.15 | **96.33** | **17.91** | 88.90 | **35.11** |

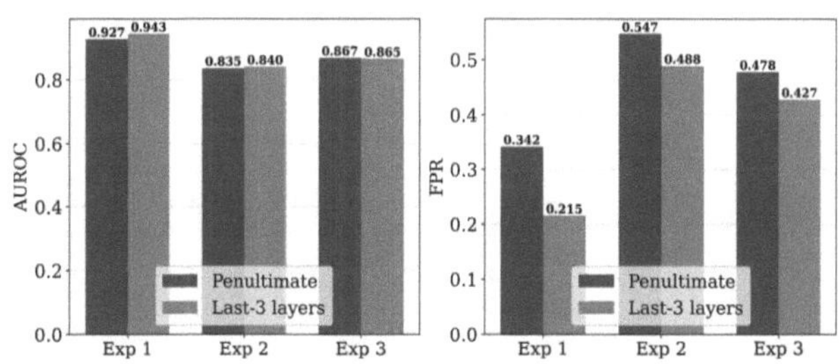


**Fig. 3.** AUROC and FPR using the penultimate layer vs. the last three layers across three ID datasets: CIFAR-10 (Exp1), CIFAR-100 (Exp2), and ImageNet-1K (Exp3).

**Table 3.** OOD detection results on the **ImageNet-1k** dataset. ↑/↓ denote that higher/lower values are better. The **first**, second, and third best results are highlighted in green, orange, and blue, respectively (values in %).

| Model | Method | SUN | | Places | | iNat | | Textures | | Average | |
|---|---|---|---|---|---|---|---|---|---|---|---|
| | | AUROC↑ | FPR↓ | AUROC↑ | FPR↓ | AUROC↑ | FPR↓ | AUROC↑ | FPR↓ | AUROC↑ | FPR↓ |
| ResNet-50 | MSP | 79.25 | 79.48 | 77.48 | 78.59 | 84.65 | 62.69 | 74.78 | 87.76 | 79.04 | 77.13 |
| | MaxLogit | 74.78 | 85.81 | 72.64 | 85.90 | 80.25 | 76.47 | 69.25 | 91.74 | 74.23 | 84.98 |
| | Energy | 50.82 | 89.72 | 50.55 | 90.23 | 51.27 | 87.51 | 48.57 | 94.28 | 50.30 | 90.44 |
| | Mahalanobis | 79.60 | **54.97** | 77.26 | **65.66** | 90.48 | 29.89 | 87.13 | 53.09 | 83.62 | 50.90 |
| | GradNorm | 30.01 | 94.09 | 29.56 | 95.31 | 30.72 | 90.89 | 39.25 | 91.06 | 32.38 | 92.84 |
| | NNGuide | **81.14** | 73.11 | **79.40** | 72.72 | **97.21** | **12.79** | 72.39 | 71.51 | 82.53 | 57.53 |
| | NECO | 25.13 | 97.40 | 23.98 | 97.67 | 23.85 | 97.13 | 51.02 | 89.28 | 31.00 | 95.37 |
| | ReAct | 24.10 | 96.45 | 23.08 | 97.37 | 23.87 | 95.26 | 35.14 | 94.52 | 26.55 | 95.90 |
| | ViM | 51.78 | 90.45 | 51.36 | 90.59 | 82.38 | 55.98 | 49.41 | 93.88 | 58.73 | 82.72 |
| | ESOOD | 83.45 | 60.67 | 75.67 | 80.38 | 79.76 | 57.87 | 68.87 | 76.89 | 76.94 | 68.95 |
| | LareX | 82.98 | 61.14 | 85.60 | 70.07 | 91.02 | 30.78 | 96.34 | 20.49 | 88.99 | 45.62 |
| | **Ours** | 80.68 | 59.27 | 77.68 | 66.02 | 93.56 | 27.25 | **96.08** | **19.55** | **87.00** | **43.02** |
| RegNet | MSP | 83.36 | 58.32 | 81.51 | 58.44 | 89.84 | 39.13 | 80.74 | 64.79 | 83.86 | 55.17 |
| | MaxLogit | 79.12 | 63.33 | 75.80 | 61.54 | 85.63 | 44.89 | 73.44 | 69.48 | 78.50 | 59.81 |
| | Energy | 84.05 | 58.25 | 82.10 | 58.52 | 90.69 | 39.10 | 81.46 | 64.85 | 84.57 | 55.18 |
| | Mahalanobis | 50.94 | 85.99 | 52.84 | 85.01 | 57.14 | 79.51 | 84.98 | 53.62 | 61.48 | 76.03 |
| | GradNorm | 78.64 | 74.42 | 71.62 | 81.55 | 87.02 | 54.17 | 81.86 | 67.50 | 79.78 | 69.41 |
| | NECO | 83.17 | 63.49 | 81.02 | 66.29 | 90.30 | 37.00 | 84.19 | 47.98 | 84.67 | 53.69 |
| | ReAct | **92.78** | **32.86** | **90.18** | **41.80** | **97.02** | **14.43** | 91.69 | 35.21 | **92.92** | **31.08** |
| | ViM | 53.48 | 84.06 | 55.05 | 83.37 | 60.16 | 77.40 | 89.62 | 44.16 | 64.58 | 72.25 |
| | NNGuide | 74.55 | 79.44 | 83.72 | 55.52 | 92.08 | 33.31 | 89.60 | 35.76 | 84.99 | 51.01 |
| | ESOOD | 81.01 | 65.34 | 80.98 | 59.33 | 91.98 | 35.21 | 80.23 | 40.65 | 83.55 | 50.13 |
| | LareX | 83.03 | 60.92 | 89.75 | 63.72 | 91.43 | 29.14 | 97.76 | 15.61 | 90.49 | 42.35 |
| | **Ours** | 83.83 | 56.28 | 80.52 | 63.79 | 94.76 | 22.45 | **98.48** | **7.20** | 89.40 | 37.43 |

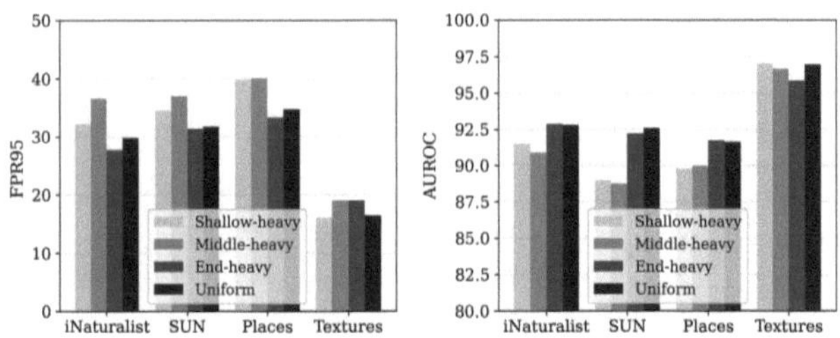


**Fig. 4.** Impact of weighting schemes across layers on overall detection performance.

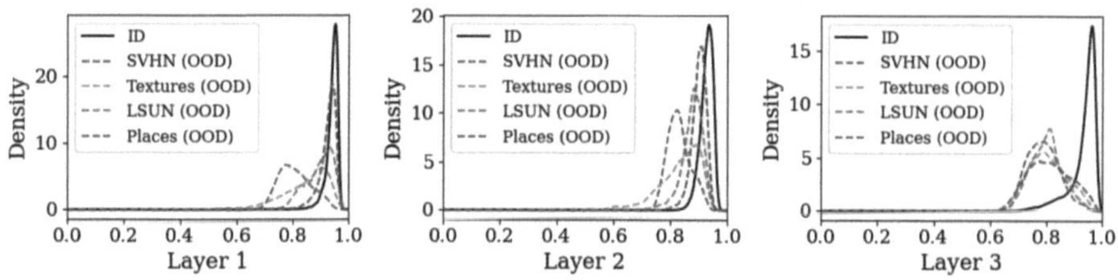


**Fig. 5.** Per-layer cosine similarity scores for the last three layers of ResNet-18 on ID data (CIFAR-10) and four OOD datasets (SVHN, Textures, LSUN, Places).

**Weightage of Backbone Stages.** We evaluate the contribution of different backbone stages under two configurations:

1. ***Feature source:*** We compare multi-stage features—the final activations from the last three residual stages (before global average pooling)—against using only the penultimate (top) stage. As shown in Fig. 3, aggregating features from the last three stages consistently yields lower FPR and higher AUROC than relying solely on the penultimate stage across all three benchmarks.
2. ***Stage weighting:*** Within the multi-stage setting, we vary the relative weights assigned to each stage. Figure 4 evaluates four weighting schemes: *shallow-heavy* (largest weight on the earliest stage), *middle-heavy* (largest weight on the middle stage), *top-heavy* (largest weight on the deepest stage), and *uniform* (equal weights). The uniform weighting consistently achieves the strongest performance across datasets.

## 5 Conclusion and Future Work

We proposed a novel post-hoc, training-free OOD detection method that leverages internal network representations and class labels from ID data. By constructing class-specific prototypes from intermediate features and comparing test samples via cosine similarity, our approach captures discriminative cues across multiple layers. Extensive experiments show consistent improvements over existing post-hoc detectors across diverse OOD datasets and ID benchmarks, including large-scale settings such as ImageNet-1K, where it improves AUROC by 4.14% and reduces FPR by up to 13.85%. We hope our work underscores the value of internal feature geometry for OOD detection and inspires research that further integrates representation learning with reliable uncertainty estimation.

## References


1. Ammar, M.B., Belkhir, N., Popescu, S., Manzanera, A., Franchi, G.: NECO: neural collapse based out-of-distribution detection. arXiv (2023)
2. Arnez, F.: Latent representation entropy density for distribution shift detection. In: Proceedings of the Conference on Uncertainty in Artificial Intelligence (UAI) (2024)
3. Dong, X., Guo, J., Li, A., Ting, W.M., Liu, C., Kung, H.T.: Neural mean discrepancy for efficient out-of-distribution detection. In: Proceedings of the IEEE/CVF Conference on Computer Vision and Pattern Recognition (CVPR), pp. 19195–19205 (2021)
4. Guan, X., Chen, J., Bu, S., Zhou, Y., Zheng, W., Wang, R.: Exploiting discrepancy in feature statistic for out-of-distribution detection. In: Proceedings of the AAAI Conference on Artificial Intelligence (AAAI), vol. 38, pp. 19858–19866 (2024)
5. Guglielmo, M.: Leveraging intermediate representations for better out-of-distribution detection. arXiv (2025)
6. Gul, S., Elmahallawy, M., Madria, S., Tripathy, A.: FisherMask: enhancing neural network labeling efficiency in image classification using fisher information. In: 2024 IEEE International Conference on Big Data (BigData), pp. 1302–1308. IEEE (2024)
7. Gul, S., Elmahallawy, M., Madria, S., Tripathy, A.: LPLgrad: optimizing active learning through gradient norm sample selection and auxiliary model training. In: 2024 IEEE International Conference on Big Data (BigData), pp. 900–909. IEEE (2024)
8. Harun, M.Y., Gallardo, J., Kanan, C.: Controlling neural collapse enhances out-of-distribution detection and transfer learning. arXiv (2025)
9. Hendrycks, D., et al.: Scaling out-of-distribution detection for real-world settings. arXiv (2019)
10. Hendrycks, D., Gimpel, K.: A baseline for detecting misclassified and out-of-distribution examples in neural networks. arXiv (2016)
11. Huang, R., Geng, A., Li, Y.: On the importance of gradients for detecting distributional shifts in the wild. In: Advances in Neural Information Processing Systems (NeurIPS), vol. 34, pp. 677–689 (2021)
12. Jelenić: Out-of-distribution detection by leveraging between-layer transformation smoothness. arXiv (2023)
13. Lambert, B., Forbes, F., Doyle, S., Dojat, M.: Multi-layer aggregation as a key to feature-based OOD detection. In: International Workshop on Uncertainty for Safe Utilization of Machine Learning in Medical Imaging, pp. 104–114 (2023)
14. Liu, L., Qin, Y.: Detecting out-of-distribution through the lens of neural collapse. In: Proceedings of the IEEE/CVF Conference on Computer Vision and Pattern Recognition (CVPR), pp. 15424–15433 (2025)
15. Liu, W., Wang, X., Owens, J., Li, Y.: Energy-based out-of-distribution detection. In: Advances in Neural Information Processing Systems (NeurIPS), vol. 33, pp. 21464–21475 (2020)
16. Park, J., Jung, Y.G., Teoh, A.B.J.: Nearest neighbor guidance for out-of-distribution detection. In: Proceedings of the IEEE/CVF International Conference on Computer Vision (ICCV), pp. 1686–1695 (2023)
17. Rodriguez-Opazo, C., Teney, D., Ranasinghe, D., Abbasnejad, E.: Mysteries of the deep: Role of intermediate representations in out-of-distribution detection. In: Advances in Neural Information Processing Systems (NeurIPS) (2025). to appear

18. Sehwag, V., Chiang, M., Mittal, P.: SSD: a unified framework for self-supervised outlier detection. arXiv (2021)
19. Sun, Y., Guo, C., Li, Y.: React: out-of-distribution detection with rectified activations. In: Advances in Neural Information Processing Systems (NeurIPS), vol. 34, pp. 144–157 (2021)
20. Sun, Y.: Out-of-distribution detection with deep nearest neighbors. In: Proceedings of the 39th International Conference on Machine Learning (ICML), pp. 20827–20840. Proceedings of Machine Learning Research (PMLR) (2022)
21. Wang, H., Li, Z., Feng, L., Zhang, W.: ViM: out-of-distribution detection with virtual-logit matching. In: Proceedings of the IEEE/CVF Conference on Computer Vision and Pattern Recognition (CVPR), pp. 4921–4930 (2022)
22. Wang, H., Zhao, C., Chen, F.: Efficient out-of-distribution detection via layer-adaptive scoring and early stopping. Front. Big Data **7**, 1444634 (2024)
23. Wu, Y.: Pursuing feature separation based on neural collapse for out-of-distribution detection. In: International Conference on Learning Representations (ICLR) (2025)
24. Yang, G.: EOOD: entropy-based out-of-distribution detection. arXiv (2025)

# Author Index





R. C.-W. Wong et al. (Eds.): PAKDD 2026, LNAI 16597, pp. 613–615, 2026.
https://doi.org/10.1007/978-981-92-1300-9

MIX
Papier aus verantwortungsvollen Quellen
Paper from responsible sources
FSC® C105338

If you have any concerns about our products,
you can contact us on
**ProductSafety@springernature.com**

In case Publisher is established outside the EU,
the EU authorized representative is:
**Springer Nature Customer Service Center GmbH**
**Europaplatz 3, 69115 Heidelberg, Germany**

Printed by Libri Plureos GmbH
in Hamburg, Germany